Biosynthesis and Biodegradation of Wood Components

Biosynthesis and Biodegradation of Wood Components

Edited by

TAKAYOSHI HIGUCHI

Wood Research Institute
Kyoto University
Uji, Kyoto, Japan

1985

ACADEMIC PRESS, INC.

(Harcourt Brace Jovanovich, Publishers)

Orlando San Diego New York London
Toronto Montreal Sydney Tokyo

ACADEMIC PRESS, INC.
Orlando, Florida 32887

United Kingdom Edition published by
ACADEMIC PRESS INC. (LONDON) LTD.
24/28 Oval Road, London NW1 7DX

Library of Congress Cataloging in Publication Data

Main entry under title:

Biosynthesis and biodegradation of wood components.

Includes bibliographies and index.
1. Wood--Chemistry--Congresses. 2. Wood--Deterioration--Congresses. 3. Biosynthesis--Congresses.
I. Higuchi, Takayoshi.
TS932.B56 1985 582.1'5041 84-12298
ISBN 0-12-347880-4 (alk. paper)

PRINTED IN THE UNITED STATES OF AMERICA

85 86 87 88 9 8 7 6 5 4 3 2 1

Contents

4. Metabolism and Synthetic Function of Cambial Tissue

GRAEME P. BERLYN and YORK C. BATTEY

5. Cell Organelles and Their Function in Biosynthesis of Cell Wall Components: Control of Cell Wall Assembly during Differentiation

D. H. NORTHCOTE

6. Biosynthesis of Plant Cell Wall Polysaccharides

HIDETAKA HORI and ALAN D. ELBEIN

7. Biosynthesis of Lignin

TAKAYOSHI HIGUCHI

8. Biosynthesis of Cutin, Suberin, and Associated Waxes

P. E. KOLATTUKUDY and KARL E. ESPELIE

9. Occurrence of Extractives in Wood Tissue

W. E. HILLIS

10. Biosynthesis and Metabolism of Phenolic Acids and Monolignols

G. G. GROSS

11. Occurrence and Biosynthesis of Quinones in Woody Plants

E. LEISTNER

12. Biosynthesis of Flavonoids

HANS GRISEBACH

13. Biosynthesis of Tannins

W. E. HILLIS

14. Biosynthesis of Stilbenes

H. KINDL

15. Biosynthesis of Terpenoid Wood Extractives

RODNEY CROTEAU and MARK A. JOHNSON

20. Degradative Pathways of Lignin Model Compounds

TAKAYOSHI HIGUCHI

21. Lignin Biodegradation: The Microorganisms Involved, and the Physiology and Biochemistry of Degradation by White-Rot Fungi

T. KENT KIRK and MIKIO SHIMADA

22. Biodegradation of Aromatic Extractives of Wood

WOLFGANG BARZ and KLAUS-MICHAEL WELTRING

Contributors

Numbers in parentheses indicate the pages on which the authors' contributions begin.

FERNAND BARNOUD (441), Centre de Recherches sur les Macromolécules Végétales, CNRS et Université de Grenoble I, St. Martin d'Heres 38402, France

WOLFGANG BARZ (607), Lehrstuhl für Biochemie der Pflanzen, Westfälische Wilhelms Universität, D-4400 Münster, Federal Republic of Germany

YORK C. BATTEY (63), Yale University, School of Forestry and Environmental Studies, Greeley Laboratory, New Haven, Connecticut 06511

GRAEME P. BERLYN (63), Yale University, School of Forestry and Environmental Studies, Greeley Laboratory, New Haven, Connecticut 06511

HOU-MIN CHANG (535), Department of Wood and Paper Science, North Carolina State University at Raleigh, Raleigh, North Carolina 27695

CHEN-LOUNG CHEN (535), Department of Wood and Paper Science, North Carolina State University at Raleigh, Raleigh, North Carolina 27695

W. A. CÔTÉ, JR. (1), N. C. Brown Center for Ultrastructure Studies, State University of New York, College of Environmental Science and Forestry, Syracuse, New York 13210

RODNEY CROTEAU (379), Institute of Biological Chemistry, and Biochemistry/Biophysics Program, Washington State University, Pullman, Washington 99164

ROBERT F. H. DEKKER (505), CSIRO, Division of Chemical and Wood Technology, Clayton, Victoria 3168, Australia

ALAN D. ELBEIN (109), Department of Biochemistry, University of Texas Health Science Center, San Antonio, Texas 78284

K.-E. ERIKSSON (469), Swedish Forest Products Research Laboratory, S-11486 Stockholm, Sweden

KARL E. ESPELIE (161), Institute of Biological Chemistry, Washington State University, Pullman, Washington 99164

D. A. I. GORING (51), Pulp and Paper Research Institute of Canada, and Department of Chemistry, McGill University, Montreal, Quebec H3A 2A7, Canada

HANS GRISEBACH (291), Lehrstuhl für Biochemie der Pflanzen, Biologischen Institut II, Universität Freiburg, D-7800 Freiburg, Federal Republic of Germany

G. G. GROSS (229), Universität Ulm, Abteilung Allgemeine Botanik, D-7900 Ulm, Federal Republic of Germany

HIROSHI HARADA (1), Department of Wood Science and Technology, Faculty of Agriculture, Kyoto University, Kyoto 606, Japan

TAKAYOSHI HIGUCHI (141, 557), Wood Research Institute, Kyoto University, Uji, Kyoto 611, Japan

W. E. HILLIS (209, 325), CSIRO, Division of Chemical and Wood Technology, Highett, Victoria 3190, Australia

HIDETAKA HORI (109), Department of Biochemistry, University of Texas Health Science Center, San Antonio, Texas 78284

MARK A. JOHNSON (379), Institute of Biological Chemistry, and Biochemistry/Biophysics Program, Washington State University, Pullman, Washington 99164

H. KINDL (349), Fachbereich Chemie, Philipps-Universität Marburg, D-3550 Marburg, Federal Republic of Germany

T. KENT KIRK (579), Forest Products Laboratory, Forest Service, U.S. Department of Agriculture, Madison, Wisconsin 53705

P. E. KOLATTUKUDY (161), Institute of Biological Chemistry, Washington State University, Pullman, Washington 99164

E. LEISTNER (273), Institut für Pharmazeutische Biologie, Rheinischen Friedrich-Wilhelms-Universität Bonn, D-5300 Bonn 1, Federal Republic of Germany

H. MEIER (43), Institute de Biologie Végétale et de Phytochimie, Université de Fribourg, CH-1700 Fribourg, Switzerland

D. H. NORTHCOTE (87), Department of Biochemistry, University of Cambridge, Cambridge CB2 1QW, United Kingdom

KATIA RUEL (441), Centre de Recherches sur les Macromolécules Végétales, CNRS et Université de Grenoble I, St. Martin d'Heres 38402, France

SHIRO SAKA[1] (51), Pulp and Paper Research Institute of Canada, and Department of Chemistry, McGill University, Montreal, Quebec H3A 2A7, Canada

MIKIO SHIMADA (579), Wood Research Institute, Kyoto University, Uji, Kyoto 611, Japan

[1]Present address: Daicell Chemical Industries, Ltd., Himeji, Hyogo 671-12, Japan.

KLAUS-MICHAEL WELTRING (607), Lehrstuhl für Biochemie der Pflazen, Westfälische Wilhelms-Universität, D-4400 Münster, Federal Republic of Germany

T. M. WOOD (469), Rowett Research Institute, Aberdeen AB2 95B, United Kingdom

Preface

From prehistoric times wood has been used as a convenient multipurpose material for housing, fuel, furniture, and fiber products. Today the global consumption of wood exceeds 3×10^9 m^3 per year and is increasing. Wood is by far the most abundant biomass on earth, and it can be endlessly renewed.

The need to develop renewable alternatives to petroleum, coupled with the world's growing requirements for fuel and food, will guarantee the more widespread use, and more efficient conversion, of wood and other forest products. The Editor of *Science*, Philip H. Abelson, stated: "For a brief period of human history oil dominated the energy and chemical scene. Wood is in the process of resuming its ancient central role, but on a broader scale as science and technology point the way to more effective production and use."* The oil crisis of the 1970s not only focused renewed research attention on forestry and the wood sciences, but it also helped fuel the intense and still-expanding interest in biotechnology among scientists and laymen alike.

Improvements in tree growth rates, qualitative and quantitative control of the biosynthesis of wood constituents, and new and more efficient conversions of wood and its residues and by-products all fall within the potential of biotechnology. In the past decade this potential has been discussed in several international symposia and seminars and in countless small forums worldwide.

Nevertheless, this is the first comprehensive book ever prepared on wood biochemistry. In treating both the biosynthesis and biodegradation of wood, this book is intended as a basic reference for those who seek to apply biotechnology to wood production and use.

*Editorial in *Science,* 12 March 1982.

Chapters 1, 2, and 3 deal with the structure and chemistry of the major components of wood cell walls. Chapters 4 through 15 discuss the metabolism and synthetic function of cambial tissue, the function of organelles involved in biosynthesis, and the biosynthesis both of the major cell wall components of wood (cellulose, hemicelluloses, and lignins) and of wood extractives. Chapters 16 through 22 treat the microbial degradation of these components and of wood extractives.

The editor expresses his cordial thanks to Dr. T. Kent Kirk, Forest Products Laboratory, Madison, who kindly reviewed the manuscripts of Chapters 16 through 21 and contributed valuable suggestions to the book. The editor is also grateful to the authors, who are internationally renowned in their respective fields, to the publisher, and to Ms. Hitomi Kojima, Wood Research Institute, Kyoto University, for valuable secretarial assistance.

TAKAYOSHI HIGUCHI

CHAPTER 1

Structure of Wood

H. HARADA

Department of Wood Science and Technology
Faculty of Agriculture
Kyoto University
Kyoto, Japan

W. A. CÔTÉ, JR.

N. C. Brown Center for Ultrastructure Studies
State University of New York
College of Environmental Science and Forestry
Syracuse, New York

I. General

The principal sources of commercial timber are the trees of the gymnosperms and of the angiosperms. Softwood lumber is derived from coniferous trees, while hardwood is a product of broad-leaved species, the dicotyledons.

A mature tree, of either the softwood or the hardwood type, consists of the stem, branch, and root. Since the stem is the principal source of

ISBN 0-12-347880-4

woody material for the manufacture of lumber and other products, this chapter deals with the stem, particularly the wood. Proceeding from the outside to the inside of the stem, there is a layer of bark (phloem), which is separated from the major component, wood (xylem), by a thin initiating layer called the cambium. The pith is located at the center of the stem.

There are three aspects of sections of wood: the cross or transverse section, which is exposed by cutting horizontally across the stem; the tangential section, produced by cutting tangent to the bark; and the radial section or face, which results from splitting the wood radially from pith to bark.

A. *Development of Wood*

The stem of a tree increases in volume both by height growth and by thickening growth. The length growth of the tree results in the increase of the tissue originating from the apical meristem at the top of the tree, and the thickening growth arises from the lateral meristem, the vascular cambium, between the xylem and the phloem. The tissue produced by the apical meristem is primary tissue and that by the lateral meristem is secondary tissue.

1. Primary Tissue and Secondary Tissue

The apical meristem, which is composed of cells resembling bubbles, forms the new cells in the very tip of the tree stem during the growing season. Below the region where the new cells are formed, the newly formed cells change in size, shape, and function, becoming permanent tissue through differentiation. This differentiated meristematic zone consists of the protoderm, the procambium, and the ground meristem, and is called the primary tissue. Among these three zones of the primary tissue, the procambium differentiates into the primary vascular bundle, which is made up of the primary xylem, the fascicular cambium, and the primary phloem. The fascicular cambium is interconnected through the rise of the interfascicular cambium. This gives rise to the complete cylinder of the vascular cambium, generally called the cambium, and leads to formation of the secondary tissue—that is, the secondary xylem and phloem. The tree stem continues to grow in diameter year after year by addition of successive layers of the secondary xylem and phloem.

2. Cambium

The cambium contains two types of initials, fusiform initials and ray initials. In both softwoods and hardwoods, the fusiform initials produce

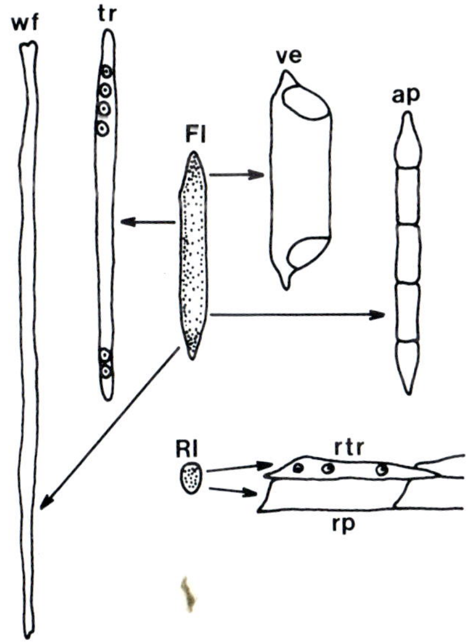

Fig. 1. Schematic diagram of the cambial initials and the axial and ray cells of wood which were derived from the cambial initials. FI, Fusiform initial; RI, ray initial; ve, vessel element; tr, tracheid; wf, wood fiber; ap, axial parenchyma cell; rtr, ray tracheid; rp, ray parenchyma cell. From Shimaji (1976).

all the axial cells, and the ray initials produce the ray cells of the xylem and phloem (Fig. 1). The fusiform initials are the fusiform type of cells with tapering ends. In softwoods and in the more primitive hardwoods, the fusiform initials are relatively long in length and their tapering ends overlap each other. This type of arrangement of the fusiform initials is called an unstoried cambium. The highly developed hardwoods have relatively short and nearly uniform fusiform initials. In the most highly developed hardwoods such as *Robinia pseudoacacia,* the fusiform initials are considerably arranged in horizontal rows in the tangential sections. This type of fusiform initial arrangement is known as storied cambium. The ray initials produce the ray cells such as ray tracheids or ray parenchyma cells in both the xylem and phloem.

The division of the fusiform initials takes place in two ways: (1) periclinal division, which is undertaken in the tangential plane and results in the increase of the stem diameter and (2) pseudotransverse or anticlinal division, which is undertaken in the radial plane and results in the increase of the stem circumference. The formation of new xylem and phloem cells is achieved by the periclinal division of the cambial initials. One of the two new cells formed by such a division continues to function as a cambial initial. The other cell of the two becomes a mother cell, capable of limited further periclinal division. The axial cells in xylem and phloem, so formed from the fusiform initials, are aligned in radial rows. The radial arrangement of the cells is generally maintained in softwoods, while hardwoods show no such evidence because of the postcambial enlargement of the vessel elements.

On the other hand, the ray initials that produce ray cells are elongated radially. Subsequently, the number of the ray initial divisions is limited, and the derived cells mature without any further cell division. The region that is composed of undifferentiated cells, such as the cambial initials and the mother cells, is called the cambial zone. The immature cells are situated in the outermost part of the cambial zone and differentiate into the various kinds of mature xylem and phloem cells: for example, softwood tracheids and hardwood fibers and vessel elements. As an exception, the formation of parenchyma strands and strand tracheids arises by postcambial division of the derived cells from the fusiform initials.

The developing cells mature through the following two stages: (1) the enlargement of cells and (2) the thickening and lignification of cell walls. The enlargement of the cells produces an increase both in diameter and in length. Increase in diameter is particularly evident in the earlywood tracheids of softwoods and the vessel elements of the hardwoods. Diameter increase in softwood tracheids is only in the radial direction; in contrast, that in hardwood vessel elements is both in the radial and tangential directions. Increase in length is typical of the hardwood fibers. Other types of cells elongate to a lesser extent or not at all, as in the case of earlywood vessel elements in the more highly specialized hardwoods. Therefore the length of matured vessel elements of earlywood in the more highly specialized hardwoods is thought to reflect the length of the fusiform initials.

At or near the completion of enlargement of the new cells, cell wall begins to be deposited on the inside surface of the cell. The new wall is called a secondary wall, while the wall formed during both the division and enlargement of the cell is called a primary wall. The formation of pits, spiral thickenings, and perforations takes place during cell enlargement and secondary wall formation. Lignification is initiated at the stage of wall thickening at the outer layers of the secondary wall and is accomplished after the completion of the secondary wall. Completion of secondary wall formation and the lignification of cell wall in tracheids, fibers, and vessel elements is followed by the death of the cell, and at this point these cells are mature. Parenchyma cells, however, differ from the other cells in that they are living as long as they are part of the sapwood.

B. *Earlywood and Latewood*

The transverse section of the growth-increment layer in the wood formed during one growing period is called the growth ring. There is usually one growing period every year in trees grown in temperate zones, so the term annual ring is often used instead of growth ring. According to this terminology of annual ring, the parts of wood formed in spring and in summer of a growing season should be called springwood and summerwood, respectively. The season during which spring- and summerwood are formed, however, does not always coincide with the actual spring and summer seasons, depending on the regions where the tree grows. Therefore, it is more appropriate to use the terms early- and latewood instead of spring- and summerwood.

The distinction of earlywood from latewood is generally obvious both in softwoods and in ring-porous wood of hardwoods but is not always clear in diffuse-porous wood of hardwoods. In softwoods, the wood that developed wide rings is generally low in specific gravity and inferior in quality, because wood quality depends on the percentage of latewood within a growth ring with the constant width of the latewood. On the other hand, in ring-porous wood of hardwoods, the wood with wide growth rings is superior in quality, because the wood quality depends on whether the growth ring is wide with the constant width of the pore zone, which is occupied by the rows of vessel pores. In hardwoods, the wood in which the growth ring is conspicuously narrow, as appears often in oak, zelkova, and ash, is called "Nukame" wood in Japan, and the wood is soft. In tropical species, the growth rings of the wood are not readily seen in the trees because of continuous growth during all seasons of the year. However, the boundary of the growth increment, which is very similar to that of the growth ring, is usually found, and this can be the result of wet and dry seasons.

C. *Sapwood and Heartwood*

The part of the wood core in the tree in which some kinds of xylem cells are living, and hence the part that is physiologically active, is called sapwood, while the physiologically dead part of the xylem is called heartwood. In some instances, the wood of an intermediate zone between the sapwood and the heartwood is called intermediate wood. The transition of sapwood to heartwood is accompanied by the formation of various organic substances such as extractives and extraneous materials, and in hardwoods frequently also by the formation of tyloses in vessels. Sapwood is generally distinguished from heartwood by contrast in color, but in species of *Picea, Abies, Tsuga,* and *Populus,* sapwood can not be distinguished from heartwood on this basis. The recognized characteristics of heartwood, such as the existence of so-called heartwood substances, pit aspiration in softwood tracheids, and the occurrence of tyloses in hardwood vessels, are considered to be advantages because the wood is made more resistant to fungal and insect attack than is sapwood. On the negative side, heartwood is less permeable to liquids and gases and therefore more difficult to treat with wood preservatives.

False heartwood is often found in trees such as beech, birch, and maple, which do not normally exhibit colored heartwood. The colored region of false heartwood is formed not along the growth ring but rather in irregular patterns.

II. The Cells of Wood

A. *Softwood Cells*

The principal types of cell in softwood are listed in Table I. Only two types of cells—tracheids and ray parenchyma cells—always occur in softwoods. The shapes of the principal softwood cells are illustrated in Fig. 1, and a scanning electron micrograph of macerated tracheids and ray parenchyma cells from softwoods is also shown in Fig. 2. In addition, the arrangement of these cells in softwoods can be seen in Fig. 3.

The regular alignment of the tracheids in radial rows, which is seen in the transverse section, is a characteristic feature of softwoods. The tangential diameter of tracheids is quite constant within the growth ring. The distinct difference in tangential width of some rows of tracheids in transverse section (the row shown by the arrow in Fig. 3) is related to the fact that the tracheids were cut across near their ends which are narrower tangentially than those of the central portions. The proportion of volume of wood occupied by tracheids in softwoods is over 90%.

Table I
Cells of Softwood

Axial Cells	Transverse Cells
Tracheid	Ray tracheid
Strand tracheid	Ray parenchyma cell
Axial parenchyma cell	Ray epithelial cell
Epithelial cell	

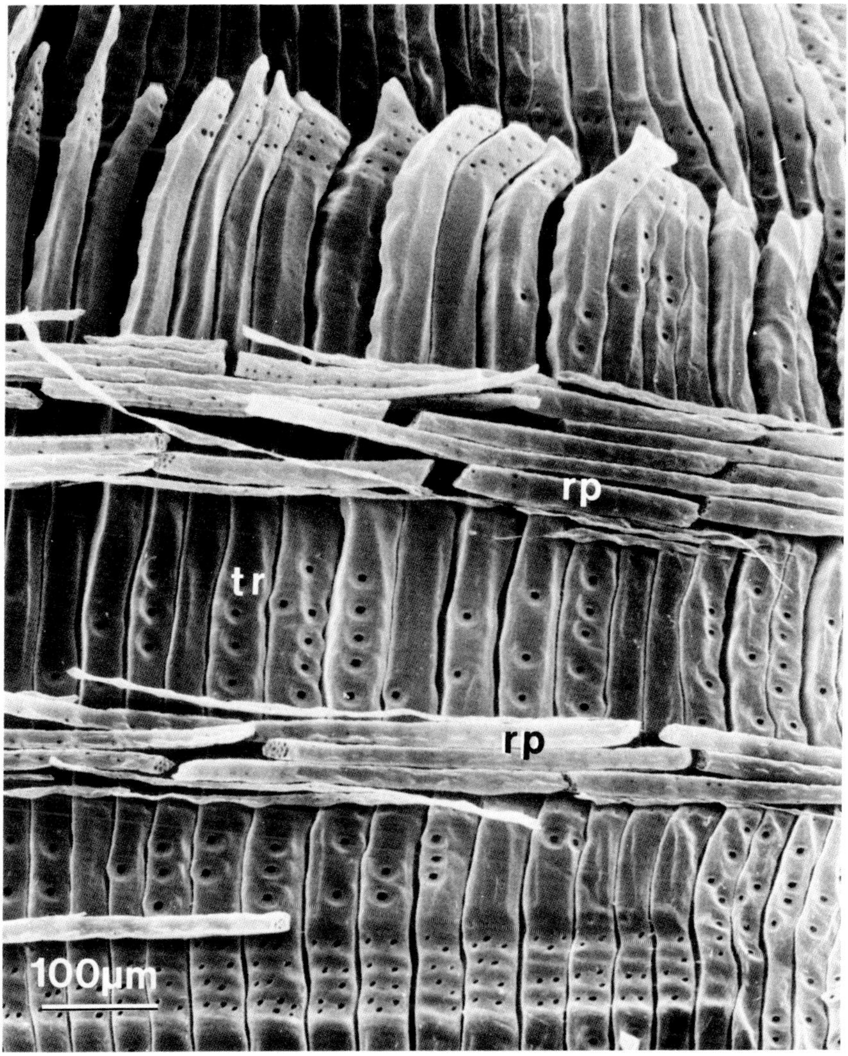

Fig. 2. Scanning electron micrograph of pulped or macerated *Pseudotsuga menziesii*, showing the tracheids (tr) with their closed ends and ray parenchyma cells (rp) arranged in ribbon-like aggregations. From Côté (1980).

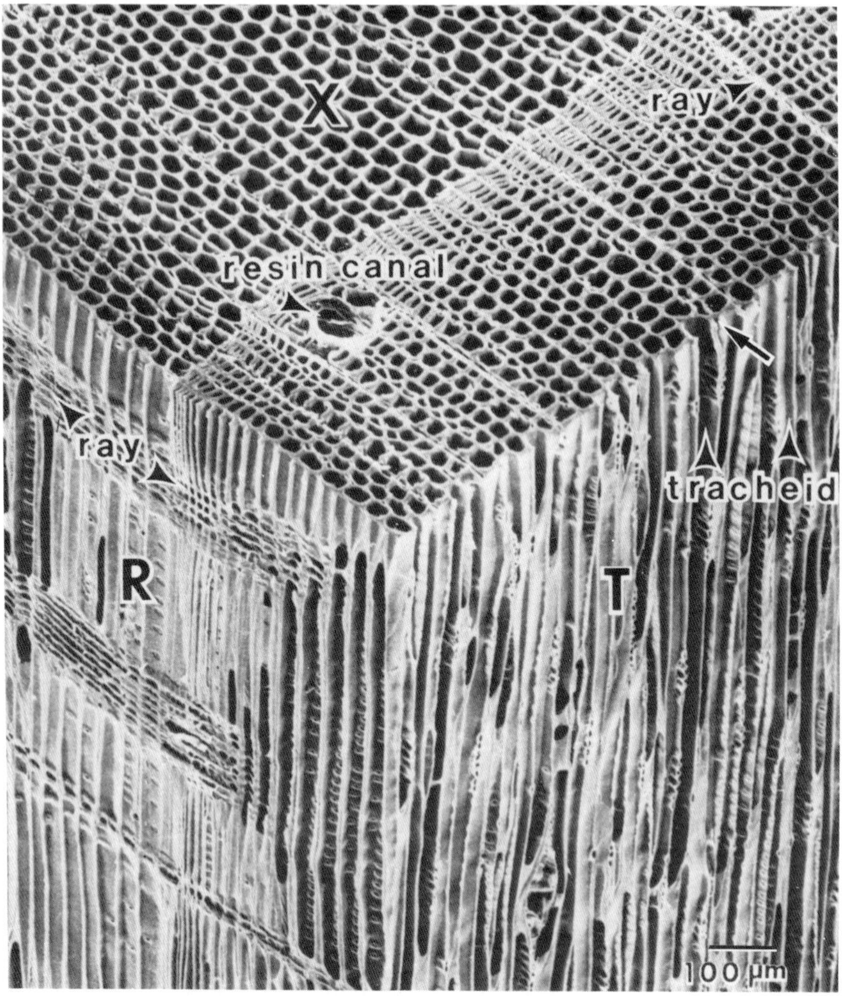

Fig. 3. Scanning electron micrograph of a typical softwood (*Pinus strobus*), showing three aspects: radial surface (R), tangential surface (T), and transverse surface (X). Note the tracheids and rays. The resin canals are found in some softwood species. From Côté (1980).

1. Tracheids

The average length of tracheids in softwoods ranges from 3.0 to 5.0 mm in softwoods (Panshin and De Zeeuw, 1980). The average tangential diameter of tracheids is from 20 to 80 μm, while the radial diameter of a tracheid varies depending on its location in the growth ring.

The tracheid wall is usually characterized by pits and, in a few species, has spiral thickenings on the inside wall lining the lumen. Structures such as trabeculae or callitrisoid thickenings are present in some species of softwoods.

The pit of softwood tracheids is the bordered pit, and it exists (1) between adjacent tracheids, (2) between tracheids and ray parenchyma cells, and (3) between tracheids and ray tracheids, respectively. The bordered pits between adjacent tracheids are numerous and large in diameter on the radial walls of the earlywood, while they are less numerous and smaller in breadth on the same walls of the latewood. Ultrastructure of the bordered pits is described in Section III,D.

2. Strand Tracheids and Axial Parenchyma Cells

Strand tracheids are the short cells arranged in axial strands and differ from the tracheids in being shorter and in possessing end walls. Strand tracheids are regarded as transitional cells between tracheids and axial parenchyma cells. Axial parenchyma cells in softwoods occur in strands and are thinner walled with simple pits, in contrast to the neighboring tracheids.

3. Ray Tracheids and Ray Parenchyma Cells

Ray tracheids are provided with smaller bordered pits than those of tracheids and are usually present on the upper and lower margins of the ray. The inner wall of ray tracheids in hard pines possesses dentations.

Ray parenchyma cells are thin-walled with simple pits. For the identification of softwoods, the half-bordered pit pair between tracheids and ray parenchyma cells is very important. The wall areas between cells of ray parenchyma and tracheids are called cross fields. The shape, size, and arrangement of the bordered pit on the tracheid side of the cross fields are classified into five types: (1) window-like pit, (2) pinoid pit, (3) cupressoid pit, (4) taxodioid pit, and (5) piceoid pit. Window-like pits are found in *Pinus* (hard pines), *Sciadopitys,* and *Podocarpus;* pinoid pits in *Pinus;* cupressoid pits in *Chamaecyparis, Thujopsis, Juniperus, Torreya,* and *Tsuga;* taxodioid pits in *Cryptomeria, Abies,* and *Thuja;* and piceoid pits in *Picea, Larix,* and *Pseudotsuga.*

4. Epithelial Cells and Ray Epithelial Cells

The epithelial or ray epithelial cells are thin-walled parenchyma cells that surround axial or radial resin canals. The resin canals are not wood cells but a space surrounded by epithelial or ray epithelial cells (see Fig.

Table II
Cells of Hardwood

Axial Cells	Transverse Cells
Vessel element	Ray parenchyma cell
Tracheid	Ray epithelial cell
Vascular tracheid	
Vasicentric tracheid	
Fiber	
Fiber tracheid	
Libriform fiber	
Axial parenchyma cell	
Epithelial cell	

3). Normal resin canals are a constant feature in *Pinus, Picea, Larix,* and *Pseudotsuga* and exist in both axial and radial directions in softwoods.

B. *Hardwood Cells*

The structural features of hardwoods that are different from those of softwoods are summarized by Panshin and De Zeeuw (1980) as follows: (1) Hardwoods possess vessel elements, while softwoods do not. (2) The radial alignment of axial cells that is evident in softwoods is lacking or obscured in hardwoods. (3) Hardwoods have more kinds of cells than those of softwoods, and these cells show more variation in size, shape, and arrangement within a growth ring. (4) Hardwood rays are more variable in width than those of softwoods.

The principal types of cells in hardwood are listed in Table II. The shapes of the main hardwood cells were illustrated in Fig. 1, and a scanning electron micrograph of a few macerated cells of a hardwood is shown in Fig. 4. The arrangement of the cells in hardwoods may be seen in Fig. 5.

1. Vessel Elements

The shape of vessel elements varies from drum-shaped to oblong with or without tail-like ends (see Fig. 4). The length of vessel elements varies considerably in different woods, from 0.19 to 1.33 mm in average length (Panshin and De Zeeuw, 1980). The length of vessel elements may reflect the length of cambial initials. The cross sections of vessel elements are called pores, and the tangential diameter of pores varies from 20 to 300 μm (Panshin and De Zeeuw, 1980). Hardwoods are divided into three groups on the basis of pore size and distribution within a growth ring: (1)

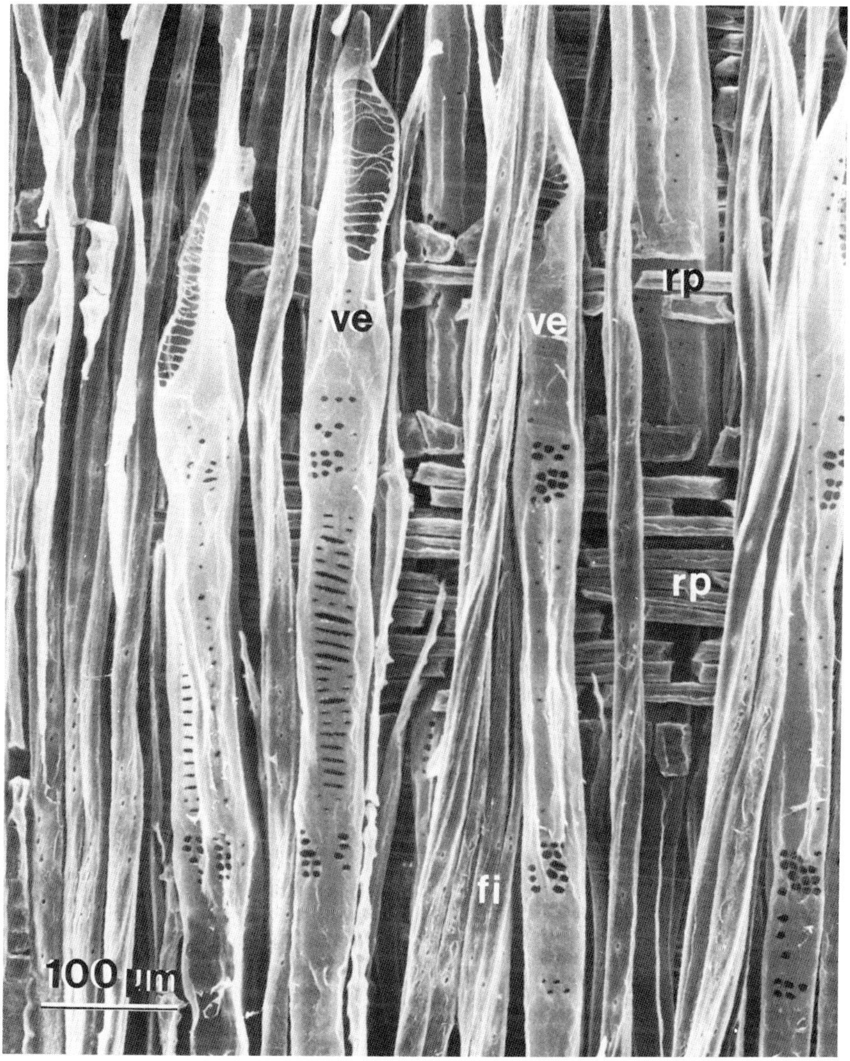

Fig. 4. Scanning electron micrograph of pulped or macerated *Liquidambar styraciflua*. The individual vessel elements (ve), fibers (fi), and the ray parenchyma cells (rp) are more distinguishable after maceration. From Côté (1980).

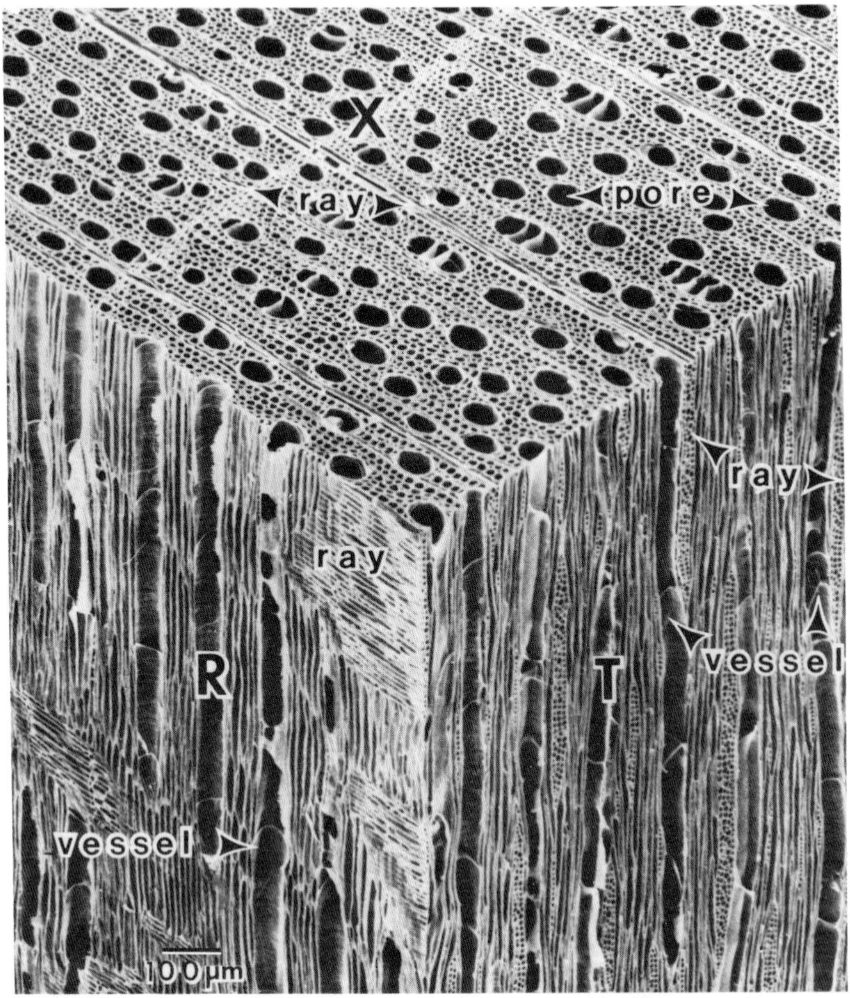

Fig. 5. Scanning electron micrograph of a typical hardwood (*Acer rubrum*), showing the radial (R); tangential (T), and transverse (X) surfaces. Note that the rays can be seen on all three surfaces. Vessels are known as pores when viewed on transverse surface. From Côté (1980).

ring-porous wood, (2) diffuse-porous wood, and (3) semi-diffuse-porous wood. Perforation plates, which are found at the junction of end walls of two vessel elements in the axial direction, are classified into simple perforation plates and multiple perforation plates. The latter are further divided into scalariform perforation plates and reticulate perforation plates.

The vessel element wall is provided with bordered pits, and the nature and the extent of the pits vary depending on the kind of cell in contact

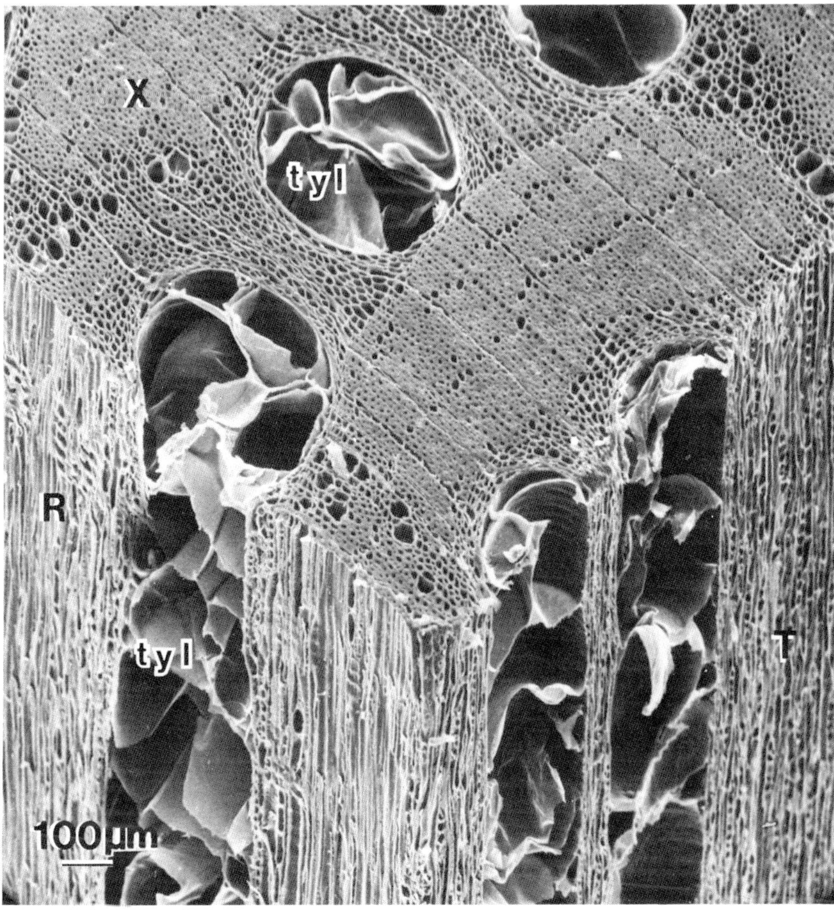

Fig. 6. Scanning electron micrograph of *Quercus alba*. The earlywood vessels are blocked by tyloses (tyl), which proliferate from parenchyma cells adjacent to vessels. (Micrograph by A. C. Day.)

with the vessel. The intervessel bordered pit is conspicuous and three types of arrangement are recognized: (1) opposite pitting, (2) alternate pitting, and (3) scalariform pitting.

The most typical inclusion of vessel elements is the tylosis. Tyloses are saclike structures that develop in a vessel through the proliferation of the protoplast of a parenchyma cell through a pit pair. When tyloses are developed abundantly, the vessel lumen is completely filled with tyloses (Fig. 6). Tyloses are usually formed in the inner sapwood prior to its transformation into heartwood. Tyloses are also formed, however, as a result of injury, and these are called traumatic tyloses. Tyloses may be considered as a kind of wood cell formed in the lumen of a vessel element, from the viewpoint of both its formation process and its wall structure.

2. Tracheids

Vascular tracheids are very similar in size, shape, and position to small latewood vessel elements, but are not imperforate at the ends. Vasicentric tracheids are short and irregularly shaped cells and are abundant around the large earlywood vessels of ring porous woods.

3. Fibers

Fibers of two types occur in hardwoods: (1) fiber tracheids and (2) libriform fibers. Both types usually have very thick walls. The difference between the two types is in the type of pitting. Fiber tracheids have bordered pits with slit-like pit apertures. Libriform fibers usually have simple pits. Fiber tracheids and libriform fibers vary in diameter, length, cell wall thickness, and volume. The average fiber length varies from 0.64 to 2.30 mm, and the volume percentage of fibers in hardwoods ranges from 34.7 to 75.7 (Panshin and De Zeeuw, 1980).

4. Axial Parenchyma Cells

Axial parenchyma cells are short, brick-shaped (procumbent or upright cell), or isodiametric (square cell), equipped with simple pits. The types of arrangements of axial parenchyma cells as viewed on the transverse sections of hardwoods are classified by their location or distribution as follows:

1. Apotracheal parenchyma (axial parenchyma cells arranged independent of the pores or vessels): (a) diffuse parenchyma, (b) diffuse-in-aggregate parenchyma, (c) reticulate parenchyma, (d) scalariform parenchyma, (e) terminal parenchyma, (f) initial parenchyma, and (g) banded apotracheal parenchyma.
2. Paratracheal parenchyma (axial parenchyma cells obviously associated with the vessels): (a) scanty paratracheal parenchyma, (b) vasicentric parenchyma, (c) aliform parenchyma, (d) confluent parenchyma, and (e) unilaterally paratracheal parenchyma.

5. Ray Parenchyma Cells

The radially arranged cells in hardwoods are ray parenchyma cells only. The ray parenchyma cells in hardwoods differ from those in softwoods in having a great variation in shape, width, and height. Their arrangement within the ray is also different.

Ray parenchyma cells may be classified into types: (1) procumbent cells, which are narrow and radially elongated, (2) upright cells, which are short and vertically oriented, occurring on the upper and lower margins of the ray, and (3) square cells that are nearly square in radial section. Cells that are nearly square in radial section, however, are regarded as a type of upright cell (Panshin and De Zeeuw, 1980), hence there are two basic types of parenchyma cells.

The rays in hardwoods are classified by their size and distribution, as well as by the cell shapes within the ray. Uniseriate rays are one cell wide, while multiseriate rays are more than one cell in width. Rays may be distributed in a diffuse manner, grouped together as in aggregate rays, or may be compound rays. Rays consisting of only procumbent cells are called homogeneous rays, while those consisting wholly or in part of square or upright cells are classed as heterogeneous rays.

Ray parenchyma cells may contain crystals, silica, and numerous other materials, such as gums, resins, tannins, oils, and latex.

The volume of wood made up of ray parenchyma cells in hardwoods may have an important effect on its physical properties. It varies among different wood species, the average volume of hardwood consisting of ray parenchyma cells ranging from 5.3 to 34.1%, according to data on a number of American and European species (Panshin and De Zeeuw, 1980).

III. The Wood Cell Wall

The walls of wood cells are composed of three structural substances: cellulose is the framework substance, hemicellulose is the matrix substance, and lignin is the encrusting substance (Wardrop, 1964a).

Cellulose exists in the form of microfibrils and is associated with the matrix and encrusting substances (Fig. 7).

The microcapillaries surrounding the cell wall framework are filled with the matrix and encrusting substances. Hemicelluloses and other carbohydrate materials are deposited in the cell wall as the matrix substances. Lignin deposition is not initiated with the start of cell wall formation but occurs in general as the final phase of differentiation.

The nature of lignin as the encrusting substance is shown by the distribution of the lignin skeleton in the cell wall, demonstrated by acid removal of cellulose and hemicellulose (Fig. 8).

A. *The Cellulose Microfibrils*

The crystalline nature of cellulose in wood has been revealed by the evidence provided by studies with X-ray diffraction and polarization microscopy.

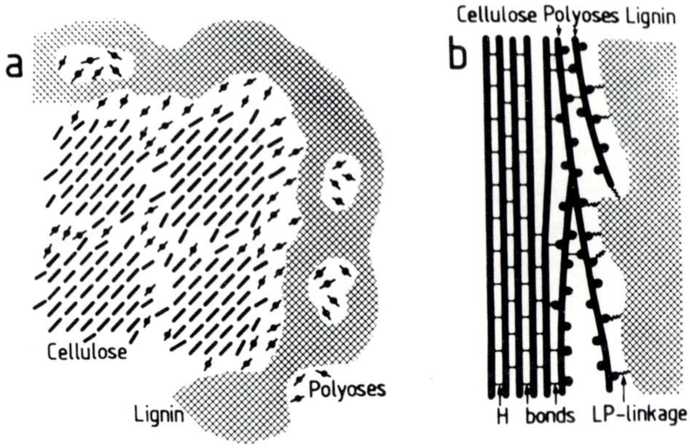

Fig. 7. Schematic diagram for the ultrastructural arrangement of cellulose, hemicellulose (polyoses), and lignin in the wood cell wall. **a,** Transverse section; **b,** longitudinal section. From Fengel and Wegener (1984).

Recently, this crystalline nature was confirmed by electron diffraction patterns of the secondary wall of wood cells in selected areas (Revol and Goring, 1982) (Fig. 9).

That the cellulose molecular chains are organized into strands as cellulose microfibrils is clearly demonstrated with the aid of electron microscopy (Fig. 10).

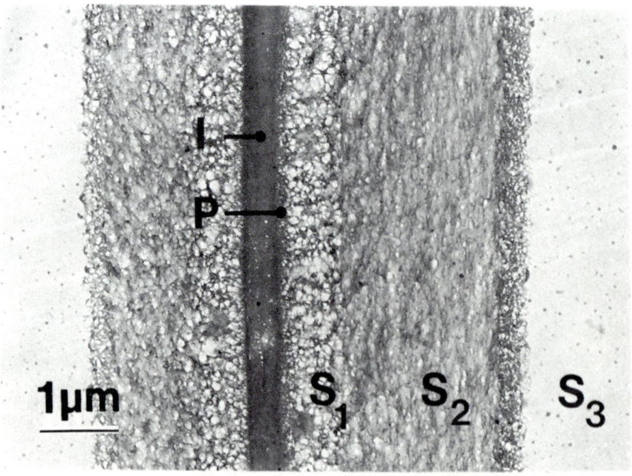

Fig. 8. Transmission electron micrograph of ultrathin longitudinal section of earlywood tracheids from *Pinus densiflora,* showing the distribution of lignin in the cell wall that was skeletonized using the HF acid technique.

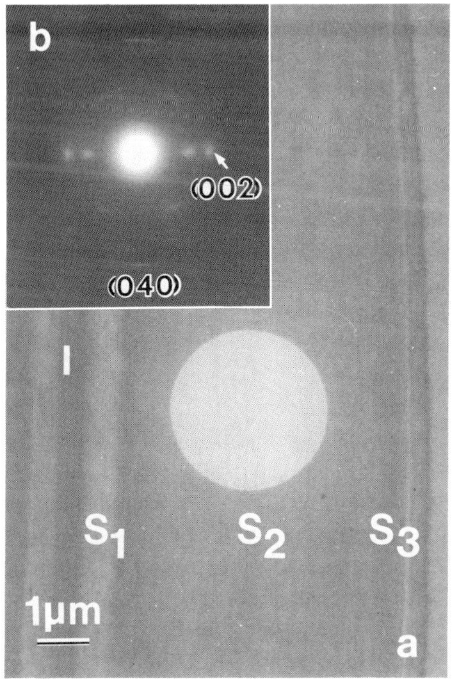

Fig. 9. **a,** Transmission electron micrograph of ultrathin longitudinal section of the latewood tracheid of *Pinus densiflora,* showing the intercellular layer (I), S_1, S_2, and S_3 layers. **b,** Electron diffraction diagram of the selected area in the S_2 layer, which is represented by the small circle. Note that the indexing of individual planes in the electron diffraction diagram refers to Meyer and Misch (1937) unit cell of cellulose I. From Sugiyama and Harada (1982).

The term elementary fibril is used for the cellulosic strands of smallest possible diameter. Mühlethaler (1960, 1965) applied this term to cellulose fibrils with a diameter approximately 3.5 nm, which can be resolved using the negative-contrast preparation technique for electron microscopy. An elementary fibril of this cross-sectional dimension could contain ~40 cellulose chains.

However, the width of cellulose microfibrils was reported to be different in various cellulose materials (Balashov and Preston, 1955). It is about 2.5 nm for wood holocellulose, about 3 nm for the gelatinous layer of hardwood tension wood fibers, about 5 nm for Whatman cellulose, and about 20 nm for *Valonia* (Goto *et al.,* 1975; Harada and Goto, 1982).

Furthermore, from the cross-sectional examination of *Valonia* cellulose fibrils by means of the negative-contrast preparation technique or by dif-

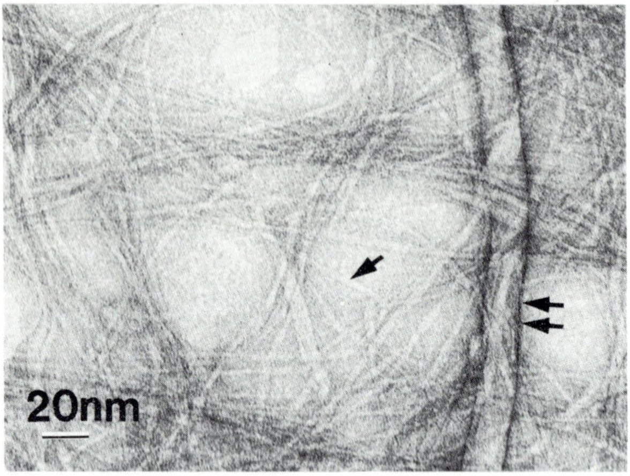

Fig. 10. Transmission electron micrograph of the holocellulose of *Pinus densiflora* (arrow) and of *Valonia macrophysa* (double arrows) (disintegration, negatively stained with uranyl acetate), showing the difference of cellulose microfibril width between wood and *Valonia*. From Sugiyama and Harada (1983).

fraction contrast in the bright-field mode of electron microscopy, it has not been possible to detect inside one cellulose microfibril any subunits corresponding to the elementary microfibrils observed after sonication (Goto *et al.*, 1982; Revol, 1982). Therefore, in this chapter, the fibrillar structure that can be detected using electron microscopy is called the cellulose microfibril.

Cellulose microfibrils are almost square in cross section in wood, are rectangular or square in *Valonia,* and are evidently of indefinite length (Goto *et al.*, 1973; Harada and Goto, 1982; Revol, 1982) (Fig. 11). They are joined laterally into lamellae that are then organized into cell wall layers.

Within the cellulose microfibrils are zones where cellulose chains are oriented to such a high order that X-ray or electron diffraction patterns can be produced as with true crystals. These regions were observed in *Valonia* by means of diffraction contrast in the bright-field mode of electron microscopy (Sugiyama and Harada, 1983) (Fig. 12). The lattice image of $(10\bar{1})$ planes (the lattice spacing of 0.54 nm) has been visualized in *Valonia macrophysa* (Sugiyama *et al.*, 1984) (Fig. 13). This observation rules out the existence of either smaller crystalline subelements in the transverse direction or periodicity in the longitudinal direction. In wood,

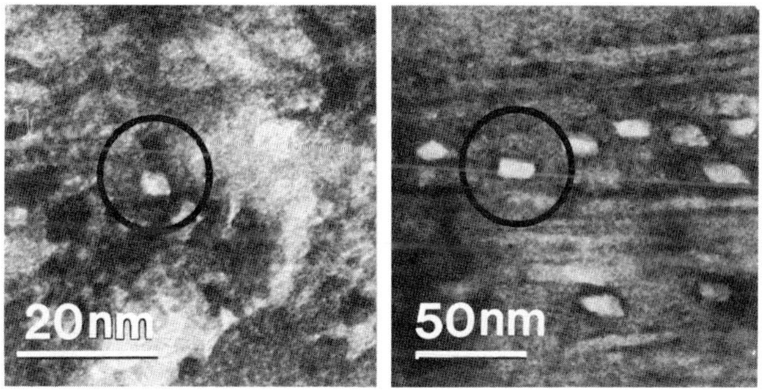

Fig. 11. Transmission electron micrograph of ultrathin sections of the cellulose microfibrils. **a,** From the gelatinous layer of *Populus euramericana* tension wood **b,** From *Valonia macrophysa* (negatively stained with uranyl acetate). From Goto *et al.* (1973, 1975).

the crystalline regions are thought to be interrupted by less ordered paracrystalline regions along the cellulose microfibrils.

Paracrystalline regions exist between cellulose microfibrils, as well as within the microfibril. In these areas, the cellulose chains may be in close

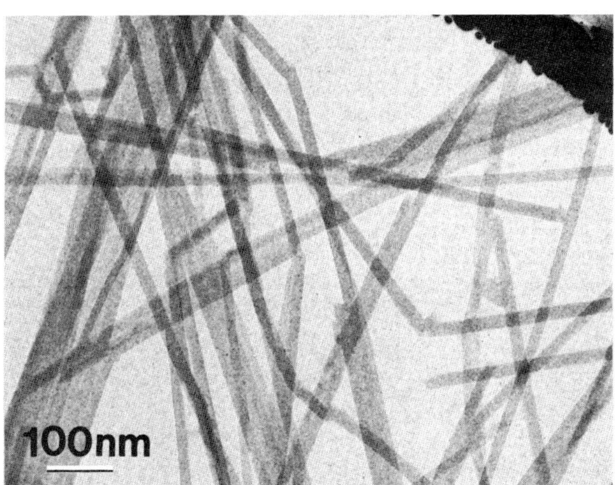

Fig. 12. Transmission electron micrograph of the cellulose microfibrils from *Valonia macrophysa* (disintegration, diffraction contrast in the bright field mode), showing the crystalline structure of cellulose microfibrils. From Sugiyama and Harada (1983).

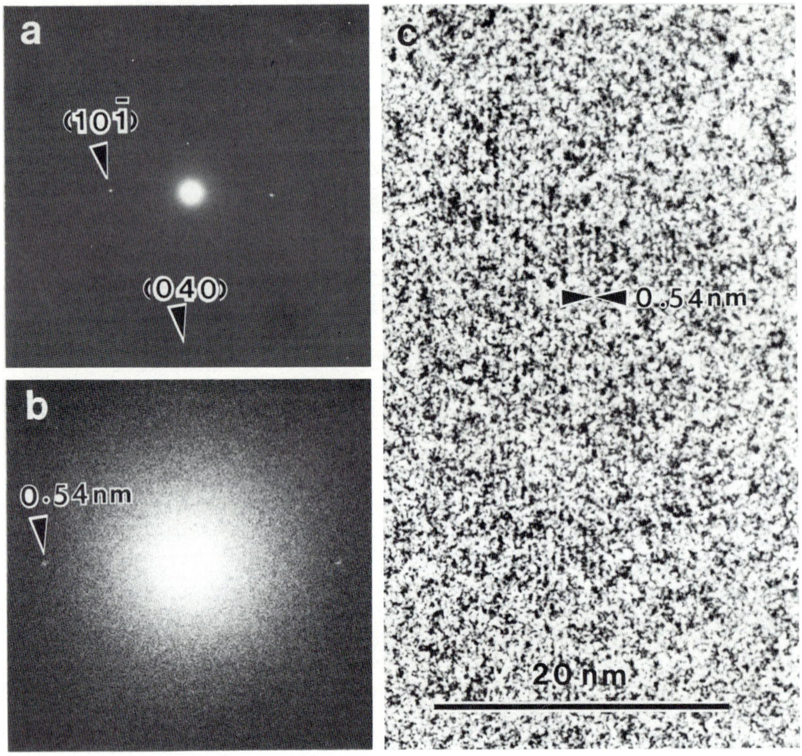

Fig. 13. Transmission electron micrograph of a cellulose microfibril in *Valonia macrophysa,* showing the 0.54 nm lattice line (**c**). Note that (**a**) is a typical electron diffraction pattern from one microfibril and (**b**) is an optical diffraction pattern of the photomicrograph. From Sugiyama *et al.* (1984).

association with matrix substances and lignin (Fengel, 1970; Kerr and Goring, 1975) (Fig. 14).

B. Individual Cell Wall Layers

1. Tracheids and Wood Fibers

When transverse sections of tracheids or wood fibers were examined under crossed Nicols with a polarizing microscope, layering was indicated by the variable birefringence in different portions of the cell wall (Fig. 15). On the basis of such evidence, Kerr and Bailey (1934) proposed a generalized structure for the secondary cell walls of softwood tracheids and hardwood fibers. According to their concept, the secondary wall of

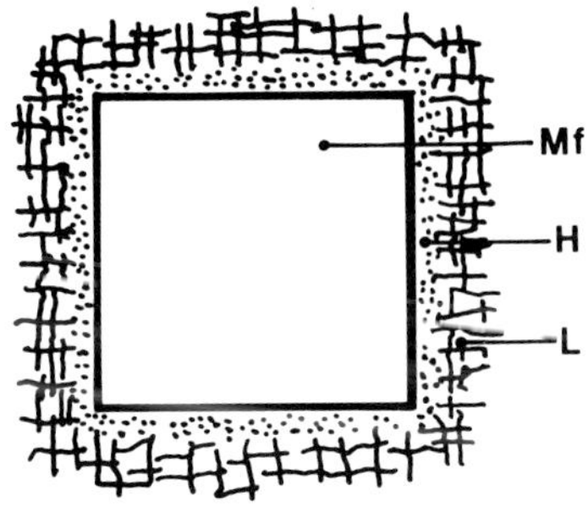

Fig. 14. Schematic diagram of the ultrastructural arrangement of a cellulose microfibril (Mf), hemicellulose (H), and lignin (L) in the wood cell wall.

Fig. 15. Polarized-light photomicrographs of transverse sections. **a,** From earlywood tracheids of *Cryptomeria japonica*. **b,** From wood fibers of *Fagus crenata*. These reveal the three-layered structure of the cell wall due to the birefringence of cellulose microfibrils. The S_1 and S_3 layers appear bright, while the S_2 is at total extinction.

normal wood cells consists of three layers: a relatively narrow or thin outer layer (S_1) and inner layer (S_3) and a relatively thick middle layer (S_2). They noted that the cellulose fibrils are oriented approximately parallel to the long axis of the cell wall in the middle layer, while the orientation in the inner and outer layers is close to 90° to the longitudinal axis. They suggested that the term middle lamella or intercellular layer be used in referring to the isotropic substance that separates the walls of adjoining cells. The term primary wall (P) is applied to the original wall of the cell, formed at the cambium, while the term secondary wall (S) is used for the layers of secondary thickening that are formed inside the primary wall. The concept and terminology are still accepted, even following extensive study with electron microscopy (Fig. 16).

The present concept of microfibrillar orientation in layers is based largely on evidence from electron microscope observations supported by X-ray diffraction and polarization microscopy.

In Fig. 17, textures to be found in cell-wall layers of typically xylary elements (softwood tracheids and hardwood fibers) are shown diagrammatically. The thin primary wall (P) consists of a loose aggregation of microfibrils oriented more or less axially on the outer surface and transversely on the inner surface to the cell axis. The microfibril orientation in the primary wall is interpreted by the multi-net growth hypothesis proposed by Roelofsen (1951) and supported for conifer cambium by

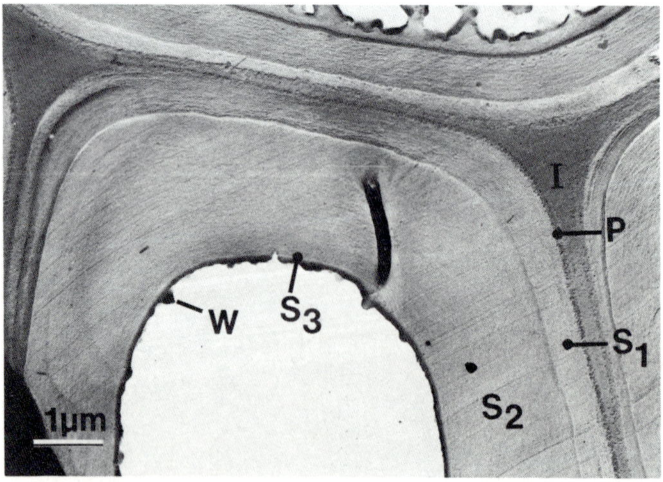

Fig. 16. Transmission electron micrograph of ultrathin transverse section from *Fagus crenata,* showing the intercellular layer (I), primary wall (P), different layers of the secondary wall (S_1, S_2, S_3), and warty layer (W) in a wood fiber.

Wardrop (1958). In this multi-net hypothesis, the microfibrils are first deposited transverse to the cell axis and are passively shifted longitudinally during cell expansion (Fig. 18). An opposite view, the ordered fibril hypothesis, is proposed by Roland *et al.* (1975) and supported by Itoh *et al.* (1975). According to this hypothesis, whether the orientation of microfibrils becomes transverse, oblique, or longitudinal is determined at the time of deposition of cell wall and may not be reoriented thereafter.

The primary walls do show lamellation with relatively broad distances between microfibrils in each lamella. The three layers of the secondary wall, designated as S_1 (outer layer), S_2 (middle layer), and S_3 (inner layer), are organized in a plywood type of construction. The microfibrillar angle to the cell axis in the individual layers of the secondary walls of conifer tracheids is shown in Table III as an example. The S_1 or S_3 with a large microfibrillar angle to the cell axis is designated as a flat helix, and the S_2 with a small angle as a steep helix. The layers themselves are of lamellae of microfibrils with varying amounts of shift in orientation, visible in electron micrographs.

The S_1 is composed of several lamellae with alternating S- and Z-helices of microfibrillar orientation (Imamura *et al.*, 1972), and this structure in the S_1 is termed "crossed fibrillar texture" (Wardrop, 1964b). The thickness of the S_1 is 0.12–0.35 μm in earlywood spruce tracheids (Jayme and Fengel, 1961), 0.46–0.62 μm in latewood tracheids of *Pinus densiflora* (Saiki, 1970), and ~0.51 μm in wood fibers of *Fagus crenata* (Harada, 1964).

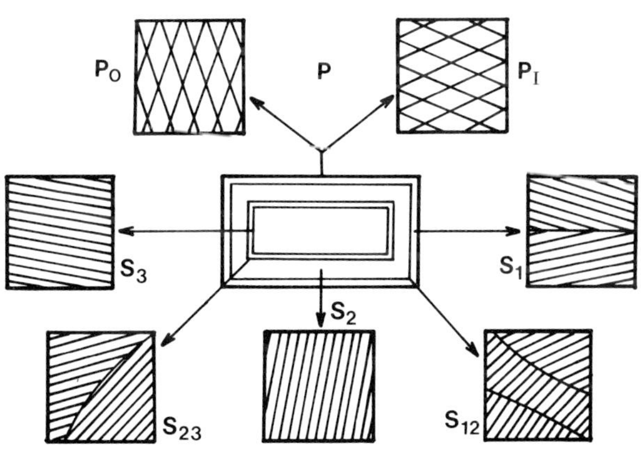

Fig. 17. Schematic diagram of the microfibril orientation in the primary wall and different layers of the secondary wall from tracheids and fibers: P_O, P_I, outer and inner surfaces in the primary wall; S_{12}, S_{23}, intermediate layers between S_1 and S_2 and between S_2 and S_3, respectively. Modified from the model of Wardrop and Harada (1965).

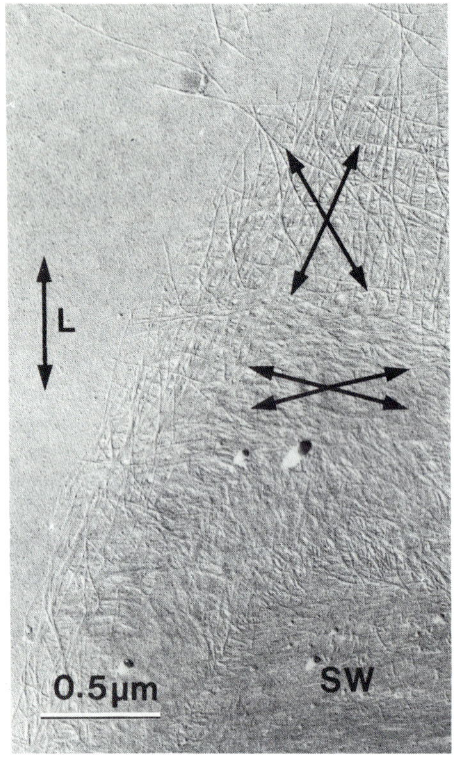

Fig. 18. Transmission electron micrograph of ultrathin longitudinal oblique section of a wood fiber from *Populus koreana* (acidified $NaClO_2$–5% KOH, shadowing), showing the microfibril orientation in the primary wall indicated by crossed arrows. Note that the microfibril orientation is a rather flat helix in the inner surface and rather steep in the outer surface. SW, Secondary wall; L, wood fiber axis. From Fujii (1981).

Table III

Microfibril Angle of the Secondary Wall Layers in the Tracheid[a]

	Earlywood			Latewood		
Wood species	S_1	S_2	S_3	S_1	S_2	S_3
Pinus densiflora (45 yr)						
Radial wall	71	26.5	83	77	8.0	79
Tangential wall	68	20	83	75	5.5	83
Cryptomeria japonica (55 yr)						
Radial wall	68	14	81	83	3.5	85
Tangential wall	64	3.5	85	80	3.0	82

[a] From Saiki (1970). Microfibril angles are given in degrees.

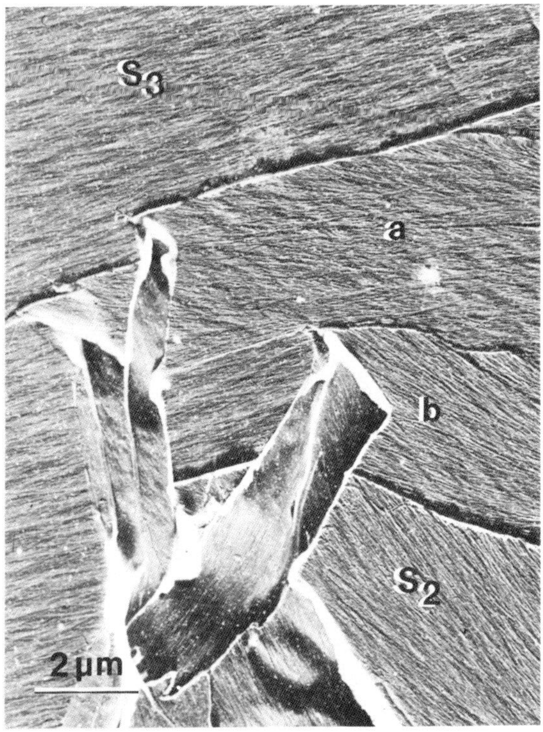

Fig. 19. Transmission electron micrograph of a replica of a torn earlywood tracheid from *Picea jezoensis*, showing the intermediate layer (a and b) between the S_2 and S_3. From Harada *et al.* (1958).

Although the S_2 exhibits a microfibrillar orientation with steep helices, there are transition lamellae on its inner and outer faces. The several lamellae in these regions show a gradual shift of microfibrillar angles between S_1 and S_2 and between S_2 and S_3 (Harada *et al.*, 1958; Harada, 1965) (Fig. 19). The gradual shift of microfibril angles is more abrupt between S_2 and S_3 than between S_1 and S_2. The transition lamellae in the secondary wall are not detected in electron micrographs of ultrathin sections, since this lamella is relatively thin compared with the S_1 and S_2.

The middle layer of the secondary wall has a thickness of 1.66 μm in earlywood tracheids and of 6.94 μm in latewood tracheids of *Pinus densiflora* (Saiki, 1970). Therefore the S_2 contributes most to the bulk of the cell wall material, as well as to its physical properties. It is a compact region in which a high degree of parallelism of the microfibrils exists, except in the transition lamellae. The number of lamellae composing the S_2 may vary from 30 or 40 in thin-walled or earlywood cells to 150 or more in latewood elements. However, Chafe (1974) reported that in the S_2 cellulose microfibrils are distributed in nonlamellate fashion, since the lamellation at the level of the microfibril may be due to a misinterpretation of structure observed in sectioned material.

The S_3 is a thin layer of flat helices of microfibril orientation like the S_1. It is estimated that this layer is 0.10 μm thick both in softwood tracheids and in hardwood fibers (Saiki, 1970; Harada, 1965). As opposed to the highly oriented S_2, the S_3 is loose-textured. The S_3, birefringent to a somewhat lesser degree than the S_1, is poorly developed. The enzymatically produced cavities made by soft-rot fungi in the S_2 layer are oriented at an angle to the cell axis (Fig. 20) and provide evidence for the orientation of cellulose microfibrils in that layer. This feature was recognized by Bailey and Vestal (1937). The microfibrillar orientation is a Z-helix in the S_2 and an S-helix in the S_3, although the S_1 is the crisscrossed arrangement of S- and Z-helices. Preston (1974) suggests that the structure with various microfibrillar angles in the secondary wall passes through only one cycle, but this may be the brief duration of wall thickening in higher plant cell walls compared with that in the algae.

2. Vessel Elements

Microfibrillar orientation in the walls of specialized cells such as vessel elements cannot be described quite as simply as for coniferous tracheids and hardwood fibers. The thin primary wall (P) in general has microfibrillar orientation similar to that of tracheids and wood fibers. Kishi *et al.* (1981a) report that the microfibrils in the primary wall (P) extend straight and are arranged parallel to one another, and that the layer consists of three parts, P-outer, P-middle, and P-inner, each showing different microfibrillar orientation. The microfibrils are oriented transversely with respect to the vessel axis in the P-outer and are oriented at random in the P-middle. The P-inner consists of crossed polylamellate structure. Kishi *et al.* (1977, 1979) report from the examination of vessel elements from about 30 species of Japanese hardwoods with polarizing and electron microscopy that the layered structure of the secondary wall can be classified into three categories: the typical three-layered structure, the undetectable layering structure, and the multilayered structure (Fig. 21). The typical three-layered structure consists of S_1, S_2, and S_3 similar to those of tracheids and wood fibers, although the S_1 and S_3 are thicker than those of tracheids and wood fibers. This type appears in the vessel walls of species such as *Populus, Juglans, Alnus, Betula, Fagus, Prunus, Acer,* and *Tilia*. The undetectable or unlayered structure has only microfibrils with the orientation of a flat helix, but there are some variations in microfibrillar angles in the secondary walls. This type is observed in the earlywood vessel walls

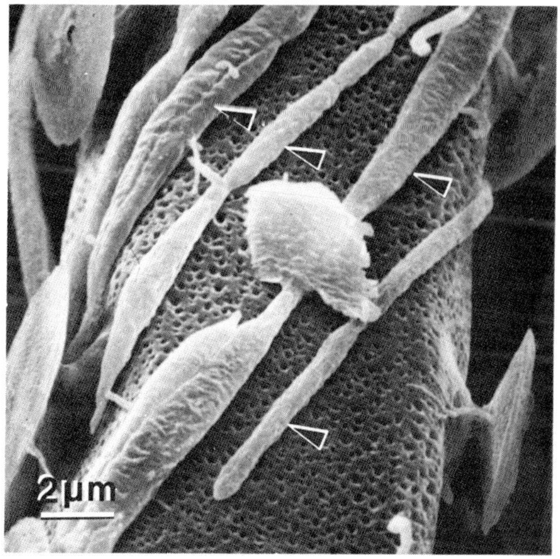

Fig. 20. Scanning electron micrograph of the stylen-molding of enzymatically produced cavities made by soft-rot fungi in the S_2 layer of an earlywood tracheid from *Thuja orientalis*, showing the cavities running helically around the tracheid axis. From Okumura *et al.* (1976).

of species in genera such as *Castanea, Quercus, Ulmus, Zelkova,* and *Morus*. The multilayered structure has more than four layers in which microfibril angles to vessel axis change and contain the so-called bow-shaped pattern (Bouligand, 1972; Yamanaka and Harada, 1968). This type is seen in species of *Cinnamomum, Pterocarya,* and *Machilus,* and in the latewood of *Firmiana* and *Fraxinus*.

The bow-shaped pattern can be seen in electron micrographs of transverse sections of the cell walls and results from the progressive changes of microfibrillar orientation in the wall between 0 and 90°.

3. Parenchyma Cells

It is evident from recent studies that ray and axial parenchyma cells both in softwoods and in hardwoods have variations or complexities in their wall structure, structures that are not observed in the cell wall of tracheids and wood fibers.

In the softwoods, the cell wall organization of the ray parenchyma cells is fundamentally classified into two types: the first type consists of the primary wall and the protective layer, and the second type consists of the primary wall, the secondary wall, and the protective layer (Fujikawa and Ishida, 1975; Harada *et al.*, 1976).

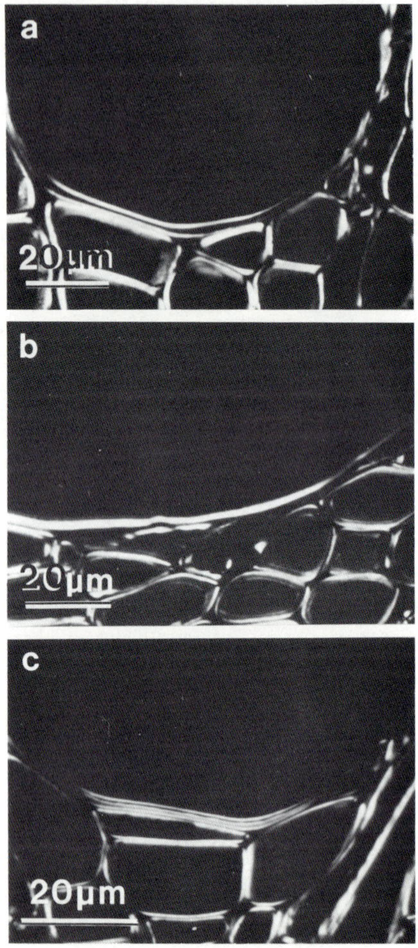

Fig. 21. Polarized-light photomicrographs of transverse oblique sections. **a,** From *Ostrya japonica*. **b,** From *Morus bombycis*. **c,** From *Pterocarya rhoifolia*. Note the different layered structure in the secondary wall of hardwood vessels. Note that **a, b,** and **c** are the three-layered, undetectable layering, and multilayered structures, respectively. From Kishi *et al.* (1977).

The primary wall is composed of P_1 with microfibrils of almost parallel orientation to the ray cell axis, of P_2 with the network appearance of microfibrils, and of P_3 with several crossed lamellae at microfibrillar angles of 30–60° (Fig. 22). The secondary wall organization is very similar to that of softwood tracheids. Fujikawa and Ishida (1975) report that the ray parenchyma cells in the Taxaceae, Podocarpaceae, Cephalotaxaceae, and

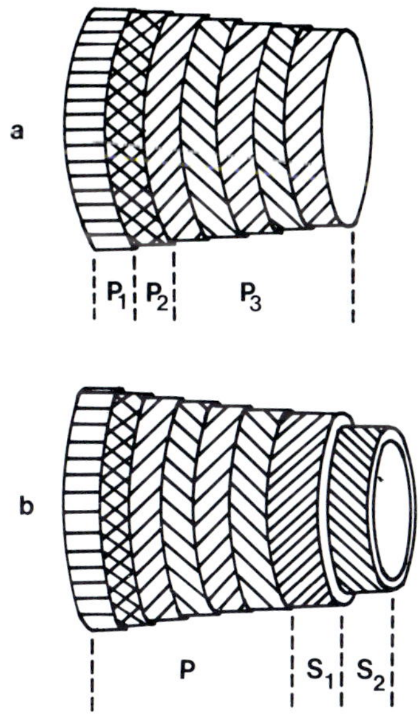

Fig. 22. Schematic diagram of the microfibril orientation in the cell wall of softwood ray parenchyma cell. Note that **a** and **b** are the first and the second types, respectively. (The protective layer is omitted in the diagram.) From Fujikawa and Ishida (1975).

Cupressaceae belong to the first type, and those in *Abies, Pseudotsuga, Tsuga, Picea, Larix,* and haploxylon of *Pinus* are of the second type.

However, the diploxylon of *Pinus* differs in that the walls develop in two stages; that is, the primary wall and inner protective layer form in the sapwood, and, just before the heartwood forms, the secondary wall and the protective layer are deposited. In the axial parenchyma cells of softwood, the cell wall organization is very similar to that of ray parenchyma cells, except that microfibrils are arranged in a flat helix with respect to the cell axis in the P_1.

In hardwoods, the primary wall of ray parenchyma cells have the so-called polylamellated structure proposed by Chafe and Chauret (1974). The microfibrils are organized in the form of an oriented mesh with an angle of 45–60° to the cell axis on the inner part. They shift toward axial orientation moving outward, and are nearly parallel to the cell axis on the outer surface of the wall (Fig. 23). In contrast, in the primary wall of axial parenchyma cells, microfibrils are more transversely oriented on the outer part of the wall than on the inner part.

A characteristic feature of parenchyma cell walls in hardwoods is the existence of an amorphous layer in the secondary wall, as proposed by Fujii *et al.* (1981). The amorphous layer corresponds to the isotropic layer

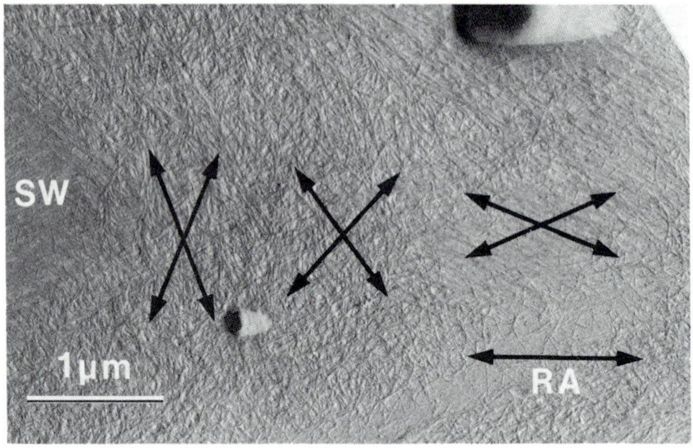

Fig. 23. Transmission electron micrograph of ultrathin radial oblique section of the ray parenchyma cell from *Castanea crenata* (delignification with 5% KOH, shadowing), showing the microfibril orientation in the primary wall indicated by crossed arrows. Note that the microfibril orientation changes from flat to steep toward the outer part of the cell wall (right). SW, Secondary wall; RA, radial axis. From Fujii (1981).

or the protective layer pointed out by Chafe and Chauret (1974). The amorphous layer is rich in hemicellulose and contains some pectic substances, along with some scattered cellulose microfibrils, and is well lignified, although this layer is poor in lignin at the first stage of cell wall development.

Fujii *et al.* (1980, 1981) report, from the examination of ray and axial parenchyma cell wall in about 50 species of Japanese hardwoods, that the secondary wall is composed of a lignified cellulosic layer (CL) and an amorphous layer (AL) and that the cell wall structure can be classified into three types according to the presence and organization of these two kinds of layers: 3CL type, 3CL + AL type, and 3CL + AL + ICL type (Fig. 24), where ICL refers to the inner cellulosic layer, which is located inside the amorphous layer. The 3CL-type wall structure may be considered as the standard structure of parenchyma cells in hardwoods, while the 3CL + AL-type wall structure occurs in cells that have extensive pit contact with vessels.

C. Distribution of Chemical Constituents in the Cell Wall

The chemical constituents of the cell wall are not uniformly distributed within it (see Chapters 2 and 3, this volume). Quantitative data on chemi-

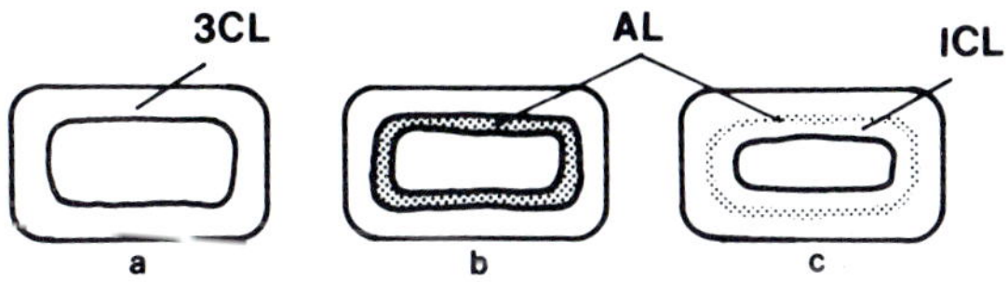

Fig. 24. Schematic diagram of the cell wall organization of hardwood ray parenchyma cell, showing the three types of earlywood structure. **a,** 3CL type. **b,** 3CL + AL type. **c,** 3CL + AL + ICL type. (CL, Lignified cellulosic layer; AL, amorphous layer; ICL, inner lignified cellulosic layer.) From Fujii *et al.* (1980).

cal constituents for each cell wall layer are limited because of difficulty in separating the various cell wall layers for analysis. Therefore, techniques utilizing light and electron microscopy are important analytical approaches.

1. Lignin

There are three techniques available for investigating lignin distribution in the wood cell wall. Details for each technique may be found in the indicated references: (1) transmission electron microscopy (TEM) observation of the lignin skeleton that is left after removal of polysaccharides with HF acid (Côté *et al.,* 1966; Côté, 1967, 1977); (2) ultraviolet absorption spectroscopy using ultraviolet (UV) microscopy (Fergus *et al.,* 1969; Fergus and Goring, 1970); and (3) the analysis of brominated lignin with scanning electron microscopy (SEM) and energy-dispersive X-ray analysis (EDXA) (Saka and Thomas, 1982; Saka *et al.,* 1982). An example obtained by the first technique is shown in Fig. 25.

In softwood tracheids, lignin is located at a higher packing density in the intercellular layer and primary wall than in the secondary wall. The S_1, S_2, and S_3 are lignified to approximately the same extent. Lignin distribution in the cell wall of softwood parenchyma cells and in wood fibers, vessel elements, and parenchyma cells in hardwoods is similar to that in softwood tracheids. The secondary wall of vessel elements is highly lignified compared with wood fibers, much like that in softwood tracheids.

Fergus *et al.* (1969) and Fergus and Goring (1970) reported on the lignin distribution of *Picea mariana* and *Betula papyrifera* following the study of thin transverse sections with a UV microscope. The average lignin concentration in the compound middle lamella of earlywood and latewood tracheids of *Picea mariana* was about twice that in the secondary wall. The lignin concentration of the intercellular layer at the cell corners of adjacent tracheids is nearly four times that in the secondary wall, but the

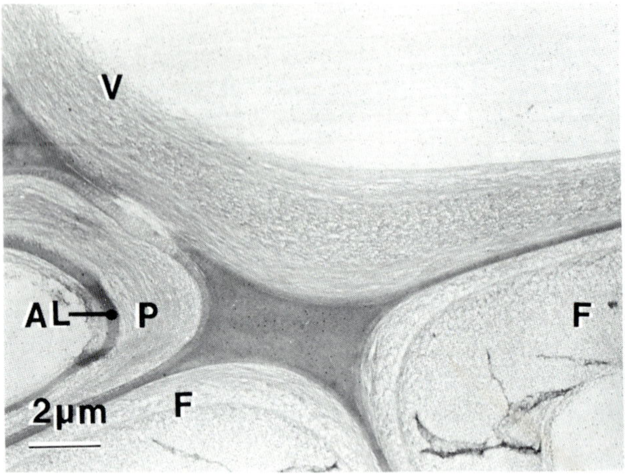

Fig. 25. Transmission electron micrograph of ultrathin transverse section of *Acer mono,* which was skeletonized using HF acid, showing the lignin distribution in the individual cell walls of vessel (V), wood fiber (F) and axial parenchyma cell (P), respectively. (AL, Amorphous layer.) From Kishi *et al.* (1982).

volume of the secondary wall is much greater than the volume of the intercellular layer. Thus, for earlywood, 72% of the total lignin is in the secondary wall, leaving only 28% in the compound middle lamella and cell corner intercellular layer regions. The corresponding values for latewood are 82 and 18%, respectively. The lignin concentration in the intercellular layer is about twice that in the secondary wall.

For the cells of *Betula papyrifera,* the lignin contents of wood fibers and vessels are 0.19–0.22 g/g and 0.24–0.28 g/g, respectively. However, because of their greater abundance, wood fibers contain 77% of the total lignin. The wood fiber secondary wall is the dominant morphological region of the xylem tissue. It contains lignin at 0.16–0.19 g/g and accounts for 77% of lignin in wood fibers and 60% of lignin in the whole wood. The total of wood fiber, vessel, and ray-cell secondary walls contains over 81% of lignin, less than 19% being in the primary wall and intercellular layer. The weight concentration of lignin in the wood fiber intercellular layer is high, ranging from 0.72 to 0.85 g/g (see Chapter 3).

2. Cellulose and Hemicelluloses

The most detailed analysis of cellulose and hemicelluloses in wood cell walls was published by Meier (1964). He used specimens of *Pinus sylvestris* and *Betula verrucosa* in a very demanding microtechnique and analyzed the constituent sugars in different cell wall layers of the woods.

Arabinan is almost completely confined to M (middle lamella) and P (primary wall), and galactan to M + P + S_1. On the other hand, glucomannan either increases (primary wall) from M + P to S_3 (softwoods) or remains approximately constant at a low level (hardwoods), while glucuronoxylan has a higher concentration in S_1, S_2, and S_3 than in M + P (hardwoods). Cellulose content is greater throughout in the secondary wall layers than in the primary wall (see Chapter 2, Section III).

D. *Sculpturing of the Cell Wall*

While cellulosic fibers of plants such as cotton are relatively simple smooth-walled composites of lamellae, in wood the cell walls are almost invariably interrupted by gaps (pits) and by thickenings or other sculptured features. In fact, microfibrillar orientation is not easily measured because the microfibrils often follow streamlined paths around wall cavities or anastamose into complex structures.

1. Pit Structure

a. Softwood Bordered Pit: Pit Border. Between the crossed Nicols of a polarizing microscope, the pit border of earlywood tracheids appears as shown as in Fig. 26. When it is observed using the sensitive

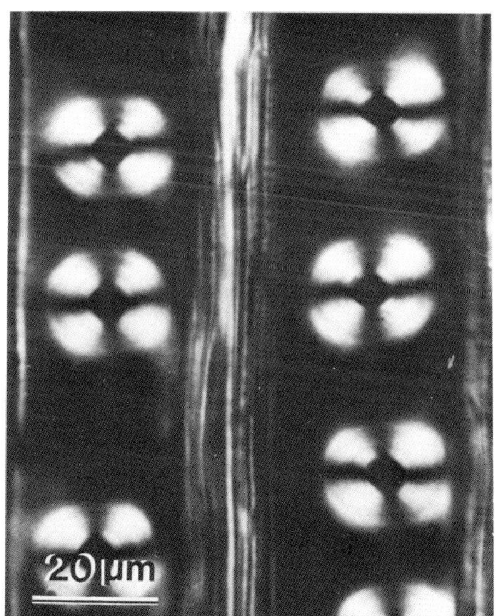

Fig. 26. Polarized-light photomicrograph of radial section from the earlywood of *Pinus densiflora,* showing the circular microfibril orientation of the bordered pit region in the double cell wall between crossed Nicols.

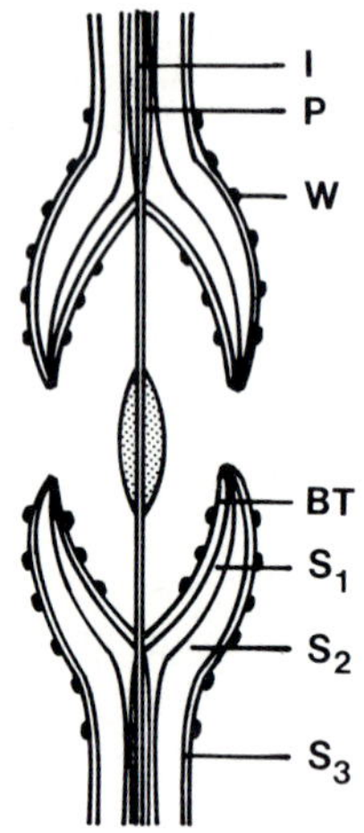

Fig. 27. Schematic diagram of a bordered pit pair in softwood tracheids. BT, Border thickening; W, warty layer. From Harada and Côté (1967).

test plate in addition to the observation between crossed Nicols, the pit border shows a Maltese cross separating sectors, with alternations of addition and subtraction colors. It is clear that cellulose microfibrils of the pit border must therefore circle around the border. From examination by the major extinction position, the orientation of iodine crystals deposited within the border, and replicas examined via TEM, the pit border region of the cell wall is composed of the border thickening (BT), S_1, S_2, and S_3 from the outer part of the cell wall (Wardrop and Dadswell, 1957; Harada and Côté, 1967). The presence of the BT and the thicker S_1 are features of the pit border wall that differ from those of the unpitted wall area (Fig. 27). The microfibrils circle only at the BT, but sweep around the pit at the individual layers, S_1, S_2, and S_3.

b. Pit Membrane. In softwood pit pairs, the pit membranes are evidently specialized to operate in a valve-like manner (Fig. 28). Many species have a thickening added to the center of the pit membrane. This is disk-like and is called a torus. The torus is suspended from fine cellulosic strands that form an open membrane or margo around the torus. Liquid flow is possible from tracheid to tracheid through this membrane, provided that the torus is not pressed against one aperture or the other. The margo consists of an open net of radially oriented microfibril bundles superimposed on an unoriented primary-wall network, and it extends from the torus to the pit border.

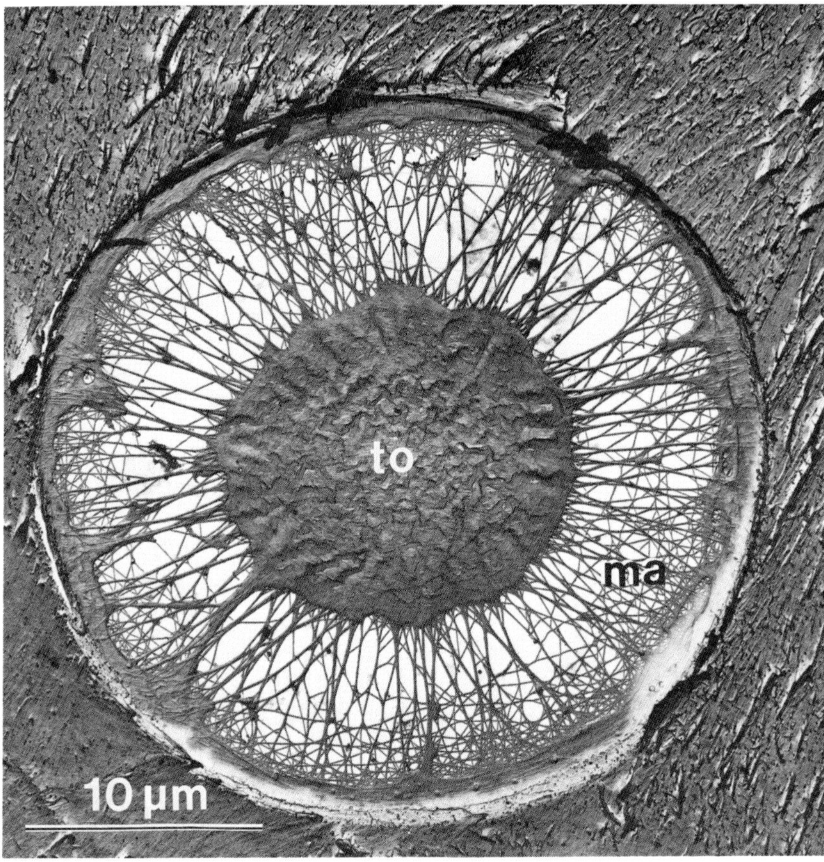

Fig. 28. Transmission electron micrograph of radial surface replica from *Tsuga canadensis,* showing a bordered pit membrane in earlywood tracheid. Note the openings in the margo (ma) and the disk-like torus (to). From Côté (1977).

The torus of softwood tracheids is usually convex lens-shaped in cross section and fairly circular in outline. The torus is usually thickened by the deposition of well-oriented cellulose microfibrils in addition to the primary wall microfibrils, and the outline is irregularly extended over the margo in some species. On the other hand, the torus is little thickened in other cases. Harada (1964) called the thickened torus and the unthickened torus the pine type and the cryptomeria type, respectively (Fig. 29). The former type is found in species of the families Pinaceae and Sciadopitaceae, and the latter type involves species of Ginkgoaceae, Taxaceae,

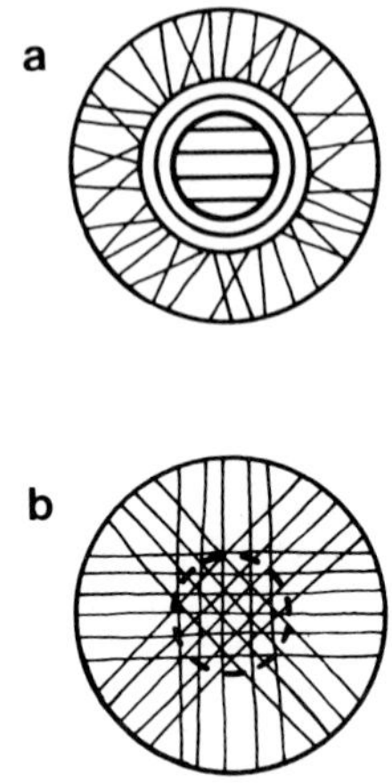

Fig. 29. Schematic diagrams of the bordered pit membrane structure of earlywood tracheids in softwood. **a,** The thickened torus (the pine type). **b,** The unthickened torus (the cryptomeria type).

Chephalotaxaceae, Cupressaceae, Podocarpaceae, and Araucareaceae. According to Harada *et al.* (1958), the margo of pit membranes is classified into two types by whether the densely packed microfibrils are present. The first type of margo has more small openings than that of the second one; that is, the former type consists of more densely packed microfibrils, radially arranged and unoriented like the primary wall, than does the latter. The first type is named the dense margo and generally exists in the earlywood tracheids of the species with the cryptomeria-type torus and in the latewood tracheids and the tracheids to ray tracheids. The second type is called the rough margo and is found in the earlywood tracheids of the species with the pine-type torus. The openings of margo are occluded with heartwood extractives or are encrusted with lignin-like materials.

The formation of the pit membrane, particularly that of the openings in the margo, is described by Imamura and Harada (1973). The microfibrillar structure of the margo and torus is detected in the later stages of primary wall formation. The margo and torus have the same microfibrillar structure as in mature pits, except for being embedded with hemicellulose even at this point in this differentiating stage. The embedded hemicellulose is removed by enzyme action immediately after the secondary wall formation is completed in the tracheid, although the torus is made resistant to enzyme action by the abundance of pectin materials in the development stage.

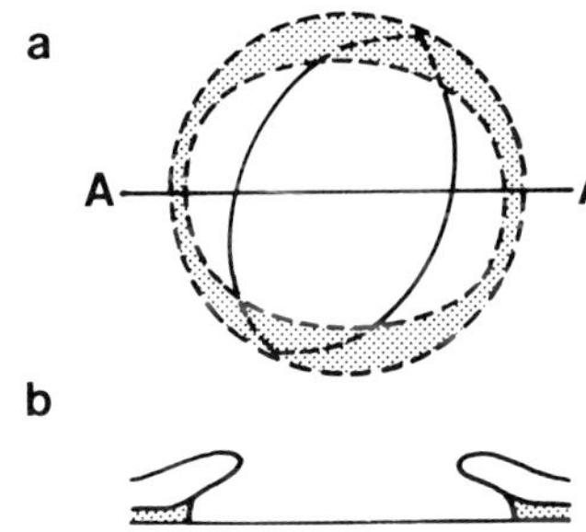

Fig. 30. Schematic diagram of the cell-wall organization of pit border region in a half-bordered pit pair between a tracheid and a ray parenchyma cell. **a,** Surface view. **b,** Transverse sectional view cut through the plane A–A. The dotted area is created by the S_1, S_2, and S_3 layers, but the empty area except for the pit aperture is formed only by the S_2 and S_3 layers. From Furukawa (1978).

c. Softwood Half-Bordered Pit and Simple Pit. The cell wall organization in the pit-border region of the cross-field pits between tracheids and ray parenchyma cells differs from that of the intertracheid pits with respect to the relative thickness and position of each layer found in the pit border. Furukawa (1978) reports that, in *Cryptomeria japonica,* the S_1 is located on both sides of the pit annulus but not in the overhanging border (Fig. 30). On the other hand, the overhanging border region consists mainly of S_2. Therefore, the shape of the pit aperture is determined by the deposition of microfibrils of the S_2.

The pit membrane structure of a half-bordered pit pair between tracheids and ray parenchyma cells is represented diagrammatically in Fig. 31 (Harada *et al.,* 1976). The pit membrane is quite thick. There is no torus in the center of the pit membrane, and no openings can be seen even at high magnification with TEM. Radiating microfibrils are recognized on the tracheid lumen surface of the pit membrane. On the ray parenchyma cell side of the pit membrane, four successive layers with different textures are present, including several lamellae alternating in their criss-crossed microfibril orientations. The membrane structure between tracheids and axial parenchyma cells is similar to that of the cross-field pits of ray parenchyma cells, except for transversely oriented microfibrils in the P_1 layer, as shown in Fig. 31. The remarkable feature of the membrane structure of simple pit pairs in the interparenchymatous pits is the presence of plasmodesmatal pores (Fig. 32).

d. Hardwood Pit. The cell wall of pit border consists of BT, P, S_1, S_2, and S_3 in tracheids and fiber tracheids of hardwood, like that of softwood tracheids. However, the pit border of vessels has a complicated

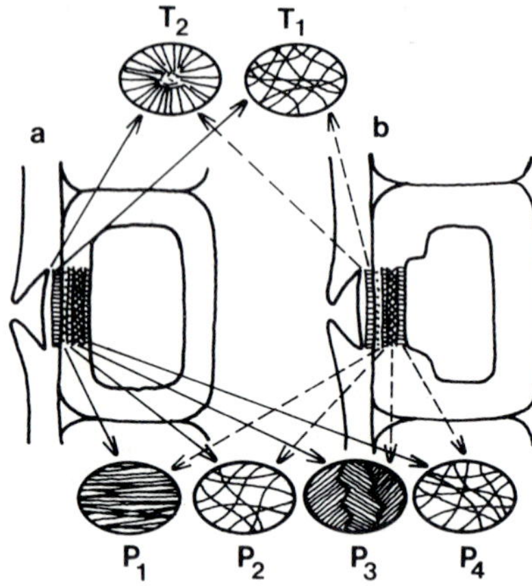

Fig. 31. Schematic diagram of the cell-wall layers of the crossed-field pit membrane between a tracheid and a ray parenchyma cell in softwood. **a,** Thin-walled ray cell. **b,** Thick-walled ray cell. From Harada *et al.* (1976). Copyright © 1976. Reprinted from *Applied Polymer Symposia,* No. 28, by permission of John Wiley & Sons, Inc.

structure, because it lacks not only BT but also S_1 in some part of the pit border region (Kishi *et al.,* 1981b). The pit membrane of the bordered, half-bordered, and simple pit pairs in hardwoods is a uniform structure exhibiting the primary wall texture, and there is usually no evidence of a torus or of radiating strands of microfibrils (Fig. 33). However, Ohtani and Ishida (1978) reported the existence of a torus in the intervessel pit

Fig. 32. Transmission electron micrograph of the inner surface replica of a transverse wall in a mature ray parenchyma cell from *Pinus densiflora,* showing the plasmodesmatal pores in a simple pit membrane. From Harada *et al.* (1976). Copyright © 1976. Reprinted from *Applied Polymer Symposia,* No. 28, by permission of John Wiley & Sons, Inc.

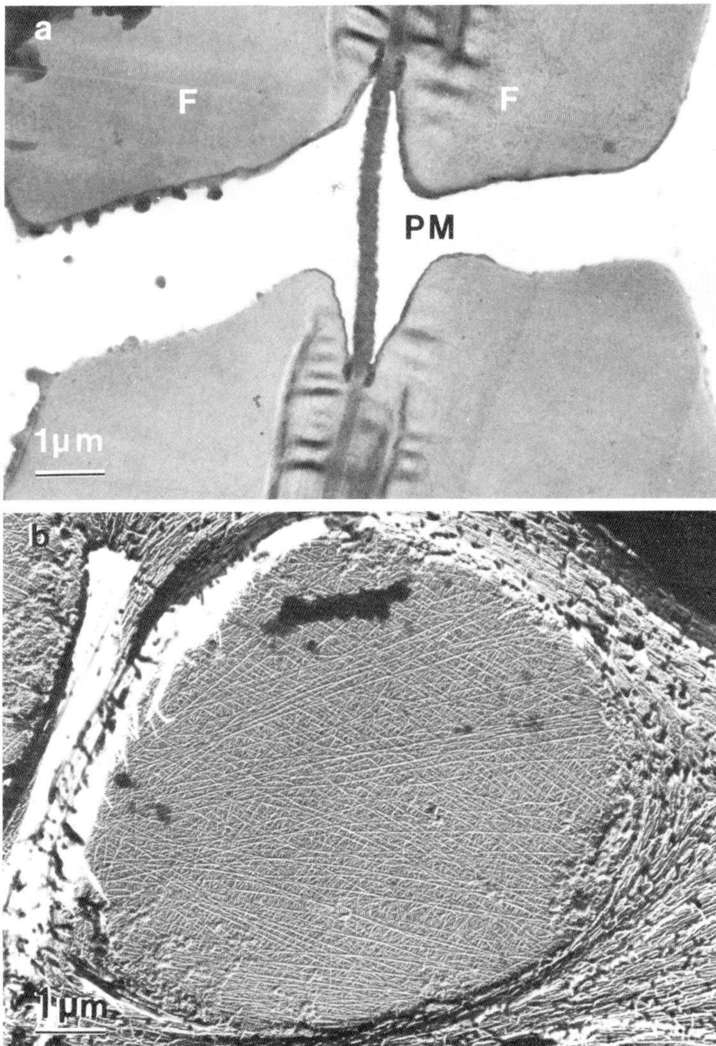

Fig. 33. Transmission electron micrograph of ultrathin transverse section (**a**) from *Fagus crenata* and split surface replica (**b**) from *Populus euramericana,* showing the pit membrane structure of intertracheid (**a**) and intervessel (**b**) pitting. F, Fiber tracheids; PM, pit membrane.

membrane of several species of hardwoods. The pit membrane of simple pit pairs has plasmodesmatal pores, as shown in Fig. 32.

2. Wart Structure and Vestured Pit

The warty layer is one of the major structural features of wood cells found by electron microscopy (Kobayashi and Utsumi, 1951). It was first found in softwood tracheids and later in the tracheids, vessels, and wood fibers of hardwoods (Fig. 34). The wart measures from 0.05 to 0.5 μm in diameter both in softwood tracheids and in vessels and wood fibers. Warts usually have a rounded cone shape but may be branched in hardwood vessels. The wart is larger in the cell corners than in the other parts of a softwood tracheid. The major chemical constituents of warts are reported to be lignin and hemicellulose, according to examination by component-removal treatments of ultrathin wood sections (Mori *et al.*, 1983). The warts are believed to arise from the extra wall materials and the remains of cytoplasm that are deposited on the S_3 layer through the plasma membrane. The warty layer is continuous over the inner surface of the cell wall and interior surface of the pit chamber.

The warty layer does not exist in all species of softwoods and hardwoods. Harada *et al.* (1958) reported that the warty layer of softwood tracheids exists in the species of *Abies, Agathis, Cedrus, Chamaecyparis, Cryptomeria, Cunninghamia, Ginkgo, Juniperus, Keteleria, Sciadopitys, Sequoia, Taxodium, Thuja, Thujopsis, Tsuga,* and diploxylon of *Pinus*. In addition, Ohtani (1979) reports that the warty layer of hardwood vessels is

Fig. 34. Transmission electron micrograph of ultrathin transverse section from *Fagus crenata*, showing the warty layer of a fiber tracheid.

present in the species of *Acacia, Acer, Albizia,* and *Fagus,* and that it is present in the genus *Fagus* in hardwood fiber tracheids.

In hardwood, the pit chamber and pit apertures that are decorated by outgrowths of wall material are known as vestured pits. The outgrowths of vestured pits are chemically constructed of lignin, hemicellulose, and a little pectin, and they do not contain cellulose different from the secondary wall. The shape and size of the outgrowths of vestured pits are variable. Ohtani and Ishida (1976) classified 15 types, depending on the shape, from a simple to a complicated one. The development of vestured pit outgrowths is regarded to be similar to that of the warts (Côté and Day, 1962).

References

Bailey, I. W., and Vestal, M. R. (1937). *J. Arnold Arbor.* **18,** 261–272.
Balashov, V., and Preston, R. D. (1955). *Nature (London)* **176,** 64–65.
Bentum, A. L. K., Côté, W. A., and Timell, T. E. (1969). *Wood Sci. Technol.* **3,** 218–231.
Bouligand, Y. (1972). *Tissue Cell* **4,** 189–217.
Chafe, S. C. (1974). *Protoplasma* **79,** 145–158.
Chafe, S. C., and Chauret, G. (1974). *Protoplasma* **80,** 129–147.
Côté, W. A. (1967). "Wood Ultrastructure. An Atlas of Electron Micrographs." Washington Univ. Press, Seattle and London.
Côté, W. A. (1977). *In* "Recent Advances in Phytochemistry" (F. A. Loewus and V. C. Runeckles, eds.), Vol. XI, pp. 1–44. Plenum, New York.
Côté, W. A. (Ed.) (1980). "Papermaking Fibers," pp. 22, 24, 25, 31. Syracuse Univ. Press, Syracuse, New York.
Côté, W. A., and Day, A. C. (1962). *Tappi* **45,** 906–910.
Côté, W. A., Timell, T. E., and Zabel, R. A. (1966). *Holz Roh-Werkst.* **24,** 432–438.
Fengel, D. (1970). *Tappi* **53,** 497–503.
Fengel, D., and Wegener, G. (1984). "Wood—Chemistry, Ultrastructure, Reactions," Fig. 8-7, p. 237. de Gruyter, Berlin.
Fergus, B. J., and Goring, D. A. I. (1970). *Holzforschung* **28,** 118–124.
Fergus, B. J., Procter, A. R., Scott, J. A. N., and Goring, D. A. I. (1969). *Wood Sci. Technol.* **3,** 117–138.
Fujii, T. (1981). "Cell Wall Structure of Xylem Parenchyma in Angiosperm Species." Ph. D. Thesis, Kyoto University.
Fujii, T., Harada, H., and Saiki, H. (1980). *Mokuzai Gakkaishi* **26,** 373–380.
Fujii, T., Harada, H., and Saiki, H. (1981). *Mokuzai Gakkaishi* **27,** 149–156.
Fujikawa, S., and Ishida, S. (1975). *Mokuzai Gakkaishi* **21,** 445–456.
Furukawa, I. (1978). *Bull. Fac. Agric. Tottori Univ.* **30,** 171–179.
Goto, T., Harada, H., and Saiki, H. (1973). *Mokuzai Gakkaishi* **19,** 463–468.
Goto, T., Harada, H., and Saiki, H. (1975). *Mokuzai Gakkaishi* **21,** 537–542.
Harada, H. (1962). *Mokuzai Gakkaishi* **8,** 252–258.
Harada, H. (1964). *Mokuzai Gakkaishi* **10,** 221–225.
Harada, H. (1965). *In* "Cellular Ultrastructure of Woody Plants" (W. A. Côté, Jr., ed.), pp. 215–233. Syracuse Univ. Press, Syracuse, New York.
Harada, H., and Côté, W. A. (1967). *Holzforschung* **21,** 81–85.
Harada, H., and Goto, T. (1982). *In* "Cellulose and Other Natural Polymer Systems" (R. M. Brown, Jr., ed.), pp. 383–401. Plenum, New York.

Harada, H., Imamura, Y., and Saiki, H. (1976). *Appl. Polym. Symp.* **28,** 1239–1253.
Harada, H., Miyazaki, Y., and Wakashima, T. (1958). *Bull. Govt. For. Exp. Stn.* **104,** 1–115.
Imamura, Y., and Harada, H. (1973). *Wood Sci. Technol.* **7,** 189–205.
Imamura, Y., Harada, H., and Saiki, H. (1972). *Bull. Kyoto Univ. For.* **44,** 183–193.
Itoh, T. (1975). *Bot. Mag. Tokyo* **88,** 145–156.
Jayme, G., and Fengel, D. (1961). *Holz Roh-Werkst.* **19,** 50–55.
Kerr, T., and Bailey, I. W. (1934). *J. Arnold Arbor.* **15,** 327–349.
Kerr, A. J., and Goring, D. A. I. (1975). *Cellul. Chem. Technol.* **9,** 563–573.
Kishi, K., Harada, H., and Saiki, H. (1977). *Bull. Kyoto Univ. For.* **49,** 122–126.
Kishi, K., Harada, H., and Saiki, H. (1979). *Mokuzai Gakkaishi* **25,** 521–527.
Kishi, K., Harada, H., and Saiki, H. (1981a). *J. Soc. Mater. Sci. Jpn.* **30,** 673–678.
Kishi, K., Harada, H., and Saiki, H. (1981b). *Mokuzai Gakkaishi* **27,** 343–349.
Kishi, K., Harada, H., and Saiki, H. (1982). *Bull. Kyoto Univ. For.* **54,** 209–216.
Kobayashi, K., and Utsumi, N. (1951). Unpublished.
Meier, H. (1964). *In* "The Formation of Wood in Forest Trees" (M. H. Zimmermann, ed.), pp. 137–151. Academic Press, New York.
Meyer, K. H., and Misch, L. (1937). *Helv. Chim. Acta* **20,** 232–244.
Mori, N., Fujita, M., Saiki, H., and Harada, H. (1983). *Bull. Kyoto Univ. For.* **55,** 299–306.
Mühlethaler, K. (1960). *Beih. Z. Schweiz. Forstv.* **30,** 55–65.
Mühlethaler, K. (1965). *In* "Cellular Ultrastructure of Woody Plants" (W. A. Côté, Jr., ed.), pp. 191–198. Syracuse Univ. Press, Syracuse, New York.
Ohtani, J. (1979). *Bull. College Exp. For. Hokkaido Univ.* **36,** 585–607.
Ohtani, J., and Ishida, S. (1976). *Res. Bull. Coll. Exp. For. Hokkaido Univ.* **33,** 407–435.
Ohtani, J., and Ishida, S. (1978). *Mokuzai Gakkaishi* **24,** 673–675.
Okumura, S., Harada, H., and Saiki, H. (1976). Unpublished.
Onaka, F. (1936). *Ringaku Kaishi* **18,** 588–602.
Panshin, A. J., and De Zeeuw, C. (1980). "Textbook of Wood Technology," 4th Ed., pp. 134, 161–164, 170–172, 181, 197. McGraw-Hill, New York.
Preston, R. D. (1974). "The Physical Biology of Plant Cell Walls," p. 302. Chapman and Hall, London.
Revol, J. F. (1982). *Carbohydr. Polym.* **2,** 123–134.
Revol, J. F., and Goring, D. A. I. (1982). *Wood Sci.* **14,** 120–126.
Roelofsen, P. A. (1951). *Biochim. Biophys. Acta* **7,** 43–53.
Roland, J. C., Vian, B., and Reis, D. (1975). *J. Cell. Sci.* **19,** 239–259.
Saiki, H. (1970). *Mokuzai Gakkaishi* **16,** 244–249.
Saka, S., and Thomas, R. J. (1982). *Wood Sci. Technol.* **16,** 167–179.
Saka, S., Whiting, P., Fukazawa, K., and Goring, D. A. I. (1982). *Wood Sci. Technol.* **16,** 269–277.
Shimaji, K. (1976). *In* "Mokuzai-no Soshiki" (K. Shimaji, S. Sudo, and H. Harada, eds.), p. 35. Morikita Shuppan, Tokyo.
Sugiyama, J., and Harada, H. (1982). Unpublished.
Sugiyama, J., and Harada, H. (1983). Unpublished.
Sugiyama, J., Harada, H., Fujiyoshi, Y., and Uyeda, N. (1984). *Mokuzai Gakkaishi* **30,** 98–99.
Wardrop, A. B. (1958). *Aust. J. Bot.* **6,** 299–305.
Wardrop, A. B. (1964a). *In* "The Formation of Wood in Forest Trees" (M. H. Zimmermann, ed.), pp. 87–88. Academic Press, New York.
Wardrop, A. B. (1964b). *Holzforschung* **11,** 102–110.
Wardrop, A. B., and Dadswell, H. E. (1957). *Holzforschung* **11,** 33–41.
Wardrop, A. B., and Harada, H. (1965). *J. Exp. Bot.* **16,** 356–371.
Yamanaka, K., and Harada, H. (1968). *Bull. Kyoto Univ. For.* **40,** 293–300.

CHAPTER **2**

Localization of Polysaccharides in Wood Cell Walls

H. MEIER

Institut de Biologie Végétale et de Phytochimie
Université de Fribourg
Fribourg, Switzerland

I. Introduction

Wood (secondary xylem) from an angiospermous tree consists of three main types of cells: libriform fibers, vessels (trachеary elements) and parenchymatous cells. In softwoods (gymnosperms), there are usually two main types of cells: tracheids and parenchymatous cells (see Chapter 1, Section II).

Only in a very few species has the overall chemical composition of each main cell type been studied in some detail. Studies of the ultrastructural distribution of wall components within the walls have always been restricted to the quantitatively most important wood cells, the libriform fibers and the tracheids.

II. Polysaccharides of Different Cell Types

The usual wood analyses give average values of the different components in all cell types of a wood sample. In most cases, 80–95%[1] of the dry

[1] Although in birch the percentage volume of vessels is about 25% and that of parenchyma cells is about 10%, the percentage weights are only about 10 and 5%, respectively, with regard to the polysaccharides (Perilä, 1962).

Biosynthesis and Biodegradation of Wood Components

ISBN 0-12-347880-4

Table I

Polysaccharide Composition of Different Xylematic Cell Elements Isolated from Holocellulose[a]

	Pinus silvestris L.		*Betula verrucosa* Ehrh.		
Polysaccharide	Vertical tracheids (%)	Parenchyma cells and ray tracheids (%)	Libriform fibers (%)	Vessels (%)	Parenchyma cells (%)
Cellulose	56	50	51	53	14
O-Acetylgalacto-glucomannan	25	20	—	—	—
Glucomannan	—	—	2	—	1
Arabino-4-*O*-methyl-glucuronoxylan	17	28	—	—	—
O-Acetyl-4-*O*-methyl-glucuronoxylan	—	—	46	45	84
Others	2	2	1	2	1

[a] Values calculated from Perilä (1961).

weight of wood is due to one single cell type: the vertical tracheids in coniferous woods, and the libriform fibers in deciduous woods. The global analyses then reflect more or less the composition of the tracheids and fibers, but one should be aware that the composition of other cell types in the sample might be very different. Especially in deciduous woods with a high percentage of vertical and ray parenchyma cells, which have quite a different composition than the libriform fibers and vessels, the overall analyses may give values that are very different from those of any specific cell type in the sample. Table I illustrates the differences between vertical tracheids and parenchyma cells (including ray tracheids) from holocellulose of pine and libriform fibers, vessels, and parenchyma cells from holocellulose of birch. As a general rule, the hemicellulose content in parenchyma cells is higher than in the other cell types. In parenchyma cells of birch, the xylan content is more than 80% of all the polysaccharides present (Table I).

As far as the author is aware, nothing is known about possible differences in the composition of vertical tracheids and ray tracheids in conifers or between vertical and ray parenchyma cells in dicotyledonous trees.

Differences have also been shown to occur in pine between earlywood and latewood (Table II), which differ mainly in having thin and thick S_2 layers, respectively, in their tracheid cell walls. There is more mannan and less xylan in latewood than in earlywood.

Table II
Polysaccharide Composition of Earlywood and Latewood of Pine (*Pinus silvestris* L.)[a]

Polysaccharide	Earlywood (%)	Latewood (%)
Cellulose	56.7	56.2
O-Acetylgalactoglucomannan	21.3	26.0
Arabino-4-*O*-methylglucuronoxylan	18.6	14.1
Others	3.4	3.7

[a] From Meier (1964).

III. Polysaccharides in Different Cell Wall Layers of Tracheids and Libriform Fibers

Several fundamentally different experimental approaches have been used to study the ultrastructural localization of polysaccharides in wood cell walls:

1. The *additive* approach, in which cells at different developmental stages of increasing age are analyzed during the deposition of their wall layers.
2. The *microscopic* approach, in which the cell walls are studied by light or electron microscopy before and after fractional extraction of certain polysaccharides or after contrasting them specifically.
3. The *radioautographic* approach, in which certain polysaccharides in the walls are labeled by feeding a radioactive precursor.
4. The *direct* approach, in which fully developed cells are dissected mechanically into single cell wall layers that can then be analyzed separately.

The additive approach was applied by Meier and Wilkie (1959) to pine tracheids and by Meier (1961) to spruce tracheids and birch fibers. During the aging of the young tracheids and fibers in the neighborhood of the cambial zone, successive polysaccharide layers are deposited next to one another by the living protoplasts. In the additive approach, tracheids or fibers are isolated and analyzed at various stages of maturation by total hydrolysis with sulfuric acid. From a knowledge of the monosaccharides in the hydrolysates, the percentages of the different polysaccharides with roughly known chemical structures can be calculated. An attempt is then made to deduce the composition of the different cell wall layers. This approach to the problem entails the assumption that once a cell wall layer has been deposited, its polysaccharide composition does not subsequently alter.

The young tracheids or fibers in radial sections through the outermost part of the xylem bordering the cambial zone were separated into four categories of increasing age: (M + P), (M + P + S_1), (M + P + S_1 + $S_{2\ \text{outer part}}$), and (M + P + S_1 + S_2 + S_3) tracheids or fibers. These separations were carried out with a micromanipulator under the polarizing microscope, where the different layers exhibit different optical properties. For the calculation of the composition of each layer (from sugar analysis of the hydrolysates of the four cell categories), it was necessary to know the weights of the different layers. These weights were deduced from the volumes of the layers whose thicknesses were measured on light and electron micrographs. As the densities are certainly not the same in all the layers, the percentage weights used for the calculations (2% for P, 10% for S_1, 78% for S_2, and 10% for S_3 in the case of the pine tracheids) are not very accurate. Hence, the compositions as shown in Table III are relatively reliable for the M + P and the S_1 layers but are less accurate for the innermost layers. For all three woods investigated, the cellulose content is lowest in the (M + P) layer, which contains a high percentage of pectic material (pectic acid, galactan, arabinan). There is, however, a marked difference between birch and the two coniferous woods in the arabinan content of the (M + P) layer: in birch it is only half as high as in pine. For the xylan content, the reverse is true. In birch, the cellulose content is highest in the inner part of the S_2 and in the S_3 layer. The S_1 and the outer part of the S_2 layer have a very high content of 4-*O*-methylglucuronoxylan.

In spruce and pine tracheids, the composition of the different cell wall layers is similar. The glucomannan content increases from the outer parts of the cell walls to the inner parts. The arabino-4-*O*-methylglucuronoxylan content is very high in the S_3 layer. There seems to be practically no galactan and arabinan in the secondary wall layers of pine and spruce tracheids.

The additive approach has also been used by Côté *et al.* (1968) to study the tracheids of balsam fir compression wood (see below) and by Takabe *et al.* (1983) to investigate *Cryptomeria* tracheids. A variation of the method has been employed by Larson (1969): $^{14}CO_2$ was fed to photosynthesizing young red pines some days before the analysis of the differentiating xylem in a way analogous to that of Meier and Wilkie (1959).

In the microscopic approach, the hemicelluloses either are removed or are specifically marked before the observation of sections. This approach has been tried by Boutelje and Hollmark (1972), who degraded hemicelluloses by enzymes and determined the loss of substance by interference

Table III

Polysaccharide Composition of the Different Cell Wall Layers of Birch Fibers and of Tracheids from Pine and Spruce[a]

Polysaccharide	Cell wall layer			
	(M + P)[b] (%)	S_1 (%)	$S_{2\,outer}$ (%)	$S_{2\,inner} + S_3$ (%)
Betula verrucosa Ehrh.				
Galactan	16.9	1.2	0.7	0.0
Cellulose	41.4	49.8	48.0	60.0
Glucomannan	3.1	2.8	2.1	5.1
Arabinan	13.4	1.9	1.5	0.0
O-Acetyl-4-*O*-methyl-glucuronoxylan	25.2	44.1	47.7	35.1
Pinus silvestris L.				
Galactan[c]	16.4	8.0	0.0	0.0
Cellulose	33.4	55.2	64.3	63.6
O-Acetylglucomannan	7.9	18.1	24.4	23.7
Arabinan	29.3	1.1	0.8	0.0
Arabino-4-*O*-methyl-glucuronoxylan	13.0	17.6	10.7	12.7
Picea abies Karst.				
Galactan[c]	20.1	5.2	1.6	3.2
Cellulose	35.5	61.5	66.5	47.5
O-Acetylglucomannan	7.7	16.9	24.6	27.2
Arabinan	29.4	0.6	0.0	2.4
Arabino-4-*O*-methyl-glucuronoxylan	7.8	15.7	7.4	19.4

[a] From Meier (1964). M, Middle lamella; P, primary wall; S_1, S_2, and S_3 represent the outer, middle, and inner layer of the secondary wall, respectively.

[b] Contains also pectic acid, which has not been taken into account.

[c] The galactan content indicated is somewhat too high, since some galactose is part of the glucomannan.

microscopy. Although interference microscopy is a very powerful method for detecting mass losses as low as 1% of the original mass, not much information about the distribution of hemicellulose was gained. The reason was essentially that the enzmes used were not pure and/or active enough to hydrolyze one specific polysaccharide to completion.

Sinner *et al.* (1973) used electron microscopy for the detection of enzymatic degradation of partially delignified beech wood (*Fagus silvatica* L.) by specific xylanases. These authors found that the outer layer of the secondary wall (S_1) showed a strong degradation. The S_2 layer was attacked only in the outer part and near the S_3 but to a lesser extent than S_1. Since only 30% of the total xylan was removed, this study cannot show the distribution of xylan through the cell wall.

Hoffmann and Parameswaran (1976) made another attempt to detect the polysaccharide distribution in the cell walls of delignified spruce tracheids by a microscopic method. They localized the total hemicelluloses in spruce holocellulose and pulp fibers by contrasting the oxidized polysaccharides with heavy metal ions and by electron microscopic examination. The highest concentration of hemicelluloses was found in the S_1 layer.

A radioautographic approach was tried by Mullis *et al.* (1976). Radioactive arabinose was administered through the basal end of cut first-year ramets of *Populus tremuloides*. It was found that in the young xylem, essentially arabinose and xylose residues were radioactively marked, demonstrating that the L-arabinose fed was quite specifically incorporated into pentosans. Autoradiography of thin sections prepared for electron microscopy revealed that pentosans were deposited by apposition throughout the entire secondary cell wall.

The most direct approach for the study of polysaccharide distribution has been tried by Hardell and Westermark (1981). These authors dissected tracheids from spruce wood (*Picea abies*) after a slight chlorite delignification. They scratched off, with specially formed tweezers under a stereomicroscope, the outer cell wall layers and so obtained three fractions, (M + P), (M + P + S_1), and (S_2 + S_3), which were separately hydrolyzed and analyzed by gas chromatography. Although the (M + P) fraction had a substantially higher content of arabinose and galactose residues than the other fractions, the percentage content of xylose and mannose residues varied very little (xylose: 10.1–11.6% and mannose: 15.5–17.6%) in the three cell wall fractions. The authors concluded that mannans and xylans are evenly distributed throughout the wall and that the hemicellulose content in the different wall layers is proportional to the cellulose and not to the lignin content. The disadvantage of this most direct method seems to be that a preliminary slight delignification is necessary, during which a partial dissolution and also a redistribution of hemicelluloses might occur. The arabinose and galactose content in the (M + P) layer, which are indicated to be 7.3 and 7.6%, respectively, are most likely substantially lower than in the nondelignified wood.

In conclusion, it must be admitted that all the methods used have serious drawbacks and that the problem of the distribution of the different polysaccharides through the cell wall cannot be considered as settled.

IV. Polysaccharides of Reaction Wood Cell Walls

There are important differences in the polysaccharide content of normal and reaction wood (Table IV). A very complete review of the chemistry and topochemistry of compression wood has recently been published by Timell (1982). Reaction wood (compression wood) tracheids from *Picea abies* contain much more galactan and less galactoglucomannan and cellulose than tracheids of normal wood. Brodzki (1972) and Waterkeyn *et al.* (1982) have shown that in the helical cavities, which are characteristic for the secondary walls of compression wood tracheids, there are depositions of callose. In 1970 and 1972, Hoffmann and Timell had already reported the isolation of callose ("laricinan") from compression wood of larch, red spruce, and balsam fir (2–4% of the wood), but without knowing its localization. The extra galactan of compression wood is localized essentially in the S_1 and in the outer part of the S_2 layer of the tracheids (Côté *et al.*, 1968). Reaction wood (tension wood) fibers from *Betula verrucosa* have a higher galactan and cellulose content, but they have less 4-*O*-methylglucuronoxylan than normal wood. A number of comparative analyses of tension and normal wood of different species have been published by Timell (1969).

The high cellulose content of tension wood is certainly due to the presence of the "gelatinous layer" (Norberg and Meier, 1966). The extra galactan of tension wood is obviously localized in the S_1 and S_2 layers of the secondary wall.

Table IV

Polysaccharide Composition of Normal Wood and Reaction Wood[a]

	Picea abies Karst.		*Betula verrucosa* Ehrh.	
Polysaccharide	Normal wood (%)	Compression wood (%)	Normal wood (%)	Tension wood (%)
Cellulose	62	50[b]	50	62
O-Acetylgalactoglucomannan	22	14	—	—
Glucomannan	—	—	4	1
Arabino-4-*O*-methylglucuronoxylan	13	17	—	—
O-Acetyl-4-*O*-methylglucuronoxylan	—	—	44	24
Others	3	19[c]	2	13[c]

[a] From Meier, 1964.
[b] Includes also some callose.
[c] Mainly galactan.

V. Concluding Remarks

It is evident that further efforts are necessary for obtaining a clear picture of the distribution of polysaccharides in wood cell walls. Whereas microscopic methods (in connection with enzymatic degradation, chemical extraction, radioactive labeling, specific contrasting, etc.) can only yield qualitative results, the most promising approach seems to be microdissection of wood tissue and of single cells (preferably without prior delignification) followed by total hydrolysis of the different fractions and microanalysis of the monosaccharides formed. A second, but much more pretentious, step would be the methylation analysis of the polysaccharides of each fraction obtained by microdissection.

References

Boutelje, J. B., and Hollmark, B. H. (1972). *Holzforschung* **26,** 76–81.
Brodzki, P. (1972). *Acta Soc. Bot. Pol.* **41,** 321–327.
Côté, W. A., Kutscha, N. P., Simson, B. W., and Timell, T. E. (1968). *Tappi* **51,** 33–40.
Hardell, H.-L., and Westermark, U. (1981). *In* "The Ekman-Days 1981," Vol. 1. International Symposium on Wood and Pulping Chemistry, Stockholm.
Hoffmann, G. C., and Timell, T. E. (1970). *Wood Sci. Technol.* **4,** 159–162.
Hoffmann, G. C., and Timell, T. E. (1972). *Sven. Papperstidn.* **75,** 135–141.
Hoffmann, P., and Parameswaran, N. (1976). *Holzforschung* **30,** 62–70.
Larson, P. R. (1969). *Holzforschung* **23,** 17–26.
Meier, H. (1961). *J. Polym. Sci.* **51,** 11–18.
Meier, H. (1964). *In* "Formation of Wood in Forest Trees" (M. Zimmermann, ed.), pp. 137–151. Academic Press, New York.
Meier, H., and Wilkie, K. C. B. (1959). *Holzforschung* **13,** 177–182.
Mullis, R. H., Thompson, N. S., and Parham, R. A. (1976). *Planta* **132,** 241–248.
Norberg, P. H., and Meier, H. (1966). *Holzforschung* **20,** 174–178.
Perilä, O. (1961). *J. Polym. Sci.* **51,** 19–26.
Perilä, O. (1962). *Suomen Kemistilethi* **B35,** 176–178.
Sinner, M., Parameswaran, N., Dietrichs, H. H., and Liese, W. (1973). *Holzforschung* **27,** 36–42.
Takabe, K., Fujita, M., and Harada, H. (1983). *Mokuzai Gakkaishi* **29,** 183–189.
Timell, T. E. (1969). *Sven. Papperstidn.* **72,** 173–181.
Timell, T. E. (1982). *Wood Sci. Technol.* **16,** 83–122.
Waterkeyn, L., Caeymaex, S., and Decamps, E. (1982). *Bull. Soc. R. Bot. Belg.* **115,** 145–155.

CHAPTER 3

Localization of Lignins in Wood Cell Walls

S. SAKA[1]
D. A. I. GORING

Pulp and Paper Research Institute of Canada
and Department of Chemistry
McGill University
Montreal, Quebec, Canada

I. Introduction

Many methods have been used to study the distribution of lignin in wood. One of the oldest is selective staining followed by examination under the light microscope (Brauns, 1952). Staining with potassium permanganate (Crocker, 1921) has been used extensively for detection of lignin by electron microscopy (Hepler *et al.*, 1970; Bland *et al.*, 1971; Parham, 1974; Saka *et al.*, 1979), although some doubt exists as to its specificity for lignin (Hoffmann and Parameswaran, 1976; Kishi *et al.*, 1982).

The distribution of lignin has also been studied by examination of lignin skeletons created by the removal of the carbohydrates by brown-rot fungi (Meier, 1955) or concentrated hydrofluoric acid (Sachs *et al.*, 1963; Bentum *et al.*, 1969; Parham and Côté, 1971; Fujii *et al.*, 1981). Although some alteration of the lignin through condensation reactions may take place and the presence of the residual carbohydrates may obscure the data, overall

[1] Present address: Research Center, Daicel Chemical Industries, Ltd., Himeji, Hyogo, Japan 671-12.

ISBN 0-12-347880-4

the results obtained by this method are in reasonable agreement with those from potassium permanganate staining (Parham, 1974).

Useful as the above methods are in elucidating the presence of lignin in the various morphological regions of wood, they do not lend themselves to the quantitative estimation of the lignin distribution across the cell wall. For quantitative assay of the lignin distribution, the method based on ultraviolet (UV) microscopy of thin sections of wood is one of the best procedures. This approach was initiated by Lange (1954), who estimated the weight concentration of lignin in the secondary wall and compound middle lamella of Norway spruce tracheids to be 16 and 73%, respectively. This result was in excellent agreement with the value of 71% for the Douglas fir middle lamella fractions obtained with a microdisection technique by Bailey (1936). Previous to Bailey's work, Ritter (1925) had concluded that approximately 75% of the lignin in wood is located in the middle lamella and that the other 25% is located in the secondary wall. Considerable confusion then appeared in the literature because several authors did not make a clear distinction between the Bailey–Lange result and that of Ritter. No doubt some of this confusion has arisen by use of the percent symbol to denote both the percentage fraction of the total wood lignin contained in a particular morphological region and the lignin content of that region. Perhaps instead of % it would be less confusing to use g/g to denote lignin concentrations: i.e., g lignin/g cell wall substance. The symbol % would then be reserved for the proportion of the total lignin in a particular tissue area.

Later, Goring and co-workers (Scott *et al.*, 1969; Scott and Goring, 1970a; Wood and Goring, 1974) refined the UV microscopy method by preparing the thin section (0.5 μm) and thus avoiding errors caused by nonparallel illumination. Goring and co-workers could then determine the quantitative distribution of lignin in wood (Fergus *et al.*, 1969; Scott and Goring, 1970b; Fergus and Goring, 1970a,b; Wood and Goring, 1971; Musha and Goring, 1975), and they proved that the Bailey–Lange result is correct. They also proved that Berlyn and Mark (1965) were correct when they pointed out that because the volume fraction of the middle lamella was so small, it could not contain more than 40% of the total lignin in wood.

Other methods have also been proposed. Lange and Kjaer (1957) introduced the use of interference microscopy for the quantitative assay of lignin, a technique further refined by Boutelje (1972). More recently, Saka *et al.* (1978, 1981; Saka and Thomas, 1982a,b) have developed a new technique for a quantitative determination of the distribution of lignin in wood. The method involves a specific bromination for lignin in a nonaqueous system ($CHCl_3$) followed by the measurement of bromine concentrations in the various morphological regions of wood by scanning electron microscopy (SEM) coupled with energy-dispersive X-ray analysis (EDXA). From knowing the relative reactivity of bromine to different types of lignins, the distribution of lignin in the different parts of the cell can be determined.

II. The Distribution of Lignin in Conifers

Table I shows the distribution of lignin in black spruce (*Picea mariana* Mill.) tracheids as determined by UV microscopy (Fergus *et al.*, 1969). The results show that the concentration of lignin in the middle lamella is considerably higher than that in the secondary wall. However, because the secondary wall makes up a much larger proportion of the total tissue volume, the majority of the lignin (72% in earlywood and 82% in latewood) is found in the secondary wall. As seen in Fig. 1, the lignin is uniformly distributed across the secondary wall of black spruce tracheids.

In Table II, the distribution of lignin in the xylem of Douglas fir [*Pseudotsuga menziesii* (Mirb.) Franco] is given (Wood and Goring, 1971). Here the ray cells are included. The concentration and distribution of lignin in the various morphological regions of the tracheids do not differ much from those shown for black spruce in Table I.

Table III shows the distribution of lignin in loblolly pine (*Pinus taeda* L.) tracheids as determined by bromination with SEM–EDXA (Saka and Thomas, 1982b). It is interesting to note that with this method it is possible to resolve the S_1, S_2, and S_3 layers in the secondary wall quite clearly. Such resolution is often difficult with UV microscopy. Apparently, in

Table I

The Distribution of Lignin in Black Spruce Tracheids as Determined by Ultraviolet Microscopy[a]

Wood	Morphological region	Tissue volume (%)	Lignin: % Total	Lignin: Conc. (g/g)
Earlywood	S	87	72	0.23
	ML	9	16	0.50
	ML_{cc}	4	12	0.85
Latewood	S	94	82	0.22
	ML	4	10	0.60
	ML_{cc}	2	8	1.00

[a] From Fergus *et al.* (1969).

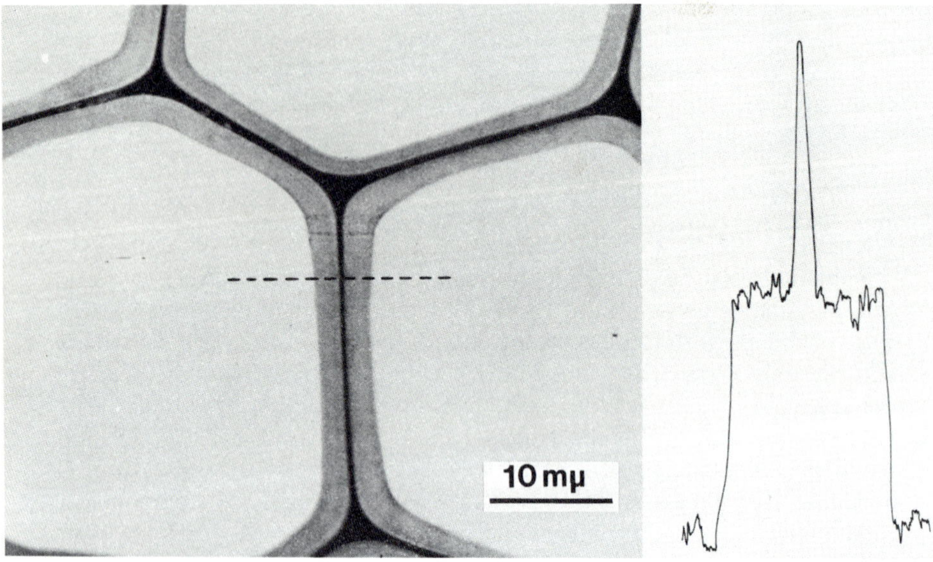

Fig. 1. Ultraviolet photomicrograph, taken at 240 nm, of the cell walls of black spruce earlywood tracheids. The densitometer tracing was taken along the dotted line. From Fergus *et al.* (1969).

Table II

The Distribution of Lignin in Douglas Fir Xylem as Determined by Ultraviolet Microscopy[a]

Wood	Morphological region	Tissue volume (%)	Lignin % Total	Lignin Conc. (g/g)
Earlywood	Tracheid S	74	58	0.25
	Tracheid ML	10	18	0.56
	Tracheid ML_{cc}	4	11	0.83
	Ray parenchyma S	8	10	0.40
	Ray tracheid S	4	3	0.28
Latewood	Tracheid S	90	78	0.23
	Tracheid ML	4	10	0.60
	Tracheid ML_{cc}	2	6	0.90
	Ray parenchyma S	3	4	—
	Ray tracheid S	1	2	—

[a] From Wood and Goring (1971).

Table III

The Distribution of Lignin in Loblolly Pine Tracheids as Determined by Bromination with SEM–EDXA[a]

Wood	Morphological region	Tissue volume (%)	Lignin	
			% Total	Conc. (g/g)
Earlywood	S_1	13	12	0.25
	S_2	60	44	0.20
	S_3	9	9	0.28
	ML	12	21	0.49
	ML_{cc}	6	14	0.64
Latewood	S_1	6	6	0.23
	S_2	80	63	0.18
	S_3	5	6	0.25
	ML	6	14	0.51
	ML_{cc}	3	11	0.78

[a] From Saka and Thomas (1982b).

loblolly pine, the lignin concentration in the S_2 layer is lower than in either S_1 or S_3. These differences are confirmed in the line profile of the bromine X rays shown in Fig. 2. Interestingly, Fukazawa and Imagawa (1981) examined in the ultraviolet microscope the tracheids of Japanese fir (*Abies sachalinensis* Fr. Schm.) and found a high UV absorbance near the

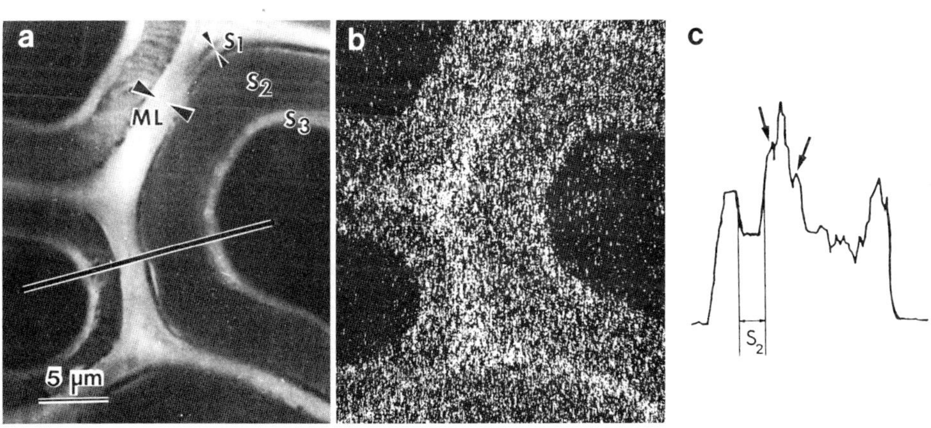

Fig. 2. **a,** Scanning electron micrograph of brominated loblolly pine latewood tracheids (0.5 μm section). **b,** The distribution map of Br-L X-rays was taken of the same area as the SEM micrograph. **c,** The distribution of bromine was taken along the line across the double cell wall (Saka *et al.,* 1981). Arrows indicate S_1 layer. ML, Middle lamella; S_1, S_2, S_3, secondary wall layers.

lumen–wall interface of the juvenile wood tracheids. Comparison of Tables I, II, and III shows that, minor differences notwithstanding, the trends in the distribution in the tracheids of the three softwoods are similar.

The ray parenchyma cells, which constitute about 5% by weight of the xylem in conifers, have been shown to possess significantly higher lignin contents than the whole wood. Bailey (1936) determined by a direct ana-

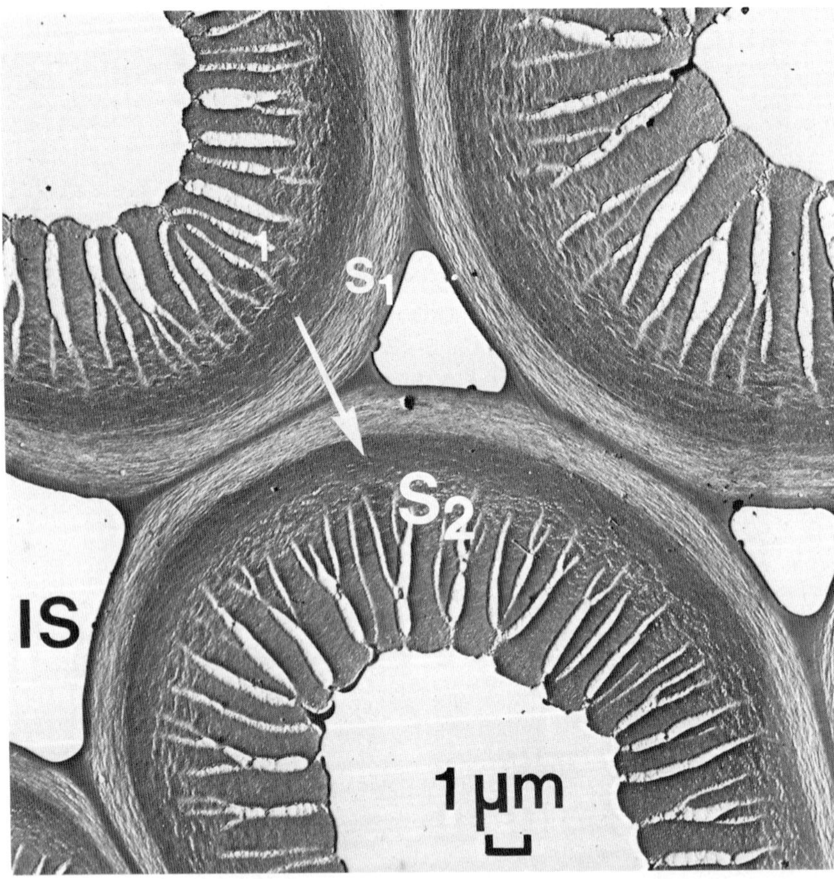

Fig. 3. Electron micrograph of a lignin skeleton of tracheids in the compression wood of larch, prepared by treatment of a cross section with hydrofluoric acid. Arrow indicates the highly lignified S_2 (L) layer. IS Indicates an intercellular space (T. E. Timell, private communication, 1984). The authors are grateful to Dr. Timell for providing this electron micrograph.

lytical method a lignin content of 0.41 g/g for the segregated ray parenchyma cells of Douglas fir. Harada and Wardrop (1960) found a lignin content of 0.44 g/g for the ray parenchyma of Japanese cedar (*Cryptomeria japonica* D. Don). Using UV microscopy, Fergus *et al.* (1969) estimated a lignin concentration of 0.40 g/g in the ray parenchyma cells of black spruce. The data for Douglas fir earlywood shown in Table II are in good agreement with these results. Furthermore, Fergus *et al.* (1969) found the lignin concentration in the secondary walls of the ray parenchyma to be about the same in both earlywood and latewood portions.

Several papers have been published on the distribution of lignin in the compression wood of softwoods. Timell (1982) has given an excellent review of this work. An electron micrograph of the lignin skeleton from the compression wood of larch [*Larix laricina* (Du Roi) K. Koch] is shown in Fig. 3 (T. E. Timell, private communication, 1984). Although the morphology is different, the distribution of lignin in compression wood tracheids is about the same as in normal wood, except that the concentrations of lignin are higher and that no lignin is present at the cell corner middle lamella. A repeatedly observed peculiarity in compression wood is a ring in the S_2 layer (S_2(L)] in which the lignin concentration is about the same as it is in the middle lamella. Table IV shows a comparison made by Timell (1982) of the lignin concentrations measured by Wood and Goring (1971) on Douglas fir and by Fukazawa (1974) on Japanese fir (*Abies sachalinensis* Fr. Schm.). The overall trends are the same, although the lignin contents in Japanese fir are lower in most of the morphological regions.

Table IV

Concentration of Lignin in the Morphological Regions of the Compression Wood of Douglas Fir[a] and of Japanese Fir[b]

Morphological region	Lignin conc. (%)	
	Douglas fir	Japanese fir
S_1	40	29
S_2(L)	54	42
S_2	36	26
ML	49	49
ML_{cc}	75	65

[a] From Wood and Goring (1971).
[b] From Fukazawa (1974).

III. The Distribution of Lignin in Hardwoods

Hardwood lignins consist mainly of guaiacyl and syringyl residues. It seems that the guaiacyl : syringyl ratio changes from one morphological region to another. Fergus and Goring (1970a,b) attempted to measure the distribution of lignin in white birch (*Betula papyrifera* Marsh.) by UV spectral analysis. However, the syringyl and guaiacyl residues have markedly different UV absorptivities. Thus, a correct measurement of the lignin distribution by UV microscopy requires a knowledge of the relative amounts of syringyl and guaiacyl residues in the various morphological regions. More recently, Saka *et al.* (1984; Saka and Goring, 1984) have combined bromination–EDXA with UV microscopy to measure the guaiacyl : syringyl ratio and estimate the concentration of lignin in the different parts of the cell wall.

Table V shows the ratio of guaiacyl and syringyl residues in various morphological regions of white birch wood as determined by UV microscopy and bromination–EDXA (UV–EDXA) (Saka and Goring, 1984). For comparison, the results obtained earlier by UV spectral analysis (Fergus and Goring, 1970b) are included. It is apparent by both methods that the guaiacyl : syringyl ratio of lignin is different in the different morphological regions investigated. The fiber secondary wall (S_2) contains predominantly syringyl residues, while the vessel secondary wall (S_2) consists mostly of guaiacyl residues.

The study by UV spectral analysis (Fergus and Goring, 1970b) revealed

Table V

Distribution of Guaiacyl and Syringyl Residues in Lignin in White Birch

Morphological region[a]	Guaiacyl : Syringyl	
	Bromination with TEM–EDXA[b]	UV Spectral analysis[c]
Fiber S_2	12 : 88	Syringyl
Vessel S_2	88 : 12	Guaiacyl
Ray parenchyma S	49 : 51	Syringyl
$ML_{cc(F/F)}$	91 : 9	50 : 50
$ML_{cc(F/V)}$	80 : 20	Guaiacyl
$ML_{cc(F/R)}$	100 : 0	50 : 50
$ML_{cc(R/R)}$	88 : 12	50 : 50

[a] F/F, Fiber/fiber; F/V, fiber/vessel; F/R, fiber/ray; R/R, ray/ray.
[b] From Saka and Goring (1984).
[c] From Fergus and Goring (1970a).

that the ray parenchyma cell consists predominantly of syringyl-type lignin as found in the fiber secondary wall. However, a guaiacyl : syringyl ratio of about 1 : 1 was found by UV–EDXA (Saka and Goring, 1984).

For the cell corner areas of the middle lamella (ML_{cc}), about 80–100% of the lignin was guaiacyl residue with the remaining 0–20% being syringyl residue. This result is somewhat at variance with the previous proposal (Fergus and Goring, 1970b) that the middle lamella lignin around fibers and ray cells contains about equal proportions of guaiacyl and syringyl residues but supports the later suggestion of Musha and Goring (1975) that the middle lamella lignin consists entirely of guaiacylpropane units.

The results shown in Table V indicate that syringyl and guaiacyl groups are distributed differently in hardwoods. Work by others has supported these findings. Wolter *et al.* (1974) have shown that a pure guaiacyl lignin is associated with the vessels in aspen callus cultures. Kirk *et al.* (1975) found that the topochemistry of the fungal degradation of lignin in birch wood was consistent with the presence of syringyl-rich lignin in the fiber walls. A syringyl-rich lignin has been isolated from several hardwoods by Yamasaki *et al.* (1978). Perhaps the most convincing evidence comes from investigations in which the wood is fractionated and the guaiacyl : syringyl ratio determined for the lignins in the various tissue regions isolated. In this manner, Hardell *et al.* (1980) have shown that the middle lamella and vessel wall lignins of birch wood are rich in guaiacyl units, whereas the syringyl : guaiacyl ratio is high in the fiber and ray cell walls. Cho *et al.* (1980) studied the chemical characteristics of the film-like substance that they isolated from the fines of birch. This material was recognized as containing a high proportion of the compound middle lamella and was shown to possess a low ratio of syringyl to guaiacyl units.

Table VI shows the distribution of lignin in white birch wood as determined by UV–EDXA (Saka and Goring, 1984). For the fiber and vessel secondary walls, the lignin is uniformly distributed across the S_1, S_2, and S_3 layers. The lignin concentration of the vessel wall is, however, about 1.9 times higher than that of the fiber, while ray parenchyma cells have a rather lower lignin concentration than that of the fiber. The lignin concentration of the middle lamella between cell corner areas (ML) was found to be 10–30% lower than that of the cell corner middle lamella (ML_{cc}).

For comparison, the data of lignin distribution in white birch wood, as determined by UV microscopy (Fergus and Goring, 1970a), are listed in Table VI. It is of interest to note that lignin concentrations in fiber and vessel secondary walls determined by the two different methods are in agreement. However, the lignin concentration in the ray parenchyma cells as measured by UV microscopy (Fergus and Goring, 1970a) was

Table VI

The Distribution of Lignin in White Birch[a]

Element	Morphological region[b]	Tissue volume (%)	Lignin conc. (g/g)	
			UV–EDXA[c]	UV Only[d]
Fiber	S_1	11.4	0.14	—
	S_2	58.5	0.14	0.16
	S_3	3.5	0.12	—
	ML	5.2	0.36	0.34
	$ML_{cc(F/F)}$	2.4	0.45	0.72
Vessel	S_1	1.6	0.26	—
	S_2	4.3	0.26	0.22
	S_3	2.3	0.27	—
	ML	0.8	0.40	0.35
	$ML_{cc(F/V)}$	≃0	0.58	—
Ray parenchyma	S	8.0	0.12	0.22
	ML	2.0	0.38	—
	$ML_{cc(F/R)}$	≃0	0.47	—
	$ML_{cc(R/R)}$	≃0	0.41	—

[a] From Saka and Goring (1984) and Fergus and Goring (1970a).
[b] F/F, Fiber/fiber; F/V, fiber/vessel; F/R, fiber/ray; R/R, ray/ray.
[c] Saka and Goring (1984).
[d] Calculated using xylem lignin content of 0.199 g/g, from Fergus and Goring (1970a).

nearly twice as high as the results obtained by UV–EDXA (Saka and Goring, 1984). The middle lamella between two cell corners (ML), both for fibers and vessels, showed similar values by the two techniques. In contrast, the concentration in the cell corner middle lamella (ML_{cc}) was lower by UV–EDXA (Saka and Goring, 1984). The observed discrepancies are probably due to the uncertainty in estimating the guaiacyl : syringyl ratio when the measurement is made by UV microscopy alone.

The results obtained by the new method involving both UV microscopy and bromination–EDXA (Saka and Goring, 1984) confirmed, in general the previous data by UV microscopy (Fergus and Goring, 1970a,b). However, some discrepancy was found for the ray parenchyma and the middle lamella region. A good method for resolving such a discrepancy would be to separate physically various types of tissues without introducing any chemical changes (Hardell *et al.*, 1980; Whiting *et al.*, 1981). Analysis of the separated tissues could then provide definitive information on the distribution of different types of lignin in the various morphological regions of hardwoods.

References

Bailey, A. J. (1936). *Ind. Eng. Chem. Anal. Ed.* **8,** 52–55.

Bentum, A. L. K., Côté, Jr., W. A., Day, A. C., and Timell, T. E. (1969). *Wood Sci. Technol.* **3,** 218–231.

Berlyn, G. P., and Mark, R. E. (1965). *For. Prod. J.* **16,** 140–141.

Bland, D. E., Foster, R. C., and Logan, A. F. (1971). *Holzforschung* **25,** 137–143.

Boutelje, J. B. (1972). *Sven. Papperstidn.* **75,** 683–686.

Brauns, F. E. (1952). "The Chemistry of Lignin," Chapter IV. Academic Press, Inc., New York.

Cho, N. S., Lee, J. Y., Meshitsuka, G., and Nakano, J. (1980). *Mokuzai Gakkaishi* **26,** 527–533.

Crocker, E. C. (1921). *Ind. Eng. Chem.* **13,** 625–627.

Fergus, B. J., and Goring, D. A. I. (1970a). *Holzforschung* **24,** 118–124.

Fergus, B. J., and Goring, D. A. I. (1970b). *Holzforschung* **24,** 113–117.

Fergus, B. J., Procter, A. R., Scott, J. A. N., and Goring, D. A. I. (1969). *Wood Sci. Technol.* **3,** 117–138.

Fujii, T., Harada, H., and Saiki, H. (1981). *Mokuzai Gakkaishi* **27,** 149–156.

Fukazawa, K. (1974). *Res. Bull. Coll. Exp. For. Hokkaido Univ.* **31**(1), 87–114.

Fukazawa, K., and Imagawa, H. (1981). *Wood Sci. Technol.* **15,** 45–55.

Harada, H., and Wardrop, A. B. (1960). *Mokuzai Gakkaishi* **6,** 34–41.

Hardell, H.-L., Leary, G. J., Stoll, M., and Westermark, U. (1980). *Sven. Papperstidn.* **83,** 71–74.

Hepler, P. K., Fosket, D. E., and Newcomb, E. H. (1970). *Am. J. Bot.* **57,** 85–96.

Hoffmann, P., and Parameswaran, N. (1976). *Holzforschung* **30,** 62–70.

Kirk, T. K., Chang, H.-m., and Lorenz, L. F. (1975). *Wood Sci. Technol.* **9,** 81–86.

Kishi, K., Harada, H., and Saiki, H. (1982). *Bull. Kyoto Univ. For.* **54,** 209–216.

Lange, P. W. (1954). *Sven. Papperstidn.* **57,** 525–537, 563–567.

Lange, P. W., and Kjaer, A. (1957). *Norsk Skogind.* **11,** 425–432.

Meier, H. (1955). *Holz Roh-Werkst.* **13,** 323–338.

Musha, Y., and Goring, D. A. I. (1975). *Wood Sci. Technol.* **9,** 45–58.

Parham, R. A. (1974). *Wood Sci.* **6,** 305–315.

Parham, R. A., and Côté, Jr., W. A. (1971). *Wood Sci. Technol.* **5,** 49–62.

Ritter, G. J. (1925). *Ind. Eng. Chem.* **17,** 1194–1197.

Sachs, I. B., Clark, I. T., and Pew, J. C. (1963). *J. Polym. Sci. Part C* **2,** 203–212.

Saka, S., and Goring, D. A. I. (1984). In preparation.

Saka, S., and Thomas, R. J. (1982a). *Wood Sci. Technol.* **16,** 1–18.

Saka, S., and Thomas, R. J. (1982b). *Wood Sci. Technol.* **16,** 167–179.

Saka, S., Thomas, R. J., and Gratzl, J. S. (1978). *Tappi* **61,** 73–76.

Saka, S., Thomas, R. J., and Gratzl, J. S. (1979). *Wood Fiber* **11,** 99–108.

Saka, S., Thomas, R. J., and Gratzl, J. S. (1981). *Proc. ISWPC, June 1981, Stockholm, Sweden,* Vol. 1, SPCI Report 38, pp. 35–42.

Saka, S., Whiting, P., Fukazawa, K., and Goring D. A. I. (1982). *Wood Sci. Technol.* **16,** 269–277.

Saka, S., Hosoya, S., St-Germain, F. G. T., and Goring, D. A. I. (1984). In preparation.

Scott, J. A. N., and Goring, D. A. I. (1970a). *Wood Sci. Technol.* **4,** 237–239.

Scott, J. A. N., and Goring, D. A. I. (1970b). *Cellul. Chem. Technol.* **4,** 83–93.

Scott, J. A. N., Procter, A. R., Fergus, B. J., and Goring, D. A. I. (1969). *Wood Sci. Technol.* **3,** 73–92.

Timell, T. E. (1982). *Wood Sci. Technol.* **16,** 83–122.
Whiting, P., Favis, B. D., St-Germain, F. G. T., and Goring, D. A. I. (1981). *J. Wood Chem. Technol.* **1,** 29–42.
Wolter, K. E., Harkin, J. M., and Kirk, T. K. (1974). *Physiol. Plant* **31,** 140–143.
Wood, J. R., and Goring, D. A. I. (1971). *Pulp Paper Mag. Can.* **72,** T95–T102.
Wood, J. R., and Goring, D. A. I. (1974). *J. Microsc.* **100,** 105–111.
Yamasaki, T., Hata, K., and Higuchi, T. (1978). *Holzforschung* **32,** 44–47.

CHAPTER 4

Metabolism and Synthetic Function of Cambial Tissue

GRAEME P. BERLYN
YORK C. BATTEY

Yale University
School of Forestry and Environmental Studies
Greeley Laboratory
New Haven, Connecticut

I. Introduction

Forests are the most productive areas of the earth, both in terms of total energy collected per annum and in terms of energy collected per unit area per annum (Lehninger, 1965; Whittaker, 1975). The primary transducer that changes solar energy into chemical energy in the form of stable chemical bonds is the process of photosynthesis. Humans call certain useful aggregates of these stable chemical bonds wood.

Most forest growth studies deal primarily with height growth, diameter growth, and processes that contribute to wood formation, but little attention has been paid to cambial physiology at the tissue level (Ford, 1981). If we were to base this chapter on direct experimental evidence, it would be one of the world's shortest, because the cambium is a single layer of cells

ISBN 0-12-347880-4

that are not necessarily contiguous (Berlyn, 1982), and the cambium has never been isolated and studied biochemically, although some have tried (see Berlyn, 1964, 1970, 1979). On the other hand, if we were to base our chapter on inferred logical constructs, it would include this entire volume and then some. Therefore we shall define the cambium and its functions and then conclude with a discussion of the physiological factors that affect or control cambial function. Thus we will try to present a holistic setting for the more specific biochemical, cytological, and cytochemical chapters in this book.

The cambium forms in the primary plant body and gives rise to the secondary vascular tissues (secondary xylem and phloem), the so-called secondary plant body. Normally, the cambium differentiates in the newly formed part of the primary plant body, where it first appears in the vascular bundles and then extends circumferentially through the interfascicular tissue (group meristem) to form a hollow cylinder around the stem. Under proper stimulus, however, cambium can form over wounds and can originate from older parenchyma cells in cortex or phloem. In most cases the procambium of the primary plant body forms a continuum with the cambium and connects it to the leaves and shoot meristems that provide the cambium with the assimilates that it requires as an energy source for its principle function of making cells. The leaves and meristems also provide hormones and time signals that instruct the cambium what to do and when to do it. Similarly, there is a structural and functional continuum between the stem cambium and the root meristems. The roots provide water and mineral nutrients that are required for cambial function and also provide hormones and perhaps time signals that serve as information systems for the cambium. For example, the transition from earlywood to latewood formation by the cambium is often correlated with a sharp increase in water stress. Thus water stress may serve as (at least) the initiator of a time signal to the cambium. It is not beyond comprehension that the cambium requires a combination or synthesis of time signals before it can initiate a metabolic action.

The single tier of cells that comprises the vascular cambium consists of two types of initials: the fusiform initials, which give rise to the longitudinal (axial) elements of the vascular tissues, and the ray initials, which give rise to the radially directed elements. The sole function of these initials is to make cells, and these cells produced by the cambium comprise three kinds of tissues: xylem, phloem, and cambium (see Chapter 1, Section I,A). In order to produce these three tissues, the cambium undergoes three different types of cell division: multiplicative, additive, and transformative (Berlyn, 1982). Additive divisions are those tangential, longitudinal divisions that add xylem and phloem tissue to the wood body, and both fusiform initials and ray initials function in this most fundamental type of division. Multiplicative divisions occur when ray initials or fusiform initials divide to form two new initials, producing the extra cambium that is necessary to keep pace with the increase in stem circumference produced by the additive divisions. Transformative divisions occur when a fusiform initial is transformed or partly transformed into a ray initial or ray initials. This type of division appears to be intimately associated with the internal metabolism of the cambium. For example, when a multiplicative division occurs in a fusiform initial, it is quite common that only one of the daughter initials remains in the cambium, and the one that remains is usually the one that is closest to a ray initial. The ratio of fusiform initials to ray initials (and thus rays) decreases with age (10.3 at age 1 year to 2.6 at age 60 in *Pinus strobus*), and this implies that the rays are important sources of food materials for the cambium and that these energy demands increase with increasing tree and fusiform initial size. Both tree size and fusiform initial size increase with age. This is reflected in increasing fiber length with age.

It is not really sufficient to say that the three types of division (additive, multiplicative, and transformative) are a result of altered planes of the mitotic spindle. The three types are integrated into a tissue system, and the products of these divisions differentiate in a complex pattern after the division is completed. Thus there is an information system that tells the initial to divide, and another one that specifies the plane of the division, and still another one that specifies the developmental path of the daughter cells produced by the division.

At this point we can ask, what are the metabolic and synthetic functions that the cambium needs to perform the three types of cell divisions and the subsequent differentiation of the derivatives? While it is clear that events in the cambium itself control the fate of the products of the multiplicative and transformative divisions, it is not certain what role the cambium has in the differentiation of the products of the additive divisions, the xylem and phloem. Nevertheless, the main function of the cambium is additive division. The other kinds of divisions are internal or housekeeping, in that they only exist because they permit the additive divisions to continue unabated throughout the life of the tree. Furthermore, the tracheids derived from a single fusiform initial in *Picea sitchensis* show autocorrelation of morphological traits (Ford and Robards, 1976), which suggests some influence of the cambium or its xylem derivatives. Berlyn (1961) noted that in *Populus deltoides* the fusiform initials in close proximity to each functioned as a group in producing fiber tracheids with similarities in regard to the proportion of the fibers that contain a gelatinous layer. These morphogenetic clumps or domains suggest that, at the least, the same specific biochemical milieu that exists in the fusiform initials also exists in their derivatives at a given space–time point.

For our purposes we will focus on the cambium itself and on the developing xylem (cambial zone), from the initiation of the xylem mother cell (XMC) to the fully differentiated xylem cell. The cambium itself is concerned with the synthesis of the primary cell wall, cytoplasm, the mitotic apparatus, and chromosome replication. These activities are largely a function of directed protein and nucleic acid synthesis. The XMC, on the other hand, are largely concerned with secondary cell-wall formation, and this is largely a function of the synthesis of cellulose, hemicellulose, and lignin. There is some evidence (e.g., Bethel, 1964), however, that the amount of secondary as well as primary wall material of a tracheid is determined when it is formed by the fusiform initial. Some cell division also occurs in the cambial zone, and the frequency may even exceed that in the cambium during periods of rapid growth. The developing vascular tissues are heavily dependent on photosynthesis and respiration and the allocation mechanisms that permit them to utilize current and stored energy sources. Phloem mother cells (PMC) are organized primarily for cellulose and hemicellulose synthesis and the production of transport proteins, P-protein, and cytoskeletal elements.

If the cambium itself does function both as a cell organizer and as a cell maker, it is heavily dependent on bioenergetics. Rays are energy storage depots, and the cambium insures that the PMC and XMC contain sufficient ray tissue by controlling the frequency of transformative divisions. In addition to transporting carbohydrates from the phloem into the xylem, the rays also release sugar to vessels from specialized ray cells, termed "contact cells," that are connected to the vessels by large pits (Sauter, 1972; Sauter *et al.*, 1973). Sauter (1972) found that starch began to disappear from xylem parenchyma cells in the spring at the same time that sugars appeared in the vessel sap of sugar maple. The process is too fast for diffusion, and respiratory activity is required for release of sucrose into vessels. The cambium produces wood of a genetic type and yet is capable of responding to environmental stimuli to manifest phenotypic plasticity in its products. The cambium could exert some control over these processes by packaging into the cytoplasm of its derivatives a system or potential system (epigenetic control) that defines the activity of the various metabolic compartments of the derivative (XMC or PMC). This might be accomplished by organelle distribution at cell division or by including activators for so-called chrono- or homoeotic genes (North, 1983). The cambium, like the apical meristems of plants, is a continuing embryonic region and subject to embryonic-like developmental controls. In general, regulated genes are those whose activities are regulated in response to environmental and/or developmental signals (Struhl, 1982).

Conceptually, they can be construed to have three elements, namely, structure, promoter, and regulator. The regulatory elements are responsible for the change in level of expression in response to time signals from the environment or intrinsic developmental chronometers.

Whatever the genetic mechanism is, it must function in energy transport, storage, and utilization, because photosynthesis and respiration are central processes in cambial function.

II. Photosynthesis

A. *Introduction*

The management of forests today is focused on maximizing the amount of net photosynthesis (P_N) per unit of land area (S) and on optimizing the conversion of P_N into usable wood substance. Forests are of global importance in the carbon fixation cycle, because forests occupy one-third of the land surface of the earth and produce two-thirds of the annual dry matter production (Lehninger, 1965; Leith, 1972, 1975). This amounts to about 6×10^{10} metric tons of dry matter produced annually by the world's forests and from which our forest products must come. Because of this massive amount of carbon fixation, forests are a key component in maintaining the carbon dioxide balance of the earth. Largely because of the process of transpiration, which is a consequence of carbon fixation, forests are also thought to exert a significant influence on the weather patterns of the world.

Basically, photosynthesis is a process whereby light energy is transformed into chemical energy, some of which is manifested in aggregates of stable chemical bonds such as wood and food. Life on earth is, of course, totally dependent on this process, and until now, anthropogenic effects on photosynthesis were minor.

Since the industrial revolution some 200 years ago, however, detrimental effects on carbon fixation have increased steadily due to the rapid rise in the levels of phytotoxic chemicals in the environment. Extensive clearing of forests has also been a factor; however, all of these effects are overshadowed by the relatively recent development of thermonuclear weapons, which have the potential not only to inhibit carbon fixation but also to terminate it. Even in a so-called limited nuclear war, the damage to the ozone layer would be severe, and the subsequent influx of ultraviolet radiation would have a distinctly negative effect on photosynthesis (Brandle *et al.*, 1977; Westing, 1977; Trocine *et al.*, 1981). This effect would be combined with cloud, smog, and radionuclides, however, the result of such interactions would certainly be a reduction in carbon fixation.

Photosynthesis is initiated when photosynthetically active photons (PAP, 400–700 nm) strike the photosynthetic event horizon—the chlorophyll molecules in the grana thylakoids of the chloroplast. Thus, photosynthesis has structural or spatial coordinates as well as a time axis. For this reason, we subdivide photosynthesis into four complex and interdependent space–time compartments: (1) pigment excitation; (2) electron transfer; (3) product formation; and (4) product translocation, allocation, storage, and utilization (including cambial activity and wood formation).

B. *Pigment Excitation*

Pigment excitation occurs in 10^{-15} to 10^{-6} sec. Chlorophyll absorbs the appropriate photons in a quantized manner and is projected from the ground state into the excited singlet state or in some cases the triplet state (see Parson and Ke, 1982). Of a pair of valence electrons in the outer shell, one is raised to the higher energy level. If its spin is opposite the paired electron in the ground state (as it was before excitation), the excited state is referred to as a singlet (or some subdivision, like first or second singlet) state. If the spin is reversed, so that the ground-state and excited-state electron have the same spin, then the triplet state is attained. Once in the excited state, the electron may be ejected from its orbital and used in a photochemical reaction, leaving a "green hole" (the oxidized chlorophyll molecule) in its wake. Other possible fates include a return to ground state by fluorescence (or phosphorescence, resonance transfer, or heat emission). The direct result of the photochemical reaction is the formation of ATP and the reduction of NADP, and when this is occurring, energy losses due to fluorescence are quenched. The photochemical reaction only occurs in special portions of the chlorophyll, termed reaction centers, which are contained in groups of pigments called photosystems. The rest of the chlorophyll and the accessory pigments of a particular photosystem transfer the light energy they collect to the reaction center by a process called resonance or exciton transfer. Thus, these collector molecules serve as antennae for the reaction center. Each antenna pigment transfers a portion of its absorbed energy (exciton) to another antenna pigment with a longer wavelength absorption maximum. The transfer continues until the reaction center is reached; thus the reaction center of the photosystem has the longest wavelength absorption maximum. Higher plants have two types of photosystems (PSI and PSII) linked together in series (see Emmerson 1958; Parson and Ke, 1982; Govindjee and Govindjee, 1975). Thus the light energy absorbed over a rather large number of wavelengths is channeled, in $\sim 10^{-9}$ sec, into the long-wavelength trap, which either utilizes the energy to make ATP and reduce

NADP or releases it as fluorescence. This is why it is the number of photons absorbed and not the energy per photon that is important in photosynthesis. Apparently, a similar amount of energy per photon arrives at the reaction center no matter what the energy of the photon was at the point of contact with the event horizon (photon equivalence).

The structural compartment or event space for photosynthesis is the chloroplast. The light-gathering pigments are situated in the thylakoid membranes of the grana; according to Miller (1982), these photosynthetic membranes are assemblages of subunits. Each subunit contains a reaction center, polypeptides, antennae pigments, and electron transport components. Miller's subunit is about 12 nm in diameter and 5 nm thick and consists of a large central element that protrudes from the membrane, surrounded by six smaller lobes. Other authors (Dyson, 1978; Armond and Arntzen, 1977) identify two classes of membrane particles. The external, smaller, and more densely packed particles appear to be associated with PSI activity, while the larger, more sparsely distributed particles, associated with the interior of the membrane, show PSII activity. Ribulose bisphosphate carboxylase/oxygenase particles are also thought to be loosely attached to the outer surface of the thylakoid membrane. It should be noted that the reaction centers of green plants have never been isolated from their antenna complex, and thus the photosystems may not have the separability of the single photosystem of prokaryotes. The chlorophyll is wholly contained within the thylakoids of the grana, at least to the detection limits of fluorescence microscopy (G. P. Berlyn, unpublished).

The following calculation indicates the magnitude of the light-gathering capacity of a hypothetical leaf, 6 in. × 4 in. (15.24 cm × 10.16 cm = 154.84 cm^2):

$$1.6 \times 10^5 \frac{\text{cells}}{\text{cm}^2} \times 154.84 \frac{\text{cm}^2}{\text{leaf surface}} = 2.48 \times 10^7 \frac{\text{cells}}{\text{leaf surface}}$$

$$15 \frac{\text{thylakoids}}{\text{granum}} \times 60 \frac{\text{grana}}{\text{chloroplast}} \times 50 \frac{\text{chloroplasts}}{\text{cell}} = 4.5 \times 10^4 \frac{\text{thylakoids}}{\text{cell}}$$

$$2.48 \times 10^7 \frac{\text{cells}}{\text{leaf surface}} \times 4.5 \times 10^4 \frac{\text{thylakoids}}{\text{cell}} = 1.12 \times 10^{12} \frac{\text{thylakoids}}{\text{leaf surface}}$$

Each thylakoid is $\sim 5 \times 10^{-5}$ cm in diameter or 2×10^{-9} cm^2; therefore $1.12 \times 10^{12} \times 2 \times 10^{-9} = 2000$ cm^2 of thylakoid surface in this one-cell-thick leaf surface. If the leaf contains four layers of chlorenchyma (excluding epidermis), there would be ~ 8000 cm^2 of thylakoid surface in the leaf. However, our studies have indicated that the light-gathering capacity of leaves of *Populus deltoides* is sufficient to absorb all the light that impinges on the nonveinous portion of the leaf. The light that is actually transmitted through the leaf is that portion that passes through the veins. When an intact leaf is observed in a microfluorospectrophotometer, even the finest veins of the aereoles transmit the incident light, while only fluorescent light is emitted from the rest of the leaf. It is conceivable that the veins also function as light pipes and thereby increase total photon collection.

C. *Electron Transfer*

After the light energy is transferred to high-energy electrons, it is moved from molecule to molecule until the extra energy has been depleted. The bulk of this energy is not lost to entropy but is transferred to make ATP and NADPH, which can then be used for metabolism and biosynthesis. The path of movement of the photoexcited electrons is termed a photosystem, and we have mentioned that green plants have two of these. Prokaryotes have only PSI and therefore do not evolve O_2. A unique feature of photosynthesis is that electrons are induced to flow away from water (against the free-energy gradient), releasing molecular oxygen.

The primary electron donor for PSI is P700, and the primary electron donor for PSII is P680. The pigment–protein complex of the reaction centers is the primary electron donor for the photosystems. Electrons pass from PSII to the acceptor Q (quencher), which is most probably a quinone. From there they pass through possibly three acceptors (A_i), which may include at least one *b*-type cytochrome. The electrons are then moved through a plastoquinone pool (PQ), through cytochrome *f* to plastocyanin, which is able to pass electrons to PSI. During this noncyclic electron flow, ATP is generated (photophosphorylation) by a mechanism believed to be similar to the chemiosmotic process postulated by Mitchell (1966) for mitochondria. The difference is that while mitochondria eject protons during oxidation, chloroplasts accumulate them (ΔpH $\sim$3.5), causing the external medium to become alkaline. The pH gradient gives rise to a voltage gradient, and these gradients are potential sources of energy. In the chloroplast, the ATPase particles are situated on the outer side of the membrane, instead of the inside as in mitochondria. This corresponds to the reverse gradient in pH. The energy of the gradient is utilized for ATP synthesis in the process of photophosphorylation. In photosynthesis the weak oxidant, CO_2, must oxidize a weak reductant, H_2O, producing a strong oxidant, O_2, and a strong reductant, carbohydrate. This requires a great amount of ATP energy, i.e., 112 kcal/mol CO_2 reduced or 672 kcal/mol glucose produced.

D. *Path of Carbon*

To date three biochemical pathways of carbon fixation have been found in plants: the reductive pentose phosphate (Calvin) cycle, also known as the C_3 pathway; the C_4 cycle (Hatch–Slack–Kortschak); and the CAM cycle (crassulacean acid metabolism). These pathways are associated with different taxa and, on a biochemical level, differ in the enzymes and intermediates involved in the primary carboxylation steps. Structurally, there are differences in the arrangement and organization of the photosynthetic tissue.

Only the C_3 pathway is known to function in woody plants, and our discussion will focus on this cycle. It was the first pathway discovered, and the steps were determined by Calvin and co-workers (Benson and Calvin, 1947; Bassham and Calvin, 1957). At the heart of this cycle is the enzyme ribulose 1,5-bisphosphate carboxylase/oxygenase (rubisco), which catalyzes the first step in the pathway: the carboxylation of the phosphorylated pentose, ribulose 1,5-bisphosphate. The six-carbon intermediate formed during this reaction is unstable and rapidly decomposes to form the three-carbon intermediate, 3-phosphoglyceric acid. A total of 11 enzymes are known to be involved in this reductive pentose phosphate pathway. Five of these enzymes, including rubisco, are localized exclusively in the chloroplast stroma; the other six are found in both the stroma and the cytoplasm. The stromal enzymes are thought to be the primary regulators of the overall rate of the C_3 cycle. This regulatory process is quite complex and involves, among other things, the concentrations of inorganic ions, the levels of NADPH and ATP generated during the electron-transfer phase, and feedback or allosteric control of enzymes early in the cycle by metabolites and by intermediates produced further along (Bassham, 1971; Walker, 1976). The C_3 cycle is ubiquitous in all plants, even those that can utilize the other two fixation cycles, and thus it is apparently the universal cycle. It appears probable that the C_4 and CAM pathways directly or indirectly evolved from it (Smith and Robbins, 1974). Only the C_3 cycle is autocatalytic, in that more CO_2 acceptor molecules can be generated by the cycle than are initially present at the start of the process (Kelly and Latzko, 1976).

III. Respiration Processes

A. *Respiration*

Through the intricate and complex pathways of photosynthesis, woody plants utilize the energy of the sun to synthesize all the major macromolecules necessary for their existence. Tapping the energy locked within these compounds, and within others obtained from exogenous sources, is ultimately controlled by the complex energy-harnessing process, respiration.

Broadly, this process can be defined as the biochemical reactions whereby large organic molecules are broken down into smaller molecules, which are then oxidized with the concomitant release of energy. Some of the energy released by the oxidations is then utilized to maintain the plant in a beneficial equilibrium with its environment (maintenance respiration), to provide energy for the synthesis of new macromolecules and tissue (growth respiration), and to power the processes of assimilation and allocation of nutrients within the plant.

At the biochemical level, the harnessing of the energy locked within organic compounds is a complex process, occurring in two stages: an anaerobic phase (glycolysis) and an aerobic phase (respiration). The aerobic phase can be further subdivided into two stages: a catabolic stage (the tricarboxylic acid cycle) and an energy-conservation stage (electron transport with concomitant oxidative phosphorylation).

The anaerobic phase of the energy harnessing process is one of several catabolic pathways that are grouped under the heading anaerobic fermentations. These are ancient metabolic pathways that seem most likely to have developed in the first organisms that evolved in the initially anaerobic, reducing environment of the earth of several billion years ago. Today these pathways are found in all organisms, both in the obligate anaerobes (those organisms, principally bacteria, that can not use or even survive in the presence of oxygen) and in the facultative anaerobes (those organisms, unicellular eukaryotes and most higher plants, that can function in either the presence or absence of oxygen).

Many different organic compounds can serve as substrates for anaerobic fermentations, including lipids, polysaccharides, and proteins. However, the most common substrates are the monomers of the polysaccharides, particularly glucose. Of those anaerobic metabolic pathways that utilize glucose as a substrate, the most common are glycolysis and alcoholic fermentation. The former occurs in many microorganisms and the cells of most higher plants, while the latter pathway is found principally in yeasts and, interestingly, in the root systems of trees during flooding. (The surfeit of water produces anaerobic conditions in the soil, inhibiting aerobic metabolism.)

The metabolic pathway of glycolysis is localized in the cytosol. The overall reaction can be summarized by the following equation:

$$\text{glucose} + 2\ P_i + 2\ \text{ADP} + 2\ \text{NAD}^+ \longrightarrow$$
$$2\ \text{lactate} + 2\ \text{ATP} + 2\ H_2O + 2\ \text{NADH} + 2\ H^+ \qquad \Delta G^{\circ\prime} = -47\ \text{kcal mol}^{-1} \qquad (1)$$

The lactate formed via the glycolytic degradation of glucose cannot be catabolized any further, and here the anaerobic phase of the energy-harnessing process halts.

When compared with the standard free-energy change for the complete oxidation of glucose, summarized in the equation

$$\text{glucose} + 6\ O_2 \longrightarrow 6\ CO_2 + 6\ H_2O \qquad \Delta G^{\circ\prime} = -686.0\ \text{kcal mol}^{-1} \tag{2}$$

it is obvious that anaerobic fermentation is capable of liberating only a very small percentage of the chemical energy locked within the glucose molecule. Utilization of the full chemical energy potentially available in organic compounds is accomplished in the next step in the energy-harnessing process, the aerobic phase, respiration.

As mentioned earlier, the aerobic phase, which is localized in specialized organelles (the mitochondria), can be further subdivided into two stages: a catabolic stage and an energy-conservation stage. The catabolic stage of respiration is known as the tricarboxylic acid (TCA) cycle. It is the common central pathway for the ultimate catabolism of not only carbohydrates but also fatty and amino acids. The TCA cycle is apparently universal and has been found in the cells of all aerobic organisms investigated to date.

The joining point of the anaerobic and aerobic phases of the energy harnessing process is the compound acetyl-CoA. This compound is produced by the oxidation and decarboxylation of pyruvate. The formation of acetyl-CoA from pyruvate, while not a direct part of the TCA cycle, is nonetheless the prerequisitive step for the catabolism of all carbohydrates via the TCA cycle. In fact, acetyl-CoA formation is the obligatory step for the degradation not only of carbohydrates but of all substrates, including amino, fatty, and nucleic acids.

The acetyl-CoA formed from pyruvate then enters the TCA cycle via an enzymatic reaction with oxaloacetate to form citrate. The citrate, in a series of linked steps, is then converted back to oxaloacetate, with the concomitant release, during the cycle, of two molecules of CO_2 and four pairs of hydrogen atoms.

In contrast to the linear nature of the anaerobic phase of the energy-harnessing process, the aerobic phase is cyclic and catalytic. In a full turn around the TCA cycle, the two-carbon acetyl group from acetyl-CoA combines with the four-carbon compound oxaloacetate to form the six-carbon citrate. The citrate is then decarboxylated twice to again form a four-carbon compound, which is converted in the next series of steps back to oxaloacetate. The oxaloacetate can then accept another two-carbon acetyl group and commence the cycle again.

In the catabolic stage of respiration, glucose is further degraded; however, the chemical energy released is not captured and stored in a usable form at this point. This is accomplished in the next stage of respiration: electron transport and concomitant oxidative phosphorylation. In this stage, the four pairs of hydrogen atoms produced during one full turn of the TCA cycle are shunted down a chain of oxidation–reduction enzymes embedded in the inner membranes of the mitochondria. These enzymes (which can be classified into four different types: cytochromes, pyridine-linked dehydrogenases, flavin-linked dehydrogenases, and iron–sulfur proteins) transfer electrons, ultimately from glucose and other organic substrates, to molecular oxygen.

The standard free-energy change that occurs during the transfer of a pair of electron equivalents from NADH ($E_0' = -0.32$ V) to molecular oxygen ($E_0' = +0.82$ V) via the electron-transport chain is considerable. A decrease in free energy of almost 53 kcal mol^{-1} occurs, which is more than sufficient energy to generate several molecules of ATP from ADP and inorganic phosphate ($\Delta G^{\circ\prime} = +7.3$ kcal mol^{-1}). In fact, a total of 15 molecules of ATP are formed for each molecule of pyruvate oxidized to completion via the TCA cycle and the respiratory chain. As there are two molecules of pyruvate produced from each molecule of glucose that passes through the glycolytic pathway, this amounts to a total of 30 molecules of ATP for each molecule of glucose oxidized to completion:

$$2 \text{ pyruvate} + 30 \text{ ADP} + 30 \text{ P}_i + 5 \text{ O}_2 \longrightarrow 30 \text{ ATP} + 34 \text{ H}_2\text{O} + 6 \text{ CO}_2 \quad (3)$$

The amount of energy stored in the form of ATP from the catabolism of a molecule of glucose is the sum of the overall equations for glycolysis [Eq. (1)] and respiration [Eq. (2)] and amounts to a net gain of 32 molecules of ATP per molecule of glucose. This, however, is not the total amount, for several additional molecules of ATP can be generated from the oxidation of the two molecules of NADH formed during glycolysis. The pair of electrons obtained from the oxidation of NADH are shunted down the respiratory electron transport chain. Depending on the point of entry into the chain, two or three molecules of ATP can be synthesized for each pair of electrons. Thus, four or six additional molecules of ATP are obtained, resulting in a total of 36 or 38 molecules of ATP for each molecule of glucose catabolized. This represents an impressive energy-harnessing efficiency of over 38%.

The TCA cycle is found in virtually all facultative anaerobes and aerobic organisms (i.e., microorganisms and higher plants and animals). There is, however, a modified form of the TCA cycle, known as the glyoxylate cycle, that is found only in microorganisms and many higher plants. This cycle bypasses the decarboxylation steps of the TCA cycle by utilizing enzymes that are located in another specialized organelle—the glyoxysome. This cycle allows carbohydrates to be synthesized from fatty acids or acetyl-CoA as the sole carbon source. The cycle provides a source of succinate, a key precursor in gluconeogenesis. The overall reaction of the glyoxylate cycle is

$$2 \text{ acetyl-CoA} + NAD^+ + 2\,H_2O \longrightarrow \text{succinate} + 2\,\text{CoA} + NADH + H^+ \quad (4)$$

An additional oxidative pathway in plants, and indeed many other organisms, is the pentose phosphate pathway, also known as the hexose monophosphate shunt. It is an intricate and complex pathway, having multiple functions. The overall reaction of the pathway is

$$\text{glucose-6-phosphate} + 12\,NADP^+ + 7\,H_2O \longrightarrow 6\,CO_2 + 12\,NADPH + 12\,H^+ + P_i \quad (5)$$

Depending on the requirements and energy status of the cell or organism, the pentose phosphate pathway is utilized in several different ways. Most importantly, it serves, in modified form, as a participant in the CO_2 assimilatory pathways for glucose formation in the dark reactions of photosynthesis. Additional information on the topic can be found in Davies (1980) and Goodwin and Mercer (1972).

B. *Photorespiration*

An additional pathway for CO_2 evolution exists in plants but not animals. It is termed photorespiration, to distinguish it from "dark" respiration and to indicate the unique difference between the two: that is, CO_2 evolution via this metabolic pathway occurs when plants are illuminated. Many studies have shown that this pathway is separate and distinct from the TCA cycle and electron transport, being unaffected by many of the factors (e.g., metabolic inhibitors, environmental conditions, etc.) that inhibit "dark" respiration.

The principle substrate for the photorespiratory pathway appears to be glycolic acid. This compound is oxidized to yield glyoxylic acid, which is then catabolized, depending on the species, to other compounds (e.g., formate or glycine), with the concomitant release of CO_2.

The rate of photorespiration (based on CO_2 evolution) varies in different plants. In herbaceous and woody C_3 plants it is at its greatest, while in C_4 plants, due to differences in the arrangement of the biochemical and cellular photosynthetic apparatus, it is greatly reduced or absent.

As yet, the exact metabolic function of photorespiration is unknown. It appears on the surface to be a wasteful process, as it results in the net loss of photosynthate, ATP, and NADPH. However, given its almost universal occurrence in higher plants, it is hard to imagine that it serves no useful metabolic purposes. Additional information on the topic can be found in Zelitch (1971) and Hatch *et al.* (1971). An excellent review of the biochemistry of the process can be found in Lorimer (1981).

There is little direct information on the respiration of the cambium. Goodwin and Goddard (1940) showed that on a fresh-weight basis, the cambium is the tissue showing the highest respiration rate. However, when respiration was analyzed on a per nitrogen atom basis, the recently formed xylem was more active than the cambium. This is probably due to the fact that respiration on a fresh-weight basis in the xylem includes much hydrophyllic cell wall substance. Differentiating xylem performs a good deal of cellular work in the biosynthesis of cell wall substance, and thus a high respiration rate is to be expected. However, as autolysis begins in tracheary cells, a climacteric-type rise in respiration is also a possibility. Another type of respiration, the alternate pathway, is involved in the climacteric of fruits and might function in tracheary tissue too. This alternate oxidase pathway is also involved in thermogenesis of arums, a process that attracts insects and thus facilitates pollination. The mitochondria from the arums have been shown to have two parallel electron-transport pathways from substrate to oxygen. The second or alternative pathway has a terminal oxidase that is cyanide insensitive and possesses the properties of a nonheme iron–sulfur protein. This pathway is deficient in the energy-conservation phase and yields very little ATP through oxidative phosphorylation.

IV. Carbohydrate Allocation and Utilization

The process of photosynthesis may be summarized by the following equation:

$$6\ CO_2 + 18\ ATP + 12\ NADPH + 12\ H_2O \longrightarrow C_6H_{12}O_6 + 18\ ADP + 18\ P_i + 12\ NADP^+ + 6\ H_2O + 6\ O_2$$

The energy differential between the hexose product (2870 kJ mol^{-1}) and the reactants (3210 kJ mol^{-1}) is −340 kJ mol^{-1} ($\Delta G°$), and thus this reaction can proceed spontaneously. However, less than half of the energy incorporated into the hexose product is converted to cell walls such as wood. Larcher (1975) estimates that in *Fagus sylvatica* the allocation of photosynthate is 35% for growth (largely cell walls), 45% for respiration, and 20% lost by shedding, leaching, and root exudates. These uses are not uncorrelated: for example, respiration provides energy necessary for the biosyntheses required in growth and maintenance of living tissue. Some of the photosynthate is put into storage, and this storage is stratified into ready reserves and deep reserves (Glerum and Balatincez, 1980). In some cases, photosynthate is translocated to other individuals in the forest through root grafts. The amount of photosynthate allocated to the various categories will vary with genetics, climate, season, species, age, forest type and structure, degree of predation, pollution, and soil characteristics. For example, in the tropics a tree may be induced to channel more of its photosynthate into respiration due to high temperature, but this may be compensated for by higher light intensity and longer growing season.

In general, the allocation of photosynthate is a function of both innate developmental pattern and response to stress (stimuli). Bethel (1964) proposed that the amount of cell wall material allocated to a given conifer tracheid is established at the time of the cell division giving rise to that particular tracheid and is independent of the amount of enlargement that the cell undergoes. It is not clear whether this determinative cell division is that of the fusiform initial or the XMC, but the inference is that it is the division of the fusiform initial. Bethel's data further suggest that every cell within an annual ring that originates from a single fusiform initial will have the same cell wall volume. These data may be related to the autocorrelation within radial files observed by Ford and Robards (1976) and the autocorrelated physiological clumping reported by Berlyn (1961). These data also find some (but not complete) support in those of Mark (1967) and Kennedy (1961) but are contradicted by those of Beckwith (1969). If cell wall area is constant (or nearly so) radially along an annual ring, it would mean that the cambium is not a weak sink in the spring (as is generally supposed because of the "thin" cell walls of earlywood) but in fact is a strong sink throughout the annual cycle. This would also mean that rate of cell division and not auxin is the key factor determining carbohydrate utilization and therefore, to some degree, rate of photosynthesis. The major function of auxin is assumed to be control of cell enlargement; if wall volume is constant, then auxin does not control carbohydrate allocation to the cell wall. In some cases (e.g., cotton) fiber elongation is promoted by gibberellic acid and not affected by auxin (Beasley and Ting, 1973). The literature also suggests that the utilization of reserves stimulates assimilation rate, but the accumulation of reserves like starch may not itself curtail assimilation rate (Little and Loach, 1973). This may be due to the fact that the photosynthate is translocated away from the chloroplasts and compartmentalized; however, rubisco is inhibited by the first stable product of the carboxylation reaction, phosphoglyceric acid, and also by citrate (Bidwell, 1974). The citrate inhibition is thought to be due to citrate accumulation as a result of limited growth respiration and higher photosynthate availability.

While auxins may not function in controlling the amount of cell wall substance in a given cell, they can affect this process indirectly by stimulating photosynthesis. Also, auxin supply can be utilized artificially to produce "long-day latewood," that is, tracheids with large radial diameter and thick walls (Larson, 1969). Thus, in this case, the allocation of carbohydrate to cell-wall structure is amplified by auxin (and possibly other hormones, as well as a somewhat increased supply of carbohydrates due to the longer day lengths used in these experiments). Auxin can also induce compression wood in conifers, and in this case there is increased cell division and thick walls; however, compression-wood tracheids are shorter, and the actual increase in wall volume per cell may not be as great as it appears. Cytokinins are generally associated with cell division, and if cell division determines carbohydrate allocation to tracheids, then cytokinins may function in this process too.

The factors influencing cambial function can be summarized as physical (water, light, temperature, wind, gravity, fire, pressure, stress, bioelectrical); mineral (structural, stabilizing, coenzymes); hormonal (influence of cell division, cell size, and cell qualities such as chemical composition, volume, and density of cell walls); stress (mechanical, physiological, elastic, plastic); genetic (regulatory, developmental, and perhaps chrono- and homoeotic type genes, specific environmental sensing systems, self-assembly systems, feedback regulators); intrabiotic (source–sink effects on carbohydrate translocation, allocation, and utilization); interbiotic (competition, infection, predation, allelopathy, antibiosis, phytoalexins, epiphytism, symbiosis, parasitism); and anthropogenic (herbicides, pesticides, industrial chemicals and gases, acid rain, radiation, radionuclides).

Most of these agents can be considered *stressors,* environmental factors (external or internal) that at some dosage interval can produce stress in a given organism or tissue. Physiological stress is generally defined as stress that produces a change (strain) toward a physiologically limiting condition. Plants vary in their ability to resist stress, and this stress resistance generally involves stress avoidance and/or stress tolerance. Mechanical stress and biological stress are not completely identical, and the concepts and terminology can only be interchanged within limits. However, the induction of stress at a space–time point in the organism may induce the formation of a sink for nutrients and thus influence structural allocation (Gordon and Larson, 1968). For example, if a tree is stressed (within limits) by wind, it will often respond by increased wood production. This could reduce the photosynthate allocation to other structural and metabolic compartments if photosynthesis is limiting. In cases where accumulation of photosynthetic products has inhibited (if it does) photosynthesis (Sweet and Wareing, 1966; King *et al.,* 1967; Upmeyer and Koller, 1973), however, the new structural allocation might stimulate photosynthesis and thus increase the total carbohydrate supply. An important question is how stress is translated into metabolic and synthetic functions. Using Selye's (1976) general adaptive syndrome (GAS) developed for animal physiology, we have developed a hypothetical scheme for a plant stress syndrome (PSS). Several sources have been used in developing this scheme (Lang, 1961; Salisbury and Ross, 1978; Berlyn, 1979, 1982). The inclusion of the possibility of environmental modification is based on work with flax genotrophs (Cullis, 1973; Durrant and Timmis, 1973; Timmis and Ingle, 1973). Additional sources of information include Dhillon *et al.* (1978) and Grime and Mowforth (1982).

0. Ground state (GS)

External	Internal
Stressors, Zeitgebers →	Stress

↓

I. Stage of alarm (SA)
Messenger networks activated

↓

→ II. Stage of resistance (SR)
Withdrawal from ready reserves
Stress metabolism responses:
Direct, triggered, modulated, conditioned, carry-over, canalization (homeostasis), genome modification

↓

III. Stage of exhaustion (SE)

↙ ↘

Withdrawal from deep reserves ⇌ Senescence metabolism → . . . → Death

↓

→ IV. Stage of compensation/overcompensation (SC)
Debilitated tissue repaired
New tissue synthesis, (e.g., wood formation, compression wood, tension wood)
Resetting of stress threshold

Note that stress is not always detrimental but can increase growth and strength (Newcomb, 1895). Trees that are guyed to prevent wind sway are weak and easily blown down once the guy wires are removed. Similarly, dense pine plantations are subject to heavy wind loss if they are thinned late and/or too heavily.

The environmental stressors affect the total annual net carbon fixation and the allocation of the assimilates to the various organs and parts of organs within the tree (Gordon and Larson, 1968, 1970). Each species has its physiological niche and has developmental adaptation to the habitats that satisfy its niche specifications. Thus, trees distribute growth between roots, stems, and leaves in ways that insure maximum survival and reproduction. For example, in high-altitude environments or on minimal or deficient sites, there is often a greater investment in roots as opposed to stems or even leaves (Tranquillini, 1979; Keyes and Grier, 1981). Salisbury and Ross (1978, p. 162) state that in alpine and arctic areas, photosynthesis so exceeds respiration that carbohydrate accumulation is not a problem despite the short growing season. However, Billings (1973, 1974) points out that the total energy available for plant growth as measured in degree days above 0°C is greatly reduced in alpine habitats, and this is often further constrained by frequent cloud cover and fog. Nevertheless, even if sufficient photosynthate is produced, carbohydrate utilization is under time constraint in high-elevation environments, because as elevation increases, the completion of a number of critical developmental processes is curtailed. For example, DeLucia and Berlyn (1983) found that in balsam fir (*Abies balsamea*), mean cuticle thickness decreased with increasing elevation while mean cuticular water loss increased. In fact, the altitudinal limit of tree growth may be determined by the inability of tissue to fully mature in terms of the production of lignin, cuticle, and epicuticular wax (Wardle, 1971; Tranquillini, 1979).

Denne and Dodd (1981) concluded that the net influence of environmental factors on fiber dimensions of wood is small, even though it may be economically important. They reason that environmental effects on rates of differentiation are often offset by effects on duration of differentiation. Furthermore, it was reported that effects on substrate availability are offset by opposite effects on photosynthetic efficiency and on substrate allocation. Worrall (1968) also found that early initiation of cambial activity was correlated with early cessation, but the correlation coefficient was only 0.24. Northern races of *Picea abies* ceased growth first, and this would support a climate control effect. Date of cessation of cambial activity largely controlled latewood width. Wood density depends on the relative amounts of earlywood and latewood and on their respective densities.

The amount of substrate for wood formation depends on both current photosynthate and reserves. Deciduous species attain their peak content of reserves in autumn. The substrate decreases slightly during the winter months and then rapidly in the spring as new growth begins. Conifers, on the other hand, show a rapid increase in reserves in the late summer and early autumn and then a slight increase over the winter months, followed by a rapid decrease in the spring. These reserves are primarily in the form of starch and lipids. Lipid synthesis is reported to show an autumnal increase in the xylem parenchyma and pith (Glerum, 1977; Glerum and Balatinecz, 1980). Timell (1980) also shows that the cambium and XMC have heavy accumulation of starch and lipids in the dormant condition. Interestingly, Glerum found that the autumnal photosynthate was primarily allocated to respiration rather than the build-up of reserves; however, this respiration may supply energy required for synthesis of compounds important in the induction and maintenance of cold hardiness. Wood formation in the following spring was largely a function of current photosynthate, and the reserves were channeled into other components and respiration. In fact, according to Möller *et al.* (1954) as well as Larcher (1975), trees only allocate about one-third of the total photosynthate to structural growth. The respiration component is not entirely lost to entropy, however, and in fact Ledig *et al.* (1976) found that respiration and growth were episodic: growth was always preceded by a pulse of respiration. The cambium has the highest respiration of the stem tissues, exclusive of the apical meristems (Goodwin and Goddard, 1940), and it would be interesting to know if respiration of the initials and daughter-cell formation are also episodic.

The reactivation of the cambium after dormancy has been intensively studied, and many recent reviews of this topic are available (Berlyn, 1982; Little and Wareing, 1981; Savidge and Wareing, 1981). Most of the literature deals with trees growing in the north temperate zone, but it has been noted that generalized concepts of cambial growth rhythms should include a consideration of all of the cambiumiferous plants in the various climatic zones of the planet, including the complex patterns of trees growing in the tropics (Berlyn, 1982; Bormann and Berlyn, 1981). It has long been noted that cambium tends to reactivate basipetally below buds and developing leaves, structures that produce auxin. The addition of exogenous auxin was shown to be capable of initiating and stimulating cambial activity in nondormant or temperature-dormant tissue. One problem with this so-called hormone theory is that the cambium reactivates faster (42 mm hr^{-1}) than polar auxin transport [for a fuller discussion see Berlyn (1982)]. This suggests that (1) auxin moves by bulk flow in the phloem translocation stream; (2) the currently accepted rate of polar auxin transport (9–10 mm hr^{-1}) is too low; (3) the cambium synthesizes or releases auxin from bound form in response to a faster moving signal; or (4) the cambium changes in sensitivity to auxin.

Little and Wareing (1981) found that free and diffusable IAA remained high all winter in *Picea sitchensis* and that IAA concentration was always substantially higher than ABA (which also showed little seasonal fluctuation). They concluded that the primary factor in the metabolic reactivation of the cambium was a change in sensitivity. Since IAA is present in the cambial zone all year, as is sucrose (Parker, 1958), and the first cell divisions after reactivation generally occur in the XMC rather than in the initials themselves, Savidge and Wareing postulate that cambial reactivation may be dependent on amino acids arriving in the transpiration stream from roots and leaves. These compounds increase rapidly in the spring; thus, when the temperature is suitable, they supposedly function in promoting protein synthesis, which in turn sets off the reactivation metabolism. This could be accomplished in a number of ways and might, for example, cause the postulated change in hormone sensitivity.

References

Armond, P. A., and Arntzen, C. J. (1977). Localization and characterization of photosystem-2 in grana and stroma lamellae. *Plant Physiol.* **59,** 398–404.

Bassham, J. A. (1971). The control of photosynthetic carbon metabolism. *Science,* **172,** 526–534.

Bassham, J. A., and Calvin, M. (1957). *In* "The Path of Carbon in Photosynthesis." Prentice-Hall, New Jersey.

Beasley, C. A., and Ting, I. P. (1973). The effects of plant growth substances on *in vitro* fiber development from fertilized cotton ovules. *Am. J. Bot.* **60,** 130–139.

Beckwith, J. R. (1969). "Crown growth and wood development in young loblolly pine." Unpublished doctoral dissertation, Yale University School of Forestry and Environmental Studies, New Haven, Connecticut.

Benson, A. A., and Calvin, M. (1947). The dark reduction of photosynthesis. *Science* **105,** 648–649.

Berlyn, G. P. (1961). Factors affecting the incidence of reaction tissue in *Populus deltoides* Bartr. *Iowa State J. Sci.* **35,** 367–424.

Berlyn, G. P. (1964). Recent advances in wood anatomy: The cell walls in secondary xylem. *For. Prod. J.* **14,** 467–476.

Berlyn, G. P. (1970). Ultrastructural and molecular concepts of cell-wall formation. *Wood Fiber* **2,** 196–227.

Berlyn, G. P. (1979). Physiological control of differentiation of xylem elements. *Wood Fiber* **11,** 109–126.

Berlyn, G. P. (1982). Morphogenic factors in wood formation and differentiation. *In* "New Perspectives in Wood Anatomy" (P. Baas, ed.), pp. 123–150. Nijhoff, The Hague.

Bethel, J. S. (1964). The developing architecture of secondary xylem in conifers. *For. Sc.* **10,** 89–91.

Bidwell, R. G. S. (1974). *In* "Plant Physiology." Macmillan, New York.

Billings, W. D. (1973). Arctic and alpine vegetation: Similarities, differences, susceptibility to disturbance. *Bioscience* **23,** 697–704.

Billings, W. D. (1974). Adaptations and origins of alpine plants. *Arctic Alpine Res.* **6,** 129–142.

Bormann, F. H., and Berlyn, G. P. (eds.) (1981). *In* "Age and growth rate of tropical trees" (Bull. No. 94). Yale Univ. School of For. and Environmental Studies, New Haven, Connecticut.

Brandle, J. R., Campbell, W. F., Sisson, W. B., and Caldwell, M. M. (1977). Net photosynthesis electron transport capacity, and ultrastructure of *Pisum sativum* L. exposed to ultraviolet-B radiation. *Plant Physiol.* **60,** 165–169.

Cullis, C. A. (1973). DNA differences between flax genotrophs. *Nature* (*London*) **243,** 515–516.

Davies, D. D. (1980). *In* "The Biochemistry of Plants, Vol. 2: Metabolism and Respiration." Academic Press, New York.

DeLucia, E., and Berlyn, G. P. (1984). The effect of increasing elevation on leaf cuticle thickness and cuticular transpiration in balsam fir. *Can. J. Bot.* in press.

Denne, M. P., and Dodd, R. S. (1981). The environmental control of xylem differentiation. "Xylem Cell Development" (J. R. Barnett, ed.), pp. 236–255. Castle House, Kent.

Dhillon, S. S., Berlyn, G. P., and Miksche, J. P. (1978). Nuclear DNA content in populations of *Pinus rigida*. *Am. J. Bot.* **65,** 192–196.

Durrant, A., and Timmis, J. N. (1973). Genetic control of environmentally induced changes in *Linum*. *Heredity* **30,** 367–379.

Dyson, R. D. (1978). "Cell Biology," 2nd Ed. Allyn and Bacon, Boston.

Emerson, R. (1958). The quantum yields of photosynthesis. *Ann. Rev. Plant Physiol.* **9,** 1–24.

Ford, E. D. (1981). Can we model xylem production by conifers? *Stud. For. Suec.* **160,** 19–29.

Ford, E. D., and Robards, A. W. (1976). Short term variation in tracheid development in the early wood of *Picea sitchensis*. *In* "Wood Structure in Biological and Technological Research" (P. Baas, A. J. Bolton, and D. M. Catling, eds.). Leiden Bot. Series No. 3. Leiden Univ. Press, The Hague.

Glerum, C. (1977). "Formation and distribution of food reserves during autumn and their subsequent utilization in jack pine." Ph.D. thesis, University of Toronto, Ontario.

Glerum, C., and Balatinecz, J. J. (1980). Formation and Distribution of Food Reserves During Autumn and Their Subsequent Utilization in Jack Pine. *Can. J. Bot.* **58,** 40–54.

Goodwin, R. H., and Goddard, D. R. (1940). The oxygen consumption of isolated woody tissues. *Am. J. Bot.* **27,** 234–237.

Goodwin, T. W., and Mercer, E. I. (1972). "Introduction to Plant Biochemistry." Pergamon Press, New York.

Gordon, J. C., and Larson, P. R. (1968). Seasonal course of photosynthesis, respiration and distribution of ^{14}C in young *Pinus resinosa* trees as related to wood formation. *Plant Physiol.* **43,** 1617–1621.

Gordon, J. C., and Larson, P. R. (1970). Redistribution of ^{14}C-labelled reserve food in young red pines during shoot elongation. *For. Sci.,* 14–20.

Govindjee, and Govindjee, R. (1975). Introduction to photosynthesis. *In* "Bioenergetics of Photosynthesis" (Govindjee, ed.), pp. 2–50. Academic Press, Inc., New York.

Grime, J. P., and Mowforth, M. A. (1982). Variation in genome size—an ecological interpretation. *Nature* (*London*) **299,** 151–153.

Hatch, M. D., Osmond, C. B., and Slatyer, R. O. (1971). "Photosynthesis and Photorespiration." Wiley, New York.

Kelly, G. J., and Latzko, E. (1976). Regulatory aspects of photosynthetic carbon metabolism. *Ann. Rev. Plant Physiol.* **27,** 181–205.

Kennedy, R. W. (1961). Variation and periodicity of summerwood in some second growth Douglas fir. *Tappi* **44,** 161–166.
Keyes, M. R., and Grier, C. C. (1981). Above- and below-ground net production in 40-year-old Douglas fir stands on low and high productivity sites. *Can. J. For. Res.* **11,** 599–605.
King, R. W., Wardlaw, I. F., and Evans, L. T. (1967). Effect of assimilate utilization on photosynthetic rate in wheat. *Planta* **77,** 261–276.
Lang, A. (1961). Unpublished outline—elaborated in "Plant Physiology," (F. B. Salibury and C. Ross, eds.), pp. 690–691. Wadsworth, Belmont, California.
Larcher, W. (1975). "Physiological Ecology." Springer-Verlag, New York.
Larson, P. R. (1969). "Wood Formation and the Concept of Wood Quality (Bulletin No. 74). Yale University School of Forestry and Environmental Studies, New Haven, Connecticut.
Ledig, F. T., Drew, A. P., and Clark, J. G. (1976). Maintenance and constructive respiration, photosynthesis and net assimilation rate in seedlings of pitch pine (*Pinus rigida* Mill.). *Ann. Bot.* **40,** 289–300.
Lehninger, A. L. (1965). "Bioenergetics." Benjamin, New York.
Leith, H. (1972). Über die Primärproduktion der Erde. *Z. Angew. Bot.* **46,** 1–37.
Leith, H. (1975). Primary productivity in ecosystems: Comparative analysis of global patterns. *In* "Unifying Concepts in Ecology" (W. H. vanDobben and R. H. Lowe-McConnell, eds.), pp. 67–88. Junk, The Hague.
Little, C. H. A., and Loach, K. (1973). Effect of changes in carbohydrate concentration on the rate of net photosynthesis in mature leaves of *Abies balasamea*. *Can. J. Bot.* **51,** 751–758.
Little, C. H. A., and Wareing, P. F. (1981). Control of cambial activity and dormancy in *Picea sitchensis* by indol-3-ylacetic and abscisic acids. *Can. J. Bot.* **59,** 1480–1492.
Lorimer, G. H. (1981). The carboxylation and oxygenation of ribulose 1,5-bisphosphate: The primary events of photosynthesis and photorespiration. *Ann. Rev. Plant Physiol.* **32,** 349–384.
Mark, R. E. (1967). "Cell Wall Mechanics of Tracheids." Yale University Press, New Haven, Connecticut.
Miller, K. R. (1982). Three-dimensional structure of a photosynthetic membrane. *Nature (London)* **200,** 53–55.
Mitchell, P. (1966). Chemiosmotic coupling in oxidative and photosynthetic phosphorylation. *Biol. Rev.* **41,** 445–502.
Möller, C. M., Müller, D., and Nielsen, J. (1954). Graphic presentation of dry matter production of european beech. *Forstl. Forsoegsvaes. Dan.* **21,** 327–335.
Newcomb, F. C. (1895). The regulatory formation of mechanical tissue. *Bot. Gaz.* **20,** 441–448.
North, G. (1983). Genes and development: Cloning the genes that specify fruit flies. *Nature (London)* **303,** 134–136.
Parker, J. (1958). Changes in sugars and nitrogenous compounds of tree barks from summer to winter. *Naturwissenschaften* **45,** 139.
Parson, W. W., and Ke, B. (1982). Primary photochemical reactions. *In* "Photosynthesis: Energy Conversion by Plants and Bacteria" (Govindjee, ed.), Vol. 1, pp. 331–385. Academic Press, New York.
Salisbury, F. B., and Ross, C. W. (1978). "Plant physiology," 2nd Ed. Wadsworth, Belmont, California.
Sauter, J. J. (1972). Respiratory and phosphatase activities in contact cells of wood rays and their possible role in sugar secretion. *Z. Pflanzenphysiol.* **67,** 135–145.

Sauter, J. J., Iten, W., and Zimmermann, M. H. (1973). Studies on the release of sugar into the vessels of sugar maple (*Acer saccharum*). *Can. J. Bot.* **51,** 1–8.
Savidge, R. A., and Wareing, P. F. (1981). Plant-growth regulators and the differentiation of vascular elements. *In* "Xylem Cell Development" (J. R. Barnett, ed.), pp. 192–235. Castle House, Kent.
Selye, Hans. (1976). "The Stress of Life." McGraw-Hill, New York.
Smith, B. N., and Robbins, M. J. (1974). Evolution of C_4 photosynthesis: An assessment based on $^{13}C/^{12}C$ ratios and kranz anatomy. *In* "Proc. Third Internat. Cong. Photosyn" (M. Ausan, ed.), pp. 1579–1587. Elsevier, Amsterdam.
Struhl, K. (1982). Regulatory sites for *his*-3 gene expression in yeast. *Nature* (*London*) **300,** 284–287.
Sweet, G. B., and Wareing, P. F. (1966). Role of plant growth in regulating photosynthesis. *Nature* (*London*) **210,** 77–79.
Timell, T. E. (1980). Organization and ultrastructure of the dormant cambial zone in compression wood of *Picea abies*. *Wood Sci. Tech.* **14,** 161–179.
Timmis, J. N., and Ingle, J. (1973). Environmentally induced changes in r-RNA redundancy. *Nature New Biol.* **244,** 235–236.
Tranquillini, W. (1979). "Physiological Ecology of the Alpine Timberline." Springer-Verlag, Berlin.
Trocine, R. P., Rice, J. D., and Wells, G. N. (1981). Inhibition of seagrass photosynthesis by ultraviolet-B radiation. *Plant Physiol.* **68,** 74–81.
Upmeyer, D. J., and Koller, H. R. (1973). Diurnal trends in net photosynthetic rate and carbohydrate levels in soybean leaves. *Plant Physiol.* **51,** 871–874.
Walker, D. A. (1976). Regulatory mechanisms in photosynthetic carbon metabolism. *Curr. Top. Cell Regul.* **11,** 203–241.
Wardle, P. (1971). An explanation for alpine timberline. *N. Z. J. Bot.* **9,** 371–402.
Westing, A. H. (1977). "Weapons of Mass Destruction and the Environment." Taylor and Francis, London.
Whittaker, R. H. (1975). "Communities and Ecosystems." 2nd Ed. Macmillan, New York.
Worrall, J. F. (1968). "Interrelationships Among Some Phenological and Wood Property Variables in Norway Spruce." Ph.D. Thesis, Yale University, New Haven, Connecticut.
Zelitch, I. (1971). "Photosynthesis, Photorespiration, and Plant Productivity." Academic Press, Inc., New York.

CHAPTER **5**

Cell Organelles and Their Function in Biosynthesis of Cell Wall Components: Control of Cell Wall Assembly during Differentiation

D. H. NORTHCOTE

Department of Biochemistry
University of Cambridge
Cambridge, United Kingdom

I. Introduction

Since the cell wall is outside the plasma membrane, the synthesis of the wall necessitates a complex mechanism by the cell for coordinated synthesis and secretion. The secretory material is held at the cell surface and is modified and added to, not only during the life of the cell, but even after its death, by contributions from neighboring cells. The cell wall plays many parts in the function of the whole plant, such as protection against predators, storage of nutrients, skeletal structure, and formation of conducting and transport vessels within the plant. Since the wall is formed by the cytoplasm and its structure and composition are continually altered during the development of the cell and also in response to environmental

Biosynthesis and Biodegradation of Wood Components

ISBN 0-12-347880-4

changes, there must be an exchange of information between the developing wall and the cytoplasm. This allows control of the synthesis and allows secretion to be maintained (Northcote, 1972, 1977).

It is now apparent that in eukaryotic cells there is a continuous transport of membranes and material contained in membrane-bound vesicles outward to the plasma membrane and cell wall from the Golgi apparatus. There is also inward movement from the plasma membrane to endosomes and lysosomes (vacuoles in higher plants) and movement of membrane-bound vesicles from the plasma membrane to the Golgi apparatus. The membrane is recycled at various stages in the transport chain by a shuttle of vesicles, especially during the transport from plasma membrane to endosomes (Farquhar, 1983; Helenius *et al.,* 1983). In this way, although the membrane material is preserved and reassembled at the source during the transport, the material that the vesicles carry is progressively moved and possibly modified on its route to the destination or sink. Within the Golgi apparatus, a major part of the synthesis of the polysaccharides of the cell wall occurs. There is thus a communication system from the plasma membrane surface to the endomembrane system of the cell (Willingham and Pastan, 1982).

Receptors on the plasma membrane surface could hold the growth factors that may control the cells' development, while receptors on the lumen side of the membranes of the Golgi apparatus may allow the partitioning of polysaccharides to that part of the membrane system that forms the export vesicles (Sly and Fischer, 1982; Pastan and Willingham, 1983). After dissociation of the ligands, it is possible that the receptors are recirculated to the respective surfaces.

This chapter is divided into two sections dealing with (1) the formation and location of precursors of polysaccharides and lignin and (2) the polymerization systems and location of polymers.

The cell wall will be considered as a composite consisting of a matrix phase, composed of hemicellulose, pectin, water, and lignin, and a microfibrillar phase, consisting of cellulose microfibrils. Other materials, such as proteins, waxes, cutin, and suberin, will be considered in conjunction with these main divisions as appropriate (Northcote, 1972).

II. Formation and Location of Precursors

A. *Precursors of Polysaccharides*

The initial precursors of all the glycosylation reactions are the nucleoside diphosphate sugar compounds that are formed by pyrophosphory-

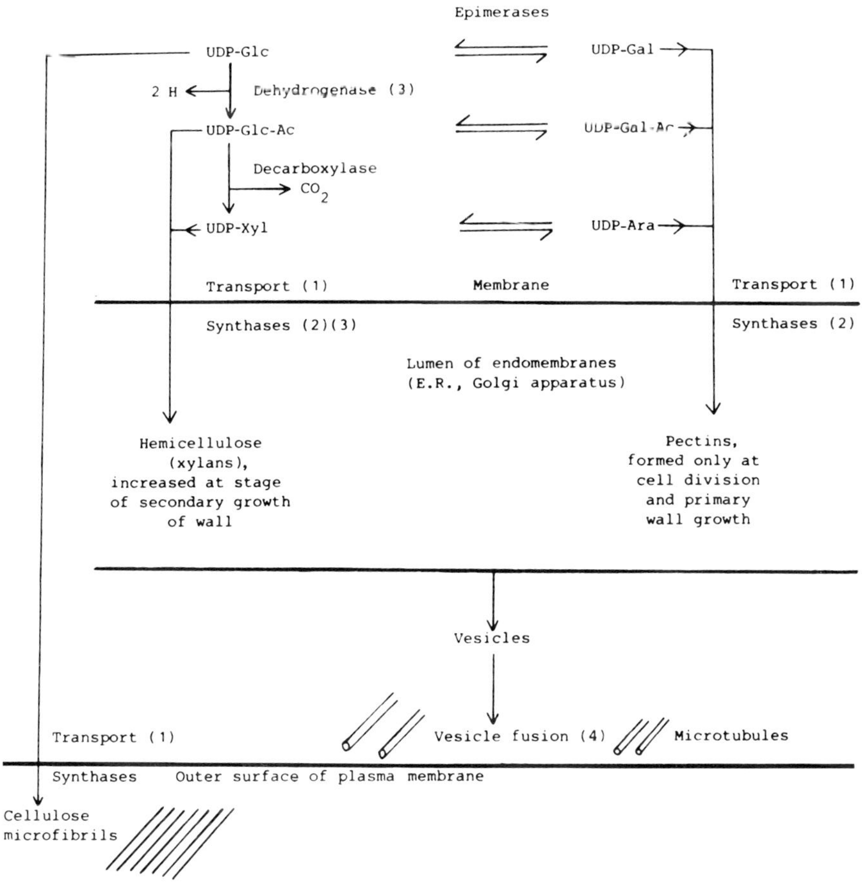

Fig. 1. Interconversions of nucleoside diphosphate sugars for synthesis of cell wall polysaccharides. The synthesis of the polysaccharides occurs at the endomembrane system of the cell, and possible control points for the formation of the polymers and their incorporation into the cell wall are indicated by numbers (1)–(4). (1) Transport systems at membrane. (2) Possibly by control of enzyme synthesis: synthases. (3) Biochemical modulation: UDP-Glc-dehydrogenase by level of UDP-Xyl; synthases by levels of nucleoside mono- and diphosphates (energy status of cell). (4) Vesicle fusion: conditions such as presence of particular proteins at the membrane and Ca^{2+}.

lases and that may be interconverted one to the other by dehydrogenases, decarboxylases, and epimerases (Fig. 1). The sugar moiety of the compounds is derived mainly from glucose, sucrose, or inositol (Northcote, 1969a). The enzymes for the reactions are soluble and probably occur free in the cytoplasm outside the endomembrane system (Dalessandro and

Northcote, 1977c). However, some epimerase activity and some of the other enzymes are found loosely bound to the membranes, probably at the cytoplasmic surface, although there is no evidence for their exact location (Feingold *et al.,* 1960; Axelos and Peaud-Lenöel, 1978; Dalessandro and Northcote, 1981a). In recent years it has been shown that many soluble enzymes found experimentally within the cytoplasm are loosely bound, either to membranes in an organized manner or by forming organized aggregates (Masters, 1981). In this way, sequences of directed reactions can be achieved in a concerted way with minimal dispersal of substrates. The control of these sequences of reactions can be maintained, modulated, and monitored by direct biochemical feedback control. Such a biochemical modulation is found for the interconversion reactions that lead to UDP-Xyl formation, since it is found that the level of this nucleoside diphosphate sugar influences the activity of the UDP glucose dehydrogenase, which is one of the initial steps in the production of the xylose derivative (Fig. 1). Thus the production of UDP-Xyl is controlled in response to the demand for the formation of xylans (Dalessandro and Northcote, 1977c).

During the initial stages of wall formation and also after the application of exogenous auxin to growing plant cells, it can be shown that the composition of pectin deposited in the wall changes (Rubery and Northcote, 1970; Hanke and Northcote, 1974; Boffey and Northcote, 1975). This is especially noticeable in relation to the proportion of polymers that contain arabinose in the pectin. The composition of the pectin at these early stages of wall formation and rapid cell expansion is important because the texture of the wall is made up in part by the gel properties of the pectin, which is dependent on its composition. Pectin makes up about 10–20% (w/w) of the matrix polysaccharides of the primary wall (Northcote, 1969b). The association of the pectin with water accounts for the fluid nature of the wall at these early stages and allows movement of microfibrils within the matrix as the wall expands. During the growth of the cell wall, there are small variations in the epimerase activities that interconvert the nucleoside diphosphate sugars (Fig. 1). These are minor changes, and although they probably serve to control, in part, the composition of the pectin, the epimerases are not the site at which the overall control of wall synthesis is exerted (Dalessandro and Northcote, 1977a). This is especially clear at the onset of secondary thickening, and subsequently during the thickening when pectin synthesis is shut down but the epimerase activities remain unchanged, so that precursors of pectin (UDP-Gal, UDP-Gal-Ac, and UDP-Ara) are still capable of being formed and the cell probably holds metabolic pools of these compounds (Dalessandro and Northcote, 1977a,b,c).

B. Lignin Precursors and Their Formation

The building units of lignin are formed from carbohydrate via the shikimic acid pathway to give aromatic amino acids (see Chapter 10, Section II). Once the aromatic amino acids are formed, a key enzyme for the control of lignin precursor synthesis is phenylalanine ammonia-lyase (PAL). This enzyme catalyzes the production of cinnamic acid from phenylalanine (Fig. 2). It is very active in those tissues of the plant that become lignified, although it is also a central enzyme for the production of other phenylpropanoid-derived compounds such as flavonoids and coumarins, which can occur in many parts of the plant and in many different organelles. Radioactive phenylalanine and cinnamic acid are directly incorporated into lignin in vascular tissue (Rubery and Northcote, 1968).

The induction of PAL activity at the onset of vascular differentiation can be shown by the use of plant tissue cultures (Haddon and Northcote, 1975; Kuboi and Yamada, 1978; Fukuda and Komamine, 1982). Xylem cells with secondary thickened and lignified walls are differentiated over a time course of 3–14 days by the application of the plant growth factors naphthylene acetic acid (NAA) and kinetin in the ratio 5 : 1 (1.0 mg/liter NAA, 0.2 mg/liter kinetin) to bean cells (*Phaseolus vulgaris*), either to a solid callus or to a liquid suspension (Haddon and Northcote, 1975; Bevan and Northcote, 1979a). The time for differentiation varies with the type of culture, solid or suspension, and with the frequency and duration of subculture, but for any one culture it is relatively constant (Haddon and Northcote, 1975; Bevan and Northcote, 1979b). For a fast-growing liquid suspension it occurred after 3–4 days from subculture to the induction medium (Jones and Northcote, 1981). At the time of differentiation when the xylem vessels are forming, the activity of PAL rises to a maximum (Fig. 3). The rising phase of the enzyme activity was inhibited by actinomycin D and by D-2-(4-methyl-2,6-dinitroanilino)-*N*-methylpropionamide (MDMP). These two inhibitors of protein synthesis were applied under carefully controlled conditions. Their concentrations and the period of time of exposure of the cells to the drugs were adjusted so that during the period when the production of enzyme activity was inhibited, the cells were active and, as judged by various criteria, were relatively normal (Jones and Northcote, 1981). That the rise in enzyme activity was inhibited by the two drugs, one of which inhibited transcription (actinomycin D) and the other translation (MDMP), indicated that both these processes were necessary for the response to the hormones.

The induction of PAL activity by the two growth factors can be separated in time so that they may act at different sites within the cell to bring

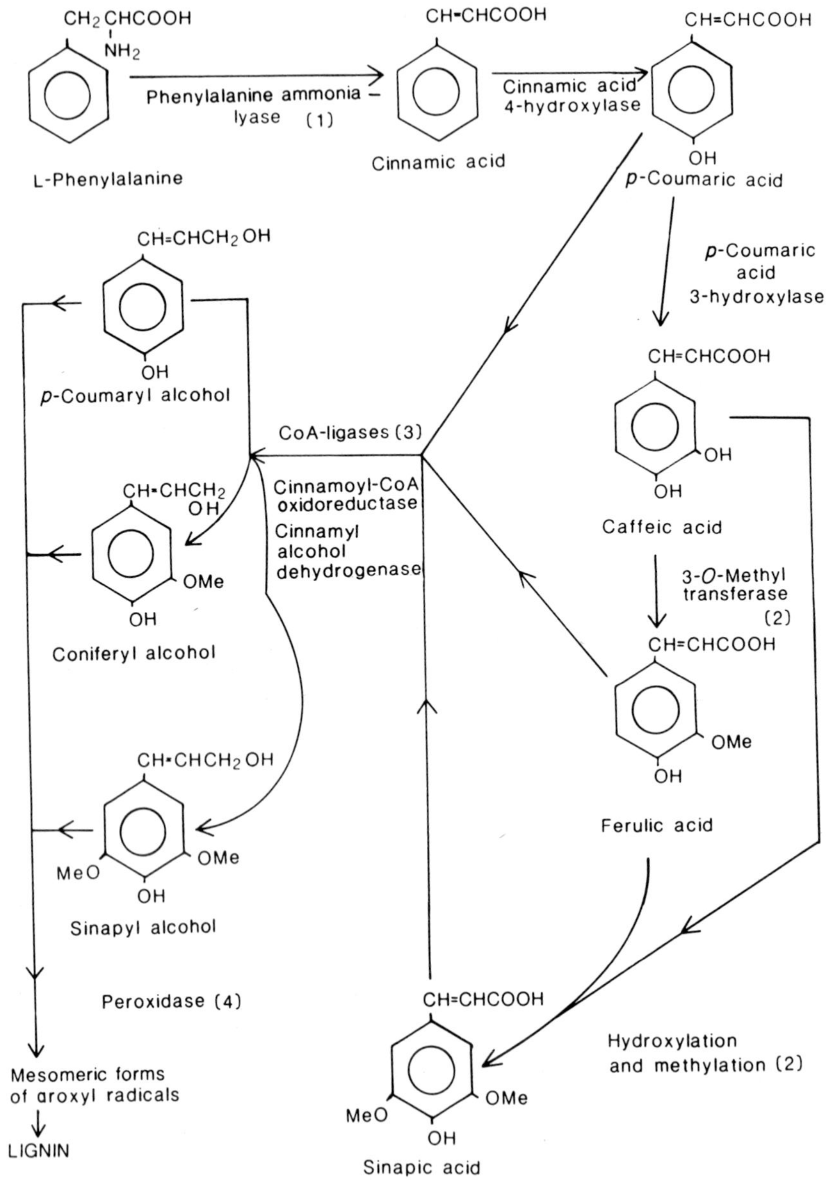

Fig. 2. The enzymatic steps involved in the formation of the precursors of lignin. Possible control points for the formation and type of lignin synthesized are indicated by numbers (1)–(4). See text for discussion.

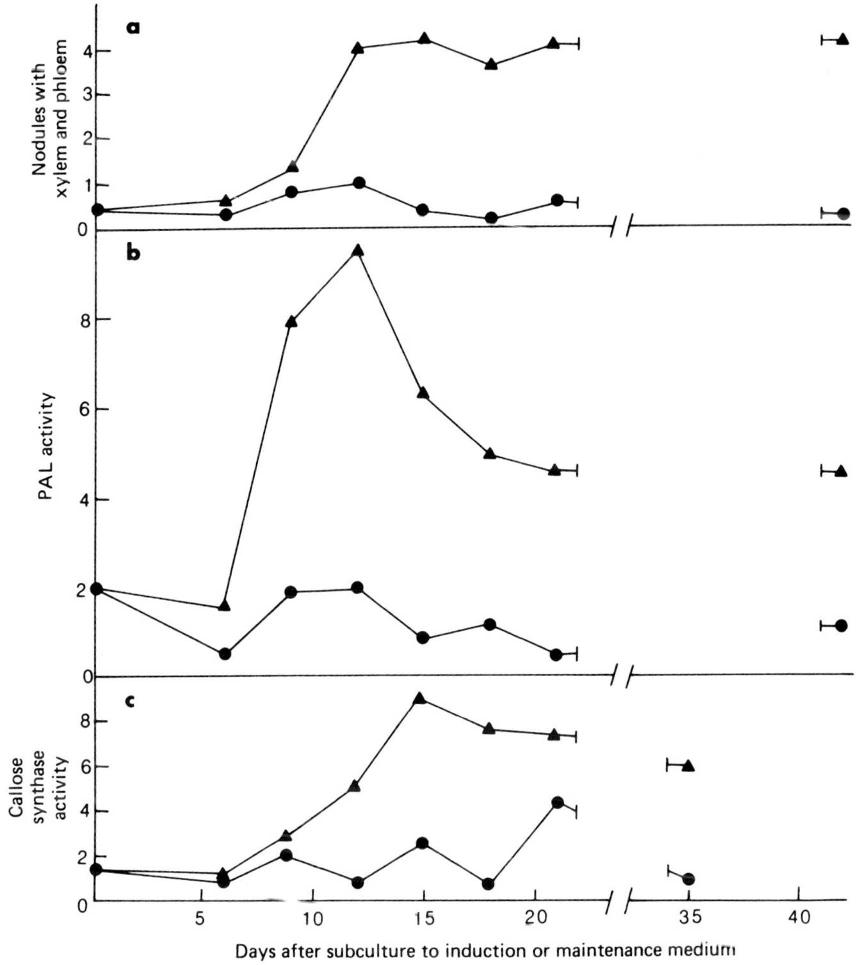

Fig. 3. The time course of differentiation in bean (*Phaseolus vulgaris*) callus after transfer from maintenance medium (sucrose 2%; 2,4-dichlorophenoxyacetic acid 2 mg/liter; kinetin 0.0 mg/liter) either to induction medium (▲) (sucrose 3%; 1-naphthylacetic acid 1 mg/liter; kinetin 0.2 mg/liter) or to maintenance medium (●). (**a**) Number of nodules containing xylem and phloem. (**b**) Phenylalanine ammonia-lyase activity. (**c**) Callose synthase activity. The amounts in each case are expressed per gram fresh weight of tissue in arbitrary units.

about the response (Bevan and Northcote, 1979a). Auxin added at the time of subculture of the tissue changes the pattern of protein synthesis of the cells by changing the transcription pattern of the mRNA after 2 hr (Bevan and Northcote, 1981a). Kinetin does not have this effect (Bevan and Northcote, 1981b).

The pathway for the production of the lignin building units is shown in Fig. 2. It has been demonstrated that the methylations are brought about by *S*-adenosylmethionine : caffeic acid, 3-*O*-methyl transferase. This transferase is meta specific and can also methylate 5-hydroxyferulic acid and 3,4,5-trihydroxycinnamic acid to sinapic acid (Poulton, 1981). The induction of this enzyme in bean cultures during differentiation is coincident with the rise in PAL activity (Haddon and Northcote, 1976). The hydroxylation of the aromatic ring of cinnamic acid is brought about by cinnamic acid 4-hydroxylase, and a further hydroxylase, *p*-coumaric acid 3-hydroxylase, also occurs to give caffeic acid (Butt and Lamb, 1981). The 4-hydroxylase activity, in some tissue, is induced at the same time as the PAL activity (Amrhein and Zenk, 1970; Hahlbrock and Wellmann, 1973). Therefore some coordinated induction of gene expression for the production of lignin precursors during differentiation is possible.

The PAL activity that is necessary for lignin formation occurs in the cytoplasm or bound to the cytoplasmic surface of the endoplasmic reticulum membranes. The cinnamic acid produced is probably carried on the lipid surface of the membranes, since it is lipophilic, and it is sequentially hydroxylated by the membrane-bound hydroxylases (Hanson and Havir, 1981; Butt and Lamb, 1981). In this way there is the possibility of at least a two-step channeling route from phenylalanine to *p*-coumaric acid. The transmethylases then direct the methyl groups to the meta positions. There is a difference between the transmethylases from angiosperms and those from gymnosperms, since with the latter preparations the enzyme is relatively inactive on 5-hydroxyferulate. Thus it is possible that these transmethylases may have some part in the control of the type of lignin formed (see Chapter 7, Section VI). The guaiacyl type (3-methoxy-4-hydroxy) is found in gymnosperms and the syringyl–guaiacyl type (3,5-dimethoxy-4-hydroxy) in dicotyledons (Grisebach, 1981).

The acid building units are reduced to the corresponding alcohols before polymerization to lignin (Fig. 2). This reduction occurs as a two-step process involving the CoA ester of the acid and using NADPH as cofactor. The enzymes CoA ligases, cinnamoyl-CoA : NADPH oxidoreductase, and cinnamyl alcohol dehydrogenase are involved, and they give finally the three building units of lignin: *p*-coumaryl alcohol, coniferyl alcohol, and sinapyl alcohol (Grisebach, 1981). Various isoenzymes of the CoA ligases may also control the type of lignin that is formed, since the isoenzymes have different affinities and activities for *p*-coumaric acid, ferulic acid, and sinapic acid and these isoenzymes occur in different proportions in different plants and in different tissues of the same plant (Grisebach, 1981).

The activation of the phenylpropionic acids and their subsequent reduction may occur in vesicles that fuse with the plasma membrane and empty the precursors of lignin into the wall (Alibert *et al.*, 1977). The NADPH for the reduction is provided by the pentose phosphate pathway (Pryke and ap Rees, 1976, 1977). The control for the final steps that produce the cinnamyl alcohols, which are the immediate building units of the lignin, is therefore dependent on the energy status of the cell, since ATP is necessary for the ligase activity, and it is also dependent on the distribution of carbohydrate metabolism between the pentose phosphate pathway and glycolysis. It is within the wall that the polymerization process to form a complex lignin cage occurs.

III. Polymerization Systems and Location of Polymers

A. *Polysaccharide Polymerization*

The nucleoside diphosphate sugars used for the synthesis of pectin and hemicellulose are polymerized within the lumen of the endoplasmic reticulum and the Golgi apparatus (Northcote, 1979a, 1982a). A major site of control therefore could be the transport of these glycosylation donor compounds from the cytoplasm into the lumen of the membrane system.

It is now well established in animal tissues that the glycosylation of some proteins is carried out by building up a core oligosaccharide by glycosylation of a polyisoprenoid diphosphate compound (dolichyl pyrophosphate) (Spiro and Spiro, 1982). This glycolipid is formed by transfer of sugars from nucleoside diphosphate sugar precursors directly, or indirectly from polyisoprenoid monophosphate sugars, which are in turn formed by transfer of sugars from nucleoside diphosphate sugars. The core oligosaccharide is assembled within the lumen of the endoplasmic reticulum. In some cases it has been shown that the glycosylation of polyisoprenoid phosphate and the transport across the membrane to the lumen occur as a coordinated process (Lennarz, 1982). This glycosyltransferase–transport reaction on the lipid therefore occurs at the cytoplasmic face of the endoplasmic membrane, by an enzyme that may be an integral protein of the membrane, and it functions with the lipid for the transport of the sugar moiety into the lumen. The oligosaccharide attached to the polyisoprenyl diphosphate, when fully assembled and in the lumen of the endoplasmic reticulum, is transferred to the protein. This second glycosyltransferase is at the lumen face of the membrane (Snider and Robbins, 1982; Snider *et al.*, 1982). After assembly on the protein, the oligosaccharide is processed by trimming of some sugar residues and modified by further transglycosylations. These secondary glycosylations take place within the lumen of the Golgi apparatus and are usually made directly from nucleoside diphosphate sugars. At this part of the endomembrane system, the nucleoside diphosphate sugars may be transported across the membrane by specific transporters (Brandan and Fleischer, 1981; Sommers and Hirschberg, 1982; Eggens and Dallner, 1982; Kuhn and White, 1976; Snider, 1982).

In plant tissue, although glycoproteins similar to those found in animals are formed in a manner similar to that described for animals, it is by no means clear whether polyisoprenylphosphate and protein act as intermediates during the formation of the polysaccharides of the wall. No transfer directly from a polyisoprenyl diphosphate oligosaccharide derivative to polysaccharide (as distinct from the carbohydrate of a glycoprotein) in plants has yet been demonstrated (Northcote, 1979a). Some large polysaccharide molecules have been shown to be formed via glycoproteins, however, and in these cases polyisoprenyl diphosphate derivatives carrying oligosaccharides, which are related to the composition and structure of the polysaccharide, have been found. In the root-tip cells of maize, a slime is produced at the outer root-cap cells. This slime is made up of polysaccharide material that contains a high content of fucose (about 20–25%). This characteristic feature of the polysaccharide makes it possible to study the mechanism and site of its formation within the cell. Using radioactive sugars fed to the root, it can be shown by radioautography that the slime is secreted across the plasma membrane by vesicle fusion via an exocytotic mechanism similar to that by which the hemicelluloses and pectins are deposited in the wall (Northcote and Pickett-Heaps, 1966). The vesicles are derived from the Golgi apparatus, where the slime is concentrated before secretion. It is probable that a major part of the synthesis of the polysaccharide occurs in this organelle. Analysis of membrane fractions of the cells provides evidence that both the endoplasmic reticulum and Golgi apparatus function for the production and transport of the slime from the cytoplasm (Northcote, 1982b; Harris and Northcote, 1971; Bowles and Northcote, 1972, 1976).

Analysis of the polymers extracted from isolated membrane systems of the root tip shows that they contain glycoproteins. The carbohydrate composition of two of these glycoproteins resembles that of the slime polysaccharides, and pulse–chase radioactive experiments indicate that the glycoproteins could be intermediates and carriers of the carbohydrate that finally appears in the free slime polysaccharide secreted from the root tip (Green and Northcote, 1978). The membranes of the cells of the root tip contain polyisoprenyl diphosphate oligosaccharides, some of which contain fucose and are made up of 13–14 glycosyl monomers. The composition of these oligosaccharides resembles that of the composition of the glycoproteins and the slime polysaccharide (Dixon and Northcote, 1983). There are at least two types of fucosyl transglycosylases found in the membrane system of the cells of the root tip. One transfers fucose to polyisoprenyl phosphate, and the other transfers fucose to polysaccharide or glycoprotein. The GDP-Fuc : polyisoprenyl phosphate transfucosylase is present in the endoplasmic reticulum, while the second transferase, although it occurs mainly in the Golgi apparatus, is present in significant amounts in the endoplasmic reticulum (Green and Northcote, 1979a,b). Whether the transfucosylase that glycosylates the polymer uses a lipid intermediate as a substrate at either location is not known. However, the evidence does make it possible to suggest that the polysaccharide is assembled by the transfer of a regular sequence of sugars in the form of an oligosaccharide of at least 13 sugar residues from a polyisoprenyl phosphate derivative. The synthesis of the lipid oligosaccharide probably starts at the endoplasmic reticulum, and transfer of oligosaccharide onto nascent protein may occur in the endoplasmic reticulum. The polysaccharide is then elaborated on the protein primer, extended within the Golgi apparatus, and finally secreted.

In any one root, simultaneously, different polymers are synthesized within the endomembrane system of cells that are at different positions in the root. Thus the outer root-cap cells produce slime, while further up the root no slime is formed, the cells become differentiated to form vascular tissue, and the walls are secondarily thickened so that there is an increase in the amount of xylose-containing polymers that are secreted into the wall (Bowles and Northcote, 1972). In contrast, at the root tips just behind the cap cells, the meristematic tissue is actively dividing, and the cells are enlarging. At this time, relatively large amounts of pectin containing arabinose, galactose, and galacturonic acid are laid down into the wall. Even within the same cell, different polymers such as slime, membrane glycoproteins, and cell wall polysaccharides are synthesized within the endomembrane system at the same time, and these products are transported to different sites. Thus the transglycosylases or synthases that produce the different polymers and that are found at the membrane can be varied both qualitatively and quantitatively, and these systems are therefore control steps whereby one aspect of differentiation is brought about.

The bulk of the transglycosylation reactions for hemicellulose and pectin polysaccharides of the cell wall occurs in the Golgi apparatus, but it is possible that initial glycosylations may occur in the endoplasmic reticulum if protein and polyisoprenoid carriers are involved. The transport processes for the nucleoside diphosphate sugars are not known, but, as in animal tissues, there may be at least two systems. The mechanism of transport used would thus depend on the site, either endoplasmic reticulum or Golgi apparatus, and also on whether polyisoprenoids are involved in a transglycosylation–transport process or whether protein transporters of the nucleotide sugars are present. Whatever the mechanism, the transport may be an important control point for polysaccharide production (Fig. 1), since although the enzymes for polymerization are developed within the lumen of the membranes, the synthases will not function unless the substrates are available to them. Their full potential and activity can be restricted in any part of the endomembrane system by the transport, both quantitatively and qualitatively, of nucleoside diphosphate sugars. These considerations may account for the ambiguities encountered between experiments on *in vitro* and *in vivo* systems. It is possible, for instance, to detect polysaccharide synthase systems in characterized parts of the endomembranes isolated from broken cells, and it can be suggested from a knowledge of this enzyme complement that the membranes *in vivo* will synthesize particular polysaccharides. When intact cells are used and fed radioactive monosaccharides, however, these particular polysaccharides sometimes cannot be found at the membrane site where the synthase activities are known to occur (Bolwell and Northcote, 1983a). Thus, although a particular part of the endomembrane system has the potential for the polysaccharide synthesis, this does not always occur *in vivo*. One of the reasons for this difference may be the control of the transport of the precursors. This will be true not only for parts of the system such as the endoplasmic reticulum and Golgi apparatus but also for the plasma membrane, which can receive enzymes via the vesicles that fuse with it from the Golgi apparatus. This latter consideration may well account for the usual appearance of cellulose synthesis only at the plasma membrane and not within the internal cell membranes (Bowles and Northcote, 1972; Northcote, 1974a,b). In this regard, it is interesting to note that some systems for cellulose synthesis are thought to occur via polyisoprenoid diphosphate oligosaccharides and glycoproteins (Brett and Northcote, 1975; Hopp *et al.*, 1978a,b; Datema *et al.*, 1983), while the nucleoside diphosphate glucose donor is provided from the cytoplasmic side of the plasmamembrane.

The activities of the synthases for hemicellulose and pectin formation are directly altered during the differentiation of the cell, and this provides one of the major control steps of polysaccharide formation during differentiation. These synthetic systems can be investigated by dissecting cells at various stages of differentiation (cambium to xylem) from intact stems or by induction of differentiation in tissue culture cells by the application of plant growth factors. The synthase activities are membrane bound and can be estimated by preparation of the membranes from the broken cells and using radioactive nucleoside diphosphate sugars to measure incorporation into polymers. During the active cell division that is present in the vascular cambium of stems and meristematic tissues of roots, or during the rapid cell divisions of tissue cultures when they are subcultured, the synthases for pectin polysaccharides, such as arabinan synthase and polygalacturonic acid synthase, increase. During differentiation, when the secondary wall is formed and pectin synthesis ceases, these synthases for pectin decrease in activity, while xylan synthase activity for hemicellulose production increases (Table I) (Bolwell and Northcote, 1981; Dalessandro and Northcote, 1981b). When phloem tissue is induced in bean tissue culture, callose synthase, which produces the characteristic $\beta(1 \rightarrow 3)$ glucan at the sieve-plate pores is increased (Fig. 3) (Haddon and Northcote, 1975). Generally, the activity of these synthases reflects the amount of polysaccharide deposited in the wall at the particular stage of differentiation.

Table I
Synthase Activities in Differentiating Cells

Source of enzyme	Xylan synthase activity[a]	Arabinan synthase activity[a]
Sycamore trees		
Cambium cells	0.7	—
Differentiating xylem cells	1.6	—
Differentiated xylem cells	4.2	—
Hypocotyls of bean		
6 Days after imbibition	0.07	0.15
6–7 Days after imbibition (rapid extension growth)	0.17	0.60
9–10 Days after imbibition (stele formation)	0.45	0.20
Bean callus tissue		
Subcultured on maintenance medium		
At subculture	0.03	0.02
5–8 Days after subculture (period of cell division)	0.10	0.52
Subcultured on induction medium		
At subculture	0.02	0.02
5–8 Days after subculture	0.12	0.20
9–12 Days after subculture (period of xylem formation)	0.40	0.10

[a] Activity data are given nmol/min/mg/protein; —, not estimated.

There are two methods for the control of these synthase activities, and both have been shown to operate for particular enzymes. The xylan synthase of sycamore stems can be biochemically modulated by the energy status of the cell, since its activity is decreased in the presence of increased levels of nucleoside mono- and diphosphates, in particular UDP (Dalessandro and Northcote, 1981b). In addition to this, during differentiation the total activity of the synthase is increased (Bolwell and Northcote, 1981; Dalessandro and Northcote, 1981b). Similarly, during the induced differentiation of bean callus tissue, the synthase activity also increases (Bolwell and Northcote, 1981) (Table I). The callus tissue can be grown in the presence of inhibitors of protein synthesis in a manner identical to that used in experiments to investigate the increase in PAL activity. It can be shown that both transcription and translation are involved for control of the activities of the xylan and arabinan synthases during cell division and induced differentiation (Bolwell and Northcote, 1983b). There is also an indication using monoclonal antibodies to plant membrane proteins that the decrease in arabinan synthase when pectin synthesis ceases is correlated with a decrease in the amount of enzyme present at the membranes (Bolwell and Northcote, 1984).

In *Acetabularia mediterranea,* the control of the synthesis of cell wall polysaccharide (mannan) seems to be regulated, as in higher plants, by the activity of the synthase, in this case mannan synthase. On termination of growth of the algal cell, the activity of the synthase stops, while during normal growth, high activities are found in the apical regions of the cell where growth and mannan synthases occurs (Bachmann and Zetsche, 1979).

1. Microfibril Formation

The synthesis of cellulose must be considered from two points of view: (1) the biochemical formation of $\beta(1 \rightarrow 4)$ glucan chains; and (2) the aggregation, almost crystallization, of the chains into discrete and fundamental units, the microfibrils, and their orientation within the wall in definite directions.

In some algae, cellulose microfibrils can be formed within the vesicles of the Golgi apparatus and exported as part of sculptured scales to the outside of the cell (Allen and Northcote, 1975; Romanovicz, 1982). Even some cells of higher plants synthesize $\beta(1 \rightarrow 4)$ glucan chains within the Golgi apparatus. It is probable, for instance, that the glucan chains of xyloglucans are formed within the internal endomembrane system (Northcote, 1979b). The glucan chains at the core of certain plant slimes (e.g., maize root slime) are also synthesized within the Golgi apparatus before the slime is secreted (Wright and Northcote, 1976). The bulk of the unsubstituted longer $\beta(1 \rightarrow 4)$ glucan chains that make up the microfibrils of higher plant cell walls are formed only at the plasma membrane, however, and the microfibrils are spun out from this membrane directly into the wall. Electron-microscopic observations and radioautographic and radioactive incorporation experiments all indicate that the plasma membrane is the locus of cellulose synthesis (Romanovicz, 1982; Mueller and Brown, 1980; Wooding, 1968; Willison and Klein, 1982). Thus, although the enzyme system for the formation of cellulose may be synthesized and incorporated into the membranes within the cytoplasm, it normally becomes operational only when the membrane is incorporated into the plasma membrane.

Freeze-etch studies of the plasma membrane of higher plants and some algae have revealed that an organized group of particles are present at one end of microfibrils developed on the membrane surface (Mueller and Brown, 1980; Giddings *et al.,* 1980). These terminal complexes are thought to traverse the membrane, and the particles making up the complex can be seen as complimentary clusters at both the E and P faces of the freeze-fractured membrane. The particles may therefore provide a transmembrane connection between the microfibrils and the cytoplasm. Each terminal complex is thought to generate a single microfibril (except in *Micrasterias,* where several rosettes of particles forming rows are thought to contribute to a single microfibril, the size of which depends on the number of rosettes in the row). There is at present no evidence that demonstrates that the terminal complex is an enzyme system for spinning out the microfibril, although this is the general assumption that is made. There is much evidence to suggest generally that the particles seen on the plasma membrane faces after freeze-fracture are proteins (Nicolson, 1976; Pinto da Silva, 1981; Sequest *et al.,* 1974; Vail *et al.,* 1974). Some caution over the significance of organized arrays of particles localized at the plasma membrane must, however, be observed, since these patterns can arise from what were previously randomly distributed particles by experimental conditions such as plasmolysis applied to the cell before fracture (Wilkinson and Northcote, 1980).

If the terminal complex does spin out the 35- to 250-glucose chains that make up the microfibrils, then presumably the chains are all elongated in the same direction. Thus, the glucan chains in the microfibril should be parallel, as distinct from antiparallel. It is necessary to postulate that the terminal complexes move in the membrane so that the microfibril is spun out in an oriented manner around the cell in the wall. If the microfibrils are oriented in definite directions, then this movement must be controlled, and mechanisms to bring about both the movement and to control its direction are necessary. One of the ways by which the movement of the terminal complexes could be brought about and also directed is by means of microtubules and other cytoskeletal components that occur in the cytoplasm just under the plasma membrane and that could be connected to the transmembrane particles of the complex (Heath and Seagull, 1982). It has been observed many times that the orientation of the microtubules in the cytoplasm just under the plasma membrane is in the same direction as that of the microfibrils deposited in the wall (Wooding and Northcote, 1964; Hepler and Newcomb, 1964; Pickett-Heaps and Northcote, 1966; Robinson and Quader, 1982). Alternatively, the movement of the complexes might be caused by the crystallization of the glucan chains into the microfibrils as they arise from the complexes, and the microtubules might act only as guides to direct the movement in a definite direction (Herth, 1980). In this way, although the direction is controlled by the microtubules in the cytoplasm, the movement is generated by the formation of the microfibrils rather than the cytoskeletal elements of the cytoplasm. No definite evidence to distinguish the various theories is at present available, although there is a great deal of evidence that correlates the orientation of the microtubules with the orientation of the microfibrils (Heath and Seagull, 1982).

2. Vesicle Fusion

Studies on wall regeneration on protoplasts and in plasmolyzed cells and also on the new cell walls formed at the cell plate during cytokinesis have indicated the sequence of wall formation. Usually the matrix is deposited first by vesicle fusion, and into this matrix the microfibrils are woven (Northcote, 1977, 1979b). The early matrix is rich in pectin, and its composition varies during these initial stages.

All the polysaccharide deposited into the wall depends on vesicle fusion, either for the direct transfer of the polymers formed within the endomembrane system, or for the incorporation of enzymes into the plasma membrane or into the wall. The bulk of the vesicles arise from the Golgi apparatus and they are transported to particular areas of the cell surface. The secreted material from the Golgi apparatus is almost certainly sequestered to particular cisternae and perhaps even to particular parts of a cisternum (Quinn *et al.*, 1983; Tartakoff, 1982; Griffiths *et al.*, 1983). By analogy with model systems suggested for animal cells, the sequestering of material within the Golgi cisternae and the fusion of the vesicle with preexisting membrane at the cell surface is dependent on receptors (Sly and Fischer, 1982; Pastan and Willingham, 1983; Farquhar, 1983). These are capable of recognition of ligands either on or contained within the membranes. In some cells, vesicle fusion is dependent on a characteristic distribution of granules (proteins) within the membrane (Satir *et al.*, 1973; Burwen and Satir, 1977; Da Silva and Nogueira, 1977). In this way, the production and fusion of the vesicle at different but definite sites of the cell may occur. The vesicles produced from the Golgi apparatus may also be directed to particular sites at the cell surface by microtubules (Northcote, 1971). This is very clearly illustrated during the development of organized secondary thickening that occurs in xylem elements. A spiral or reticulate secondary wall is laid down, and during its formation the microtubules are distributed just under the plasma membrane at the sites of the thickening (Wooding and Northcote, 1964; Pickett-Heaps and Northcote, 1966).

Part of the control of vesicle fusion at the surface is mediated by the ionic atmosphere at the membrane, and Ca^{2+} is necessary for the fusion to occur (Loister and Layter, 1973; Dahl *et al.*, 1978; Baydoun and Northcote, 1980a,b, 1981). The rate of vesicle fusion can be a limiting process for the rate of cell-wall formation, since at any one time the number of vesicles ready for fusion exceeds the number that are fusing and depositing material into the wall (Morris and Northcote, 1977). In this way the composition and amount of wall material deposited may respond very quickly to a stimulus at the cell surface, which allows the rate of vesicle fusion to vary. A new steady state would then be achieved that produced the requisite number of vesicles necessary to maintain the new rate of fusion. The turnover of the Golgi apparatus can be very fast in plant cells, and times of 5–40 min have been calculated (Robinson and Kristen, 1982).

B. *Polymerization of Lignin and Formation of Covalent Cross-Linkages*

The cinnamyl alcohols may reach the wall as the free alcohols or as β-glucosides formed by glucosyltransferases with UDP-Glu (Ibrahim and Grisebach, 1976). For polymerization, the free alcohol is necessary, and β-glucosidases occur in the walls of tissues that are lignified (Freudenberg, 1965). Glucosides may be important for the transport of the alcohols to the walls, but they are not obligatory for lignin synthesis. They may act as reservoirs of the lignin precursors. The precursors arise in cells that are undergoing lignification, or they may arise from neighboring cells of young differentiating xylem, which themselves are not at the stage of massive lignification (Rubery and Northcote, 1968). Lignification is brought about by the oxidation of the alcohols to yield mesomeric phenoxy radicals with half-lives of about 45 sec, so that rapid polymerization occurs (see Chapter 7, Section II). At the same time, linkages of these radicals, and hence the lignin, to carbohydrate can take place (Freudenberg, 1965, 1968). It is probable that the final oxidation of the phenolic hydroxy group to give the free radicals is brought about by a specific isoenzyme of peroxidase that occurs in the walls of plant cells (Yung and Northcote, 1975; Mader, 1976; Mader *et al.*, 1977). This isoenzyme is an anodic-migrating component. In *Zinnia elegans,* the activity of wall-bound peroxidase increases during the onset of lignification, although the total soluble peroxidase activity of the cell may decrease (Fukuda and Komamine, 1982; Haddon and Northcote, 1976). Another isoenzyme of peroxidase might be required to produce the hydrogen peroxide on which the oxidative polymerization process depends (Mader and Amberg-Fisher, 1982). The synthesis and activity of some isoenzymes of peroxidase are therefore possible control sites for lignification (see Chapter 7, Section III).

In the rigid secondary wall of woody tissue, the lignin replaces the water of the growing cell wall and forms a hydrophobic matrix around the microfibrils. Strong hydrogen bonds occur between the polysaccharides at the microfibrillar–matrix interface and between components of the matrix. These, together with the covalent bonds formed between carbohydrate and lignin, make the wall a composite in which the linear polysaccharide polymers are enclosed in a cross-linked polymer cage. The wall has great tensile strength because of the microfibrils and a rigid structure because of the lignified matrix (Northcote, 1972).

The peroxidase of the wall may also establish other covalent linkages between wall polymers. The wall normally contains protein, which may consist in part of one or all of three types of hydroxyproline-rich glycoproteins: (1) insoluble glycoprotein associated with cellulose (Heath and Northcote, 1971), (2) soluble lectins and agglutinins (Allen *et al.*, 1978; Leach *et al.*, 1982; Muray and Northcote, 1978) and (3) soluble high molecular weight arabinogalactan glycoprotein (Clark *et al.*, 1979; Fincher *et al.*, 1983). In addition to these, various enzymes such as peroxidases, phosphatases, and carbohydrases are present. The tyrosine residues of the proteins may be oxidized by peroxidase to give cross-linkages of isodityrosine between the polypeptides (Fry, 1982; Cooper and Varner, 1983). The existence of ferulic acid in the cell walls of some plants allows the possibility of oxidation to give diferulic acid ester linkages joining polysaccharide chains (Fry, 1983; Tanner and Morrison, 1983).

C. *Transport and Formation of Lipid*

Many cell walls have layers in the outer regions of the wall that carry lipid material. These are cutin, suberin, and waxes (Gould and Northcote,

1984). How these are transported to the outside of the cell and wall is not known. Pores have not been found, nor has a volatile lipid solvent been detected that would carry the lipid through the hydrophilic wall.

The fatty acids are synthesized in chloroplasts or proplastids and moved into the cytoplasm and the endomembrane system for further modification and synthesis of neutral fats, phospholipids, and other compounds (Stumpf and Shimakata, 1983). The fatty acids could be carried by proteins by a process similar to the way in which serum albumin binds fatty acid in the bloodstream of mammals. Other types of lipid might be formed into complexes analogous to low-density lipoproteins of the type found in animal tissues, where the lipid core of the lipoprotein is surrounded by a hydrophilic cortex made up of protein, phospholipid, and cholesterol (Nilsson-Ehle *et al.*, 1980). This allows the lipid to be moved in an aqueous environment. The protein or glycoprotein shell could also act as possible ligands for particular receptors at the membrane of the cell at which the export occurs. The lipoproteins, if they are present, would probably be formed within the endomembrane lumen and would receive the proteins at the endoplasmic reticulum.

References

Alibert, G., Ranjera, R., and Boudet, A. M. (1977). *Physiol. Veg.* **15,** 279–301.
Allen, A. K., Desai, N. N., Neucherger, A., and Creeth, J. M. (1978). *Biochem. J.* **171,** 665–674.
Allen, D. M., and Northcote, D. H. (1975). *Protoplasma* **83,** 389–412.
Amrhein, N., and Zenk, M. H. (1970). *Naturwissenschaften* **57,** 312.
Axelos, M., and Peaud-Lenöel, C. (1978). *Biochimie* **60,** 35–44.
Bachmann, P., and Zetsche, K. (1979). *Planta* **145,** 331–337.
Baydoun, E. A. H., and Northcote, D. H. (1980a). *J. Cell Sci.* **45,** 147–167.
Baydoun, E. A. H., and Northcote, D. H. (1980b). *J. Cell Sci.* **45,** 169–186.
Baydoun, E. A. H., and Northcote, D. H. (1981). *Biochem. J.* **193,** 781–791.
Bevan, M., and Northcote, D. H. (1979a). *Planta* **147,** 77–81.
Bevan, M., and Northcote, D. H. (1979b). *J. Cell Sci.* **39,** 339–353.
Bevan, M., and Northcote, D. H. (1981a). *Planta* **152,** 32–35.
Bevan, M., and Northcote, D. H. (1981b). *Planta* **152,** 24–31.
Boffey, S. A., and Northcote, D. H. (1975). *Biochem. J.* **150,** 433–440.
Bolwell, G. P., and Northcote, D. H. (1981). *Planta* **152,** 225–233.
Bolwell, G. P., and Northcote, D. H. (1983a). *Biochem. J.* **210,** 497–507.
Bolwell, G. P., and Northcote, D. H. (1983b). *Biochem. J.* **210,** 509–515.
Bolwell, G. P., and Northcote, D. H. (1984). *Planta* **162,** 139–146.
Bowles, D. J., and Northcote, D. H. (1972). *Biochem. J.* **130,** 1133–1145.
Bowles, D. J., and Northcote, D. H. (1976). *Biochem. J.* **142,** 139–144.
Brandan, E., and Fleischer, B. (1981). *Fed. Proc., Proc. Am. Soc. Expl. Biol.* **40,** 1660.
Brett, C. T., and Northcote, D. H. (1975). *Biochem. J.* **148,** 107–117.
Burwen, S. J., and Satir, B. H. (1977). *J. Cell Biol.* **73,** 660–671.
Butt, V. S., and Lamb, C. J. (1981). *In* "The Biochemistry of Plants: Secondary Plant Products" (E. E. Conn, ed.), Vol. 7, pp. 627–665. Academic Press, London.

Clark, A. E., Anderson, R. L., and Stone, B. A. (1979). *Phytochemistry* **18,** 521–540.
Cooper, J. B., and Varner, J. E. (1983). *Biochem. Biophys. Res. Commun.* **112,** 161–167.
Dahl, G., Schudt, C., and Gratzl, M. (1978). *Biochem. Biophys. Acta* **514,** 105–116.
Dalessandro, G., and Northcote, D. H. (1977a). *Planta* **134,** 39–44.
Dalessandro, G., and Northcote, D. H. (1977b). *Biochem. J.* **162,** 281–288.
Dalessandro, G., and Northcote, D. H. (1977c). *Biochem. J.* **162,** 267–279.
Dalessandro, G., and Northcote, D. H. (1981a). *Planta* **151,** 53–60.
Dalessandro, G., and Northcote, D. H. (1981b). *Planta* **151,** 61–67.
Da Silva, P. P., and Nogueira, L. (1977). *J. Cell Biol.* **73,** 161–181.
Datema, R., Schwarz, R. T., Rivas, L. A., and Pont Lezica, R. (1983). *Plant Physiol.* **71,** 76–81.
Dixon, W. T., and Northcote, D. H. (1983). In preparation.
Eggens, I., and Dallner, G. (1982). *Biochimica Biophys. Acta* **686,** 77–93.
Farquhar, M. G. (1983). *Fed. Proc., Proc. Am. Soc. Expl. Biol.* **42,** 2407–2413.
Feingold, D. S., Neufeld, E. F., and Hassid, W. Z. (1960). *J. Biol. Chem.* **235,** 910–913.
Fincher, G. B., Stone, B. A., and Clark, A. E. (1983). *Annu. Rev. Plant Physiol.* **34,** 47–70.
Freudenberg, K. (1965). *Science* **148,** 595–600.
Freudenberg, K. (1968). *In* "Constitution and Biosynthesis of Lignin" (Mol. Biology, Biochemistry and Biophysics Series) (A. Kleinzeller, ed.), Vol. 2, pp. 45–122. Springer-Verlag, Berlin.
Fry, S. C. (1982). *Biochem. J.* **204,** 449–455.
Fry, S. C. (1983). *Planta* **157,** 111–123.
Fukuda, H., and Komamine, A. (1982). *Planta* **155,** 423–430.
Giddings, T. H., Brower, D. L., and Staehelin, L. A. (1980). *J. Cell Biol.* **84,** 327–339.
Gould, J., and Northcote, D. H. (1984). *In* "Bacterial Adhesion: Mechanisms and Physiological Significance" (M. M. Fletcher and D. C. Savage, eds.), Plenum, New York.
Green, J. R., and Northcote, D. H. (1978). *Biochem. J.* **170,** 599–608.
Green, J. R., and Northcote, D. H. (1979a). *Biochem. J.* **178,** 661–671.
Green, J. R., and Northcote, D. H. (1979b). *J. Cell Sci.* **40,** 235–244.
Griffiths, G., Quinn, P., and Warren, G. (1983). *J. Cell Biol.* **96,** 835–850.
Grisebach, H. (1981). *In* "The Biochemistry of Plants Secondary Plant Products" (E. E. Conn, ed.), Vol. 7, pp. 457–478. Academic Press, London.
Haddon, L. E., and Northcote, D. H. (1975). *J. Cell Sci.* **17,** 11–26.
Haddon, L., and Northcote, D. H. (1976). *Planta* **128,** 255–262.
Hahlbrock, K., and Wellmann, E. (1973). *Biochim. Biophys. Acta* **304,** 702–706.
Hanke, D. E., and Northcote, D. H. (1974). *J. Cell Sci.* **14,** 29–50.
Hanson, K. R., and Havir, E. A. (1981). *In* "The Biochemistry of Plants: Secondary Plant Products" (E. E. Conn, ed.), Vol. 7, pp. 577–625. Academic Press, London.
Harris, P. J., and Northcote, D. H. (1971). *Biochim. Biophys. Acta* **237,** 56–64.
Heath, I. B., and Seagull, R. W. (1982). *In* "The Cytoskeleton in Plant Growth and Development" (C. W. Lloyd, ed.), pp. 163–182. Academic Press, London.
Heath, M. F., and Northcote, D. H. (1971). *Biochem. J.* **125,** 953–961.
Helenius, A., Mellman, I., Wall, D., and Hubbard, A. (1983). *Trends Biochem. Sci.* **8,** 245–250.
Hepler, P. K., and Newcomb, E. H. (1964). *J. Cell Biol.* **20,** 529–533.
Herth, W. (1980). *J. Cell Biol.* **87,** 442–450.
Hopp, H. E., Romero, P. A., Daleo, G. R., and Pont Lezica, R. (1978a). *Eur. J. Biochem.* **84,** 561–571.
Hopp, H. E., Daleo, G. R., Romero, P. A., and Pont Lezica, R. (1978b). *Plant Physiol.* **61,** 248–251.

Ibrahim, R. K., and Grisebach, H. (1976). *Arch. Biochem. Biophys.* **176,** 700–708.
Jones, D. H., and Northcote, D. H. (1981). *Eur. J. Biochem.* **116,** 117–125.
Kuboi, T., and Yamada, Y. (1978). *Biochim. Biophys. Acta* **542,** 181–190.
Kuhn, N. J., and White, A. (1976). *Biochem. J.* **154,** 243–244.
Leach, J. E., Cantrell, M. A., and Sequeira, L. (1982). *Plant Physiol.* **70,** 1353–1359.
Lennarz, W. J. (1982). *Philos. Trans. R. Soc. London* **B300,** 129–144.
Loister, Z., and Layter, A. (1973). *J. Biol. Chem.* **248,** 422–432.
Mader, M. (1976). *Planta* **131,** 11–15.
Mader, M., and Amberg-Fisher, V. (1982). *Plant Physiol.* 1128–1131.
Mader, M., Nessel, A., and Bopp, M. (1977). *Z. Pflanzenphysiol.* **82,** 247–260.
Masters, C. J. (1981). *CRC Crit. Rev. Biochem.* **11,** 105–143.
Morris, M. R., and Northcote, D. H. (1977). *Biochem. J.* **166,** 603–618.
Mueller, S. C., and Brown, R. M. (1980). *J. Cell Biol.* **84,** 315–326.
Muray, R. H. A., and Northcote, D. H. (1978). *Phytochemistry* **17,** 623–629.
Nicolson, G. L. (1976). *Biochim. Biophys. Acta* **457,** 57–108.
Nilsson-Ehle, P., Garfinkel, A. S., and Schotz, M. C. (1980). *Ann. Rev. Biochem.* **49,** 667–692.
Northcote, D. H. (1969a). *Essays Biochem.* **5,** 90–137.
Northcote, D. H. (1969b). *Symp. Soc. Gen. Microbiol.* **19,** 333–349.
Northcote, D. H. (1971). *Symp. Soc. Exp. Biol.* **25,** 51–69.
Northcote, D. H. (1972). *Ann. Rev. Plant Physiol.* **23,** 113–132.
Northcote, D. H. (1974a). *Philos. Trans. R. Soc. London* **B268,** 119–128.
Northcote, D. H. (1974b). *In* "Plant Carbohydrate Biochemistry" (J. P. Pridham, ed.). Academic Press, London.
Northcote, D. H. (1977). *In* "Cell Surface Reviews" (G. Poste and G. L. Nicholson, eds.), Vol. 4, pp. 717–739. North-Holland, Amsterdam.
Northcote, D. H. (1979a). *In* "Biomembranes" (L. A. Manson, ed.), Vol. 10, pp. 51–76. Plenum, London.
Northcote, D. H. (1979b). *In* "Polysaccharides in Food" (J. M. V. Blanshard and J. R. Mitchell, eds.), pp. 3–13. Butterworths, London.
Northcote, D. H. (1982a). *In* "Nucleic Acids and Proteins in Plants. I. Encyclopaedia of Plant Physiology" (D. Boulter and B. Parther, eds.), New Series Vol. 14A, pp. 637–655. Springer-Verlag, Berlin.
Northcote, D. H. (1982b). *Philos. Trans. R. Soc. Lond.* **B300,** 195–206.
Northcote, D. H., and Pickett-Heaps, J. D. (1966). *Biochem. J.* **98,** 159–167.
Pastan, I., and Willingham, M. C. (1983). *Trends Biochem. Sci.* **8,** 250–254.
Pickett-Heaps, J. D., and Northcote, D. H. (1966). *J. Exp. Bot.* **17,** 20–26.
Pinto da Silva, P. (1981). *Science* **213,** 230–233.
Poulton, J. E. (1981). *In* "The Biochemistry of Plants: Secondary Plant Products" (E. E. Conn, ed.), Vol. 7, pp. 667–723. Academic Press, London.
Pryke, J. A., and ap Rees, T. (1976). *Planta* **132,** 279–289.
Pryke, J. A., and ap Rees, T. (1977). *Phytochemistry* **16,** 557–560.
Quinn, P., Griffiths, G., and Warren, G. (1983). *J. Cell Biol.* **96,** 851–856.
Robinson, D. G., and Kristen, U. (1982). *Int. Rev. Cytol.* **77,** 89–127.
Robinson, D. G., and Quader, H. (1982). *In* "The Cytoskeleton in Plant Growth and Development" (C. W. Lloyd, ed.), pp. 109–126. Academic Press, London.
Romanovicz, D. K. (1982). *In* "Cellulose and other Natural Polymer Systems" (R. M. Brown, ed.), pp. 127–147. Plenum, London.
Rubery, P. H., and Northcote, D. H. (1968). *Nature (London)* **219,** 1230–1234.
Rubery, P. H., and Northcote, D. H. (1970). *Biochim. Biophys. Acta* **222,** 95–108.

Satir, B., Schooley, C., and Satir, P. (1973). *J. Cell Biol.* **56,** 153–176.
Sequest, J. P., Gulik-Krzywicki, T., and Sandet, C. (1974). *Proc. Natl. Acad. Sci.* **71,** 3294–3298.
Sly, W. S., and Fischer, H. D. (1982). *J. Cellul. Biochem.* **18,** 67–85.
Snider, M. D. (1982). *Nature (London)* **298,** 117–118.
Snider, M. D., and Robbins, P. W. (1982). *J. Biol. Chem.* **257,** 6796–6801.
Snider, M. D., Huffaker, T. C., Couto, J. R., and Robbins, P. W. (1982). *Philos. Trans. R. Soc. London* **B300,** 207–233.
Sommers, L. W., and Hirschberg, C. B. (1982). *J. Biol. Chem.* **257,** 10811–10817.
Spiro, R. G., and Spiro, M. J. (1982). *Philos. Trans. R. Soc. Lond.* **B300,** 117–127.
Stumpf, P. K., and Shimakata, T. (1983). *In* "Biosynthesis and Function of Plant Lipids" (W. W. Thomson, J. B. Mudd, and M. Gibbs, eds.), Proc. 6, Symp. University of California, Riverside. American Society Plant Physiologists, Maryland.
Tanner, G. R., and Morrison, I. M. (1983). *Phytochemistry* **22,** 1433–1439.
Tartakoff, A. M. (1982). *Philos. Trans. R. Soc. London* **B300,** 173–184.
Vail, W. J., and Papahadjopoulos, D., and Moscarello, M. A. (1974). *Biochim. Biophys. Acta* **345,** 463–467.
Wilkinson, M. J., and Northcote, D. H. (1980). *J. Cell Sci.* **42,** 401–415.
Willingham, M. C., and Pastan, L. H. (1982). *J. Cell Biol.* **94,** 207–212.
Willison, J. H. M., and Klein, A. S. (1982). *In* "Cellulose and other Natural Polymer Systems" (R. M. Brown, ed.), pp. 61–85. Plenum, New York.
Wooding, F. B. P. (1968). *J. Cell Sci.* **3,** 71–79.
Wooding, F. B. P., and Northcote, D. H. (1964). *J. Cell Biol.* **23,** 327–337.
Wright, K., and Northcote, D. H. (1976). *Protoplasma* **88,** 225–239.
Yung, K-H., and Northcote, D. H. (1975). *Biochem. J.* **151,** 141–144.

CHAPTER 6

The Biosynthesis of Plant Cell Wall Polysaccharides

HIDETAKA HORI
ALAN D. ELBEIN

Department of Biochemistry
University of Texas Health Science Center
San Antonio, Texas

I. Introduction

The discovery of sugar nucleotides by Leloir and co-workers in the early 1950s was an important landmark for future studies on the biosynthesis of polysaccharides and complex carbohydrates (Caputto *et al.*, 1950; Leloir, 1953). Since the time of those studies, there has been an explosive increase in our understanding of the mechanism of polymerization of sugars, as will be discussed in this chapter.

ISBN 0-12-347880-4

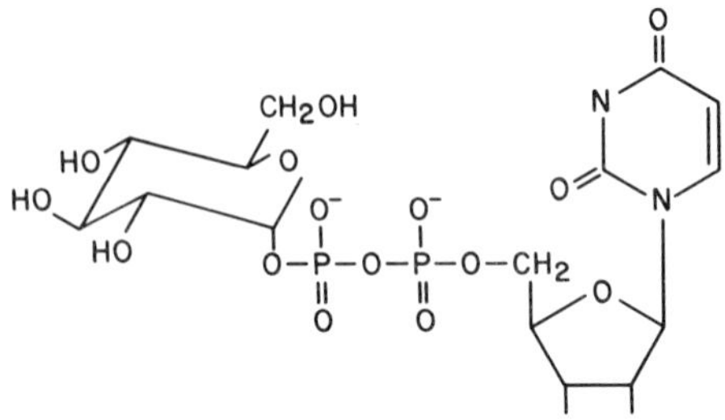

Fig. 1. Structure of uridine 5′-(α-glucopyranosylpyrophosphate) or UDP-glucose.

It is now quite clear that, in most cases, the individual monosaccharides are first converted to their sugar nucleotide derivatives, and these high-energy intermediates are then utilized in the biosynthetic polymerization reactions. Figure 1 presents the structure of a typical sugar nucleotide, uridine 5′-(α-D-glucopyranosyl pyrophosphate), more frequently referred to as UDP-glucose, or UDP-Glc. The next section of this chapter discusses the enzymatic reactions involved in the formation and the interconversions of the various sugar nucleotides in plants. Other sections focus on the intermediates in the biosynthesis of various plant polysaccharides.

II. Formation of Sugar Nucleotides

A. *Activation of Monosaccharides*

In order for sugars to be converted to their sugar nucleotide derivatives, they must first be activated by phosphorylation. The major source of many of the phosphorylated sugars of plants is the Calvin cycle, which gives rise to D-fructose 6-phosphate. As shown in Fig. 2, fructose 6-phosphate can then be transformed to D-mannose 6-phosphate, D-glucose 6-phosphate, *N*-acetyl-D-glucosamine 6-phosphate, and so on (Walker and Robinson, 1980; Feingold and Avigad, 1980). These sugar 6-phosphates can then be transformed via mutase reactions to the corresponding sugar 1-phosphate (see Fig. 2), and then to the sugar nucleotide.

In addition to these reactions, plants contain a number of specific kinases that can phosphorylate monosaccharides to provide sugar phosphates as substrates for the sugar nucleotide pyrophosphorylases. Thus, in germinating seedlings, for example, storage polysaccharides may be hydrolyzed to the free monosaccharides, which can then be reutilized for the formation of cell wall polysaccharides (Loewus, 1974). In a large

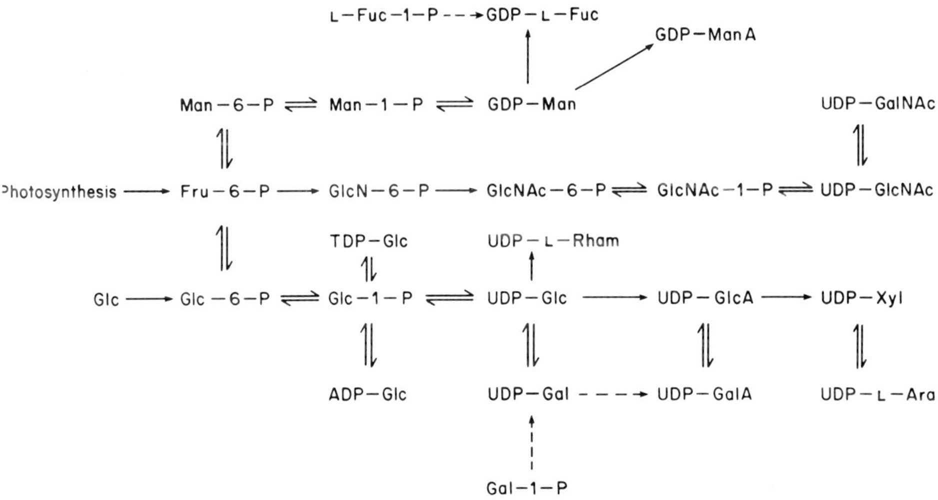

Fig. 2. Summary of reactions involved in the metabolism of hexoses and pentoses and their conversion to sugar nucleotides.

number of different plants, kinases have been demonstrated that can phosphorylate D-glucose, D-mannose, D-fructose, and D-glucosamine at the C-6 position, and L-arabinose, D-galactose, and D-galacturonate at the C-1 position. These include such enzymes as hexokinase (Saltman, 1953), L-arabinokinase (Neufeld *et al.*, 1960), D-galactokinase (Chan and Hassid, 1975), D-glucuronokinase (Leibowitz *et al.*, 1977), and D-galacturonokinase (Neufeld *et al.*, 1961).

Sugar 6-phosphates, such as glucose 6-phosphate or mannose 6-phosphate, have to be converted to the sugar 1-phosphates before they can be utilized as substrates for sugar nucleotide formation. Most plant tissues contain appropriate mutases capable of catalyzing these reactions. For example, phosphoglucomutase catalyzes the interconversion of glucose 6-phosphate and glucose 1-phosphate (Najjar, 1955); phosphomannomutase catalyzes the interconversion of mannose 6-phosphate and mannose 1-phosphate (Sutherland *et al.*, 1949); phospho-*N*-acetylglucosamine mutase catalyzes the interconversion of *N*-acetylglucosamine 6-phosphate and *N*-acetylglucosamine 1-phosphate (Brown, 1953), and so on.

B. *Formation of Nucleotide Sugars*

The major type of reaction by which sugar nucleotides are formed *de novo* is by the action of sugar nucleotide pyrophosphorylases. The general reaction for these enzymes is as follows (where N is a nucleotide):

$$\text{NTP} + \text{glycose 1-P} \rightleftharpoons \text{NDP-glycose} + \text{inorganic pyrophosphate}$$

A specific example of a pyrophosphorylase reaction is that of UDP-glucose pyrophosphorylase (UDP: α-D-glucose-1-phosphate uridylyltransferase, EC 2.7.7.9), which catalyzes the formation of UDP-glucose via the following reaction (Leloir, 1953; Buchanan, 1953):

$$\text{UTP} + \text{D-glucose 1-P} \rightleftharpoons \text{UDP-glucose} + \text{PP}$$

Although pyrophosphorylases that catalyze the formation of many different sugar nucleotides have been demonstrated in plant tissues, those catalyzing the formation of D-glucose-containing nucleotides probably predominate. Thus, in addition to UDP-glucose, plants also contain ADP-D-glucose (Dankert *et al.*, 1964), GDP-D-glucose (Sanwal and Preiss, 1969), TDP-D-glucose (Katan and Avigad, 1966), and perhaps even CDP-D-glucose (Axelos and Peaud-Lenoel, 1969). These sugar nucleotides are formed from specific pyrophosphorylases.

In addition to glucose nucleotides, a large number of other sugar nucleotides have also been identified in plants. Many of these are probably also formed by specific nucleotide pyrophosphorylases, including GDP-D-mannose (DeAsua *et al.*, 1966), UDP-D-galactose[1] (Neufeld *et al.*, 1957), UDP-*N*-acetyl-D-glucosamine (Solms and Hassid, 1957), ADP-*N*-acetyl-D-glucosamine (Passeron *et al.*, 1964), GDP-L-fucose (Lin and Hassid, 1966), UDP-D-glucuronic acid (Selvendran and Isherwood, 1967), UDP-D-galacturonic acid (Neufeld and Feingold, 1961), UDP-D-xylose (Ginsburg *et al.*, 1956), and UDP-L-arabinose (Franz and Meier, 1969). Some of these sugar nucleotides, however, may also be formed by transformation of other sugar nucleotides (see next section), and there are undoubtedly many other sugar nucleotide pyrophosphorylases yet to be described.

III. Interconversions of Sugar Nucleotides

As already indicated, many sugar nucleotides are synthesized directly from their sugar 1-phosphates and the appropriate nucleotide triphosphate via a pyrophosphorylase reaction. However, a number of other sugar nucleotides may be formed by transformations of already existing sugar nucleotides.

[1] There is a pyrophosphorylase for this sugar nucleotide in animal cells, but in plants it may only be formed by an epimerase reaction (see below).

A. *4-Epimerases*

Probably the best known and most widely studied enzymes involved in sugar nucleotide interconversions are the 4-epimerases. These enzymes have been known since Leloir's initial studies on sugar nucleotides (Leloir, 1951), when the interconversion of UDP-D-glucose and UDP-D-galactose was demonstrated. This enzyme, UDP-D-glucose-4-epimerase, has been demonstrated in all forms of life; in plants, it may represent the only means to synthesize UDP-galactose. UDP-glucose-4-epimerase has become the prototype for this class of enzymes, and its mechanism of action has been well studied (Fan and Feingold, 1969). The enzyme requires NAD^+ as a cofactor, and the reaction mechanism involves UDP-4-ketohexose as an enzyme-bound intermediate (Adair *et al.*, 1973). Figure 3 shows the typical reaction for this 4-epimerase, and several excellent reviews give the details of these reactions (Gabriel *et al.*, 1975; Adams, 1976).

In addition to the D-glucose : D-galactose pair, other 4-epimeric pairs are also found in nature, such as D-glucuronic acid : D-galacturonic acid and L-arabinose : D-xylose. Mung bean seedlings were shown to contain UDP-derivatives of all of the 4-epimer pairs (Neufeld *et al.*, 1957; Dalessandro and Northcote, 1977). In addition, extracts of this tissue catalyzed

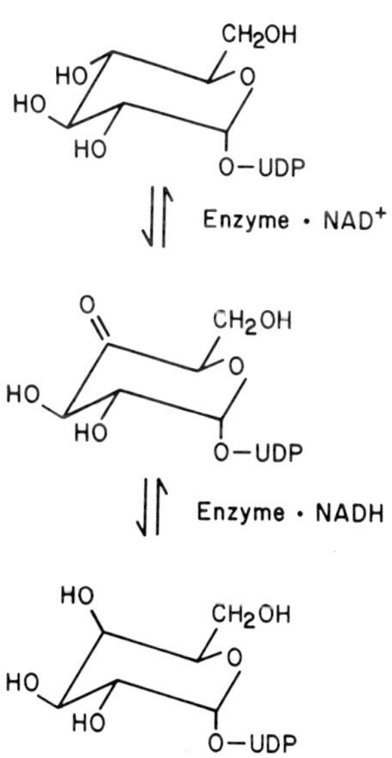

Fig. 3. Interconversion of UDP-galactose and UDP-glucose by the 4-epimerase, showing formation of 4-keto intermediate.

the interconversion of UDP-L-arabinose and UDP-D-xylose (Axelos and Peaud-Lenoel, 1978) as well as of UDP-glucuronic acid and UDP-galacturonic acid (Feingold *et al.*, 1960). Particulate preparations of mung bean seedlings could also cause the interconversion of TDP-galactose and TDP-glucose (Neufeld, 1962). Although the mechanism of the reaction has not been worked out in all of these 4-epimerases, it is probably similar to that of the UDP-glucose-4-epimerase (Fig. 3) and involves a 4-keto intermediate.

B. *Other Epimerases*

Suggestive evidence from several sources indicate the presence of 2-epimerases in some tissues that can interconvert D-glucose and D-mannose as the sugar nucleotide derivatives. Thus, in particulate enzyme preparations of mung bean seedlings, a glucomannan synthesized from GDP-[^{14}C]mannose contained radioactivity in both the D-glucose and D-mannose moieties (Elbein, 1969). The interconversion of TDP-D-glucose and TDP-D-mannose was also demonstrated in an enzyme preparation from *Streptomyces griseus* (Baddiley *et al.*, 1961). In addition to these epimerases, other epimerases appear to be involved in the biosynthesis of deoxyhexoses and other unusual sugars (see below).

C. *Formation of Uronic Acids and Pentoses*

Because of the structural similarities between D-glucose, D-glucuronic acid, and D-xylose, it was postulated, many years ago, that D-glucuronic acid could arise by oxidation of D-glucose and also that decarboxylation of D-glucuronic acid would give D-xylose. These reactions have now been shown to occur at the sugar nucleotide level in plants and other organisms.

The enzyme that catalyzes the oxidation of UDP-D-glucose, that is, UDP-D-glucose : NAD oxidoreductase (UDP-D-glucose dehydrogenase, EC 1.1.1.22) was first demonstrated in mammalian liver (Strominger *et al.*, 1954; Atlermatt and Neish, 1956) and was later partially purified from pea seedlings (Strominger and Mapson, 1957). It has since been demonstrated in a number of other plant tissues. The reaction, as shown in Fig. 4, involves a four-electron oxidation at the C-6 position of the D-glucose moiety of UDP-D-glucose with the concomitant reduction of 2 mol of NAD^+. The mammalian enzyme has been purified to homogeneity (Zalitis and Feingold, 1969), and its mechanism of action has been studied (Ordman and Kirkwood, 1977). Presumably, the enzyme forms a Schiff base

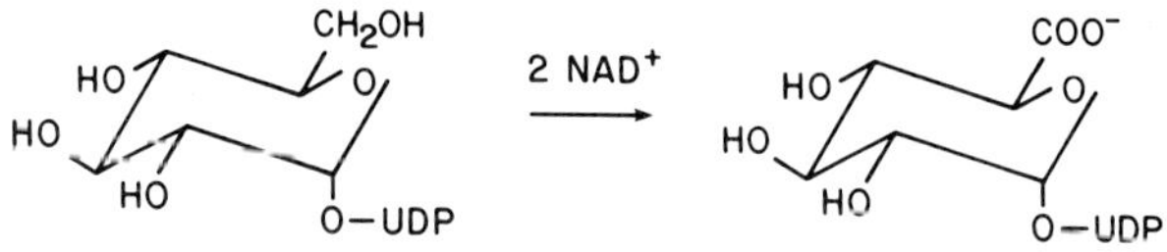

Fig. 4. Reaction for the conversion of UDP-glucose to UDP-glucuronic acid utilizing 2 mol NAD^+.

between C-6 of the glucose and an ε-amino group of an appropriately placed lysine on the enzyme. This leads to formation of a hemithioacetal, which is then further oxidized to the carboxylic acid. In many plant tissues, the UDP-D-glucose dehydrogenase displays complex inhibition by UDP-D-xylose (Neufeld and Hall, 1965). This inhibition may represent a control in the synthesis of pentose nucleotides.

Particulate enzyme preparations from mung bean seedlings and other plant tissues were shown to convert UDP-D-glucuronic acid to a mixture of UDP-D-galacturonic acid, UDP-D-xylose, and UDP-L-arabinose (Neufeld *et al.*, 1958). This early study indicated that these extracts contained a 4-epimerase for the uronic acids (see above) as well as UDP-uronic acid decarboxylase. However, the origin of the pentose nucleotides was not clear. Kinetic studies with partially purified extracts from which the UDP-D-glucuronic acid-4-epimerase had been removed showed that the first detectable product arising from UDP-D-glucuronic acid was UDP-D-xylose. As the reaction progressed, however, UDP-L-arabinose accumulated (Feingold *et al.*, 1960; Ankel and Feingold, 1965). The results suggested that UDP-D-xylose resulted from decarboxylation of UDP-D-glucuronic acid and that this pentose nucleotide was subsequently epimerized to UDP-L-arabinose (Fig. 5). The enzyme involved in decarboxylation, UDP-D-glucuronate decarboxylase (E.C. 4.1.1.35), is widely distributed in nature and has been detected in every plant tissue examined. This is probably a reflection of the important and ubiquitous presence of D-xylose and L-arabinose in plant polysaccharides. The decarboxylase was highly purified from wheat germ and shown to convert UDP-D-glucuronic acid to UDP-D-xylose and carbon dioxide only (Ankel and Feingold, 1966). The partially purified enzyme from *Cryptococcus laurentii* showed an absolute requirement for NAD^+ (Schutzbach and Feingold, 1970). The mechanism of the decarboxylation reaction, involving a carbonyl intermediate, was proposed as shown in Fig. 5. As indicated above, the UDP-D-xylose can be transformed to UDP-L-arabinose by a 4-epimerase.

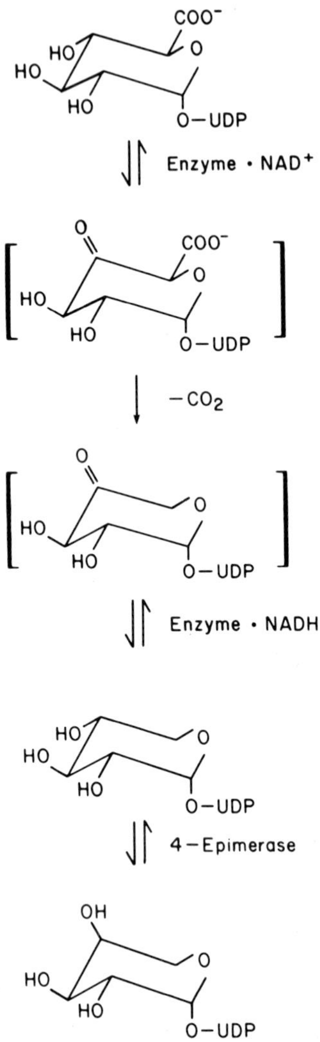

Fig. 5. The reactions involved in the formation of UDP-xylose and UDP-arabinose. The mechanism of decarboxylation of UDP-glucuronic acid involves a 4-keto intermediate, as does the 4-epimerase reaction.

D. Biosynthesis of Deoxyhexoses

Plant, animal, and bacterial cells contain a number of different kinds of 6-deoxyhexoses such as L-fucose (6-deoxy-L-galactose), L-rhamnose (6-deoxy-L-mannose), and quinovose (6-deoxy-D-glucose). In addition, vari-

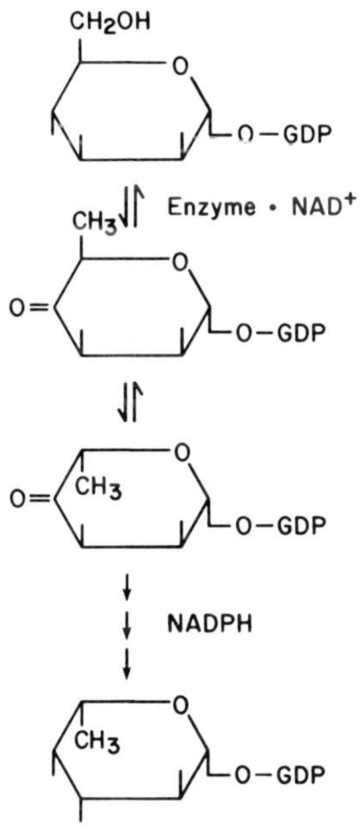

Fig. 6. The reactions involved in the biosynthesis of L-fucose. The 6-deoxyhexoses and 3,6-dideoxyhexoses are synthesized by similar series of reactions involving 4-keto intermediates and epimerases.

ous dideoxyhexoses such as digitoxose (2,6-dideoxy-D-glucose), colitose (3,6-dideoxy-L-galactose), and tyvelose (3,6-dideoxy-D-mannose) are found in certain organisms. With few exceptions, these sugars are formed from preexisting sugar nucleotides by a rather complex series of oxidation–reduction and epimerization reactions.

The first deoxyhexose to be studied was L-fucose, whose biosynthesis was examined in *Aerobacter aerogenes* (Ginsburg, 1961). Cell-free extracts of this bacterium catalyzed the formation of GDP-L-fucose from GDP-D-mannose via an NADPH-dependent series of reactions. Cell-free extracts of other bacteria catalyze the conversion of TDP-D-glucose to TDP-L-rhamnose (Glaser and Kornfeld, 1961) or GDP-D-mannose to GDP-colitose (Elbein and Heath, 1965). In plants, the transformation of UDP-D-glucose to UDP-L-rhamnose (Barber, 1962) and GDP-D-mannose to GDP-D-rhamnose occur by a rather complex series of reactions that are similar in mechanism to those observed for bacterial systems. Figure 6 demonstrates the general series of reactions that account for the biosynthesis of the deoxyhexoses.

The enzyme that catalyzes the first reaction in Fig. 6, called GDP-D-mannose 4-6-hydro-lyase (EC 4.2.1.47), has been purified from several sources, including seedlings of *Phaseolus aureus* (Liao and Barber, 1972). Although no cofactor was found for this enzyme, other oxidoreductases have been shown to contain tightly bound NAD^+. This oxidoreductase, or a similar type of enzyme, is also involved in the biosynthesis of dideoxyhexoses such as colitose (Gabriel and Van Lenten, 1978). The mechanism of this reaction has been studied in detail. For example, during conversion of TDP-D-glucose to TDP-4-keto-6-deoxy-D-glucose, there is a quantitative transfer of the C-4 hydrogen atom to C-6. In addition, a proton from the medium is incorporated at C-5 of the product (Gabriel, 1973). Enzyme-bound NADH is formed upon oxidation of the substrate at C-4.

While the first step in deoxyhexose formation is well understood, there is considerably less information about other steps in the pathway, that is, conversion of the 4-keto-6-deoxyhexose intermediate to the final product. Since the products differ in configuration from the substrates (for example, in configuration at C-3 and C-5 in Fig. 6), several epimerases and a reductase must be involved in subsequent stages of the pathway, in addition to a requirement for NADPH.

IV. The Role of Lipid-Linked Saccharides in the Biosynthesis of Complex Carbohydrates

The involvement of lipid-linked oligosaccharides in the biosynthesis of the oligosaccharide portion of the N-linked glycoproteins is clearly established in both plant and animal cells. However, the question as to whether these intermediates are also involved in the biosynthesis of cell wall polysaccharides is still unanswered. In this section, the mechanism of biosynthesis of lipid-linked saccharides in plants will be briefly reviewed and their role in plant glycoprotein biosynthesis will be outlined. For more details on these intermediates in plant and animal cells, the reader is referred to several reviews on this subject (Elbein, 1979; Struck and Lennarz, 1980; Staneloni and Leloir, 1982).

A. *In Cellulose Synthesis*

Probably the first suggestion for the presence of lipid-linked saccharides as potential intermediates in polysaccharide synthesis came from studies with the cellulose-producing bacterium *Acetobacter xylinum* (Colvin, 1959). In these studies, an ethanol extract from the organism was capable of causing the formation of cellulose microfibrils when incubated with an extract from the same bacterium. These workers subsequently isolated "glucolipids" following incubation of cell-free extracts with UDP-[^{14}C]glucose. These glucolipids were reported to contain acid labile glucose, cellobiose, and perhaps higher polymers of glucose (Kjosbakken and Colvin, 1973). Whether these intermediates are really precursors to cellulose remains unclear, however, since the sugar portions of these lipids have not been shown to be transferred to cellulose. Furthermore, other workers in this area have also not found evidence for the involvement of lipid-linked saccharides in cellulose synthesis (Delmer, 1976). Clearly, glucosylphosphorylpolyprenols are formed from UDP-glucose in many systems, but this lipid intermediate may be a precursor for the glucosyl moieties of glycoproteins (see below), or for glucose residues in bacterial cell wall polymers.

When membrane preparations from *Pisum sativum* were incubated with UDP-[^{14}C]glucose, radioactivity was incorporated into various compounds that were extracted with butanol, phenol, or alkali. In addition, radioactivity was incorporated into insoluble material that was solubilized by digestion with proteolytic enzymes. The radioactivity in each of the fractions (i.e., butanol, phenol, and alkali) was shown to be in glucose following complete acid hydrolysis and paper chromatography. Based on Smith degradation and other studies, the authors suggest that the butanol fraction contains an oligosaccharide–lipid having an oligosaccharide of 14 glucose units linked in $\beta(1\rightarrow3)$ and $\beta(1\rightarrow4)$ glycosidic bonds. Both the phenol-extractable and alkali-extractable materials gave glucose and erythritol upon Smith degradation, suggesting the presence of $(1\rightarrow3)$ and $(1\rightarrow4)$ linkages. These workers suggest that glucose from UDP-glucose is transferred to lipid and/or protein intermediates to produce an oligoglucan–lipid or oligoglucan–protein. The oligoglucan is then postulated to be transferred or converted to polyglucan (Brett and Northcote, 1975). Unfortunately, in these experiments, no kinetics were done, so that transfer of radioactivity from one component to another was not demonstrable.

In cultured cotton ovules (Delmer, 1976), glucose from UDP-glucose is also incorporated into lipid. More than 95% of the glucose was incorporated into steryl glucosides, however, and only small amounts were associated with glucosylphosphorylpolyprenol. At least with mung bean seedlings, the glucolipid (Elbein, 1979) appeared to be involved in glycoprotein biosynthesis by serving as a precursor in the formation of $Glc_3Man_9GlcNAc_2$-pyrophosphoryldolichol (see below).

Glucolipids have been claimed to be precursors to cellulose biosynthesis in the alga *Prototheca zopfii* (Hopp *et al.*, 1978). In those studies, a particulate enzyme from the alga incorporated 5% of the radioactive sugar nucleotide into butanol-soluble material. Three polar glucolipids were isolated by chromatography on DEAE-cellulose and were partially identified as glucosylphosphorylpolyprenol, glucosylpyrophosphorylpolyprenol, and the triglucosyl lipid, Glc-β-Glc-β-Glc-pyrophosphoryl lipid. In addition, extraction of the reaction mixtures with $CHCl_3$: CH_3OH : H_2O (10 : 10 : 3) gave oligosaccharide–lipids. These oligosaccharides were composed of varying numbers of glucose residues to give a spectrum from disaccharides to decasaccharides. Since only glycerol and erythritol were detected upon Smith degradation, the glycosidic linkages were most likely (1→4) bonds. In addition, radioactive glucose was incorporated into water-soluble polymers that eluted in the void volume from Sephadex G-50, and the formation of this polymer was inhibited by UMP or UDP. On the other hand, the formation of the glucolipids and the water-soluble polymer were enhanced by the addition of exogenous ficaprenylphosphate plus Triton X-100. The incorporation of glucose from glucosylphosphoryllipid into the shorter chain lipid-linked oligosaccharides and into the water-soluble polymer was enhanced by the addition of 0.4 mmol UDP-glucose to the incubation mixtures, suggesting that UDP-glucose was the glucosyl donor for the shorter oligosaccharides and glucosylphosphoryllipid for the longer chains. The water-soluble polymer could be partially digested with pronase or almost completely digested with cellulase or β-glucosidase, but not by α-glucosidase, amylase, or dextranase. The authors also reported that addition of GDP-glucose to the reaction mixtures leads to the formation of water-insoluble and alkali-insoluble polymers.

Based on their results, these workers postulated the scheme for cellulose biosynthesis shown in Fig. 7. In this pathway, both GDP-glucose and UDP-glucose are cellulose precursors, with UDP-glucose being utilized to form the lipid intermediates and protein intermediates and GDP-glucose for elongation of polysaccharide chains. While this sequence of reactions is intriguing and could account for many of the controversies regarding cellulose synthesis, thus far, similar results have not been reported in any other systems. Nor have these authors reported any further studies in this system. Thus, we will have to await further substantiation in this or other systems to determine the mechanisms of cellulose biosynthesis.

B. *In Glycosylation of Plant Proteins*

The role of lipid-linked saccharides in the biosynthesis of the oligosaccharide portion of the N-linked glycoproteins in animal cells is quite well established, and their role in the biosynthesis of plant glycopro-

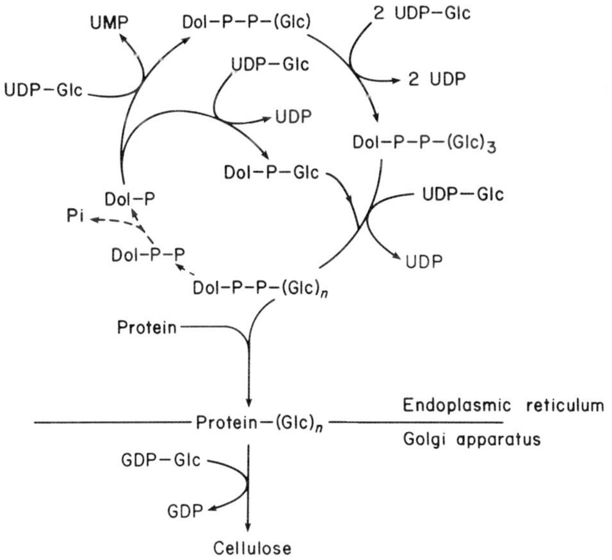

Fig. 7. Postulated series of reactions for the biosynthesis of cellulose in algae. Both UDP-glucose and GDP-glucose, as well as lipid-linked saccharides, are believed to be involved.

teins has also become reasonably well documented. For example, tunicamycin is an antibiotic that inhibits the formation of GlcNAc-pyrophosphoryldolichol, the first step in the formation of the lipid-linked oligosaccharides (Tkacz and Lampen, 1975). In animal cells, this antibiotic prevents glycosylation of those proteins having N-linked oligosaccharides (Elbein, 1981). Tunicamycin also inhibits the formation of the GlcNAc-lipid in plant extracts (Ericson *et al.*, 1977), and incubation of plant cells with this inhibitor prevents glycosylation of N-linked proteins (Hori and Elbein, 1981).

The reactions involved in the formation of the lipid-linked oligosaccharides are outlined in Fig. 8. These reactions involve the transfer of GlcNAc from UDP-GlcNAc, mannose from GDP-mannose, and glucose from UDP-glucose to various lipid intermediates and lipid-linked oligosaccharides to give the final lipid-linked oligosaccharide, Glc_3Man_9-$GlcNAc_2$-pyrophosphoryldolichol (Li *et al.*, 1978; Elbein, 1979).

Initially it was demonstrated in a number of laboratories that mannose from GDP-mannose was incorporated into material that was soluble in chloroform–methanol. This "mannolipid" was negatively charged, since it bound to DEAE-cellulose, and the sugar was released as free mannose by mild acid hydrolysis (Alam and Hemming, 1973; Beevers and Mense,

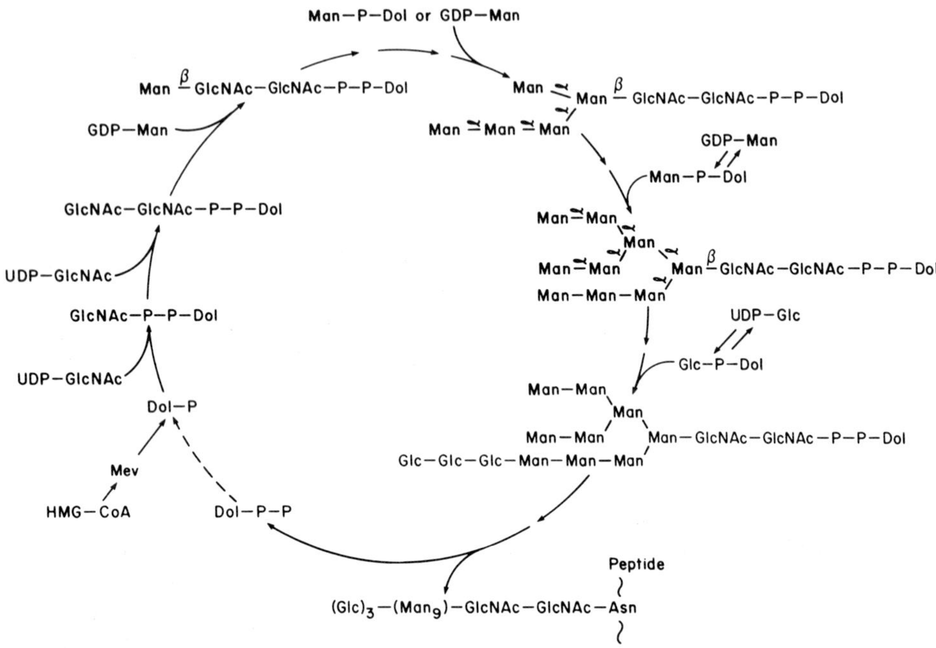

Fig. 8. The pathway of biosynthesis of the oligosaccharide chains of N-linked glycoproteins. The lipid pathway outlined here gives rise to a $Glc_3Man_9GlcNAc_2$-lipid, which undergoes a number of processing reactions after transfer to protein.

1978; Ericson and Delmer, 1977; Forsee and Elbein, 1972; Smith *et al.*, 1976; Waechter and Lennarz, 1976). These properties were consistent with those of mannosylphosphorylpolyprenol identified in animal systems (Waechter and Lennarz, 1976). Figure 9 shows the structure of mannosylphosphoryldolichol. Subsequent studies showed that the mannose was also incorporated into more hydrophilic lipids, which were not extractable with chloroform : methanol but which could subsequently be extracted with chloroform : methanol : H_2O (10 : 10 : 3) (Forsee and Elbein, 1975; Brett and Leloir, 1977; Lehle *et al.*, 1976). These lipids proved to be a series of oligosaccharide–lipids that contained various sized oligosaccharides of mannose linked to lipid in a pyrophosphoryl bond. The oligosaccharides could be released from the lipid by mild acid hydrolysis and separated into various oligosaccharides by gel filtration on Biogel P-4 (Forsee and Elbein, 1975; Brett and Leloir, 1977; Lehle *et al.*, 1976; Hori and Elbein, 1982). Later studies identified the lipid protion of the mannolipid as a dolichol by mass spectrometry (Delmer *et al.*, 1978).

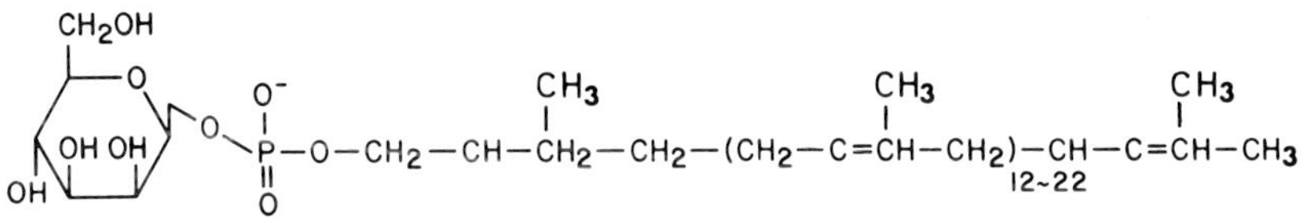

Fig. 9. The structure of mannosylphosphoryldolichol, a lipid intermediate in the glycosylation reactions.

Since the mannose-containing oligosaccharide–lipids were found to have GlcNAc at the reducing end of the oligosaccharides, the incorporation of GlcNAc from UDP-GlcNAc into lipid-soluble material was examined in several plant systems (Hori and Elbein, 1982). GlcNAc was also incorporated into charged lipids by several plant extracts; however, these lipids bound more tightly to DEAE-cellulose than the mannosylphosphoryldolichols. Subsequently, it was determined that the GlcNAc-lipids extractable by chloroform : methanol were a mixture of GlcNAc-pyrophosphorylpolyprenol and GlcNAc-GlcNAc-pyrophosphorylpolyprenol (Lehle *et al.*, 1976; Hori and Elbein, 1982; Delmer *et al.*, 1978; Forsee *et al.*, 1976). The radioactive GlcNAc was also found in the various mannose-containing oligosaccharide–lipids that had been identified earlier. A variety of chase experiments and other studies established that two GlcNAc residues were first transferred to the polyprenol to form the GlcNAc-GlcNAc-pyrophosphorylpolyprenol and then a number of mannose residues were added. In these *in vitro* studies, no evidence for glucose incorporation into the lipid-linked oligosaccharides was obtained, although the presence of glucose in the animal lipid-linked oligosaccharide had already been established.

Studies with several different plant cell culture systems, however, demonstrated that glucose was incorporated into lipid-linked oligosaccharides (Staneloni *et al.*, 1980; Lehle, 1981; Hori *et al.*, 1982). Although the complete structure of the oligosaccharide has not been established, it has been demonstrated to be a $Glc_3Man_9GlcNAc_2$, and some structural studies have suggested that it is probably similar to, if not identical to, the $Glc_3Man_9GlcNAc_2$ of animal cells (Hori *et al.*, 1982). In animal cells, the $Glc_3Man_9GlcNAc_2$ is transferred from its lipid to the protein while the protein is being synthesized on membrane-bound polysomes. Although this has not been established in plants, there is some preliminary evidence for the transfer of oligosaccharide to protein (Staneloni *et al.*, 1981). In fact, the oligosaccharides containing glucose were transferred better than those without glucose.

Although the oligosaccharide transferred to protein contains glucose, the N-linked oligosaccharides of these glycoproteins do not contain glucose. Thus, it appeared that the glucose residues must be removed after transfer to protein. Subsequently, it was demonstrated that the animal glycoproteins undergo a number of processing reactions as they are transported from the endoplasmic reticulum through the Golgi apparatus. These reactions involve removal of all three glucose residues by two membrane-bound glucosidases, and the removal of up to six mannose residues by at least two different mannosidases (Tulsiani *et al.*, 1982). In animal cells, these processing reactions allow the cell to utilize the same oligosaccharide precursor to form high-mannose, hybrid, and complex oligosaccharides. Several of the processing glycosidases of animal cells have been purified and studied in some detail.

In plants, processing reactions have not been shown directly. In one study, however, soybean cells were labeled for short times with [^{3}H]mannose or [^{3}H]glucose and then chased for various times in unlabeled medium. Glycopeptides were prepared from the membrane fraction of these cells, and these glycopeptides were sized on columns of Biogel P-4. At very early times, the glycopeptides were larger in size, but as the chase proceeded, they became smaller and smaller. A combination of data indicated that the glycoproteins initially contained glucose and that they were subsequently processed by the removal of glucose (Hori and Elbein, 1983). In this plant system, or in others, it is not clear how far the processing goes. That is, are mannose residues also removed, and if so, how many? And are other sugars added to these proteins, and at what stage? These questions remain to be answered.

V. Biosynthesis of Cellulose

Cellulose is the most abundant carbohydrate in nature and the most abundant compound of plant cell walls. It makes up the basic structural material of the cell walls of all higher land plants and is also found in some algae. In addition, it is synthesized by a few bacteria. Cellulose is a $\beta(1\rightarrow4)$-linked D-glucan having a degree of polymerization of about 10,000. The glucan chains line up in a parallel arrangement to form elementary fibrils. The cellulose microfibril has a well-defined crystalline structure, but how the chains associate to give the crystalline arrays and how the microfibrils are deposited to give a coherent cell wall are poorly understood. In addition, almost nothing is known about how cellulose chains are initiated or terminated, nor whether the synthesis of primary cell wall cellulose is mediated by the same enzyme system and/or the same mechanism as secondary cell wall cellulose.

Essentially, three main approaches have been used to try to answer fundamental questions about cellulose synthesis, orientation, and termination. In one case, autoradiographic techniques with labeled sugars have been coupled with electron microscopy. While those techniques can show some synthetic reactions as well as ultrastructure and localization, it cannot tell us anything about the mechanism of polymerization or sorting. Furthermore, when one uses labeled sugars as precursors, the label is not specific for cellulose, but will label many cell-wall polysaccharides as well as other materials. Another approach to studying cellulose biosynthesis and translocation is by cell-fractionation technique. If cell fractionation is coupled with labeling studies and chase experiments, it should be possible to follow the movement of cellulose or some of its precursors or intermediates. The problem in this case is that the amount of material that can be isolated from various fractions or the amount of radioactivity that can be incorporated is quite small, and therefore this method has its limitations. The third approach is the biochemical approach. The problem with this methodology is that it frequently involves cell homogenates, so that the cell integrity is destroyed. Nevertheless, this is a powerful method and has been utilized in many studies.

In some early autoradiographic studies, the deposition of [^{14}C]glucose was shown to be localized exclusively along the plasma membrane of xylem and phloem in sycamore seedlings. The deposition was not seen in and around the Golgi body (Wooding, 1968). It was also shown that radioactive glucose was deposited on the secondary wall before it appeared within the cells (Northcote and Lewis, 1968). In other experiments designed to follow hemicellulose synthesis, however, [^{3}H]glucose was found to be deposited over the Golgi bodies in a 10-min pulse using wheat root cap cells (Pickett-Heaps, 1969). Although some effort was made to stain the carbohydrate, it was not possible to determine the exact structure of the labeled materials.

An important approach to determine the synthetic site of cellulose formation was an electron-microscopic study. By freeze-fracture techniques and electron microscopy, a particle was located on the plasmalemma surface that was thought to be an enzyme-containing particle involved in orienting the deposit of the cellulose microfibrils. The dimension of the particle was ~500 Å. These particles, as well as their function, however, have not been established, as a result of incomplete evidence (Robinson, 1977). Thus, in algae, cellulose microfibrils have been seen to be associated with particles at the plasma membrane, while in higher plants the evidence for cellulose associated with such particles is poor. However, it does seem likely that the synthesis of cellulose occurs near or at the plasma membrane. In yeast, chitin microfibrils were found to be associated with particles that were about 350–1000 Å in size (Ruiz-Herrera *et al.*, 1975). These particles were shown to incorporate GlcNAc, from UDP-GlcNAc, into polymer thought to be chitin.

The deposition of microfibrils of cellulose at the plasma membrane becomes oriented from a random arrangement into a parallel orientation. Thus, the primary cell wall at inception usually possesses randomly oriented microfibrils, but when the cells begin to elongate, the innermost and newly formed microfibrils tend to become parallel at 90° to the ground axis (Roelofson, 1965). Microtubules, originally described by Ledbetter and Porter (1963), appear to play a role in microfibril arrangement and orientation. Thus, microfibrils of cellulose have been demonstrated to run parallel to microtubules in higher plants (Hepler and Palevitz, 1974). Furthermore, the drug colchicine inhibits microtubule formation, and this drug when given to algae disrupts the normal orientation of the microfibrils (Robinson *et al.*, 1976; Sachs *et al.*, 1977; Grimm *et al.*, 1977; Hogetsu and Shibaoka, 1978). In *Oocystis solitaria* treated with 10^{-2} *M* colchicine, the regular alternating pattern of microfibrils is modified. After 24 hr, however, the drug was removed, and the cells were able to regain the normal orientation of cellulose fibers. Although others have claimed that cytoplasmic streaming may be responsible for the orientation of microfibrils (Colvin, 1980), this phenomenon is also believed to be controlled by microtubules.

A review by Delmer (1976) has pointed out that no one has yet succeeded in achieving the *in vitro* synthesis of cellulose. At best, various investigators have been able to incorporate small amounts of radioactive glucose from various sugar nucleotides into polymers that appear to be $\beta(1\rightarrow4)$ glucans. The initial *in vitro* studies on cellulose synthesis were done in *Acetobacter xylinium* by Glaser (1958). Cell-free extracts of this bacterium were shown to incorporate [^{14}C]glucose from UDP-[^{14}C]glucose into alkali-insoluble material that had the properties of a $\beta(1\rightarrow4)$ glucan. On the other hand, in plants, glucose from both GDP-glucose and UDP-glucose has been reported to be incorporated into cellulose in various experiments with cell-free plant extracts. Thus, GDP-glucose was shown to be incorporated into alkali-insoluble polymer by extracts of mung bean seedlings, cotton fibers, and other tissues (Elbein *et al.*, 1964). The radioactive product was shown to give the typical radioactive cellodextrin series on partial acid hydrolysis, suggesting that it was constructed of $\beta(1\rightarrow4)$ bonds. Subsequent studies with this system, however, have shown that [^{14}C]mannose from GDP-[^{14}C]mannose is also incorporated into alkali-insoluble polymer and that the incorporation of glucose, from GDP-glucose, is enhanced by the presence of GDP-mannose (Elbein and Hassid, 1966). Another study has also demonstrated the incorporation of mannose and glucose from their guanosine nucleotides into glucomannan (Villemez, 1971). Thus, it seems likely that GDP-glucose and GDP-mannose are both precursors for a $\beta(1\rightarrow4)$-linked glucomannan (Section

VII). However, GDP-glucose could still be a precursor for cellulose biosynthesis, although this remains to be definitively established. As indicated in the section on lipid-linked saccharides, GDP-glucose has been reported to be involved in postlipid-mediated reactions of cellulose formation.

A number of studies have implicated UDP-glucose as the glucosyl donor for cellulose (Delmer, 1976). While some of these reports are probably well founded, in other cases, it seems likely that the workers were studying either $\beta(1\rightarrow3)$ glucan synthesis or the formation of a mixed polymer of $\beta(1\rightarrow3)$ and $\beta(1\rightarrow4)$ bonds. The problem in some of these cases is that the characterization of the product was based largely on its insolubility in alkali, and this is clearly not a characteristic of cellulose alone. Thus, $\beta(1\rightarrow3)$ glucans, formed from UDP-glucose by cell-free extracts, may also be insoluble in hot 2% NaOH (Ericson and Elbein, 1980). Furthermore, some of these polymers may be glucans having both $\beta(1\rightarrow3)$ and $\beta(1\rightarrow4)$ glycosidic bonds. In some studies, however, the product did show properties that were indicative of cellulose. Thus, Brummond and Gibbons (1964) reported that when UDP-[^{14}C]glucose was used at concentrations of 1 to 6 m*M*, 3–7% of the total radioactivity in the incubation mixture was incorporated into cellulose, while at GDP-[^{14}C]glucose concentrations of 0.65 to 5.7 m*M*, 2–8% of the total radioactivity was incorporated into cellulose (Brummond and Gibbons, 1964). In another study, however, particulate extracts from *Lupinus albus* incorporated glucose from UDP-glucose into insoluble glucan, which contained 80% of the glucose as $\beta(1\rightarrow4)$ glucan (Stafford and Brummond, 1970). Ordin and Hall (1968) tested various concentrations of UDP-glucose with particulate extracts of *Avena* and examined the products formed in these cases (Ordin and Hall, 1968). They found that at concentrations of 4 μM, glucose, from UDP-glucose, was incorporated into cellulose since digestion with a *Streptomyces* cellulase gave rise to radioactive cellobiose. At 1 m*M* UDP-glucose, however, digestion of the product gave mostly the $\beta(1\rightarrow3)$ disaccharide laminaribiose. These results and those of others indicate that UDP-glucose is a likely precursor of cellulose, but the concentrations of this nucleotide in *in vitro* studies are critical in terms of the product(s) formed. It seems obvious that much more work is needed to decipher the mechanism of cellulose synthesis. In addition, it seems likely that it will take a new approach and perhaps a different system to achieve synthesis of this molecule.

A considerable amount of progress has been made in biosynthetic studies with *Acetobacter xylinum* (Aloni and Benziman, 1982). In these studies, membrane preparations from the bacterium were able to catalyze high rates of $\beta(1\rightarrow4)$ glucan synthesis, in contrast to earlier studies from various other laboratories. The secret in these more recent studies was the finding that the "cellulose synthase" was an enzyme with a rather complex regulatory system. Thus, the enzyme was activated specifically by GTP. This GTP activation was found to depend on the presence of a protein factor that easily dissociates from the enzyme and therefore was apparently missed by earlier workers. When the membrane preparations were prepared in the presence of PEG 4000, however, they retained the protein factor and then were activated by GTP. The factor is thought to be a regulatory protein subunit of the enzyme. These workers have now solubilized "cellulose synthase" (Aloni *et al.*, 1983), and therefore we may expect to hear more about the regulation and mechanism of this interesting system.

In cotton hairs, maturation can be separated into two distinguishable phases. In the early phase, primary walls are produced, whereas in the late phase, the cells produce the secondary wall (Franz, 1969). In the early stage, cellulose production is very low, but cellulose synthesis is very rapid in the secondary stage and gives nearly all the cellulose produced (Marx-Figini, 1966). Interestingly enough, the degree of polymerization of cellulose produced in the secondary stage is quite constant, suggesting that secondary wall cellulose synthesis is controlled by templates. Many investigators have attempted to define the templates involved in this process (Machlachlan, 1982). In cotton fibers, the formation of $\beta(1\rightarrow3)$ glucan was enhanced by cellulose at concentrations around 10 m*M*. This enhancement was especially effective at low concentrations of UDP-glucose, that is, $\sim 40\ \mu M$ (Delmer, 1976). Pea-stem membrane preparations catalyzed the incorporation of glucose from both GDP-glucose and UDP-glucose into various soluble and insoluble fractions. Cellobiose and cellulose accelerated glucose incorporation only from GDP-glucose into alkali-soluble and -insoluble fractions (Spencer *et al.*, 1971). Carboxymethylcellulose (CMC) also enhanced the incorporation in buffer-soluble material. The authors postulated that the enhancement by CMC and cellulose was due to nonspecific adsorption of radioactive materials to the polymer surface. However, it has been pointed out that this postulation does not explain the fact that cellobiose was only effective toward GDP-glucose (Machlachlan, 1982).

VI. Biosynthesis of Noncellulosic β-Glucans

A number of other β-glucans having either $\beta(1\rightarrow3)$-linked D-glucose units, or a mixture of $\beta(1\rightarrow3)$- and $\beta(1\rightarrow4)$-linked glucose units are found in plants. Some of these β-glucans may belong to the hemicellulose groups, and, as already indicated, it seems probable that some of these polymers were actually being synthesized rather than cellulose. However, there is one β(1→3) glucan that is not a member of the hemicellulose group, and whose biosynthesis has been thoroughly studied. This is the polysaccharide callose, which is generally considered to be a wound polysaccharide; it is also found in sieve plates of phloem and other higher plant tissues.

In a classic study, Feingold and co-workers (1958) showed that a soluble enzyme preparation from mung bean seedlings, as well as particulate enzyme preparations from parsley, zucchini, cabbage, and spinach, was capable of incorporating D-glucose from UDP-[^{14}C]glucose into a polymer that was immobile on paper chromatograms. The authors characterized the product as a β(1→3) glucan on the basis of partial acid hydrolysis and identification of the products, as well as by lead tetraacetate oxidation of β-glucosidase digestion (Feingold *et al.*, 1958). Laminaribiose, and to a lesser extent D-glucose and other sugars, stimulated the incorporation of [^{14}C]glucose into polymer. Although these workers were not able to determine the exact role of these activators in the biosynthetic reaction, it was evident from incorporation studies and from the relatively high concentrations of activator required that these compounds were not functioning as primers in the reaction. Since this study, many researchers have also noted the stimulation of β-glucan synthesis by laminaribiose or cellobiose. For example, the synthesis of paramylon, a storage β(1→3) glucan, by extracts of *Euglena gracilis* was stimulated 20–30% by laminaribiose (Marechal and Goldemberg, 1969). In that study, it was shown that laminaribiose did not act as a primer, since incubation of labeled laminaribiose with unlabeled UDP-glucose did not give rise to labeled polymer.

VII. Synthesis of Hemicellulosic Polysaccharides

The major hemicellulosic polysaccharide in the cell wall of dicotyledonous plants is a xyloglucan. Xyloglucan has the ability to hydrogen bond to cellulose chains, because it bears a structural relationship to cellulose, and this hydrogen bonding serves to connect the cellulose fibers to pectin polysaccharides. The xyloglucan is actually misnamed, since this polysaccharide also contains L-fucose and D-galactose as well as small amounts of L-arabinose. The structure of the polymer is a β(1→4) glucan core to which single α(1→6)-linked xylose units are attached. The sugars L-fucose and/or D-galactose may be linked to the D-xylose residues.

Xyloglucans are found in seeds and in cell walls of various dicots. They are composed of a β(1→4)-glucan backbone to which short side chains of D-xylose are attached via an α(1→6) linkage:

```
          α-D-Xylp                    R-α-D-Xylp
             1                             1
             ↓                             ↓
             6                             6
-β-D-Glcp-(1→4)β-D-Glcp-(1→4)β-D-Glcp-(1→4)β-D-Glcp-
```

where R = Gal or Fuc(1→2)Gal.

The biosynthesis of xyloglucan has been studied in elongating pea stems (Ray, 1975, 1980). Cell-free extracts of this tissue catalyzed the incorporation of D-glucose from UDP-[^{3}H]glucose and D-xylose from UDP-[^{14}C]xylose into a water-soluble xyloglucan. This polysaccharide gave rise to the disaccharide xylosylglucose upon partial acid hydrolysis. Both glycosyl transferases [i.e., UDP-glucose : β(1→4) glucan glucosyl transferase and the UDP-xylose : xyloglucan xylosyl transferase] were found to be localized in the Golgi apparatus of the cell. The addition of unlabeled UDP-glucose to the incubation mixtures greatly stimulated the incorporation of xylose, from UDP-xylose, into the polymer. The authors provided the following pieces of evidence to indicate that the stimulation was due to glucan synthetase activity rather than to any other cause: (1) the maximum stimulation of xylose incorporation by UDP-glucose occurred at pH 7–8, where glucan synthesis was also optimal; (2) the stimulation occurs at UDP-glucose concentrations that result in the formation of amounts of glucan that are equal to or greater than the amount of xylose incorporated, (3) at higher concentrations of UDP-xylose, more xylose is incorporated into the polymers and xylose incorporation becomes more dependent on the presence of UDP-glucose; (4) when the glucan synthetase is inhibited by Cd^{2+} UDP-glucose no longer stimulates xylose incorporation. These results suggest that UDP-glucose stimulates xylose incorporation by forming the glucan backbone that then serves as an acceptor for xylose. Thus, in the presence of xylose, the glucan chain becomes modified to yield a heteropolysaccharide of D-xylose and D-glucose units in which the D-xylose residues appear to be attached mainly as nonreducing termini onto the glucan core.

Xylans are mainly secondary wall components, but in monocots they are found in the primary cell wall. They are the most abundant noncellulosic polysaccharide in many angiosperms and may account for 20–30% of the dry weight of the woody tissue. The xylans are linear polysaccharides composed of β(1→4)-linked D-xylopyranose units that are frequently branched with variable amounts of 4-*O*-methylglucuronic acid and/or

L-arabinofuranose:

$$\begin{array}{ccc} \alpha\text{-L-Ara}f & & \alpha\text{-D-4-MeGlcA}p \\ 1 & & 1 \\ \downarrow & & \downarrow \\ 3 & & 2 \end{array}$$
$$\text{-Xyl}p\text{-}(1\rightarrow4)\text{-}\beta\text{-D-Xyl}p\text{-}(1\rightarrow4)\text{-}\beta\text{-D-Xyl}p\text{-}(1\rightarrow4)\text{-}\beta\text{-D-Xyl}p\text{-}$$

As expected, the biosynthesis of arabinoxylan involved the participation of the sugar nucleotides UDP-D-xylose and UDP-L-arabinose. Thus, when particulate enzyme preparations from immature corn cobs were incubated with UDP-[^{14}C]xylose, the radioactive product resembled native arabinoxylan and contained radioactivity in both xylose and arabinose (Bailey and Hassid, 1966). This indicated that the particulate enzyme contained the UDP-xylose : UDP-arabinose 4-epimerase. The incorporation of 4-*O*-methyl-D-glucuronic acid into xylan has also been demonstrated in extracts of corn cobs (Kauss and Hassid, 1967). In these studies, glucuronic acid from UDP-D-glucuronic acid was incorporated into polymer, and then the glucuronic acid residues became methylated by the transfer of methyl groups from *S*-adenosyl-L-methionine to the 4 position of glucuronic acid (Kauss, 1969). However, the products have not been well enough characterized to determine whether they are typical hemicellulosic xylans.

Glucomannans represent another hemicellulosic polysaccharide that is found in the cell walls of various higher plants along with galactoglucomannans. These two polysaccharides are the major cell wall components of gymnosperms, whereas in angiosperms, glucomannan is a minor component. The general structure of the glucomannan and galactoglucomannan is

$$\begin{array}{c} \text{D-Gal}p \\ 1 \\ \downarrow \\ 6 \end{array}$$
$$\text{-}\beta\text{-D-Man}p\text{-}(1\rightarrow4)\text{-}\beta\text{-D-Glc}p\text{-}(1\rightarrow4)\text{-}\beta\text{-D-Man}p\text{-}(1\rightarrow4)\text{-}\beta\text{-D Glc}p\text{-}$$

In mung bean particulate enzyme preparations, it was shown that glucose from GDP-[^{14}C]glucose was incorporated into a $\beta(1\rightarrow4)$glucan and that this incorporation was greatly stimulated by the addition of unlabeled GDP-mannose to the incubation mixtures (Elbein and Hassid, 1966). Subsequent studies showed that mannose from GDP-[^{14}C]mannose was incorporated into an alkali-insoluble polymer (Elbein, 1969). In this case, however, the addition of unlabeled GDP-glucose to reaction mixtures containing GDP-[^{14}C]mannose did not result in a stimulation in the incorporation of radioactivity; instead the incorporation of [^{14}C]mannose was markedly decreased. The radioactive glucomannan formed from GDP-[^{14}C]mannose was compared to that produced from GDP-[^{14}C]glucose in the presence of unlabeled GDP-mannose. Partial acid hydrolysis of these two products, or enzymatic hydrolysis with a β-mannanase, liberated a similar series of radioactive oligosaccharides from both biosynthetic products, and these labeled oligosaccharides contained both D-glucose and D-mannose, indicating that both glucose and mannose were incorporated into the same polymer. Since both mannose and glucose became labeled when incubation mixtures contained only GDP-[^{14}C]mannose, it seemed likely that the particulate enzyme contained a 2-epimerase capable of interconverting GDP-glucose and GDP-mannose. The labeled oligosaccharides produced by hydrolysis were also characterized by methylation analysis and periodate oxidation. These experiments indicated that the polymer was a linear (1→4) glycan composed of D-mannose and D-glucose. Thus, one of the major disaccharides found in the hydrolysates was characterized as 4-*O*-β-D-mannopyranosyl-β-D-glucopyranoside, but 4-*O*-β-D-glucopyranosyl-β-D-mannopyranose was also found. These results have been confirmed by other workers in several plant systems (Heller and Villemez, 1972).

VIII. Synthesis of Pectic Substances

The term pectic substances refers to a group of closely associated polysaccharides isolated from the primary cell walls and intercellular regions of higher plants. Pectins, which are the major components, are esterified galacturonans or, more commonly, rhamnogalacturonans in which β(1→4)-linked D-galacturonan chains are interrupted at various intervals by the insertion of L-rhamnose residues. In pectins, varying proportions of the galacturonic acid residues are present as methyl esters. Pectic substances also include diesterified pectic acid and its salts, and other neutral polysaccharides.

The biosynthesis of pectins occurs by the transfer of monosaccharides from their sugar nucleotide donors to an acceptor polysaccharide. The initial studies, using a particulate enzyme preparation from mung bean seedlings, demonstrated the incorporation of [^{14}C]galacturonic acid from UDP-[^{14}C]galacturonic acid into a polygalacturonic acid product that appeared to be labeled throughout the chain (Villemez *et al.*, 1965). UDP-galacturonic acid was the most active glycosyl donor in this reaction, although TDP-galacturonic acid and CDP-galacturonic acid could also serve as galacturonic acid donors to a lesser extent (Lin *et al.,* 1966). Thus, the glycosyl transferase involved in this reaction is apparently not completely specific for the uridine derivative. In this regard, it is interesting to point out that sugar beets contain an enzyme system for converting TDP-glucose to TDP-galacturonic acid (Katan and Avigad, 1966).

The addition of methyl groups to form the typical methyl esters of galacturonic acid takes place either during chain polymerization or once the polygalacturonic acid chain has been formed. This conclusion is based on several different studies. In the first place, UDP-galacturonic acid chemically methylated at position 6 of the sugar will not serve as a glycosyl donor for the formation of polygalacturonic acid, strongly implying that methylation occurs after sugar transfer (Kauss and Hassid, 1967). Second, *in vivo* studies with methyl-labeled L-methionine demonstrated that this amino acid was the methyl donor in these reactions (Villemez *et al.*, 1966). The mechanism of methylation was examined with cell-free extracts of mung bean seedlings and *S*-adenosyl-L-methionine (SAM) as the methyl donor. SAM was shown to be the direct methyl donor in the reaction. The methyl groups were shown to be attached by ester bonds rather than as ethers by the fact that they were labile to alkaline digestion. The pH optimum for the methyl transferase was ~6.8, and the apparent K_m for SAM was $\sim 5.9 \times 10^{-5}$ *M*. No requirement for a divalent cation could be shown with this enzyme. On the other hand, there is little information available on the incorporation of acetyl groups into these polymers, although the likely acetyl donor is acetyl-CoA. Since acetylated sugar nucleotides have not been found in plants, it is probable that acetylation occurs after sugar transfer.

The synthesis of the polygalacturonic acid backbone and its subsequent methylation depends on distinct intracellular compartmentalization. Thus, exogenously added polygalacturonic acid could not be methylated by the particulate methyl transferase. This enzyme would act only on polygalacturonic acid that had been synthesized by the galacturonosyl transferase also present in the particulate enzyme preparation. In addition, these workers found no difference in the amount of methyl-esterified pectin formed in the reactions, either in the presence or absence of added methyl esterase. This suggested that the product formed in these reactions was not accessible to methyl esterase, perhaps because it was compartmentalized. If these workers first liberated the methyl-labeled pectin from the particles by treatment with detergent or with phospholipase A and then treated with methyl esterase, however, the product was completely demethylated within seconds (Kauss *et al.*, 1969). This enzyme, pectin methyl esterase, has been proposed to have a role in cell wall loosening that precedes plant cell growth. However, the exact relationship of the enzyme to cell wall growth remains to be established.

A number of *in vitro* studies have attempted to examine the mechanism of incorporation of neutral sugars into pectins, either as side chains or as blocks within the polygalacturonic acid chain. The problems in these experiments are (1) radioactivity from the sugar nucleotide usually ends up in a number of neutral polysaccharides and it is frequently difficult to determine whether pectin or a hemicellulose, or both, have been formed, and (2) the incorporation of radioactivity is usually quite low and therefore characterization of the product is difficult. In spite of these difficulties, the incorporation of radioactive sugars into various polymers has been examined in a number of *in vitro* systems. For example, apiose is transferred from UDP-apiose into an apiogalacturonan by a particulate enzyme preparation from duckweed (Kindel, 1973). A particulate enzyme from mung bean seedlings incorporated L-arabinose from UDP-arabinose into an arabinan as well as D-xylose from UDP-xylose into xylan (Odzuck and Kauss, 1972) and also into xyloglucan (Villemez and Hinman, 1975). These mung bean extracts also incorporated D-galactose from UDP-galactose into galactan (Panayotato and Villemez, 1973). Again, it should be mentioned that it is not clear that these studies really represent pectin synthesis rather than synthesis of hemicellulose components or other polysaccharides. While the apiogalacturonan most probably is pectin-related, the arabinan and galactan may represent neutral polymers that are frequently found associated with pectins, or they could represent the neutral blocks on their way to becoming associated with pectins (see below).

In a different approach to pectin synthesis, Stoddard and Northcote (1976) did labeling and chase studies with suspension-cultured sycamore cells in order to follow pectin synthesis and assembly (Stoddart and Northcote, 1976). When the cells were labeled for 1 hr with [^{14}C]glucose and then chased for 1 hr with unlabeled glucose, the first radioactive pectin-type material was a partially methylated galacturonan. Upon longer incubations in unlabeled medium, the radioactivity became associated with neutral blocks, with the pectic arabinan–galactan, and with neutral sugars, especially L-arabinose and D-galactose. At the same time, the label disappeared from the initially labeled component. Quantitative measurements showed that galacturonan was moving from the first-labeled material into acidic (type I) pectinic acids by deesterification and into the branched (type II) pectinic acids by the addition of blocks of neutral sugars. Similarly, it appeared that radioactivity could be passing from the arabinan–galactan to the neutral blocks of type II pectinic acids in the form of arabinose and galactose. Pulse–chase experiments with [^{14}C]-L-arabinose as the label confirmed that sugars were passing from the neutral blocks into pectin. Based on these studies, the authors proposed a scheme whereby the primary galacturonorhamnan chain is methylated and esterified to give type I pectinic acid. Then xylose from UDP-xylose is added, and the product of this reaction is converted to type II pectinic acid by the addition of arabinogalactan.

The mucilage or slime polysaccharides that are secreted by the roots of many terrestrial plants also appear to be pectinic in nature. In wheat and corn, these polymers are formed and secreted from peripheral root-cap cells using an integrated membrane system composed of endoplasmic reticulum and Golgi apparatus (Northcote, 1973). The polymer isolated from wheat was a water-soluble acidic polysaccharide that had a high proportion of D-galactose and L-fucose (Harris and Northcote, 1974). The slime secreted by corn has many of the same properties as that secreted by wheat. Thus, it has a molecular weight in excess of 2×10^6, a density of 1.63 g/cm^3, and contains the neutral sugars L-fucose (39%) and D-galactose (30%) as the major components and smaller amounts of L-arabinose, D-mannose, D-xylose, D-glucose, and D-galacturonic acid (Paull *et al.*, 1975).

The initial biosynthesis of slime polysaccharides was demonstrated by *in vivo* studies. Thus, when corn roots were incubated with [^{14}C]glucose, the sugar units of the polysaccharide (fucose, galactose, xylose, arabinose, glucose, mannose, and galacturonic acid) all became labeled (Bowles and Northcote, 1972). Also, when corn roots were incubated with [^{3}H]-L-fucose, three components became labeled, two of which were glycoproteins while the third was the slime polysaccharides (Green and Northcote, 1978). A fucosyl transferase was solubilized from the endoplasmic reticulum and characterized (James and Jones, 1979). This enzyme catalyzed the transfer of L-fucose from GDP-L-fucose to the slime polysaccharide. The enzyme activity was optimal at pH 7.0 in the presence of 6.8% (v/v) Triton X-100 and did not require divalent cations for activity. In fact, the activity was inhibited at concentrations of $MgCl_2$ and $MnCl_2$ above 1 m*M*. Exogenously added slime polysaccharide, isolated from corn roots and defucosylated by mild acid treatment, could serve as a fucose acceptor. The rate of release of L-fucose by acid was the same for the biosynthetic product as for the native corn root slime.

Acknowledgments

We are grateful to Pamela Lenow for an excellent job of editing and typing. Studies from the authors' laboratory were supported by grants from the National Institutes of Health (AM21300 and HL17783) and from the Robert A. Welch Foundation.

References

Adair, W. L., Gabriel, O., Stathakos, D., and Kalckar, H. M. (1973). *J. Biol. Chem.* **248,** 4640–4648.

Adams, E. (1976). *Adv. Enzymol.* **44,** 69–138.

Alam, S. S., and Hemming, I. W. (1973). *Phytochemistry* **12,** 1641–1649.

Aloni, Y., and Benziman, M. (1982). *In* "Cellulose and Other Natural Polymer Systems" (R. M. Brown, Jr., ed.), pp. 341–361. Plenum, New York.

Aloni, Y., Cohen, R., Benziman, M., and Delmer, D. (1983). *J. Biol. Chem.* **258,** 4419–4423.

Ankel, H., and Feingold, D. S. (1965). *Biochemistry* **4,** 2468–2475.

Ankel, H., and Feingold, D. S. (1966). *Biochemistry* **5,** 182–189.

Atlermatt, H. A., and Neish, A. C. (1956). *Can. J. Biochem. Physiol.* **34,** 405–413.

Axelos, M., and Peaud-Lenoel, C. (1969). *Bull. Soc. Chim. Biol.* **51,** 261–273.

Axelos, M., and Peaud-Lenoel, C. (1978). *Biochimie* **60,** 35–44.

Baddiley, J., Blumson, N. L., DiGirolamo, M. (1961). *Biochim. Biophys. Acta* **50,** 391–393.

Bailey, R. W., and Hassid, W. Z. (1966). *Proc. Natl. Acad. Sci U.S.A.* **56,** 1586.

Barber, G. A. (1962). *Biochem. Biophys. Res. Commun.* **8,** 204–208.

Beevers, L., and Mense, R. M. (1978). *Plant Physiol.* **60,** 703–708.

Brown, D. H. (1953). *J. Biol. Chem.* **204,** 877–889.

Brummond, D. O., and Gibbons, A. P. (1964). *Biochem. Biophys. Res. Commun.* **17,** 156–159.

Buchanan, J. G. (1953). *Arch. Biochem. Biophys.* **44,** 140–149.

Bowles, D. J., and Northcote, D. H. (1972). *Biochem. J.* **130,** 1133–1145.

Brett, C. T., and Leloir, L. F. (1977). *Biochem. J.* **161,** 93–101.

Brett, C. T., and Northcote, D. H. (1975). *Biochem. J.* **148,** 107–117.

Caputto, R., Leloir, L. F., Cardini, E. C., and Paladini, S. C. (1950). *J. Biol. Chem.* **184,** 338–350.

Chan, P. H., and Hassid, W. Z. (1975). *Anal. Biochem.* **64,** 372–379.

Colvin, R. (1959). *Nature (London)* **183,** 1135–1136.

Colvin, J. R. (1980). *In* "The Biochemistry of Plants" (J. Preiss, ed.), Vol. 3, pp. 543–570. Academic Press, New York.

Dalessandro, G., and Northcote, D. H. (1977). *Biochem. J.* **162,** 267–279.

Dankert, M., Passeron, S., Recondo, E., and Leloir, L. F. (1964). *Biochem. Biophys. Res. Commun.* **14,** 358–362.

DeAsua, L. J., Carminatti, H. T., and Passeron, S. (1966). *Biochem. Biophys. Acta.* **128,** 582–585.

Delmer, D. P. (1976). *In* "Recent Advances in Phytochemistry" (F. A. Loewus and V. C. Runeckles, eds.), Vol. 3, pp. 45–76. Plenum, New York.

Delmer, D. P., Kulow, C., and Ericson, M. C. (1978). *Plant Physiol.* **61,** 25–29.

Elbein, A. D. (1969). *J. Biol. Chem.* **244,** 1608–1619.

Elbein, A. D. (1979). *Annu. Rev. Plant Physiol.* **30,** 239–272.

Elbein, A. D. (1981). *Trends Biochem. Sci.* **6,** 219–221.

Elbein, A. D., and Hassid, W. Z. (1966). *Biochem. Biophys. Res. Commun.* **23,** 311–318.

Elbein, A. D., and Heath, E. C. (1965). *J. Biol. Chem.* **240,** 1919–1925.

Elbein, A. D., Barber, G. A., and Hassid, W. Z. (1964). *J. Am. Chem. Soc.* **86,** 309–310.

Ericson, M. C., and Delmer, D. P. (1977). *Plant Physiol.* **59,** 341–347.

Ericson, M. C., and Elbein, A. D. (1980). *In* "The Biochemistry of Plants" (J. Preiss, ed.), Vol. 3, pp. 589–615. Academic Press, New York.

Ericson, M., Gafford, J., and Elbein, A. D. (1977). *J. Biol. Chem.* **252,** 7431–7433.

Fan, D. F., and Feingold, D. S. (1969). *Plant Physiol.* **44,** 599–604.

Feingold, D. S., and Avigad, G. (1980). *In* "The Biochemistry of Plants," (J. Preiss, ed.), Vol. 3, pp. 101–170. Academic Press, New York.

Feingold, D. S., Neufeld, E. F., and Hassid, W. Z. (1958). *J. Biol. Chem.* **233,** 783–788.

Feingold, D. S., Neufeld, E. F., and Hassid, W. Z. (1960). *J. Biol. Chem.* **235,** 910–913.

Franz, G., and Meier, H. (1969). *Biochim. Biophys. Acta* **184,** 658–659.
Forsee, W. J., and Elbein, A. D. (1972). *Biochem. Biophys. Res. Commun.* **49,** 930–939.
Forsee, W. T., and Elbein, A. D. (1975). *J. Biol. Chem.* **250,** 9283–9293.
Forsee, W. T., Valkovich, G., and Elbein, A. D. (1976). *Arch. Biochem. Biophys.* **174,** 469–479.
Franz, G. (1969). *Phytochemistry* **8,** 737–741.
Gabriel, O. (1973). *Adv. Chem. Ser.* **117,** 387–410.
Gabriel, O., and Van Lenten, L. (1978). *In* "Biochemistry of Carbohydrates II" (D. Manners, ed.), Vol. 16, pp. 1–36. Univ. Park Press, Baltimore.
Gabriel, O., Kalckar, H. M., and Darrow, R. A. (1975). *In* "Subunit Enzymes: Biochemistry and Function" (R. Ebner, ed.), pp. 85–135. Dekker, New York.
Ginsburg, V. (1961). *J. Biol. Chem.* **236,** 2389–2393.
Ginsburg, V., Stumpf, P. K., and Hassid, W. Z. (1956). *J. Biol. Chem.* **223,** 977–983.
Glaser, L. (1958). *J. Biol. Chem.* **232,** 627–636.
Glaser, L., and Kornfeld, S. (1961). *J. Biol. Chem.* **236,** 1795–1797.
Green, J. R., and Northcote, D. H. (1978). *Biochem. J.* **170,** 599–608.
Grimm, I., Sachs, H., and Robinson, D. G. (1977). *Cytobiologie* **14,** 61–74.
Harris, P. J., and Northcote, D. H. (1970). *Biochem. J.* **120,** 479–491.
Heller, J. S., and Villemez, C. L. (1972). *Biochem. J.* **129,** 645–655.
Hepler, P. K., and Palevitz, B. A. (1974). *Ann. Rev. Plant Physiol.* **25,** 309–362.
Hogetsu, T., and Shibaoka, H. (1978). *Planta* **140,** 15–18.
Hopp, H. E., Romero, R. A., Daleo, G. R., and Pont Lezica, R. (1978). *Eur. J. Biochem.* **84,** 561–571.
Hori, H., and Elbein, A. D. (1981). *Plant Physiol.* **67,** 882–886.
Hori, H., and Elbein, A. D. (1982). *Plant Physiol.* **70,** 12–20.
Hori, H., and Elbein, A. D. (1983). *Arch. Biochem. Biophys.* **215,** 12–21.
Hori, H., James, D. W., and Elbein, A. D. (1982). *Arch. Biochem. Biophys.* **215,** 12–21.
James, Jr., D. W., and Jones, R. L. (1979). *Plant Physiol.* **64,** 914–918.
Katan, R., and Avigad, G. (1966). *Biochem. Biophys. Res. Commun.* **24,** 18–24.
Kauss, H. (1969). *Phytochemistry* **9,** 985–988.
Kauss, H., and Hassid, W. Z. (1967). *J. Biol. Chem.* **242,** 1680–1684; 3449–3453.
Kauss, H., Swanson, A. L., and Odzuck, W. (1969). *Biochem. Biophys. Acta* **192,** 55–61.
Kindel, P. A. (1973). *In* "Biogenesis of Plant Cell Wall Polysaccharides" (F. Loewus, ed.), pp. 85–94. Academic Press, New York.
Kjosbakken, J., and Colvin, R. (1973). *In* "Biogenesis of Plant Cell Walls" (F. Loewus, ed.), pp. 361–371. Academic Press, New York.
Ledbetter, M. C., and Porter, K. R. (1963). *J. Cell Biol.* **19,** 239–250.
Lehle, L. (1981). *FEBS Lett.* **123,** 63–66.
Lehle, L., Fartaczek, F., Tanner, W., and Kauss, H. (1976). *Arch. Biochem. Biophys.* **175,** 419–426.
Leibowitz, M. D., Dickinson, M. D. B., Loewus, F. A., and Loewus, M. A. (1977). *Arch. Biochem. Biophys.* **179,** 559–564.
Leloir, L. F. (1951). *Arch. Biochem. Biophys.* **33,** 186–190.
Leloir, L. F. (1953). *Adv. Enzymol.* **14,** 193–218.
Li, E., Tabas, I., and Kornfeld, S. (1978). *J. Biol. Chem.* **253,** 7762–7770.
Liao, T. H., and Barber, G. A. (1972). *Biochim. Biophys. Acta* **276,** 85–93.
Lin, T. Y., and Hassid, W. Z. (1966). *J. Biol. Chem.* **241,** 3283–3293.
Lin, T. Y., Elbein, A. D., and Su, J. C. (1966). *Biochem. Biophys. Res. Commun.* **22,** 650–657.

Loewus, F. A. (1974). *Rec. Adv. Phytochem.* **8,** 179–207.
Machlachlan, G. A. (1982). *In* "Cellulose and Other Natural Polymer Systems" (R. M. Brown, Jr., ed.), pp. 327–339. Plenum, New York.
Marechal, L. R., and Goldemberg, S. H. (1969). *J. Biol. Chem.* **239,** 3163–3167.
Marx-Figini, M. (1966). *Nature* **210,** 754–755.
Najjar, V. A. (1955). "Methods in Enzymology" (S. P. Colowick and N. O. Kaplan, eds.), pp. 294–299. Academic Press, New York.
Neufeld, E. F. (1962). *Biochem. Biophys. Res. Commun.* **7,** 461–466.
Neufeld, E. F., and Feingold, D. S. (1961). *Biochim. Biophys. Acta* **53,** 589–590.
Neufeld, E. F., and Hall, C. W. (1965). *Biochem. Biophys. Res. Commun.* **19,** 456–461.
Northcote, D. H., and Lewis, D. R. (1968). *J. Cell Sci.* **3,** 199–206.
Neufeld, E. F., Ginsburg, V., Putnam, E. W., Fanshier, D., and Hassid, W. Z. (1957). *Arch. Biochem. Biophys.* **69,** 602–616.
Neufeld, E. F., Feingold, D. S., and Hassid, W. Z. (1958). *J. Am. Chem. Soc.* **80,** 4430–4433.
Neufeld, E. F., Feingold, D. S., and Hassid, W. Z. (1960). *J. Biol. Chem.* **235,** 906–909.
Neufeld, E. F., Feingold, D. S., Ilves, S. M., Kessler, G., and Hassid, W. Z. (1961). *J. Biol. Chem.* **236,** 3102–3105.
Northcote, D. H. (1973). in "Plant Carbohydrate Biochemistry" (J. B. Pridham, ed.), pp. 165–181. Academic Press, New York.
Odzuck, W., and Kauss, H. (1972). *Phytochemistry* **11,** 2489–2494.
Ordin, L., and Hall, M. A. (1968). *Plant Physiol.* **43,** 473–476.
Ordman, A. B., and Kirkwood, S. (1977). *J. Biol. Chem.* **252,** 1320–1326.
Panayotato, N., and Villemez, C. L. (1973). *Biochem. J.* **133,** 263–271.
Passeron, S., Recondo, E., and Dankert, M. (1964). *Biochim. Biophys. Acta* **89,** 372–374.
Paull, R. E., Johnson, C. M., and Jones, R. L. (1975). *Plant Physiol.* **56,** 300–306.
Pickett-Heaps, J. D. (1969). *J. Histochem. Cytochem.* **15,** 442–445.
Ray, P. M. (1975). *Plant Physiol.* **56,** (Suppl. 84).
Ray, P. M. (1980). *Biochem. Biophys. Acta* **629,** 431–444.
Robinson, D. G. (1977). *In* "Advances in Botanical Research" (H. W. Woodhouse, ed.), Vol. 5, pp. 89–151. Academic Press, New York.
Robinson, H., Sachs, H., and Mayer, F. (1976). *Planta* **140,** 75–82.
Roelofson, P. A. (1965). *In* "Advances in Botanical Research" (R. D. Preston, ed.), Vol. 2, pp. 69–149. Academic Press, New York.
Ruiz-Herrera, J., Sing, V. O., Van Der Woude, W. J., and Bartnicki-Garcia, S. (1975). *Proc. Natl. Acad. Sci. U.S.A.* **72,** 2706–2710.
Sachs, H., Grimm, I., and Robinson, D. G. (1977). *Cytobiologie* **14,** 49–60.
Saltman, P. (1953). *J. Biol. Chem.* **200,** 145–154.
Sanwal, G. G., and Preiss, J. (1969). *Phytochemistry* **8,** 707–723.
Schutzbach, J. S., and Feingold, D. S. (1970). *J. Biol. Chem.* **245,** 2476–2482.
Selvendran, K. R., and Isherwood, F. A. (1967). *Biochem. J.* **105,** 723–738.
Smith, M. M., Axelos, M., and Peaud-Lenoel, C. (1976). *Biochemie* **58,** 1195–1211.
Solms, J., and Hassid, W. Z. (1957). *J. Biol. Chem.* **228,** 357–364.
Spencer, F. S., Shore, G., and Maclachlan, G. A. (1971). *Arch. Biochem. Biophys.* **152,** 311–317.
Stafford, L. D., and Brummond, D. O. (1970). *Phytochemistry* **9,** 253–258.
Staneloni, R. J., and Leloir, L. F. (1982). *CRC Crit. Rev. Biochem.* **12,** 289–326.
Staneloni, R. J., Tolmasky, M. E., Petriella, C., Ugalde, R. A., and Leloir, L. F. (1980). *Biochem. J.* **191,** 257–260.

Staneloni, R. J., Tolmasky, M. E., Petriella, C., and Leloir, L. F. (1981). *Plant Physiol.* **68,** 1175–1179.
Stoddart, R. W., and Northcote, D. H. (1976). *Biochem. J.* **105,** 45–59.
Strominger, J. L., and Mapson, L. W (1957). *Biochem. J.* **66,** 567–572.
Strominger, J. L., Kalckar, H. M., Axelrod, J., and Maxwell, E. S. (1954). *J. Am. Chem. Soc.* **76,** 6411–6412.
Struck, D. P., and Lennarz, W. J. (1980). *In* "The Biochemistry of Glycoproteins and Proteoglycans" (W. J. Lennarz, ed.), p. 35–83. Plenum, New York.
Sutherland, E. W., Cohn, M., Posternak, T., and Cori, C. F. (1949). *J. Biol. Chem.* **180,** 1285–1295.
Tkacz, J., and Lampen, J. O. (1975). *Biochem. Biophys. Res. Commun.* **65,** 248–257.
Tulsiani, D. R. P., Hubbard, S. C., Robbins, P. W., and Touster, O. (1982). *J. Biol. Chem.* **257,** 3660–3668.
Villemez, C. L. (1971). *Biochem. J.* **121,** 151–157.
Villemez, C. L., and Hinman, M. (1975). *Plant Physiol.* **56** (Suppl. 79).
Villemez, C. L., Lin, Y. T., and Hassid, W. Z. (1965). *Proc. Natl. Acad. Sci. U.S.A.* **54,** 1626–1632.
Villemez, C. L., Swanson, A. L., and Hassid, W. Z. (1966). *Arch. Biochem. Biophys.* **116,** 446–452.
Waechter, C. J., and Lennarz, W. J. (1976). *Ann. Rev. Biochem.* **45,** 95–112.
Walker, D., and Robinson, S. P. (1980). *In* "The Biochemistry of Plants" (P. K. Stumpf and S. Hatch, ed.), Vol 8, pp. 193–236. Academic Press, New York.
Wooding, F. B. P. (1968). *J. Cell Sci.* **3,** 71–80.
Zalitis, J., and Feingold, D. S. (1969). *Arch Biochem. Biophys.* **132,** 457–465.

CHAPTER 7

Biosynthesis of Lignin

TAKAYOSHI HIGUCHI

Wood Research Institute, Kyoto University,
Kyoto, Japan

I. Introduction

Lignin occurs in cell walls of true vascular plants, ferns, and club mosses but not in those of mosses, algae, and microorganisms (Kawamura and Higuchi, 1964). Lignin is generally distributed as a matrix component with hemicelluloses in the spaces of intercellulose microfibrils in primary and secondary walls, and in middle lamellae, and functions to connect cells to one another and to harden the cell walls of xylem tissues. The lignified cell walls are resistant to microbial attack and contribute to the compressive strength of the stem. Lignified cell walls are considered necessary to the stems of woody plants and the conductive xylem tissues for water transport because they strengthen the tissues and decrease the permeability of cell walls. The lignin content of woody stems of arboreus gymnosperms and angiosperms ranges from 15 to 36%.

The process of lignification of cell walls can be observed microscopically by taking advantage of the ultraviolet (UV) absorption of lignins at 280 nm (see Chapter 3). It has been shown that lignification at the cell corner, at the intercellular layer, and at the secondary wall proceeds

Biosynthesis and Biodegradation of Wood Components

ISBN 0-12-347880-4

separately (Fujita *et al.*, 1978): lignification is initiated in the primary walls adjacent to the corners of the cell undergoing cellulose deposition in the S_1 layers near the cambium and then proceeds in the intercellular layers and primary walls (Wardrop and Bland, 1959). The lignification of the secondary walls occurs separately, after cellulose deposition in the S_3 layer has been completed (see, Chapter 1, Section III,C). Lignin occurs in intimate association with cell-wall polysaccharides, and it has been shown that some hemicelluloses are linked by covalent bonds with lignin (Fengel and Wegener, 1983). It was also suggested that cell-wall proteins could be connected to lignin by covalent linkages during polymerization of monolignols in primary walls (Whitmore, 1978).

Chemical and spectrometric studies of softwood lignin (Adler, 1977; Sakakibara, 1983) indicate that the lignin is an aromatic polymer in which the monomeric guaiacylpropane units are connected by both ether and carbon–carbon linkages; several substructures[1] involved in lignin macromolecules were elucidated, and it was found that guaiacylglycerol-β-aryl ether (β-*O*-4′ substructure) (**1**) is the most abundant interphenylpropane linkage (40–60%) in lignin, and followed by phenylcoumaran (β-5′ substructure, 10%) (**2**), diarylpropane (β-1′ substructure, 5–10%) (**3**), pinoresinol (β-β′ substructure, 5%) (**4**), biphenyl (5-5′ substructure, 10%) (**5**), diphenyl ether (4-*O*-5′ substructure, 5%) (**6**), etc.

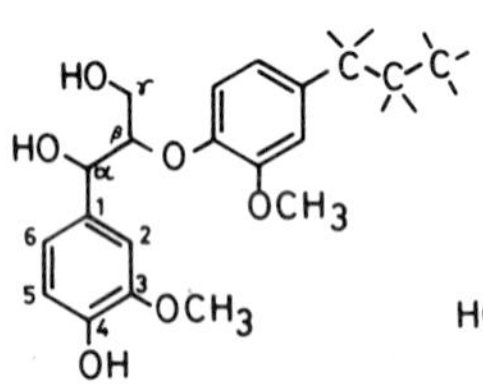

(1)

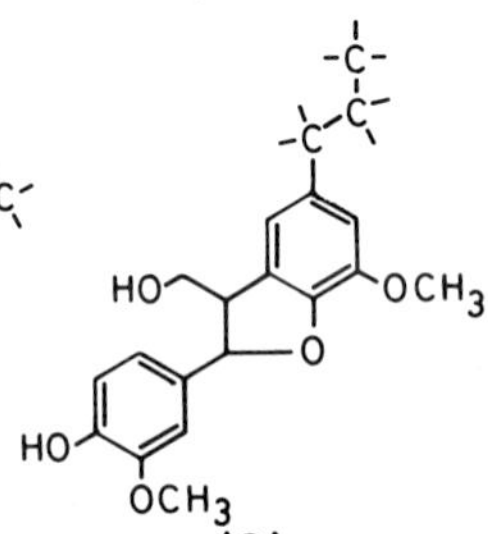

(2)

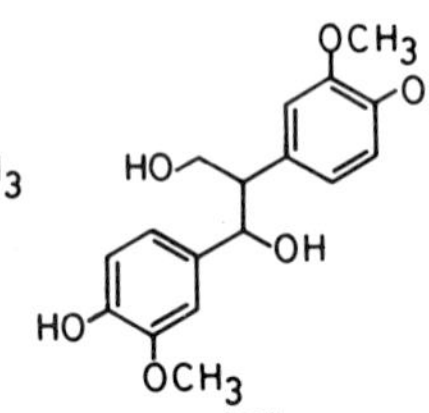

(3)

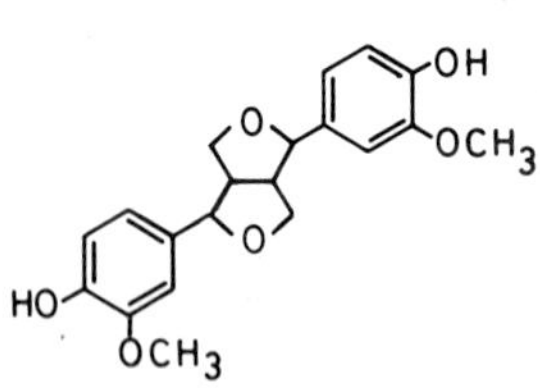

(4)

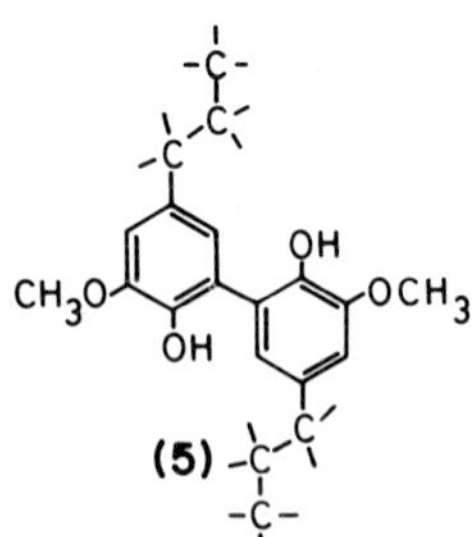

(5)

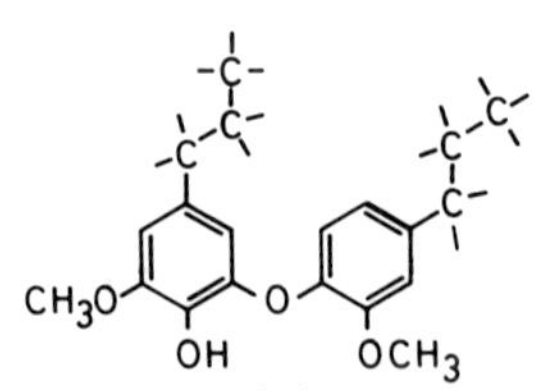

(6)

[1] Dimeric structures containing the principal linkage mode between monomeric phenylpropane units in lignin macromolecules.

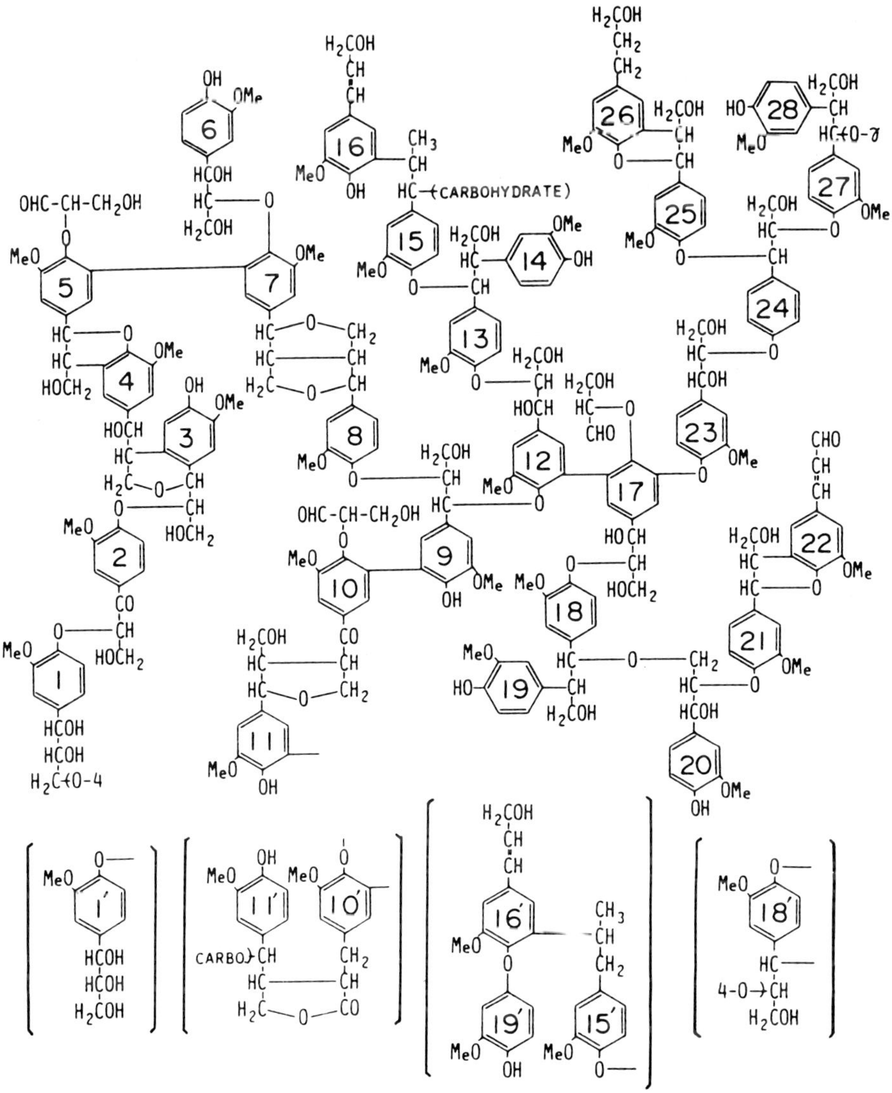

Fig. 1. A structural model of softwood lignin.

The lignin of hardwoods such as beech is composed of approximately equal amounts of guaiacyl- and syringylpropane units connected by linkages similar to those found in spruce lignin. Grass lignin, such as bamboo lignin, is considered to be composed of guaiacyl-, syringyl-, and *p*-hydroxyphenylpropane units, also connected by similar linkages, and *p*-coumaric acid is esterified to the terminal hydroxyl groups of the side chains. Figure 1 shows the structural model of softwood lignin proposed on the basis of chemical and spectrometric studies (Sakakibara, 1983).

Recent studies on the lignin biosynthesis indicated that *p*-hydroxycinnamyl alcohols such as coniferyl, sinapyl, and *p*-coumaryl alcohols, which are derived from L-phenylalanine via the corresponding cinnamic acids, are direct precursors of lignins: these *p*-hydroxycinnamyl alcohols were named monolignols by Freudenberg (1965).

II. Dehydrogenative Polymerization of Monolignols to Lignin Macromolecules

L-Phenylalanine, an essential amino acid in primary metabolism, is converted to *trans*-cinnamic acid by phenylalanine ammonia-lyase (PAL), which is widely distributed in lignifying tissues and is involved as a key enzyme in the synthesis of various phenolic secondary products, including lignin. *trans*-Cinnamic acid thus formed is converted to *p*-coumaric, ferulic, and sinapic acids via successive hydroxylation and methylation reactions. These hydroxycinnamic acids are then converted to their CoA esters, which are reduced to monolignols, that is, *p*-coumaryl alcohol (**7**), coniferyl alcohol (**8**), and sinapyl alcohol (**9**), via the respective hydroxycinnamaldehydes (see Chapter 10, Sections II, III, IV).

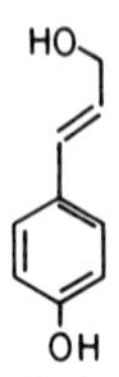

(7)

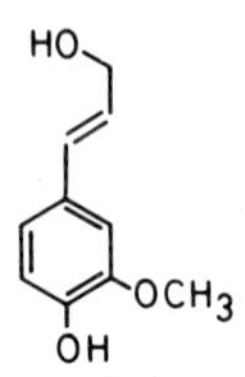

(8)

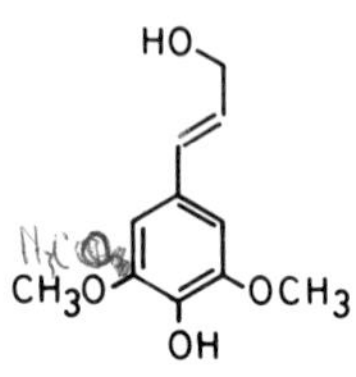

(9)

In 1939, Erdtman postulated that lignin is formed by a coupling reaction of phenoxy radicals of *p*-hydroxyphenylpropane compounds with unsaturated side chains, and he subsequently suggested that coniferous lignin is formed by enzymatic dehydrogenation of coniferyl alcohol (Erdtman, 1950). Freudenberg and associates later confirmed that coniferyl alcohol is dehydrogenatively polymerized to a high polymer material (dehydrogenation polymer, DHP) whose properties are closely related to those of coniferous lignin, chemically and spectrometrically (Freudenberg, 1965). They isolated several oligolignols as intermediates in the dehydrogenation reaction, determined their chemical structures, and found that these oligolignols are further polymerized enzymatically to DHP. These results indicated that coniferous lignin could be formed by dehydrogenative polymerization of coniferyl alcohol.

It was shown by Freudenberg *et al.* that coniferyl alcohol is dehydrogenated by mushroom laccase/O_2 or peroxidase/H_2O_2 to yield phenoxy radicals in mesomeric forms, as shown in Fig. 2. The radicals of the alcohol couple nonenzymatically in a random fashion to yield quinone methides, which play a central role in the formation of lignin macromolecules as reactive intermediates. The dimeric quinone methides thus formed are converted to various dilignols by the addition of water or by intramolecular nucleophilic attack on the benzyl carbons by primary alcohol or quinone groups, as shown in Fig. 3. The dilignols are further dehydrogenated by the enzymes to their radicals, which are finally converted to lignin and lignin–carbohydrate complexes (LCC) via radical couplings followed by nucleophilic attack on the benzyl carbons of the oligomeric quinone methides by water, by the aliphatic and phenolic hydroxyl groups of lignols, and by hydroxyl groups of sugar residues of cell wall polysaccharides (Fig. 4).

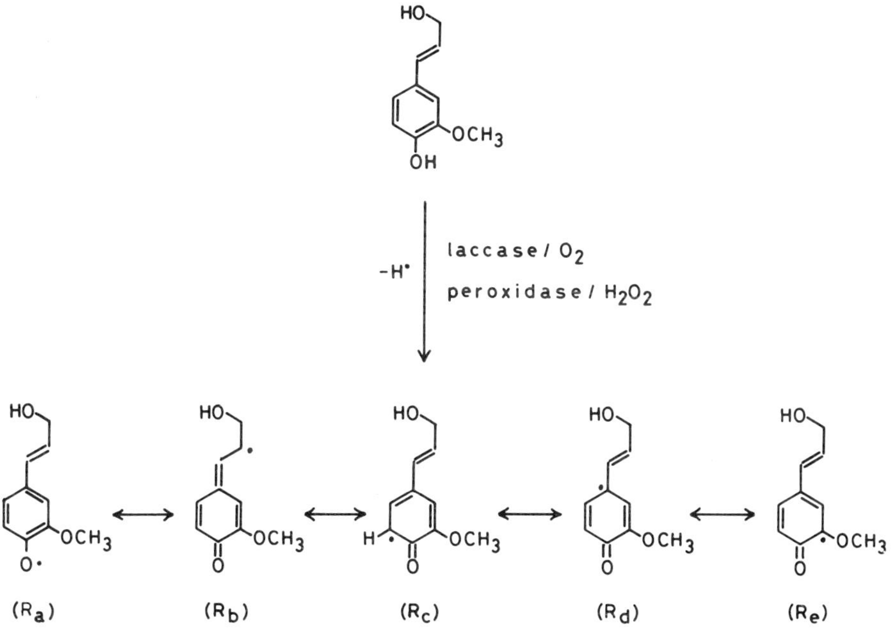

Fig. 2. Mesomeric forms of the phenoxy radical of coniferyl alcohol formed by the mediation of laccase/O_2 or peroxidase/H_2O_2.

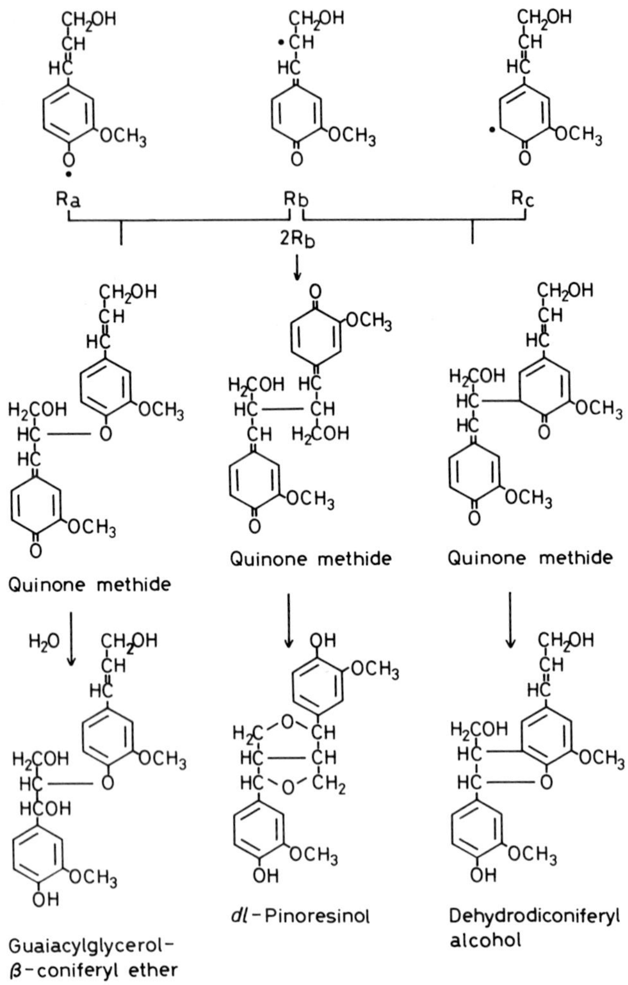

Fig. 3. Formation of dilignols via the coupling of monolignol radicals and subsequent nucleophilic attack by hydroxyl and quinone groups on the benzyl carbons of the quinone methides.

Based on the extensive investigations on dehydrogenative polymerization of respective monolignols by peroxidase/H_2O_2 (Freudenberg, 1965) and analyses of natural lignins (Adler, 1977; Sakakibara, 1983), it has been established that gymnosperm lignin is mainly formed by dehydrogenative polymerization of coniferyl alcohol, angiosperm lignin by the dehydrogenative polymerization of a mixture of coniferyl and sinapyl alcohols, and grass lignin by the dehydrogenative polymerization of a mixture of coniferyl, sinapyl and *p*-coumaryl alcohols.

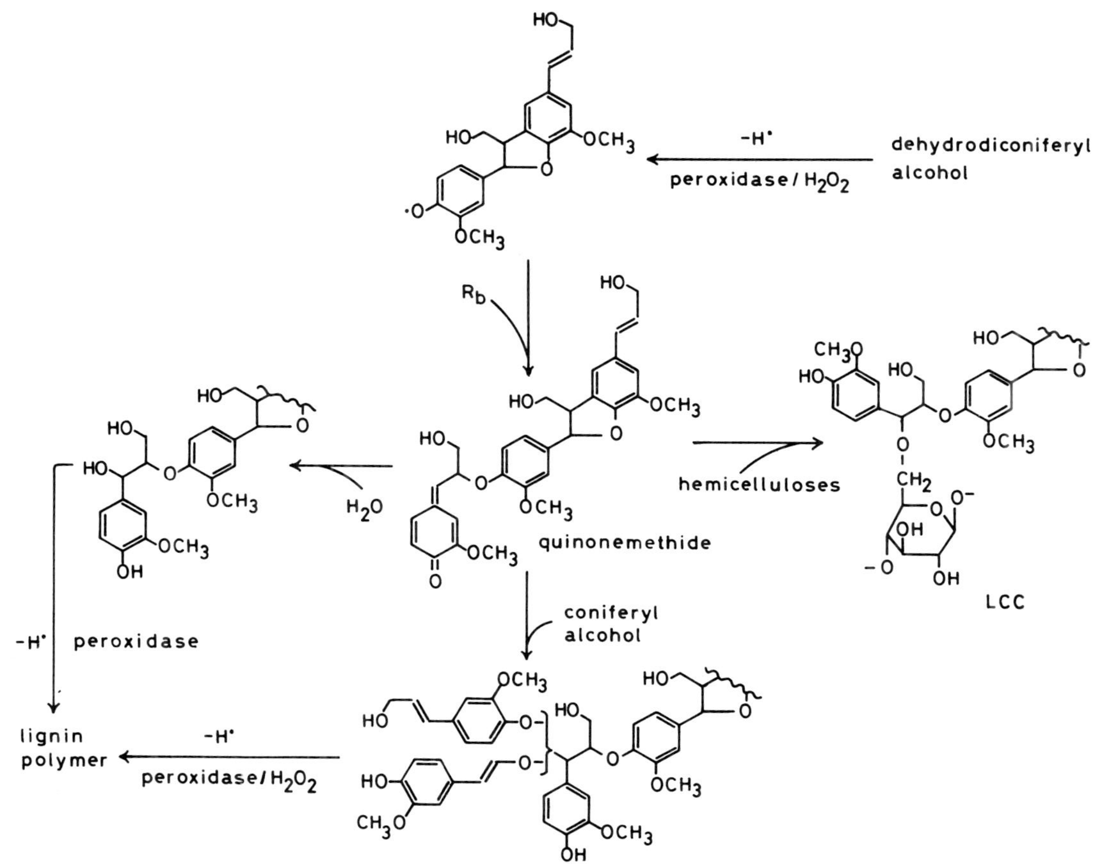

Fig. 4. Formation of lignin polymers and lignin–carbohydrate complexes (LCC) via oligolignol quinone methide.

III. Role of Peroxidase in the Dehydrogenative Polymerization of Monolignols to Lignins

Freudenberg (1965) found that coniferyl alcohol is dehydrogenatively polymerized to a pale-brown precipitate by an enzyme from a mushroom (*Psalliota campestris*) under aeration, and that the precipitate (DHP) is closely related to spruce milled wood lignin (MWL). Higuchi (1958) found later that the enzyme used in earlier experiments by Freudenberg is a mixture of laccase (*p*-diphenol oxidase) and tyrosinase (*o*-diphenol oxidase), and that the former enzyme/O_2 in the mushroom is involved only in the dehydrogenative polymerization of coniferyl alcohol. Further investigations (Higuchi and Ito, 1958; Freudenberg *et al.*, 1958) showed that peroxidase/H_2O_2 yields a DHP of coniferyl alcohol, whose chemical properties resemble closely those of the DHP obtained with mushroom laccase/O_2. Thereafter, Nakamura (1967) found that a purified plant laccase from a Japanese lacquer is incapable of catalyzing the dehydrogenation of coniferyl alcohol, but that pure bamboo shoot peroxidases dehydrogenate coniferyl alcohol to yield a DHP. Evidence for the involvement of peroxidase but not laccase in the dehydrogenative polymerization of monolignols in higher plants was obtained by Harkin and Obst (1973): syringaldazine solution applied to cross sections of the stubs of angiosperms and gymnosperms produced an intense purple color, due to tetramethoxyazo-*p*-methylene quinone in the xylem tissue adjacent to the cambium, only when H_2O_2 was added.

The origin of H_2O_2 required for the peroxidase reaction was investigated using horseradish cell walls (Elstner and Heupel, 1976) and cell walls isolated from *Forsythia* xylem (Gross *et al.*, 1977). They found that H_2O_2 is formed in a complex reaction that involves a dismutation reaction of superoxide radical ($O_2^{\cdot -}$) generated by the reduction of O_2 with $NAD^{\cdot}$. The $NAD^{\cdot}$ was postulated to be formed via oxidation of NADH, which is provided by the oxidation of malate with NAD : malate oxidoreductase bound to cell walls, by the phenoxy radicals generated in an oxidase reaction of some phenols with a peroxidase–Mn^{2+} complex (Fig. 5). The formation of H_2O_2 was stimulated by various monophenols, especially by coniferyl alcohol, which indicated that the monolignols could be directly involved in the regulation of H_2O_2 required for the dehydrogenation. Gross *et al.* (1977) found that coniferyl alcohol is dehydrogenated, without addition of H_2O_2, in the presence of added malate by lignifying *Forsythia* cell walls, which contain a bound malate dehydrogenase and a peroxidase. It seems that the scheme reasonably explains the dehydrogenative polymerization of lignols as well as the formation of H_2O_2 at the

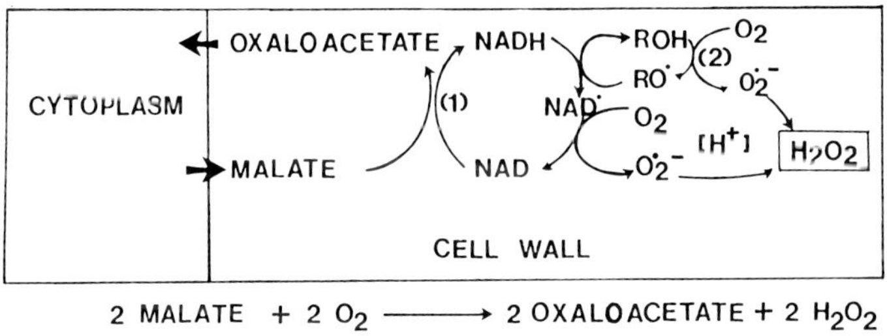

Fig. 5. A hypothetical scheme for the formation of hydrogen peroxide in the plant cell walls. (1) NAD : malate oxidoreductase; (2) peroxidase–Mn^{2+} complex.

lignification site without requiring transport of toxic H_2O_2 from the cytoplasm to the cell wall. It is considered that oxaloacetate formed from malate is transported into cytoplasm and converted to malate again by malate–oxaloacetate shuttle and then supplied to the wall repeatedly.

If peroxidase is really involved in lignification, the enzyme should be localized in tracheary elements, especially in the secondary walls, which become heavily lignified. A positive correlation of the localization of peroxidase with lignified walls has been found by several workers (Higuchi, 1957; Van Fleet, 1959; Gagnon, 1968). Helper *et al.* (1972) found that peroxidase activity in wound vessel elements of *Coleus* is located in the reticulate secondary wall and in the primary wall where secondary thickenings have been progressing. They also found that the enzyme activity is located in the plasmalemma of differentiating vessel elements, especially where it overlies the secondary thickening, and in the dictyosomes and their associated vesicles. These results strongly support the thesis that peroxidase is involved in dehydrogenative polymerization of monolignols into lignin in higher plants.

Westermark (1982) proposed a possible participation of superoxide rad icals ($O_2^{\cdot -}$) in the dehydrogenative polymerization of coniferyl alcohol in lignifying cell walls, based on the following facts.

1. Superoxide radicals, which are known to be generated by several biological systems, are relatively stable in the presence of Ca^{2+} located in cell walls.
2. The radicals stabilized would penetrate cell walls by diffusion and catalyze the dehydrogenation of coniferyl alcohol to the radicals that yield lignin.
3. The thickness of lignin-rich lamella (35 Å) estimated by electron-microscopic observations in the middle secondary layers of cell walls is less than the diameter of horseradish peroxidase (55–60 Å), which may

not be consistent with the peroxidase-inclusive lignification within the lamella.

4. For large molecules of peroxidase to penetrate to the capillaries in cell walls and catalyze the formation of network structures of lignified cell walls would be difficult.

The validity of the proposal could be demonstrated by characterization of oxidized products of monolignols by $O_2^{\cdot-}$, characterization and distribution of $O_2^{\cdot-}$-generating enzymes (especially in the secondary wall), determination of accurate dimensions of the amorphous region of cellulose microfibrils, and elucidation of ultrastructural localization of lignin.

Dehydrogenation experiments on monolignols have mostly been made with commercially available horseradish peroxidase. Many isozymes of peroxidase have been found in plants, however, and in lignification of plant cell walls, cell wall-bound peroxidases would be involved in the dehydrogenation of monolignols (Gross *et al.*, 1977; Helper *et al.*, 1972) (see Chapter 5, Section III,B). It seems not unlikely that peroxidases bound with cell membrane remain in plasmalemma with cell walls during formation of secondary cell walls and participate in the dehydrogenative polymerization of monolignols provided by the synthetic machinery such as Golgi vesicles in the cytoplasm, because many enzymes—especially polysaccharidases, which are involved in construction of cell walls in growth and thickening—have been found in cell walls.

IV. Structural Differences in Dehydrogenation Polymers

Freudenberg and Neish (1968) found that the yields of dehydrodiconiferyl alcohol (β-5′ dimer) and *dl*-pinoresinol (β-β′ dimer) are considerably higher than the yield of guaiacylglycerol-β-coniferyl ether (β-*O*-4′ dimer) (see Fig. 3) when coniferyl alcohol solution is added at once to the peroxidase/H_2O_2 solution (zulaufverfahren), whereas the yield of guaiacylglycerol-β-coniferyl ether increases when the substrate is added dropwise over long periods of time to the enzyme solution (zutropverfahren). The difference of the polymerization processes in the two methods was discussed in detail by Sarkanen and Ludwig (1971). They postulated that in the former method the coupling of monolignol radicals, of which the concentration is higher, would be predominant, and the dilignols thus formed are dehydrogenatively polymerized to tetramers and finally to a large lignin molecule (bulk polymer) via mainly 4-*O*-5′ and 5-5′ linkages (Fig. 6). In the latter method, the coupling of monolignol radicals occurs initially to give oligolignols, and hence monolignol radicals are succes-

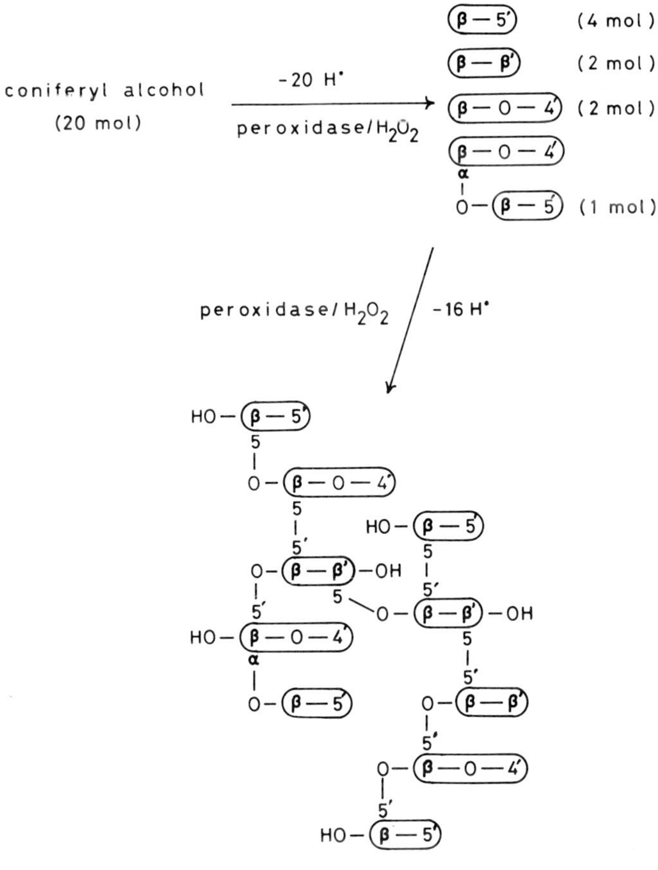

Fig. 6. Formation of a bulk polymer of coniferyl alcohol by the "zulauf" method. The radicals of coniferyl alcohol (20 molecules) couple to form dilignols, which comprise β-5′, β-β', and β-*O*-4′ substructures. The dilignols are dehydrogenatively polymerized to tetralignols and finally to a polymeric lignin that contains considerable amounts of double bonds in the side chains via mainly 5-*O*-4′ and 5-5′ linkages.

sively coupled to the oligolignol radicals initially formed, via mainly β-*O*-4′ linkages, to produce an endwise polymer, because the concentration of monolignol radicals is lower (Fig. 7). It seems that in the actual lignification of plant cell walls, the mode of polymerization could be controlled by the amount of monolignols supplied to cell walls.

The molecular weight (M_n) of coniferyl alcohol DHPs by zulauf- and zutropverfahren are usually 1000–1200 and 2500, far lower than the molecular weight of natural lignins. Tanahashi and Higuchi (1981) found that DHPs with higher molecular weight, which are less soluble in organic

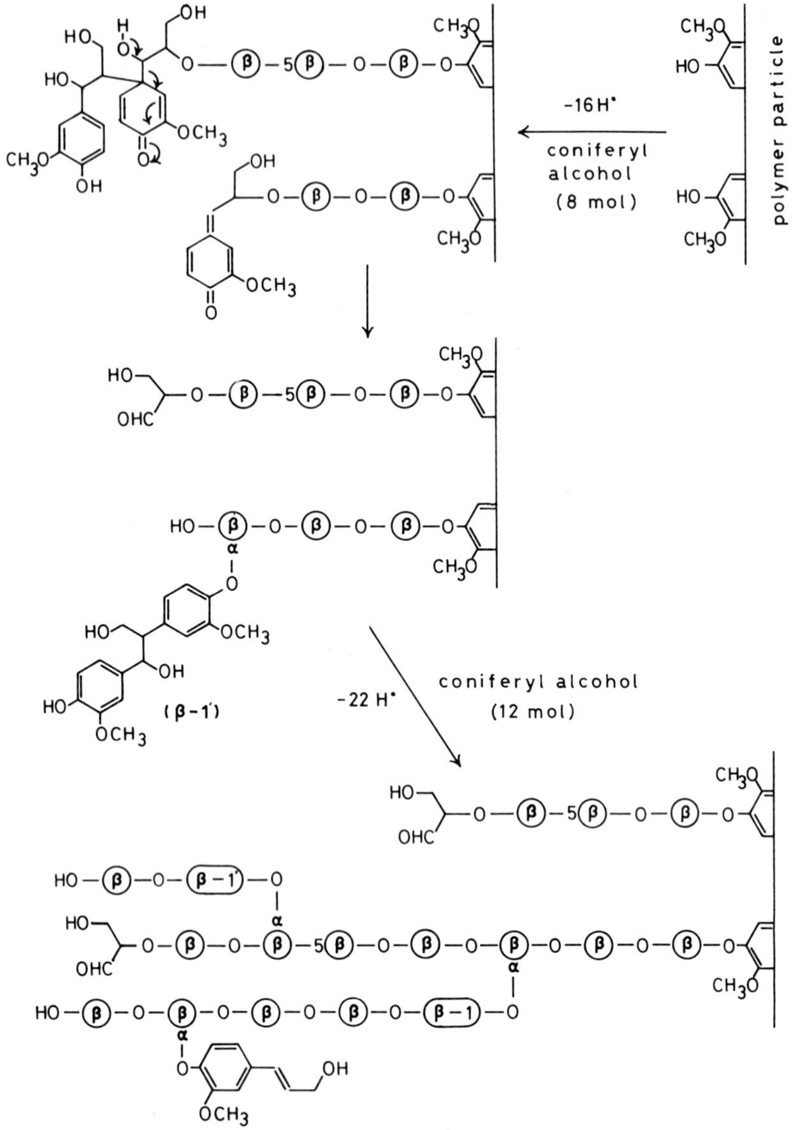

Fig. 7. Formation of an end-wise polymer of coniferyl alcohol by the "zutropf" method. Coniferyl alcohol is added dropwise for long periods to a system consisting of polymer particles and peroxidase/H_2O_2. Coniferyl alcohol radicals couple one by one with phenoxy radicals present on the surface of the polymer particles via mainly *β*-*O*-4′ linkages. Branchings of the polymer molecule occur via *α*-*O*-4′ linkages formed by the attack of phenolic hydroxyl groups of *β*-1′ dilignols on benzyl carbons of the quinone methides at the end of the linear polymer. The proportion of double bonds in the side chains and phenolic hydroxyl groups would be considerably lower than in the bulk polymer.

lignin solvent, were formed when a cellulosic dialysis tube containing horseradish peroxidase was added to a flask containing H_2O_2 and coniferyl alcohol solutions. The M_w is about 6000, which is higher than that of bamboo MWL (M_w 3500). The results suggest that the concentration of polymer radicals in the tube is high; smaller molecules of oligolignols diffuse through the cellulose membrane. Couplings between polymer radicals and monomer radicals in the tube would proceed in preference to monolignol–monolignol coupling, resulting in higher molecular weight polymers. In the lignification of plant cell walls, polymerization of lignins may proceed in a similar way, involving diffusion of lignols across cell membrane and cell walls.

V. Formation of Syringyl Lignin in Angiosperm Woods

Freudenberg and Neish (1968) indicated that a mixture of coniferyl and sinapyl alcohols in equal amounts is dehydrogenatively polymerized by mushroom enzyme/O_2 to a mixed DHP similar to guaiacyl–syringyl lignin in angiosperms, but that sinapyl alcohol alone does not yield a sinapyl DHP. However, it was later found that DHP of sinapyl alcohol is formed with horseradish peroxidase/H_2O_2, and the chemical and spectral properties resemble those of syringyl lignin (Yamasaki *et al.*, 1976; Tanahashi *et al.*, 1976). On the other hand, it was shown by UV-spectroscopic observation of birch wood that syringyl or syringyl-rich lignin is located in secondary walls of wood fibers and parenchyma cells (Musha and Goring, 1975; see this volume, Chapter 3, Section III). Chemical evidence for the occurrence of syringyl lignin in angiosperm woods has also been obtained; MWL of *Myrica rubra* Sieb *et* Zucc. was separated into syringyl-rich lignin with 25.2% OCH_3, which corresponds to 85% syringyl unit in the lignin (Yamasaki *et al.*, 1978).

The difference of the coupling mode between coniferyl and sinapyl radicals was ascribed to the steric hindrance of two methoxyl groups located at C-5 and C-3 of aromatic rings of the latter radicals; the bulkiness of two methoxyl groups was considered to inhibit the β-*O*-4′ coupling to give syringylglycerol-β-sinapyl ether. Tanahashi *et al.* (1976), however, found that several 3,5-disubstituted *p*-coumaryl alcohols, where the substituted groups are bulkier than the methoxyl group, were efficiently dehydrogenatively polymerized to dimers with ferric chloride. They further found that the yields of β-*O*-4′ dimers mostly depend on the electronic effect of the substituted groups but not the steric hindrance. Sinapyl alcohol produced syringylglycerol-β-sinapyl ether in a high yield (85%) upon oxidation with ferric chloride in dioxane.

These results indicate that sinapyl alcohol could be polymerized to a syringyl lignin in plant cell walls, and that angiosperm lignins are not a uniformly copolymerized guaiacyl–syringyl lignin but are heterogeneously comprised in part by a syringyl-rich fraction and in part by a guaiacyl-rich fraction, in addition to syringyl–guaiacyl copolymer.

VI. Differences between Gymnosperms and Angiosperms in Lignin Biosynthesis

[^{14}C]Ferulic acid administered to gymnosperms is converted to only guaiacyl lignin, but when it is administered to angiosperms it is converted to guaiacyl–syringyl lignin. Reasons for the formation of guaiacyl lignin in gymnosperms and guaiacyl–syringyl lignin in angiosperms have been investigated by Higuchi and associates (Higuchi, 1983), and the following factors were found to be involved in differentiation of guaiacyl and guaiacyl–syringyl lignins in both classes of woods:

1. *O*-Methyltransferase (OMT) of gymnosperms primarily catalyzes the ferulate formation from caffeate, while angiosperm OMTs catalyze not only the ferulate formation but also sinapate formation from 5-hydroxyferulate (see Chapter 10, Section II,C).
2. Ferulate-5-hydroxylase, which has been isolated from poplar stems and characterized as a cytochrome *P*-450-type monooxygenase (Grand, 1984), occurs only in angiosperms and participates as a key enzyme in the differentiation of lignin biosynthesis into guaiacyl and syringyl lignins.
3. *p*-Hydroxycinnamate : CoA ligase of gymnosperms is active for only ferulate, while some of the angiosperm enzymes are active for both ferulate and sinapate; it seems that the occurrence of sinapate : CoA ligase in angiosperms is not common, and an additional pathway in the synthesis of sinapyl alcohol via successive reactions of 5-hydroxyferulate, 5-hydroxyferuloyl-CoA, 5-hydroxyconiferaldehyde, and sinapaldehyde as illustrated in Fig. 8 was postulated (Higuchi, 1983). It was found that 5-hydroxyferulate was well converted to its CoA ester by both angiosperm and gymnosperm ligases and that 5-hydroxyconiferaldehyde added to the reaction medium methylated efficiently to sinapaldehyde by angiosperm OMTs (Kutsuki *et al.*, 1982).
4. *p*-Hydroxycinnamyl alcohol oxidoreductase, which catalyzes the last step of the formation of monolignols, has different substrate specificity; gymnosperm enzymes primarily catalyze the formation of coniferyl alcohol from coniferaldehyde, while angiosperm enzymes catalyze not

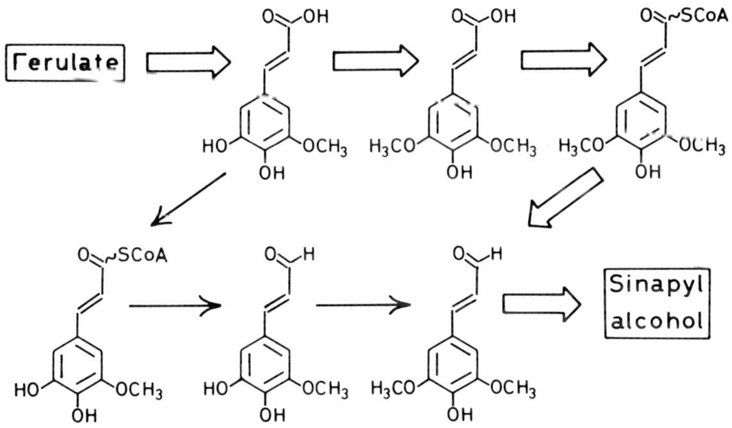

Fig. 8. Biosynthetic pathways of sinapyl alcohol in angiosperms. Broad arrows show main pathway; fine arrows show additional pathway.

only the formation of coniferyl alcohol but also the formation of sinapyl alcohol from sinapaldehyde.

It has been suggested that the occurrence of sinapate:CoA ligase is closely related to the differentiation and lignification of angiosperm tissues; poplar and cherry shoots converted ferulic acid to coniferyl alcohol, and sinapic acid to sinapyl alcohol, respectively, but the callus of angiosperms reduced ferulic acid to coniferyl alcohol but not sinapic acid to sinapyl alcohol. Angiosperm callus that is not differentiated to xylem tissue makes primarily guaiacyl and not syringyl lignin (Higuchi and Barnoud, 1964).

For the synthesis of guaiacyl lignin in gymnosperms, and guaiacyl–syringyl lignin in angiosperms except for grasses, L-phenylalanine only is used as substrate. For the synthesis of grass lignin, which is composed of guaiacyl-, syringyl-, and *p*-hydroxyphenylpropane units and contains *p*-coumaric acid esterified to terminal hydroxyl groups of the propane side chains, L-tyrosine, as well as L-phenylalanine, is used as substrate; [^{14}C]tyrosine is incorporated into *p*-hydroxyphenyl, guaiacyl, and syringyl units of the lignin polymer, and also into the esterified *p*-coumaric acid by the mediation of L-tyrosine ammonia-lyase, which is only found in grasses (Fig. 9).

The lignin of compression wood formed in the lower parts of the stem of gymnosperms has been known to contain a higher amount of lignin that consists increased amounts of condensed units and *p*-hydroxyphenyl component in comparison with normal wood lignin. Kutsuki and Higuchi

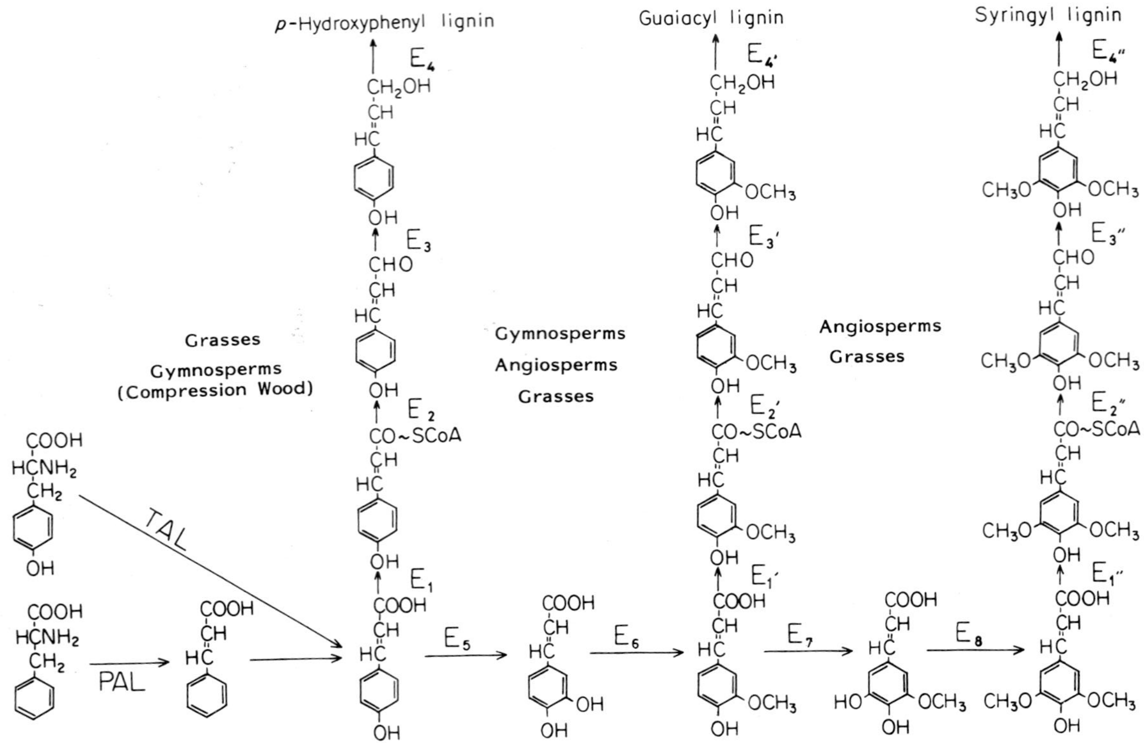

Fig. 9. Biosynthetic pathways of *p*-hydroxyphenyl, guaiacyl, and syringyl lignins. Enzymes: (E1), (E1′), and (E1″), hydroxycinnamate : CoA ligase; (E2), (E2′), and (E2″), hydroxycinnamoyl-CoA reductase; (E3), (E3′), and (E3″), cinnamyl alcohol dehydrogenase; (E4), (E4′), and (E4″), peroxidase; (E5), *p*-coumarate 3-hydroxylase; (E6) and (E8), hydroxycinnamate *O*-methyltransferase; (E7), ferulate 5-hydroxylase; PAL, phenylalanine ammonia-lyase; TAL, tyrosine ammonia-lyase.

(1981) found that the activities of phenylalanine ammonia-lyase, *O*-methyltransferase, *p*-hydroxycinnamate : CoA ligase, and cinnamyl alcohol oxidoreductase, which are involved in lignin biosynthesis, are severalfold higher in the compression wood than in opposite wood. The results suggest that in compression wood the supply of monolignols is raised by the increased enzyme activities, and this leads to the dehydrogenative polymerization by zulaufverfahren to result in a bulk lignin composed of large amounts of condensed units (Section IV).

VII. Regulation of Lignin Biosynthesis

The sequential conversion of L-phenylalanine to *p*-coumaroyl-CoA is generally involved in the biosynthesis of lignin, flavonoids, and other phenylpropanoids. Nevertheless, most of the enzymes involved in lignin biosynthesis are specific for lignin precursors and could be extracted mainly from actively lignifying tissues. PAL activity of an aspen tree and bamboos was found to be much higher in the tissues with increasing lignification than in the less lignifying tissues. The activities of cinnamate 4-hydroxylase and caffeate : *O*-methyltransferase were also higher in the increasingly lignifying bamboos (Higuchi, 1969). It was further found that hydroxycinnamate : CoA ligase, cinnamoyl-CoA reductase, and cinnamyl alcohol dehydrogenase could be extracted only from vascular bundles, especially from xylem, and not from parenchymatous tissues of celery (Gross *et al.*, 1975). A great increase in the activities on hydroxycinnamoyl-CoA reductase and cinnamyl alcohol dehydrogenase were observed during the lignification of spruce seedlings (Hahlbrock and Grisebach, 1979).

These results suggest that the lignification could be controlled by the rate of synthesis and degradation of appropriate enzymes, the substrate specificity, and the compartmentalization of the enzymes in tissues during xylem differentiation.

O-Methyltransferases isolated from the seedlings and young stems of Japanese black pine, poplar, and bamboos were found to be involved only in the formation of ferulate and sinapate as intermediates in lignin biosynthesis and not in the formation of intermediates in flavonoid synthesis. In soybean and parsley cell cultures, isozymes of OMT and hydroxycinnamate : CoA ligase, which are involved in preferential biosynthesis of lignin and flavonoids, respectively, were found (Hahlbrock and Grisebach, 1979); the respective isozymes of hydroxycinnamate : CoA ligase were competitively but differently inhibited by AMP, suggesting an important role of the enzymes in controlling the rate of formation of lignin and flavonoids. A sigmoidal dependence of enzyme activity on ATP and CoASH was found for the CoA ligase from parsley, indicating an allosteric character for this enzyme (Knobloch and Hahlbrock, 1975).

In relation to hormonal regulation of lignification, stimulation of lignin biosynthesis and of PAL and OMT activities by gibberellin (Cheng and Marsh, 1968) and kinetin (Rubery and Fosket, 1969, Yamada and Kuboi, 1976) was found. Ethylene was shown to stimulate lignin biosynthesis in swede roots (Rhodes and Wooltorton, 1973a,b), with a dramatic increase of the activities of PAL, cinnamate hydroxylase, and hydroxycinnamate : CoA ligase. Feruloyl-CoA reductase and OMT were moderately stimulated. Siegel (1954) showed that polymerization of lignin precursors by peroxidase in *Elodea densa* is inhibited by IAA, and suggested that high levels of IAA in meristematic tissues would suppress peroxidase activity, and hence lignin formation, whereas with decline of IAA concentration during growth and differentiation, lignification would increase. Higuchi (1957) suggested that a level of reducing system such as glutathion–ascorbic acid is involved in controlling the dehydrogenative polymerization of coniferyl alcohol, and with decline of the level of the reducing system lignification increases.

Recent investigation (Mizuno and Komamine, 1978) of xylogenesis in tissue cultures of carrot roots has indicated that xylem formation is induced by the cooperation of auxin and zeatin, the formation of which is mediated by the levels of cyclic AMP, which in turn is affected by illumination. The stimulation of the enzyme activities in both lignin and flavonoid synthesis by illumination was also found in the cultured cells of parsley and soybean (Grisebach and Hahlbrock, 1974).

In relation to the compartmentalization and multienzyme complexes involved in lignification, Alibert *et al.* (1972) found that the microsomal fraction from seedlings of *Quercus penduculata* catalyzes three steps from L-phenylalanine to caffeic acid. It was also found that the microsomal fraction of green shoots of *Sorghum vulgare* contains an enzyme system converting aromatic amino acids to cinnamic acids (Stafford, 1974). Electron-microscopic observations (Pickett-Heaps, 1968; Fujita *et al.,* 1983) showed that [^{3}H]phenylalanine and cinnamaic acid added to wheat plant and *Cryptomeria japonica* seedlings were first incorporated into both endoplasmic reticulum and Golgi bodies in developing xylem cells, and then into xylem thickenings. These results seem to support the idea that lignin precursors are synthesized in Golgi bodies and/or endoplasmic reticulum, transported to the wall via vesicles, and dehydrogenatively polymerized to lignin by peroxidase bound to cell walls (see Chapter, Section II,B).

References

Adler, E. (1977). *Wood Sci. Technol.* **11,** 169–218.
Alibert, G., Ranjeva, R., and Boudet, A. (1972). *Physiol. Plant.* **27,** 240–243.
Cheng, C. K. C., and Marsh, H. V. (1968). *Plant Physiol.* **43,** 1755–1759.
Elstner, E. F., and Heupel, A. (1976). *Planta* **130,** 175–180.
Erdtman, H. (1939). *Sven. Papperstidn.* **42,** 115–122.
Erdtman, H. (1950). *Research (London)* **3,** 63–67.
Fengel, D., and Wegener, G. (1983). *Proc. 1983 Int. Symp. Wood Pulping Chem.* **1,** 144–148.
Freudenberg, K. (1965). *Science* **148,** 595–600.
Freudenberg, K., and Neish, A. C. (1968). *In* "Constitution and Biosynthesis of Lignin," pp. 78–122. Springer-Verlag, Berlin, Heidelberg.
Freudenberg, K., Harkin, J., Reichert, M., and Fukuzumi, T. (1958). *Chem. Ber.* **91,** 581–590.
Fujita, M., Saiki, H., and Harada, H. (1978). *Mokuzai Gakkaishi* **24,** 158–163.
Fujita, M., Saiki, H., and Harada, H. (1983). *Proc. 1983 Int. Symp. Wood Pulping Chem.* **1,** 14–19.
Gagnon, C. (1968). *Can. J. Bot.* **46,** 1491–1494.
Grand, C. (1984). *FEBS Lett.* **169,** 7–11.
Grisebach, H., and Hahlbrock, K. (1974). *Rec. Adv. Phytochem.* **8,** 21–52.
Gross, G. G., Mansell, R. L., and Zenk, M. H. (1975). *Biochem. Physiol. Pflanz.* **168,** 41–51.
Gross, G. G., Janse, C., and Elstner, E. F. (1977). *Planta* **136,** 271–276.
Hahlbrock, K., and Grisebach, H. (1979). *Ann. Rev. Plant Physiol.* **30,** 105–130.
Harkin, J. M., and Obst, T. R. (1973). *Science* **180,** 296–298.
Helper, P. K., Rice, R. M., and Terranova, W. A. (1972). *Can. J. Bot.* **50,** 977–983.
Higuchi, T. (1957). *Physiol. Plant* **10,** 365–372, 621–632.
Higuchi, T. (1958). *J. Biochem. (Tokyo)* **45,** 515–528.
Higuchi, T. (1966). *Agric. Biol. Chem.* **30,** 667–673.
Higuchi, T. (1969). *Wood. Res.* (*Kyoto*) **48,** 1–14.
Higuchi, T. (1983). *In* "New Frontiers in Plant Biochemistry" (T. Akazawa, T. Asahi, and H. Imaseki, eds.), pp. 23–46. Japan Science Soc. Press, Tokyo.
Higuchi, T., and Barnoud, F. (1964). *In* "Chimie Biochimie de la Lignine, Cellulose, Hemicelluloses," pp. 255–274. Les Imprimeries Reunies de Chambery.
Higuchi, T., and Ito, Y. (1958). *J. Biochem. (Tokyo)* **45,** 575–579.
Kawamura, I., and Higuchi, T. (1964). *In* "Chimie Biochimie de la Lignin, Cellulose, Hemicelluloses," pp. 439–456. Les Imprimeries Reunies de Chambery.
Knobloch, K. H., and Hahlbrock, K. (1975). *Eur. J. Biochem.* **52,** 311–320.
Kutsuki, H., and Higuchi, T. (1981). *Planta* **152,** 365–368.
Kutsuki, H., Shimada, M., and Higuchi, T. (1982). *Phytochemistry* **21,** 267–271.
Mizuno, K., and Komamine, A. (1978). *Planta* **138,** 59–62.
Musha, Y., and Goring, D. A. I. (1975). *Wood Sci. Technol.* **9,** 45–58.
Nakamura, W. (1967). *J. Biochem. (Tokyo)* **62,** 54–60.
Pickett-Heaps, J. D. (1968). *Protoplasma* **65,** 181–205.
Rhodes, M. J. C., and Wooltorton, L. S. C. (1973a). *Phytochemistry* **12,** 107–118.
Rhodes, M. J. C., and Wooltorton, L. S. C. (1973b). *Qualitas Plant* **23,** 145–155.
Rubery, P. H., and Fosket, D. E. (1969). *Planta* **87,** 54–62.
Sakakibara, A. (1983). *In* "Recent Advances in Lignin Biodegradation Research" (T. Higuchi, H.-m. Chang, and T. K. Kirk, eds.), pp. 12–33. UNI Publisher, Tokyo.

Sarkanen, K. V., and Ludwig, C. H. (1971). *In* "Lignin, Occurrence, Formation, Structure and Reactions," pp. 150–155. Wiley (Interscience), New York.
Siegel, S. M. (1954). *Physiol. Plant.* **7,** 41–49.
Stafford, H. A. (1974). *Rec. Adv. Phytochem.* **8,** 53–79.
Tanahashi, M., and Higuchi, T. (1981). *Wood Res. (Kyoto)* **67,** 29–42.
Tanahashi, M., Takeuchi, H., and Higuchi, T. (1976). *Wood Res. (Kyoto)* **61,** 44–53.
Van Fleet, D. S. (1959). *Can. J. Bot.* **37,** 449–458.
Wardrop, A. B., and Bland, D. E. (1959). *In* "Biochemistry of Woods" (K. Kratzl and G. Billek, eds.), pp. 92–116. Pergamon, New York.
Westermark, U. (1982). *Wood Sci. Technol.* **16,** 71–78.
Whitmore, F. W. (1978). *Plant Sci. Lett.* **13,** 241–245.
Yamada, Y., and Kuboi, T. (1976). *Phytochemistry* **15,** 395–396.
Yamasaki, T., Hata, K., and Higuchi, T. (1976). *Mokuzai Gakkaishi,* **22,** 582–588.
Yamasaki, T., Hata, K., and Higuchi, T. (1978). *Holzforschung* **32,** 44–47.

CHAPTER **8**

Biosynthesis of Cutin, Suberin, and Associated Waxes

P. E. KOLATTUKUDY
KARL E. ESPELIE

Institute of Biological Chemistry
Washington State University
Pullman, Washington

ISBN 0-12-347880-4

I. Introduction

Three types of noncarbohydrate polymers are associated with wood. The major one, lignin, is a polymer derived entirely from aromatic monomers, whereas cutin is a polyester derived almost entirely from aliphatic monomers. Suberin, although poorly understood, appears to be a polymer containing both aromatic and aliphatic domains. The presence of aliphatic chains in suberin was first recognized in 1797, when LaGrange obtained suberic acid by treating cork with nitric acid. The term suberin was apparently used for the first time by Chevreul in 1815 to describe the insoluble material in bottle cork. The term cutin arose from "cutose," which was coined to describe the acid-resistant insoluble material obtained from plant cuticle. With the advent of modern analytical instrumentation, particularly combined gas–liquid chromatography and mass spectrometry, much progress has been made in our understanding of such polymers. We define cutin and suberin as insoluble polymeric materials, and both of these polymers are associated with soluble waxes. Progress in our knowledge of the biosynthesis of cutin, suberin, and associated waxes is reviewed in this chapter.

II. Location of Cutin, Suberin, and Waxes

A. *Ultrastructural Characteristics of Cutin, Suberin, and Waxes*

Robert Hooke was examining suberized bark of the cork oak tree when he first described the cell in 1665. More than 200 years later, von Höhnel examined cells from the same source and described the lamellar structure characteristic of suberin. In modern times, suberized regions in the bark of several tree species have been examined ultrastructurally. These studies include cork oak (Sitte, 1955), birch (Jensen, 1972; Schönherr and Ziegler, 1980), acacia (Wattendorff, 1974), spruce (Parameswaran *et al.*, 1975; Godkin *et al.*, 1983), pine (Patel, 1975), Douglas fir (Litvay and Krahmer, 1977), balsam fir, and eastern hemlock (Grozdits *et al.*, 1982; Grozdits, 1982). Such studies show that the suberized region exhibits a lamellar structure and is laid down outside the plasma membrane within the cell wall (Fig. 1). The lamellar appearance is thought to be due to alternating layers of polymer and suberin-associated wax (Meyer, 1938; Sitte, 1957; Litvay and Krahmer, 1977), and the observation that specific inhibition of wax biosynthesis prevented the development of the light bands in the periderm of wound-healing potato tissue slices provided evidence that these bands represent wax (Soliday *et al.*, 1979).

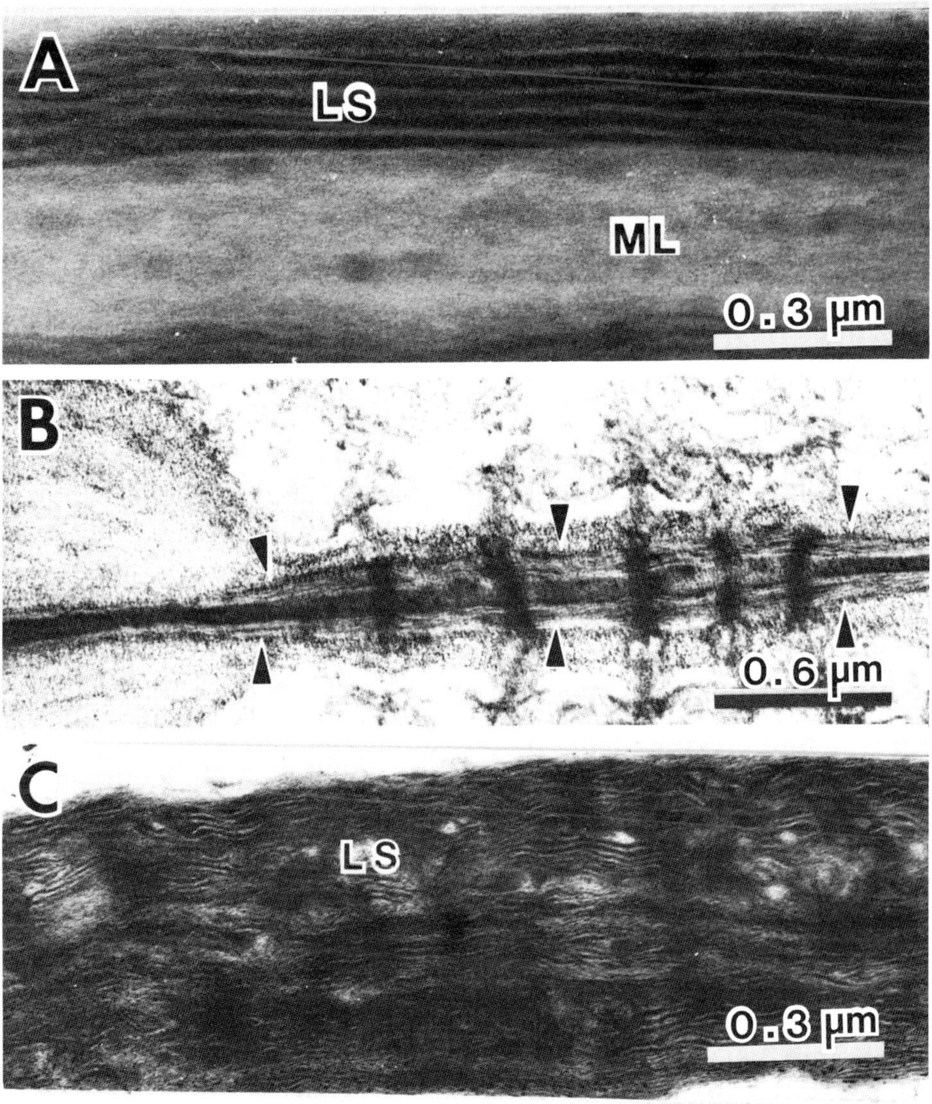

Fig. 1. Electron micrographs of suberized cell walls. (A) Periderm layer from potato tuber showing lamellar suberin (LS) and middle lamella (ML). (B) Lamellar suberin (arrowheads) in the walls of adjacent bundle sheath cells of *Zea mays*. Courtesy of Dr. T. P. O'Brien. (C) Lamellar suberin in the wall of a heavily suberized cell in the chalazal region of the inner seed coat of grapefruit.

In contrast to suberin, cutin is deposited on the outside of the cell wall and often appears as an amorphous layer, but in some cases it may also have a lamellar appearance, either throughout or at the outer portion of the cuticular layer (Martin and Juniper, 1970; Kolattukudy, 1980a,b; Wattendorff and Holloway, 1982).

Cutin-associated wax, in addition to being an embedding medium for the polymer, is often laid down on the outer surface of the plant in a crystalline structure characteristic of the plant source and the anatomical location (Martin and Juniper, 1970; Kolattukudy and Walton, 1973; Tulloch, 1976; Kolattukudy, 1980b).

B. *Anatomical Location of Suberin and Cutin*

Cutin is deposited on the outer surface of the epidermal cells of aerial organs of plants. In many cases, ambiguity about the nature of the polymer has been especially prevalent regarding internal layers that have been observed microscopically and have been defined as cutin or suberin or even "cutin–suberin-like" (Scott, 1966; Scott and Bystrom, 1970). Chemical analysis of the polymeric materials from such layers would assist in distinguishing between cutin and suberin and would allow identification of the anatomical layers observed. Such chemical analysis combined with ultrastructural studies has resolved some of the ambiguities. For example, it has been shown that a cuticle covers such plant parts as embryonic shoots in seeds (Kolattukudy, 1974), stigmata (Shaykh *et al.*, 1977), some seed coats (Espelie *et al.*, 1979; Velcheva *et al.*, 1981), and *Citrus* juice sacs (Espelie *et al.*, 1980b).

Suberin is deposited in the walls of phellem or cork cells in the periderm layer. In tree bark, the periderm is located between the inner and outer bark layers (Esau, 1977; Hemingway, 1981). Periderm layers are also deposited in several other regions of the plant for protective purposes. Such layers have been characterized ultrastructurally as suberin in plant roots (Torrey and Clarkson, 1975; Wilson and Peterson, 1983). Suberin layers have been seen in either the epidermal or hypodermal cells of the roots of most plants that have been studied (Ferguson and Clarkson, 1976; Olesen, 1978; Peterson *et al.*, 1978; Robards *et al.*, 1979), as well as in the walls of endodermal cells at varying distances from the root tip (Robards *et al.*, 1973; Ferguson and Clarkson, 1975; Tippet and O'Brien, 1976; Scott and Peterson, 1979). Ultrastructural studies also revealed suberin in the Casparian band in the endodermis (Bonnett, 1968; Haas *et al.*, 1976) and in the hypodermis (Peterson *et al.*, 1982) of plant roots. These ultrastructural studies have been confirmed by only a few chemical studies, which examined the aliphatic components of the periderm of underground storage organs (Kolattukudy and Agrawal, 1974; Kolattukudy *et al.*, 1975a; Brieskorn and Binnemann, 1975), roots of very young *Vicia faba* (Espelie *et al.*, 1979), outer cork layers from roots of apple trees (Holloway, 1982a), and both the hypodermis and endodermis of *Zea mays* roots (Pozuelo *et al.*, 1984). Suberin has also been chemically identified in an endodermal fraction from *Sorghum* stems (Espelie and Kolattukudy, 1979a).

Suberization appears to be the general response to wounding. Although it had been postulated that wounded leaf resynthesized cutin, chemical studies showed that such conclusions were erroneous and that wounding induced suberization irrespective of the composition of the natural protective polymer of the organ (Dean and Kolattukudy, 1976). Suberized layers are also found in the mestome sheath and bundle sheath cell walls of grasses (Fig. 1) (O'Brien and Carr, 1970; O'Brien and Kuo, 1975; Eleftheriou and Tsekos, 1979; Botha *et al.*, 1982) and have been identified chemically in the case of *Zea mays* (Espelie and Kolattukudy, 1979b).

Internal suberin layers are deposited around oil-gland cells (Amelunxen and Gronau, 1969; Platt-Aloia *et al.*, 1983), epidermal cells adjacent to silica-containing cells (Lawton, 1980), the base of trichomes (Thomson *et al.*, 1979), calcium oxalate idioblasts (Wattendorff, 1976a,b), and in the "hollow heart" disorder of potato (Dean *et al.*, 1977). The suberin found in idioblasts was characterized chemically and ultrastructurally (Espelie *et al.*, 1982). Suberin has been shown to be deposited in the pigment strand of wheat (Zee and O'Brien, 1970) and rice (Oparka and Gates, 1982) and in the chalazal region of the inner seed coat of *Citrus* (Espelie *et al.*, 1980b).

III. Structure and Function of Cutin, Suberin, and Waxes

A. *Isolation of Cutin and Suberin*

Cutin can be removed from many plant organs by enzymatic disruption of the pectin layer that attaches the cuticle to the epidermal wall or by chemical disruption of the layer by heating the tissue with solutions of ammonium oxalate plus oxalic acid or $ZnCl_2$ plus HCl (Orgell 1955; Matic 1956; Eglinton and Hunneman, 1968; Walton and Kolattukudy 1972b). Although this procedure works well for the isolation of well-developed thick cuticles, the very thin cuticular layers found on some plant parts might be too fragile to isolate by this method. The isolated cuticle is ground to a powder and extracted thoroughly with organic solvents to

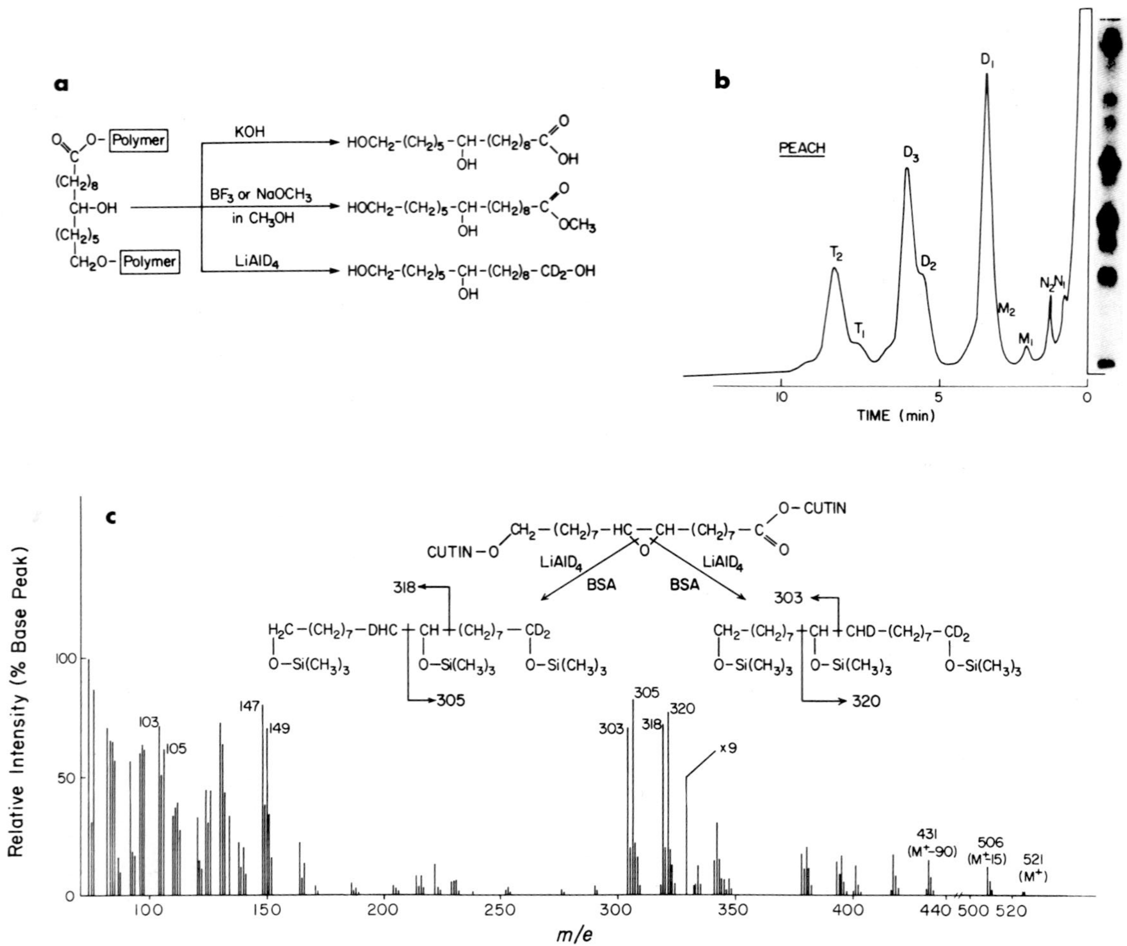

a
Polymer
KOH
BF3 or NaOCH3
in CH3OH
LiAlD4
HOCH2-(CH2)5-CH-(CH2)8-C(=O)OH
HOCH2-(CH2)5-CH-(CH2)8-C(=O)OCH3
HOCH2-(CH2)5-CH-(CH2)8-CD2-OH
b
PEACH
T2
T1
D3
D2
D1
M2
M1
N2
N1
TIME (min)
c
CUTIN-O
O-CUTIN
LiAlD4
BSA
318
305
303
320
O-Si(CH3)3
Relative Intensity (% Base Peak)
m/e
103
105
147
149
x9
431 (M+-90)
506 (M+-15)
521 (M+)

remove the soluble waxes, and the carbohydrates contained in this cutin preparation are removed by extensive treatment with cellulase and pectinase.

Since suberin is located within the cell wall, it can be isolated only as a suberin-enriched preparation, which consists of essentially suberized cell walls. Often the process of isolation starts with the physical removal of the suberized region of specific plant parts, such as the periderm (Kolattukudy and Agrawal, 1974; Kolattukudy and Dean, 1974). Treatment of the isolated material with hydrolytic enzymes such as pectinase and cellulase removes considerable amounts of carbohydrates, and this causes further enrichment of suberized walls. The remaining material is ground to a fine powder and then thoroughly extracted with organic solvents to remove the wax associated with suberin. The final preparation always contains cell wall carbohydrates. External periderm layers such as those found in underground storage tissues can be quite readily isolated in a very enriched form and then examined chemically (Kolattukudy and Agrawal, 1974; Kolattukudy *et al.*, 1975a). On the other hand, less enrichment is achieved when a similar approach is used to isolate internal suberin polymers such as those found in the bundle sheath cells of C_4 plants, the Casparian band, and the endodermal wall (Espelie and Kolattukudy, 1979a,b; Pozuelo *et al.*, 1984).

B. *Composition of Cutin*

Since cutin is an insoluble polyester composed of hydroxy and hydroxy–epoxy fatty acids, the depolymerization techniques that can be used involve ester cleavage methods such as alkaline hydrolysis (Legg and Wheeler, 1925; Matic, 1956; Huelin, 1959; Crisp, 1965; Eglinton and Hunneman, 1968), transesterification with sodium methoxide plus methanol or boron trifluoride plus methanol (Holloway, 1973; Holloway and Deas, 1973; Kolattukudy *et al.*, 1973b, 1975a; Kolattukudy and Agrawal, 1974), and hydrogenolysis with $LiAlH_4$ in tetrahydrofuran (Kolattukudy, 1970c; Kolattukudy and Walton, 1972; Walton and Kolattukudy, 1972b, Jones, 1978) (Fig. 2). The method of choice may depend on the type of

Fig. 2. (a) Chemical methods used to depolymerize cutin and suberin. (b) Thin-layer and gas–liquid chromatograms of the mixture of monomers generated by $LiAlD_4$ treatment of peach fruit cutin. The four major spots in the thin-layer chromatogram are, from the bottom, C_{18} tetraol, C_{18} triol, $C_{18:1}$ triol and diols. The components in the gas chromatogram are trimethylsilyl ethers of N_1, C_{16} alcohol; N_2, C_{18} alcohol; M_1, C_{16} diol; M_2, C_{18} diol; D_1, C_{16} triol; D_2, $C_{18:1}$ triol; D_3, C_{18} triol; T_1, $C_{18:1}$ tetraol; T_2, C_{18} tetraol. (c) Mass spectrum of component D_3 in the gas chromatogram. BSA, bis(*N*,*O*-trimethylsilyl)acetamide.

monomers present in the polymer and the amount of sample available for analysis. For example, the presence of functional groups such as epoxides can be conveniently identified by the deuterium labeling resulting from depolymerization with $LiAlD_4$ (Fig. 2). This deuterium labeling method led to the first recognition that ω-hydroxy-9,10-epoxyoctadecanoic acid is a major component of cutin (Kolattukudy *et al.*, 1971). Epoxides can also be identified from the methanolysis product of the oxirane ring (Kolattukudy and Agrawal, 1974; Holloway, 1974). The epoxy acid can be isolated by careful depolymerization with anhydrous methanol containing sodium methoxide (Kolattukudy *et al.*, 1973b). The different depolymerization methods yield monomers containing carboxyl groups and/or hydroxyl groups, and these monomers can be separated into classes by thin-layer chromatography. The method of choice is to derivatize the monomer mixture and subject the mixture to combined gas chromatography and mass spectrometry.

The major monomers of cutin in most plants have 16 or 18 carbon atoms (Fig. 3). The most common members of the C_{16} family are hexadecanoic acid, ω-hydroxyhexadecanoic acid, and dihydroxyhexadecanoic acid. Dihydroxyhexadecanoic acid is the most characteristic and dominant cutin monomer and is usually found as a mixture of positional isomers, with the midchain hydroxyl found on either C-7, C-8, C-9, or C-10, with the latter two isomers being the most prevalent (Kolattukudy 1977b; Brieskorn, 1978). The dihydroxy acids from papaya fruit cutin (predominantly 9,16-dihydroxyhexadecanoic acid) and tomato fruit cutin (10,16-dihydroxy C_{16} acid) both showed a plain positive rotation, and therefore the L configuration has been assigned to this major cutin monomer (Espelie and Kolattukudy, 1978). Various monomers that could be derived by further oxidation or reduction of the common C_{16} monomers are found as major

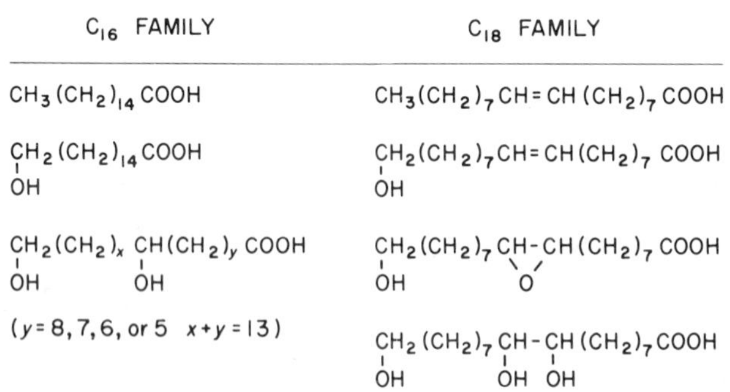

Fig. 3. The major monomer acids commonly found in cutin. In the C_{18} family, Δ^{12} unsaturated analogs also occur.

components in rare cases or as minor cutin components. These include 16-oxo-9- or 10-hydroxyhexadecanoic acid (Kolattukudy, 1972, 1974), ω-hydroxy-9- or 10-oxohexadecanoic acid (Deas *et al.*, 1974), 9,16-dihydroxy-10-oxohexadecanoic acid (Espelie *et al.*, 1983), and 1,8,16-trihydroxyhexadecane (Caldicott *et al.*, 1975).

The major members of the C_{18} family of cutin monomers are octadecanoic, octadecenoic, ω-hydroxyoctadecenoic, ω-hydroxy-9,10-epoxyoctadecanoic, and 9,10,18-trihydroxyoctadecanoic acids, together with their Δ^{12} unsaturated analogs (Kolattukudy *et al.*, 1981). Minor and/or rare monomers in the C_{18} family include 18-oxo-9,10-epoxyoctadecanoic acid (Kolattukudy, 1973), 9,10,18-trihydroxy-12,13-epoxyoctadecanoic acid, and 9,10,12,13,18-pentahydroxyoctadecanoic acid (Brieskorn and Kabelitz, 1971; Croteau and Kolattukudy, 1974a). A more comprehensive list of the monomers found so far in cutin is given in Table I. Many of the components are present in very small amounts, and in many cases these components have been found in only a few plants.

Cutin can be isolated fairly easily from fruits and leaves, and therefore the polymers from such sources have been most frequently utilized for chemical analysis (Kolattukudy *et al.*, 1981; Holloway, 1982b). Even though cutin composition shows species specificity, chemotaxonomic application of such information should be done cautiously, because the composition of the polymer can vary with anatomical location. Thus the fruit, leaf, stigma, and flower petal of *Malus pumila* each contain different proportions of C_{16} and C_{18} monomers (Espelie *et al.*, 1979). The polymer from these organs contained 73, 35, 14, and 12%, respectively, hydroxy and hydroxyepoxy C_{18} monomers. This observation is in agreement with the tentative generalization that the C_{18} family of monomers is more prevalent in cutin that covers slow-growing organs and that the C_{16} family of monomers dominates in polymers covering fast-growing organs (Kolattukudy, 1975).

C. *Composition of Suberin*

The chemical composition of suberin is not as well understood as that of cutin, in part due to the difficulty in obtaining suberin in pure form. Suberin-enriched preparations have been isolated and chemically examined from relatively few plants and from relatively few anatomical locations within the plant (Kolattukudy, 1981). On the basis of early analysis of suberin from the periderm of underground storage organs, a tentative generalization was drawn concerning suberin composition (Kolattukudy and Agrawal, 1974; Kolattukudy *et al.*, 1975a), and subsequent studies have, for the most part, confirmed the early conclusion. According to the

Table I

Fatty Acids with One or More Additional Functional Groups That Have Been Reported as Components of Cutin or Suberin[a]

Monomer		Source	Percentage of total aliphatics	Reference
Monohydroxy acids				
8-Hydroxy C_8		*Psilotum nudum* stem	0.7	Caldicott *et al.* (1975)
9-Hydroxy C_9		*Solanum tuberosum* leaf	0.5	Brieskorn and Binnemann (1975)
12-Hydroxy C_{12}		*Pinus sylvestris* leaf	9	Hunneman and Eglinton (1972)
9-Hydroxy $C_{14:1}$		*Coffea arabica* leaf	4.5	Holloway *et al.* (1972)
14-Hydroxy C_{14}		*Encephalartos altensteinii* leaf	4	Hunneman and Eglinton (1972)
9-Hydroxy C_{15}[b]		*Coffea arabica* leaf	1	Holloway *et al.* (1972)
2-Hydroxy C_{16}		*Conocephalum conicum* leaf	4.8	Caldicott and Eglinton (1976)
15-Hydroxy C_{16}		*Astarella lindenbergiana* leaf	72	Caldicott and Eglinton (1976)
16-Hydroxy C_{16}	S	*Populus tremula* bark	22	Holloway (1983)
2-Hydroxy C_{18}		*Conocephalum conicum* leaf	3.3	Caldicott and Eglinton (1976)
10-Hydroxy C_{18}[b]		*Rosmarinus officinalis* leaf	1.3	Brieskorn and Kabelitz (1971)
12-Hydroxy $C_{18:1}$		*Rosmarinus officinalis* leaf	2.3	Brieskorn and Kabelitz (1971)
18-Hydroxy C_{18}	S	*Cupressus leylandi* bark	8	Holloway (1983)
18-Hydroxy $C_{18:1}$	S	*Solanum tuberosum* storage organ	33	Brieskorn and Binnemann (1975)
18-Hydroxy $C_{18:2}$		*Spinacia oleracea* leaf	0.1	Holloway (1974)
20-Hydroxy C_{20}	S	*Beta vulgaris* tuber	2.9	Kolattukudy *et al.* (1975a)
22-Hydroxy C_{22}	S	*Gossypium hirsutum* green fiber	70	Yatsu *et al.* (1983)
20-Hydroxy C_{23}		*Conocephalum conicum* leaf	2	Caldicott and Eglinton (1976)
20-Hydroxy C_{24}		*Conocephalum conicum* leaf	10	Caldicott and Eglinton (1976)
24-Hydroxy C_{24}	S	*Euonymus alatus* "cork wings"	14	Holloway (1983)
26-Hydroxy C_{26}	S	*Quercus ilex* bark	2	Holloway (1983)
28-Hydroxy C_{28}	S	*Fraxinus excelsior* bark	0.9	Holloway (1983)

Table I (*Continued*)

Monomer	Source	Percentage of total aliphatics	Reference
Dihydroxy acids			
9,15-Dihydroxy C_{15}	*Araucaria imbricate* leaf	1.7	Hunneman and Eglinton (1972)
10,15-Dihydroxy C_{16}	*Astarella lindenbergiana* leaf	3.9	Caldicott and Eglinton (1976)
7,16-Dihydroxy C_{16}	*Pisum sativum* seed coat	4.1	Espelie *et al.* (1979)
8,16-Dihydroxy C_{16}	*Hordeum vulgare* leaf	8	Espelie *et al.* (1979)
9,16-Dihydroxy C_{16}	*Malabar papaiarnarum* fruit	73	Walton and Kolattukudy (1972b)
10,16 Dihydroxy C_{16}	*Ribes grossularia* fruit	83	Hunneman and Eglinton (1972)
10,17-Dihydroxy C_{17}	*Pinus radiata* stem	0.1	Franich and Volkman (1982)
10,18-Dihydroxy C_{18}	*Pinus sylvestris* leaf	1.0	Hunneman and Eglinton (1972)
10,18-Dihydroxy $C_{18:1}$	*Vaccinium macrocarpon* fruit	1.1	Croteau and Fagerson (1972)
Tri- and pentahydroxy acids			
6,7,16-Trihydroxy C_{16}	*Rosmarinus officinalis* leaf	17	Brieskorn and Kabelitz (1971)
9,10,16-Trihydroxy C_{16}	*Citrus paradisi* fruit	1.9	Espelie *et al.* (1983)
9,10,17-Trihydroxy C_{17}	*Rosmarinus officinalis* leaf	2.9	Brieskorn and Kabelitz (1971)
9,10,17-Trihydroxy $C_{17:1}$	*Rosmarinus officinalis* leaf	3.0	Brieskorn and Kabelitz (1971)
9,10,18 Trihydroxy C_{18}	*Agave americana* leaf	52	Espelie *et al.* (1982)
9,10,18-Trihydroxy $C_{18:1}$	*Citrus paradisi* seed coat	23	Espelie *et al.* (1980b)
9,10,12,13,18-Pentahydroxy C_{18}	*Rosmarinus officinalis* leaf	3.2	Brieskorn and Kabelitz (1971)
Epoxy and oxo acids			
16-Hydroxy-10-oxo C_{16}	*Citrus limon* fruit	34	Baker and Procopiou (1975)
9-Hydroxy-16-oxo C_{16}[b]	*Vicia faba* embryonic stem	32	Kolattukudy (1974)
9,16-Dihydroxy-10-oxo C_{16}	*Citrus paradisi* fruit	4.2	Espelie *et al.* (1983)
9,10-Epoxy-18-hydroxy C_{18}	*Citrus paradisi* seed coat	37	Espelie *et al.* (1980b)
9,10-Epoxy-18-hydroxy $C_{18:1}$	*Vitis vinifera* fruit	30	Walton and Kolattukudy (1972b)
9,10-Epoxy-18-oxo C_{18}	*Malus pumila* young fruit	—	Kolattukudy (1973)

(*continued*)

Table I (*Continued*)

Monomer		Source	Percentage of total aliphatics	Reference
Dicarboxylic acids				
C_9 Diacid		*Solanum tuberosum* leaf	1.7	Brieskorn and Binnemann (1975)
C_{14} Diacid		*Pinus radiata* stem	0.5	Franich and Volkman (1982)
C_{15} Diacid		*Pinus radiata* stem	0.7	Franich and Volkman (1982)
6-Hydroxy C_{15} diacid		*Gnetum gnemom* leaf	7	Hunneman and Eglinton (1972)
7-Hydroxy C_{15} diacid		*Sapindus saponaria* leaf	1.3	Hunneman and Eglinton (1972)
8-Hydroxy C_{15} diacid		*Sphagnum cuspidatum* leaf	0.6	Caldicott and Eglinton (1976)
C_{16} Diacid	S	*Citrus paradisi* seed coat	13	Espelie *et al.* (1980b)
$C_{16:1}$ Diacid		*Vaccinium macrocarpon* fruit	0.1	Croteau and Fagerson (1972)
7-Hydroxy C_{16} diacid		*Welwitschia mirabilis* leaf	15	Hunneman and Eglinton (1972)
8-Hydroxy C_{16} diacid		*Sphagnum cuspidatum*	7	Caldicott and Eglinton (1976)
C_{17} Diacid		*Pinus radiata* stem	5.2	Franich and Volkman (1982)
8,9-Dihydroxy C_{17} diacid		*Vaccinium macrocarpon* fruit	0.2	Croteau and Fagerson (1972)
C_{18} Diacid	S	*Ribes nigrum* bark	2.8	Holloway (1983)
$C_{18:1}$ Diacid	S	*Solanum tuberosum* tuber	31	Brieskorn and Binnemann (1975)
$C_{18:2}$ Diacid		*Vaccinium macrocarpon* fruit	0.02	Croteau and Fagerson (1972)
9,10-Dihydroxy C_{18} diacid	S	*Acer griseum* bark	17	Holloway (1983)
9,10-Epoxy C_{18} diacid	S	*Quercus suber* bark	16	Holloway (1983)
$C_{19:1}$ Diacid		*Pinus radiata* stem	8	Franich and Volkman (1982)
C_{20} Diacid	S	*Cupressus leylandi* bark	3.0	Holloway (1983)
C_{22} Diacid	S	*Gossypium hirsutum* green fiber	25	Yatsu *et al.* (1983)
C_{24} Diacid	S	*Citrus paradisi* seed coat	4.8	Espelie *et al.* (1980b)
C_{26} Diacid	S	*Euonymus alatus* "cork wings"	0.1	Holloway (1983)

[a] Monomers from suberin are indicated by S.
[b] Positional isomers also found.

current working hypothesis, suberin consists of aliphatic polyester domains covalently attached to aromatic domains which are in turn attached to the cell wall (Fig. 4).

The aliphatic portion of the polymer has been examined by utilizing the depolymerization methods already described for cutin (Holloway 1972a,b, 1982a, 1983; Rodríguez-Miguens and Ribas-Marqués, 1972a,b; Kolattukudy and Agrawal, 1974; Brieskorn and Binnemann, 1975; Kolat-

Major monomers $CH_3(CH_2)_mCOOH$

$CH_3(CH_2)_mCH_2OH$

$CH_2(CH_2)_nCOOH$ (with OH on the terminal CH_2)

$HOOC(CH_2)_nCOOH$

Phenolics (m = 18–30; n = 14–20)

Polymer

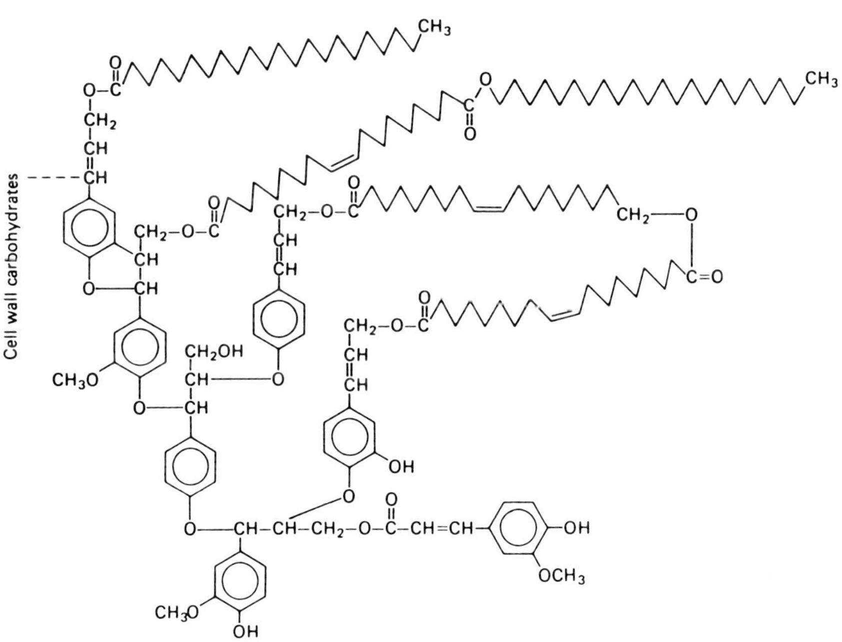

Fig. 4. A tentative model depicting the structure of suberin polymer. The model is to illustrate mainly the general features.

tukudy *et al.*, 1975a, 1981). ω-Hydroxy acids and dicarboxylic acids often constitute a very high proportion of the aliphatic portion of the polymer. For example, these two classes of acids comprised 50–90% of the aliphatic portion of the suberin from the periderm of underground storage organs (Kolattukudy *et al.*, 1975a), the chalazal region of grapefruit seed coat (Espelie *et al.*, 1980b), and the idioblasts of *Agave* leaf (Espelie *et al.*, 1982). In a number of total bark samples recently examined, which may have contained some cutin polymers, ω-hydroxy acids and dicarboxylic acids ranged from 23 to 83% of the total aliphatic components (Table II). The most common chain lengths of these two classes of acids are C_{16} and $C_{18:1}$, but monomers of longer chain length (C_{20} to C_{26}) are frequently found, with C_{22} and C_{24} being the most common of the very long chain monomers (Kolattukudy *et al.*, 1981). For example, ω-hydroxydocosanoic acid and docosanedioic acid were recently reported to comprise 70% and 25%, respectively, of the aliphatic components of the suberin polymer found in the walls of green cotton fibers (Yatsu *et al.*, 1983). Very long ($>C_{28}$) fatty acids and fatty alcohols are also characteristic components of the aliphatic portion of suberin (Kolattukudy, 1978). Table I lists both common and rare monomers found in the aliphatic portion of suberin.

Table II

Total ω-Hydroxy Acid and Dicarboxylic Acid Content of the Aliphatic Portion of Monomers Released by Depolymerization of Total Bark Samples[a]

Source	ω-Hydroxy acid (C_{16}–C_{28})	Dicarboxylic acid[b] (C_{16}–C_{26})
Quercus robur	21.3	15.1
Quercus ilex	23.9	10.5
Fagus sylvatica	18.7	11.9
Castanea sativa	13.4	9.9
Betula pendula	21.6	8.0
Quercus suber	40.7	32.7
Acer griseum	39.6	25.9
Fraxinus excelsior	24.3	40.4
Acer pseudoplatanus	45.3	20.5
Ribes nigrum	49.7	20.2
Populus tremula	54.5	27.8
Sambucus nigra	50.5	24.6
Laburnum anagyroides	48.6	21.8
Cupressus leylandii	51.2	31.6

[a] From Holloway (1983).
[b] Includes midchain substituted diacids.

The phenolic portion of suberin has not been well characterized. Combined histochemical and ultrastructural studies indicate that suberized layers show the presence of phenolic components (Tippet and O'Brien, 1976; Peterson *et al.*, 1978; Scott and Peterson, 1979; Pearçe and Rutherford, 1981). Early chemical studies also indicated that the suberin polymer contains phenolics. Saponification of suberin from Douglas fir bark released phenolic acids (43% of extractive-free cork) and aliphatic acids (35%) as the two major classes of polymer components (Hergert and Kurth, 1952). Selective extraction of a mixture obtained by partial saponification of Douglas fir suberin yielded a fraction (22% of extractive-free cork) that was partially characterized as a phenolic acid–hydroxy acid ester. Similarly, ferulic acid esterified to ω-hydroxydocosanoic acid was isolated by saponification of the suberin from oak cork (Guillemonat and Traynard, 1963). Ferulic acid was also identified as a component (6%) of suberin from white fir bark, while sinapic acid was identified as a component of *Quercus suber* suberin (Hergert, 1958). More recently, suberin in the periderm layer of underground storage organs has been shown to contain small amounts of esterified ferulic acid (Riley and Kolattukudy, 1975). Depolymerization of suberin releases colored, water-soluble phenolic components and also leaves behind a large proportion of insoluble phenolic-containing residue (Holloway, 1972a; Kolattukudy and Agrawal, 1974; Kolattukudy *et al.*, 1975a; Kolattukudy, 1981). However, the nature of the phenolic material contained in such fractions is unknown.

Alkaline nitrobenzene oxidation similar to that used on lignin (Hartley, 1971) was recently used as a technique to release aromatic components from suberin. Nitrobenzene oxidation of suberized potato tissue released *p*-hydroxybenzaldehyde and vanillin (Cottle and Kolattukudy, 1982a). That these aromatic aldehydes originated from suberin was suggested by the observation that the time course of deposition of the material that gave rise to the aldehydes coincided with the time course of suberization during the wound healing of potato tissue slices (Cottle and Kolattukudy, 1982a). Furthermore, induction of suberization of abscisic acid in potato tissue culture also was associated with deposition of the insoluble phenolic components that generated the aromatic aldehydes upon nitrobenzene oxidation. Nitrobenzene oxidation of radioactive potato periderm isolated from tuber slices that had metabolized [U-^{14}C]cinnamic acid released radioactive components, which were fractionated as shown in Fig. 5. The majority of activity was located in the four numbered fractions: 1, 25%; 2, 9%; 3, 15%; 4, 20% (Cottle and Kolattukudy, 1982a). Thin-layer chromatography (TLC) and radiolabeled gas chromatography of fraction 1 indicated that *p*-hydroxybenzaldehyde and vanillin contained most of the label of this fraction, but the labeled components in the other fractions

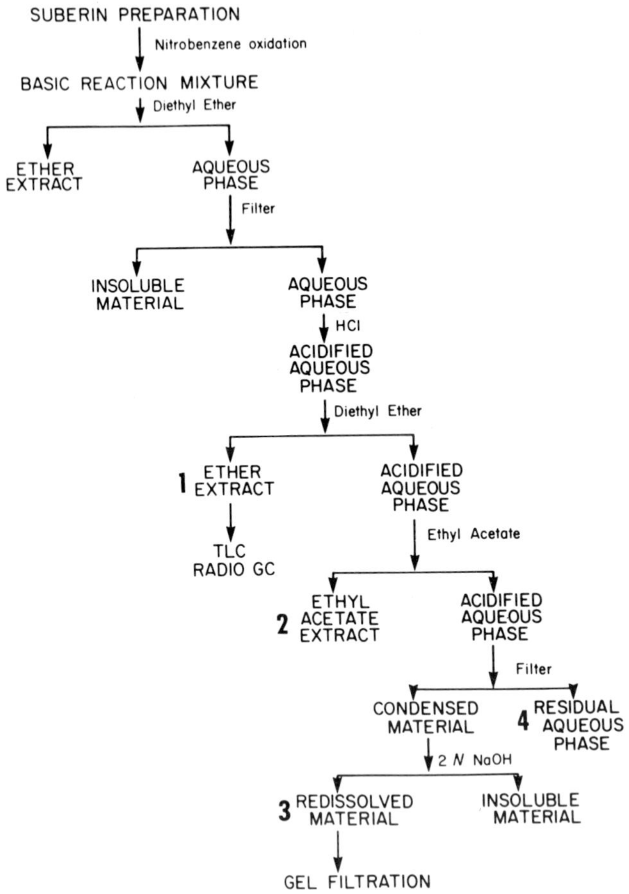

Fig. 5. Flowchart of the fractionation of the products generated by alkaline nitrobenzene oxidation of a radiolabeled preparation of potato suberin. From Cottle and Kolattukudy (1982a).

remain unidentified. It seems clear that nitrobenzene oxidation is releasing only a portion of the suberin phenolics contained in potato suberin. Oxidation of suberin from *Agave* idioblasts (Fig. 6) (Espelie *et al.*, 1982) and of suberin from the hypodermis and endodermis of *Zea mays* roots (Pozuelo *et al.*, 1984) also generated *p*-hydroxybenzaldehyde and vanillin. Nitrobenzene oxidation of several suberin and cutin-enriched samples indicates that *p*-hydroxybenzaldehyde and vanillin are released from most suberin samples, whereas syringaldehyde is more likely to be released

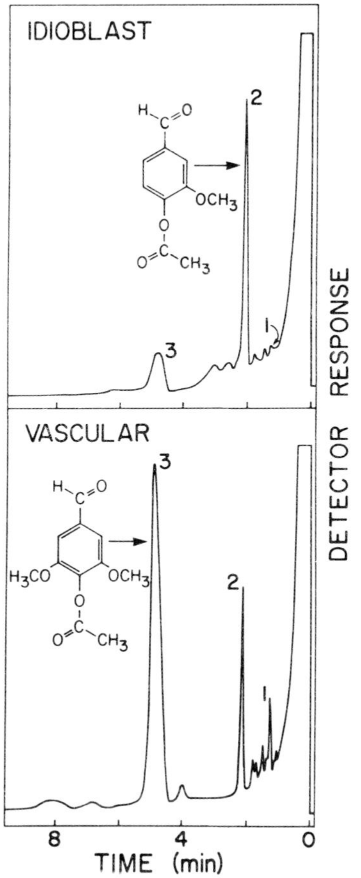

Fig. 6. Gas–liquid chromatograms of acetylated aromatic aldehydes generated by alkaline nitrobenzene oxidation of suberized idioblast cell walls and lignified vascular tissue from *Agave americana* leaves. Compounds 1, 2, and 3 were identified by their mass spectra to be acetylated *p*-hydroxybenzaldehyde, vanillin, and syringaldehyde, respectively.

from lignin-containing tissue (e.g., pecky rot lignin and *Agave* vascular tissue) (Table III).

Since the phenolic domains of suberin could possibly have some similarities to lignin, we attempted to solubilize a portion of the suberin-enriched cell wall preparation from potato periderm with dioxane after extensive ball-milling, resulting in 4–5% solubilization of the material (K. E. Espelie and P. E. Kolattukudy, unpublished results), giving rise to a fraction analogous to Björkman lignin (Björkman, 1956). However, this

Table III

Aromatic Aldehydes Released by Nitrobenzene Oxidation of Cutin- and Suberin-Enriched Samples[a]

Tissue	*p*-Hydroxybenzaldehyde	Vanillin	Syringaldehyde	Total
Potato periderm	3.28	16.51	0.53	20.3
Potato periderm LAH residue	1.77	6.66	0.20	8.6
Chalazal region of grapefruit seed coat	1.19	15.63	1.80	18.6
Grapefruit inner seed coat	3.45	1.84	0.16	5.5
Grapefruit outer seed coat	0.40	38.72	1.09	40.2
Cotton seed coat	0.37	6.44	1.07	7.9
Pecky rot lignin	—	34.90	—	34.90
Agave idioblasts	0.02	0.26	0.12	0.4
Agave vascular leaf tissue	0.36	1.95	10.91	13.2
Wound-healed rutabaga	0.93	1.59	—	2.5
Parsnip periderm	—	0.25		0.3
Carrot periderm	—	0.51	—	0.5
Quercus suber bark	3.00	1.32	1.17	5.9
Zea mays root	2.49	2.13	0.13	4.8
Zea mays root (low Mg^{2+})	7.38	5.51	4.98	17.9
Tomato fruit cutin	5.07	—	—	5.1
Apple fruit cutin	—	—	—	—

[a] Values are given as μg/mg dry tissue.

fraction was not enriched in either aliphatic or aromatic components over the insoluble residue. Other procedures, including treatment at high pressure and 160°C in dioxane or refluxing in HCl plus dioxane or HCl plus dimethylformamide resulted in 20–50% solubilization of the suberin-enriched preparation, but in each case the residue contained the majority of aliphatics and aromatics.

D. *Composition of Plant Waxes*

Waxes are normally extracted from the cuticle by dipping the intact organ in organic solvents (e.g., chloroform or hexane) for a short period of time (e.g., 30–60 sec) at room temperature. Longer wash periods are required to remove wax embedded in the cuticular matrix. X-Ray diffraction of powdered apple cutin preparation, which had been extracted in a Soxhlet extractor for 96 hr, showed the presence of residual crystalline wax (V. P. Agrawal and P. E. Kolattukudy, unpublished results), demonstrating the difficulty of completely removing wax embedded in the polymeric matrix. The lamellar structure of suberin and its location within the cell wall make the waxes associated with this polymer more difficult to extract (Dean and Kolattukudy, 1977; Espelie *et al.*, 1980a). Lengthy (15- and 21-hr) successive Soxhlet extractions of powdered *Pinus sylvestris* bark gave hydrocarbon fractions with markedly different compositions (Streibl *et al.*, 1978), indicating the difficulty of effecting complete wax extraction. Thus, suberin-associated waxes, which can be obtained only by a very thorough extraction, are likely to contain lipids from other locations within the tissue.

Thin-layer chromatography and column chromatography are the normal procedures utilized to separate waxes into separate chemical classes (Kolattukudy and Walton, 1973; Tulloch, 1973, 1976; Kolattukudy, 1980b; Baker, 1982). Gas–liquid chromatography plus mass spectrometry of isolated fractions (or their appropriate derivatives) is the preferred method of characterization. The waxes associated with cutin and suberin are usually a complex mixture of several classes of aliphatic compounds, each of which may contain many homologous series of compounds. Some of the most common wax components are listed in Table IV. Wax composition is species dependent; however, anatomical location, age, and environmental conditions can also significantly affect the composition. Hydrocarbons, the most ubiquitous wax components, vary from <1% (e.g., in carnauba wax; Downing *et al.*, 1961) to >90% (e.g., in *Solandra grandiflora;* Herbin and Robins, 1968). Cuticular hydrocarbons are predominantly odd-chain-length alkanes, often dominated by n-C_{29} or n-C_{31}. Fatty alcohols are often the dominant class in cuticular wax, with C_{26} and C_{28} as the most common chain lengths. Esters and fatty acids are very common but are often relatively minor components.

Table IV

Cutin- and Suberin-Associated Wax Components[a]

Compound type	General structure	Chain-length range	Comments
Hydrocarbons			
n-Alkanes	$CH_3(CH_2)_nCH_3$	C_{21} to C_{35}	Most common wax component; high proportion of even chain length in suberin-associated waxes
Isoalkanes	$CH_3CH(CH_3)R$	C_{25} to C_{35}	Not as widespread as *n*-alkanes

(*continued*)

Table IV (*Continued*)

Compound type	General structure	Chain-length range	Comments
Anteisoalkanes	$CH_3CH_2CH(CH_3)R$	C_{24} to C_{36}	
Alkenes	$R^1CH{=}CHR^2$	C_{17} to C_{33}	In algae, dienes and trienes also found; branched alkenes might also be minor components
Terpenoid hydrocarbons	Farnasene, pristane, and cyclic hydrocarbons, phyllocladene, isophyllocladene (+) kaurene, and isokaurene, etc.		
Aromatic hydrocarbons	Anthracene and phenanthrene type with an attached alkane chain		
Ketones	R^1COR^2	C_{25} to C_{33}	Not as common as alkanes
α-Ketols	$R^1CH(OH)COR^2$		Minor and rare component
Secondary alcohols	$R^1CH(OH)R^2$	C_9 to C_{33}	Hydroxyl usually near the middle but toward the end also found; 2-ols found in suberin-associated waxes and in cuticular waxes that have β-diketones
β-Diketones	$R^1COCH_2COR^2$		Carbonyls at 12,14, at 14,16-, at 16,18 positions; in some cases major wax components
Hydroxy- or oxo-β-diketones	$R^1CH(OH)(CH_2)_nCOCH_2COR^2$		Usually minor components
Monoesters	R^1COOR^2	C_{30} to C_{60}	Composed of even-chain-length saturated acids and alcohols; esters of aromatic acids found in carnauba
Phenolic esters	1,16-Dioxo, 1-hydroxy-16-oxo-, and 1,16-dihydroxy-hexadecan-7-yl-*p*-coumarate; ferulic acid esters of C_{18} to C_{28} primary alcohols.		Commonly found in suberin-associated wax

Table IV *(Continued)*

Compound type	General structure	Chain-length range	Comments
Diesters	$R^1—C(=O)—OCH_2(CH_2)_nCH_2—OC(=O)—R^2$	Diol C_9 to C_{12}; acids C_{22} and C_{24} trans Δ^2	Rare and minor components
Polyesters	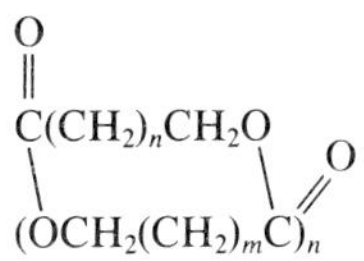		Cyclic and noncyclic; major components only in gymnosperms
Primary alcohols	RCH_2OH	C_{12} to C_{36}	Even chain lengths predominate
Aldehydes	RCHO	C_{14} to C_{34}	Not as common as alcohols, polymeric forms also found; even chain length and saturated
Acids			
Alkanoic acids	RCOOH	C_{12} to C_{36}	Very common; even-chain-length saturated chains predominate
Dicarboxylic acids	$HOOC(CH_2)_nCOOH$	C_{16} to C_{30}	Uncommon; found in suberin-associated wax
γ-Lactones	$RCHCH_2CH_2CO$ (ring closed through O)	C_{22} to C_{28}	Rare and minor
ω-Hydroxy acids	$CH_2(OH)(CH_2)_nCOOH$	C_{10} to C_{34}	Uncommon; found in suberin-associated wax
Diols			
α,ω-Diols	$CH_2(OH)(CH_2)_nCH_2OH$	C_{20} to C_{32}	Probably occur as esters (see diester above), not very common in waxes
α,β-Diols	$CH_2(OH)CH(OH)(CH_2)_nCH_3$	C_{14} to C_{32}	Rare and minor
Terpenes and steroids	Ursolic acid, oleanolic acid, betulin, β-sitosterol		Probably common in suberin-associated wax

[a] From Kolattukudy (1980b).

In contrast to cuticular waxes, suberin-associated waxes have not been examined thoroughly. An analysis of the wax from the periderm of several underground storage organs indicates that suberin-associated waxes contain the same classes of components as cuticular waxes but that there may be some characteristic differences (Espelie *et al.*, 1980a). The hydrocarbons in suberin wax have a larger percentage of even-chain-length homologs, a shorter average chain length, and a broader distribution of homologs. These differences were also seen in a study of hydrocarbons from the bark and leaf of several trees (Fig. 7). Fatty acids may comprise a higher proportion of suberin-associated waxes (10–50%) than is usually found in cuticular wax (Espelie *et al.*, 1980a). In the two comparative analyses reported, the chain lengths of major fatty acids and fatty alcohols are shorter in suberin-associated waxes (e.g., C_{22}) than in a cuticular wax (C_{26} and C_{28}) from the same plant (Espelie *et al.*, 1982; Yatsu *et al.*, 1983). In both of these cases, *Agave* idioblasts and green cotton fibers, C_{22} was a major homolog in wax components and in the polymer. Similarly, ω-hydroxy and dicarboxylic acids, characteristic components of the suberin

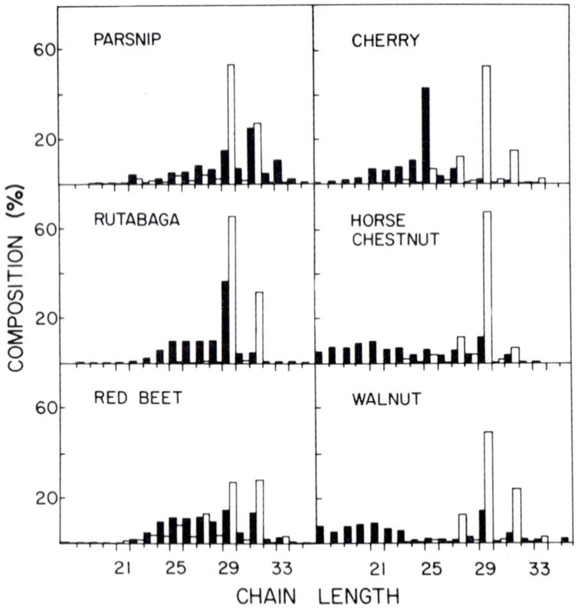

Fig. 7. Chain-length distributions of the alkane hydrocarbons present in the wax from periderm (solid bars) and from the cuticle of leaves (open bars). For each chain length, the periderm component is on the left and the cuticular component is on the right. The periderm wax was extracted from underground storage organs of the vegetables (Espelie *et al.*, 1980a) and from the bark of the trees (Streibl *et al.*, 1978).

polymer, were also found in the free form in suberin-associated waxes (Loveland and Laver, 1972a,b; Espelie *et al.*, 1980a). Another example of the occurrence of possible precursors of suberin in wax are the ferulic acid esters of long-chain (C_{18} to C_{28}) alcohols found in Douglas fir bark and potato tuber periderm (Adamovics *et al.* 1977). A wider variety of waxes needs to be studied before any firm conclusions can be drawn about the nature of suberin-associated waxes.

E. Function of Cutin, Suberin, and Waxes

The primary function of cuticular layers and suberized cell walls is to serve as a physical barrier to prevent water loss. Diffusion experiments with isolated cuticular layers showed that the waxes associated with cutin provided the major diffusion barrier to moisture. Extraction of the waxes from cuticular layers increased the water permeability by a factor of 300 to 2000 (Schönherr, 1976, 1982; Schönherr and Mérida, 1981; Schmidt *et al.*, 1981). Similar experiments demonstrated that wax is the major barrier in restricting gas diffusion across cuticular layers (Lendzian, 1982). Suberin also serves as an efficient barrier to water loss, but its location within the cell wall prohibits diffusion resistance studies on isolated polymer. Extraction of wax from suberin layers does, however, cause decreased diffusion resistance (Schönherr and Ziegler, 1980; Espelie *et al.*, 1980a). On the basis of the observation that specific inhibition of wax biosynthesis prevented development of diffusion resistance without affecting the synthesis of the suberin polymer in the periderm of wound-healing potato tuber tissue, it was concluded that waxes constituted the major diffusion barrier of suberin (Soliday *et al.*, 1979). This conclusion brought into question the report that waxes did not play a major role in providing diffusion resistance in *Betula* periderm (Schönherr and Ziegler, 1980). However, a more recent report from the same laboratory (Vogt *et al.*, 1983) has confirmed the conclusions of Soliday *et al.* (1979). It is possible that the relative contribution of waxes to the overall diffusion resistance of suberized walls varies with factors which are not presently known.

Suberin serves as a barrier to solute movement within the plant, especially in the roots (Torrey and Clarkson, 1975). Suberin also functions as a diffusion barrier to seal off certain regions of the plant at certain developmental stages (Zee and O'Brien, 1970; Espelie *et al.*, 1980b; Fussel and Dwarte, 1980; Oparka and Gates, 1982). Suberin also is deposited in the walls of idioblasts and thus serves to seal off crystals of calcium oxalate, probably as a disposal mechanism (Wattendorff, 1976a,b, 1980). In a similar fashion, an internal suberin layer is laid down to seal off resin-containing trichomes in some plants (Thomson *et al.*, 1979).

Table V

Effect of Mg^{2+} Concentration in the Growth Medium on the Composition of α,ω-Diols Generated by $LiAlD_4$ Treatment of Roots of *Zea mays*[a]

Mg^{2+} concentration (m*M*)	Diols (μg/g dry tissue)					
	C_{16}	C_{18}	C_{22}	C_{24}	C_{26}	Total
0.03	364.1	216.7	2.9	12.8	5.8	602.3
0.06	243.6	95.0	0.1	26.3	0.1	364.1
2.0	165.2	4.1	—	0.9	—	170.2

[a] From Pozuelo *et al.* (1984).

Both cutin and suberin serve as barriers to the entry of pathogens into plants. Although both polymers are susceptible to degradation by fungal enzymes (Kolattukudy and Köller, 1983), there is some evidence that fungal attack may result in the release of antifungal components (such as phenolics) from cutin or suberin polymers (Grambow, 1978; Grambow and Grambow, 1978). Constituents in the wax associated with either polymer may also affect the growth and differentiation of fungi and thereby serve a protective function in the plant (Blakeman and Sztejnberg, 1973; Wynn, 1976; Harmon *et al.*, 1980).

There are many indications that suberization is one response that the plant utilizes in an effort to cope with various stress situations. Trees respond to wound healing by laying down a series of "barrier zones" (Shigo and Marx, 1977), some of which have been described as suberized layers that function effectively to stop the spread of fungal infection (Pearce and Rutherford, 1981; Tippet *et al.*, 1982, 1983). There are numerous reports of "wound lignin" forming in response to injury (Vance *et al.*, 1980), but in many instances it seems quite probable that these protective polymers are actually suberin (Kolattukudy, 1981). The deposition of suberin during the wound healing of underground storage organs, leaves, and fruits has been well documented, and this suberization results in the restoration of diffusion resistance (Kolattukudy and Dean, 1974; Dean and Kolattukudy, 1976). Mechanical stress in the form of physical impedance to root growth led to increased suberization in the endodermal cell walls of barley (Wilson and Robards, 1978). Stress caused by mineral deficiency can also lead to changes in suberization. For example, deficiency of magnesium led to increased suberization in both the hypodermal and endodermal cells of the roots of *Zea mays* (Pozuelo *et al.*, 1984) (Table V), and iron deficiency caused a dramatic decrease in suberization of the roots of *Phaeseolus* (P. C. Sijmons, H. F. Bienfait, K. E. Espelie, and P. E. Kolattukudy, manuscript in preparation). Cold stress can also lead to increased deposition of aliphatic polymeric layers. Electron-microscopic and chemical examination showed that cold-hardened rye plants have an increased amount of aliphatic polyester-containing layers in the epidermis and in the mestome sheaths compared to nonhardened plants. (Griffith *et al.*, 1981, 1985). Cork cells in tree bark serve as an efficient insulation (Esau, 1977).

IV. Biosynthesis of Cutin

Biosynthetic pathways for the major monomers of cutin have been established by (a) the incorporation of radiolabeled precursors into the predicted components of cutin in tissue slices, (b) incorporation of postulated intermediates into the subsequent members of the biochemical pathway, and (c) demonstrating each postulated biosynthetic reaction in cell-free preparations. Although some express concern that experiments with tissue slices and cell-free preparations involve wounding of the tissue (Holloway, 1982b; Lendzian and Schönherr, 1983), the above approaches are generally considered to be reliable methods for establishing biochemical pathways.

A. *Biosynthesis of the C_{16} Family of Cutin Acids*

1. Pathway for Biosynthesis of C_{16} Monomers

The structural relationship among the three major components of the C_{16} family of cutin acids suggested that hexadecanoic acid is hydroxylated first at C-16 and then at either C-10, C-9, C-8, or C-7. With the finding that labeled hexadecanoic acid was readily incorporated into the three major C_{16} monomers of the cutin covering rapidly expanding *V. faba* leaves, it became possible to examine the biochemical pathway for cutin biosynthesis (Kolattukudy, 1970a,b). Synthetic 16-hydroxy[G-^{3}H]hexadecanoic acid and biosynthetically prepared 16-hydroxy[1-^{14}C]hexadecanoic acid were incorporated without degradation into the dihydroxyhexadecanoic acid of cutin by *V. faba* leaf disks (Kolattukudy and Walton, 1972). Since neither [10-^{14}C]palmitoleic nor [10-^{14}C]palmitelaidic acids were incorporated into cutin, the possibility of a dehydrogenation/hydration mechanism for the introduction of the midchain hydroxyl moiety was ruled out. Direct hydroxylation was suggested when it was shown that one of the

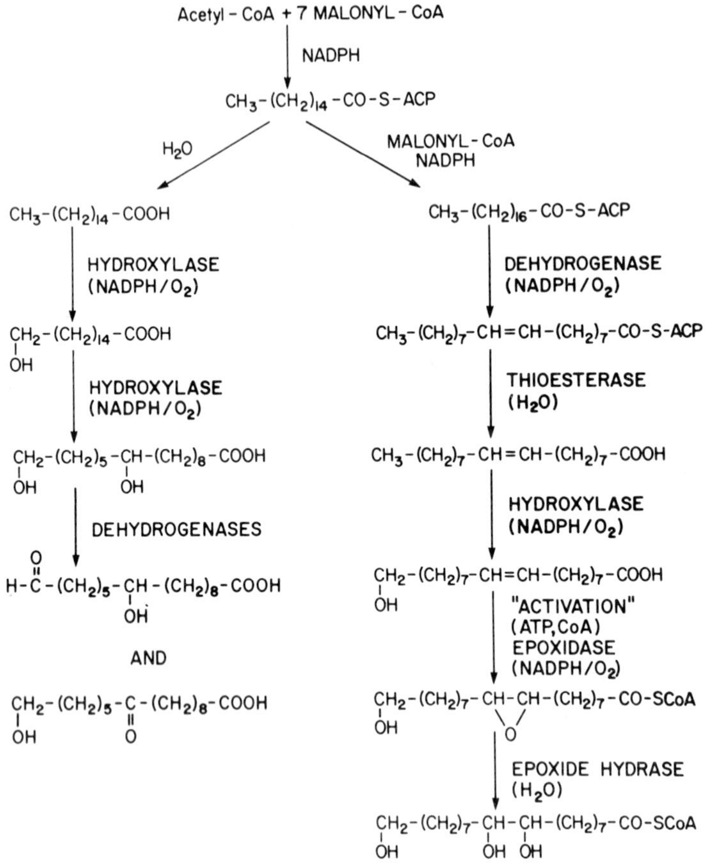

Fig. 8. Biosynthetic pathways for the major cutin monomers.

four hydrogen atoms at C-9 and C-10 of [10-^{14}C, 9,10-^{3}H]hexadecanoic acid was lost during conversion of this acid into dihydroxyhexadecanoic acid. On the basis of the results of such radiolabeling experiments, the pathway shown in Fig. 8 was proposed for the biosynthesis of the C_{16} family of cutin acids. Subsequently, each of the postulated reactions was demonstrated in cell-free preparations.

2. ω-Hydroxylation of Fatty Acids

An endoplasmic reticulum fraction from embryonic shoots of *V. faba* catalyzed ω-hydroxylation of hexadecanoic acid with NADPH and oxygen as the required cofactors (Soliday and Kolattukudy, 1977). Free fatty acid appeared to be the true substrate for this hydroxylation, as all attempts to hydroxylate the CoA ester of this acid failed. Hydroxylation rates were maximal at pH 8, and hexadecanoic acid gave a sigmoidal substrate saturation pattern. The microsomal preparations catalyzed ω-hydroxylation of other fatty acids, notably octadecenoic acid. Mixed-function oxidase inhibitors, such as metal ion chelators, sodium azide, and thiol-directed reagents, inhibited the ω-hydroxylation catalyzed by the microsomal preparations. The involvement of a cytochrome (cyt) *P*-450-type protein in the ω-hydroxylation was suggested by the inhibition of this reaction by CO. However, this ω-hydroxylase showed an unusually high sensitivity to CO, and light at 420–460 nm failed to reverse the inhibition caused by CO. Thus, if a cyt *P*-450-type protein is involved in this system, it must have an unusually high affinity for CO, and possibly other unique properties.

3. Midchain Hydroxylation of ω-Hydroxyhexadecanoic Acid

Midchain hydroxylation of ω-hydroxyhexadecanoic acid was demonstrated in a cell-free preparation from *V. faba* epidermis (Walton and Kolattukudy, 1972a). After a reverse-phase high-performance liquid chromatographic technique was developed to assay for midchain hydroxylation, it was shown that the C-10 hydroxylase was located in the endoplasmic reticulum fraction isolated from embryonic shoots of *V. faba* (Soliday and Kolattukudy, 1978). This midchain hydroxylation required O_2 and NADPH and showed an optimal pH of 7.5, and the apparent K_m for ω-hydroxyhexadecanoic acid was 50 μM. This reaction was inhibited by NaN_3 and by metal ion chelators, as well as by thiol-directed reagents. Carbon monoxide inhibited this midchain hydroxylation, and, unlike the ω-hydroxylation catalyzed by the same preparation, the inhibition of midchain hydroxylation caused by CO was photoreversible. Therefore, a typical cyt *P*-450 appears to be involved in this hydroxylation.

B. *Biosynthesis of the C_{18} Family of Cutin Acids*

1. Pathway for Biosynthesis of C_{18} Monomers

Exogenous labeled precursors administered to skin slices from developing apple fruits were incorporated into the C_{18} family of cutin acids (Kolattukudy *et al.*, 1971). [1-^{14}C]Acetate was incorporated into both C_{16}

and C_{18} monomers, and [1-^{14}C]hexadecanoic acid was incorporated into only the C_{16} family of monomers. [1-^{14}C]Octadecanoic acid was not incorporated into any monomers with midchain substitution, suggesting that the direct hydroxylation pathway is not utilized in the C_{18} family. Thus octadecenoic acid was suggested to be the precursor of the C_{18} cutin monomers, and, in fact, [1-^{14}C]octadecenoic acid was readily converted to 18-hydroxyoctadecenoic acid, 18-hydroxy-9,10-epoxyoctadecenoic acid, and 9,10,18-trihydroxyoctadecanoic acid in several plant tissues (Kolattukudy *et al.*, 1973b). Exogenous 18-hydroxy[18-^{3}H]octadecenoic acid was converted to 18-hydroxy-9,10-epoxyoctadecanoic acid and 9,10,18-trihydroxyoctadecanoic acid in several tissues (Croteau and Kolattukudy, 1974b). Similarly synthetic 18-hydroxy-9,10-epoxy[18-^{3}H]octadecanoic acid was directly converted to 9,10,18-trihydroxyoctadecanoic acid in leaf and fruit skin disks of apple. These results suggested a pathway that involves an ω-hydroxylation, followed by epoxidation of the C-9 double bond and a subsequent epoxide hydration. A similar sequence of reactions for linoleic acid and linolenic acid was also demonstrated in apple skin slices (Kolattukudy *et al.*, 1973b). Even though in most tissues the epoxidation and hydration is specific for and limited to the C-9 double bond, at least in some plants such as *Rosmarinus officinalis,* subsequent to the epoxidation and hydration of the C-9 double bond, the C-12 double bond can also undergo epoxidation followed by hydration, resulting in a pentahydroxy acid (Croteau and Kolattukudy, 1974a). In this tissue, exogenous [1-^{14}C]linoleic acid was converted to 9,10,12,13,18-pentahydroxy C_{18} acid, and all of the expected intermediates were also labeled.

2. Enzymes Involved in the Biosynthesis of C_{18} Monomers

The epoxidation enzyme was discovered in cell-free preparations from spinach, a plant in which cutin is composed of the epoxy acid as a major component. A 3000-*g* particulate preparation from young spinach leaves catalyzed the conversion of 18-hydroxy[18-^{3}H]octadecenoic acid to 18-hydroxy-9,10-epoxyoctadecanoic acid (Croteau and Kolattukudy, 1975a). This reaction required O_2 and NADPH as cofactors and occurred optimally at pH 9. In addition, ATP and CoA were also required, suggesting that activation of the carboxyl group seven methylene groups away from the epoxidation site was required for this reaction. The apparent K_m for 18-hydroxyoctadecenoic acid was 75 μM, and the enzyme showed a fairly stringent substrate specificity in that the cis-Δ^9 and a free hydroxyl group at C-18 were required. The epoxidation was inhibited by chelators, and this inhibition was reversed by Fe^{2+}. Photoreversible inhibition of the epoxidation by CO strongly suggested that a cyt *P*-450-type protein was involved in the enzymatic epoxidation.

Enzymic epoxide hydration was catalyzed by a 3000-*g* particulate fraction from the skin of young apple fruit. This preparation catalyzed the conversion of labeled 18-hydroxy-*cis*-9,10-epoxyoctadecanoic acid to *threo*-9,10,18-trihydroxyoctadecanoic acid (Croteau and Kolattukudy, 1975b). This epoxide hydrase showed a pH optimum of 6.5, required no cofactors, and was located exclusively in the 3000-*g* particulate fraction containing fragments of cuticular membranes and cell walls. This biosynthetic epoxide hydrase showed a fairly stringent substrate specificity, unlike the catabolic epoxide hydrases found in the microsomes of animal tissues (Oesch *et al.*, 1971; Lu *et al.*, 1975). For example, styrene oxide, a commonly used substrate for the catabolic epoxide hydrase, was not hydrated by the plant enzyme, and any modification of the functional groups in the 18-hydroxy-9,10-epoxyoctadecanoic acid rendered it a poor substrate for the plant enzyme. Trichloropropene oxide, which is a potent inhibitor of the animal enzyme, was a poor inhibitor of the plant enzyme. Thiol-directed reagents strongly inhibited the biosynthetic epoxide hydrase from plants. All these differences are not necessarily associated with the biosynthetic function, as a soluble catabolic epoxide hydrase from a fungus also showed properties similar to those of the higher plant enzymes, except for a very much higher pH optimum (Kolattukudy and Brown, 1975).

C. *Biosynthesis of Cutin from Monomers*

Labeled monomer acids were incorporated into an insoluble polymer by a 3000-*g* particulate fraction isolated from the epidermis of rapidly expanding *V. faba* leaves (Croteau and Kolattukudy, 1973). This incorporation required ATP and CoA, suggesting that cutin biosynthesis involves a transfer of a hydroxyacyl group from CoA to a free hydroxyl group of cutin. Although two pH optima near 7.0 and 8.5 were observed when free acid was used as the substrate, a single optimum near 8.5 was observed with hexadecanoyl-CoA. The transacylation occurred optimally at pH 8.5 (Croteau and Kolattukudy, 1974c). Apparent K_m values with the free acids were 20, 67, and 100 μM for hexadecanoic acid, 16-hydroxyhexadecanoic acid, and 10,16-dihydroxyhexadecanoic acid, respectively. The enzyme preparation from *V. faba* showed a preference for the C_{16} family of acids, reflecting the composition of *V. faba* cutin. However, fatty acids from C_{10} to C_{18} and C_{18} hydroxy acids were also incorporated into cutin by the *V. faba* enzyme preparation.

The enzyme(s) catalyzing the esterification process was dissociated from endogenous primer by mild ultrasonic treatment in order to characterize the nature of the endogenous polymer to which the monomers were esterified (Croteau and Kolattukudy, 1974c). The soluble preparation thus obtained required exogenous cutin as a primer and the most active primer was found to be cutin isolated from very young leaves of *V. faba*. The primer-dependent synthesis of cutin from monomers showed the same cofactor requirements and substrate specificity as those already noted for the particulate preparation. Experiments with chemically modified primers showed that the carboxyl groups of the incoming monomers were esterified to free hydroxyl groups present in the monomer. Thus, chemical or enzymatic treatments that increased the number of hydroxyl groups present in cutin or opened the structural matrix of the polymer increased the priming efficiency of cutin. All the available evidence strongly suggests that hydroxyacyl moieties, which are possibly generated in close proximity to the cuticular matrix or transported to the cuticle, are transferred from their CoA esters to the growing polymer by a transacylase(s) which is associated with the cuticular matrix.

V. Biosynthesis of Suberin

A. *Pathway for Biosynthesis of Suberin Aliphatic Monomers*

Wound-healing tissue slices of potato were chosen as a system for a biosynthetic study of suberin, in part because the composition of the aliphatic components of the suberin deposited in the wound periderm was found to be identical to that of the natural periderm (Kolattukudy and Dean, 1974). The time course of the appearance of the aliphatic components of suberin in the wound-healing potato disks correlated with the time course of the incorporation of [1-^{14}C]octadecenoic acid into 18-hydroxyoctadecenoic acid and the corresponding dicarboxylic acid, which are the two major aliphatic components of potato suberin (Dean and Kolattukudy, 1977). The extent of incorporation of exogenous [1-^{14}C]acetate into the other characteristic suberin components, very long chain fatty acids and alcohols, was also consistent with the time course of suberization. On the basis of these results and what was known about the biosynthesis of similar compounds in other systems, the pathway shown in Fig. 9 was proposed for the biosynthesis of the aliphatic monomers of suberin (Kolattukudy, 1977a). The unique reaction in this pathway is the

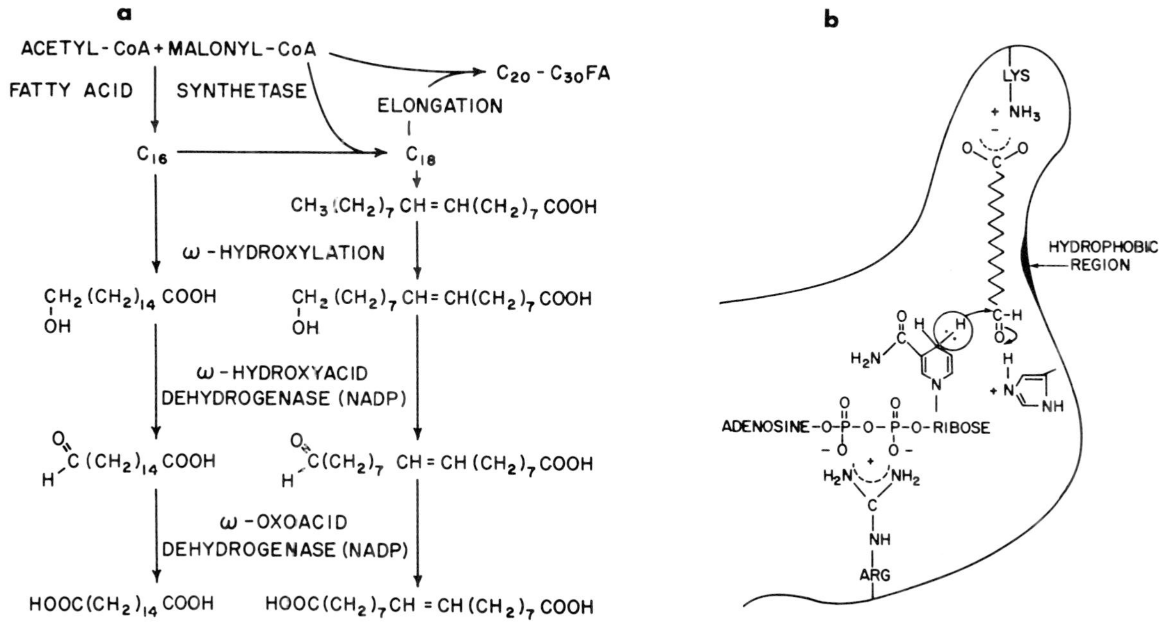

Fig. 9. **(a)** Biosynthetic pathways for the major aliphatic components of suberin. **(b)** A representation of the active site of ω-hydroxy acid dehydrogenase involved in the synthesis of the dicarboxylic acids characteristic of suberin. From Agrawal and Kolattukudy (1978b).

oxidation of ω-hydroxy acid to dicarboxylic acid, a monomer quite characteristic of suberin (Kolattukudy, 1978).

B. *Enzymes Involved in the Biosynthesis of the Aliphatic Components of Suberin*

The oxidation of the ω-hydroxyl moiety to yield a dicarboxylic acid is a key step in suberin biosynthesis. This oxidation has been demonstrated in cell-free preparations from the excised epidermis of *V. faba* leaves (Kolattukudy *et al.*, 1975b). This dehydrogenase showed maximal activity near pH 8, had a high degree of preference for NADP, and was located mainly in the 100,000-*g* supernatant. Modification of the substrate, by esterification of the carboxyl group, replacement of the carboxyl group by a methyl group, or introduction of another hydroxyl group at C-10, rendered it a poor substrate. Thiol-directed reagents strongly inhibited the oxidation of the ω-hydroxy group. The ω-oxo acid, the suspected intermediate in the conversion of the ω-hydroxy acid to the dicarboxylic acid, could be trapped as the dinitrophenylhydrazone, and the enzyme preparation also catalyzed oxidation of synthetic ω-oxo acid to the dicarboxylic acid.

Since oxidation of ω-hydroxy acid to the dicarboxylic acid is a rather unique process involved in suberization, the extracts of wound-healing potato slices were examined for ω-hydroxy acid dehydrogenase activity. These extracts catalyzed the oxidation of ω-hydroxyhexadecanoic acid to the ω-oxo acid, which was in turn converted to C_{16} dicarboxylic acid (Agrawal and Kolattukudy, 1977). The two oxidation steps were catalyzed by distinct dehydrogenases, which were separable by gel filtration (Agrawal and Kolattukudy, 1978b). The ω-hydroxy acid dehydrogenase but not the ω-oxo acid dehydrogenase was induced during suberization. The ω-hydroxy acid dehydrogenase, which appears to be induced for suberization, was purified to homogeneity from acetone powder extracts of wound-healing potato slices. The purified enzyme was found to be a dimer with a protomer molecular weight of 30,000. The pH optima for the reaction in the forward (formation of oxo acid) and reverse directions were 9.5 and 8.5, respectively, and values of V in forward and reverse directions were 140 and 3200 nmol min^{-1} mg^{-1}, respectively. Apparent K_m values for NADP, 16-hydroxyhexadecanoic acid, NADPH, and 16-oxohexadecanoic acid were 100, 20, 5, and 7 μM, respectively. The equilibrium constant of the reaction at pH 9.5 and 30°C was found to be 1.4×10^{-9} M. Surprisingly, hydride from the A side of NADPH was transferred to 16-oxohexadecanoic acid by the purified enzyme at nearly the same rate as from the B side.

Initial velocity and product inhibition studies with the purified ω-hydroxy fatty acid dehydrogenase indicated an ordered sequential mechanism where NADPH was added first, followed by the ω-oxo acid, and the hydroxy acid was released before NADP (Agrawal and Kolattukudy, 1978a). Substrate inhibition by NADPH was observed at concentrations of >0.2 mM, and the inhibitory NADPH molecule appeared to compete with the oxo acid, indicating that it formed a dead-end complex with the E-NADPH form of the enzyme. Based on chemical modification studies with phenylglyoxal, an arginine residue was suggested to be involved in the binding of NADPH. Reversible inhibition of the reaction by a Schiff's base formation with pyridoxal 5′-phosphate and pH dependence of the K_m of the oxo acid but not of hexadecanal strongly suggested that a lysine residue was involved in binding the distal carboxyl group of the oxo acid. Inactivation of the dehydrogenase by diethylpyrocarbonate and reversal of this inhibition by hydroxylamine suggested the involvement of the imidazole group of a histidine residue in the reaction. This suggestion was supported by the fact that the enzyme was protected from inactivation by NADPH but not by AMP. With C_3 to C_8 alkanals as model substrates, the K_m for the substrate decreased drastically from 7000 to 90 μM as the chain length increased. Additional increases in chain length from C_8 to C_{20} caused only a small decrease in K_m, and homologs $>C_{20}$ gave extremely low rates. These results suggested that the substrate binding involves a hydrophobic pocket that cannot accommodate aliphatic chains $>C_{20}$. This preference for C_{18} and shorter chain lengths is consistent with the fact that the longer chain dicarboxylic acids are not major components of potato suberin even though the corresponding ω-hydroxy acids may be present in high proportions (Kolattukudy and Agrawal, 1974). This enzyme specificity might be more generally applicable to the ω-hydroxy fatty acid dehydrogenases involved in suberin biosynthesis, as the presence of shorter homologs in the dicarboxylic acids noted in potato suberin appears to be common in the suberin from other plants (Kolattukudy and Dean, 1974; Espelie *et al.*, 1982; Yatsu *et al.*, 1983; Holloway, 1983). Based on the results just discussed, the active site of the dehydrogenase was proposed to be as indicated in Fig. 9.

C. *Biosynthesis of the Aromatic Components of Suberin*

The biosynthesis of the aromatic components of suberin has not been studied in detail although phenolic metabolism in wound-healing plant tissues has been studied extensively (Rhodes and Wooltorton, 1978). Although these studies deal primarily with the metabolic changes that occur

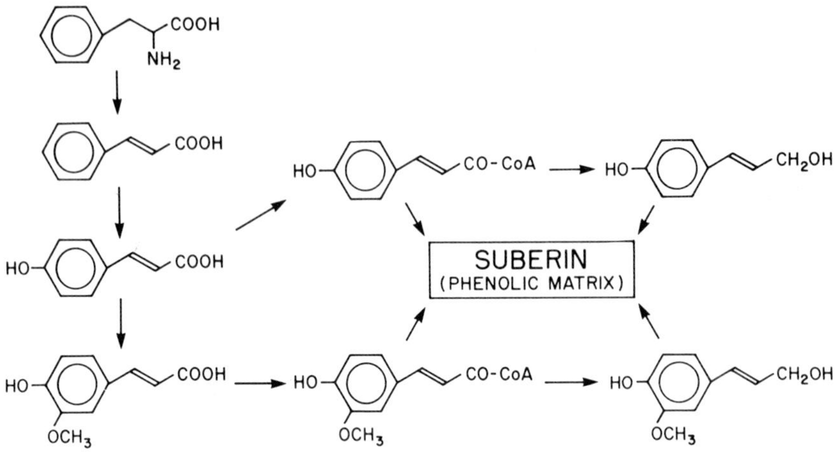

Fig. 10. Biosynthetic pathway for the aromatic components of suberin. Some phenolic acids and alcohols could be esterified to fatty alcohol or fatty acids prior to incorporation into the polymer by peroxidases.

24–48 hr after wounding, well before the onset of suberization, it is probable that the phenolic metabolism observed during these early periods provides the monomers utilized for suberization. The sequence of reactions that may be involved in this process is depicted in Fig. 10. The individual reactions involved in this process have been elucidated with cell-free preparations and in some cases purified enzymes (Gross, 1977; Rhodes and Wooltorton, 1978; Zenk, 1979; Hanson and Havir, 1979) (see Chapter 10).

The deposition of aromatic components into the phenolic matrix of suberin in wound-healing potato slices was measured by determining the amount of *p*-hydroxybenzaldehyde and vanillin generated by alkaline nitrobenzene oxidation (Cottle and Kolattukudy, 1982a). The deposition of such phenolics into the polymeric material showed a lag period of about 3 days, followed by several days of rapidly increasing phenolic deposition and subsequent cessation. Both L-[U-^{14}C]phenylalanine and [U-^{14}C]cinnamic acid generated labeled insoluble polymeric material in suberizing potato disks, although the latter acid was far more efficient than the former (Cottle and Kolattukudy, 1982a). Nitrobenzene oxidation of the polymeric material derived from labeled cinnamic acid generated labeled vanillin and *p*-hydroxybenzaldehyde (see Fig. 6). The incorporation of the aromatic precursors into the polymeric material in suberizing potato tissue, the deposition of aliphatic monomers into the polymeric material, the deposition of waxes characteristic of suberin-associated waxes, and the development of diffusion resistance all showed very similar time courses (Fig. 11) (Cottle and Kolattukudy, 1982b).

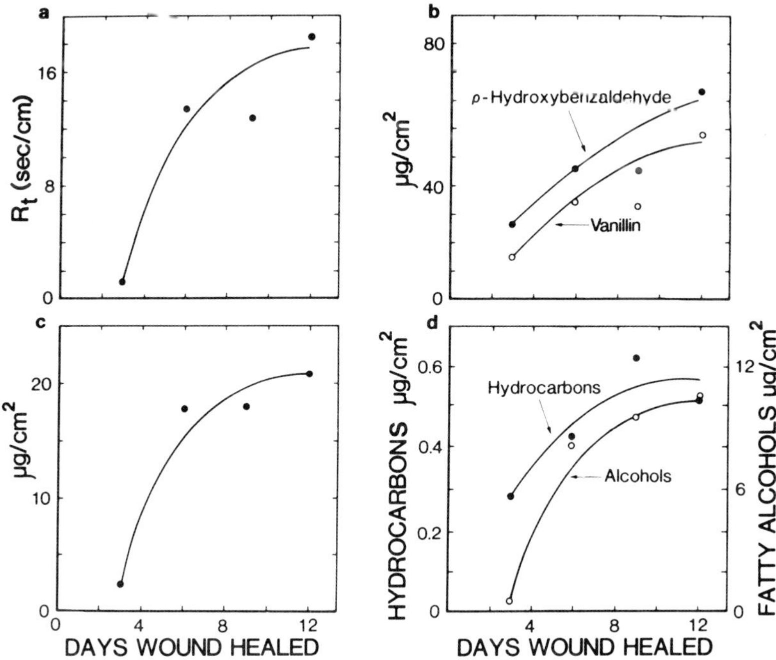

Fig. 11. Time course of development of diffusion resistance and deposition of suberin-associated components during the wound healing of potato tuber tissue slices. (a) Diffusion resistance. (b) Phenolics measured as aromatic aldehydes generated by alkaline nitrobenzene oxidation. (c) Polymeric aliphatics measured as diols generated by $LiAlH_4$ reduction. (d) Waxes recovered by solvent extraction. From W. Cottle and P. E. Kolattukudy (unpublished results).

D. Biosynthesis of Suberin from Monomers

The biosynthesis of suberin from monomers has not been studied in cell-free preparations. The aromatic monomers may undergo polymerization in a manner analogous to that suggested for lignin synthesis (Gross, 1977; Grisebach, 1981) (see Chapter 7). Peroxidase present in cell walls can catalyze polymerization of phenolic monomers and covalent bond formation between cell walls and the phenolic polymer (Gross, 1977; Whitmore, 1978; Grisebach, 1981). The following observations suggest that the phenolic matrix of suberin may be formed by a similar process involving peroxidases. During the wound healing of potato tuber, a specific peroxidase was found to be induced, and it was localized in the suberizing cells (Borchert, 1974). Both the time course and the spatial distribution of wound-induced peroxidase activity in this tissue correlated with that of suberization (Borchert, 1978). In potato tissue culture, the time course of suberization induced by abscisic acid was correlated with that of the induction of the suberization-linked isoperoxidase (Cottle and Kolattukudy, 1982b). Peroxidase activity in roots appears to be localized in the epidermal and endodermal cell walls (Mueller and Beckman, 1978; Smith and O'Brien, 1979), which are known to be areas which undergo suberization. The polymeric material generated by potato parenchyma *in vitro* from *p*-coumaric and ferulic acids appeared to be highly condensed, as it released only small amounts of aromatic aldehydes upon nitrobenzene oxidation (Bland and Logan, 1965). The suberization-linked peroxidase might generate a similar high degree of condensation in the phenolic matrix of suberin, and thus nitrobenzene oxidation of potato suberin also gives relatively low yields of *p*-hydroxybenzaldehyde and vanillin (Cottle and Kolattukudy, 1982a).

The H_2O_2 needed for the peroxidase-catalyzed polymerization process could be provided in a manner similar to that suggested for lignification (Gross, 1977). A cell wall-bound malate dehydrogenase (Elstner and Heupel, 1976; Gross and Janse, 1977; Gross *et al.*, 1977) might produce sufficient amounts of NADH to allow generation of adequate amounts of H_2O_2 by the wall-bound peroxidase (see Chapter 7, Section III).

The aliphatic components of suberin are presumably esterified to the functional groups in the phenolic polymer. In polymers that have a large proportion of dicarboxylic acids and a small amount of midchain hydroxylated acids, there would not be enough hydroxyl moieties for esterification of all of the carboxyl groups in the aliphatic components, and therefore the phenolic moieties are also probably involved in esterification of the aliphatics. Activated ω-hydroxy acids and dicarboxylic acids are probably involved in the esterification process in a manner analogous to that shown for cutin biosynthesis. It was suggested (Kolattukudy, 1981) that fatty alcohols might be incorporated into suberin by the peroxidase-catalyzed polymerization of phenypropanoic acid esters of the alcohols, such as the feruloyl esters of C_{18} to C_{28} alcohols found in jack pine and Douglas fir bark and potato tuber periderm (Rowe *et al.*, 1969; Fang, 1974; Adamovics *et al.*, 1977). However, no direct evidence is yet available to support this suggestion.

E. *Regulation of Suberization*

The fact that wounding causes suberization even on tissue normally protected by cutin (Dean and Kolattukudy, 1976) indicates that this process may be a general response to wounding in plants. The finding that thorough washing of the wound surface of potato tuber inhibited wound-healing suggested that some chemical produced by the wound might initiate the suberization process (Soliday *et al.*, 1978). Abscisic acid (ABA) was removed by this washing, and addition of ABA to the washed tissue partially restored the suberizing ability of the tissue, suggesting that this plant hormone might trigger suberization. In support of this conclusion was the finding that in potato tissue culture suberization was induced by ABA in a dose-dependent manner. This ABA-induced suberization was demonstrated by the observation that ABA treatment of the tissue culture caused a threefold increase in the deposition of polymeric aliphatics, a fourfold increase in polymeric aromatics, and a fivefold increase in suberin-associated waxes (Cottle and Kolattukudy, 1982b). The role of ABA in suberization is presumably indirect, because suberization could be inhibited by washing the tissue even 3 days after wounding, although ABA was not removed by this washing (Soliday *et al.*, 1978). Experiments with actinomycin D and cycloheximide suggested that the transcriptional and translational processes directly related to the synthesis of the aliphatic components of suberin occurred between 72 and 96 hr after wounding (Agrawal and Kolattukudy, 1977). Washing of the tissue any time prior to this period (72 hr) inhibited suberization. Thus it appears that ABA, generated during the first day after wounding, triggers a process that results in the production of the suberization-inducing factor that causes the induction of the enzymes involved in suberization.

The mechanism of induction of suberization in potato tissue culture by ABA was studied using two key enzymes uniquely involved in suberization: ω-hydroxy fatty acid dehydrogenase and the suberization-linked isoperoxidase. Even though the former, which is not present in fresh potato tuber tissue, is induced during wound-healing of tissue slices, the tissue culture already had a fairly high level of this enzyme, and ABA treatment resulted in only a relatively small increase in its activity. The isoperoxidase, on the other hand, showed a dramatic increase as a result of ABA treatment of the tissue culture (Cottle and Kolattukudy, 1982b). This isoperoxidase was isolated and purified from both wound-healing potato tissue slices and ABA-treated tissue culture. The enzyme from both sources appeared to be immunologically similar, if not identical (K. E. Espelie and P. E. Kolattukudy, unpublished results).

VI. Biosynthesis of Waxes

A. *Biosynthesis of Very Long Fatty Acids*

Plant waxes characteristically contain very long aliphatic chains (C_{20} to C_{36}) that are formed by chain elongation of fatty acids generated by fatty acid synthetase. Chain elongation was demonstrated in several plant tissues by incorporation of C_2 to C_{24} acids into long acids (Kolattukudy and Walton, 1973; Kolattukudy, 1980b). These long chain acids were also synthesized by cell-free preparations from different plants (Macey and Stumpf, 1968; Kolattukudy *et al.*, 1972; Lessire, 1973; Cassagne and Lessire, 1979). Chain elongation, localized in the microsomal fraction, was shown by these studies to require malonyl-CoA, NADPH, and an endogenous acyl moiety, and it was reported that the endogenous acyl moiety may be CoA ester rather than an acyl carrier protein derivative (Cassagne and Lessire, 1978). Variable effects of inhibitors and cofactors on chain elongation (Liu, 1972; Kolattukudy, 1975; Kolattukudy and Brown, 1975; Kolattukudy *et al.*, 1976; Cassagne and Lessire, 1978; von Wettstein-Knowles, 1979) and genetic evidence (von Wettstein-Knowles, 1974) suggest that different chain-elongating enzyme systems, each with separate specificities, are involved in the generation of the wide variety of chain length classes found in plant waxes.

B. *Biosynthesis of Fatty Alcohols*

The incorporation of exogenous C_2 to C_{20} radioactive fatty acids into very long fatty alcohols in plant tissues suggested that these alcohols are produced by the reduction of elongated fatty acids (Mazliak, 1963; Kolattukudy, 1966, 1980b; Cassagne, 1970). The first cell-free preparation capable of catalyzing the reduction of fatty acid to alcohol was obtained from *Euglena gracilis* (Kolattukudy, 1970c). Acyl-CoA reductase is located primarily in the microsomal membranes in *Euglena* (Khan and Kolattukudy, 1973). Since in the membranes acyl-CoA reduction is coupled to aldehyde reduction, free aldehydes are not found. However, acyl-CoA reductase activity extracted from an acetone powder of young *Brassica oleracea* leaves was fractionated into two components: one catalyzed NADPH-dependent reduction of the acyl-CoA to aldehyde, and the other fraction catalyzed an NADPH-dependent reduction of the aldehyde to alcohol (Kolattukudy, 1971). Such enzyme systems are probably present in plants such as *B. oleracea* that have free aldehydes in their wax.

C. *Biosynthesis of Wax Esters*

An acyl transfer mechanism is probably involved in the biosynthesis of wax esters (Kolattukudy, 1967). A partially purified protein fraction obtained from the acetone powder prepared from rapidly expanding leaves of *B. oleracea* catalyzed the esterification of fatty alcohols with acyl-CoA as the acyl donor. The extract also contained enzymes that catalyzed the

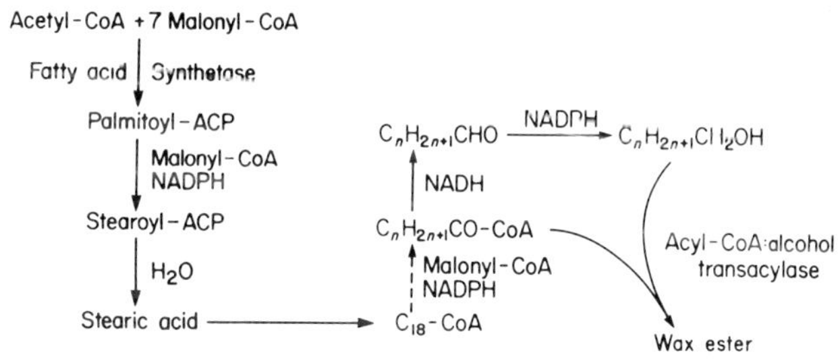

Fig. 12. Biosynthetic pathway for very long acids, aldehydes, alcohols, and wax esters.

esterification of fatty alcohols with free fatty acids and phospholipids as acyl donors. Nevertheless, acyl-CoA fatty alcohol transacylase probably is the physiologically significant route for wax ester synthesis. On the basis of current evidence, the biosynthesis of wax esters can be summarized as shown in Fig. 12.

D. Biosynthesis of Hydrocarbons and Derivatives

Head-to-head condensation between two fatty acids was ruled out as a possible mechanism for the biosynthesis of hydrocarbons on the basis of the results of incorporation experiments with specifically labeled precursors, as discussed elsewhere (Kolattukudy and Walton, 1973; Kolattukudy, 1975). Consequently, an elongation–decarboxylation mechanism was proposed for the biosynthesis of hydrocarbons (Kolattukudy, 1966, 1967). This hypothesis suggests that a C_{16} or C_{18} fatty acid is elongated to the appropriate chain length and then decarboxylated (Fig. 13). A large amount of evidence obtained from the experiments with higher plant tissues supports this hypothesis, as summarized elsewhere (Kolattukudy *et al.*, 1976; Kolattukudy, 1980b). Such evidence includes the direct conversion of very long exogenous acids into alkanes containing one less carbon than the precursor acid (Kolattukudy *et al.*, 1972; Cassagne and Lessire, 1974; Kolattukudy *et al.*, 1974). The demonstration that cell-free preparations from *Pisum sativum* catalyzed the conversion of [9,10,11-^{3}H]C_{32} acid into C_{31} and C_{30} alkanes was perhaps the most convincing evidence in support of the elongation–decarboxylation mechanism (Khan and Kolattukudy, 1974). This conversion required O_2 and ascorbate and was inhibited by such thiol compounds as dithioerythritol. More recent results suggest that the elongating enzyme system is localized in the cytoplasmic side of the endoplasmic reticulum, while the hydrocarbon-generating enzyme system is found in the luminal side of the membrane (Bognar *et al.*,

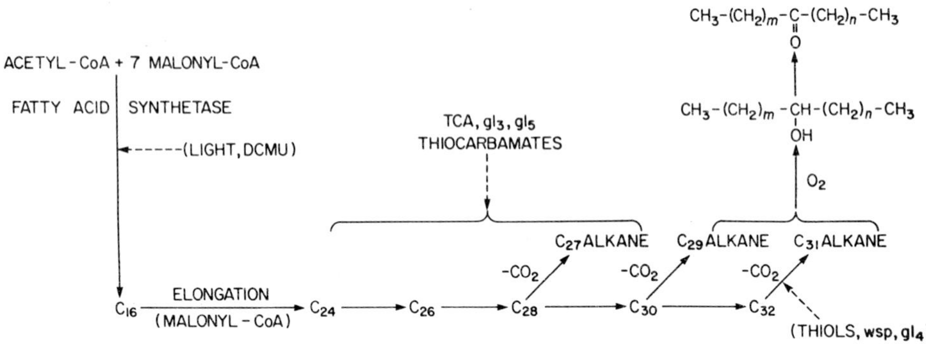

Fig. 13. A biosynthetic pathway for alkanes, secondary alcohols, and ketones. Steps that are probably affected by light, chemicals and mutations are indicated; gl_3, gl_4, and gl_5 are mutants of *B. oleracea,* and wsp is a mutant of *P. sativum.*

1984). The hydrocarbons produced in the lumen might then be excreted via vesicles fused with the plasma membrane.

The hydrocarbon-generating particulate enzyme preparation from *P. sativum* utilized C_{18}, C_{22}, C_{24}, and C_{32} fatty acids as substrates, suggesting that the specificity of the chain elongation process may determine the chain lengths of the alkanes generated by a specific organism. Odd-chain fatty acids were generated from the even-chain fatty acid substrates by the α-oxidation activity contained in the *P. sativum* microsomes, and the hydrocarbon-producing enzyme preferentially used these odd-chain products and thus generated alkanes with two carbon atoms less than the parent acid. The α-oxidation intermediate from which alkane originated had one carbon less than the parent acid. Time-course studies with particulate preparations from *P. sativum* suggested that the alkane-generative intermediate was aldehyde (Bognar *et al.,* 1984). In fact, fatty aldehyde was shown to be converted to the corresponding alkane by decarbonylation, a novel biochemical reaction (Cheesbrough and Kolattukudy, 1984). Thus the mechanism of biosynthesis of alkane is best depicted as

$$\text{RCOSCoA} \longrightarrow \text{R}\overset{\text{O}}{\overset{\|}{\text{C}}}\text{H} \longrightarrow \text{RH}$$

Hydroxylation of alkanes and subsequent oxidation of the resulting secondary alcohols generate the secondary alcohols and ketones, which are often found in waxes (Fig. 13). The experimental results supporting these conclusions have been previously summarized (Kolattukudy, 1975). Direct evidence for this hypothesis was obtained when exogenous alkane was converted to the secondary alcohol and when exogenous secondary alcohol was oxidized to the ketone in *B. oleracea* leaves (Kolattukudy and Liu, 1970; Kolattukudy *et al.*, 1973a).

E. *Biosynthesis of β-Diketones*

Modification of the elongation–decarboxylation mechanism probably results in the production of β-diketones (Fig. 14). This mechanism was supported by the observed distribution of radioactivity in the hentriacontan-14,16-dione generated from specifically labeled [^{14}C]acetate and by the fact that exogenous labeled C_{12}, C_{14}, and C_{16} (but not C_{18}) acids were incorporated into the aliphatic chains of the β-diketone moiety (von Wettstein-Knowles, 1979; Mikkelsen, 1979). Genetic evidence (Lundquist and von Wettstein-Knowles, 1982) and differential inhibition of the incorporation of labeled precursors into various classes of wax components are also consistent with the pathway shown in Fig. 14. β-Diketones are probably hydroxylated to yield hydroxy-β-diketones and alkan-2-ols, which are found in plant waxes containing β-diketones and are probably derived from the same β-keto intermediates as the β-diketones.

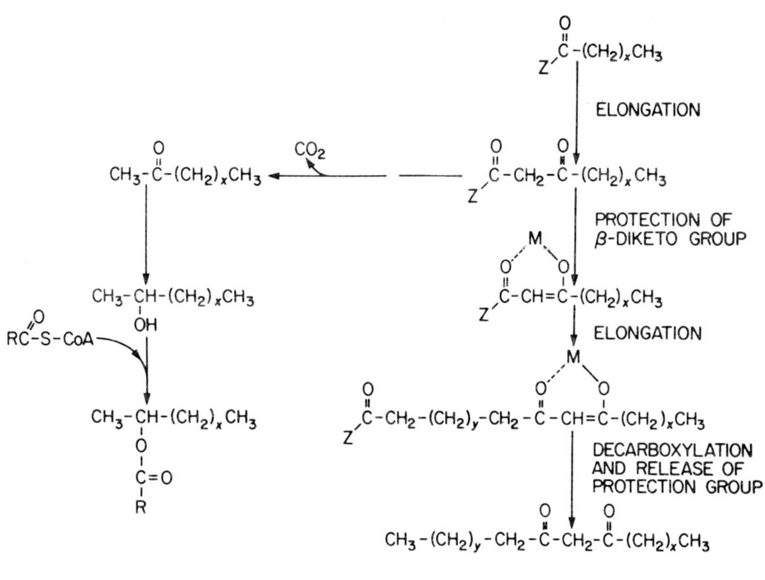

Fig. 14. Proposed pathways for the biosynthesis of β-diketones and esters of alkan-2-ols. Based partly on the proposal of Mikkelsen and von Wettstein-Knowles (1978).

VII. Conclusion

The nature of the waxes associated with the phytopolymers cutin and suberin has been examined using combined gas chromatography and mass spectrometry, and as a result, considerable progress has been made in an understanding of the chemistry of such waxes. The broad outlines of the biochemical pathways involved in the biosynthesis of the major components of plant waxes have become available during the past two decades. Although cutin and suberin have attracted much less attention than the familiar biopolymer lignin, the composition, biosynthesis, and biodegradation of cutin have been investigated using modern approaches. Much progress has been made in our understanding of these aspects of cutin, although much remains to be done.

The chemistry and biochemistry of suberin is poorly understood. Because of the insoluble nature and the complexity of structure of suberin, modern techniques such as ^{13}C NMR obtained on solid samples with magic angle spinning are needed to elucidate the nature of suberin. With such structural information, it will be possible to make further progress in the biochemistry of this polymer.

Acknowledgment

The work from the authors' laboratory was supported by grants GM-18278 from the U.S. Public Health Service and PCM-8306835 from the National Science Foundation.

References

Adamovics, J. A., Johnson, G., and Stermitz, F. R. (1977). *Phytochemistry* **16,** 1089–1090.
Agrawal, V. P., and Kolattukudy, P. E. (1977). *Plant Physiol.* **59,** 667–672.
Agrawal, V. P., and Kolattukudy, P. E. (1978a). *Arch. Biochem. Biophys.* **191,** 452–465.
Agrawal, V. P., and Kolattukudy, P. E. (1978b). *Arch. Biochem. Biophys.* **191,** 466–478.
Amelunxen, F., and Gronau, G. (1969). *Z. Pflanzenphysiol.* **60,** 156–168.
Baker, E. A. (1982). *In* "The Plant Cuticle" (D. F. Cutler, K. L. Aluin, and C. E. Price, eds.), pp. 139–165. Academic Press, New York.
Baker, E. A., and Procopiou, J. (1975). *J. Sci. Food Agric.* **26,** 1347–1352.
Björkman, A. (1956). *Sven. Papperstidn.* **59,** 477–485.
Blakeman, J. P., and Sztejnberg, A. (1973). *Physiol. Plant Pathol.* **3,** 269–278.
Bland, D. E., and Logan, A. F. (1965). *Biochem. J.* **95,** 515–520.
Bognar, A. L., Paliyath, G., Rogers, L., and Kolattukudy, P. E. (1984). *Arch. Biochem. Biophys.* in press.
Bonnett, Jr., H. T. (1968). *J. Cell Biol.* **37,** 199–205.
Borchert, R. (1974). *Dev. Biol.* **36,** 391–399.
Borchert, R. (1978). *Plant Physiol.* **62,** 789–793.
Botha, C. E. J., Evert, R. F., Cross, R. H. M., and Marshall, D. M. (1982). *Protoplasma* **112,** 1–8.
Brieskorn, C. H. (1978). *Fette, Seifen Anstrichm.* **80,** 15–20.

Brieskorn, C. H., and Binnemann, P. H. (1975). *Phytochemistry* **14,** 1363–1367.
Brieskorn, C. H., and Kabelitz, L. (1971). *Phytochemistry* **10,** 3195–3204.
Caldicott, A. B., and Eglinton, G. (1976). *Phytochemistry* **15,** 1139–1143.
Caldicott, A. B., Simoneit, B. R. T., and Eglinton, G. (1975). *Phytochemistry* **14,** 2223–2228.
Cassagne, C. (1970). Les hydrocarbures végétaux: biosynthesis et localisation cellulaire. Ph.D. Thesis, University of Bordeaux, France.
Cassagne, C., and Lessire, R. (1974). *Arch. Biochem. Biophys.* **165,** 274–280.
Cassagne, C., and Lessire, R. (1978). *Arch. Biochem. Biophys.* **191,** 146–152.
Cassagne, C., and Lessire, R. (1979). *In* "Advances in the Biochemistry and Physiology of Plant Lipids" (L. O. Appelqvist and C. Liljenberg, eds.), pp. 393–398. Elsevier, New York.
Cheesbrough, T. M., and Kolattukudy, P. E. (1984). *Proc. Natl. Acad. Sci. U.S.A.* in press.
Cottle, W., and Kolattukudy, P. E. (1982a). *Plant Physiol.* **69,** 393–399.
Cottle, W., and Kolattukudy, P. E. (1982b). *Plant Physiol.* **70,** 775–780.
Crisp, C. E. (1965). The biopolymer cutin. Ph.D. Thesis, University of California, Davis.
Croteau, R., and Fagerson, I. S. (1972). *Phytochemistry* **11,** 353–363.
Croteau, R., and Kolattukudy, P. E. (1973). *Biochem. Biophys. Res. Commun.* **52,** 863–869.
Croteau, R., and Kolattukudy, P. E. (1974a). *Arch. Biochem. Biophys.* **162,** 458–470.
Croteau, R., and Kolattukudy, P. E. (1974b). *Arch. Biochem. Biophys.* **162,** 471–480.
Croteau, R., and Kolattukudy, P. E. (1974c). *Biochemistry* **13,** 3193–3202.
Croteau, R., and Kolattukudy, P. E. (1975a). *Arch. Biochem. Biophys.* **170,** 61–72.
Croteau, R., and Kolattukudy, P. E. (1975b). *Arch. Biochem. Biophys.* **170,** 73–81.
Dean, B. B., and Kolattukudy, P. E. (1976). *Plant Physiol.* **58,** 411–416.
Dean, B. B., and Kolattukudy, P. E. (1977). *Plant Physiol.* **59,** 48–54.
Dean, B. B., Kolattukudy, P. E., and Davis, R. W. (1977). *Plant Physiol.* **59,** 1008–1010.
Deas, A. H. B., Baker, E. A., and Holloway, P. J. (1974). *Phytochemistry* **13,** 1901–1905.
Downing, D. T., Kranz, Z. H., and Murray, K. E. (1961). *Aust. J. Chem.* **14,** 619–627.
Eglinton, G., and Hunneman, D. H. (1968). *Phytochemistry* **7,** 313–322.
Eleftheriou, E. P., and Tsekos, I. (1979). *Protoplasma* **100,** 139–153.
Elstner, E. F., and Heupel, A. (1976). *Planta* **130,** 175–180.
Esau, K. (1977). *In* "Anatomy of Seed Plants" 2nd ed., pp. 183–198. Wiley, New York.
Espelie, K. E., and Kolattukudy, P. E. (1978). *Lipids* **13,** 832–833.
Espelie, K. E., and Kolattukudy, P. E. (1979a). *Plant Physiol.* **63,** 433–435.
Espelie, K. E., and Kolattukudy, P. E. (1979b). *Plant Sci. Lett.* **15,** 225–230.
Espelie, K. E., Dean, B. B., and Kolattukudy, P. E. (1979). *Plant Physiol.* **64,** 1089–1093.
Espelie, K. E., Sadek, N. Z., and Kolattukudy, P. E. (1980a). *Planta* **148,** 468–476.
Espelie, K. E., Davis, R. W., and Kolattukudy, P. E. (1980b). *Planta* **149,** 498–511.
Espelie, K. E., Wattendorff, J., and Kolattukudy, P. E. (1982). *Planta* **155,** 166–175.
Espelie, K. E., Köller, W., and Kolattukudy, P. E. (1983). *Chem. Phys. Lipids* **32,** 13–26.
Fang, H. H.-L. (1974). Ph.D. Thesis, Oregon State University, Corvallis.
Ferguson, I. B., and Clarkson, D. T. (1975). *New Phytol.* **75,** 69–79.
Ferguson, I. B., and Clarkson, D. T. (1976). *New Phytol.* **77,** 11–14.
Franich, R. A., and Volkman, J. K. (1982). *Phytochemistry* **21,** 2687–2689.
Fussel, L. K., and Dwarte, D. M. (1980). *J. Exp. Bot.* **31,** 645–654.
Godkin, S. E., Grozdits, G. A., and Keith, C. T. (1983). *Wood Sci. Technol.* **17,** 13–30.
Grambow, H. J. (1978). *Z. Pfanzenphysiol.* **88,** 369–372.
Grambow, H. J., and Grambow, G. E. (1978). *Z. Pflanzenphysiol.* **90,** 1–9.
Griffith, M., Brown, G. N., Espelie, K. E., and Kolattukudy, P. E. (1981). *Plant Physiol. Suppl.* **67,** 317.

Griffith, M., Huner, N. P. A., Espelie, K. E., and Kolattukudy, P. E. (1985). *Protoplasma* in press.
Grisebach, H. (1981). *In* "The Biochemistry of Plants" (E. E. Conn, ed.), Vol. 7, pp. 301–316. Academic Press, New York.
Gross, G. G. (1977). *Rec. Adv. Phytochem.* **11,** 141–184.
Gross, G. G., and Janse, C. (1977). *Z. Pflanzenphysiol.* **84,** 447–452.
Gross, G. G., Janse, C., and Elstner, E. F. (1977). *Planta* **136,** 271–276.
Grozdits, G. A. (1982). *Wood Sci.* **15,** 110–118.
Grozdits, G. A., Godkin, S. E., and Keith, C. T. (1982). *Wood Sci. Technol.* **16,** 305–316.
Guillemonat, A., and Traynard, J.-C. (1963). *Bull. Soc. Chim. Fr.* 142–144.
Haas, D. L., Carothers, Z. B., and Robbins, R. R. (1976). *Am. J. Bot.* **63,** 863–867.
Hanson, K. R., and Havir, E. A. (1979). *Rec. Adv. Phytochem.* **12,** 91–137.
Harmon, G. E., Mattick, L. R., Nash, G., and Nedrow, B. L. (1980). *Can. J. Bot.* **58,** 1541–1547.
Hartley, R. D. (1971). *J. Chromatogr.* **54,** 335–344.
Hemingway, R. W. (1981). *In* "Organic Chemicals from Biomass" (I. S. Goldstein, ed.), pp. 189–248. CRC Press, Boca Raton, Florida.
Herbin, G. A., and Robins, P. A. (1968). *Phytochemistry* **7,** 257–268.
Hergert, H. L. (1958). *For. Prod. J.* **8,** 335–339.
Hergert, H. L., and Kurth, E. F. (1952). *TAPPI* **35,** 59–66.
Holloway, P. J. (1972a). *Chem. Phys. Lipids* **9,** 158–170.
Holloway, P. J. (1972b). *Chem. Phys. Lipids* **9,** 171–179.
Holloway, P. J. (1973). *Phytochemistry* **12,** 2913–2920.
Holloway, P. J. (1974). *Phytochemistry* **13,** 2201–2207.
Holloway, P. J. (1982a). *Phytochemistry* **21,** 2517–2522.
Holloway, P. J. (1982b). *In* "The Plant Cuticle" (D. F. Cutler, K. L. Alvin, and C. E. Price, eds.), pp. 45–85. Academic Press, Inc., New York.
Holloway, P. J. (1983). *Phytochemistry* **22,** 495–502.
Holloway, P. J., and Deas, A. H. B. (1973). *Phytochemistry* **12,** 1721–1735.
Holloway, P. J., Deas, A. H. B., and Kabaara, A. M. (1972). *Phytochemistry* **11,** 1443–1447.
Huelin, F. E. (1959). *Aust. J. Biol. Sci.* **12,** 175–180.
Hunneman, D. H., and Eglinton, G. (1972). *Phytochemistry* **11,** 1989–2001.
Jensen, W. (1972). *Quimica* **68,** 871–878.
Jones, J. H. (1978). *Plant Physiol.* **62,** 831–832.
Khan, A. A., and Kolattukudy, P. E. (1973). *Biochemistry* **12,** 1939–1948.
Khan, A. A., and Kolattukudy, P. E. (1974). *Biochem. Biophys. Res. Commun.* **61,** 1379–1386.
Kolattukudy, P. E. (1966). *Biochemistry* **5,** 2265–2275.
Kolattukudy, P. E. (1967). *Biochemistry* **6,** 963–975.
Kolattukudy, P. E. (1970a). *Biochem. Biophys. Res. Commun.* **41,** 299–305.
Kolattukudy, P. E. (1970b). *Plant Physiol.* **46,** 759–760.
Kolattukudy, P. E. (1970c). *Biochemistry* **9,** 1095–1102.
Kolattukudy, P. E. (1971). *Arch. Biochem. Biophys.* **142,** 701–709.
Kolattukudy, P. E. (1972). *Biochem. Biophys. Res. Commun.* **49,** 1040–1046.
Kolattukudy, P. E. (1973). *Lipids* **8,** 90–92.
Kolattukudy, P. E. (1974). *Biochemistry* **13,** 1354–1363.
Kolattukudy, P. E. (1975). *In* "Recent Advances in the Chemistry and Biochemistry of Plant Lipids" (T. Galliard and E. I. Mercer, eds.), pp. 203–246. Academic Press, Inc., New York.
Kolattukudy, P. E. (1977a). *In* "Lipids and Lipid Polymers in Higher Plants" (M. Tevini and H. K. Lichtenthaler, eds.), pp. 271–292. Springer-Verlag, Berlin.

Kolattukudy, P. E. (1977b). *Rec. Adv. Phytochem.* **11,** 185–246.

Kolattukudy, P. E. (1978). *In* "Biochemistry of Wounded Plant Tissues" (G. Kahl, ed.), pp. 43–84. de Gruyter, Berlin.

Kolattukudy, P. E. (1980a). *Science* **208,** 990–1000.

Kolattukudy, P. E. (1980b). *In* "The Biochemistry of Plants, Lipids: Structure and Function" (P. K. Stumpf and E. E. Conn, eds.), Vol. 4, pp. 571–645. Academic Press, Inc., New York.

Kolattukudy, P. E. (1981). *Annu. Rev. Plant Physiol.* **32,** 539–567.

Kolattukudy, P. E., and Agrawal, V. P. (1974). *Lipids* **9,** 682–691.

Kolattukudy, P. E., and Brown, L. (1975). *Arch. Biochem. Biophys.* **166,** 599–607.

Kolattukudy, P. E., and Buckner, J. S. (1972). *Biochem. Biophys. Res. Commun.* **46,** 801–807.

Kolattukudy, P. E., and Dean, B. B. (1974). *Plant Physiol.* **54,** 116–121.

Kolattukudy, P. E., and Köller, W. (1983). *In* "Biochemical Plant Pathology" (J. A. Callow, ed.), pp. 79–100. Wiley, New York.

Kolattukudy, P. E., and Liu, T. J. (1970). *Biochem. Biophys. Res. Commun.* **41,** 1369–1374.

Kolattukudy, P. E., and Walton, T. J. (1972). *Biochemistry* **11,** 1897–1907.

Kolattukudy, P. E., and Walton, T. J. (1973). *Prog. Chem. Fats Other Lipids* **13,** 119–175.

Kolattukudy, P. E., Walton, T. J., and Kushwaha, R. (1971). *Biochem. Biophys. Res. Commun.* **42,** 739–745.

Kolattukudy, P. E., Buckner, J. S., and Brown, L. (1972). *Biochem. Biophys. Res. Commun.* **47,** 1306–1313.

Kolattukudy, P. E., Buckner, J. S., and Liu, T.-Y. J. (1973a). *Arch. Biochem. Biophys.* **156,** 613–620.

Kolattukudy, P. E., Walton, T. J., and Kushwaha, R. P. S. (1973b). *Biochemistry* **12,** 4488–4498.

Kolattukudy, P. E., Croteau, R., and Brown, L. (1974). *Plant Physiol.* **54,** 670–677.

Kolattukudy, P. E., Kronman, K., and Poulose, A. J. (1975a). *Plant Physiol.* **55,** 567–573.

Kolattukudy, P. E., Croteau, R., and Walton, T. J. (1975b). *Plant Physiol.* **55,** 875–880.

Kolattukudy, P. E., Croteau, R., and Buckner, J. S. (1976). *In* "Chemistry and Biochemistry of Natural Waxes" (P. E. Kolattukudy, ed.), pp. 289–347. Elsevier, New York.

Kolattukudy, P. E., Espelie, K. E., Soliday, C. L. (1981). *In* "Encyclopedia of Plant Physiology New Series, Plant Carbohydrates II" (W. Tanner and F. A. Loewus, eds.), Vol. 13B, pp. 225–254. Springer-Verlag, Berlin.

Lawton, J. R. (1980). *Bot. J. Linnean Soc.* **80,** 161–177.

Legg, V. H., and Wheeler, R. V. (1925). *J. Chem. Soc.* 1412–1421.

Lendzian, K. J. (1982). *Planta* **155,** 310–315.

Lendzian, K. J., and Schönherr, J. (1983). *Planta* **158,** 70–75.

Lessire, R. (1973). Etude de la biosynthèse des acides gras a tres longue chaine et leurs relations avec les alcanes dans la cellule d'epiderme d'*Allium porrum*. Ph.D. Thesis, University of Bordeaux, France.

Litvay, J. D., and Krahmer, R. L. (1977). *Wood Sci.* **9,** 167–173.

Liu, T. J. (1972). The biosynthetic relationships among long chain fatty acids, alkanes, secondary alcohols and ketones in plants. M.S. Thesis, Washington State University, Pullman.

Loveland, P. M., and Laver, M. L. (1972a). *Phytochemistry* **11,** 430–432.

Loveland, P. M., and Laver, M. L. (1972b). *Phytochemistry* **11,** 3080–3081.

Lu, A. Y. H., Ryan, D., Jerina, D. M., Daly, J. W., and Levin, W. (1975). *J. Biol. Chem.* **250,** 8283–8288.

Lundquist, U., and von Wettstein-Knowles, P. (1982). *Carlsberg Res. Commun.* **47,** 29–43.

Macey, M. J. K., and Stumpf, P. K. (1968). *Plant Physiol.* **43,** 1637–1647.

Martin, J. T., and Juniper, B. E. (1970). "The Cuticles of Plants." St. Martin's Press, New York.
Matic, M. (1956). *Biochem. J.* **63,** 168–178.
Mazliak, P. (1963). La cire cuticulaire des pommes (*Malus pumila* L.). Ph.D. Thesis, University of Paris.
Meyer, M. (1938). *Protoplasma* **29,** 552–586.
Mikkelsen, J. D. (1979). *Carlsberg Res. Commun.* **44,** 133–147.
Mikkelsen, J. D., and von Wettstein-Knowles, P. (1978). *Arch. Biochem. Biophys.* **188,** 172–181.
Mueller, W. C., and Beckman, C. H. (1978). *Can. J. Bot.* **56,** 1579–1587.
O'Brien, T. P., and Carr, D. J. (1970). *Aust. J. Biol. Sci.* **23,** 275–287.
O'Brien, T. P., and Kuo, J. (1975). *Aust. J. Bot.* **23,** 783–794.
Oesch, R., Kaubisch, N., Jerina, D. M., and Daly, J. W. (1971). *Biochemistry* **10,** 4858–4866.
Olesen, P. (1978). *Protoplasma* **94,** 325–340.
Oparka, K. J., and Gates, P. J. (1982). *Protoplasma* **113,** 33–43.
Orgell, W. H. (1955). *Plant Physiol.* **30,** 78–80.
Parameswaran, N., Kruse, J., and Liese, W. (1975). *Z. Pflanzenphysiol.* **77,** 212–221.
Patel, R. N. (1975). *N. Z. J. Bot.* **13,** 149–167.
Pearce, R. B., and Rutherford, J. (1981). *Physiol. Plant Pathol.* **19,** 359–369.
Peterson, C. A., Peterson, R. L., and Robards, A. W. (1978). *Protoplasma* **96,** 1–21.
Peterson, C. A., Emanuel, M. E., and Wilson, C. (1982). *Can. J. Bot.* **60,** 1529–1535.
Platt-Aloia, K. A., Oross, J. W., and Thomson, W. W. (1983). *Bot. Gaz.* **144,** 49–55.
Pozuelo, J. M., Espelie, K. E., and Kolattukudy, P. E. (1984). *Plant Physiol.* **74,** 256–260.
Rhodes, M. J. C., and Wooltorton, L. S. C. (1978). *In* "Biochemistry of Wounded Plant Tissues" (G. Kahl, ed.), pp. 243–286. de Gruyter, Berlin.
Riley, R. G., and Kolattukudy, P. E. (1975). *Plant Physiol.* **56,** 650–654.
Robards, A. W., Jackson, S. M., Clarkson, D. T., and Sanderson, J. (1973). *Protoplasma* **77,** 291–311.
Robards, A. W., Clarkson, D. T., and Sanderson, J. (1979). *Protoplasma* **101,** 331–347.
Rodríguez-Miguens, B., and Ribas-Marqués, I. (1972a). *An. Quim.* **68,** 301–308.
Rodrígues-Miguens, B., and Ribas-Marqués, I. (1972b). *An. Quim.* **68,** 1301–1306.
Rowe, J. W., Bower, C. L., and Wagner, E. R. (1969). *Phytochemistry* **8,** 235–237.
Schmidt, H. W., Mérida, T., and Schönherr, J. (1981). *Z. Pflanzenphysiol.* **105,** 41–51.
Schönherr, J. (1976). *Planta* **131,** 159–164.
Schönherr, J. (1982). *In* "Encyclopedia of Plant Physiology New Series Physiological Plant Ecology II Water Relations and Carbon Assimilation" (O. L. Lange, P. S. Nobel, C. B. Osmond, and H. Ziegler, eds.), Vol. 12B, pp. 153–179. Springer-Verlag, Berlin.
Schönherr, J., and Mérida, T. (1981). *Plant Cell Environ.* **4,** 349–354.
Schönherr, J., and Ziegler, H. (1980). *Planta* **147,** 345–354.
Scott, F. M. (1966). *Nature (London)* **210,** 1015–1017.
Scott, F. M., and Bystrom, B. G. (1970). *In* "New Research in Plant Anatomy" (N. K. B. Robson, D. F. Cutler, and M. Gregory, eds.), pp. 15–23. Academic Press, Inc., New York.
Scott, M. G., and Peterson, R. L. (1979). *Can. J. Bot.* **57,** 1063–1077.
Shaykh, M., Kolattukudy, P. E., and Davis, R. (1977). *Plant Physiol.* **60,** 907–915.
Shigo, A. L., and Marx, H. G. (1977). Compartmentalization of Decay in Trees (Agriculture Information Bulletin, No. 405) pp. 679–686. United States Department of Agriculture Forest Service.
Sitte, P. (1955). *Mikroskopie* **10,** 178–200.

Sitte, P. (1957). *In* "Die Chemie der Pflanzenzellwand" (E. Treiber, ed.), pp. 421–432. Springer, Berlin.
Smith, M. M., and O'Brien, T. P. (1979). *Aust. J. Plant Physiol.* **6,** 201–219.
Soliday, C. L., and Kolattukudy, P. E. (1977). *Plant Physiol.* **59,** 1116–1121.
Soliday, C. L., and Kolattukudy, P. E. (1978). *Arch. Biochem. Biophys.* **188,** 338–347.
Soliday, C. L., Dean, B. B., and Kolattukudy, P. E. (1978). *Plant Physiol.* **61,** 170–174.
Soliday, C. L., Kolattukudy, P. E., and Davis, R. W. (1979). *Planta* **146,** 607–614.
Streibl, M., Stransky, K., and Hérout, V. (1978). *Collect. Czech. Commun.* **43,** 320–326.
Thomson, W. W., Platt-Aloia, K., and Koller, D. (1979). *Bot. Gaz.* **140,** 249–260.
Tippet, J. T., and O'Brien, T. P. (1976). *Aust. J. Bot.* **24,** 619–632.
Tippet, J. T., Bogle, A. L., and Shigo, A. L. (1982). *Eur. J. For. Pathol.* **12,** 357–364.
Tippet, J. T., Shea, S. R., Hill, T. C., and Shearer, B. L. (1983). *Aust. J. Bot.* **31,** 197–210.
Torrey, J. G., and Clarkson, D. T. (eds.) (1975). "The Development and Function of Roots." Academic Press, Inc., New York.
Tulloch, A. P. (1973). *Phytochemistry* **12,** 2225–2232.
Tulloch, A. P. (1976). *In* "Chemistry and Biochemistry of Natural Waxes" (P. E. Kolattukudy, ed.), pp. 235–287. Elsevier, New York.
Vance, C. P., Kirk, T. K., and Sherwood, R. T. (1980). *Annu. Rev. Phytopathol.* **18,** 259–288.
Velcheva, M. P., Espelie, K. E., and Ivanov, C. P. (1981). *Phytochemistry* **20,** 2225–2227.
Vogt, E., Schönherr, J., and Schmidt, H. W. (1983). *Planta* **158,** 294–301.
von Wettstein-Knowles, P. (1979). *In* "Advances in the Biochemistry and Physiology of Plant Lipids" (L.-O. Appelquist and C. Liljenberg, eds.), pp. 1–26. Elsevier, New York.
Walton, T. J., and Kolattukudy, P. E. (1972a). *Biochem. Biophys. Res. Commun.* **46,** 16–21.
Walton, T. J., and Kolattukudy, P. E. (1972b). *Biochemistry* **11,** 1885–1897.
Wattendorff, J. (1974). *Z. Pflanzenphysiol.* **73,** 214–225.
Wattendorff, J. (1976a). *Planta* **128,** 163–165.
Wattendorff, J. (1976b). *Planta* **130,** 303–311.
Wattendorff, J. (1980). *Biol. Uns. Zeit* **10,** 81–90.
Wattendorff, J., and Holloway, P. J. (1982). *Ann. Bot.* **49,** 769–804.
Whitmore, F. W. (1978). *Phytochemistry* **17,** 421–425.
Wilson, A. J., and Robards, A. W. (1978). *Protoplasma* **95,** 255–265.
Wilson, C. A., and Peterson, C. A. (1983). *Ann. Bot.* **51,** 759–769.
Wynn, W. K. (1976). *Phytopathol.* **66,** 136–146.
Yatsu, L. Y., Espelie, K. E., and Kolattukudy, P. E. (1983). *Plant Physiol.* **73,** 521–524.
Zee, S.-Y., and O'Brien, T. P. (1970). *Aust. J. Biol. Sci.* **23,** 1153–1171.
Zenk, M. H. (1979). *Rec. Adv. Phytochem.* **12,** 139–176.

CHAPTER **9**

Occurrence of Extractives in Wood Tissues

W. E. HILLIS

CSIRO
Division of Chemical and Wood Technology
Highett, Victoria, Australia

I. Introduction

A. *Aims of Chapter*

The term "extractives" is convenient in describing the secondary nonstructural components of wood and bark. The term, however, implies a uniformity of properties and chemistry among the components that does not exist. The number of individual components of extractives is very

Biosynthesis and Biodegradation of Wood Components

ISBN 0-12-347880-4

large, and many classes of compounds are involved. Their constitution varies widely according to the botanical origin of the sample, the type of tissue in the sample, and the nature of the extracting solvent used.

This chapter aims to draw attention to the complexity of extractives in order to lay the basis for the detailed discussions of some of the major chemical classes to which most components of the extractives belong. It also directs attention to the selectivity with which some of these components may be formed, so that increased attention will be given to the cellular functions that control the biosynthesis of individual components. Aspects of this subject have been treated in more detail elsewhere (Hillis, 1962, 1971, 1977; Shigo and Hillis, 1973; Bauch, 1980) and are summarized here, with attention drawn to key references and more recent developments.

B. *Descriptions and Definitions*

1. Extractives

The nonstructural components of the wood and bark tissues of plants are commonly called *extractives* or "extraneous components." The largest amounts are found in the cell lumens, but they may also be present in the cell walls. The nature and amount of extractives removed varies with respect to the tissue and to the solvent used for extraction. With neutral solvents, differing amounts are retained by the substrate, particularly within the cell wall.

2. Sapwood

The wood of young trees or seedlings consists entirely of *sapwood,* which by definition (IAWA, 1964) is that "portion in the living tree which contains living cells and reserve material (e.g., starch)." The proportion of living cells is about 10%, although the percentage is variable, ranging between 5 and 40% (Hillis, 1977), and these cells consist mostly of vertical parenchyma and transversely oriented ray cells. The living cells conduct the metabolic processes, store food, and translocate water, minerals, and primary metabolites. The sapwood forms the pale-colored outer zone in stems, branches, or roots when heartwood is present. After a certain period, which is often characteristic for the species, heartwood is formed. Usually this transformation occurs within 5 to 30 years of tree growth. There are some trees, however, such as sugar maple, that contain only sapwood after 100 years' growth.

3. Heartwood

Heartwood forms "the inner layers of wood which in the growing tree have ceased to contain living cells and in which the reserve materials (e.g. starch) have been removed or converted into heartwood substances" (IAWA, 1964). It is generally darker in color than sapwood, but it is not always easy to distinguish between the two zones. Heartwood has a lower content of inorganic ions than the sapwood (Ellis, 1965; Ziegler, 1968), and the ionic content is raised if discoloration occurs. Heartwood frequently has an abrupt boundary with the adjacent sapwood and traverses growth rings, has a uniform color, and its volume in stems, branches, and roots varies widely. Heartwood formation is a function of most trees, and the formation is partly under genetic and partly under environmental control.

4. Transition Zone

In a number of species, heartwood, and sometimes woundwood, is surrounded by a *transition zone* that can be readily distinguished by a color that is paler than the color of either sapwood or heartwood. It can contain a smaller amount of extractives, other secondary components, or of water than do the surrounding tissues, and there is evidence to support the view (Hillis, 1977) that the formation of heartwood extractives is initiated in this zone.

5. Discolored Wood

Wounds or deformations that allow the entry of air and microorganisms into the living tree (e.g., logging wounds or branch stubs) can initiate the formation of *discolored wood* (see also Bauch, 1980), which does not form regularly as does heartwood. Just as in heartwood, the cells die at the same time as discoloration occurs, and the reserve materials are converted to extractives. Discolored wood appears as a very irregular column of wood extending up and down the interior of the tree from the point of injury, but rarely in a concentric pattern (Shigo and Hillis, 1973; Shigo *et al.*, 1977). When not associated with obvious injuries, this type of wood has been described as facultative or false heartwood (Bosshard, 1968) and is found in genera such as *Acer, Fagus,* and *Fraxinus*. Heartwood formation is a regular or inherent occurrence in most trees, but discolored wood is an additional change of sapwood or of heartwood and does not react further to renewed wounding.

Whereas the color of heartwood is relatively uniform, with an increase in depth of color from the pith to the outer heartwood, discolored wood frequently is characterized by its dark-colored streaks. The extractives of discolored wood have received relatively little attention, appear largely to be highly polymerized materials, and usually have a very moderate toxicity to fungi. Prolonged extraction with neutral solvents will not remove the dark color observed in the streaks of discolored wood.

II. Distribution of Extractives

A. *In Trees of Different Families, Genera, and Species*

1. Differences in Type

The many classes of extractives found in wood and bark tissues include polyphenols (or benzenoid types), terpenoids and tropolones, fats and waxes, salts of organic acids, complex polysaccharides, and nitrogenous compounds. Polyphenols are the most common extractives and are present in all heartwoods, from trace to large amounts, and in many sapwoods.

Polyphenols include the monoaryl products, such as gallic acid derivatives, stilbenes, and the phenylpropanoids, including coumarins, lignans, and the large group of flavonoids. Some of these components exist in the glycosidic form.

Almost always, extractives contain a mixture of more than one group of monomeric compounds with preponderantly polymerized materials. The ratio of amounts of monomeric compounds to each other in tissue extracts may vary, but the general composition remains the same. This situation is the basis for chemotaxonomy, and the resolution of the mixtures by chromatographic and other methods provides patterns facilitating the identity of plant material into families (e.g., the Cupressaceae is the sole source of tropolones), genera, and even in some cases into species.

Polyphenols are the most widely used for chemotaxonomy. Heartwood extractives have a "conservative" composition in comparison with leaf extracts, which can reveal biochemical or physiological variants of a species. Some variants with differing heartwood constituents, however, are known (Hillis, 1966). Distinctive compositions of heartwood polyphenols have been used to group species in genera such as *Pinus* (and other conifers) (Erdtman, 1963), *Prunus* (Hasegawa, 1958), *Acacia* (Tindale and Roux, 1974), and *Eucalyptus* (Hillis, 1972). Two *Nothofagus* species with an almost identical wood anatomy contain extractives with markedly different compositions (Hillis and Orman, 1959).

The increasing sensitivity and precision of analytical methods has shown some complex components that may be formed selectively (see Section III,C), a possible existence of a hierarchy of secondary components, an order in formation, a varying sensitivity to changes in the direction of biosynthesis, and a specialized function of some parenchyma.

2. Differences in Amount

The amounts of extractives in tissues vary widely between families, species, and tissues and within tissues, as exemplified in Table I. The amount of extractives present in heartwood is at least partly under genetic control, which in the case of *Pinus elliottii* is strong with regard to oleoresin yield and the composition of turpentine (Hillis, 1977). In other pine species, the inheritance of yield of oleoresin appears to be less genetically controlled. Despite the importance of extractives in affecting wood quality and in providing tannin extracts for commercial purposes, there have been few studies of genetic and environmental influences on their formation. The effect of site on the color of furniture woods such as black walnut indicates the possibilities for its influence (Phelps and McGinnis, 1980; Nelson *et al.*, 1969).

B. In Different Tissues

The difficulties in collecting samples from the cambial zone has hindered studies of biosynthesis in the region controlling the growth of woody plants and their maintenance. The intensive enzymatic activity during the major growth periods rapidly changes a number of extractives during sample collection. The amount of gallic acid derivatives appears to be high in relation to cell wall material in the cambial region of eucalypts during these periods (Hillis and Carle, 1958). High amounts of procyanidins are found in the bark of *Pinus* spp. and low amounts of flavonoids and stilbenes in the heartwood, but there is little information on the nature of the cambial extractives in these species.

The composition of the cambial extractives can differ from that of either the bark or sapwood or heartwood extractives of the same species. In *Eucalyptus marginata,* gallic and ellagic acid derivatives preponderate in the cambium, whereas in the phloem they are accompanied by proanthocyanidins, and in the heartwood proanthocyanidins are the major extractives components. On the other hand, the major polyphenols in all tissues of *E. delegatensis, E. regnans* F. Muell., and *E. sieberi* L. Johnson are ellagitannins, but kino (formed in damaged cambium) is almost entirely polymerized prodelphinidins (Hillis and Carle, 1962; Hillis, 1977).

Table I

Content of Extractives in Different Zones[a]

Family	Species	Tissue	Extractives content (%)	Type[b] (main component)
Anacardiaceae	*Schinopsis balansae* Engl.[1]	Heartwood	20–25	Polyphenol
	Rhus coriaria L.[1]	Bark	6	Polyphenol
		Wood	0.4	Polyphenol
Cupressaceae	*Thuja plicata* D. Don[6]	Heartwood	0.6	Tropolone
Fagaceae	*Quercus cinerea* Michx.[1]	Wood	1.3	Polyphenol
	Quercus robur L.[1]	Wood	9.2	Polyphenol
Mimosaceae	*Acacia mearnsii* deWild[3]	Bark	35–40	Polyphenol
		Sapwood	<0.5	Polyphenol
		Heartwood	1–5	Polyphenol
Pinaceae	*Pinus ponderosa* Douglas[1,2]	Bark	11	Polyphenol
		Wood	1	Polyphenol
		Sapwood	3.2–4.8	Lipid and terpenoid
		Heartwood	2.7–9.9	Lipid and terpenoid
	Pinus radiata D. Don[3,4,5]	Bark	15–25	Polyphenol
		Sapwood	<0.5	Polyphenol
		Heartwood	1–2	Polyphenol
		Sapwood	0.8–1.9	Lipid and terpenoid
		Heartwood	1.6–3.8	Lipid and terpenoid
Salicaceae	*Populus grandidentata* Michx.[2]	Sapwood	1.02	Lipid and terpenoid
		Heartwood	1.03	Lipid and terpenoid
Myrtaceae	*Eucalyptus astringens* Maiden[1,3]	Bark	40–75	Polyphenol
		Sapwood	<1	Polyphenol
		Heartwood	<5	Polyphenol
	E. delegatensis R. T. Bak.[7] (syn. *E. gigantea*)	Sapwood	10.7	Polyphenol
		Heartwood, outer	18.3	Polyphenol
		Heartwood, inner	10.1	Polyphenol
	E. marginata Donn. *ex*. Sm.[1]	Sapwood	<1	Polyphenol
		Heartwood	23–31	Polyphenol
Rhizophoraceae	*Rhizophora mucronata* (Lam.) Merr.[1]	Bark	16–40	Polyphenol
		Sapwood	0.5	Polyphenol
		Heartwood	5	Polyphenol

[a] Superscript numerals refer to the following references: 1, Hillis (1962); 2, Mutton (1962); 3, Hillis from various sources; 4, Sands (1975); 5, Hemingway and Hillis (1971); 6, Gardner (1962); 7, Hillis and Carle (1960).

[b] The content of polyphenols is frequently recorded in the literature as "tannin," the compounds that convert animal hides and skins to leather.

III. Variations in Extractives

A. *In Amounts*

1. Age of the Tree

The amount of extractives in the sapwood of some genera such as *Eucalyptus, Robinia,* and *Acacia* increases only slightly with distance from the cambial region and as the age of tissue increases. On the other hand, there is descriptive evidence that in other species (Hillis, 1977) a significant increase occurs.

The progressive increase in extractives content from the pith outward in dark-colored heartwoods can be readily seen, and determinations of extractives components in light-colored species confirm the same pattern (Hillis, 1962). Environmental conditions as a whole contribute to variations in extractives content, but the age of *Pinus echinata* trees has more influence than any other variable (Posey and Robinson, 1969). The lower amounts of extractives in heartwood of young trees is well recognized in wood utilization, with some properties improved and others reduced. The lower levels of extractives in the heartwood of young trees reduces their durability, so that the onset of decay in ground contact or the development of discoloration and decay in the living tree takes place more readily.

2. Position in the Tree

In some species, the formation of heartwood usually commences at about chest height, and, in mature trees, its cross-sectional area (relative to disk area) at this level is greater than at the stump [as in *Pinus* spp., *Pseudotsuga menziesii* (Mirb.) Franco (Hillis, 1977)]. The area of heartwood usually tapers from the level of initiation toward the crown, and with the younger tissues at higher levels in the tree, the extractives content is lower.

3. Growth Rate of the Tree

Contradictory observations exist concerning the effect of growth rate on the amount of extractives formed. Between-tree variations in the amount of total heartwood extractives in *Quercus rubra* L. and *Juglans nigra* L. failed to show any relationship with growth rate (Nelson, 1975). On the other hand, from a number of observations, the quantity of heartwood extractives is negatively correlated with rate of diameter growth (Hillis, 1977).

B. *In Damaged Wood*

1. Included Sapwood

When the cambium is killed, particularly at the peak of the growth season, the formation of further volumes of heartwood in a radial direction from the injury is prevented. Subsequently, after the sapwood in adjacent regions has been transformed to heartwood, there remains a zone of "included sapwood," which can be distinguished by its paler color. The amount of extractives in this zone, which is found in a number of species, will be close to that of normal sapwood, so that the zone will degrade more readily than heartwood.

"Target-ring pattern" is a related phenomenon, in which a number of concentric, pale-colored bands of sapwood appear in the heartwood. These bands, which contain low amounts of extractives, may be uniform and regular in appearance, or irregular and incompletely circular. The pattern has been observed in diverse species such as *Thuja plicata, Pseudotsuga menziesii,* and *Eucalyptus marginata.*

2. Extracellular Deposits

If the cambium of some species is injured mechanically or by insect attack in the spring season or is affected by extremes of heat or severe cold, extracellular secondary components will form within 3 weeks. These components are retained in the wood structure or exude through openings in the bark to dry on the outside of the tree as soft or brittle solids. If the injury is sufficiently severe, parenchyma will not bridge across the lesion, callus tissue will form on the developing sapwood, included sapwood will result, and the secondary components will be stored in the "pockets" that form over the injured region. These pockets can contain up to 2 liters of fluid. If the injury is less severe, parenchyma bridges will form radially between the cambium and woody tissues, but the formation of fibers is delayed. The resultant "veins" will be filled with small amounts of the extracellular components.

These components or exudates have compositions characteristic of the particular genus or species, and are often distinctly different from the extractives in the surrounding tissues. The "gums" are clear glassy masses with colors from pale yellow to brown, composed largely of polysaccharides and formed by *Acacia, Prunus,* and other species. "Resins" formed by *Pinus* spp., *Pseudotsuga menziesii,* and others are pale-colored materials, more or less soluble in various organic solvents but not in water, with a variable composition, usually of a terpenoid nature. When the cambium of eucalypts and a few other species is injured, an abrupt reorganization of biosynthesis results in the formation of kino, a polyphenolic extracellular fluid. Immediately after formation, kinos are colorless, but after an hour they become increasingly red or red-brown in color. Some kinos are almost entirely polymerized proanthocyanidins, whereas others contain small amounts of lignans or flavonoids, which in some cases permit identification of their origin (Hillis, 1977).

3. Resin-Impregnated Wood

The butts of large-diameter stems of some *Pinus* species contain unusually large amounts of resinous material. This accumulation may be due to the rupture of parts of the anastomosing resin canal system and the escape of canal resin due to the high resin pressure found in healthy trees. The naval stores industry is affected by the decreasing availability of the stumps of these trees; as a result, in the past decade, the resin content of young pines (notably *P. elliottii* Engel.) has been increased by injecting paraquat (Roberts, 1973), ethylene-forming, and other compounds (Lightwood Research Conference, 1975–1977) into the sapwood. "Resin tapping" was formerly practiced on a large scale on dipterocarps such as *Hopea* and *Shorea* spp. growing in tropical regions to provide damars for the varnish industry (Mantell *et al.*, 1942; Howes, 1949).

4. Shake Deposits

Heart shakes, frost checks, ring shakes, etc. are found as star-shaped or radial splits in the stems of living trees. They sometimes contain deposits, some of which are pure compounds which may not be present in the extractives from the surrounding heartwood. Lignans, terpenoids, flavonoids, and their glycosides have been reported, as well as aluminum succinate (Hillis, 1977; Bauch, 1980).

5. Discolored Wood

Discolored wood usually results from damage to the youngest tissues, whereas the oldest tissues are the first to be transformed into heartwood. The differences between sapwood–heartwood and sapwood–discolored wood are mainly in the amount and distribution of inorganic elements, the pH, the amount and composition of extractives and moisture content.

The amount of extractives in damaged or injured wood is higher than in the adjacent sapwood, and the amorphous deposits of melanistic substances are more abundant than in heartwood. The extractable materials frequently differ quantitatively in different parts of the discolored region. Discolored woods from different trees of the same species, and apparently resulting from the same cause, can contain different ratios of extractable components. The components in discolored and fungally affected wood can be of classes different from those found in unaffected heartwood (Shigo and Hillis, 1973; Hillis, 1977). Certain types of wounds in some species result in the accumulation of materials inhibiting to decay fungi; in other cases, the decay resistance of the tissues is not increased.

When the heartwood of living trees is injured, discoloration can occur, but there is no evidence that new or additional monomeric compounds are formed. The discoloration is apparently due to increased polymerization or oxidation of existing extractives and is the first stage of decay (Shigo and Hillis, 1973).

A significant characteristic of discolored wood is the increased amount of inorganic materials compared with the amount in the surrounding tissues. This is in contrast to heartwood, which yields lower amounts of ash than the sapwood of the same specimen or species. Detailed studies of some species, such as *Robinia pseudoacacia* L. (Hart, 1968), show complex situations, in that selective increases in the amounts of potassium, calcium, and magnesium, but not phosphorus, ions occur (see also Shigo and Hillis, 1973). In addition to these changes, increases in the pH and moisture content of discolored wood relative to sapwood have been observed (Shigo and Hillis, 1973; Ellis, 1965; Safford *et al.*, 1974; Bauch, 1980).

C. *In Different Cell Types*

Extractives are found in all the anatomical elements of those woods containing high amounts of extractives, but only in the ray cells of those woods with small amounts.

1. Parenchyma

The function of some parenchyma is solely to form crystals of calcium oxalate or of calcium carbonate. These inorganic salts exist as amorphous powders or characteristically shaped crystals, sometimes of large size in comparison with adjacent fibers. In other cases, the parenchyma can subsequently form organic compounds after the inorganic deposits, such as silica conglomerates, appear (Hillis, 1977).

Parenchyma can form, with considerable selectivity, complex polyphenols such as bayin (a *C*-glucosyl flavone) in *Castanospermum australe* A. Cunn., which is also the major component of the deposits in the vessels of this species. In a detailed study of the formation of heartwood in *Juglans nigra* and *Quercus rubra,* it was found that physiological conditions near the heartwood periphery had a much greater influence on the rate and duration of biosynthesis of benzenoid compounds than did the number or volume of active parenchyma units (Nelson, 1975).

2. Vessels

In *Intsia bijuga* O. Ktze., some vessels may be completely filled with pure robinetin (3,7,3′,4′,5′-pentahydroxy flavone), whereas adjacent vessels may be empty or filled with a mixture of C-5 OH flavones, stilbenes, and polymerized procyanidin (Hillis and Yazaki, 1973a; Hillis, 1977). Most of the parenchyma adjacent to the vessels filled with deposits can be devoid of colored material.

3. Fibers and Tracheids

The contents of different tracheids in the small clusters of cells that comprise floccosoids in *Tsuga heterophylla* (Raf.) Sarg. have different appearances. Some predominantly contain the lignan conidendrin, others hydroxymatairesinol, and others an unidentified lignan (Krahmer *et al.*, 1970; Parker *et al.*, 1974).

4. Cell Walls

Several studies have shown that the cell wall matrix of heartwood contains extractives that would enter the cell wall when the tonoplast ruptures and releases its contents. The few studies of cell wall in its original undried form indicate water occupies the 25% "free" spaces [in the case of *Pinus resinosa* Ait. sapwood (Berlyn, 1969)] existing in the form of capillaries with average cross sections of 160–600 nm in width (Hillis, 1977). This width would accommodate a number of molecules, as the widest part of a glucose molecule is 63 nm and a flavonoid monomer ~150 nm. Proanthocyanidin monomers or polymers would be able to penetrate these capillaries, perhaps interact with the structural components of the cell wall, and become trapped in the collapsed capillaries when the cell wall dries. Finely ground, dark-colored woods such as red-colored eucalypts or *Acacia* spp. probably retain components in the cell wall after prolonged extraction with methanol. Extracted eucalypt wood yields red anthocyanidin-like colors or ellagic acid when heated with butanol–hydrochloric acid, and the extracted wood of *Acacia* spp. continues to fluoresce yellow under ultraviolet light.

In contrast, the extractives formed in the parenchyma cells of sapwood during its change to discolored wood do not penetrate the fiber walls of some species (Bosshard, 1968, 1974). This could be due to polymerization/oxidation reactions resulting in molecules too large to penetrate the capillaries of the cell wall or to be extracted by neutral solvents on prolonged extraction.

5. Specialized Cell Types

Specialized cell types are found in a few species; examples include vertical intercellular canals in members of the Dipterocarpaceae, tanniferous tubules in *Horsfeldia* spp., and oil cells in the Lauraceae. In the conifers there are certain genera, namely *Picea, Pinus, Pseudotsuga,* and *Larix,* in which there are typical "resin canals." These occur as an anastomosing canal system, surrounded by epithelial cells capable of secreting resinous material under pressure into the cavities of the canal. This resin has a composition different from that of the heartwood or the sapwood parenchyma of the same species (Mutton, 1962).

IV. Extractives Formation

A. *The Specificity of Formation of Extractives Components*

Numerous anatomical observations indicate the existence of a considerable degree of specificity in the formation of extractives. Examination by sensitive quantitative techniques developed relatively recently will provide greater understanding of extractives formation. These techniques should provide information on the possible sequential formation of extractives in pathways that may be interrupted, blocked, or diverted when cellular conditions change.

The parenchyma that form tyloses (see Section IV,B) in *Eucalyptus botryoides* Sm. appear to occupy a particular position in the rays (Necesany, 1973). The parenchyma in the young sapwood of *Syncarpia glomulifera* (Sm.) Niedenz form silica conglomerates, and several years later at the heartwood periphery the same parenchyma become filled with polyphenolic extractives (De Silva and Hillis, 1976). The outer edge of the transition zone surrounding the heartwood of *E. polyanthemos* Schau has a strong fluorescence under ultraviolet light. The fluorescence is due to 3,4,3′-tri-*O*-methylellagic acid-4′-glucoside, which is also present in the inner layers of the transition zone and heartwood, where its fluorescence is masked by procyanidins (Hillis and Yazaki, 1973b). It is noteworthy this ellagic acid derivative is a distinctive component that characterizes a clear-cut group of eucalypt species to which *E. polyanthemos* belongs. In another example, crystals of ellagic acid are formed in specially developed parenchyma cells of the Myrtaceae (Chattaway, 1953).

Wound wood surrounding the tunnels made by the ovipositor of the wood wasp (*Sirex noctilio*) in newly formed *Pinus radiata* sapwood contains a characteristic mixture of polyphenols. It lacks the flavonoids of the heartwood but contains two fungicidal stilbenes (pinosylvin and pinosylvin monomethyl ether) in ratios different from those of the heartwood (Hillis and Inoue, 1968). Mechanical injury does not have the same effect. The sapwood of *Prunus domestica* L. affected by the fungus *Stereum purpureum* contains considerable amounts of the coumarin scopoletin, which is present in only very small amounts in sapwood and absent from the normal heartwood of other *Prunus* species (Hillis and Swain, 1959; Hillis, 1977). Other examples of the diversion of biosynthetic pathways are known.

It would be expected that all biosynthesis would occur in living cells with the full complement of organelles. However, as has been mentioned, pure robinetin has been found in vessels of *Intsia bijuga* but not in the adjacent parenchyma, which would have been the only living cells (Hillis and Yazaki, 1973a). Also, specific lignans have been found as major components in the tracheids of *Tsuga heterophylla,* but not in the adjacent parenchyma (Krahmer *et al.*, 1970).

B. Changes Associated with Extractives and Heartwood Formation

1. Moisture Changes

In softwoods, the moisture content of the heartwood is considerably lower than that of the sapwood (Hillis, 1962; Kozlowski, 1971), and the abrupt change is concomitant with an increase in extractives content. The decrease in moisture content on the formation of heartwood of some hardwoods, however, is small. A steady loss of water may take place across the stem, and in the case of *Fagus sylvatica* L., the sapwood–heartwood boundary has a 60% moisture content. The transition zone observed between sapwood and heartwood of some species can have lower moisture contents than the surrounding tissues (Hillis, 1977).

Higher moisture contents in heartwood than in sapwood have been recorded for a number of hardwood species (Kozlowski, 1971; Hillis, 1977). "Wetwood," containing more water than the inner sapwood zone, occurs independently of normal heartwood and can exist as small or large zones in the sapwood or heartwood in a large number of softwoods and hardwoods (Hartley *et al.*, 1961). Although microorganisms are frequently associated with wetwood, they may not be the direct cause of its formation. In some cases the extractives content of wetwood may be higher than that of the heartwood.

2. Anatomical Changes

a. Aspiration of Pits and Formation of Tyloses. When the osmotic pressure of the living parenchyma is reduced through injury or reduction of moisture in adjacent cells, different changes can occur. In conifers, reduction of moisture content below a certain level in one tracheid will result in the aspiration (closure) of the pit (interconnecting opening for fluid movement) to the adjacent tracheid. Aspiration occurs simultaneously with heartwood formation, and movement of fluid is hindered considerably (Harris, 1954).

With hardwoods, when the osmotic pressure is lowered in injured sapwood or at the heartwood periphery, the inner surfaces of the horizontal ray parenchyma extend through the pit aperture into the lumen of vessels to initiate tylosis formation. These balloon-like structures hinder the movement of fluids through the vessels, but are formed only when the pit opening is more than 8–10 μm. Tyloses vary considerably in structure and may have thin or thick lignified secondary walls. Tyloses can form abruptly in the transition zone (Bamber, 1976), and in eucalypts they form before extractives appear (Necesany, 1973).

b. Extractives Formation. The marked increase in the amount of extractives at the heartwood periphery appears first in the ray parenchyma and then, as the extractives content increases in the vertical parenchyma, in the vessels and fibers.

3. Biochemical Changes

a. Cytological Aspects. Cytological examinations have shown that the living parenchyma alter along the gradient from the cambium to the pith or to the periphery of heartwood or discolored wood (Bosshard, 1974; Nobuchi *et al.*, 1982). These alterations are often associated with the slight increase of extractives in the sapwood (Hillis, 1977; Bauch, 1980). The amount of starch declines between the middle sapwood and the transition zone or heartwood periphery, where it is completely metabolized. Some lipids show a similar distribution and most polyphenolic glycosides are hydrolyzed in a similar way.

The amount of starch or lipids in the inner sapwood is insufficient to form the amount of extractives found in the heartwood—particularly in some species. The mechanism by which extractives increase in quantity at the periphery of heartwood or woundwood has not been fully established. For some species, aging of the sapwood cells proceeds with the gradual degeneration and disintegration of the nuclei in the parenchyma, and these cells progressively lose their organelles and vitality (Bosshard, 1974). On the other hand, the activity of some enzymes that are necessary for the biosynthesis of extractives has been found to increase in the inner sapwood or at the heartwood periphery. These include dehydrogenases, an amylase, enzymes controlling starch hydrolysis, peroxidases, and phenol oxidases (Hillis, 1977). Histological tests on the transition zone of *Pinus radiata* and *Cryptomeria japonica* have indicated that the activity of the pentose phosphate shunt increased more than that of the citric acid cycle (Shain and Mackay, 1973; Higuchi *et al.*, 1967). There are numerous gaps in information on the expected increases in biological activity required for extractives formation at or adjacent to the periphery of heartwood, due to the difficulties in collecting samples at the period when the extractives are formed. There is increasing evidence that heartwood formation is a seasonal activity that differs for different species (Hillis, 1977; Bauch, 1980; Fukazawa *et al.*, 1980; Nobuchi *et al.*, 1982).

b. The Role of Ethylene. Recently attention has been given to the role that the phytohormone ethylene may play in the formation of extractives at the heartwood periphery. In other vegetative tissues, ethylene can be produced as a result of various stresses such as dehydration and protoplasmic disorganization, and it acts as a regulatory hormone in a variety of physiological changes occurring at many stages in the ontogeny of plants (Abeles, 1972). Very small amounts (1–5 ppm and less) of ethylene effectively trigger a wide range of changes dependent on the tissues involved. These include increase in the activity, or *de novo* synthesis, of a number of enzymes such as phenylalanine ammonia-lyase and other enzymes involved in phenol synthesis, polyphenoloxidase and particularly peroxidase (Hillis, 1977).

The amount of ethylene present has been related to the concentration of polyphenols formed in vegetative tissues, and there is an association between the amount of ethylene produced and the formation of polyphenols in radiata pine. Characteristic heartwood polyphenols have been formed when blocks of living pine sapwood, which were partly dried, were placed in a stream of ethylene (Shain and Hillis, 1972b). In another case they were formed when the pine sapwood was injured and stored in a manner that could result in the formation of ethylene (Jorgensen and Balsillie, 1969). In addition, the transition zone surrounding the heartwood of radiata pine, or the *Sirex* lesions in young radiata pine sapwood, yield larger amounts of ethylene than surrounding tissues, and its formation clearly preceded phenol synthesis (Shain and Hillis, 1972a). The transition zone of *Eucalyptus tereticornis* Sm. produces more ethylene than sapwood and ethylene in the sapwood close to the heartwood boundary in *Juglans nigra* and *Prunus serotina* Ehrh. reaches a maximum at certain periods (Hillis, 1977). More recent studies by Nelson *et al.* (1981), however, indicate the complexity of the situation affecting xylem production of ethylene.

The injection of the ethylene-releasing chemical chloroethyl phosphonic acid into sapwood has resulted in the formation of the characteristic heartwood polyphenols of *Rhus* (Hillis, 1975), resin enrichment in pines (Wolter, 1977; Nelson and Hillis, 1976), copious exudation of a carbohydrate gum from a *Prunus* sp. (Hillis, 1975), and enhanced yields of rubber from *Hevea braziliensis* Muell. Arg. (Abraham *et al.*, 1968). Injection into the cambial region of *Eucalyptus sieberi, E. regnans* F. Muell., *E. calophylla* R. Br. ex Lindl., *E. sideroxylon* A. Cunn. ex Woolls, and *E. astringens,* however, resulted in the copious formation of kinos containing predominantly polymerized proanthocyanidins (Hillis, 1975; Nelson and Hillis, 1976). The kinos had compositions similar to those produced naturally with none of the ellagitannins present in undamaged cambial tissues.

V. General Features of the Formation of Extractives and of Heartwood

A. *Extractives*

The monomeric components of extractives have received the overwhelming attention for study, and they convey numerous properties to wood, whether they are located in the lumens or the cell walls of the various wood elements. Further studies on their biosynthesis and their biodegradability will lead to a greater understanding of the properties of wood and its improvement.

The proportion of monomeric components in an extract is partly dependent on the solvent and is often high when ether or benzene is used to extract those woods characterized by the presence of lipids or terpenoids. On the other hand, the major proportion of methanol extracts of wood containing various classes of polyphenols is polymerized materials, and frequently it is greater than 90% of the total extract.

There is limited information on the polymerized components of extractives. Recent work shows significant proportions can be stereospecific bi-

and trimolecular proanthocyanidins (Drewes *et al.*, 1967; Hemingway *et al.*, 1981; Fletcher *et al.*, 1977; Haslam, 1982). Polymerized proanthocyanidins are commonly known as condensed tannins (see Chapter 13). There is evidence the oligomers of proanthocyanidins can conglomerate into molecules of considerable size (Yazaki and Hillis, 1977, 1980), but so far they have not been separated into their smaller component parts.

The heartwoods of some eucalypts contain up to 25% extractives, of which 10% are galloyl- and hexahydroxydiphenoylglucose derivatives (Seikel and Hillis, 1970). The dark-colored polymeric material accompanying the monomers has not been fractionated. These polymers could be complex, due to the ready oxidizability of the pyrogallol moieties and subsequent polymerization. The small yield of ellagic and gallic acids from the hydrolyzed polymers supports this view. Extracts containing these mixtures are classified as hydrolyzable tannins (see Chapter 13).

B. *Heartwood Formation*

With the range of types of heartwood, woundwood, and discolored woods (as well as barks), there are likely to be several mechanisms for their formation and the extractives they contain. It is not surprising that there are different theories for the formation of heartwood and, indirectly, the heartwood extractives. Because of the relatively constant width of the sapwood zone surrounding heartwood within a species, it is commonly assumed that the parenchyma cells gradually age and degrade and cease to be active in the inner zones of the sapwood. In this zone it has been proposed that semianaerobic conditions exist and the phenols are changed by oxidative polymerization reactions (Frey-Wyssling and Bosshard, 1959).

It has also been suggested that air enters tree trunks through broken branch stubs and other wounds and accumulates in the interior of stems. The resultant reduction of the water content of the wood in turn affects the metabolism of the living parenchyma cells, so that heartwood and its extractives are formed (Hartig, 1888; Hillis, 1977).

Other theories propose the translocation of secondary metabolites from the cambial tissues and the parenchyma to the heartwood periphery as a type of internal excretion of biochemical "waste" products (Stewart, 1967). It was considered that these "excretory substances" are translocated in nontoxic concentrations, and, in trees that form heartwood, they accumulate in amounts that rise to lethal levels, so that the innermost parenchyma die and further translocation inward ceases. On the basis of the premise that heartwood transformation is a development process rather than a deterioration of cell function with age, it has also been suggested (Bamber, 1976) that a heartwood-inducing substance moves centripetally along the rays from the cambium. This substance reaches the threshold level for heartwood initiation in cells adjacent to the impermeable heartwood core.

Most evidence supports the general scheme that the amount of extractives in heartwood, and in damaged and discolored wood will be influenced by the amount of primary metabolites transported centripetally from the cambial zone to their peripheries, or, in the case of bark, the amount transported centrifugally. The amount of primary metabolites would be dependent on the amount in the translocation stream needed for cell wall formation in the cambium. An increase in metabolic activity at the periphery, or in the transition zone, possibly triggered by ethylene, would result in the formation of extractives characteristic for the species and for the conditions. Subsequently, further and notably oxidative changes may take place over a period of time.

References

Abeles, F. B. (1972). "Ethylene in Plant Biology." Academic Press, New York.

Abraham, P. D., Wycherly, P. R., and Pakianathan, S. W. (1968). *J. Rubber Res. Inst. Malaya* **20,** 291–297.

Bamber, R. K. (1976). *Wood Sci. Technol.* **10,** 1–8.

Bauch, J. (1980). *In* "Natural Variations in Wood Properties" (J. Bauch, ed.), No. 131, pp. 69–98, Mitteileingen, Bund für Forst u. Holzwirtschaft, Hamburg-Reinbek.

Berlyn, G. P. (1969). *Am. J. Bot.* **56,** 498–503.

Bosshard, H. H. (1968). *Wood Sci. Technol.* **2,** 1–12.

Bosshard, H. H. (1974). "Holzkunde" Bd 2: "Zur Biologie, Physik and Chemie des Holzes." Birkhauser, Basel und Stuttgart.

Chattaway, M. M. (1953). *Aust. J. Bot.* **1,** 27–38.

De Silva, D., and Hillis, W. E. (1976). Unpublished manuscript.

Drewes, S. E., Roux, D. G., Egger, S. H., and Feeney, J. (1967). *J. Chem. Soc. Ser. C.* 1217–1227.

Ellis, E. L. (1965). *In* "Cellular Structure of Wood Plants" (W. A. Cote, ed.), pp. 181–189. Syracuse Univ. Press, New York.

Erdtman, H. (1963). *In* "Chemical Plant Taxonomy" (T. Swain, ed.), pp. 89–125. Academic Press, London.

Fletcher, A. C., Porter, L. J., Haslam, E., and Gupta, R. K. (1977). *J. Chem. Soc. Perkin Trans. I* 1628–1643.

Frey-Wyssling, A., and Bosshard, H. H. (1959). *Holzforschung* **13,** 129–137.

Fukazawa, K., Yamamoto, K., and Ishida, S. (1980). *In* "Natural Variations of Wood Properties" (J. Bauch, ed.), No. 131, pp. 113–131. Mitteilungen der BFH, Hamburg-Reinbek.

Gardner, J. A. F. (1962). *In* "Wood Extractives" (W. E. Hillis, ed.), pp. 317–330. Academic Press, New York.

Harris, J. M. (1954). *New Phytol.* **53,** 517–524.
Hart, J. H. (1968). *For. Sci.* **24,** 334–338.
Hartig, J. H. (1888). *Allg. Forst-u. Jagdztg* **64,** 52–56.
Hartley, C., Davidson, R. W., and Crandell, B. S. (1961). Report No. 2215. U.S. Dept. Agriculture Forest Service, Forest Prod. Lab.
Hasegawa, M. (1958). *J. Jpn. For. Soc.* **40,** 111–119.
Haslam, E. (1982). *In* "The Flavonoids: Advances in Research" (J. B. Harborne and T. J. Mabry, eds.), pp. 417–448. Chapman and Hall, London.
Hemingway, R. W., and Hillis, W. E. (1971). *APPITA* **21,** 439–443.
Hemingway, R. W., Foo, L. Y., and Porter, L. J. (1981). *J. Chem. Soc. Chem. Commun.* 320–322.
Higuchi, T., Fukazawa, K., and Shimada, M. (1967). *Hokkaido Univ. Coll. Exp. For. Res. Bull.* **25,** 167–192.
Hillis, W. E. (1962). *In* "Wood Extractives" (W. E. Hillis, ed.), pp. 60–132. Academic Press, New York.
Hillis, W. E. (1966). *Phytochemistry* **5,** 541–556.
Hillis, W. E. (1971). *Wood Sci. Technol.* **5,** 272–289.
Hillis, W. E. (1972). *APPITA* **26,** 113–122.
Hillis, W. E. (1975). *Phytochemistry* **14,** 2559–2562.
Hillis, W. E. (1977). *In* "Recent Advances in Phytochemistry" (F. A. Loewus and V. C. Runeckles, eds.), Vol. II, pp. 247–309. Plenum, New York.
Hillis, W. E., and Carle, A. (1958). *Holzforschung* **12,** 136–141.
Hillis, W. E., and Carle, A. (1960). *Biochem. J.* **74,** 607–615.
Hillis, W. E., and Carle, A. (1962). *Biochem. J.* **82,** 435–439.
Hillis, W. E., and Inoue, T. (1968). *Phytochemistry* **7,** 13–22.
Hillis, W. E., and Orman, H. R. (1959). *J. Linn. Soc. (Bot.)* **58,** 175–184.
Hillis, W. E., and Swain, T. (1959). *J. Sci. Food Agric.* **10,** 533–537.
Hillis, W. E., and Yazaki, Y. (1973a). *Phytochemistry* **12,** 2491–2495.
Hillis, W. E., and Yazaki, Y. (1973b). *Phytochemistry* **12,** 2969–2979.
Howes, F. N. (1949). "Vegetable Gums and Resins." Chronica Botanica, Waltham, Massachusetts.
IAWA (1964). "Multilingual Glossary of Terms Used in Wood Anatomy." Mitteilungen der Schweizerischen Anstalt für das Versuchswesen 40, Part 1.
Jorgensen, E., and Balsillie, D. (1969). *Can. J. Bot.* **47,** 1015–1016.
Kozlowski, T. T. (1971). "Growth and Development of Trees," Vol. I. Academic Press, New York.
Krahmer, R. L., Hemingway, R. W., and Hillis, W. E. (1970). *Wood Sci. Technol.* **4,** 122–139.
Lightwood Research Conference (1975–1977). "Proc. Annual Lightwood Res. Coord. Council." USDA Southeastern Forest Expt. Sta., Asheville, North Carolina.
Mantell, C. L., Kopf, C. W., Curtis, J. L., and Rogers, E. M. (1942). "The Technology of Natural Resins," Wiley, New York.
Mutton, D. B. (1962). *In* "Wood Extractives" (W. E. Hillis, ed.), pp. 331–363. Academic Press, New York.
Necesany, V. (1973). *Holzforschung* **27,** 73–79.
Nelson, N. D. (1975). *Can. J. For. Res.* **5,** 291–301.
Nelson, N. D. (1978). *Can. J. Bot.* **56,** 626–634.
Nelson, N. D., and Hillis, W. E. (1976). Unpublished data.
Nelson, N. D., Maeglin, R. R., and Wahlgren, H. E. (1969). *Wood Fiber* **1,** 29–37.
Nelson, N. D., Rietveld, W. J., and Isebrands, J. G. (1981). *For. Sci.* **27,** 537–543.

Nobuchi, T., Kuroda, K., Iwata, R., and Harada, H. (1982). *Mokuzai Gakkaishi* **28,** 669–676.
Parker, M. L., Barton, G. N., and Jozsa, L. A. (1974). *Wood Sci. Technol.* **8,** 229–232.
Phelps, J. E., and McGinnes, E. A. (1980). *In* "Natural Variations in Wood Properties" (J. Bauch, ed.), No. 131, pp. 169–192. Mitteilunger der BFH, Hamburg-Reinbek.
Posey, C. E., and Robinson, D. W. (1969). *TAPPI* **52,** 110–114.
Roberts, D. R. (1973). Research Note S.E.-191. Forest Service, U.S. Dept. Agriculture.
Safford, L. O., Shigo, A. L., and Ashley, M., (1974). Gradients of cation concentration in discolored and decayed wood of red maple. *Can. J. For. Res.* **4,** 435–440.
Sands, R. (1975). *N.Z. J. Sci.* **5**(1), 74–86.
Seikel, M. K., and Hillis, W. E. (1970). *Phytochemistry* **9,** 1115–1128.
Shain, L., and Hillis, W. E. (1972a). *Phytopathology* **61,** 841–845.
Shain, L., and Hillis, W. E. (1972b). *Can. J. Bot.* **51,** 1331–1335.
Shain, L., and Mackay, J. F. G. (1973). *Can. J. For. Res.* **51,** 737–739.
Shigo, A. L., and Hillis, W. E. (1973). *Ann. Rev. Phytopathol.* **11,** 197–222.
Shigo, A. L., Shortle, W. G., and Garrett, P. (1977). *J. Arbori.* **3,** 114–118.
Stewart, C. M. (1967). *Nature (London)* **214,** 138–141.
Tindale, M. D., and Roux, D. G. (1974). *Phytochemistry* **13,** 829–839.
Wolter, K. E. (1977). *Proc. Lightwood Res. Coord. Counc.* 90–99.
Yazaki, Y., and Hillis, W. E. (1977). *Holzforschung* **31,** 20–25.
Yazaki, Y., and Hillis, W. E. (1980). *Holzforschung* **34,** 125–130.
Ziegler, H. (1968). *Holz Roh-Werkst.* **26,** 61–68.

CHAPTER **10**

Biosynthesis and Metabolism of Phenolic Acids and Monolignols

G. G. GROSS

Universität Ulm
Abteilung Allgemeine Botanik
Ulm, Federal Republic of Germany

I. Introduction

Extensive interpretation of the term "phenolic acid" would mean that every compound being characterized by both a carboxyl function and at least one phenolic hydroxyl group falls into this category. In common practice, however, this designation is used for only a limited number of natural products, particularly cinnamic acids (phenylacrylic acids) and

Biosynthesis and Biodegradation of Wood Components

ISBN 0-12-347880-4

benzoic acids (phenylcarboxylic acids). In this connection, a second inaccuracy of the above definition is apparent: compounds like the unsubstituted parent acids as well as derivatives with substituted hydroxyls, both thus lacking phenolic characteristics, represent well established members of this group of secondary products on grounds of obvious biogenetic relations—an aspect that certainly is of higher significance in the classification of most natural products than mere consideration of structural properties.

A wide variety of differently substituted phenolic acids is known to occur, although frequently in low concentration, as typical components of many plants and plant organs. The main importance of these compounds, however, is their obvious role as precursors of a host of secondary metabolites. As summarized in Fig. 1, these pathways involve reduction of the carboxyl function, formation of esters, amides, or lactones, side-chain

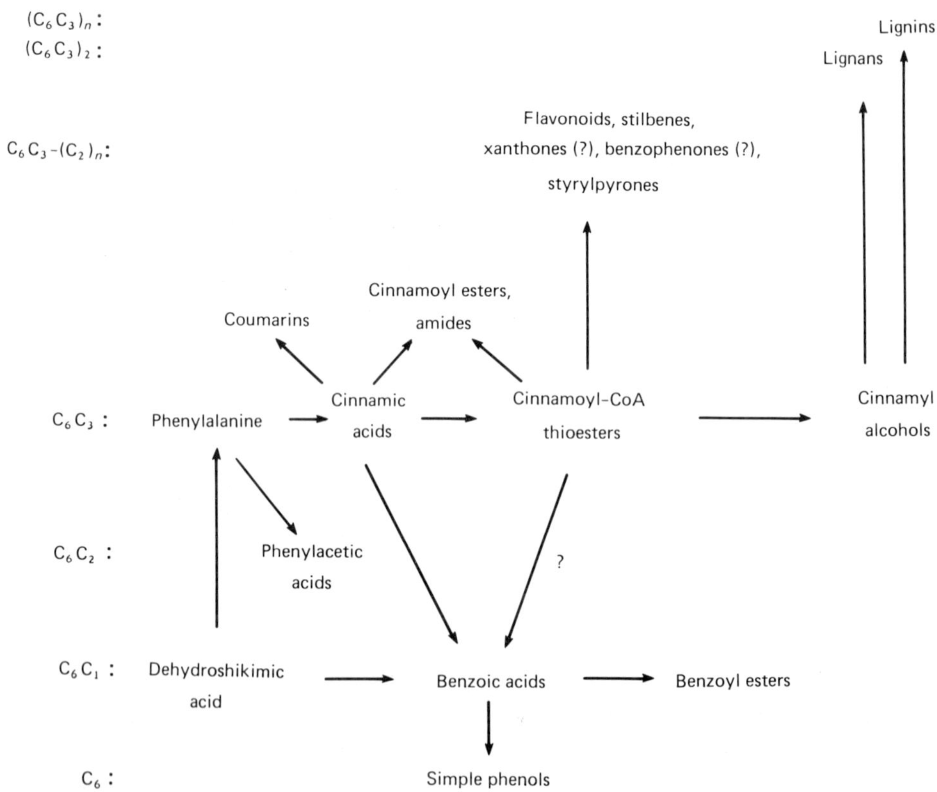

Fig. 1. Metabolic grid of major biosynthetic pathways involving phenolic acids and their derivatives. Modified after Kindl (1971).

elongation and polymerization reactions, and degradation processes. The latter aspect is of particular significance with respect to the biogenesis of phenylacetic acids. In addition to these reactions, which primarily concern the basic carbon skeleton of the individual classes of compounds, secondary alterations affecting the substitution pattern of their aromatic nuclei are also frequently encountered.

In this chapter, the pathways will be discussed that, as outlined above, are involved in the formation of phenolic acids and in their subsequent conversion to smaller or more complex molecules. Initial investigations in this area, two or three decades ago, were based mainly on isotope-incorporation studies using labeled presumed precursors. By this technique, many excellent results were obtained that provided a fruitful basis for more detailed subsequent research. It must be recognized, however, that such procedures find their limitations when biochemical reaction mechanisms are to be tackled or when a strict discrimination between precursors, intermediates, and obligate intermediates is required (only the latter is mandatory for the biosynthesis of a particular compound) (Brown, 1966). Since, from pragmatic considerations, it is almost impossible to work with higher plant mutants, enzymology is the only practicable tool for solving such questions. This method has indeed developed to an important and almost indispensable procedure when analyzing the details of biochemical events, and many gaps in our knowledge have been filled by this means. As a consequence, this chapter will emphasize mainly the results that have been achieved by enzymatic studies on the biosynthesis and metabolism of phenolic acids and their derivatives.

Before discussing these pathways in detail, it appears appropriate to insert a few comments on the nomenclature of phenolic acids which will frequently be mentioned in the subsequent sections. Table I summarizes the trivial names of a variety of common cinnamic and benzoic acids, which still are widely preferred instead of the often lengthy systematic nomenclature. Derivatives of these acids (e.g., esters) should always be characterized by the ending *-oyl* to avoid confusion with the corresponding alcohol derivatives ending with *-yl* (e.g., benzoyl versus benzyl).

II. Biosynthesis of Cinnamic Acids: The Phenylalanine–Cinnamic Acid Pathway

It is well documented that the C_6C_3 carbon skeleton characteristic of cinnamic acids originates from L-phenylalanine or, in certain cases, from its hydroxylated analog L-tyrosine. These two aromatic amino acids are

Table I
Nomenclature of Common Cinnamic and Benzoic Acids

Substitution pattern 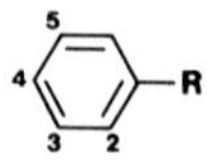	Cinnamic acids (R = CH=CH-COOH)	Benzoic acids (R = COOH)
—	Cinnamic	Benzoic
2-OH	*o*-Coumaric	Salicylic
3-OH	*m*-Coumaric	*m*-Hydroxybenzoic
4-OH	*p*-Coumaric	*p*-Hydroxybenzoic
4-OMe	*p*-Methoxycinnamic	Anisic
3-OH, 4-OH	Caffeic	Protocatechuic
3-OMe, 4-OH	Ferulic	Vanillic
3-OH, 4-OMe	Isoferulic	Isovanillic
3-OCH_2O-4	3,4-Methylenedioxycinnamic	Piperonylic
3-OMe, 4-OMe	3,4-Dimethoxycinnamic	Veratric
3-OH, 4-OH, 5-OH	3,4,5-Trihydroxycinnamic	Gallic
3-OMe, 4-OH, 5-OH	5-Hydroxyferulic	—
3-OMe, 4-OH, 5-OMe	Sinapic	Syringic

the end products, together with tryptophan, of the so-called shikimic acid pathway. This biogenetic route, which was originally identified in bacteria, characterizes both microorganisms and plants but does not operate in animals, with the consequence that such amino acids represent essential nutrient factors for these organisms. As a constituent of primary metabolism, the shikimate pathway lies beyond the scope of this article and will not be discussed here (for relevant reviews, see Floss, 1979; Gilchrist and Kosuge, 1980).

Among these amino acids, phenylalanine is of outstanding importance, as this compound represents the common precursor from which the majority of phenolic natural products is derived. The starting point of these biosynthetic routes is the deamination of phenylalanine to cinnamic acid, a principal step that marks the transition from primary to secondary metabolism. The aromatic ring of the resulting parent compound cinnamic acid is further modified by subsequent hydroxylation and methylation steps, yielding the derivatives depicted in Fig. 2. The entire sequence is known as the "phenylalanine–cinnamic acid pathway." The three principal reaction types involved in this sequence (i.e., deamination, hydroxylation, and methylation) will be discussed below in more detail.

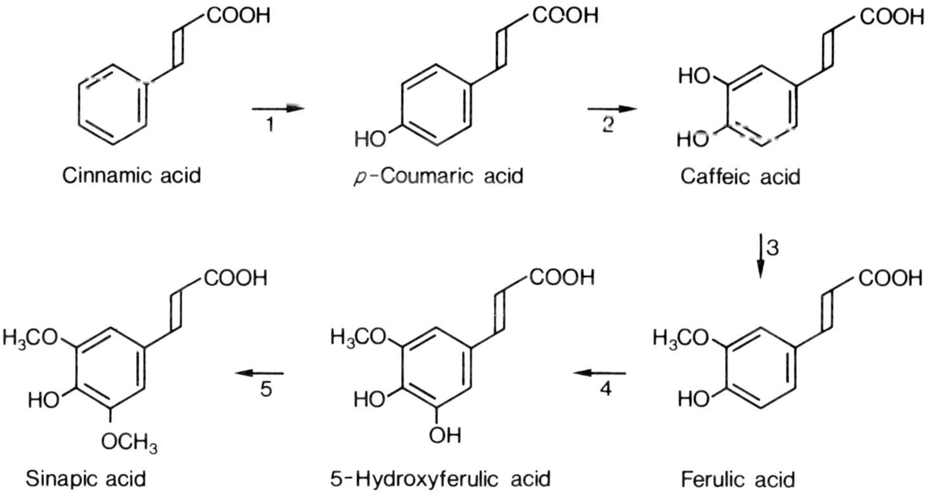

Fig. 2. The cinnamic acid pathway. The enzymes involved are (1) cinnamate 4-hydroxylase, (2) *p*-coumarate 3-hydroxylase, (3 and 5) catechol *O*-methyltransferase, and (4) ferulate 5-hydroxylase. A less probable pathway, which, however, cannot be excluded totally to date, is the sequence caffeic → 3,4,5-trihydroxycinnamic → 5-hydroxyferulic acid.

A. *Ammonia-Lyases*

Although it had long been known from feeding experiments that phenylalanine can be converted to cinnamic acids, the mechanism of this reaction remained unclear. Initial assumptions on the intermediacy of phenylpyruvate and phenyllactate (Brown *et al.*, 1959), based on conceivable analogies with the metabolism of fatty acids, could not be verified. The question was settled by the detection of a previously unknown enzyme, phenylalanine ammonia-lyase (PAL), which catalyzed the direct nonoxidative deamination of L-phenylalanine to cinnamic acid in one step (Koukol and Conn, 1961). Simultaneously, an analogous tyrosine ammonia-lyase (TAL) was isolated (Neish, 1961). Today, the existence of such enzymes has been proved not only with a wide variety of higher plants but also with bacteria, fungi, and algae, which clearly demonstrates the general importance of this conversion in phenolic metabolism. Consequently, PAL has become the most intensively studied enzyme in this field, resulting in a vast number of publications dealing with its occurrence, structure, reaction mechanisms, catalytic properties, synthesis, and activation or inhibition, and only some essentials of these widespread aspects can be discussed here.

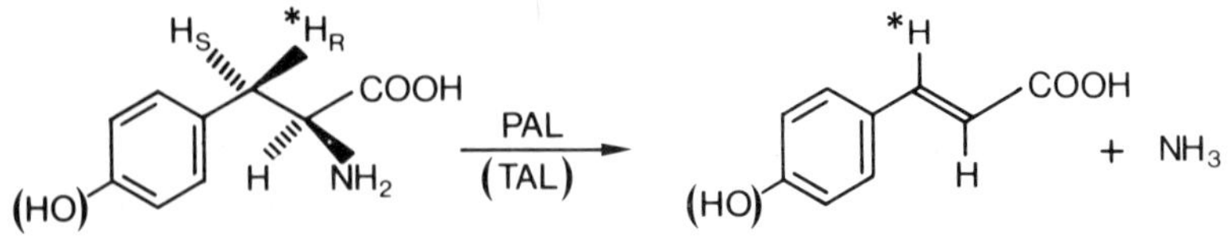

Fig. 3. Stereochemistry of the nonoxidative deamination of L-phenylalanine (L-tyrosine) by phenylalanine (tyrosine) ammonia-lyase to *trans*-cinnamic (*trans-p*-coumaric) acid. Modified after Ellis *et al.* (1973).

The reaction catalyzed by PAL or TAL does not require cofactors. With stereospecifically tritium-labeled substrates, it has been shown that the pro-3*S* hydrogen is lost in the formation of cinnamic or *p*-coumaric acid, respectively (Fig. 3) (Hanson *et al.*, 1971; Ife and Haslam, 1971; Ellis *et al.*, 1973). The release of the elements of ammonia thus represents an anti elimination, which leads directly to the natural trans configuration of cinnamic acids. The reactions occurring at the active site of PAL, which is thought to contain a dehydroalanine imine derivative

$$[\rangle C{=}N{-}C_{\alpha}({=}C_{\beta}H_2)CO{-}]$$

as prosthetic group, have been discussed in detail by Hanson and Havir (1979, 1981).

A yet unsolved problem is the question of the existence of a genuine tyrosine-specific ammonia-lyase. Although many enzyme preparations have been reported to exhibit only PAL activities, TAL (which is found mainly in grasses) never occurs independently and is always accompanied by high PAL levels. One explanation would be that TAL is only the result of certain less specific PAL species accepting tyrosine as substrate also, in contrast to the phenylalanine-specific enzymes that eventually underwent small structural changes, thus excluding tyrosine from the active site (Hanson and Havir, 1979). This view is supported by studies of Havir *et al.* (1971), which found a constant PAL : TAL ratio in the purification of the maize enzyme and a parallel loss of both activities in inactivation experiments. These considerations, however, are not necessarily contradictory to the existence of two separate enzymes in other organisms. Arguments for the latter opinion have included varying PAL : TAL ratios during purification, different pH optima for both substrates, or independent stimulatory effects [$MnCl_2$ treatment of castor bean seedlings, for instance, resulted in a pronounced stimulation of PAL, but not of TAL (Gregor, 1976)]. It is apparent that settling of these questions requires future work with a variety of highly purified enzyme preparations.

Structural studies with PAL from various sources revealed that the enzymes isolated so far from plants or fungi have almost the same molecular weight, about 300,000–330,000 and are composed of four subunits that are either identical or closely similar in size. Only the enzyme from the procaryote *Streptomyces*, an actinomycete, is significantly smaller (molecular weight 226,000) (see Hanson and Havir, 1981). Since PAL operates at an important metabolic branching point, this tetrameric structure offers an effective opportunity for regulating the activity of this key enzyme of phenolic metabolism via allosteric mechanisms. There are numerous examples of inhibitors that affect the catalytic activity of PAL. Moreover, the levels of PAL are controlled by a variety of internal or external factors (e.g., developmental status, wounding, infection, hormones, or light), and it appears that these effects involve *de novo* synthesis of the enzyme. In addition, PAL levels can be regulated by reversible association with specific macromolecular inhibitors whose existence has been predicted from *in vivo* experiments with the protein synthesis inhibitor cycloheximide. Such inactivators have been isolated from sunflower leaves (Creasy, 1976) and gherkin hypocotyls (French and Smith, 1975; Billett *et al.*, 1978), and the latter was characterized as a hydrophobic protein having a molecular weight of 19,000.

In this context, the powerful PAL-specific inhibitor 2-aminooxy-3-phenylpropionic acid (AOPP) commands interest. This hydroxylamino analog of phenylalanine was found to block PAL activity *in vivo* at the nanomole level without apparent side effects (Amrhein *et al.*, 1976; Amrhein and Gödeke, 1977; Amrhein, 1979; Holländer *et al.*, 1979), thus allowing studies on the physiology of phenolic metabolism by a strategy similar to research with bacterial mutants. In the presence of this inhibitor, relatively recent studies (Amrhein *et al.*, 1983) revealed the complete absence of lignin in mung bean seedlings, resulting in collapsed xylem vessels, as shown by scanning electron microscopy. Biochemical proof of the importance of lignin for the resistance against tensile forces in conducting xylem cells has thus been provided. For a detailed treatment of the various aspects outlined above, see the reviews of Camm and Towers (1977) and Hanson and Havir (1979, 1981).

B. *Hydroxylases*

The second reaction type encountered in the cinnamic acid pathway involves two consecutive hydroxylation steps catalyzing the sequence

cinnamate ⟶ *p*-coumarate ⟶ caffeate

(reactions 1 and 2 in Fig. 2). Whereas numerous details of these reactions have been elucidated during the past decade, the subsequent ferulate 5-

hydroxylase (reaction 4 in Fig. 2) has been identified only recently (Grand, 1984). The eventual conversion of caffeate to 3,4,5-trihydroxycinnamate, in contrast, is still a matter of hypothetical disputation. It should be mentioned in this context that a further hydroxylation step, the formation of tyrosine from phenylalanine, is known with certainty from animal tissues only and appears not to occur in plants (Hasegawa and Maier, 1972; see also Camirand *et al.*, 1983). Concerning the first-mentioned hydroxylases, which are usually designated as cinnamate 4-hydroxylase and *p*-coumarate 3-hydroxylase, respectively, it is obvious that both enzymes follow the same general equation

$$RH + O_2 + AH_2 = ROH + H_2O + A$$

and thus are classified as monooxygenases ("mixed-function oxidases"). In contrast to oxidases utilizing O_2 as a terminal electron acceptor, these enzymes incorporate half of the O_2 molecule into the substrate and reduce the other half to H_2O with the aid of an external electron donor AH_2. The details of the reaction mechanisms of these two hydroxylases differ so significantly, however, that a separate discussion of their characteristics is justified. For an extensive treatment of these enzymes, see the review articles of Butt (1979, 1980), Butt and Lamb (1981), Hanson and Havir (1979), Malmström (1982), West (1980), and White and Coon (1980).

1. Cinnamate 4-Hydroxylase

This multienzyme system specifically introduces a hydroxyl group to the para position of *trans*-cinnamate and is completely inactive with the cis isomer of the substrate (Pfändler *et al.*, 1977). The hydroxylase is a tightly membrane-bound hemoprotein and is linked with the cytochrome *P*-450 system, which receives reducing equivalents through a short electron transport chain, with $NADPH_2$ as the preferred cosubstrate. The involvement of *P*-450 has been studied in seedlings from pea (Russel, 1971) or *Sorghum* (Potts *et al.*, 1974), in swede root (*Brassica napobrassica*) disks (Hill and Rhodes, 1975), and in slices of potato (Rich and Lamb, 1977) or of *Helianthus tuberosus* (Benveniste *et al.*, 1977; Reichhart *et al.*, 1980). It is characteristic of the *P*-450 system that it is inactivated by the binding of carbon monoxide, resulting in a complex exhibiting an absorption maximum at 450 nm. Photodissociation of this complex with light of 450 nm restores the enzyme activity (Potts *et al.*, 1974). It should be mentioned in this context that 1-aminobenzotriazole has recently been reported as a selective, and perhaps specific, suicide substrate for cinnamate 4-hydroxylase in plant tissues (Reichhart *et al.*, 1982).

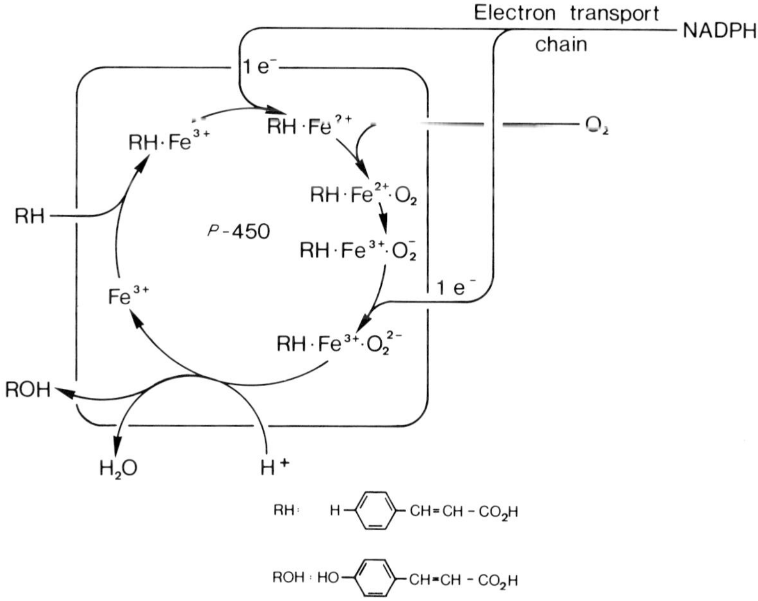

Fig. 4. Principles of the reaction catalyzed by cinnamate 4-hydroxylase.

Besides para hydroxylation, hydroxylation at the ortho position yielding *o*-coumaric acid has also been observed. Such reactions have been studied in sweet clover (*Melilotus alba;* Gestetner and Conn, 1974), potato (Czichi and Kindl, 1975a), the green alga *Dunaliella marina* (Czichi and Kindl, 1975b), soybean cell cultures (Postius and Kindl, 1977), or cucumber cotyledons (Czichi and Kindl, 1977). These enzymes, which again were found to be closely associated with membranes (endoplasmic reticulum or, particularly, chloroplast lamellae), appear to be involved in coumarin biosynthesis.

Some details of the *P*-450–catalyzed hydroxylation reaction are depicted in Fig. 4. After binding of substrate RH to the oxidized enzyme, two one-electron reduction steps and binding of O_2 occur, resulting in the formation of a Fe^{3+} peroxide. Finally, cleavage of the oxygen–oxygen bond leads to the liberation of the products ROH and H_2O, thus regenerating the reoxidized resting enzyme form. This last step is only partially understood. According to the "oxenoid hypothesis," a peracid $RCO(OO)^-$ could oxidize the aromatic substrate to an aryl epoxide, which then rearranges in a nonenzymatic reaction to the hydroxylated derivative (Fig. 5). Trapping experiments with the nucleophilic agent glutathione provided indirect evidence for the actual existence of this presumed ex-

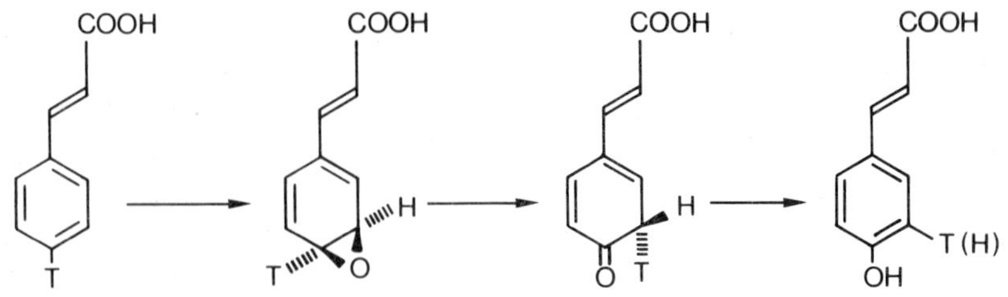

Fig. 5. The hydroxylation of cinnamic acid via an intermediary arene oxide and the NIH shift.

tremely labile intermediate (Diesperger and Sandermann, 1978). This reaction has long been known to involve a so-called NIH-shift, i.e., the transfer of the 4 hydrogen to the adjacent 3 position. With tritium-labeled substrates, a pronounced retention of this isotope has been observed—85% in experiments with *Catalpa* leaves, for instance (Zenk, 1967). The preferential loss of protium in such experiments is explained as the result of a large kinetic isotope effect.

2. *p*-Coumarate 3-Hydroxylase

This enzyme activity is also known by a variety of other names, including phenolase, polyphenol oxidase, tyrosinase, and catechol oxidase. There are several reasons for this rather confusing nomenclature. First, such enzymes often exhibit very low degrees of substrate specificity. Second, phenolases normally catalyze two different reactions. One is the *hydroxylase* or *monophenolase* reaction, by which a monophenol is hydroxylated according the equation

$$\text{monophenol} + O_2 + AH_2 = o\text{-diphenol} + H_2O + A$$

where AH_2 represents a variety of external reductants such as $NAD(P)H_2$, ascorbate, tetrahydropteridine, or even the product itself. The second reaction, the *oxidase* reaction, leads to the formation of quinones:

$$o\text{-diphenol} + \tfrac{1}{2}O_2 = \text{orthoquinone} + H_2O$$

The situation is further complicated by the fact that enzyme preparations are known that preferentially catalyze the hydroxylation reaction, while others are mainly or exclusively involved in the formation of quinones. Consequently, it is still a matter of dispute to what extent phenolases are related with the pathway discussed here. In this context, the discovery that tentotoxin [cyclo(*N*-methyl-L-alanyl-L-leucyl-N-methyl-*trans*-dehydrophenylalanylglycyl)], a chlorosis-inducing fungal toxin, completely eliminated phenolase activity in mung bean seedling but had virtually no effect on the quality and quantity of caffeic acid derivatives in the plants demands interest (Duke and Vaughan, 1982). It follows that phenolase cannot be the exclusive catalyst in the hydroxylation of *p*-coumaric acid, and, indeed, other systems carrying out this conversion have been reported. Peroxidase, for instance, causes hydroxylations in the presence of O_2 and dihydroxyfumarate. This reaction appears to be rather unspecific, however, and the requirement for the cofactor dihydroxyfumarate casts severe doubt on the physiological significance of this system (see West, 1980). Illuminated chloroplasts are able to hydroxylate *p*-coumarate in the abscence of additional reductants. Since this conversion is completely inhibited by superoxide dismutase, it was concluded that $O_2^{\cdot -}$ radicals generated by the chloroplast are involved in providing reductants required in the formation of caffeic acid (Halliwell, 1975). This reaction, however, is supposed to involve a phenolase, and it is evident that clarification of the entire question discussed here will require further intensive investigations. In this context, it should be noted that the function of phenolase in higher plants has been reviewed recently by Vaughn and Duke (1984).

Finally, some data on the 3-hydroxylation of *p*-coumarate will be reported here that mainly refer to studies on the phenolase from spinach beet (*Beta vulgaris*) that has been purified to apparent homogeneity by Vaughan *et al.* (1975). This copper-containing enzyme is, like cinnamate hydroxylase, bound to particles, especially to chloroplast lamellae, from which it could be released by detergents. No changes in the hydroxylase : oxidase ratio were observed during the purification of the enzyme (Vaughan and Butt, 1969), which was found to be a monomeric protein with a molecular weight of 40,000 (Vaughan *et al.*, 1975). In kinetic studies by McIntyre and Vaughan (1975), a ping-pong mechanism was proposed involving an intermediary oxidized stable form of the enzyme (Fig. 6). Evidence was presented to show that the oxidized enzyme remaining after the release of caffeic acid is subsequently reduced by catalytic amounts of that product. The resulting caffeoyl *o*-quinone is recycled by the above-mentioned external cofactors. These results are consistent with previous observations where enzyme assays showed a distinct lag period before hydroxylation reached its maximum rate, and this lag could be eliminated by the addition of caffeic acid or other *o*-diphenols (Vaughan and Butt, 1970). Further studies (Vaughan and McIntyre, 1975) revealed that treatment of the phenolase with H_2O_2, yielding an oxygenated enzyme species, resulted in a stimulation of the hydroxylating reaction but did not affect the oxidation of caffeic acid. This observation supports the

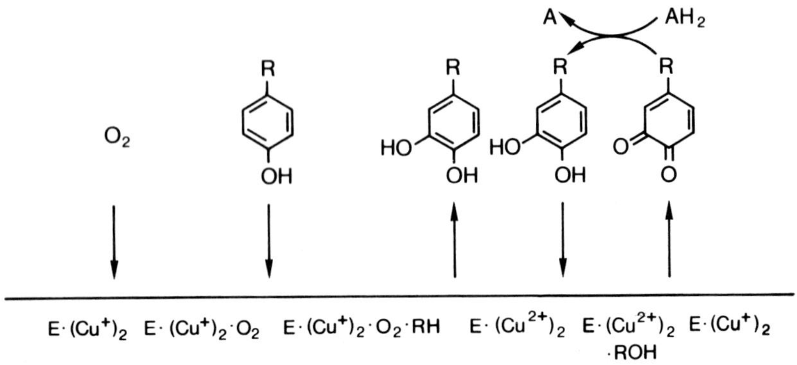

Fig. 6. Flow diagram of the reaction sequence catalyzed by *p*-coumarate 3-hydroxylase. $NAD(P)H_2$, ascorbate, reduced pteridines, or photosynthetic electron transport chains can serve as the external reductant AH_2. Modified after McIntyre and Vaughan (1975). R = CH═CH—COOH.

reaction sequence depicted in Fig. 6 and suggests that the formation of the enzyme–O_2 complex is a rate-limiting step.

3. Ferulate 5-Hydroxylase

As a result of labeling experiments, it has long been suggested that the conversion of ferulate to sinapate proceeds via 5-hydroxyferulate. In spite of many attempts, the hydroxylase catalyzing the synthesis of this intermediate could be detected only recently (Grand, 1984). The enzyme, isolated from young poplar stems, was found to be associated with the microsomal fraction. Moreover, it was characterized as an O_2- and NADPH-requiring, cytochrome *P*-450-dependent mixed-function monooxygenase that, however, is not identical with cinnamate 4-hydroxylase. This discovery of a second membrane-bound hydroxylase of the *P*-450 type within the cinnamic acid pathway lends further support to recent doubts as to the role of phenolase in relation to this pathway (see Section II,B,2).

C. *O-Methyltransferases*

The third reaction type encountered in the cinnamate pathway comprises two methylation steps yielding ferulate and sinapate, respectively (reactions 3 and 5 in Fig. 2). In these reactions, a phenolic hydroxyl of the substrate is converted to a methoxy group by a conventional transmethylation, with *S*-adenosyl L-methionine serving as the methyl donor. The resulting demethylated cofactor, *S*-adenosyl L-homocysteine, can further be split to adenosine and the amino acid by a specific hydrolase, and it is assumed that this degradation is of importance with respect to the unfavorable equilibrium of the preceding methylation step (Poulton and Butt, 1976).

It should be noted that the methyltransferase activities mentioned here require substrates characterized by a 3,4-dihydroxy substitution pattern of the aromatic ring. Consequently, these enzymes are usually found in the literature under the trivial name "catechol *O*-methyltransferase" (OMT or COMT). Such transferases have been isolated from many different plants (see Poulton, 1981). In most cases, these enzymes were found to be meta specific, but also methylation of para or both para and meta hydroxy groups has been observed. With respect to cinnamic acids, there is increasing evidence from studies with purified enzyme preparations that these compounds are methylated exclusively or at least predominantly in the meta position (e.g., Poulton and Butt, 1975; Poulton *et al.*, 1976a; Legrand *et al.*, 1976; Sütfeld and Wiermann, 1978, 1980; Sharma and Brown, 1979; Tsang and Ibrahim, 1979a,b), whereas other phenolic substrates such as benzoic acids, flavonoids, or coumarins can be methylated either in the para position (e.g., Legrand *et al.*, 1976; Tsang and Ibrahim, 1979a,b) or in the meta position (Poulton *et al.*, 1976b; Sütfeld and Wiermann, 1978). This pronounced enzyme specificity explains why the overwhelming majority of naturally occurring cinnamic acids and many of their derivatives are characterized by a common free phenolic hydroxyl function at the 4 position. This trend, of course, does not exclude other possibilities. For instance, enzymes that have been extracted from the endoplasmic reticulum of *Populus* cells catalyzed the methylation of caffeic acid not only to ferulic acid but also to the rather uncommon products isoferulic and 3,4-dimethoxycinnamic acid (Charrière-Ladreix, 1979).

Finally, some results from comparative studies on *O*-methyltransferases from gymnosperms and angiosperms are of importance (see Chapter 7, Section VI). The examples given in Table II show that these enzymes catalyze meta-directed methylations in both taxonomic groups. A marked difference, however, is encountered in their substrate specificities. The gymnosperm enzymes catalyze the almost exclusive formation of ferulate from caffeate but are rather inactive with 5-hydroxyferulate. The analogous angiosperm enzymes, in contrast, methylate both substrates, with a more or less pronounced preference for the formation of sinapate. (The moderate reaction observed in both cases with trihydroxycinnamate is irrelevant, as this acid probably does not occur naturally.) The two types of methyltransferases were thus named "monofunction"

Table II

Substrate Specificity of Catechol *O*-Methyltransferases Related to Lignification from Gymnosperms and Angiosperms[a]

Substrate	Product (found or expected)	Relative activity (%)[b] of OMT from			
		Pinus[c]	*Gingko*[c]	*Populus*[d]	*Bambusa*[d]
p-Coumarate	*p*-Methoxycinnamate	0	—	0	0
m-Coumarate	*m*-Methoxycinnamate	—	—	0	0
Caffeate	Ferulate	100	100	100	100
	Isoferulate	0	0	—	—
Isoferulate	3,4-Dimethoxycinnamate	0	—	0	5
5-Hydroxyferulate	Sinapate	5	6	320	124
3,4,5-Trihydroxycinnamate	5-Hydroxyferulate	25	25	60	50

[a] From Gross (1980).
[b] Data refer to the activity observed for the formation of ferulate = 100%.
[c] Data from Shimada (1972).
[d] Data from Higuchi (1976).

and "difunction" OMT, respectively (Shimada *et al.*, 1973). It was concluded that these different substrate specificities represent one plausible biochemical explanation for the well-known significant differences between the lignins from gymnosperms (and certainly pteridophytes, too) and those from angiosperms. In this context, the isolation of a lignin-specific OMT from differentiating xylem of 10-year-old *Populus* trunks should be mentioned (Kuroda *et al.*, 1981). The $V_{max} : K_m$ ratio of this enzyme for 5-hydroxyferulate was about five times greater than that for caffeate, thus being in full accordance with the above results, which had been obtained with extracts from young shoots.

III. Cinnamoyl Coenzyme A Thioesters

As mentioned in the introduction, cinnamic acids have been recognized as precursors of a wide variety of more or less complex phenolic natural products. This biogenetic relation is evident with certain compounds (e.g., cinnamyl alcohols, cinnamoyl esters and amides, coumarins); in other cases, it is more obscured, but often observed similarities of their substitution patterns suggest that the compounds originate from cinnamic acids. Common to all these classes of metabolites, however, is that the acrylic side chain of cinnamic acids, and particularly their carboxyl group, is primarily involved in such conversions. Moreover, for thermodynamic reasons, it is apparent that the formation of these derivatives requires the participation of carboxyl-activated intermediates. In particu-

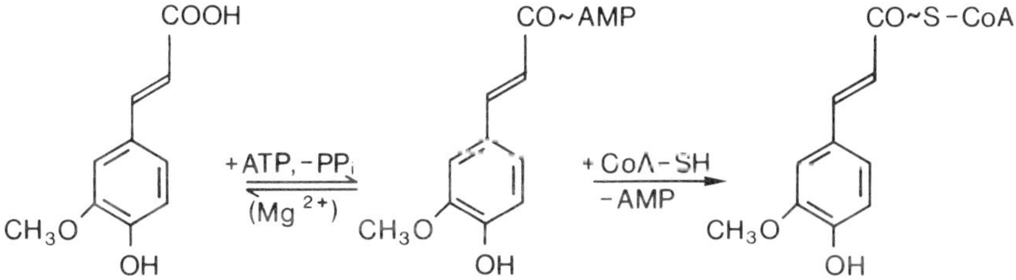

Fig. 7. Enzymatic activation of cinnamic acids by acyl-CoA ligase. Reaction shown with ferulic acid as substrate.

lar, cinnamoyl coenzyme A (CoA) thioesters have frequently been postulated as the most likely candidates in the formation of flavonoids (Grisebach, 1962), cinnamyl alcohols (Brown *et al.*, 1959; Neish, 1964), esters or amides (Gross and Zenk, 1966), and benzoic acids (Zenk, 1966). Consequently, it was postulated that cinnamoyl-CoA esters occupy a central role in the metabolism of plant phenolic compounds (Zenk, 1971; Zenk and Gross, 1972). Enzymatic studies have demonstrated unequivocally the validity of this hypothesis in many cases (see Zenk, 1979).

Initially, the most serious problem in studies related to this field was the difficulty of synthesizing CoA esters of cinnamic acids, because of their troublesome phenolic hydroxyl groups and the risk of addition reactions on the acrylic side chain. The problem was solved by using a fairly nonspecific acyl-CoA ligase from beef liver mitochondria (Gross and Zenk, 1966) that activates cinnamic acids according to the reaction depicted in Fig. 7. More recently, similar enzymes from other sources have been used, the most efficient one being a hydroxycinnamate : CoA ligase from a strain of *Pseudomonas putida* (Zenk *et al.*, 1980). Apart from these enzymatic procedures, chemical methods have been developed that are especially useful for the preparation of the CoA derivatives of those cinnamic acids that are not readily activated enzymatically (Stöckigt and Zenk, 1975).

The most prominent property of cinnamoyl-CoA esters is their characteristic UV spectrum. Besides a maximum near 260 nm due to the adenosine moiety of CoA, a second peak in the long wave UV is observed that represents the absorption of the thioester linkage (Fig. 8). The maximum of this peak depends on the substitution pattern of the cinnamate moiety, ranging from 311 nm for cinnamoyl-CoA to 352 nm for sinapoyl-CoA (Gross and Zenk, 1966; Stöckigt and Zenk, 1975). Together with the high extinction coefficient ε at these wavelengths (average about 20×10^6 cm^2/mol), these spectral characteristics provide the basis for a sensitive and convenient photometric assay that permits the direct measurement of

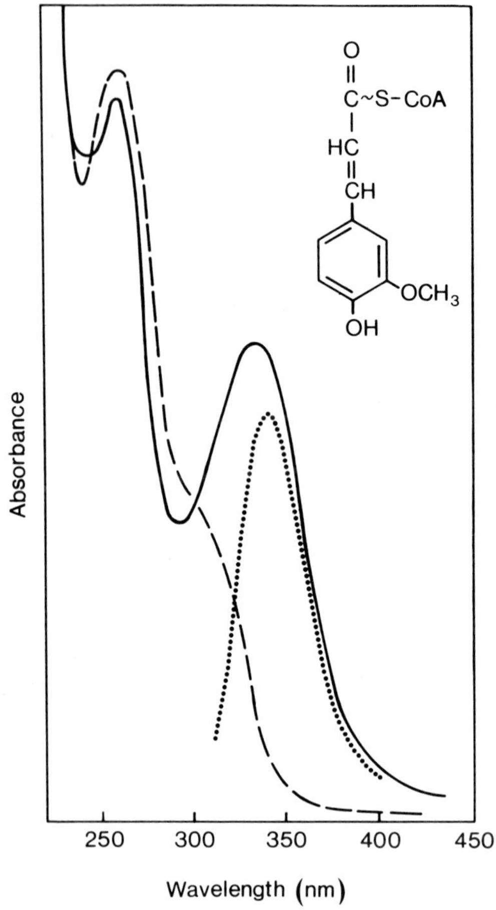

Fig. 8. Absorption spectrum of feruloyl-CoA in 0.1 *M* phosphate buffer, pH 7.0, before (——) and after (–––) hydrolysis; (·····) difference spectrum.

reactions forming or consuming cinnamoyl-CoA esters, without significant interference of the absorption of proteins, nucleic acids, or phenolic compounds (Gross and Zenk, 1966; Zenk, 1979).

The first positive results in the search for enzymes from higher plants catalyzing the formation of cinnamoyl-CoA esters were reported several years ago with extracts from spinach-beet leaves (Walton and Butt, 1970, 1971). More detailed knowledge came from intensive studies on the biosynthesis of flavonoids and lignins. Meanwhile, such enzymes have been recognized as common constituents of plants—a situation that reflects the now well-established central role of cinnamoyl-CoA thioesters in the metabolism of plant phenolics (see Gross *et al.*, 1975; see also the reviews of Gross, 1977, 1979, 1981, and Zenk, 1979). Similar enzymes have also been found in the basidiomycete *Polyporus hispidus* (Vance *et al.*, 1975) and, as already mentioned, in a bacterium (*Pseudomonas putida*, Zenk *et al.*, 1980).

Summarizing the results obtained in several laboratories with a variety of such CoA-ligases, it is evident that the basic properties of such enzymes are fairly uniform. They depend strictly on ATP as a cosubstrate that cannot be substituted by other nucleoside triphosphates, and they require Mg^{2+} (or other divalent cations) as cofactor. ATP is split in the reaction to AMP and pyrophosphate (PP_i). Stoichiometric studies revealed that the consumption of ferulate and ATP and the formation of feruloyl-CoA, AMP, and PP_i occurred in equimolar ratios. By analogy with the well-characterized activation of fatty acids, it thus appeared most likely that the reaction proceeds via an intermediate acyl adenylate, which further reacts with CoA to form the thioester (see Fig. 7). This reaction mechanism has been proven by $^{32}PP_i$–AMP exchange experiments in the absence of CoA and was further substantiated using authentic feruloyl AMP as substrate (Gross and Zenk, 1974). The pH optima were found around pH 7.5, with the exception of an isoenzyme from soybean cell cultures (pH 8.3; Knobloch and Hahlbrock, 1975) and the enzyme from *P. putida* (pH 6.7; Zenk *et al.*, 1980). Most enzyme preparations had a molecular weight of ~55,000, but higher or lower values have also been observed occasionally: pea seedlings, 75,000 (Wallis and Rhodes, 1977); parsley cell cultures, 67,000 (Knobloch and Hahlbrock, 1977); *Erythrina crista-galli*, a legume, 40,000 (Kutsuki *et al.*, 1982a).

With respect to the cinnamate substrates, it was found that the natural trans form is the preferred substrate (Knobloch and Hahlbrock, 1975). Substrate specificity studies revealed that most of the enzymes investigated so far had a pronounced affinity towards hydroxylated cinnamic acids, whereas cinnamate itself or its methoxy derivatives were not (or were only minimally) accepted as substrates. Sinapic acid, the precursor of the syringyl unit characteristic of angiosperm lignins, however, was activated only in a few cases (see Gross, 1980; Kutsuki *et al.*, 1982, and references cited therein). Such limitations cause no problem in the biosynthetic of most phenolics (it is known that secondary substitution reactions can occur at later biogenetic stages); it is highly surprising, however, that even extracts from a wide variety of actively lignifying angiosperm tissues are definitively devoid of this enzyme activity (Gross *et al.*, 1975; Kutsuki *et al.*, 1982a; Lüderitz *et al.*, 1982). This discrepancy is still an open and challenging question in our understanding of the metabolic processes involved in lignification. As a consequence, alternative routes to sinapyl alcohol must be considered, too (see Kutsuki *et al.*, 1982a; see also Chapter 7, Section VI). Another explanation, as discussed by Gross (1980), could be that enzymes activating sinapate are produced in lignifying tissues only at later developmental stages, with the consequence that disruption and extraction of fully differentiated, heavily lignified cells bears the inherent risk of extensive inactivation of such enzymes.

To date, none of these enzymes has been found to accept substrates other than cinnamic acids. This fact, together with their pronounced preference for hydroxylated cinnamic acids, has led to proposing the systematic name "hydroxycinnamate:CoA ligase (AMP)" (Gross and Zenk, 1974); alternatively, the name "*p*-coumarate:CoA ligase" is used for this class of enzymes, referring to the best substrate in most cases.

It is the purpose of these ligases to provide the activated cinnamic acids required in the formation of a wide variety of important secondary plant products. These enzymes may thus be regarded as the catalysts of the ultimate step of the so-called general phenylpropanoid metabolism (see Grisebach, 1977), which comprises the reactions from phenylalanine to cinnamoyl-CoA thioesters. Some essential characteristics of the routes emerging from this central branch point will be discussed in the subsequent sections.

IV. Natural Products Derived from Cinnamic Acids

As summarized in Fig. 1, cinnamic acids are of outstanding importance in the metabolism of phenolic compounds because of their role as common precursors of the majority of these natural products. Many of these pathways require participation of energy-rich intermediates, and enzymatic studies have shown unequivocally that CoA thioesters are involved in a variety of these conversions. It should not be overlooked, however, that other solutions of this problem have also been realized in nature. Glucose esters, for example, appear to be synthesized generally from the free acid and UDP-glucose. In the subsequent sections, some principles of the pathways leading from cinnamic acids to the various classes of phenolic acid derivatives are discussed.

A. *Cinnamic Acid Esters*

Cinnamic acid esters occur widely in the plant kingdom. A great variety of hydroxylated compounds has been shown to serve as the alcoholic moiety, including sugars, aromatic or aliphatic alcohols, hydroxyacids, sterols, terpenoids, and alkaloids (for structures of such esters, see Zenk, 1979; Gross, 1981). It has been postulated that the biosynthesis of these compounds involves the participation of activated cinnamoyl-CoA esters. The correctness of this assumption was proved for the first time by Stöckigt and Zenk (1974), who showed that cell-free extracts from cell cultures of *Nicotiana alata* catalyzed the esterification of caffeoyl-CoA and quinic acid, yielding the ubiquitously occurring phenolic plant constituent chlorogenic acid. Similar enzymes were isolated later from tomato fruits (Rhodes and Wooltorton, 1976, 1977), buckwheat hypocotyls (Ulbrich *et al.*, 1976; Ulbrich and Amrhein, 1978), potato tubers (Lamb, 1977; Rhodes and Wooltorton, 1978), and cell cultures from *Stevia rebaudiana* and various other species (Ulbrich and Zenk, 1979).

Further experiments demonstrated the free reversibility of this reaction, a fact that clearly facilitates the assay of this enzyme's activity. Substrate specificity studies revealed that several of the above enzyme preparations also exhibited high affinities toward *p*-coumaroyl-CoA, yielding *p*-coumaroyl quinate; thus, the name hydroxycinnamoyl-CoA : quinate hydroxycinnamoyl transferase was chosen. These results are also of interest with respect to the conflicting conclusions on the biosynthesis of chlorogenic acid that were drawn from tracer experiments, i.e., whether this compound is formed directly from caffeic acid or via hydroxylation of *p*-coumaroyl quinate (see Nagels and Parmentier, 1974, 1976). As pointed out by Zenk (1979), it must be assumed now that both possibilities occur, the actual pool sizes probably being the determining factor.

Meanwhile, this reaction type has been observed also in the formation of other cinnamoyl esters (Fig. 9). Extracts from *Cichorium endiva* cell suspension cultures (Ulbrich and Zenk, 1979, 1980) or of potato tubers (Rhodes *et al.*, 1979) catalyzed the synthesis of *p*-coumaroyl shikimate. An enzyme from pea seedlings acylated the sugar moiety of the flavonol kaempferol 3-triglucoside in the presence of *p*-coumaroyl-CoA (Saylor and Mansell, 1977). Finally, the formation of *p*-coumaroyl lupinine by cell-free extracts from *Lupinus luteus* seedlings was stimulated by the addition of CoA, ATP, and Mg^{2+}, the cofactors of acyl-CoA ligases (Murakoshi *et al.*, 1977).

Clear-cut evidence has thus been obtained that cinnamoyl esters can be formed via the corresponding acyl-CoA thioesters—a reaction mechanism that, however, is not exclusive. In a variety of experiments, it has been shown that 1-*O*-acyl glucose esters are formed in a completely different manner, by the reaction of UDP-glucose with the free acid. Enzymes catalyzing such an esterification of cinnamic (especially hydroxycinnamic) acids have been isolated from various plants (Corner and Swain,

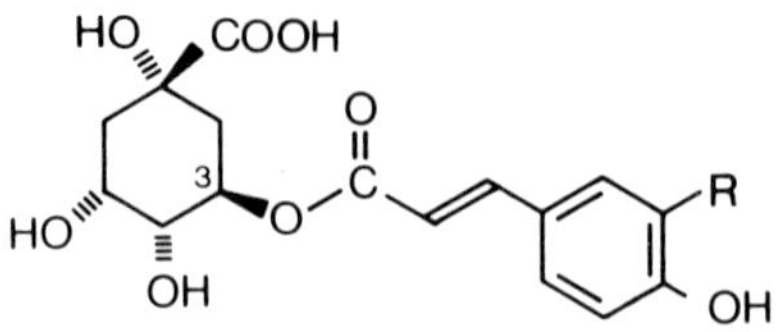

p-Coumaroylquinic acid (R=H)
Chlorogenic acid (R=OH)

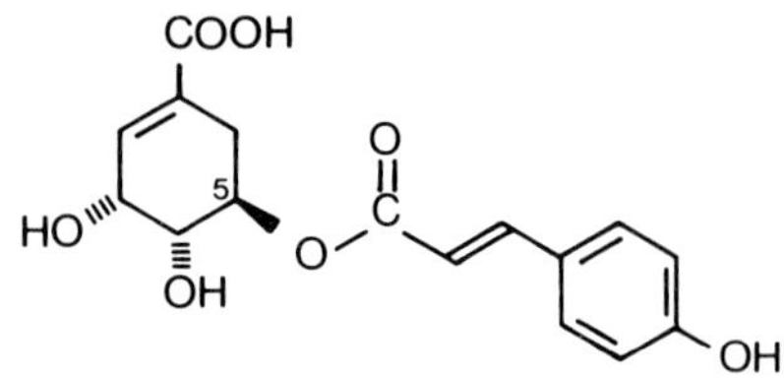

5-*O*-*p*-Coumaroylskikimic acid

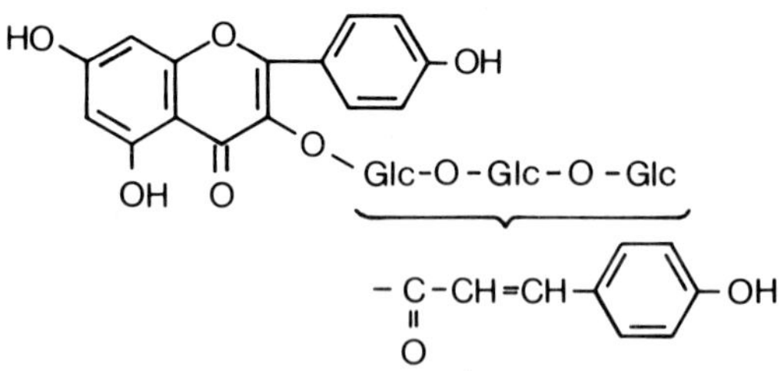

Kaempferol-3-(*p*-coumaroyltriglucoside)

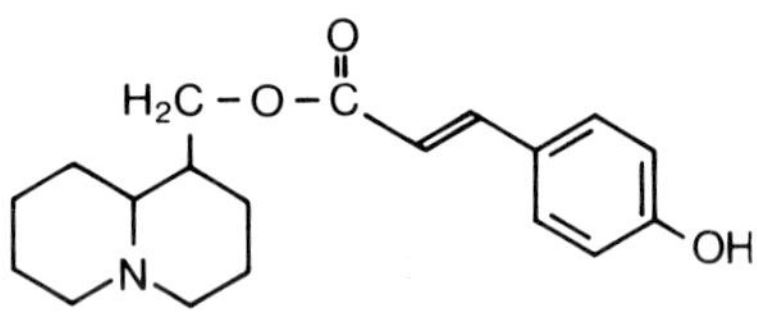

p-Coumaroyllupinin

Fig. 9. Structures of cinnamic acid esters that have been shown to be synthesized via cinnamoyl-CoA thioesters in cell-free systems.

1965; Macheix, 1977; Strack, 1980; Fleuriet *et al.*, 1980; Nurmann and Strack, 1981; Nagels *et al.*, 1981; Shimizu and Kojima, 1984). The same reaction was also found for the conjugation of glucose and benzoic acids (Corner and Swain, 1965; Strack, 1980), including gallic acid (Gross, 1982, 1983a), and of indolylacetic acid (Michalzuk and Bandurski, 1980, 1982). It thus appears that this second esterification mechanism is of general significance for the synthesis of 1-*O*-acyl glucosides. Moreover, it has recently been found that the group transfer potential of such glucose esters is sufficiently high to permit subsequent transacylation reactions (Fig. 10). Via this third reaction type, the sinapic acid esters sinapine and sinapoyl malate are formed with cell-free systems from *Raphanus sativus* (Tkotz and Strack, 1980; Strack, 1982; Strack *et al.*, 1983). Analogous enzymes catalyzing the synthesis of indolylacetyl-myo-inositol (Michalzuk and Bandurski, 1980, 1982) and of di- and trigalloylglucose (Gross, 1983b) have been reported, and it is conceivable that also other phenolic acid esters can be formed according to this newly detected pathway.

Coumarins may be regarded as a special type of cinnamic acid ester (for review, see Brown, 1979, 1981). Their basic benzopyrone structure origi-

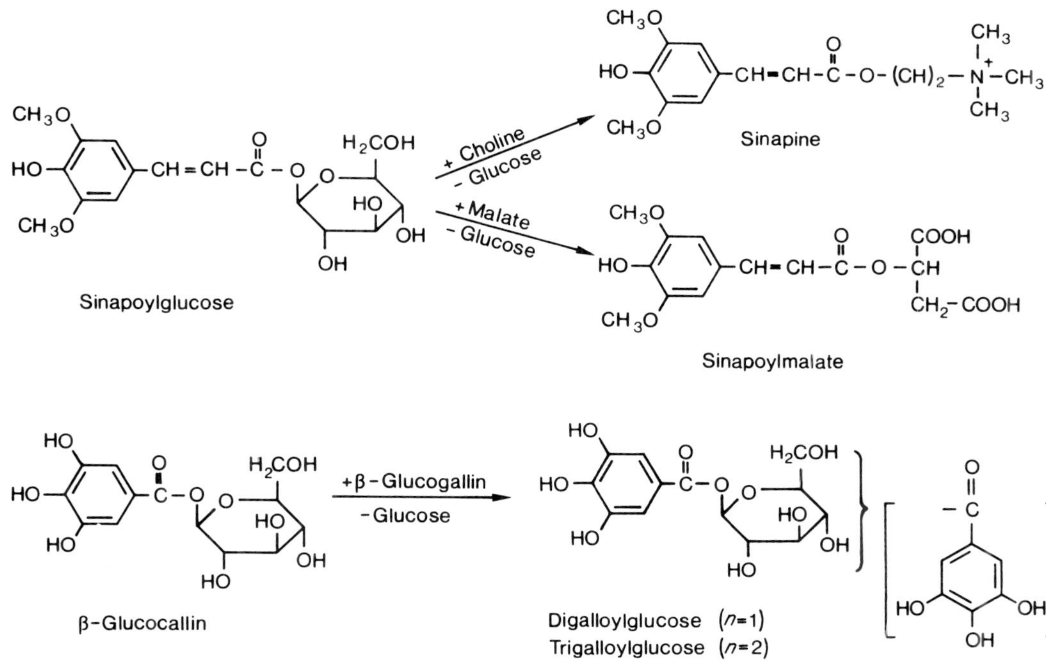

Fig. 10. Formation of phenolic acid esters by transacylation reactions involving 1-*O*-acyl glucose esters as acyl donors.

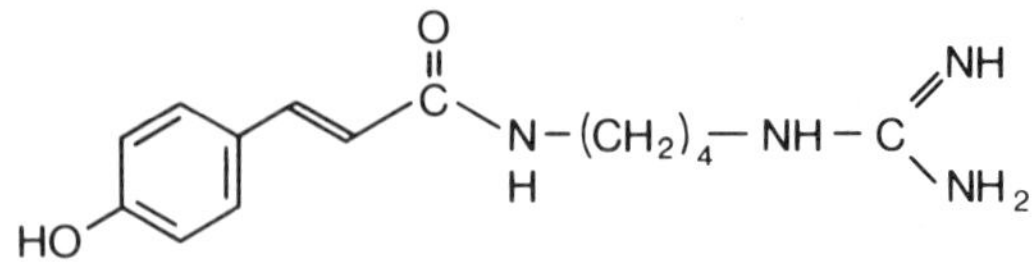

Fig. 11. Structure of *p*-coumaroyl agmatine, a cinnamoyl amide formed by the conjugation of agmatine and *p*-coumaroyl-CoA.

nates from the esterification of the carboxyl group of *cis*-cinnamic acids with a phenolic hydroxyl function in ortho position to the acrylic side chain, thus forming a lactone ring. The biosynthesis of coumarins is initiated by a specific ortho hydroxylation of *trans*-cinnamic acids (see Section II,B), followed by glucosidation of this phenolic OH group and a light- rather than enzyme-dependent trans–cis isomerization step. Under decompartmenting conditions (e.g., wounding or wilting), this "bound" coumarin is hydrolyzed by specific β-glucosidases, and the resulting unstable aglycone lactonizes spontaneously to the "free" coumarin.

B. Cinnamoyl Amides

By analogy to cinnamic acid esters, a wide variety of conjugates of cinnamic acid and its derivatives with many different amines is found in higher plants (for structures, see Zenk, 1979). Hydroxycinnamic acid amides, with putrescine, spermidine, tyramine, or spermine serving as the amine moiety, have been identified as the main phenolic constituents in the reproductive organs of a wide range of flowering plants (Martin-Tanguy *et al.*, 1978; Ponchet *et al.*, 1982), and it appears that such compounds act as natural antiviral inhibitors (reviewed by Smith, 1981). It has long been postulated (Gross and Zenk, 1966) that the biosynthesis of cinnamoyl amides involves the participation of activated CoA thioesters. This assumption has been proved recently by Bird and Smith (1981), who found that cell-free extracts from shoots of barley seedlings synthesized *p*-coumaroyl agmatine (Fig. 11) from *p*-coumaroyl-CoA and agmatine. The purified enzyme (4-coumaroyl-CoA : agmatine *N*-4-coumaroyltransferase) was shown to be specific for agmatine as the amine substrate, whereas thioesters of all the common cinnamic acids could provide the acyl moiety (Bird and Smith, 1983). Dimerization of the reaction product was also observed in these experiments, yielding hordatines, which are known to possess antifungal properties.

C. *Cinnamyl Alcohols (Monolignols) and Their Metabolism*

1. Biosynthesis of Cinnamyl Alcohols

After feeding wheat plants with radioactive ferulic acid, Higuchi and Brown (1963) were able to isolate both labeled coniferyl alcohol and traces of the corresponding aldehyde. The specific radioactivities of these products suggested that the acid had been reduced via the aldehyde to the alcohol. Any consideration of the enzymatic mechanism of such a conversion must take into account that the reduction of a carboxylic acid is a highly endergonic process that principally requires the participation of an activated intermediate. To our present knowledge, three alternatives have been realized in nature for this purpose: the formation of acyl phosphates, of acyl adenylates, or of thioesters.

The reduction of acyl phosphates represents a rather common step in numerous metabolic sequences (e.g., the reduction of 1,3-diphosphoglycerate to glyceraldehyde 3-phosphate). The participation of acyl adenylates, in contrast, appears to occur only occasionally and perhaps is restricted to fungi. This type of reaction has been demonstrated for the reduction of 2-aminoadipate in yeast (Sagisaka and Shimura, 1962; Sinha and Bhattacharjee, 1971) and for the reduction of benzoic and cinnamic acids in the ascomycete *Neurospora crassa* (Gross, 1972). Most probably, the same pattern applies to the reduction of sorbic acid in *Mucor griseo-cyanus* (Tahara *et al.*, 1979) and of aromatic acids in the basiodiomycete *Polystictus* (*Trametes*) *versicolor* (Zenk and Gross, 1965).

Aldehyde formation via the reduction of thioesters, in contrast, is obviously of particular importance. Numerous enzymes catalyzing the reduction of acyl-CoA derivatives have been isolated from bacteria, algae, animal tissues, and higher plants (for references, see Gross, 1979). That thiols other than CoA can also be utilized in such conversions is demonstrated by the reversible reduction of *S*-formylglutathione to formaldehyde (see Uotila and Koivusalo, 1974, 1979).

Concerning the reduction of cinnamic acids in higher plants, particularly with respect to their role as monomeric lignin units, Brown *et al.* suggested some years ago (1959) that the CoA esters of these acids might represent the required activated intermediates. In the meantime, intensive research has established evidence for the correctness of this assumption. This was shown for the first time by Mansell *et al.* (1972) with extracts from *Salix*, which catalyzed the reduction of ferulic acid to coniferyl alcohol in the presence of CoA. This result was confirmed by subse-

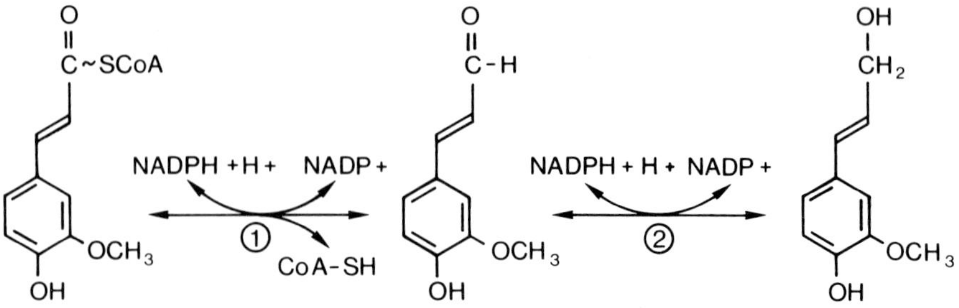

Fig. 12. The reduction of cinnamoyl-CoA thioesters to cinnamyl alcohols via the corresponding aldehyde. Reaction depicted with feruloyl-CoA, coniferyl aldehyde, and alcohol, respectively. ① Cinnamoyl-CoA reductase, cinnamoyl-CoA : NADP oxidoreductase; ② cinnamyl alcohol dehydrogenase, cinnamyl alcohol : NADP oxidoreductase.

quent reports from several laboratories on the CoA-dependent reduction of *p*-coumaric or ferulic acid with enzyme preparations from various plants (references in Gross, 1980). Subsequently, the presumed intermediacy of cinnamoyl-CoA esters was definitively proved by the successful *in vitro* reduction of authentic *p*-coumaroyl-CoA or feruloyl-CoA (Ebel and Grisebach, 1973; Gross *et al.,* 1973; Stöckigt *et al.,* 1973; Ebel *et al.,* 1974).

Apart from the alcohol that was the main product found in these experiments, some formation of the corresponding aldehydes was also observed on several occasions. These results indicated that, in accordance with the *in vivo* experiments of Higuchi and Brown (1963), the biosynthesis of cinnamyl alcohols involves two consecutive steps. The validity of this interpretation was proved for the first time by the successful isolation of two enzymes from lignifying cambial tissue of *Forsythia suspensa* that catalyzed the sequence

$$\text{feruloyl-CoA} \longrightarrow \text{coniferyl aldehyde} \longrightarrow \text{coniferyl alcohol}$$

(Fig. 12) (Gross *et al.,* 1973). The properties of these enzymes, as well as of enzymes from other plants or plant cell cultures, have since been studied in detail. Some essential characteristics resulting from these investigations are discussed here (for review, see Gross, 1977, 1979, 1980; Grisebach, 1977, 1981).

Enzymes catalyzing the first reductive step in the biosynthesis of cinnamyl alcohols—i.e., the formulation of cinnamyl aldehydes from the corresponding cinnamoyl-CoA esters—have been detected in various plants (references in Gross, 1980; Grisebach, 1981). Detailed results have been published for the enzyme from *Forsythia* xylem (Gross and Kreiten, 1975) and particularly for the highly purified preparations from soybean cell cultures and spruce (Wengenmayer *et al.*, 1976; Lüderitz and Grisebach, 1981). The enzymes from these sources are very similar with respect to size (molecular weight ~38,000; no subunits) and their requirement for NADPH as the exclusive cosubstrate. Stereospecificity studies have shown that the B-hydrogen (H_S) of NADPH is transferred to the cinnamoyl moiety. With respect to various acyl-CoA esters assayed as potential substrates it was found that these enzymes were most active with feruloyl-CoA; sinapoyl-CoA was reduced only with the enzymes from angiosperms, whereas *p*-coumaroyl-CoA was a poor substrate in all cases. These results (which are discussed in more detail at the end of this section) are consistent with the known differences between the lignins from gymnosperms and angiosperms. By analogy with hydroxycinnamate:-CoA ligases (see Section III), these enzymes appear to act exclusively on cinnamoyl substrates; no or only minimal activities were observed with the CoA derivatives of various benzoic or aliphatic acids. Since reversibility of the reaction has also been demonstrated, the systematic name "cinnamoyl-CoA : NADP oxidoreductase" is fully justified.

The final step in the biosynthesis of cinnamyl alcohols is mediated by an oxidoreductase catalyzing the reversible reduction of cinnamyl aldehydes to the alcohol (Fig. 12, reaction 2). The existence of such enzymes has been demonstrated to be a common feature of higher plants (Mansell *et al.*, 1974, 1976). Starch gel electrophoresis with extracts from a multitude of plants revealed the general presence of one (or occasionally two) NADP-linked coniferyl alcohol-specific dehydrogenases, with the exception of *Salix* species which contained multiple forms of this enzyme. Detailed studies have been carried out with purified preparations from *F. suspensa* xylem (Mansell *et al.*, 1974), *Glycine* cell cultures (Wyrambik and Grisebach, 1975, 1979, Lüderitz and Grisebach, 1981), and seedlings and cambial sap of *Picea abies* (Lüderitz and Grisebach, 1981) and of *Pinus thunbergii* (Kutsuki *et al.*, 1982b). Summarizing the results of these investigations, it is apparent that the basic properties of these enzymes are very uniform again. Gel filtration experiments revealed molecular weights of 67,000 to 80,000; after SDS gel electrophoresis, only one band of ~40,000 daltons was observed, indicating the existence of two identical subunits. Further studies demonstrated that the enzyme from soybeans contains zinc, like other alcohol dehydrogenases from various sources.

The cinnamyl alcohol dehydrogenases isolated so far specifically require NADP(H) as coenzyme. In contrast to the preceding cinnamoyl-CoA reductase, the A-hydrogen (H_R) of NADPH is transferred. In the reverse reaction, the pro-*R* hydrogen of the alcohol is removed (Klischies *et al.*, 1978; Wyrambik and Grisebach, 1979). Substrate specificity studies with the *Forsythia* enzyme (Mansell *et al.*, 1974) revealed that this enzyme catalyzes the oxidation of cinnamyl alcohol and a wide variety of its hydroxylated, methylated, methoxylated, or halogenated derivatives. Thus, this enzyme is less specific with respect to the substitution pattern of the aromatic ring than the preceding enzymes of this metabolic sequence. Further studies also showed, however, that this enzyme does not react with aliphatic substrates or with benzyl alcohols. Moreover, hydroconiferyl alcohol and homovanillyl alcohol could not serve as substrates. Consequently, this alcohol dehydrogenase seems virtually specific for substrates characterized by both a phenyl residue and an allylic side chain. The same situation appears to apply to the majority of such enzymes isolated from other sources.

According to the results summarized here, the reversible reduction of cinnamyl aldehydes is catalyzed by a rather specific dehydrogenase. The equilibrium of this reaction ($K = 2.8 \times 10^{-9}$; Wyrambik and Grisebach, 1975) greatly favors the formation of cinnamyl alcohols as the immediate precursors of lignin. Concerning this important natural product, in both qualitative and quantitative terms, it appears appropriate to discuss briefly the significance of the substrate specificities of the enzymes involved in the synthesis, activation, and reduction of cinnamic acids. In the section on *O*-methyl transferases (II,C), it was mentioned that the observation of marked differences between the substrate specificities of gymnosperm and angiosperm OMTs provides one explanation for the profound differences in the composition of softwood and hardwood lignins. Serious gaps in our understanding are encountered at the next crucial step, the formation of cinnamoyl-CoA esters (see Section III), because of the frequently observed absence of sinapate-activating enzymes even from actively lignifying angiospermous tissues—in spite of the fact that these are characterized by a high proportion (>50 mol % in dicotyledons; see Erickson *et al.*, 1973) of syringyl residues. Most probably, metabolic alterations during the development of plants must be considered. This view is supported by the finding that sinapic acid was reduced only by tissue slices of fully differentiated angiosperms but not by less differentiated tissues or cell cultures. Accordingly, almost no syringyl lignin was found in the latter tissues (Nakamura *et al.*, 1974).

No such limitations are encountered in the two subsequent reductive steps. In contrast, the results from recent comparative investigations on the substrate specificities of these enzymes in gymnosperms and angiosperms were found to reflect the respective lignin composition. *p*-Coumaroyl-CoA (which is produced in excellent yield by CoA-ligases) was a poor substrate for the reductase from both spruce and soybean, and this result is consistent with the low content of *p*-hydroxyphenyl residues in both lignin types. Feruloyl-CoA was the preferred substrate for the enzymes from both sources, whereas sinapoyl-CoA was a good substrate for the angiosperm reductase only (Lüderitz and Grisebach, 1981). A similar situation has been observed with respect to the subsequent alcohol dehydrogenase Kutsuki *et al.* (1982b) analyzed the ratio of coniferyl versus sinapyl aldehyde-reducing activities in a variety of gymnosperms and angiosperms, and it was found that gymnosperm enzymes preferentially catalyzed the formation of coniferyl alcohol. The angiosperm enzymes, in contrast, not only reduced both substrates but exhibited even higher affinities toward sinapyl aldehyde in most cases. Also, the substrate specificities of the enzymes catalyzing the synthesis and hydrolysis of cinnamyl alcohol glucosides (see below) clearly fit into this picture. In summary, convincing biochemical explanations for the characteristic constitutional differences between angiosperm and gymnosperm lignins have been elaborated by these enzymatic studies (see Chapter 7, Section VI).

2. Compounds Derived from Cinnamyl Alcohols

Cinnamyl alcohols have occasionally been observed to occur in the free state, even in plant cell cultures (e.g., *Vieitez et al.*, 1975; Andersen *et al.*, 1981). Their main function, however, is to serve as intermediates in the biosynthesis of various natural products. Undoubtedly, the most prominent among these is lignin, a macromolecule formed in higher plant cell walls by a free-radical-mediated dehydropolymerization of *p*-hydroxycinnamyl alcohols (see Chapter 7) (Gross, 1978, 1979, 1980; Grisebach, 1981). This reaction depends on cinnamyl alcohols bearing a free phenolic hydroxyl group. Lignifying tissues, however, do not normally contain these alcohols in the free form but as the corresponding 4-*O*-β-D-glucosides, where the phenolic hydroxyl is masked by the sugar moiety. Enzymes catalyzing the synthesis of such glucosides have recently been isolated and characterized (Ibrahim and Grisebach, 1976; Ibrahim, 1977; Schmid and Grisebach, 1982; Schmid *et al.*, 1982). They required UDP-glucose as donor and showed a pronounced affinity for coniferyl and syringyl alcohol as substrate, yielding both coniferin and syringin (Fig. 13), except for the enzyme from spruce, which specifically formed coniferin. *p*-Coumaryl alcohol, in contrast, was a poor substrate in all cases.

Further studies also revealed the existence of β-glucosidases that specifically hydrolyze coniferin and syringin to the free alcohols (Hösel *et al.*, 1978; Hösel and Todenhagen, 1980; Surholt and Hösel, 1981; Marcinowski and Grisebach, 1978). These enzymes were found to be attached to cell walls, presumably via ionic linkages, and their activity was correlated with the onset of lignification in spruce seedlings (Marcinowski and

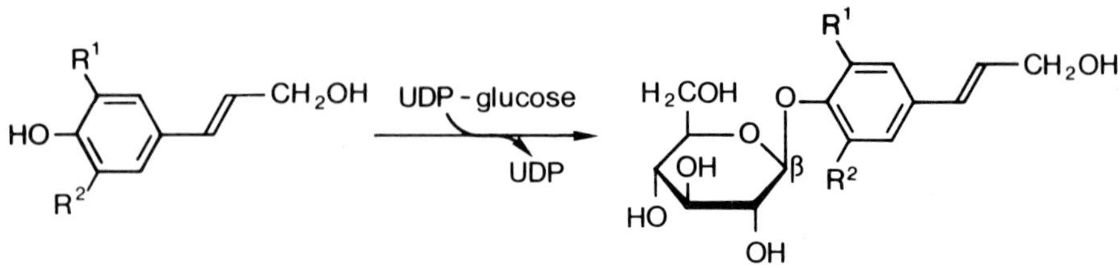

Fig. 13. Formation of 4-*O*-β-D-glucosides of cinnamyl alcohols by UDP-glucose-dependent glucosyl transferases. R^1 = H, R^2 = OCH_3 for coniferyl alcohol and coniferin; R^1 = R^2 = OCH_3 for sinapyl alcohol and syringin.

Grisebach, 1978). Thus, it must be concluded that hydroxycinnamyl alcohol glucosides are related to lignification. As lignifying cells contain the complete set of enzymes required for the synthesis and reduction of cinnamic acids it is no longer necessary to regard these glucosides as an *inter*cellular transport form; most likely, they represent an *intra*cellular storage metabolite or the form in which cinnamyl alcohols are excreted from the cytoplasm into the lignifying cell wall (Gross, 1979, 1980).

Esters of cinnamyl alcohols with various aliphatic or aromatic acids are known as constituents of many essential oils or resins, e.g., coniferyl benzoate from *Styrax benzoin* (see Schroeder, 1968), cinnamyl acetate from *Cinnamomum* bark (Angmor *et al.*, 1979), or sinapyl acetate from *Citrus bergamia* (Ehret and Montpetit, 1982). Coniferyl alcohol has also been proposed as the direct precursor of several lignans (e.g., arctiin, Fig. 14; Stöckigt and Klischies, 1977), i.e., dimeric phenylpropanes that are stereospecifically linked at the β position of the side chain. The latter suggests that their biosynthesis, in contrast to that of lignin, proceeds under strict enzymatic control.

Formally, further reduction of cinnamyl alcohols would lead to propenyl phenols or, after an additional isomerization step, to allyl phenols (Fig. 14). Tracer studies with doubly labeled coniferyl alcohol have provided evidence for the existence of such a conversion in *Ocimum basilicum,* yielding eugenol and methyleugenol with unchanged $^3H:^{14}C$ ratios (Klischies *et al.*, 1975). It should be mentioned, however, that a different pathway involving loss of the terminal side-chain carbon at the acid level has also been proposed (Manitto *et al.*, 1974; Senanayake *et al.*, 1977).

In addition, hydrogenation of the unsaturated side chain must also occur in nature, as indicated by the existence of dihydrocinnamyl alcohols. Such compounds have been proposed as intermediates in the biosynthesis of mesembrine alkaloids (Jeffs *et al.*, 1978) and of rhododendrin (Fig. 14;

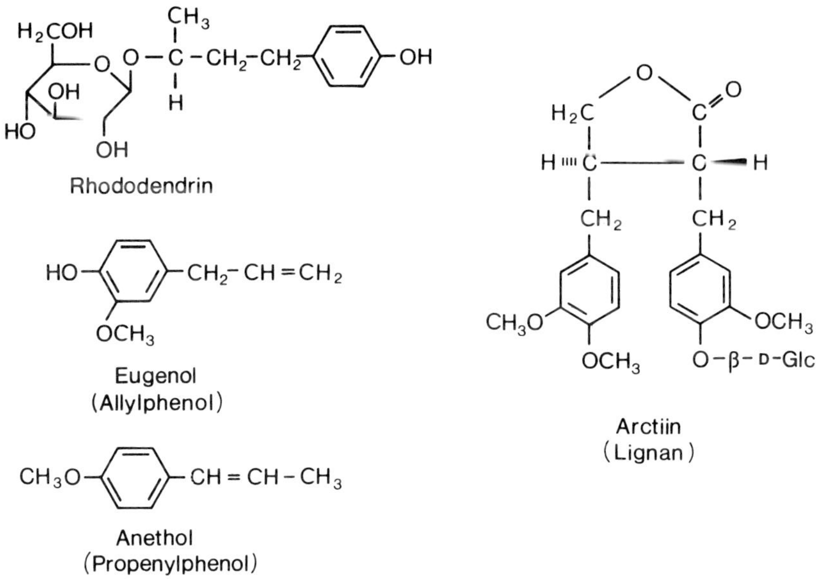

Fig. 14. Examples of low molecular weight cinnamyl alcohol derivatives.

Klischies and Zenk, 1978). Nothing is known, however, concerning the enzymology of such conversions.

D. *Side-Chain Elongation Reactions*

Grisebach postulated in 1962 that flavonoids were synthesized by the sequential condensation of a cinnamoyl-CoA ester with 3 molecules of malonyl-CoA, under the simultaneous liberation of 3 CO_2, yielding a chalcone as the first intermediate in the biosynthesis of the various flavonoid classes (see Fig. 15). This prediction was confirmed several years ago by the successful isolation of chalcone synthase, an enzyme complex catalyzing the *in vitro* synthesis of chalcones in the expected manner (see Chapter 12; Grisebach, 1979; Hahlbrock, 1981). Similar stilbene synthases have been detected in *Rheum rhaponticum* (rhubarb) and *Pinus silvestris*. Extracts from these plants were able to form various stilbenes in a reaction utilizing the same substrates, but apparently proceeding by a different cyclization mode and involving an additional decarboxylation step (see Chapter 14; Rupprich and Kindl, 1978; Schoeppner and Kindl, 1979; Rupprich *et al.*, 1980).

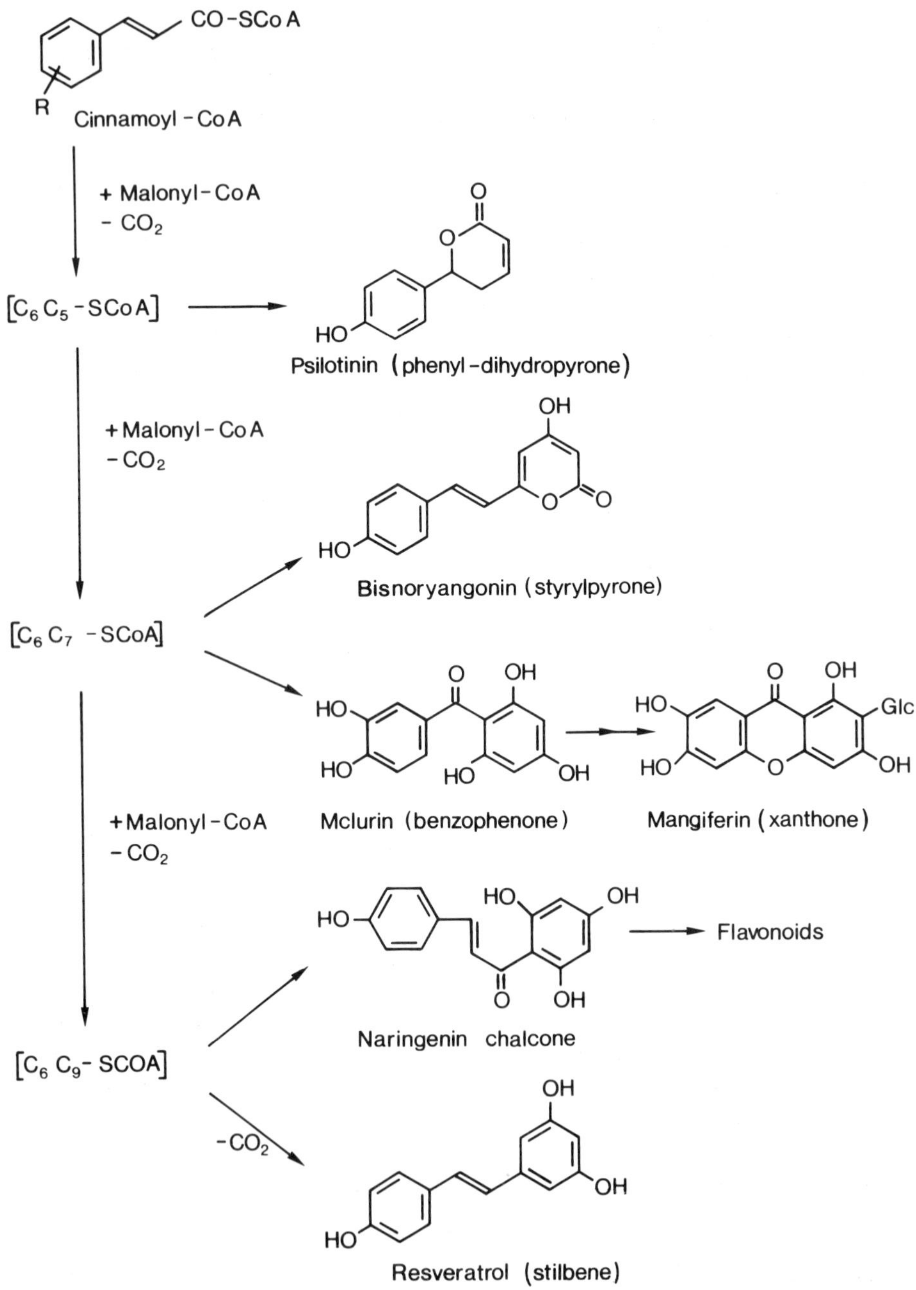

Fig. 15. Natural products that have been proved or proposed to be formed by the condensation of cinnamoyl-CoA esters and malonyl-CoA. Modified after Gross (1981).

In further studies on chalcone synthase, it was found that styrylpyrones, as well as dihydropyrones and benzalacetones, were formed under certain assay conditions (Kreuzaler and Hahlbrock, 1975; Hrazdina *et al.*, 1976). These products obviously represent short-chain condensation products of the cinnamoyl-CoA starter molecule with only one or two malonyl-CoA derived acetate units. Although these compounds must be regarded as artificial products in this *in vitro* system, it is not unlikely that such reaction mechanisms occur naturally. Psilotinin (6-phenyldihydro-α-pyrone) has recently been found to be formed from [^{13}C,^{14}C,^{3}H]-triple-labeled phenylalanine in feeding experiments with *Psilotum nudum* with retention of all the isotopes (Leete *et al.*, 1982), a result that is consistent with the assumption that this pathway involves the condensation of a C_6C_3 unit and one acetate unit. With regard to the biosynthesis of xanthones, it has been suggested that they are formed from a C_6C_1 precursor and three acetate units, yielding benzophenones, which in turn undergo cyclization to xanthones (Floss and Rettig, 1964; Gupta and Lewis, 1971). More recent experiments with the lily *Anemarrhena asphodeloides* (Asphodelaceae), however, have provided strong evidence that a C_6C_3 compound (*p*-coumaric acid) reacts with two acetate units in the initial steps of this pathway (Fujita and Inoue, 1977, 1981). Most probably, analogous condensations occur in fungi in the synthesis of hispidin, a styrylpyrone with the substitution pattern of caffeic acid (Perrin and Towers, 1973; Hatfield and Brady, 1973).

E. *Side-Chain Degradation Reactions*

The most prominent pathway belonging to this category is the one leading to benzoic acids through the removal of an acetate unit. This conversion will be discussed in Section V,A. Decarboxylation of cinnamic acids to the corresponding styrenes is known from studies on bacteria and fungi. Retention of the stereospecific configuration of the side chain in this reaction has been demonstrated with *Saccharomyces cerevisiae* (Manitto *et al.*, 1975; see also the review of Wat and Towers, 1979). No data are available about this conversion in higher plants. This applies also to the biogenesis of acetophenones, which eventually are formed in a side reaction during the β oxidation of cinnamic acids (see Zenk, 1979).

V. Benzoic Acids

A. *Biosynthesis of Benzoic Acids*

The most important mechanism for the formation of benzoic acids in plants is the side-chain degradation of cinnamic acids. Feeding experiments have shown that the removal of an acetate unit yields the corresponding benezoic acid (see Zenk, 1966), with the consequence that the substitution pattern of the individual benzoic acids is determined by their respective C_6C_3 precursors. However, secondary hydroxylations and methylations of benzoic acids similar to those of cinnamic acids are also known to occur (for detailed references to the literature, see Billek and Schmook, 1966). The situation is further complicated by demethylation reactions and particularly by the catabolism of more complex natural products (e.g., flavonoids) yielding benzoic acids as intermediates and byproducts or end products (see Ellis, 1974; Barz and Hösel, 1979; Barz and Köster, 1981). The latter reactions explain the existence of many benzoic acids characterized by substitution patterns that differ significantly from those of the commonly occurring cinnamic acids.

Concerning the mechanism of the conversion of cinnamic to benzoic acids, a reaction sequence analogous to the β-oxidation of fatty acids was proposed by Zenk (1966, 1971) (Fig. 16). This view was supported by Alibert and Ranjeva (1971) in studies on cell-free extracts from *Quercus* where the degradation of cinnamic to benzoic acids was stimulated upon the addition of ATP and CoA. More recently, a *Pseudomonas* strain growing on ferulic acid as the sole carbon source was found to possess an extremely active cinnamate : CoA ligase and several enzymes required for the initial steps in the β oxidation sequence (Zenk, 1979). Definitive proof of this pathway in plants, however, is still lacking.

In addition to these observations, there are several reports on the formation of benzoic acids from C_6C_3 precursors by cell-free systems from bacteria, fungi, algae, and higher plants that evidently were independent of added ATP and CoA, indicating the existence of a different "nonoxidative" pathway (Fig. 16). These systems were either membrane-associated enzyme complexes yielding benzoic acids (Hagel and Kindl, 1975; Löffelhardt and Kindl, 1975, 1976) or soluble enzymes producing acetate and the corresponding benzaldehyde, which was subsequently oxidized to the acid by an NAD-dependent dehydrogenase or by peroxidase and O_2 (French *et al.*, 1976; Alibert *et al.*, 1979). Summarizing these conflicting aspects, which have been discussed in detail by Zenk (1979), it must be stated that neither of the two pathways has been fully characterized to date and that it is still an open question whether one or perhaps both possibilities operate in the synthesis of benzoic acids.

The situation is further complicated by the existence of other quite different pathways to benzoic acids, particularly in microorganisms (see Gross, 1981). In higher plants, it is still unclear by what mechanisms gallic acid, the main constituent of hydrolyzable tannins, is synthesized (Fig. 17) (see Chapter 13, Section IV,B). Zenk (1964) favored a β oxidation of

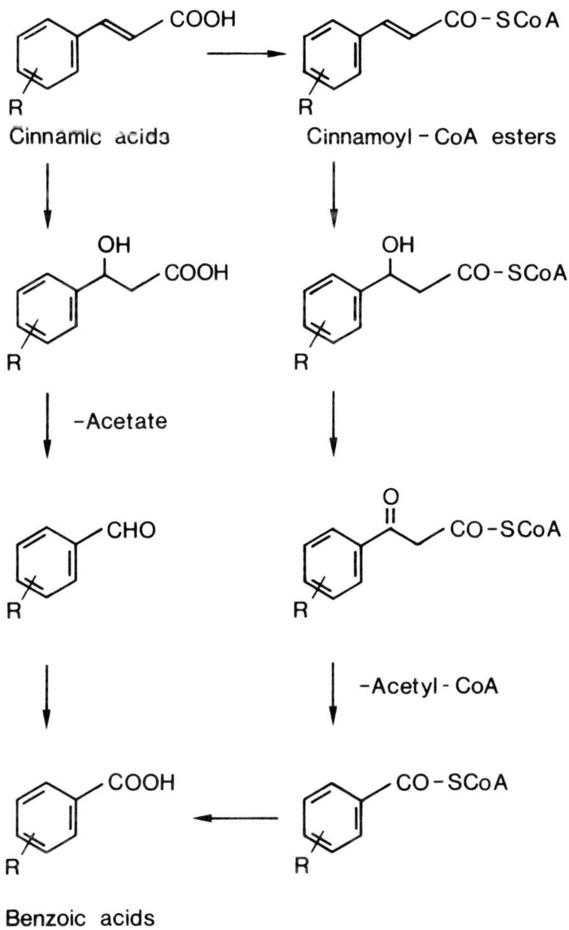

Fig. 16. Mechanisms proposed for the conversion of cinnamic to benzoic acids. Left, nonoxidative pathway; right, β oxidation pathway.

3,4,5-trihydroxycinnamic acid; it is very unlikely, however, that this precursor does really occur in plants. Other authors proposed the sequence

caffeic ⟶ protocatechuic ⟶ gallic acid

A third, attractive alternative is the formation of gallic acid by the direct aromatization of dehydroshikimic acid, with the consequence that the entire alicyclic C_6C_1 carbon skeleton is retained, whereas the pathway via phenylpropanes involves loss of the carboxyl group of the precursor (for references, see Haslam, 1981). Experiments (Amrhein *et al.*, 1984) with

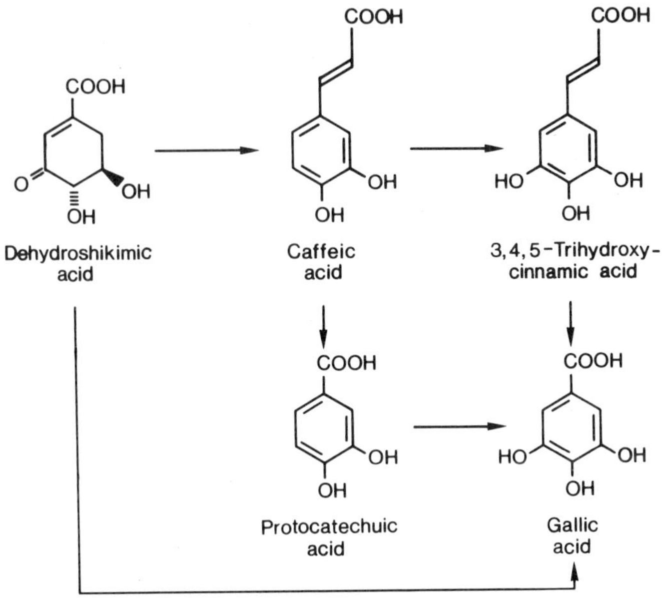

Fig. 17. Pathways proposed for the biosynthesis of gallic acid.

the inhibitors 2-aminooxy-3-phenylpropionic acid (AOPP; see Section II,A) and glyphosate (an inhibitor of shikimic acid utilization) provided strong evidence that this third pathway was operative in cell suspension cultures of *Quercus robur* and in leaves of *Rhus typhina*.

B. *Compounds Derived from Benzoic Acids*

Summarizing the results outlined in the above section, it is evident that benzoic acids to a great extent result from catabolic processes. Nevertheless, these phenolic acids can be utilized, by analogy to cinnamic acids (see Section IV), as the precursors of a wide variety of natural products. The few examples depicted in Fig. 18 show that the principal metabolic conversions that have already been discussed for cinnamic acids are encountered here again. Benzoic acids can undergo esterification (e.g., cocaine); they can be reduced via the aldehyde (e.g., vanillin) to the alcohols (e.g., salicyl alcohol), which in turn can give methyl or ethyl ethers; they can be converted to simple phenols (e.g., arbutin); and glucosylation is also of importance. In contrast to the C_6C_3 derivatives, our knowledge of the mechanisms involved in these conversions is rather limited, and this applies especially to enzymatic studies in this field.

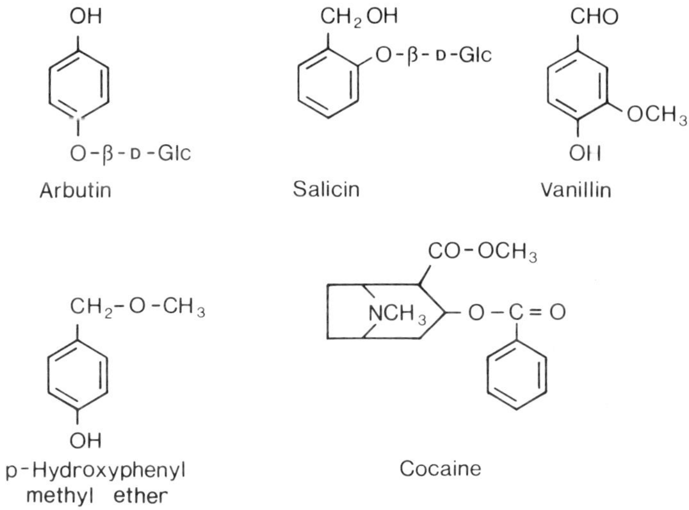

Fig. 18. Examples of naturally occurring benzoic acid derivatives.

In some of these pathways, CoA-esters were proposed as intermediates. These considerations arose from assuming the β oxidation of cinnamic acids, which should lead to the formation of benzoyl-CoA esters. These could well be the carboxyl-activated intermediates required in the formation of benzoyl esters or of benzaldehydes and alcohols (Zenk, 1966, 1971, 1979). Although such compounds are common plant constituents, little data on their synthesis in cell-free systems are available. Thus far, the *in vitro* reduction of benzoic acids has been demonstrated only with fungal enzymes, but this reaction does not require CoA-thioesters (see Section IV,C). Alternatively, benzyl alcohols could be formed from benzaldehydes as the end products of the nonoxidative cinnamic acid degradation pathway under the catalysis of NAD(P)-dependent aromatic alcohol dehydrogenases, and such enzyme activities have indeed been reported to occur in many plants (Davies *et al.*, 1973; Mansell *et al.*, 1974, 1976). Also, esterifications involving benzoyl-CoA esters still lack experimental proof. To date, only the *in vitro* synthesis of various benzoylglucose esters has been reported, but these reactions again proceeded without the participation of CoA (see Section IV,A).

Another pathway arising from benzoic acids is the route leading to ubiquinones, a sequence in which *p*-hydroxybenzoic or protocatechuic acid is prenylated with isopentenyl pyrophosphate (e.g., see Casey and Threlfall, 1978). Finally, benzoic acids can also be decarboxylated ox-

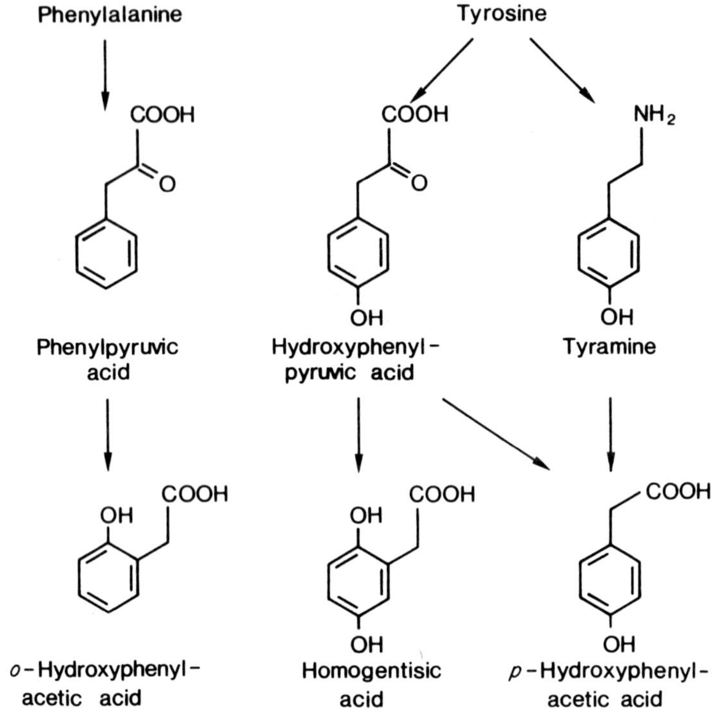

Fig. 19. Biosynthetic pathways to phenylacetic acids in plants.

idatively to yield phenols (or the corresponding quinones). *p*-Hydroxylated substrates were thus found to undergo rapid decarboxylation in the presence of peroxidase and H_2O_2, and, depending on their substitution pattern, *p*-quinones, dimers, or polymers were identified as main products (Berlin and Barz, 1975; see also Barz and Köster, 1981).

VI. Phenylacetic Acids

Phenylacetic acids occur less frequently than the structurally related cinnamic and benzoic acids. Their biosynthesis appears to involve mainly catabolic processes (see Ellis, 1974; Barz and Hösel, 1979; Barz and Köster, 1981). The principal precursors of this class of natural products are the aromatic amino acids phenylalanine and tyrosine (Fig. 19), which undergo transamination (Stafford and Lewis, 1979) or oxidative deamination to give the corresponding α-ketoacids; alternatively, the latter compounds can also be taken directly from the shikimate pathway (Fiedler *et al.*, 1982). A subsequent oxidative decarboxylation step gives rise to the formation of phenylacetic and *p*-hydroxyphenylacetic acid. An interesting alternative reaction catalyzed by the enzyme *p*-hydroxyphenylpyruvate dioxygenase involves oxidative decarboxylation, migration of the side chain, and ring oxygenation to yield homogentisic acid (see Löffelhardt and Kindl, 1979; Fiedler *et al.*, 1982). The steric course of the NIH shift encountered in this conversion has been analyzed recently by Leinberger *et al.* (1981). A similar reaction produces *o*-hydroxyphenylacetic acid from phenylpyruvate. With cell cultures from a variety of plants, a different route to phenylacetic acids was found by which tyramine (or its *N*-methyl derivatives) and dopamine were deaminated to the corresponding C_6C_2 acids (Meyer and Barz, 1978).

After introduction of a second hydroxyl group into the monohydroxylated phenylacetic acids, these compounds can be degraded further by cleavage of their aromatic ring (reviewed by Ellis, 1974; Barz and Hösel, 1979; Barz and Köster, 1981). That these catabolic pathways in fact occur in plants and are not the result of microbial contaminations has been shown by Durand and Zenk (1974a,b) with sterile plants and cell cultures from which enzymes catalyzing the above sequence could be isolated.

Although phenylacetic acids are thus formed by catabolic processes, they must not be regarded as mere degradation products in physiological terms. Phenylacetic acid and its *p*-hydroxy derivative have been shown to cause growth-regulating and morphogenetic effects in algae (Fries, 1977; Fries and Åberg, 1978), and phenylacetate has recently been identified as natural auxin in the shoots of higher plants (Wightman and Lighty, 1982). Moreover, anabolic processes can also be observed. Spinach chloroplasts have been found not only to produce homogentisic acid but also to catalyze the enzymatic prenylation of this acid with phytyldiphosphate to give 2-methyl-6-phytylquinol (plus CO_2) as the first intermediate in the synthesis of α-tocopherol and plastoquinone-9 (Fiedler *et al.*, 1982). A quite different route was detected by Bolkart and Zenk (1969) that showed that homogentisic acid, after decarboxylation to toluhydroquinone, was incorporated into the naphthoquinone chimaphilin.

VII. Concluding Remarks

Summarizing the results of a host of investigations outlined in this chapter, it is evident that the biochemical events involved in the biosynthesis and metabolism of phenolic acids are now fairly well understood. Enzymatic studies in particular, and certainly also the wide use of plant cell cultures, have significantly advanced the knowledge gained by previous tracer experiments. By this means, the reaction mechanisms involved in many metabolic sequences can now be explained in detail at the molecular level. The broad basis provided by these investigations has stimulated a growing number of studies on the more physiological aspects within this field, e.g., the questions related to compartmentation, regulation, hormonal effects or developmental processes. Our present knowledge of the enzymes catalyzing these pathways certainly facilitates future intensive and fruitful research concerning such questions.

References

Alibert, G., and Ranjeva, R. (1971). *FEBS Lett.* **19,** 11–14.
Alibert, G., Bouyssou, H., and Ranjeva, R. (1979). *C. R. Acad. Sci. Paris, Ser. D.* **288,** 681–684.
Amrhein, N. (1979). *Phytochemistry* **18,** 585–589.
Amrhein, N., and Gödeke, K. H. (1977). *Plant Sci. Lett.* **8,** 313–317.
Amrhein, N., Gerhardt, J., and Gödeke, K. H. (1976). *Planta* **131,** 33–40.
Amrhein, N., Frank, G., Lemm, G., and Luhmann, H. B. (1983). *Eur. J. Cell Biol.* **29,** 139–144.
Amrhein, N., Topp, H., and Joop, O. (1984). *Plant Physiol.* **75,** Suppl., p. 96.
Andersen, R. A., Kemp, T. R., and Vaughn, T. H. (1981). *Physiol. Plant.* **53,** 89–94.
Angmor, J. D., Dewick, P. M., and Evans, W. C. (1979). *Planta Med.* **35,** 342–347.
Barz, W., and Hösel, W. (1979). *In* "Biochemistry of Plant Phenolics" (T. Swain, J. B. Harborne, and C. F. Van Sumere, eds.), Vol. 12, pp. 339–369, Rec. Adv. Phytochem. Series. Plenum, New York.
Barz, W., and J. Köster (1981). *In* "The Biochemistry of Plants, Vol. 7, Secondary Plant Products" (E. E. Conn, ed.), pp. 35–84. Academic Press, New York.
Benveniste, I., Salaun, J. P., and Durst, F. (1977). *Phytochemistry* **16,** 69–73.
Berlin, J., and Barz, W. (1975). *Z. Naturforsch.* **30c,** 650–658.
Billek, G., and Schmook, F. P. (1966). *Österr. Chem. Ztg.* **67,** 401–409.
Billett, E. E., Wallace, W., and Smith, H. (1978). *Biochim. Biophys. Acta* **524,** 219–230.
Bird, C. R., and Smith, T. A. (1981). *Phytochemistry* **20,** 2345–2346.
Bird, C. R., and Smith, T. A. (1983). *Phytochemistry* **22,** 2401–2403.
Bolkart, K. H., and Zenk, M. H. (1969). *Z. Pflanzenphysiol.* **61,** 356–359.
Brown, S. A. (1966). *Ann. Rev. Plant Physiol.* **17,** 223–244.
Brown, S. A. (1979). *In* "Biochemistry of Plant Phenolics" (T. Swain, J. B. Harborne, C. F. Van Sumere, eds.), Vol. 12, pp. 249–286, Rec. Adv. Phytochem. Series. Plenum, New York.
Brown, S. A. (1981). *In* "The Biochemistry of Plants, Vol. 7: Secondary Plant Products" (E. E. Conn, ed.), pp. 269–300. Academic Press, New York.
Brown, S. A., Wright, D., and Neish, A. C. (1959). *Can. J. Biochem. Physiol.* **37,** 25–34.
Butt, V. S. (1979). *In* "Biochemistry of Plant Phenolics" (T. Swain, J. B. Harborne, C. F. Van Sumere, eds.), Vol. 12, pp. 433–456, Rec. Adv. Phytochem. Series. Plenum, New York.
Butt, V. S. (1980). *In* "The Biochemistry of Plants, Vol. 2: Metabolism and Respiration" (D. D. Davies, ed.), pp. 81–123. Academic Press, New York.

Butt, V. S., and Lamb, C. J. (1981). *In* "The Biochemistry of Plants, Vol. 7: Secondary Plant Products" (E. E. Conn, ed.), pp. 627–665. Academic Press, New York.
Camirand, A., Phipps, J., and Wightman, F. (1983). *Can. J. Bot.* **61,** 2302–2308.
Camm, E. L., and Towers, G. H. N. (1977). *Prog. Chem.* **4,** 169–188.
Casey, J., and Threlfall, D. R. (1978). *FEBS Lett.* **85,** 249–253.
Charrière-Ladreix, Y. (1979). *Phytochemistry* **18,** 43–45.
Corner, J., and Swain, T. (1965). *Nature (London)* **207,** 634–635.
Creasy, L. L. (1976). *Phytochemistry* **15,** 673–675.
Czichi, U., and Kindl, H. (1975a). *Planta* **125,** 115–125.
Czichi, U., and Kindl, H. (1975b). *Hoppe-Seyler's Z. Physiol. Chem.* **356,** 475–485.
Czichi, U., and Kindl, H. (1977). *Planta* **134,** 133–143.
Davies, D. D., Ugochukwu, E. N., Patil, K. D., and Towers, G. H. N. (1973). *Phytochemistry* **12,** 531–536.
Diesperger, H., and Sandermann, H. (1978). *FEBS Lett.* **85,** 333–336.
Duke, S. O., and Vaughn, K. C. (1982). *Physiol. Plant.* **54,** 381–385.
Durand, R., and Zenk, M. H. (1974a). *Phytochemistry* **13,** 1483–1492.
Durand, R., and Zenk, M. H. (1974b). *FEBS Lett.* **39,** 218–220.
Ebel, H., and Grisebach, H. (1973). *FEBS Lett.* **30,** 141–143.
Ebel, J., Schaller-Hekeler, B., Knobloch, K., Wellmann, E., Grisebach, H., and Hahlbrock, K. (1974). *Biochim. Biophys. Acta* **362,** 417–424.
Ehret, C., and Monpetit, P. (1982). *Phytochemistry* **21,** 2984–2985.
Ellis, B. E. (1974). *Lloydia* **37,** 168–183.
Ellis, B. E., Zenk, M. H., Kirby, G. W., Michael, J., and Floss, H. G. (1973). *Phytochemistry* **12,** 1057–1058.
Erickson, M., Miksche, G. E., and Somfai, I. (1973). *Holzforschung* **27,** 113–117.
Fiedler, E., Soll, J., and Schultz, G. (1982). *Planta* **155,** 511–515.
Fleuriet, A., Macheix, J. J., Suen, R., and Ibrahim, R. V. (1980). *Z. Naturforsch.* **35c,** 967–972.
Floss, H. G. (1979). *In* "Biochemistry of Plant Phenolics" (T. Swain, J. B. Harborne, C. F. Van Sumere, eds.), Vol. 12, pp. 59–89, Rec. Adv. Phytochem. Series. Plenum, New York.
Floss, H. G., and Rettig, A. (1964). *Z. Naturforsch.* **19b,** 1103–1105.
French, C. J., and Smith, H. (1975). *Phytochemistry* **14,** 963–966.
French, C. J., Vance, C. F., and Towers, G. H. N. (1976). *Phytochemistry* **15,** 564–566.
Fries, L. (1977). *Phycologia* **16,** 451–455.
Fries, L., and Åberg, S. (1978). *Z. Pflanzenphysiol.* **88,** 383–388.
Fujita, M., and Inoue, T. (1977). *Tetrahedron Lett.* **51,** 4503–4506.
Fujita, M., and Inoue, T. (1981). *Phytochemistry* **20,** 2183–2185.
Gestetner, B., and Conn, E. E. (1974). *Arch. Biochem. Biophys.* **163,** 6173–6624.
Gilchrist, D. G., and Kosuge, T. (1980). *In* "The Biochemistry of Plants, Vol 5: Amino Acids and Derivatives" (B. J. Miflin, ed.), pp. 507–531. Academic Press, New York.
Grand, C. (1984). *FEBS Lett.* **169,** 7–11.
Gregor, H. D. (1976). *Z. Pflanzenphysiol.* **77,** 372–375.
Grisebach, H. (1962). *Planta Med.* **10,** 385–397.
Grisebach, H. (1977). *Naturwissenschaften* **64,** 619–625.
Grisebach, H. (1979). *In* "Biochemistry of Plant Phenolics" (T. Swain, J. B. Harborne, and C. F. Van Sumere, eds.), Vol. 12, pp. 221–248, Rec. Adv. Phytochem. Series. Plenum, New York.
Grisebach, H. (1981). *In* "The Biochemistry of Plants, Vol. 7: Secondary Plant Products" (E. E. Conn, ed.), pp. 457–478. Academic Press, New York.

Gross, G. G. (1972). *Eur. J. Biochem.* **31,** 585–592.
Gross, G. G. (1977). *In* "The Structure, Biosynthesis and Degradation of Wood" (F. A. Loewus and V. C. Runeckles, eds.), Vol. 11, pp. 141–184, Rec. Adv. Phytochem. Series. Plenum, New York.
Gross, G. G. (1979). *In* "Biochemistry of Plant Phenolics" (T. Swain, J. B. Harborne, and C. F. Van Sumere, eds.), Vol. 12, pp. 177–220, Rec. Adv. Phytochem. Series. Plenum, New York.
Gross, G. G. (1980). *Adv. Bot. Res.* **8,** 25–63.
Gross, G. G. (1981). *In* "The Biochemistry of Plants, Vol. 7: Secondary Plant Products" (E. E. Conn, ed.), pp. 301–316. Academic Press, New York.
Gross, G. G. (1982). *FEBS Lett.* **148,** 67–70.
Gross, G. G. (1983a). *Phytochemistry* **22,** 2179–2182.
Gross, G. G. (1983b). *Z. Naturforsch.* **38c,** 519–523.
Gross, G. G., and Kreiten, W. (1975). *FEBS Lett.* **54,** 259–262.
Gross, G. G., and Zenk, M. H. (1966). *Z. Naturforsch.* **21B,** 683–690.
Gross, G. G., and Zenk, M. H. (1974). *Eur. J. Biochem.* **42,** 453–459.
Gross, G. G., Stöckigt, J., Mansell, R. L., and Zenk, M. H. (1973). *FEBS Lett.* **31,** 283–286.
Gross, G. G., Mansell, R. L., and Zenk, M. H. (1975). *Biochem. Physiol. Pflanzen* **168,** 41–51.
Gupta, P., and Lewis, J. R. (1971). *J. Chem. Soc. C,* 629–631.
Hagel, P., and Kindl, H. (1975). *FEBS Lett.* **59,** 120–124.
Hahlbrock, K. (1981). *In* "The Biochemistry of Plants, Vol. 7: Secondary Plant Products" (E. E. Conn, ed.), pp. 425–456. Academic Press, New York.
Halliwell, B. (1975). *Eur. J. Biochem.* **55,** 355–360.
Hanson, K. R., and Havir, E. A. (1979). *In* "Biochemistry of Plant Phenolics" (T. Swain, J. B. Harborne, C. F. Van Sumere, eds.), Vol 12, pp. 91–137, Rec. Adv. Phytochem. Series Plenum, New York.
Hanson, K. R., and Havir, E. A. (1981). *In* "The Biochemistry of Plants, Vol. 7: Secondary Plant Products" (E. E. Conn, ed.), pp. 577–625. Academic Press, New York.
Hanson, H. R., Wightman, R. H., Staunton, J., and Battersby, A. R. (1971). *Chem. Commun.* 185–186.
Hasegawa, S., and Maier, V. P. (1972). *Phytochemistry* **11,** 1365–1370.
Haslam, E. (1981). *In* "The Biochemistry of Plants, Vol. 7: Secondary Plant Products" (E. E. Conn, ed.), pp. 527–556. Academic Press, New York.
Hatfield, G. M., and Brady, L. R. (1973). *Lloydia* **36,** 59–65.
Havir, E. A., Reid, P. D., and Marsh, H. V. (1971). *Plant Physiol.* **48,** 621–628.
Higuchi, T. (1976). *Wood Res.* **59–60,** 180–199.
Higuchi, T., and S. A. Brown (1963). *Can. J. Biochem. Physiol.* **41,** 621–628.
Hill, A. C., and Rhodes, M. J. C. (1975). *Phytochemistry* **14,** 2387–2391.
Holländer, H., Kiltz, H. H., and Amrhein, N. (1979). *Z. Naturforsch.* **34C,** 1162–1173.
Hösel, W., and Todenhagen, R. (1980). *Phytochemistry* **19,** 1349–1353.
Hösel, W., Surholt, E., and Borgmann, E. (1978). *Eur. J. Biochem.* **84,** 487–492.
Hrazdina, G., Kreuzaler, F., Hahlbrock, K., and Grisebach, H. (1976). *Arch. Biochem. Biophys.* **175,** 392–399.
Ibrahim, R. K. (1977). *Z. Pflanzenphysiol.* **85,** 253–262.
Ibrahim, R. K., and Grisebach, H. (1976). *Arch. Biochem. Biophys.* **176,** 700–708.
Ife, R., and Haslam, E. (1971). *J. Chem. Soc. C,* 2818–2821.
Jeffs, P. W., Karle, J. M., and Martin, N. H. (1978). *Phytochemistry* **17,** 719–728.
Kindl, H. (1971). *Naturwissenschaften* **58,** 554–563.

Klischies, M., and Zenk, M. H. (1978). *Phytochemistry* **17,** 1281–1284.
Klischies, M., Stöckigt, J., and Zenk, M. H. (1975). *J. Chem. Soc. Chem. Commun.,* 879–880.
Klischies, M., Stöckigt, J., and Zenk, M. H. (1978). *Phytochemistry* **17,** 1523–1525.
Knobloch, K. H., and Hahlbrock, K. (1975). *Eur. J. Biochem.* **52,** 311–320.
Knobloch, K. H., and Hahlbrock, K. (1977). *Arch. Biochem. Biophys.* **184,** 237–248.
Koukol, J., and Conn, E. E. (1961). *J. Biol. Chem.* **236,** 2692–2698.
Kreuzaler, F., and Hahlbrock, K. (1975). *Arch. Biochem. Biophys.* **169,** 84–90.
Kuroda, H., Shimada, M., and Higuchi, T. (1981). *Phytochemistry* **20,** 2635–2639.
Kutsuki, H., Shimada, M., and Higuchi, T. (1982a). *Phytochemistry* **21,** 267–271.
Kutsuki, H., Shimada, M., and Higuchi, T. (1982b). *Phytochemistry* **21,** 19–23.
Lamb, C. J. (1977). *FEBS Lett.* **75,** 37–40.
Leete, E., Muir, A., and Towers, G. H. N. (1982). *Tetrahedron Lett.* **23,** 2635–2638.
Legrand, M., Fritig, B., and Hirth, L. (1976). *FEBS Lett.* **70,** 131–136.
Leinberger, R., Hull, W. E., Simon, H., and Rétey, J. (1981). *Eur. J. Biochem.* **117,** 311–318.
Löffelhardt, W., and Kindl, H. (1975). *Hoppe-Seyler's Z. Physiol. Chem.* **356,** 487–493.
Löffelhardt, W., and Kindl, H. (1976). *Z. Naturforsch.* **31C,** 693–699.
Löffelhardt, W., and Kindl, H. (1979). *FEBS Lett.* **104,** 332–334.
Lüderitz, T., and Grisebach, H. (1981). *Eur. J. Biochem.* **119,** 115–124.
Lüderitz, T., Schatz, G., and Grisebach, H. (1982). *Eur. J. Biochem.* **123,** 583–586.
Macheix, J. J. (1977). *C. R. Acad. Sci. Ser. D.* **284,** 53–56.
Malmström, B. G. (1982). *Ann. Rev. Biochem.* **51,** 21–59.
Manitto, P., Monti, D., and Gramatica, P. (1974). *J. Chem. Soc. Perkin Trans. I,* 1727–1731.
Manitto, P., Gramatica, P., and Ranzi, B. M. (1975). *J. Chem. Soc. Chem. Commun.* 442–443.
Mansell, R. L., Stöckigt, J., and Zenk, M. H. (1972). *Z. Pflanzenphysiol.* **68,** 286–288.
Mansell, R. L., Gross, G. G., Stöckigt, H., Franke, H., and Zenk, M. H. (1974). *Phytochemistry* **13,** 2427–2435.
Mansell, R. L., Babbel, G. R., and Zenk, M. H. (1976). *Phytochemistry* **15,** 1849–1853.
Marcinowski, S., and Grisebach, H. (1978). *Eur. J. Biochem.* **87,** 37–44.
Marcinowski, S., Falk, H., Hammer, D. K., Boyer, B., and Grisebach, H. (1979). *Planta* **144,** 161–165.
Martin-Tanguy, J., Cabanne, F., Perdrizet, E., and Martin, C. (1978). *Phytochemistry* **17,** 1927–1928.
McIntyre, R. J., and Vaughan, P. F. T. (1975). *Biochem. J.* **149,** 447–461.
Meyer, E., and Barz, W. (1978). *Planta Med.* **33,** 336–344.
Michalzuk, L., and Bandurski, R. S. (1980). *Biochem. Biophys. Res. Commun.* **93,** 588–592.
Michalzuk, L., and Bandurski, R. S. (1982). *Biochem. J.* **207,** 273–281.
Murakoshi, I., Ogawa, M., Toriizuka, K., Haginiwa, J., Ohmiya, S., and Otomasu, H. (1977). *Chem. Pharm. Bull.* **25,** 527–528.
Nagels, L., and Parmentier, F. (1974). *Phytochemistry* **13,** 2759–2762.
Nagels, L., and Parmentier, F. (1976). *Phytochemistry* **15,** 703–706.
Nagels, L., Molderez, M., and Parmentier, F. (1981). *Phytochemistry* **20,** 965–967.
Nakamura, Y., Fushiki, H., and Higuchi, T. (1974). *Phytochemistry* **13,** 1777–1784.
Neish, A. C. (1961). *Phytochemistry* **1,** 1–24.
Neish, A. C. (1964). *In* "The Formation of Wood in Forest Trees" (M. H. Zimmermann, ed.), pp. 219–239. Academic Press, New York.
Nurmann, G., and Strack, D. (1981). *Z. Pflanzenphysiol.* **102,** 11–17.

Perrin, P. W., and Towers, G. H. N. (1973). *Phytochemistry* **12,** 589–592.
Pfändler, R., Scheel, D., Sandermann, H., and Grisebach, H. (1977). *Arch. Biochem. Biophys.* **178,** 315–316.
Ponchet, M., Martin-Tanguy, J., Marais, A., and Martin, C. (1982). *Phytochemistry* **21,** 2865–2869.
Postius, C., and Kindl, H. (1977). *Z. Naturforsch.* **33C,** 65–69.
Potts, J. R. M., Weklych, R., and Conn, E. E. (1974). *Z. Biol. Chem.* **249,** 5019–5026.
Poulton, J. E. (1981). *In* "The Biochemistry of Plants, Vol. 7: Secondary Plant Products" (E. E. Conn, ed.), pp. 667–723. Academic Press, New York.
Poulton, J. E., and Butt, V. S. (1975). *Biochim. Biophys. Acta* **403,** 301–314.
Poulton, J. E., and Butt, V. S. (1976). *Arch. Biochem. Biophys.* **172,** 135–142.
Poulton, J., Hahlbrock, K., and Grisebach, H. (1976a). *Arch. Biochem. Biophys.* **176,** 449–456.
Poulton, J., Grisebach, H., Ebel, J., Schaller-Hekeler, B., and Hahlbrock, K. (1976b). *Arch. Biochem. Biophys.* **173,** 301–305.
Reichhart, D., Salaun, J. P., Benveniste, I., and Durst, F. (1980). *Plant Physiol.* **66,** 600–604.
Reichhart, D., Simon, A., Durst, F., Mathews, J. M., and Ortiz de Montellano, P. R. (1982). *Arch. Biochem. Biophys.* **216,** 522–529.
Rhodes, M. J. C., and Wooltorton, L. S. C. (1976). *Phytochemistry* **15,** 947–951.
Rhodes, M. J. C., and Wooltorton, L. S. C. (1977). *Phytochemistry* **16,** 655–659.
Rhodes, M. J. C., and Wooltorton, L. S. C. (1978). *Phytochemistry* **17,** 1225–1229.
Rhodes, M. J. C., Wooltorton, L. S. C., and Lourenço, E. J. (1979). *Phytochemistry* **18,** 1125–1129.
Rich, P. R., and Lamb, C. J. (1977). *Eur. J. Biochem.* **72,** 353–360.
Rupprich, N., and Kindl, H. (1978). *Hoppe-Seyler's Z. Physiol. Chem.* **359,** 165–172.
Rupprich, N., Hildebrand, H., and Kindl, H. (1980). *Arch. Biochem. Biophys.* **200,** 72–78.
Russel, D. W. (1971). *J. Biol. Chem.* **246,** 3870–3878.
Sagisaka, S., and Shimura, K. (1962). *J. Biochem.* **52,** 155–161.
Saylor, M. H., and Mansell, R. L. (1977). *Z. Naturforsch.* **32C,** 765–768.
Schmid, G., and Grisebach, H. (1982). *Eur. J. Biochem.* **123,** 363–370.
Schmid, G., Hammer, D. K., Ritterbusch, A., and Grisebach, H. (1982). *Planta* **156,** 207–212.
Schoeppner, A., and Kindl, H. (1979). *FEBS Lett.* **108,** 349–352.
Schroeder, H. A. (1968). *Phytochemistry* **7,** 57–61.
Senanayake, U. M., Wills, R. B. H., and Lee, T. H. (1977). *Phytochemistry* **16,** 2032–2033.
Sharma, S. K., and Brown, S. A. (1979). *Can. J. Biochem.* **57,** 986–994.
Shimada, M. (1972). *Wood Res.* **53,** 19–65.
Shimada, M., Fushiki, H., and Higuchi, T. (1973). *Mokuzai Gakkaishi* **19,** 13–21.
Shimizu, T., and Kojima, M. (1984). *J. Biochem.* **95,** 205–212.
Sinha, A. K., and Bhattacharjee, J. K. (1971). *Biochem. J.* **125,** 743–749.
Smith, T. A. (1981). *In* "The Biochemistry of Plants, Vol. 7: Secondary Plant Products" (E. E. Conn, ed.), pp. 249–268. Academic Press, New York.
Stafford, H. A., and Lewis, L. C. (1979). *Plant Physiol.* **64,** 176–181.
Stöckigt, J., and Klischies, M. (1977). *Holzforschung* **31,** 41–44.
Stöckigt, J., and Zenk, M. H. (1974). *FEBS Lett.* **41,** 131–134.
Stöckigt, J., and Zenk, M. H. (1975). *Z. Naturforsch.* **30C,** 352–358.
Stöckigt, J., Mansell, R. L., Gross, G. G., and Zenk, M. H. (1973). *Z. Pflanzenphysiol.* **70,** 305–307.
Strack, D. (1980). *Z. Naturforsch.* **35C,** 204–208.

Strack, D. (1982). *Planta* **155,** 31–36.
Strack, D., Knogge, W., and Dahlbender, B. (1983). *Z. Naturforsch.* **38C,** 21–27.
Surholt, E., and Hösel, W. (1981). *Z. Pflanzenphysiol.* **101,** 207–214.
Sütfeld, R., and Wiermann, R. (1978). *Biochem. Physiol. Pflanzen* **172,** 111–123.
Sütfeld, R., and Wiermann, R. (1980). *Z. Pflanzenphysiol.* **97,** 283–288.
Tahara, S., Kudo, M., and Mizutani, J. (1979). *Agric. Biol. Chem.* **43,** 2627–2628.
Tkotz, N., and Strack, D. (1980). *Z. Naturforsch.* **35C,** 835–837.
Tsang, Y. F., and Ibrahim, R. K. (1979a). *Z. Naturforsch.* **34C,** 46–50.
Tsang, Y. F., and Ibrahim, R. K. (1979b). *Phytochemistry* **18,** 1131–1136.
Ulbrich, B., and Amrhein, N. (1978). *Planta* **138,** 69–71.
Ulbrich, B., and Zenk, M. H. (1979). *Phytochemistry* **18,** 929–933.
Ulbrich, B., and Zenk, M. H. (1980). *Phytochemistry* **19,** 1625–1629.
Ulbrich, B., Stöckigt, J., and Zenk, M. H. (1976). *Naturwissenschaften* **63,** 484.
Uotila, L., and Koivusalo, M. (1974). *J. Biol. Chem.* **249,** 7653–7663.
Uotila, L., and Koivusalo, M. (1979). *Arch. Biochem. Biophys.* **196,** 33–45.
Vance, C. P., Nambudiri, A. M. D., Wat, C. K., and Towers, G. H. N. (1975). *Phytochemistry* **14,** 967–969.
Vaughan, P. F. T., and Butt, V. S. (1969). *Biochem. J.* **113,** 109–115.
Vaughan, P. F. T., and Butt, V. S. (1970). *Biochem. J.* **119,** 89–94.
Vaughan, P. F. T., and McIntyre, R. J. (1975). *Biochem. J.* **151,** 759–762.
Vaughan, P. F. T., Eason, R., Paton, J. Y., and Ritchie, G. A. (1975). *Phytochemistry* **14,** 2383–2386.
Vaughn, K. C., and Duke, S. O. (1974). *Physiol. Plant.* **60,** 106–112.
Vieitez, A. M., Ballester, A., and Vieitez, E. (1975). *Experientia* **31,** 1163.
Wallis, P. J., and Rhodes, M. J. C. (1977). *Phytochemistry* **16,** 1891–1894.
Walton, E., and Butt, V. S. (1970). *J. Exp. Bot.* **21,** 887–891.
Walton, E., and Butt, V. S. (1971). *Phytochemistry* **10,** 295–304.
Wat, C. K., and Towers, G. H. N. (1979). *In* "Biochemistry of Plant Phenolics" (T. Swain, J. B. Harborne, and C. F. Van Sumere, eds.), Vol. 12, pp. 371–432, Rec. Adv. Phytochem. Series. Plenum, New York.
Wengenmayer, H., Ebel, J., and Grisebach, H. (1976). *Eur. J. Biochem.* **65,** 529–536.
West, C. A. (1980). *In* "The Biochemistry of Plants, Vol. 2: Metabolism and Respiration" (D. D. Davies, ed.), pp. 317–364. Academic Press, New York.
White, R. E., and Coon, M. J. (1980). *Ann. Rev. Biochem.* **49,** 315–356.
Wightman, F., and Lighty, D. L. (1982). *Physiol. Plant.* **55,** 17–24.
Wyrambik, D., and Grisebach, H. (1975). *Eur. J. Biochem.* **59,** 9–15.
Wyrambik, D., and Grisebach, H. (1979). *Eur. J. Biochem.* **97,** 503–509.
Zenk, M. H. (1964). *Z. Naturforsch.* **19B,** 83–84.
Zenk, M. H. (1966). *In* "Biosynthesis of Aromatic Compounds" (G. Billek, ed.), pp. 45–60. Pergamon, Oxford.
Zenk, M. H. (1967). *Z. Pflanzenphysiol.* **57,** 477–478.
Zenk, M. H. (1971). *In* "Pharmacognosy and Phytochemistry" (H. Wagner and C. Hörhammer, eds.), pp. 314–346. Springer Verlag, Berlin and New York.
Zenk, M. H. (1979). *In* "Biochemistry of Plant Phenolics" (T. Swain, J. B. Harborne and C. F. Van Sumere, eds.), Vol. 12, pp. 139–176, Rec. Adv. Phytochem. Series. Plenum, New York.
Zenk, M. H., and Gross, G. G. (1965). *Z. Pflanzenphysiol.* **53,** 356–362.
Zenk, M. H., and Gross, G. G. (1972). *Rec. Adv. Phytochem.* **4,** 87–106.
Zenk, M. H., Ulbrich, B., Busse, J., and Stöckigt, J. (1980). *Anal. Biochem.* **101,** 182–187.

CHAPTER 11

Occurrence and Biosynthesis of Quinones in Woody Plants

E. LEISTNER

Institut für Pharmazeutische Biologie
Rheinischen Friedrich-Wilhelms-Universität Bonn
Bonn, Federal Republic of Germany

I. Reviews and Books on Quinones

As a consequence of the continuing progress in the chemistry, biochemistry, and physiology of quinones, several reviews and two monographs have been published on this class of natural products. The chemistry of quinones has been treated in a book edited by Patai (1974). Chemical and biological aspects such as identification, distribution, and biogenesis of quinones are dealt with in Thomson's book (Thomson, 1971), one of the most cited references in this field. Two review articles (Thomson, 1976a,b) can be considered as supplements to this book.

While many quinones are based on the benzo-, naphtho-, or anthraquinone skeleton, Thomson (1979) shows that there are many quinonoid structures that do not belong to any of these three basic types. Quinones are mainly *p*-quinones, but *o*-quinones may also occur. Their chemistry

Biosynthesis and Biodegradation of Wood Components

ISBN 0-12-347880-4

and biology have been reviewed by Pierpoint (1970). Quinones may arise by different biosynthetic pathways (Zenk and Leistner, 1968; Bentley, 1975a,b; Leistner, 1980, 1981), and there is a relationship between the occurrence of quinones in various taxa and their biosynthesis (Zenk and Leistner, 1968; Leistner 1980, 1981).

Publication of three reviews on the biosynthesis of vitamin K (Leistner and Zenk, 1971; Bentley and Meganathan, 1982, 1983) reflects the rapid progress in the understanding of the biochemistry and genetics of this biologically important naphthoquinone. Finally, it should be mentioned that a series of quinonoid compounds exhibits antibiotic activities. The biosynthesis of this type of natural products has been reviewed by Floss *et al.* (1978), while Eugster (1980) demonstrates that many quinones exist among terpenoid pigments.

II. Occurrence of Quinones in Wood and Heartwood

It is widely accepted that quinones occur in bacteria, fungi, lichens, gymnosperms, and angiosperms (Thomson, 1971; Leistner, 1981), and it has been pointed out (e.g., Leistner, 1981) that they have not yet been found in extracts of mosses and ferns. Recently, however, 3,3′-bi-(2-methyl-1,4-naphthaquinone) has been isolated from *Asplenium laccin-iatum,* a fern (Gupta *et al.,* 1979). Thus, to our knowledge, mosses are the only major taxon of the plant kingdom from which quinones have so far not been isolated.

Quinones are present in many plant families. They may occur in all organs of a plant, and they have been isolated from the wood or heart-wood of plants belonging to families in the following groups:

Group 1: Leguminosae,[1] Boraginaceae, Ebenaceae, Rhamnaceae
Group 2: Bignoniaceae, Juglandaceae, Proteaceae, Rubiaceae, Verbenaceae, Rutaceae
Group 3: Malvaceae, Sterculiaceae, Ulmaceae.

The plant families are grouped together according to the structural type of quinone.

[1] The taxonomic classification used in this chapter is that of A. Engler (1964), *Syllabus der Pflanzenfamilien,* Vol. II, Bornträger, Berlin–Nikolassee. Engler divides the family Leguminosae into the following subfamilies and genera: Mimosoideae (e.g., *Acacia*), Caesalpinoideae (e.g., *Cassia*), and Faboideae (e.g., *Dalbergia, Abrus, Cajanus*).

A. *Quinones in Group 1*

Tropical woods derived from *Dalbergia* plants (Leguminosae) contain quinones that exhibit microbiocidal and algicidal activity. Resistance to marine borer organisms is also attributed to this type of natural product (Manners and Jurd, 1977a), which may also cause allergic reactions (Hausen, 1978). *Dalbergia retusa* wood has outstanding resistance to attack by teredos, limnoria, and other marine boring organisms. The wood contains larvicidal constituents and is the most resistant among 115 different wood species. The heartwood of this plant is intensely orange in color, containing mainly two pigments: obtusaquinone (3% of undried heartwood), which is a quinonemethide and has either structure **1** or **2** (Fig. 1), and a black pigment that represents the quinhydrone of racemic 4-methoxydalbergione (**3**). Obtusaquinone is highly toxic to fish (Jurd *et al.*, 1972). A phenanthraquinone derived from *Dalbergia latifolia* (source of Indian rosewood) was named latinone (**4**). The structure of this compound was confirmed by synthesis. A biogenetic relationship between compounds I, II, III, and IV seems to be plausible (Criodain *et al.*, 1981).

Australian blackwood (*Acacia melanoxylon*, Leguminosae) is one of at least 100 wood species of commercial value that are capable of inducing an allergy of the eczematous type and allergic bronchial asthma. The sensitizers have been identified as 2,6-dimethoxybenzoquinone (**5**) and 2-methyl-6-methoxyfuranobenzoquinone (acamelin) (**6**). The wood is used for furniture (Schmalle and Hausen, 1980). The heartwood of *Bowdichia nitida* (Leguminosae) also causes allergic reactions because it also contains the benzoquinone **5**. Another compound, however, 7-hydroxy-3-(5-methoxy-1,4-benzoquinone-2-yl-4-benzopyrone (**7**), has also been isolated from this wood (Brown *et al.*, 1974). This compound is neither a typical quinone nor a typical flavonoid.

Quinones in plants belonging to the family Boraginaceae are also of mixed origin, comprising an aromatic moiety and isoprene units. Compounds of this type have been isolated from *Cordia* species (Boraginaceae) (Manners and Jurd, 1977b; Moir and Thomson, 1973a). Although biosynthetic experiments have not been carried out on these natural products, it is obvious that they are derived from geranylhydroquinone. Depending on the sequence of cyclization and oxidation processes, cordiachromes (cordiachrome A = **8**), cordiachromenes, allioquinols, or cordiols are formed. The quinonoid extractives of *Cordia alliodora* and other *Cordia* species are likely to be responsible for the durability of this wood in marine use. Compounds of this type are also present in the heartwood of *Patagonula americana* L. (Moir and Thomson, 1973b), cordiachrome G (**9**) being an example. The compound is optically active.

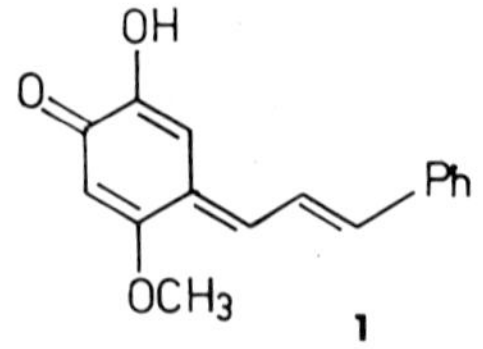

1

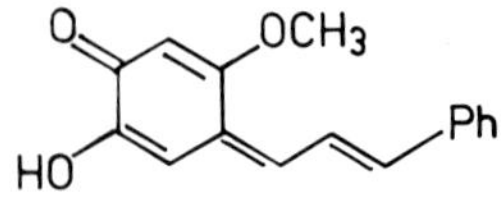

2

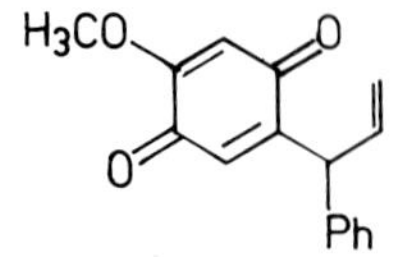

3

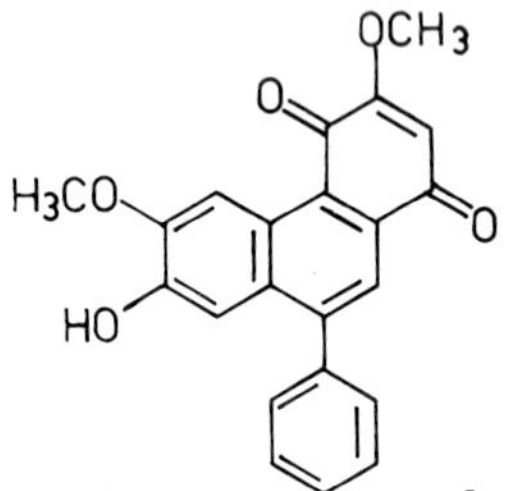

4

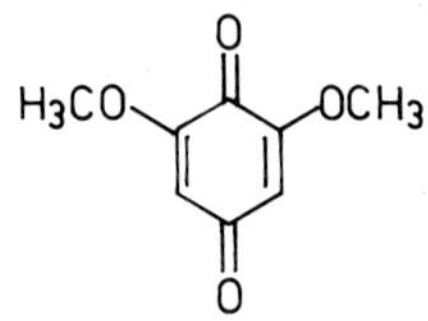

5

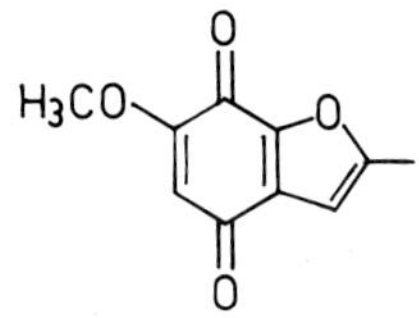

6

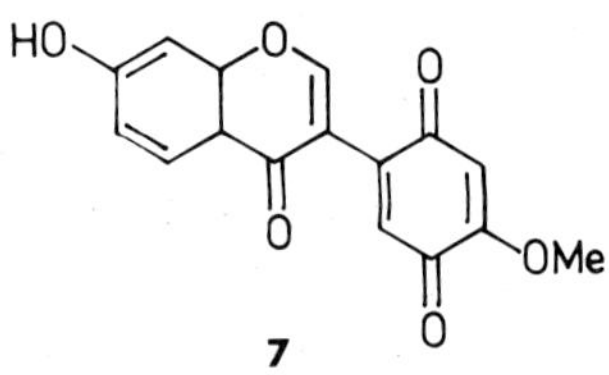

7

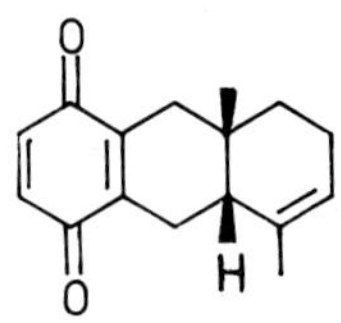

8

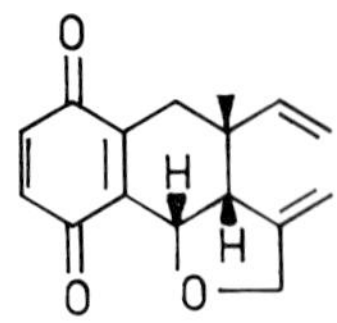

9

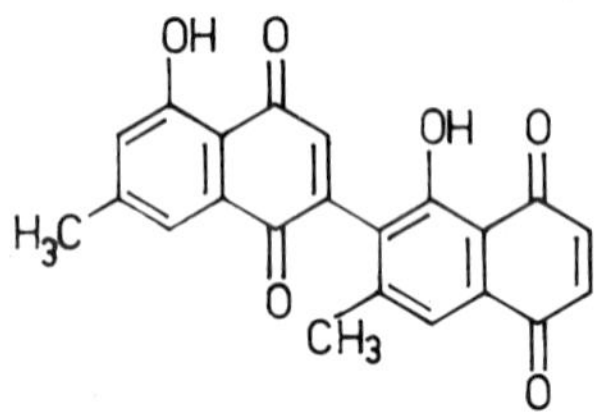

10

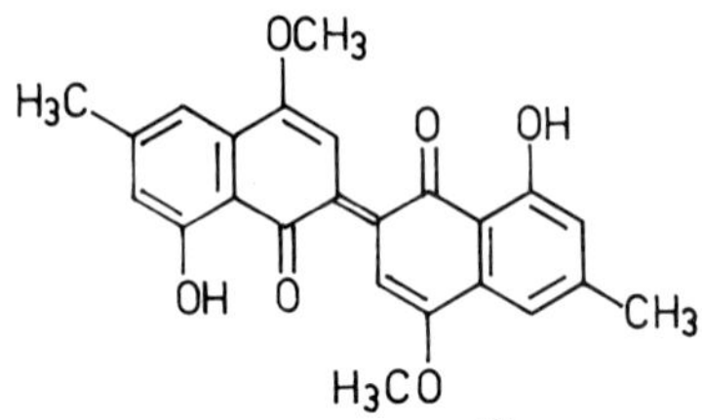

11

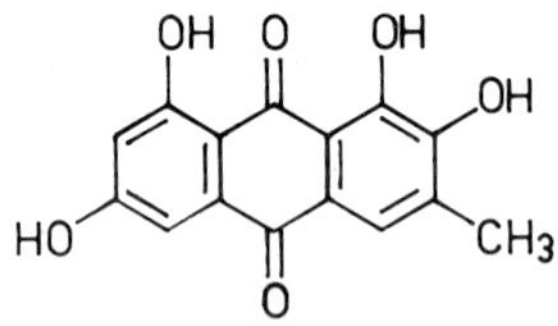

12

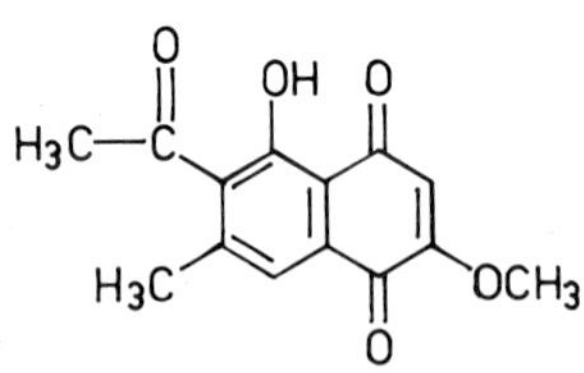

13

Fig. 1. Quinones that are derived from acetate or are of mixed biogenetic origin.

Prenylated naphthoquinones occur also in the roots of different Boraginaceae, e.g., *Lithospermum* (Yoshizaki *et al.*, 1982), *Echium* (Tabata *et al.*, 1982), *Macrotomia* (Papageorgiou, 1979), *Arnebia* (Skukla *et al.*, 1973), and *Alkanna* (Papageorgiou *et al.*, 1979). These naphthoquinones exhibit antimicrobial activity and have long been used in Japan as a medicine for wounds, burns, etc. (Fujita *et al.*, 1982).

Euclea and *Diospyros* species belong to the family Ebenaceae and are a rich source of binaphthoquinones that are responsible for antifungal, antibacterial, and termite-resistant properties. This is very important because *Diospyros* timbers are of high commercial value. Scrapings of wooden parts of the plants are used in African folk medicine for the treatment of leprosy, and the ichthyotoxic properties of the bark of some species is used in hunting (Waterman and Mbi, 1979).

Bisnaphthoquinones have been isolated from *Euclea natalensis* (King *et al.*, 1976), *E. pseudebenus* (Ferreira *et al.*, 1973), *Diospyros melanoxylon* (Sankaram *et al.*, 1981), *D. batocana* (Alves *et al.*, 1983), *D. chloroxylon* (Sidhu and Prasad, 1971), *D. tricolor* (Lillie and Musgrave, 1980), *D. bipindensis, D. cinnabarina, D. obliquifolia* (Waterman and Mbi, 1979), *D. lycioides, D. heterotricha* (Alves *et al.*, 1973), and other *Diospyros* species (Fallas and Thomson, 1968; Musgrave and Skoyles, 1974). Naphthoquinones bearing coumarin units have been isolated from *D. ismaillii* and *D. canaliculata* (Jeffreys *et al.*, 1983). The most common binaphthoquinone seems to be diospyrin (**10**), whereas a very remarkable one is diosindigo A (**11**), because it is blue and responsible for the blue color of the heartwood of some *Diospyros* species.

The family Rhamnaceae comprises woody plants (shrubs or trees). They contain anthraderivatives like emodin, chrysophanol, physcion, and the corresponding dianthrones and heterodianthrones. The compounds are present as such and as glycosides (Hegnauer, 1973). Anthra derivatives of Rhamnaceae are used as purgatives. Other physiological effects of these types of compounds have also been described (Dreyer *et al.*, 1975; Tripathi *et al.*, 1979). Examples of quinones isolated from extracts of Rhamnaceae are alternin (**12**) and 2-methoxystypandron (**13**) (Rauwald and Miething, 1982).

The fresh bark of *Frangula alnus* (synonym, *Rhammus frangula*) contains only anthrones. After storage and extraction, anthraquinones and dianthrone derivatives are isolated (Labadie, 1970). There is some doubt as to whether dianthrones and anthraquinones are components of these plants or are artifacts. The situation in *Cassia* plants (Leguminosae) is similar. Using radioimmunoassay, it has been shown that the formation of the dianthrone glycoside sennoside B is inversely related to the water loss of the leaves (Atzorn *et al.*, 1981).

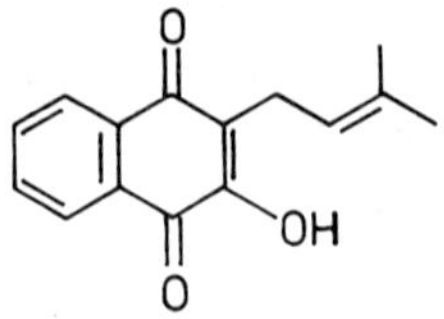

14

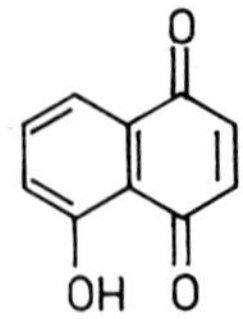

15

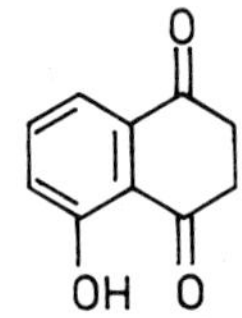

16

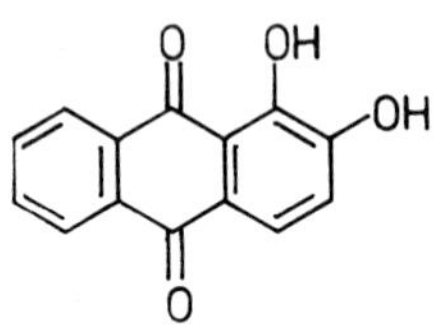

17

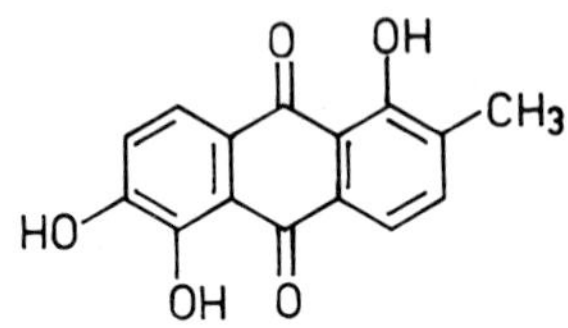

18

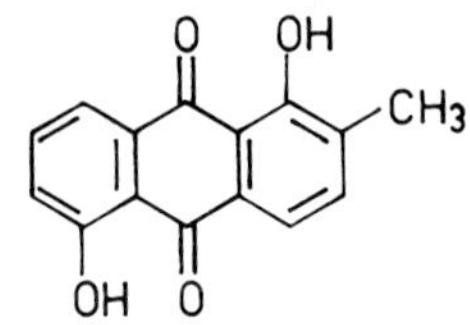

19

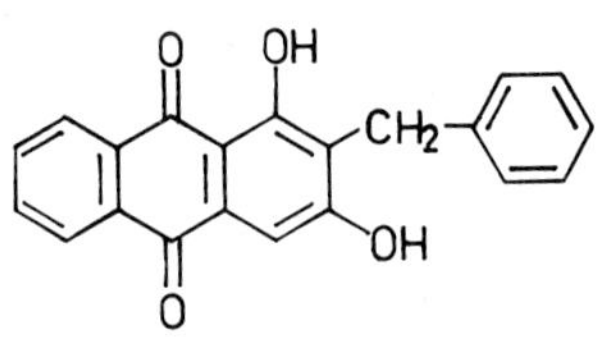

20

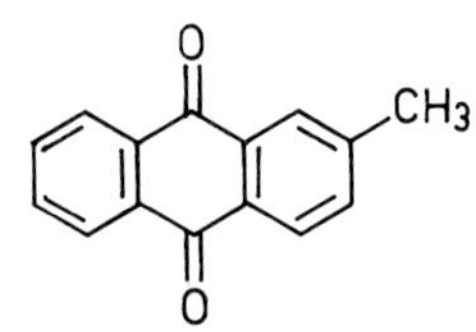

21

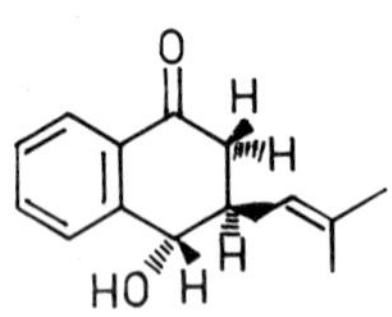

22

Fig. 2. Quinones and related compounds derived from *o*-succinylbenzoic acid.

B. Quinones in Group 2

The woody plants belonging to the Bignoniaceae contain mainly prenylated naphthalene derivatives and anthraquinones. The most common prenylated naphthoquinone seems to be lapachol (**14**) (Fig. 2). The fact that prenylated naphthoquinones co-occur with anthraquinones (Singh and Singh, 1980; Burnett and Thomson, 1967, 1968a) suggested that the former are precursors of the latter (see below).

Quinones in the Bignoniaceae have been isolated from the wood of plants belonging to the genus *Kigelia* (Inoue *et al.*, 1981a), *Radermachera*

(Inoue *et al.*, 1981b), *Markhamia* (Singh and Singh, 1980), *Haplophragma* (Joshi *et al.*, 1979), *Tecomella* (Joshi *et al.*, 1977), *Tabebuia* (Manners *et al.*, 1975; Burnett and Thomson, 1967, 1968a), *Zeyhera* (Weinberg *et al.*, 1976), *Catalpa* (Inouye *et al.*, 1975), and *Paratecoma* (Sandermann *et al.*, 1968). The constituents in the wood of these plants are responsible for their antifungal activity (Singh and Singh, 1980) and their resistance against teredos (Sandermann *et al.*, 1968).

The fact that constituents in *Catalpa* are in part prearomatic naphthalene derivatives has led Inouye *et al.* (1981, 1982) to propose a reaction sequence in which these compounds play a role as precursors of quinones (see below).

Prearomatic naphthalene derivatives may also occur in *Carya* (Hedin *et al.*, 1980) and *Juglans* (Müller and Leistner, 1978a), plants which both belong to the family Juglandaceae. *Juglans regia* is the only plant from which unsubstituted 1,4-naphthoquinone (Müller and Leistner, 1976) and the β-D-glucopyranoside of its hydroquinone (Müller and Leistner, 1978b) have been isolated. The main quinonoid constituent of this plant and of *Carya illinoensis* (Hedin *et al.*, 1980) is juglone (**15**). This compound not only occurs as a monomer, but also forms a 3,3′-dimer and a cyclotrijuglone, which are present in the root bark of *Juglans regia* (Pardhasardhi and Babu, 1978). Juglone (**15**) is a fungitoxic agent effective against *Fusicladium effusum,* a fungus that causes the most important disease of pecans in the southeastern United States (Hedin *et al.*, 1980). Juglone, its glucoside, and its congeners are washed down from the leaves by rain and are responsible for the inhibition of growth of other plants in the surroundings of *Juglans*. The German saying, "Wer unter dem Nußbaum [i.e., *Juglans regia*] schläft, taugt nichts," refers to this observation.

Compounds similar to those occurring in *Juglans* and *Carya* have also been isolated from the wood of *Lomatia* (Moir and Thomson, 1973c), *Stenocarpus* (Mock *et al.*, 1973), and *Conospermum* (Cannon *et al.*, 1975). These plants belong to the family Proteaceae. *Lomatia,* for example, contains juglone (**15**), β-hydrojuglone (**16**), and a prenylated naphthoquinone. The latter group of compounds is also found in extracts of *Stenocarpus* and *Conospermum*.

Plants of the family Rubiaceae have been extensively investigated for the presence of quinones. These compounds occur mainly in the roots but also in the bark or stem. Quinones occur in the stem of *Morinda* (Adesogan, 1973; Demagos *et al.*, 1981), in the bark of *Coprosma* (Briggs *et al.*, 1976), and in the roots of *Galium* (Heide and Leistner, 1982), *Rubia* (Dosseh *et al.*, 1981), *Asperula* (Burnett and Thomson, 1968b), *Hymenodictyon,* and *Dammacanthus* (Brew and Thomson, 1971). Plants of these genera have often been used for dyeing purposes, alizarin (**17**) being a well-known dye. Extracts of *Morinda* plants exhibit antineoplastic properties against sarcoma 180 ascites in mice (Durodola, 1974).

Many quinones in the Rubiaceae are anthraquinones. Substituents in the anthraquinone molecule are usually confined to one ring. Morindone (**18**) and soranjidiol (**19**) are exceptions. An anthraquinone with an unusual structural feature is benzylxanthropurpurin (**20**) (Brew and Thomson, 1971). A plausible hypothesis on the origin of the benzyl residue in this compound has not been advanced. Quinone constituents of the Verbenaceae are responsible for the durability of the wood. When the termite resistance of various woods and the effectiveness of different compounds was tested, tectoquinone (**21**) and 2-methoxyanthraquinone were found to be most effective against termites (Sandermann and Dietrichs, 1957). In *Tectona grandis,* the highest concentration of tectoquinone is observed in the center of the heartwood. The amount is much lower in the outer parts of the stem. Prenylated naphthalene derivatives show a different distribution in the stem. Their concentration is highest in the outer sphere of the heartwood but is relatively low in the center (Sandermann and Dietrichs, 1959). The quinones found in *Tectona* are skin-irritating (Sandermann and Simatupang, 1963, 1966). They occur predominantly in the ray cells (Sandermann, 1966). Catalponol (**22**) is a compound that is interesting because it is a prenylated naphthalene derivative that occurs in *Lippia origanoides* (Brieskorn and Pöhlmann, 1976) and is related to the biosynthesis of compounds occurring in *Catalpa* plants (see below). Tectoquinone (**21**), an anthraquinone typical of Verbenaceae, has also been isolated from a Rutaceae plant, *Clausena heptaphylla* (Chakraborty *et al.,* 1978).

C. Quinones in Group 3

Many timbers change color on exposure to daylight, and a striking example is provided by the wood of blue mahoe (*Hibiscus elatus,* Malvaceae). The freshly cut heartwood shows a dark bluish-gray color, whereas on exposure to daylight the wood color changes to a light yellow to buff-gray (Ferreira *et al.,* 1980). The wood of this tree contains sesquiterpenoids, the so-called hibiscones and hibiscoquinones. The latter are orthonaphthoquinones. Hibiscoquinone A (**23**) (Fig. 3) is an example. When dissolved in chloroform or benzene, it decolorizes rapidly on exposure to light. The photochemistry of these compounds has been investigated (Ferreira *et al.,* 1980).

Similarly, the heartwood of *Mansonia altissima* (Sterculiaceae) contains violet to yellow dyes that are mainly *o*-quinones (Galeffi *et al.,* 1969; Thomson, 1979; Eugster, 1980). The distribution of these quinones in the stem of *Mansonia* has been investigated by Sandermann (1966). Their

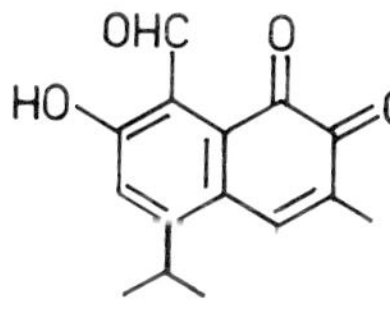

23

Fig. 3. A terpenoid quinone.

concentration is highest in the outer part of the heartwood and decreases in the center of the stem, a distribution that contrasts strongly with that found in *Tectona* (see above). *o*-Naphthoquinones that are present in *Mansonia* occur also in Ulmaceae (Rowe *et al.*, 1972; Thomson, 1979). In some *Ulmus* species they are associated with 2,6-dimethoxybenzoquinone (**5**), a compound noted for its skin-irritating properties (see above).

III. Biosynthesis of Quinones in Woody Plants and Cell Suspension Cultures

The biosynthesis of quinones was reviewed in 1981 (Leistner, 1981). In this chapter, the more recent developments will be stressed.

Roots and woody plants do not lend themselves to biosynthetic experiments or to enzyme extraction. This difficulty can be overcome by using cell suspension cultures, some of which produce quite large amounts of quinones.

A. *Biosynthesis of Quinones in Group 1*

One of the simple, naturally occurring quinones is 2,6-dimethoxybenzoquinone (**5**). Benzoquinones may in principle be derived from phenylpropanoids by way of β-oxidation (French *et al.*, 1976). This seems to be a nonoxidative process yielding the corresponding benzaldehyde or benzoic acid (see Chapter 10, Section V). The steps by which a benzoic acid can be converted to the corresponding benzoquinone have been elucidated during investigations of the biosynthesis of ubiquinone (Goewert *et al.*, 1981) and of other simple benzoquinones (Bolkart and Zenk, 1968). Compound **5** may also originate by degradation of lignin catalyzed by a peroxide/peroxidase system (Young and Stelink, 1973) (see Chapter 20, Section I). Compounds **1–4** and **7** are likely to be of a mixed phenylpropanoid/polyketide origin, whereas compound **6** is possibly derived from a prenylated benzene derivative by chain shortening.

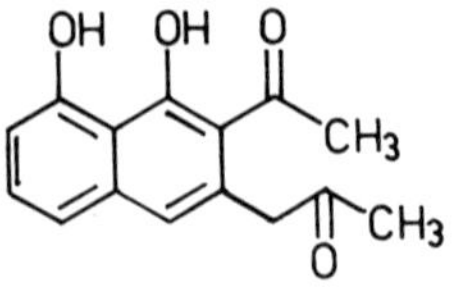

24

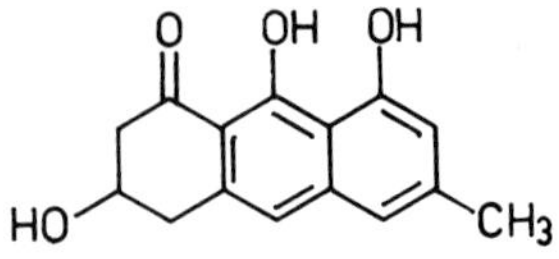

25

Fig. 4. Metabolites related to the biosynthesis of acetate-derived anthraquinones.

The remaining compounds in group 1 are certainly polyketides, although no experimental evidence exists in the case of compounds **10** and **11**. They are often referred to as juglone derivatives or juglone-type compounds. Juglone **15** itself occurs in *Carya* and *Juglans* plants and is not a polyketide (see below).

Naphthalene and anthracene derivatives such as **12** and **13** are most probably polyketides. This can be deduced from experiments on the biosynthesis of structurally related compounds such as nepodin (Bauch *et al.*, 1975), chrysophanol, and emodin (Leistner and Zenk, 1969; Leistner, 1971; Fairbairn and Muhtadi, 1972) in Rhamnaceae and Polygonaceae plants. The polyketide intermediate is still hypothetical, but some observations exist that provide some evidence concerning the mode of the ring closure and aromatization. The diketonaphthol **24** (Fig. 4), a substance regarded as a possible intermediate, was synthesized, radioactively labeled, and administered to a culture of the fungus *Penicillium islandicum*. Islandicin (1,4,8-trihydroxy-3-methyl-9,10-anthraquinone) was isolated after 14 days of incubation and found to be radioactive (incorporation 0.33–0.61%). It was concluded that the biosynthesis of acetate-derived anthraquinones, at least in microorganisms, proceeds via a bicyclic intermediate (Franck and Stange, 1981). Another substance that can be regarded as an intermediate in the biosynthesis of acetate-derived anthraquinones is germichrysone (**25**), a compound occurring in seeds but to a much larger extent in a tissue culture of *Cassia torosa* (Noguchi and Sankawa, 1982). A highly specific anthraquinone *O*-methyltransferase has been isolated from a fungus (Fujii *et al.*, 1982). Some anthra derivatives occur as *C*-glycosides, and a cell-free system catalyzing the transfer of glycose from UDP-glucose to aloeemodinathron has been obtained from *Aloe arborescens* Mill. (Grün and Franz, 1981).

B. Biosynthesis of Quinones in Group 2

Quinones in this group are derived from chorismic acid and α-ketoglutaric acid. In some cases an isoprene unit may also participate. Of the investigations on the biosynthesis of quinones in this group, those on vitamin K are most advanced. A short outline of vitamin K biosynthesis will therefore facilitate the understanding of the progress in this field.

Usually vitamin K does not occur in wood (an exception is mentioned later) but in chloroplasts (vitamin K_1) and bacteria (vitamin K_2). These systems have been used in the following work.

A bacterial enzyme system catalyzing the formation of *o*-succinylbenzoic acid (OSB, **29**) (Fig. 5) from chorismic acid (**26**) and α-ketoglutarate (**27**) has been obtained (Meganathan, 1981; Meganathan and Bentley, 1983). Two mechanisms have been proposed for the process of **26** → **29** (Bentley and Meganathan, 1983). The reaction is initiated by decarboxylation of α-ketoglutarate in the presence of thiamine pyrophosphate to give the thiamine pyrophosphate adduct of succinic semialdehyde (**28**). We have purified this enzyme system and shown that the decarboxylating enzyme can be separated from the intact enzyme system. The decarboxylating enzyme is not identical with the decarboxylase involved in the α-ketogluterate–dehydrogenase complex (Weische and Leistner, 1983). OSB (**29**) is converted to 1,4-dihydroxy-2-naphthoic acid (DHNS, **31**) in the presence of ATP, CoASH, and Mg^{2+} (Shaw *et al.*, 1982), with *o*-succinylbenzoic acid coenzyme A ester (**30**) as the intermediate (Heide *et al.*, 1982a). Subsequent prenylation (Saito and Ogura, 1981) and methylation (Azerad *et al.*, 1965) leads to vitamin K (**33**). This pathway occurs also in chloroplasts (Schultz *et al.*, 1982).

Compound **29** is also a precursor of naphthalene derivatives in *Juglans*. Callus cultures of two different *Juglans* species were established. When the medium contained naphthaleneacetic acid as a hormone rather than a combination of four hormones, good production of 1,4-naphthoquinone, juglone (**15**), and the glucosides of the corresponding hydroquinones (**37**, **38**) (Fig. 6) were formed (Müller and Leistner, 1978b). It was shown that the hydroxylation of the naphthalene derivatives occurred at the stage of the genins rather than of the glycosides, suggesting the biosynthetic relation shown in Fig. 6 (Müller and Leistner, 1978a). Furthermore, β-hydrojuglone (**16**) and α-oxotetralone were found to be present in the fruit of *Juglans* and were shown to be derived from OSB (**29**) (Müller and Leistner, 1978b).

Another cell culture that has been used for biosynthetic studies was established by Zenk *et al.* (1975). They showed that nutritional factors and hormones greatly influence anthraquinone production in *Morinda citrifolia* L. (Rubiaceae) suspension cultures. The most striking observation was that 2,4-dichlorophenoxyacetic acid (2,4-D) inhibited anthraquinone production completely, while in the presence of naphthalene acetic acid (NAA) good production was observed. The amount of anthraqui-

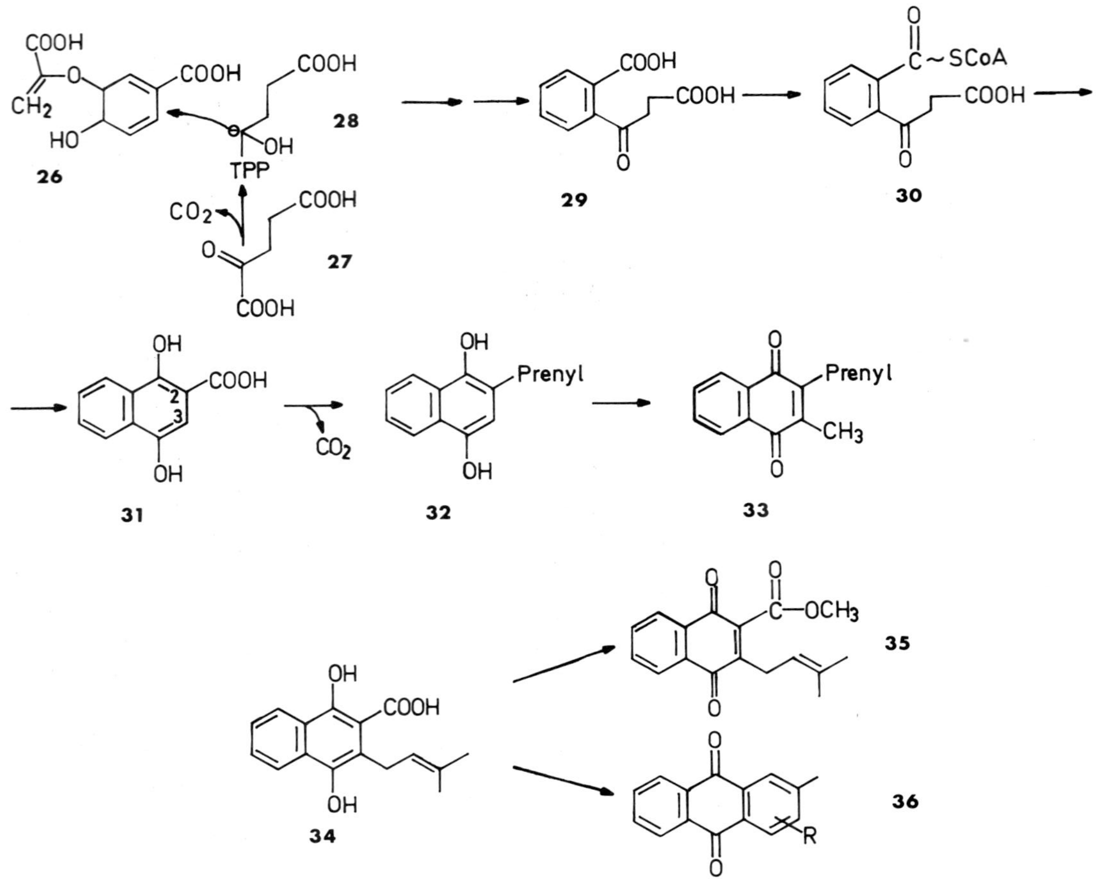

Fig. 5. Biosynthesis of bacterial menaquinones (vitamin K_2) and anthraquinones in Rubiaceae.

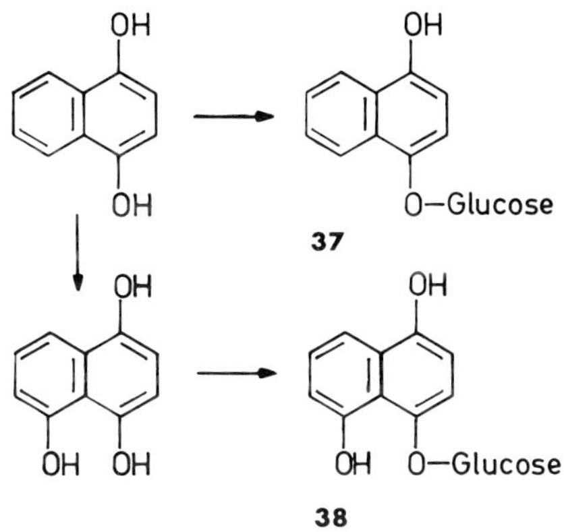

Fig. 6. Naphthalene derivatives in *Juglans*.

nones present in the tissue culture exceeded that present in the intact plant by a factor of 10 on a dry weight basis. The quinones present in the cell culture were isolated (Leistner, 1975; Inoue *et al.*, 1981c). In both cases, the same tissue culture strain was used, but the presence of different hormones resulted in a different spectrum of pigments. Using this cell culture strain, it was shown that anthraquinones from *Morinda* are derived from OSB (**29**) (Leistner, 1975), even though morindone shows a substitution pattern akin to those anthraquinones that are acetate-derived. A cell culture system similar to that of *Morinda* was established from *Galium mollugo* plants (Zenk, 1978; Bauch and Leistner, 1978). With the aid of a ^{13}C-labeling experiment and ^{13}C-NMR analysis, it was shown that OSB (**29**) is a precursor of anthraquinones in this cell culture also, and that during biosynthesis of anthraquinones (**36**), DHNA (**31**) is prenylated at C-3 rather than at C-2. This is in agreement with the fact that 2-methoxy-carbonyl-3-prenyl-1,4-naphthoquinone (**35**) is a new natural product that has been isolated from *G. mollugo* plants (Heide and Leistner, 1981, 1982). Thus, it can be assumed that the biosynthetic pathways leading to vitamin K and anthraquinones branch at DHNA (**31**). This is in agreement with the observation that an enzyme preparation (Heide and Leistner, 1983) from *G. mollugo* cell suspension cultures catalyzes the formation of OSB CoA ester and that this CoA ester is metabolized to DHNA (**31**) in the presence of naphthoatesynthase, an enzyme involved in bacterial vitamin K biosynthesis (Heide *et al.*, 1982b). This indicates that the OSB CoA ester (**30**) involved in vitamin K and in anthraquinone biosynthesis is identical.

Investigations of the biosynthesis of anthraquinones and prenylated naphthalene derivatives in *Catalpa, Tabebuia,* and *Streptocarpus* have

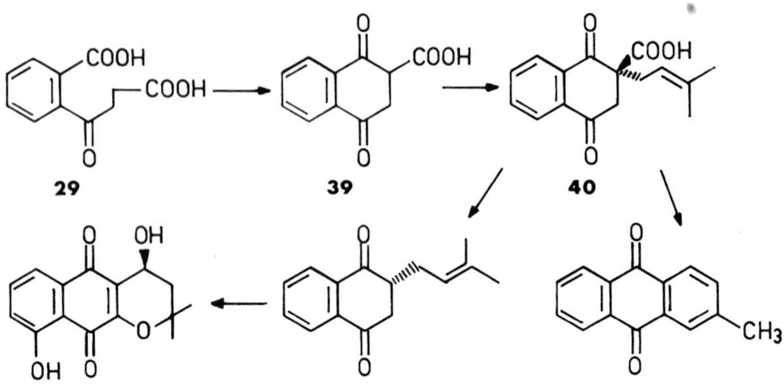

Fig. 7. Biosynthesis of quinones in *Catalpa*, *Streptocarpus*, and *Tabebuia*.

shown that yet another variation exists in the way in which OSB (**29**) can be metabolized (Inouye *et al.*, 1982) in plants. These experiments have been carried out in cell cultures (Inouye *et al.*, 1982) and wood (Inouye *et al.*, 1981). The fact that naphthalene derivatives with a hydrated ring system occur in *Catalpa* (Inoue *et al.*, 1980) has led Inouye to propose a reaction sequence by which compounds such as carboxyoxotetralone (COT, **39**) (Fig. 7) and (2-*R*)-2-prenyl-4-oxo-1-tetralone-2-carboxylic acid (**40**) are formed. These substances are indeed naturally occurring compounds, and their biosynthesis proceeds by prenylation at C-2 of a compound derived from OSB, as is the case in vitamin K biosynthesis. Vitamin K_2 (MK-1) actually co-occurs with these metabolites (Inouye *et al.*, 1981) (Fig. 7). Prenylated naphthalene derivatives are also present in *Streptocarpus* plants (Inoue *et al.*, 1983) and cell cultures (Inoue *et al.*, 1982). They occur with anthraquinones that are biosynthetically derived from a prenylated compound (**40**), as outlined in Fig. 7, and that differ from the pathway encountered in Rubiaceae (Fig. 5).

C. *Biosynthesis of Quinones in Group 3*

No experiments appear to have been carried out on the biosynthesis of terpenoid quinones.

Acknowledgments

The investigations on naphthoquinone and anthraquinone biosynthesis carried out in the author's laboratory were supported by the Deutsche Forschungsgemeinschaft and Fonds der Chemischen Industrie. Thanks are due to Dr. I. D. Spenser, Hamilton, Ontario, Canada, for his help in the preparation of the English version of this manuscript.

References

Adesogan, E. K. (1973). *Tetrahedron* **29,** 4099–4102.

Alves, A. C., Costa, M. A. C., and Ferreira, M. A. (1973). *Anais Da Faculdade De Farmacia Do Porto* **33,** 5–19.

Alves, A. C., Costa, M. A. C., and Paul, M. I. (1983). *Planta Medica* **47,** 121–124.

Atzorn, R., Weiler, E. W., and Zenk, M. H. (1981). *Planta Med.* **41,** 1–14.

Azerad, R., Bleiler-Hill, R., and Lederer, E. (1965). *Biochem. Biophys. Res. Commun.* **19,** 194–195.

Bauch, H.-J., and Leistner, E. (1978). *Planta Med.* **33,** 105–123.

Bauch, H.-J., Labadie, R. P., and Leistner, E. (1975). *J. Chem. Soc. Perkin I,* 689–692.

Bentley, R. (1975a). *Pure Appl. Chem.* **41,** 47–68.

Bentley, R. (1975b). *Biosynthesis* **3,** 181–246.

Bentley, R., and Meganathan, R. (1982). *Microbiol. Rev.* **46,** 241–280.

Bentley, R., and Meganathan, R. (1983). *Lloydia* **46,** 44–59.

Bolkart, K. H., and Zenk, M. H. (1968). *Z. Pflanzenphysiol.* **59,** 439–444.

Brew, E. J. C., and Thomson, R. H. (1971). *J. Chem. Soc.* 3001–2007.

Brieskorn, C. H., and Pöhlmann, R. (1976). *Arch. Pharm.* **76,** 829–836.

Briggs, L. H., Beachen, J. F., Cambie, R. C., Dudman, N. D. B., Steggles, A. W., and Rutledge, P. S. (1976). *J. Chem. Soc.* 1789–1792.

Brown, P. M., Thomson, R. H., Hausen, B. M., and Simatupang, M. H. (1974). *Liebigs Ann. Chem.* 1295–1300.

Burnett, A. R., and Thomson, R. H. (1967). *J. Chem. Soc.* 2100–2104.

Burnett, A. R., and Thomson, R. H. (1968a). *J. Chem. Soc.* 850–853.

Burnett, A. R., and Thomson, R. H., (1968b). *J. Chem. Soc.* 854–857.

Cannon, J. R., Joshi, K. R., McDonald, J. A., Retallack, R. W., Sierakowski, A. F., and Wong, L. C. H. (1975). *Tetrahedron Lett.* 2795–2798.

Chakraborty, D. P., Islam, A., and Roy, S. (1978). *Phytochemistry* **17,** 2043.

Criodain, R. O., O'Sullivan, M., Meegan, M. J., and Donelly, D. M. X. (1981). *Phytochemistry* **20,** 1089–1092.

Demagos, G. P., Baltus, W., and Höfle, G. (1981). *Z. Naturforsch.* **36B,** 1180–1184.

Dosseh, Ch., Tessier, A. M., and Delaveau, P. (1981). *Planta Med.* **43,** 360–366.

Dreyer, D. L., Arai, J., Buchman, C. D., Anderson, Jr., W. R., Smith, R. G., and Daves, Jr., G. D. (1975). *J. Am. Chem. Soc.* **97,** 4985–4990.

Durodola, J. I. (1974). *Planta Med.* **26,** 208–211.

Eugster, C. H. (1980). *In* "Pigments in Plants" (F.-C. Czygan, ed.), pp. 149–186. Fischer, Stuttgart, New York.

Fairbairn, J. W., and Muhtadi, F. J. (1972). *Phytochemistry* **11,** 215–219.

Fallas, A. L., and Thomson, R. H. (1968). *J. Chem. Soc. Part C,* 2279–2282.

Ferreira, M. A., Costa, M. A. C., Alves, A. C., and Lopes, M. H. (1973). *Phytochemistry* **12,** 433–435.

Ferreira, M. A., King, T. J., Ali, S., and Thomson, R. H. (1980). *J. Chem. Soc.* 249–259.

Floss, H. G., Chang, C.-J., Mascaretti, O., and Shimada, K. (1978). *Planta Med.* **34,** 345–380.

Franck, B., and Stange, A. (1981). *Liebigs Ann. Chem.* 2106–2116.

French, C. J., Vance, C. P., and Towers, G. H. N. (1976). *Phytochemistry* **15,** 564–566.
Fujii, I., Ebizuka, Y., and Sankawa, U. (1982). *Chem. Pharm. Bull.* **30,** 2283–2286.
Fujita, Y., Tabata, M., Nishi, A., and Yamada, Y. (1982). *In* "Proc. 5th Intl. Cong. Plant Tissue and Cell Culture" (A. Fujiwara, ed.) pp. 399–400. Maruzen, Tokyo.
Galeffi, C., Monache, E. M. delle, Casinovi, C. G., and Marini Bettolo, G. B. (1969). *Tetrahedron Lett.* 3583–3584.
Goewert, R. G., Sippel, C. F., Grimm, M. F., and Olson, R. E. (1981). *Biochemistry* **20,** 5611–5616.
Grün, M., and Franz, G. (1981). *Planta* **152,** 562–564.
Gupta, R. B., Khanna, R. N., and Manchanda, V. P. (1979). *Ind. J. Chem.* **18B,** 217–218.
Hausen, B. D. (1978). *Arbeitsmedizin, Sozialmedizin Präventivmedizin* **13,** 161–166.
Hedin, P. A., Collum, D. H., Langhans, V. E., and Graves, C. H. (1980). *J. Agric. Food Chem.* **28,** 340–342.
Hegnauer, R. (1973). "Chemotaxonomie der Pflanzen," Vol. 6. Birkhäuser, Basel.
Heide, L., and Leistner, E. (1981). *J. Chem. Soc. Chem. Commun.* 334–336.
Heide, L., and Leistner, E. (1982). *Z. Naturforsch.* **37C,** 354–362.
Heide, L., and Leistner, E. (1983). *Phytochemistry* **22,** 659–662.
Heide, L., Arendt, S., and Leistner, E. (1982a). *J. Biol. Chem.* **257,** 7396–7400.
Heide, L., Kolkmann, R., Arendt, S., and Leistner, E. (1982b). *Plant Cell Rep.* **1,** 180–182.
Inoue, K., Inouye, H., Taga, T., Fujita, R., Osaki, K., and Kuriyama, K. (1980). *Chem. Pharm. Bull.* **28,** 1224–1229.
Inoue, K., Inouye, H., and Chen, C.-C. (1981a). *Phytochemistry* **20,** 2271–2276.
Inoue, K., Chen, C.-C., Inouye, I., and Kuriyama, K. (1981b). *J. Chem. Soc.* 2764–2770.
Inoue, K., Nayeshiro, H., Inouye, H., and Zenk, M. H. (1981c). *Phytochemistry* **20,** 1693–1700.
Inoue, K., Ueda, S., Nayeshiro, H., and Inouye, H. (1982). *J. Chem. Soc. Chem. Commun.* 993–994.
Inoue, K., Ueda, S., Nayeshiro, H., and Inouye, H. (1983). *Phytochemistry* **22,** 737–741.
Inouye, H., Okuda, T., and Hayashi, T. (1975). *Chem. Pharm. Bull.* **23,** 384–391.
Inouye, H., Ueda, S., Inoue, K., and Shiobara, Y. (1981). *Phytochemistry* **20,** 1707–1710.
Inouye, H., Ueda, S., Inoue, K., Nayeshiro, H., and Moritome, N. (1982). *In* "Plant Tissue Culture 1982" (A. Fujiwara, ed.), pp. 375–376. Japanese Association for Plant Tissue Culture, Tokyo.
Jeffreys, J. A. D., bin Zakaria, M., Waterman, P. G., and Zhong, S. (1983). *Tetrahedron Lett.* **24,** 1085–1088.
Joshi, K. C., Singh, P., and Pardasani, R. T. (1977). *Planta Med.* **31,** 14–16.
Joshi, K. C., Singh, P., Pardasani, R. T., and Singh, G. (1979). *Planta Med.* **37,** 60–63.
Jurd, L., Stevens, K., and Manners, G. (1972). *Phytochemistry* **11,** 3287–3292.
King, T. J., Ferreira, M. A., Alves, A. C., and Thomson, R. H. (1976). *J. Chem. Soc. Chem. Commun.* 241–242.
Labadie, R. P. (1970). *Pharm. Weekblad* **105,** 473.
Leistner, E. (1971). *Phytochemistry* **10,** 3015–3020.
Leistner, E. (1975). *Planta Med.* Supplement, 214–224.
Leistner, E. (1980). *In* "Pigments in Plants" (F.-C. Czygan, ed.), pp. 352–369. Fischer, Stuttgart, New York.
Leistner, E. (1981). *In* "The Biochemistry of Plants" (E. E. Conn, ed.), Vol. 7, pp. 403–423. Academic Press, New York.
Leistner, E., and Zenk, M. H. (1969). *Chem. Commun.* 210–211.

Leistner, E., and Zenk, M. H. (1971). *In* "Methods in Enzymology" (D. B. McCormick and L. D. Wright, eds.), Vol. 18c, pp. 547–559. Academic Press, New York.
Lillie, T. J., and Musgrave, O. C. (1980). *J. Chem. Soc. Part I,* 1161–1164.
Manners, G. D., and Jurd, L. (1977a). *J. Agric. Food Chem.* **25,** 726–730.
Manners, G. D., and Jurd, L. (1977b). *J. Chem. Soc.* 405–410.
Manners, G. D., Jurd, L., Wong, R., and Palme, K. (1975). *Tetrahedron* **31,** 3019–3021.
Meganathan, R. (1981). *J. Biol. Chem.* **256,** 9386–9388.
Meganathan, R., and Bentley, R. (1983). *J. Bacteriol.* **153,** 739–746.
Mock, J., Murphy, S. T., Ritchie, E., and Taylor, W. C. (1973). *Aust. J. Chem.* **26,** 1121–1130.
Moir, M., and Thomson, R. H. (1973a). *J. Chem. Soc.* 1352–1357.
Moir, M., and Thomson, R. H. (1973b). *J. Chem. Soc.* 1556–1561.
Moir, M., and Thomson, R. H. (1973c). *Phytochemistry* **12,** 1351–1353.
Müller, W.-U., and Leistner, E. (1976). *Phytochemistry* **15,** 407–410.
Müller, W.-U., and Leistner, E. (1978a). *Phytochemistry* **17,** 1735–1738.
Müller, W.-U., and Leistner, E. (1978b). *Phytochemistry* **17,** 1739–1742.
Musgrave, O. C., and Skoyles, D. (1974). *J. Chem. Soc. Part I* 1128–1131.
Noguchi, H., and Sankawa, U. (1982). *Phytochemistry* **21,** 319–323.
Papageorgiou, V. P. (1979). *Planta Med.* **37,** 259–263.
Papageorgiou, V. P., Winkler, A., Sagredos, A. N., Digenis, G. A. (1979). *Planta Med.* **35,** 56–60.
Pardhasaradhi, M., and Babu, M. H. (1978). *Phytochemistry* **17,** 2042.
Patai, S. (1974). "The Chemistry of the Quinoid Compounds." Wiley, New York.
Pierpoint, W. S. (1970). *Rep. Rothamsted Exp. Stn. 1970,* Part 2, pp. 199–218.
Rauwald, H.-W., and Miething, H. (1982). *Z. Naturforsch.* **38C,** 17–20.
Rowe, J. W., Seikel, M. K., Roy, N. D., and Jorgensen, E. (1972). *Phytochemistry* **11,** 2513–2517.
Saito, Y., and Ogura, K. (1981). *J. Biochem.* **89,** 1445–1452.
Sandermann, W. (1966). *Naturwissenschaften* **53,** 513–524.
Sandermann, W., and Dietrichs, H. H. (1957). *Holz Roh- Werkst.* **15,** 281–297.
Sandermann, W., and Dietrichs, H. H. (1959). *Holzforschung* **13,** 137–148.
Sandermann, W., and Simatupang, M. H. (1963). *Chem. Ber.* **96,** 2182–2185.
Sandermann, W., and Simatupang, M. H. (1966). *Holz Roh- Werkst.* **24,** 190–204.
Sandermann, W., Simatupang, M. H., and Wendeborn, W. (1968). *Naturwissenschaften* **55,** 38.
Sankaram, A. V. B., Reddy, V. V. N., and Sidhu, G. S. (1981). *Phytochemistry* **20,** 1093–1096.
Schmalle, H. W., and Hausen, B. M. (1980). *Tetrahedron Lett.* **21,** 149–152.
Schultz, G., Soll, J., and Fiedler, E. (1982). "Proceedings of the 5th International Symposium on the Biochemistry and Metabolism of Plant Lipids," Groningen, The Netherlands, June 7th–10th.
Shaw, D. J., Guest, J. R., Meganathan, R., and Bentley, R. (1982). *J. Bacteriol.* **152,** 1132–1137.
Shukla, Y. N., Tandon, J. S., and Dhar, M. M. (1973). *Ind. J. Chem.* **11,** 528–529.
Sidhu, G. S., and Prasad, K. K. (1971). *Ind. J. Chem.* **9,** 767–769.
Singh, P., and Singh, A. (1980). *Pharmazie* **35,** 701–702.
Tabata, M., Tsukada, M., and Fukui, H. (1982). *Planta Med.* **44,** 234–236.
Thomson, R. H. (1971). "Naturally Occurring Quinones," 2nd ed. Academic Press, New York.

Thomson, R. H. (1976a). *In* "Chemistry and Biochemistry of Plant Pigments" (T. W. Goodwin, ed.), Vol. 2, pp. 207–231. Academic Press, New York.
Thomson, R. H. (1976b). *In* "Chemistry and Biochemistry of Plant Pigments" (T. W. Goodwin, ed.), Vol. 2, pp. 527–559. Academic Press, New York.
Thomson, R. H. (1979). *In* "Recent Advances in Phytochemistry" (T. Swain, J. B. Harborne, and C. F. Van Sumere, eds.), Vol. 12, pp. 287–312. Plenum, New York.
Tripathi, V. D., Agarwal, S. K., and Rastogi, R. P. (1979). *Ind. J. Chem.* **17B,** 89–90.
Waterman, P. G., and Mbi, C. N. (1979). *Planta Med.* **37,** 241–246.
Weinberg, M. De L. D., Gottlieb, O. R., and Oliveira, G. G. de (1976). *Phytochemistry* **15,** 570.
Weische, A., and Leistner, E. (1983). Unpublished manuscript.
Yoshizaki, F., Hisamichi, S., Kondo, Y., Sato, Y., and Nozoe, S. (1982). *Chem. Pharm. Bull.* **30,** 4407–4411.
Young, M., and Stelink, C. (1973). *Phytochemistry* **12,** 2851–2861.
Zenk, M. H. (1978). *In* "Frontiers of Plant Tissue Culture" (T. A. Thorpe, ed.), pp. 1–13. Int. Assoc. for Plant Tissue Culture, Calgary, Canada.
Zenk, M. H., and Leistner, E. (1968). *Lloydia* **31,** 275–292.
Zenk, M. H., El-Shagi, H., and Schulte, U. (1975). *Planta Med.* Supplement 79–101.

CHAPTER **12**

Biosynthesis of Flavonoids

HANS GRISEBACH

Lehrstuhl für Biochemie der Pflanzen
Biologischen Institut II
Universität Freiburg
Freiburg, Federal Republic of Germany

I. Introduction

On the basis of an analysis of the substitution pattern in rings A and B of flavonoids, Sir Robert Robinson postulated as early as 1921 that these compounds are biogenetically derived from a C_6C_3 (ring B and carbon atoms 2, 3, and 4) and a C_6 unit (ring A). This hypothesis was expanded in 1953 by Birch and Donovan, who formulated the synthesis of flavanone from three acetate units and cinnamic acid or a related compound. The first tracer experiments on biosynthesis of flavonoids were carried out in 1957 with red cabbage seedlings by the author and with buckwheat seedlings by Underhill, Watkin, and Neish. The distribution of ^{14}C in cyanidin from red cabbage and in quercetin from buckwheat with differently la-

Biosynthesis and Biodegradation of Wood Components

ISBN 0-12-347880-4

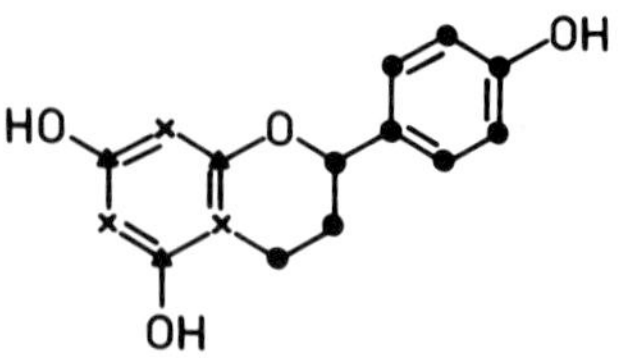

Fig. 1. Origin of carbon atoms in the flavane skeleton. ●, Phenylpropane unit; ▲, carboxyl group of acetate; ×, methyl group of acetate.

beled acetate and phenylalanine as precursors was in full accord with the Robinson–Birch hypothesis. The labeling pattern obtained is shown in Fig. 1. There followed a period of extensive tracer studies in which the importance of chalcones as key intermediates in the formation of various flavonoids and the biogenetic relationships among different flavonoids were established. The discovery of the first enzyme of the phenylpropanoid pathway, phenylalanine ammonia-lyase, by Koukol and Conn (1961) marked the beginning of enzymatic studies in the field of phenylpropanoids in general and flavonoids in particular. In this work, plant cell suspension cultures proved to be a very valuable source of enzymes. Use of parsley cell cultures led to the isolation and characterization of all but one enzyme of the flavone–glycoside pathway in this plant.

The genetics of flower pigments began with Gregor Mendel, who used peas with red and white flowers in hybridization experiments. After the structure of anthocyanins had been elucidated, mainly by the work of Willstätter, Robinson, and Karrer, extensive chemogenetic studies were started in several laboratories. The work of Wheldale, Scott-Moncrieff, Lawrence, Price, Beale, and others showed that changes in the genotype of a plant were correlated with particular structural changes in the anthocyanins. More recently, biochemical studies have supplemented the chemogenetic investigations. Use of flowers from defined genotypes proved valuable for correlations of genes with particular enzymes, as well as for the elucidation of some steps in the biosynthetic pathway. The more recent work will be described here. For the more classical studies, the reader is referred to the review by Alston (1964).

Numerous reviews on biosynthesis of flavonoids have appeared. The more recent reviews are the following: Ebel and Hahlbrock (1982), Grisebach (1979, 1982), Hahlbrock and Grisebach (1975, 1979), and Wong (1976).

In those cases where individual reactions have been studied on the enzymatic level, I shall concentrate on this aspect. I shall also cover some of the work with defined genotypes. Earlier work with labeled precursors

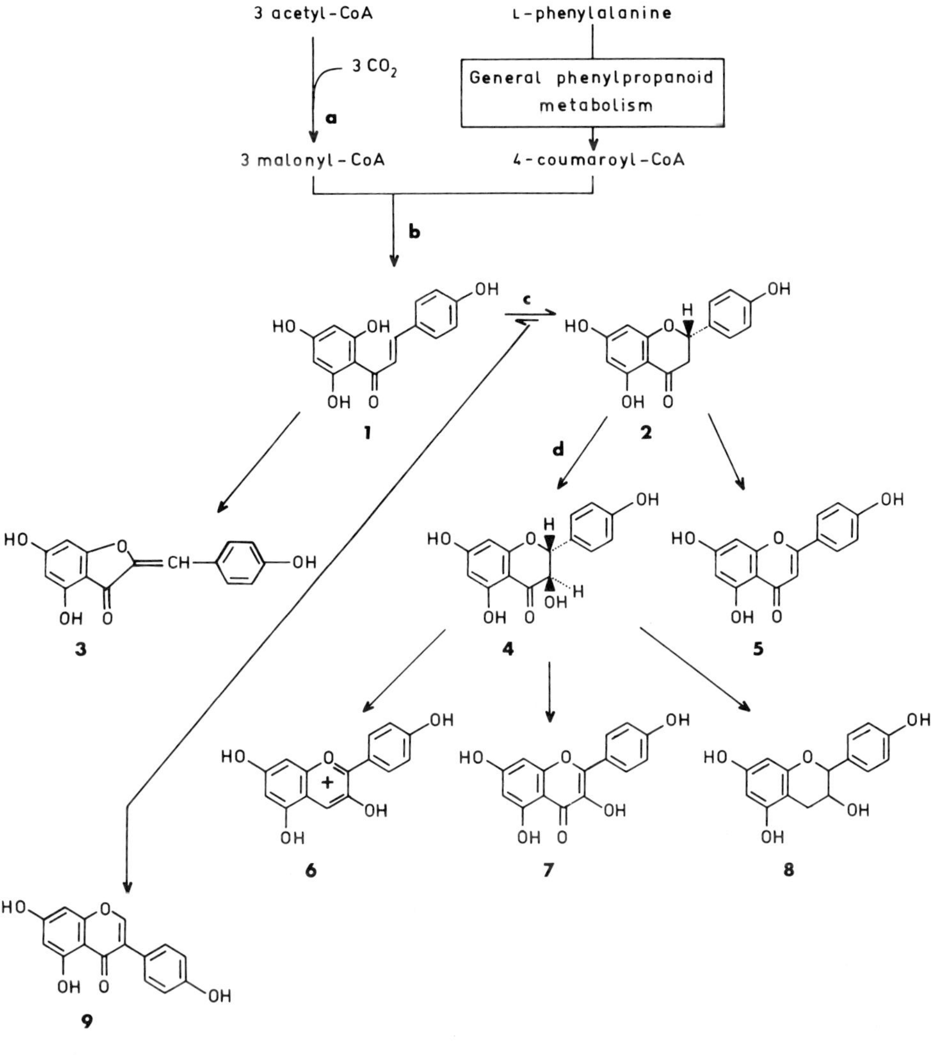

Fig. 2. Biogenetic relationships among various flavonoids. Enzymes: **a,** acetyl-CoA carboxylase; **b,** chalcone synthase; **c,** chalcone isomerase; **d,** flavanone 3-hydroxylase. Compounds: (**1**), chalcone; (**2**), flavanone; (**3**), aurone; (**4**), dihydroflavonol; (**5**), flavone; (**6**), anthocyanidin; (**7**) flavonol; (**8**) catechin; (**9**) isoflavone.

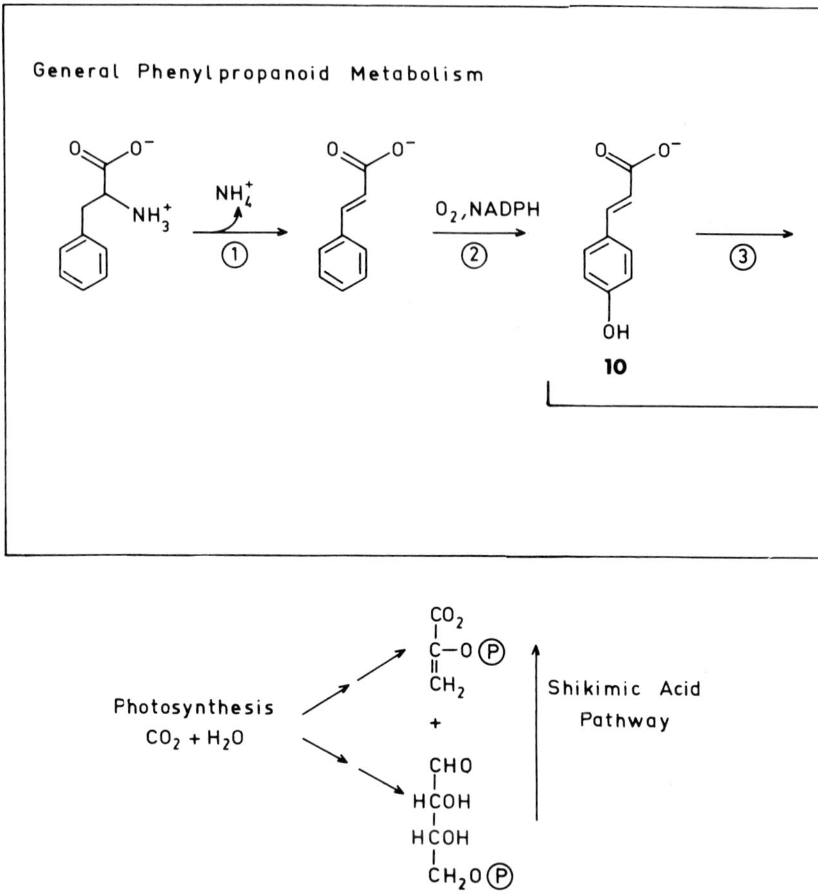

Fig. 3. General phenylpropanoid metabolism. Enzymes: ① phenylalanine ammonia-lyase; ② cinnamate 4-hydroxylase; ③ phenolase; ④ *S*-adenosylmethionine : caffeate 3-*O*-methyltransferase; ⑤ ferulate 5-hydroxylase; ⑥ 4-coumarate : CoA ligase. Compounds:

is covered in the reviews of Hahlbrock and Grisebach (1975) and Wong (1976).

II. General Overview of Flavonoid Biosynthesis

Early chemogenetic studies of flower color had indicated that in the formation of various classes of flavonoids there is competition for a common precursor (Alston, 1964). Tracer experiments and enzymatic studies

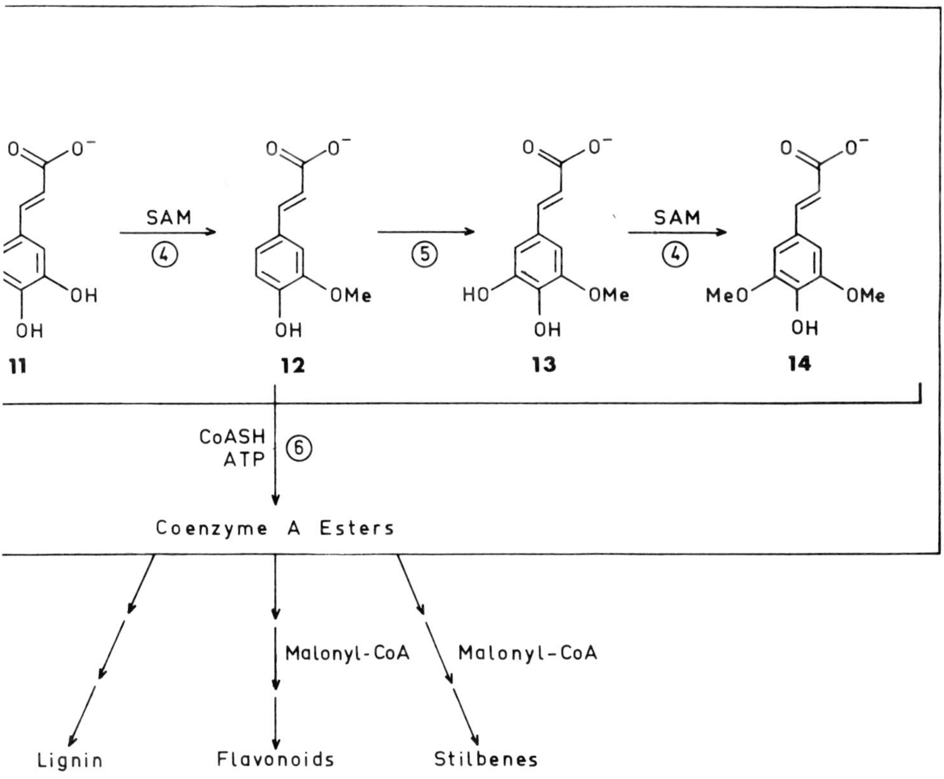

(**10**) 4-coumaric acid; (**11**) caffeic acid; (**12**) ferulic acid; (**13**) 5-hydroxyferulic acid; (**14**) sinapic acid.

proved that a common precursor for all flavonoids is a chalcone [Fig. 2 (**1**)], which is synthesized in the plant from 4-coumaroyl-CoA and three molecules of malonyl-CoA by action of the enzyme chalcone synthase (see Section IV,A). The activated 4-coumaric acid for this central reaction in flavonoid biosynthesis is supplied by the general phenylpropanoid metabolism (see Section III) and malonyl-CoA by the acetyl-CoA carboxylase reaction (Egin-Bühler *et al.*, 1980).

Further transformations of the chalcone to various flavonoids are depicted in Fig. 2. Aurone is probably synthesized directly from chalcone, whereas flavone and dihydroflavonol (see Section IV,C) are formed from the isomeric flavanone by dioxygenase type of enzymes. Flavanone originates by enzymatic cyclization of the chalcone (see Section IV,B). Dihydroflavonols are biosynthetic intermediates in the formation of anthocyanidins, flavonols, and catechins. The biosynthesis of catechins is described in Chapter 13, Section IV. Whether the rearrangement to isoflavone, involving a 2,3-aryl shift, takes place with chalcone or with flavanone is still uncertain. Isoflavones are the precursors for the large class of coumestanes.

III. General Phenylpropanoid Metabolism

General phenylpropanoid metabolism is defined as the sequence of reactions involved in the conversion of L-phenylalanine to activated cinnamic acids (Hahlbrock and Grisebach, 1975). The reaction sequence of this pathway, together with the corresponding enzymes, is shown in Fig. 3. It is not implied by this figure that only one set of enzymes is present in a particular plant. Parallel pathways could occur in different cell compartments, and metabolic channeling would be made possible by multienzyme complexes or isoenzymes.

The first enzyme of this pathway, phenylalanine ammonia-lyase, catalyzes the trans elimination of ammonia from L-phenylalanine to form *trans*-cinnamic acid. Cinnamic acid is then converted, by a sequence of hydroxylation and methylation reactions, to a number of substituted acids that can be activated to the corresponding coenzyme A (CoA) esters. These activated acids can enter different biosynthetic pathways leading to lignin, flavonoids, stilbenes, benzoic acids, and other compounds. General phenylpropanoid metabolism is therefore one of the most important pathways in higher plants, and its details are described in Chapter 10 and in the review by Ebel and Hahlbrock (1982). That a very close metabolic relationship exists among the enzymes of the general phenylpropanoid metabolism is apparent from their interdependent regulation in a number of plant tissues (Hahlbrock and Grisebach, 1979).

IV. Individual Pathways of Flavonoid Biosynthesis

A. *Chalcone Synthase*

As has already been mentioned, the key enzyme for synthesis of the central C_{15} intermediate is chalcone synthase, formerly called flavanone

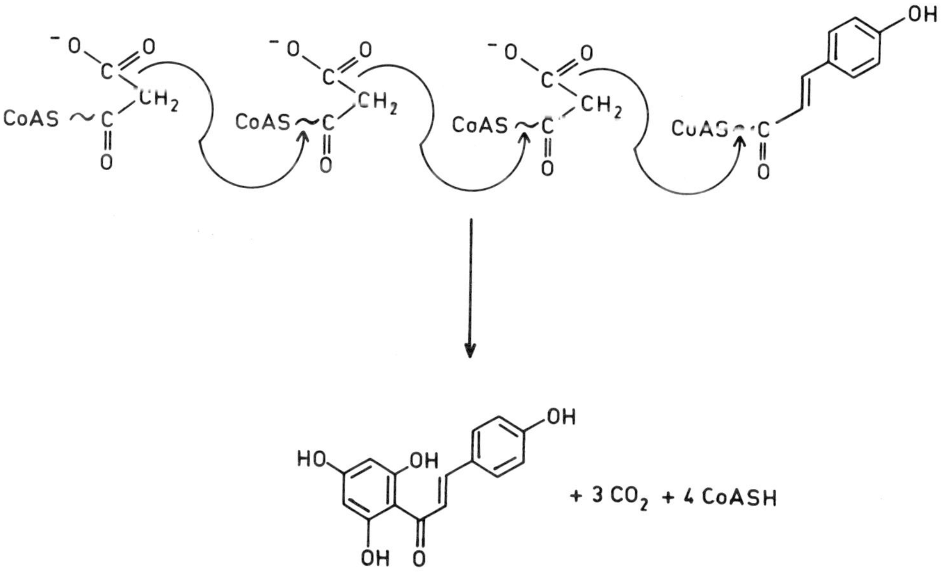

Fig. 4. Condensation of 4-coumaroyl-CoA and malonyl-CoA catalyzed by chalcone synthase.

synthase. This enzyme catalyzes the condensation of one molecule of 4-coumaroyl-CoA with three molecules of malonyl-CoA to form 4,2′,4′,6′-tetrahydroxychalcone (naringenin chalcone). The reaction shown in Fig. 4 was first believed to lead directly to the isomeric flavanone (naringenin) (Kreuzaler and Hahlbrock, 1975a). With the homogenous enzyme from parsley cell cultures, it was later demonstrated that naringenin chalcone is the immediate product of the synthase reaction (Heller and Hahlbrock, 1980; Light and Hahlbrock, 1980). Sütfeld and Wiermann (1980) came to the same conclusion from their studies with enzyme extracts from tulip anther tapetum using a biphasic enzyme assay.

In addition to 4-coumaroyl-CoA, chalcone synthase from parsley and *Haplopappus gracilis* cell cultures (Saleh *et al.*, 1978) and from *Tulipa* (Sütfeld and Wiermann, 1981) can also use caffeoyl-CoA and feruloyl-CoA as substrates, in which case the corresponding chalcones are formed. Whether these reactions lie on the pathway to 3′,4′-substituted anthocyanidins will be discussed in Section IV,D.

Further proof that chalcone is the reaction product was obtained with chalcone synthase preparations from mutants of *Dianthus caryophyllus* (genotype *ii*) and *Callistephus chinensis* (genotype *chch*), which lack chalcone isomerase (see Section IV,B) (Forkmann and Dangelmayr,

1980; Kuhn *et al.*, 1977) so that the enzymatic isomerization to naringenin is blocked.

In *Matthiola incana* (Spribille and Forkmann, 1981) and *Antirrhinum majus* (Spribille and Forkmann, 1982), chalcone synthase is controlled by the genes f^+ and niv^+, respectively. Recessive genotypes (*ff* or *nivniv*) contain neither the enzyme nor flavonoids, but accumulate cinnamic acid glucosides (Spribille and Forkmann, 1982). Chalcone synthase has been purified from cell cultures of parsley (Kreuzaler and Hahlbrock, 1975a; Kreuzaler *et al.*, 1979) and *H. gracilis* (Saleh *et al.*, 1978) and from parts of tulip and cosmos plants (Sütfeld and Wiermann, 1981).

The chalcone synthase from parsley cell cultures was purified to apparent homogeneity by a four-step procedure (Kreuzaler *et al.*, 1979). As determined by sedimentation–equilibrium measurements, the molecular weight of the enzyme is ~77,000. It is composed of two probably identical subunits of MW ~42,000. The amino acid analysis and an assay with *Lactobacillus plantarum* showed that chalcone synthase does not contain a pantetheinyl residue. These results indicate that chalcone synthase differs considerably from both 6-methylsalicylic acid synthase from *Penicillium patulum* (MW 1×10^6) and the type I fatty acid synthase from eukaryotes, in which the acyl residue is intermittently linked to a pantetheinyl arm.

More recently, the cDNA complementary to the mRNA of chalcone synthase from irradiated parsley cell cultures was cloned in *Escherichia coli*. The primary structure of the enzyme subunit was determined by sequence analysis of the cDNA. It is composed of 398 amino acids and has a molecular weight of 43,682 (Reimold *et al.*, 1983).

The substrate specificity of the chalcone synthase with respect to cinnamoyl-CoA esters is pertinent to the question whether the substitution pattern of flavonoids in the B ring is already determined at the cinnamic acid stage or whether further substitutions occur after formation of the chalcone intermediate. In all known cases, 4-coumaroyl-CoA is the most efficient substrate. Conversion of caffeoyl-CoA to eriodictyol (5,7,3′,4′-tetrahydroxyflavanone: in presence of chalcone isomerase) has been reported (Saleh *et al.*, 1978), whereas feruloyl-CoA does not seem to be a substrate for the synthase (Hrazdina *et al.*, 1976). The reaction with caffeoyl-CoA had a pH optimum between 6.5 and 7, in contrast to the reaction with 4-coumaroyl-CoA with a pH optimum at 8. In tulip anthers, it was proposed that caffeoyl-CoA and feruloyl-CoA, in addition to 4-coumaroyl-CoA, function as substrates of chalcone synthase *in vivo* (Sütfeld *et al.*, 1978).

Despite the fact that caffeoyl-CoA and feruloyl-CoA can act *in vitro* as substrates for chalcone synthase, their physiological role in flavonoid biosynthesis is doubtful. As will be discussed in Section IV,D, genetic evidence and the occurrence of a specific naringenin 3′-hydroxylase and flavonoid-specific 3′-*O*-methyltransferases point to modification of the B ring at the C_{15} level.

Chalcone synthase preparations from parsley and tulip are strongly inhibited by the reaction products, chalcone (or flavanone) and coenzyme A (Sütfeld *et al.*, 1978).

Some information on the mechanism of the chalcone synthase reaction is available. The stepwise addition of acetate units to coumaroyl-CoA can be deduced from the *in vitro* formation of side products, with the enzymes from parsley and *H. gracilis* cell cultures, that originate from the condensation with only one (benzalacetones or dihydropyrones) or two (styrylpyrones) molecules of malonyl-CoA (Kreuzaler and Hahlbrock, 1975b; Hrazdina *et al.*, 1976). The formation of these products from the putative enzyme-bound intermediates is shown in Fig. 5. While these "release products" have not been detected in parsley cell cultures and are therefore only formed under nonphysiological conditions, the occurrence of styrylpyrones in several higher plant species is well established (Mors *et al.*, 1961), and 3′-hydroxybisnoryangonin has been found, for example, in *Polyporus hispidus* (Perrin and Towers, 1973). Such compounds could therefore be synthesized by enzymes related to chalcone synthase. From the observation that chalcone synthase also catalyzes CO_2 exchange and decarboxylation of malonyl-CoA both in the presence and in the absence of 4-coumaroyl-CoA, it was postulated that the initial reaction is formation of an acetyl-CoA carbanion (Kreuzaler *et al.*, 1978). As shown in Fig. 6, the putative carbanion could condense in the normal chalcone synthase reaction with 4-coumaroyl-CoA to form the first intermediate of chain elongation; alternatively, it could react either with a proton to form acetyl-CoA (decarboxylation) or with CO_2 to regenerate malonyl-CoA (CO_2 exchange). The mechanism of naringenin chalcone formation seems therefore not to involve a concerted reaction of decarboxylation and condensation, as proposed for fatty acid biosynthesis in yeast and postulated for the synthesis of other "polyacetate" compounds derived from malonyl-CoA (Arnstadt *et al.*, 1975).

B. *Chalcone Isomerase*

Chalcone isomerase catalyzes the stereospecific isomerization of chalcone to (−)-(2*S*)-flavanone (Fig. 7). This enzyme, which has been isolated from several plants, shows specificity toward the substitution pattern of ring A of the chalcone (Hahlbrock and Grisebach, 1975). In general, the substitution pattern of the preferred substrate corresponds to that of the flavonoids found in the plant from which the enzyme was isolated. For example, chalcone isomerase from parsley is specific for 4,2′,4′,6′,-

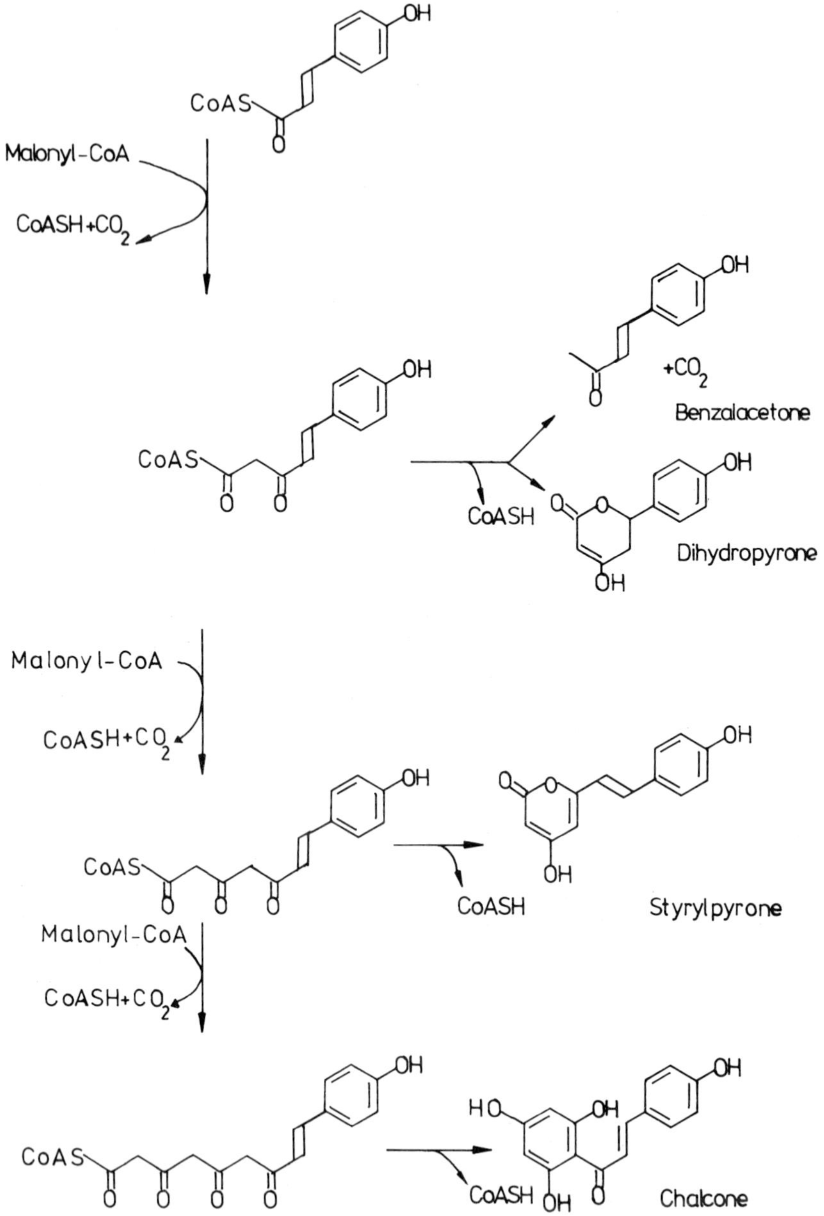

Fig. 5. Proposed mechanism for chalcone synthase from parsley. The release products (benzalacetone or others) are formed from polyketide intermediates with either one or two acetate units, respectively.

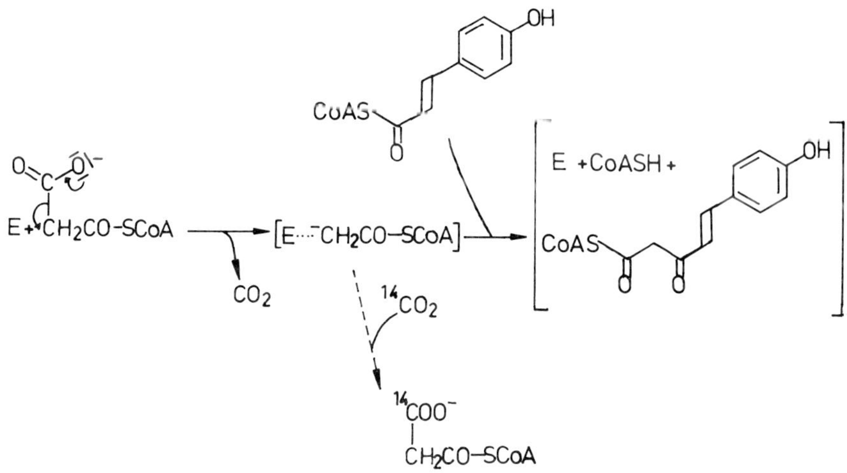

Fig. 6. Proposed mechanism for the first condensation step of chalcone synthase from parsley.

tetrahydroxychalcone, in agreement with the exclusive occurrence of 5,7-dihydroxyflavonoids in this plant (Hahlbrock *et al.*, 1970). In all cases investigated, chalcone glucosides were not substrates for isomerase.

An equilibrium constant of 37 in favor of the flavanone was determined from the kinetics of isomerization of 2′,4,4′-trihydroxychalcone to 7,4′-dihydroxyflavanone by the soybean enzyme (Boland and Wong, 1975).

Reports on the occurrence of isoenzymes of isomerase (Hahlbrock *et al.*, 1970; Grambow and Grisebach, 1971) are at least doubtful, because KCN was not added to the assay to avoid interference of peroxidase (Boland and Wong, 1975) and the reaction products were not unequivocally identified as the corresponding flavanones.

The phytoalexin kievitone [2′,4′,5,7-tetrahydroxy-8-(3,3-dimethylallyl)isoflavonone] and coumestrol were reported to act as competitive

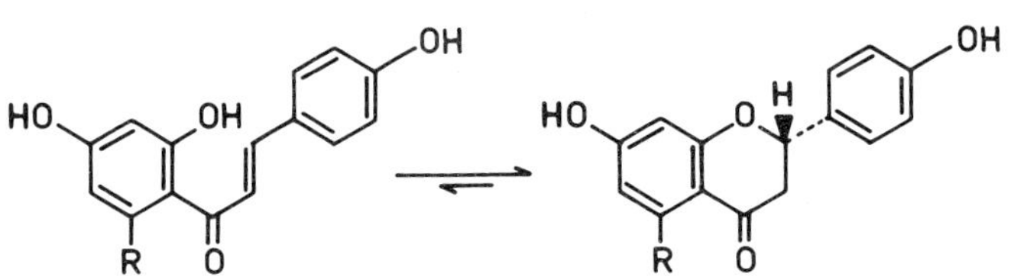

Fig. 7. Reaction catalyzed by chalcone isomerase. R can be H or OH, depending on the source of the enzyme.

inhibitors for chalcone isomerase of *Phaseolus vulgaris* with K_i values of 9.2 and 2.5 μM, respectively (Dixon *et al.*, 1982).

After establishment of the fact that chalcone and not flavanone is the first C_{15} intermediate in flavonoid biosynthesis, the important role of chalcone isomerase became evident. Further important clues about the role of this enzyme in flavonoid biosynthesis came from experiments with defined genotypes of several plants. Several anthocyanin-containing and acyanic genotypes of the china aster (*Callistephus chinensis* L.) were analyzed for their flavonoid content. Extensive crossing experiments indicated that anthocyanin synthesis is blocked in the acyanic lines by at least two complementarily acting factors, *a* and *f*. In addition, another gene (*Ch*) responsible for anthocyanin synthesis was detected. It exists in yellow lines in a recessive state and causes an accumulation of tetrahydroxychalcone 2′-glucoside (isosalipurpuroside). Measurements of chalcone isomerase activity showed that this enzyme is present in the anthocyanin- or flavonol-containing genotypes, e.g., 01 and 10*a*, but is completely absent from the yellow line 07, which accumulates the chalcone glucoside (Kuhn *et al.*, 1977). The accumulation of isosalipurpuroside should therefore be due to a lack of the isomerase. The site of the genetic block is shown in Fig. 8. The gene *Ch* could be directly responsible for synthesis of this enzyme.

These results permit the following conclusions: (1) higher amounts of anthocyanins[1] are formed only in presence of chalcone isomerase; (2) the pathway to anthocyanins proceeds via the flavanone; and (3) chalcone lies in this pathway before flavanone.

The genetic control of chalcone isomerase was also investigated in flowers of carnation (*Dianthus caryophyllus*) (Forkmann and Dangelmayr, 1980). This investigation is based on earlier chemogenetic studies on different-colored strains of carnation of Geissman and Mehlquist (1947). Table I summarizes the results of chemical analysis of four color types and the activity of chalcone isomerase. Comparison of these results with those reported by Geissman and Mehlquist suggested that in the pure yellow and in the yellow-orange strain, accumulation of tetrahydroxychalcone 2′-glucoside is caused by recessive alleles at the locus *I*, whereas wild-type alleles must be present at this locus both in the white and in the red strain. This conclusion is fully supported by the correlation of chalcone isomerase activity with genotype.

A situation similar to that in flowers of aster and carnation exists in genetically defined lines of *Petunia hybrida* differing with regard to the

[1] Lower amounts of anthocyanins can be formed even in the absence of chalcone isomerase, possibly because of chemical cyclization of chalcone.

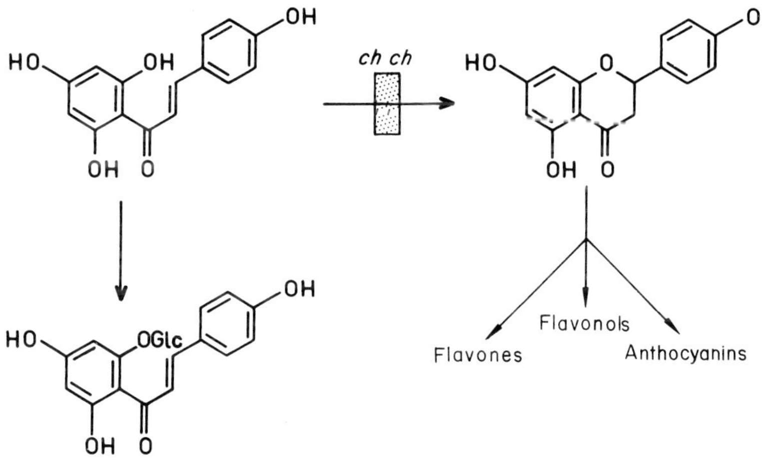

Fig. 8. Site of the genetic block in a mutant of *Callistephus chinensis* that accumulates tetrahydroxychalcone 2-glucoside.

pollen color, which is influenced by the two genes *Po* and *An4*. In genotypes with the recessive allele *po*, naringenin chalcone accumulates (De Vlaming and Kho, 1976). As shown in Table II, enzymatic investigation on anthers of four lines with different pollen coloration revealed a clear correlation between accumulation of naringenin chalcone and a lack of activity of chalcone isomerase (Forkmann and Kuhn, 1979).

Table I

Chemogenetic Characterization of Four Flower Color Types of *Dianthus caryophyllus* and Activity of Chalcone Isomerase

Strain	Color	Flavonoids	Genotype[a]	Chalcone isomerase activity
Legion d'honneur	Brick-red	Pelargonidin	*Y I A*	High
Jeanne Dionis	White	Kaempferol	*Y I aa*	High
Marie Chabaud	Yellow	Tetrahydroxychalcone 2′-glucoside, some kaempferol and quercetin	*Y ii aa*	Not detectable
Greenhouse type	Yellow-orange	Tetrahydroxychalcone 2′-glucoside, some pelargonidin	*Y ii A*	Not detectable

[a] Genotypes according to Geissmann and Mehlquist (1947); supplementation experiments (Forkmann and Dangelmayr, 1980) suggest that in recessive genotypes *aa*, anthocyanin synthesis is blocked after dihydroflavonol formation. The gene *Y* seems to influence flavonoid concentration. *I* is the gene for chalcone isomerase.

Table II

Genetically Defined Lines of *Petunia hybrida* and Activity of Chalcone Isomerase in Anthers

Line	Pollen genotype[a]	Pollen color	Pollen flavonoids	Specific activity of chalcone isomerase in anthers (ΔA min^{-1} mg protein^{-1})
V10	*Po An4*	Blue	Anthocyanin	0.572 ± 0.017
V28	*po An4*	Green	Naringenin chalcone, little anthocyanin	0.026 ± 0.014
V31	*Po an4*	White	Dihydroflavonol and flavonol	0.509 ± 0.070
V33	*po an4*	Yellow	Naringenin chalcone, some flavonol and dihydroflavonol	0.023 ± 0.020

[a] *Po* is gene for chalcone isomerase; gene *An4* controls anthocyanin synthesis in pollen.

C. *Modification of Ring C*

Tracer experiments with ^{3}H-labeled dihydrokaempferol and [1-^{14}C]-phenylalanine as an internal standard had shown that dihydrokaempferol (aromadendrin) but not kaempferol is a very efficient precursor for cyanidin and quercetin in buckwheat (*Fagopyrum esculentum* L.) seedlings (Patschke *et al.*, 1966). Similar experiments were carried out with cell suspension cultures of *H. gracilis*. In these cultures, dihydrokaempferol proved to be a much better precursor for cyanidin than 4,2′,4′,6′-tetrahydroxychalcone (administered as the 2′-glucoside) or phenylalanine (Fritsch *et al.*, 1971). From these results, it was concluded that dihydrokaempferol is an intermediate in the biosynthesis of cyanidin from the chalcone/flavanone precursor (Fig. 2).

The mechanism of the transformation of the tetrahydroxychalcone to dihydrokaempferol was investigated in *Chamaecyparis obtusa*. When the chalcone labeled with ^{3}H and ^{14}C in the β position (2 position of naringenin) was fed to this plant, the isolated dihydrokaempferol had the same ^{3}H : ^{14}C ratio as the precursor. Synthesis of dihydrokaempferol via flavone or flavonol, in which case ^{3}H would be lost, was therefore excluded, and it was assumed that dihydrokaempferol arises by direct hydroxylation of naringenin in the 3 position (Grisebach and Kellner, 1965). This hydroxylation was first demonstrated with dimethyl sulfoxide-permeabilized cells of *H. gracilis* and a microsomal preparation from these cells (Fritsch and Grisebach, 1975). A better characterization of the flavanone 3-hydroxylase was possible with flowers of defined genotypes of *Matthiola incana* (Forkmann *et al.*, 1980). For this purpose, white flowers of line 17 (genotype *ee* f^+f^+ *bb*) were used. In this line, the

Table III
Genes of *Matthiola incana*

Gene	Functional role	Corresponding enzyme
B/b	Hydroxylation of naringenin and dihydrokaempferol in the 3′ position	Microsomal 3′-hydroxylase
E/e G/g	Unknown role between dihydroflavonols and anthocyanin	
F/f	Synthesis of chalcone intermediate	Chalcone synthase

anthocyanin pathway is blocked after dihydroflavonol formation (recessive *e*), and 3′-hydroxylation also does not occur (recessive *b*) (Table III). The 3-hydroxylase from these flowers is a soluble enzyme and needs 2-oxoglutarate, ferrous ion, and ascorbate as cofactors. According to these cofactor requirements, the enzyme belongs to the 2-oxoglutarate-dependent dioxygenases (Abbott and Udenfriend, 1974). Eriodictyol (5,7,3′,4′-tetrahydroxyflavanone) was also a substrate for the enzyme, leading to the formation of dihydroquercetin.

With a soluble enzyme preparation of irradiated cell suspension cultures of parsley, the conversion of naringenin or eriodictyol to the corresponding dihydroflavonol, flavonol, and flavone was observed (Britsch *et al.*, 1981). All the reactions shown in Fig. 9 require 2-oxoglutarate, Fe^{2+}, and ascorbate as cofactors. Incubation of dihydrokaempferol with this enzyme preparation and the three cofactors leads to formation of kaempferol.

Oxidation of naringenin to apigenin had been shown before with a cell-free extract from very young primary leaves of parsley, and it was proved that flavanone and not chalcone is the substrate of this reaction (Sutter *et al.*, 1975). In addition to molecular oxygen and Fe^{2+}, the soluble enzyme preparation required another nonproteinaceous cofactor, which has now been identified as 2-oxoglutarate.

The fact that both flavanone 3-hydroxylation and formation of flavones and flavonols require the typical cofactors for 2-oxoglutarate-dependent dioxygenases was rationalized by the hypothesis that flavone and flavonol synthesis proceeds via 2-hydroxyflavanone and 2,3-dihydroxyflavanone, respectively (Britsch *et al.*, 1981). Such a reaction sequence, shown in Fig. 10, would be expected to involve three enzymes: a 2- and a 3-hydroxylase and a dehydratase. It is known that 2,3,4,7,4′-pentahydroxyflavanone (Fig. 10) and the corresponding 3′,4′-hexahydroxyflavanone

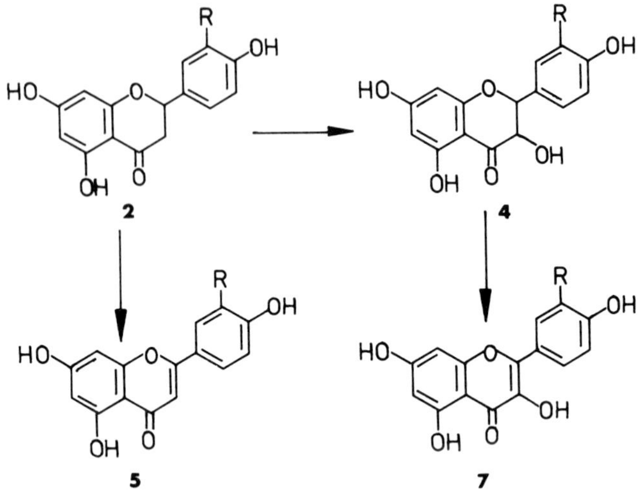

Fig. 9. Reaction products of naringenin (**2**, R = H) or eriodictyol (**2**, R = OH) with an enzyme preparation from parsley cell cultures in presence of 2-oxoglutarate, Fe^{2+}, and ascorbate: (**4**) R = H, dihydrokaempferol; (**4**) R = OH, dihydroquercetin; (**5**) R = H, apigenin; (**5**) R = OH, luteolin; (**7**) R = H, kaempferol; (**7**) R = OH, quercetin.

dehydrate under mild conditions to kaempferol and quercetin, respectively (Hauteville *et al.*, 1979). In *Populus nigra* buds (Chadenson *et al.*, 1971) and *Unona lawii* (Chopin *et al.*, 1978), substituted 2,5-dihydroxyflavanones occur together with the corresponding flavones.

Unexpectedly, the oxidation of flavanone to flavone with extracts of snapdragon flowers (*Antirrhinum majus*) was found to be localized in the microsomal fraction and to require NADPH as cofactor (Stotz and Forkmann, 1981). This is one of the rare cases in which the same product is formed by two distinctively different enzymatic reactions. This finding could be of significance in considering the evolution of flavonoid biosynthesis in plants.

Flavanone 3-hydroxylase has also been detected in flowers of the snapdragon. In the "incolorata" mutant, a correlation between recessive genotype (*incinc*) and absence of 3-hydroxylase was found (Forkmann and Stotz, 1981).

Further strong support for the role of dihydroflavonols as intermediates in anthocyanin biosynthesis has come from work with genetically defined acyanic lines of *Matthiola incana* (Forkmann, 1977), *Petunia hybrida* (Kho *et al.*, 1975, 1977), and *Antirrhinum majus* (Harrison and Stickland, 1974; Stickland and Harrison, 1974). A supplementation technique was used in these experiments that is similar to techniques used successfully

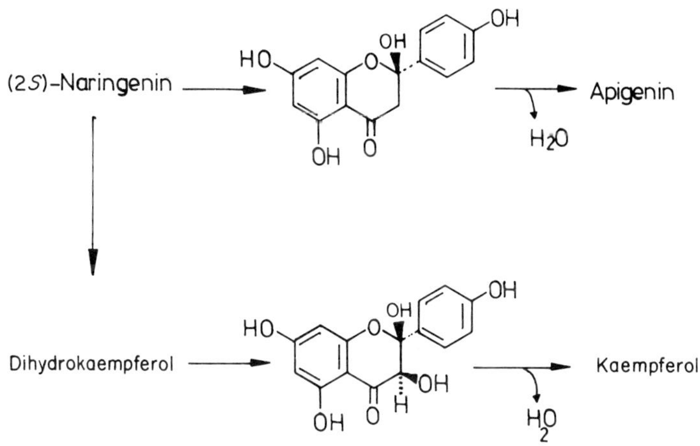

Fig. 10. Hypothetical reaction sequence from naringenin to apigenin and kaempferol via 2-hydroxyflavanones.

with mutants of microorganisms blocked in a discrete step of a biosynthetic pathway. This technique is based on the assumption that the compound lying immediately before the block accumulates and that synthesis of the end product takes place when a compound is provided that lies after the block.

Dihydroflavonols or their glycosides that accumulated in white-flowering mutants were supplied to acceptor mutants blocked in the synthesis of dihydroflavonols. This led to anthocyanin synthesis in the acceptor mutant. For example, in flower buds of the white-flowering mutant W 19 of *Petunia hybrida,* dihydroquercetin 7-glucoside, dihydroquercetin 4′-glucoside, dihydroquercetin, and dihydrokaempferol 7-glucoside accumulated. When extracts of flowers of line 19 or dihydroquercetin were supplied to *in vitro* cultured corollas of the white-flowering mutant W 18, cyanidin 3-glucoside was formed as the major pigment, together with smaller amounts of cyanidin 3-diglucoside and delphinidin 3-glucoside. These anthocyanins are the same as those detected in flowers of a red-flowering genotype, R 3, of *Petunia hybrida.*

Similar results were obtained with acyanic genotypes of *Matthiola incana* and *Antirrhinum majus.* In lines of the genotype *ff* of *Matthiola,* administration of dihydrokaempferol, dihydroquercetin, naringenin or the corresponding chalcone, or tetrahydroxychalcone 2′-glucoside permitted synthesis of anthocyanins. In contrast, hydroxycinnamic acids were ineffective. This indicates that the genetic block in these lines lies before naringenin and very probably concerns chalcone synthase. Lines with the

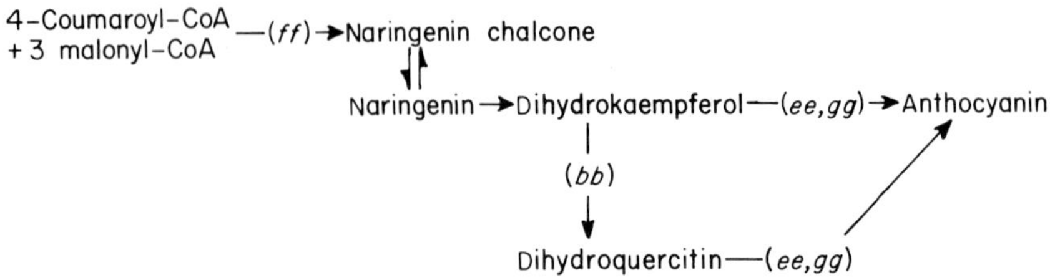

Fig. 11. Recessive alleles that have been found in *Matthiola incana*.

genotype *ee* or *gg* did not produce anthocyanins with the precursors just mentioned. This supports the assumption that the genes *e* and *g* block the pathway after dihydroflavonol synthesis but before anthocyanin formation. The probable position of the genetic blocks in *Matthiola* is shown in Fig. 11 (Forkmann, 1977). In genotype *ff*, dihydroflavonols were clearly superior to flavanone and chalcone as precursors for anthocyanin formation.

Anthocyanin biosynthesis in the aleurone tissue of maize endosperm (*Zea mays*) has also been investigated by supplementation experiments (McCormick, 1978). In this tissue, the complementary genes *C*, *C2*, *R*, *A*, *Az*, *Bz*, and *Bz2* are required for anthocyanin formation. Recessive alleles at one of the first five loci listed result in colorless aleurone, and recessive alleles at either of the bronze loci (*Bz*, *Bz2*) result in formation of a brown pigment whose structure is as yet unknown. Supplementation of seeds recessive in the locus *C2* with dihydroquercetin resulted in synthesis of cyanidin; when naringenin was supplied, the pigments formed were cyanidin and pelargonidin. From these results, it could be concluded that the *C2* block is before the flavanone intermediate.

Additional evidence for dihydroflavonols as intermediates in anthocyanin biosynthesis comes from experiments with inhibitors of phenylalanine ammonia-lyase (PAL). The most powerful inhibitor of phenylalanine ammonia-lyase is L-α-amino-oxy-β-phenylpropionic acid (L-AOPP) or its *N*-benzyloxycarbonyl derivative (Amrhein *et al.*, 1976; Amrhein and Gödeke, 1977), which inhibits this enzyme with a K_i of 1.4×10^{-9} *M*. The compound L-AOPP is also a potent inhibitor of cyanidin synthesis in buckwheat hypocotyls, in buds of *Ipomoea purpurea* and *Catharanthus roseus*, and in red cabbage seedlings (Amrhein, 1979). Inhibition of cyanidin synthesis can be fully or partially reversed by precursors of the B ring of cyanidin (*trans*-cinnamate, 4-coumarate) and by naringenin, dihydrokaempferol, or dihydroquercetin. In contrast, reversion was not observed with apigenin, kaempferol, quercetin, and dihydrofisetin (Amrhein, 1979).

Enzyme(s) for the conversion of dihydroflavonols to anthocyanidins are still unknown. In flowers of *Petunia hybrida,* wild-type alleles of both genes *An1* and *An2* are required for conversion of dihydroquercetin to cyanidin 3-glucoside. Evidence for glucosylation of dihydroquercetin was not found, and it seems most probable that dihydroquercetin is transformed to cyanidin, which is then immediately glucosylated at the 3-hydroxyl group (see Section IV,E).

D. Substitution in Ring B

1. Hydroxylation

The question of whether the oxygenation pattern of the B ring of flavonoids is determined at the cinnamic acid stage or at the stage of a C_{15} intermediate has been the subject of several investigations. With regard to anthocyanidins, the former possibility would mean that 4-coumarate is the precursor for pelargonidin, caffeate for cyanidin, ferulate for peonidin, etc. In the second case, 4-coumarate would be the precursor for all anthocyanidins, and introduction of further hydroxyl groups or methylation reactions would occur at the chalcone/flavanone stage or later in the pathway.

Tracer experiments in buckwheat seedlings with [^{14}C]4,2′,4′,6′-tetrahydroxychalcone and [^{14}C]3,4,2′,4′,6′-pentahydroxychalcone, corresponding in their substitution pattern to pelargonidin and cyanidin, respectively, showed that both chalcones were incorporated into cyanidin with about equal dilution values and incorporation rates (Patschke and Grisebach, 1965). These and other experiments with labeled chalcones or dihydrokaempferol (Patschke *et al.,* 1966) proved that the 3′ hydroxyl can be introduced at the chalcone/flavanone stage or later in the pathway.

In the search for enzymes capable of catalyzing the 3′ hydroxylation, attention was first focused on phenolases. A phenolase preparation from leaves of spinach beet (*Beta vulgaris* L. ssp. *vulgaris*) that catalyzes hydroxylation of 4-coumaric acid to caffeic acid was also able to catalyze hydroxylation of naringenin, dihydrokaempferol, and kaempferol at the 3′ position to eriodictyol, dihydroquercetin, and quercetin, respectively (Vaughan *et al.,* 1969). In contrast, a phenolase preparation from cell cultures of parsley had only weak hydroxylating activity with naringenin and none with other flavonoid substrates. Since irradiation of the cell cultures had no effect on extractable phenolase activity, it was concluded that phenolase is not specifically involved in the flavone glycoside pathway (Schill and Grisebach, 1973).

More recent results indicate that, rather than a phenolase, a microsomal mixed-function oxidase is involved in the 3′ hydroxylation reaction. Hydroxylation of naringenin and dihydrokaempferol to eriodictyol and dihydroquercetin, respectively, was found with a microsomal fraction from *H. gracilis* cell cultures (Fritsch and Grisebach, 1975) and with a microsomal fraction from flowers of defined genotypes of *Matthiola incana* (Forkmann *et al.*, 1980). The reaction is dependent on NADPH and molecular oxygen.

A microsomal fraction from cell suspension cultures of parsley also catalyzes the NADPH-dependent hydroxylation of naringenin to eriodictyol. Dihydrokaempferol, kaempferol, and apigenin could also serve as substrates for this enzyme to form dihydroquercetin, quercetin, and luteolin, respectively (Hagmann *et al.*, 1983). Because the enzyme reaction requires dioxygen and NADPH and is partly inhibited by carbon monoxide, it can be classified as a cytochrome *P*-450-dependent monooxygenase, as cinnamate 4-hydroxylase has been (Potts *et al.*, 1974). Like the latter enzyme, naringenin 3′-hydroxylase is induced by illumination of the parsley cell cultures and cannot be detected in dark-grown cell cultures. However, the fact that cinnamate 4-hydroxylase is also induced by treatment of the cells with a glucan preparation of *Phytophthora megasperma* f. sp. *glycinea* (elicitor), whereas naringenin 3′-hydroxylase is not induced by the elicitor, proves that cinnamate and flavanone hydroxylation are catalyzed by different enzymes.

Further support that hydroxylation occurs at the C_{15} level comes from work with defined genotypes. In *M. incana,* 3′ hydroxylation is found only in lines with wild-type alleles at the locus *b*. Since analytical investigations and supplementation experiments have shown that this locus is responsible for 3′ hydroxylation (Forkmann, 1980), the correlation between enzyme activity and genotype is strong support for participation of this enzyme in cyanidin biosynthesis. In the snapdragon, a correlation of wild-type allele of the gene *Eos* and the presence of 3′-hydroxylase activity was possible (Forkmann and Stotz, 1981).

Analysis of genotypes from flowers of *Petunia hybrida* (Tabak *et al.*, 1978; Doodeman *et al.*, 1982) and *Antirrhinum majus* (Stickland and Harrison, 1974; Harrison and Stickland, 1974), from aleurone tissue of maize (McCormick, 1978), and from seeds or seedlings of *Melilotus alba* (Specht *et al.*, 1976) led also to the definition of specific genes responsible for 3′ hydroxylation and allowed the conclusion that these genes act at the stage of a C_{15} intermediate.

In petals of the red campion (*Silene dioica*), however, the hydroxylation in ring B seems to be determined at the cinnamic acid stage. From chemogenetic experiments, it was concluded that gene *P* has a pleiotropic effect: it controls both formation of cyanidin and the hydroxylation pattern of the hydroxycinnamoyl group esterified with the rhamnosyl residue of cyanidin 3-rhamnosyl (1→6) glucoside 5-glucoside (Kamsteeg *et al.*, 1980a). In accord with these results, an enzyme was discovered in these petals that catalyzes hydroxylation of 4-coumaroyl-CoA to caffeoyl-CoA. Enzyme activity was only found in petal extracts of plants with the genotype *PP* (Kamsteeg *et al.*, 1981).

2. Methylation

The fact that feruloyl-CoA is a poor substrate for chalcone synthase (Section IV,A) indicates that methylation occurs after completion of the C_{15} skeleton. In several cases, *O*-methyltransferases have been found that have a pronounced specificity for the 3′-hydroxyl group of flavonoid substrates and that differ considerably from caffeate *O*-methyltransferases with respect to both substrate specificity and other properties. In all cases, *S*-adenosyl-L-methionine was found to be the methyl donor (for a review, see Ebel and Hahlbrock, 1982). Substrate specificities of methyltransferases for which Michaelis constants and maximal velocities have been determined are listed in Table IV.

In soybean cell cultures, for example, two *O*-methyltransferases are present. One enzyme has a pronounced specificity for methylation of flavonoid substrates at the 3′-hydroxyl group, whereas the second enzyme is specific for methylation of hydroxycinnamic acids in the 3 and 5 positions but does not react with flavonoids. In tobacco leaves (Legrand *et al.*, 1978), tobacco cell cultures (Tsang and Ibrahim, 1979), and *Tulipa* anthers (Sütfeld and Wiermann, 1978), the occurrence of several *O*-methyltransferases has also been reported (see Chapter 10, Section II,C).

Crude extracts from young *Robinia pseudoacacia* seedlings catalyze the para-O-methylation of apigenin to acacetin (Kuroki and Poulton, 1981). Apigenin (K_m 50 μM) was the best substrate, but significant activity was shown toward caffeic acid, 5-hydroxyferulic acid, naringenin, and quercetin, whereas 4-coumaric acid, ferulic acid, and sinapic acid were not methylated.

Recently, the methylation of anthocyanins by cell-free extracts from flower buds of *Petunia hybrida* has been reported (Jonsson *et al.*, 1982). Methylation of cyanidin, delphinidin, and petunidin in the B ring took place when the 3-(4-coumaroyl)-rutinosido-5-glucoside derivatives were used as substrates, but not with the aglyca, or with anthocyanidin 3-glucosides or 3-rutinosides. Caffeic and 4-coumaric acid were also not substrates. These results are in agreement with the finding of Wiering (1974) that in *Petunia hybrida* two pairs of duplicate genes (*Mt1*/*Mt2* and *Mf1*/*Mf2*) are responsible for methylation at the 3′ and 5′ positions, respectively, and that these genes are only expressed if two other genes (*Rf*

Table IV
Substrate Specificities of Some *O*-Methyltransferases

Source and designation	Substrate	K_m (μM)	V_{max} (μkat[b] kg^{-1})	Remarks	Reference
Parsley cell culture	Luteolin 7-*O*-glucoside	31	100[a]		Ebel *et al.* (1972)
	Luteolin	46	88[a]		
	Eriodictyol	1200	15[a]		
	Caffeic acid	1600	148[a]		
Soybean cell cultures flavonoid 3′-*O*-methyltransferase	Luteolin 7-*O*-glucoside	28	13.4		Poulton *et al* (1977)
	Luteolin	16	33.7		
	Quercetin	35	70.9		
	Eriodictyol	75	11.7		
	Dihydroquercetin	435	32.7		
	Caffeic acid	770	30.9		
Soybean cell cultures	Caffeic acid	113	72	No reaction with luteolin and quercetin	Poulton *et al* (1976)
caffeate 3-*O*-methyltransferase	5-Hydroxyferulic acid	55	142		
	3,4,5-Trihydroxycinnamic acid	100	16		
Chrysosplenium americanum					
Flavonoid 3′-*O*-methyltransferase	Quercetin	1.4	—		Ibrahim *et al* (1982)
Flavonoid 3′/5′-*O*-methyltransferase	3,7-Di-*O*-methylquercetagenin	0.5	—		
Flavonoid 6-*O*-methyltransferase	3,7-Di-*O*-methylquercetagenin	2.5	—		

[a] Relative V_{max} values.
[b] kat = mol · sec^{-1}.

and *Gf*), responsible for glucosylation, rhamnosylation, and acylation of the anthocyanin, are present in the dominant allele.

E. Glycosylation

Glycosylation seems in most cases to be the last reaction in biosynthesis of flavonoids. An exception is the methylation of acylated anthocyaninglycosides mentioned in Section IV,D. Glycosylation has recently been covered extensively in the review by Ebel and Hahlbrock (1982), and only a few representative examples will be discussed here.

The various 7-*O*-glycosides and 3,7-di-*O*-glycosides in parsley are formed by sequential glycosylation steps with UDP-glucose as the glucosyl donor. A UDP-glucose : flavone/flavonol 7-*O*-glucosyltransferase (Sutter *et al.*, 1972) and a UDP-glucose : flavonol 3-*O*-glucosyltransferase (Sutter and Grisebach, 1973) have been characterized.

Because of the instability of anthocyanidins (aglyca), it can be assumed that these are immediately glucosylated at the 3-hydroxyl group. Additional sugars may then be attached to the 3-glucosyl residue and/or to the 5-hydroxyl group.

3-*O*-Glucosyltransferases for cyanidin from red cabbage seedlings (Saleh *et al.*, 1976b), cell cultures of *H. gracilis* (Saleh *et al.*, 1976a), petals of *Silene dioica* (Kamsteeg *et al.*, 1978a), and maize (McCormick, 1978; Larson and Coe, 1977) have been described. All enzymes use UDP-glucose as the glucosyl donor.

The best substrate for the transferases from red cabbage and *H. gracilis* was cyanidin. Of the other anthocyanidins tested, pelargonidin, peonidin, and, to a lesser extent, malvidin were also glucosylated. These enzymes were not specific for anthocyanidins but could also glucosylate flavonols like kaempferol and quercetin at the 3 position. However, dihydroflavonols like dihydroquercetin did not serve as substrates.

In maize, the 3-*O*-glucosyltransferase can also glucosylate both anthocyanidins and flavonols. In this plant it could be shown that the gene *Bz* is the structural gene for this enzyme (Larson and Coe, 1977). A clear gene dosage effect was found in maize, for example, in the endosperm. Recessive alleles *Bz* cause brown pigmentation of the aleurone, which is presumably due to complexing of unstable anthocyanidins (McCormick, 1978). In studies by Dooner and Nelson (1977), a more complex genetic control of this enzyme was reported. In addition to the gene *Bz*, the genes *C* and *R* are necessary for full expression of the glucosyltransferase. The genetic control exerted by *C* and *R* seems to be indirect.

The 3-*O*-glucosyltransferase from *S. dioica* is specific for anthocyanidins. An enzyme catalyzing the transfer of the rhamnosyl moiety of UDP-L-rhamnose to the 6-hydroxyl group of the 3-*O*-bound glucose of anthocyanidin 3-*O*-glucosides was demonstrated in petal extracts of *S. dioica* (Kamsteeg *et al.*, 1980b). Enzyme activity is controlled by a single gene *N*. No rhamnosyltransferase activity was found in petals of homozygous recessive *nn* plants. The 3-*O*-glucosides of pelargononidin, cyanidin, and delphinidin could serve as acceptors. Rhamnosylation of anthocyanin 3,5-diglucoside occurred at a reduced rate.

A third transferase that catalyzes the transfer of the glucosyl moiety of UDP-glucose to the 5-hydroxyl group of cyanidin 3-rhamnosyl (1 → 6) glucoside (cyanidin 3-rutinoside) was also detected in these extracts (Kamsteeg *et al.*, 1978b). This transferase is controlled by a single gene *M*. Pelargonidin 3-rutinoside was also an acceptor. Cyanidin 3-glucoside was converted to the 3,5-diglucoside at a much lower rate than the rutinoside at pH 7.5, which is the pH maximum for glucosylation of the 3-rutinoside. At pH 6.5, however, glucosylation of the 3-glucoside proceeded at a higher rate. It seems therefore that the same transferase can use either acceptor, depending on the pH of the medium.

From these results, it can be concluded that glycosylation in *S. dioica* petals takes place in the following order: anthocyanidin 3-glucoside → 3-rutinoside → 3-rutinoside 5-glucoside, with a branch from the 3-glucoside to the 3,5-diglucoside.

F. *Acylation*

Several malonylated flavone and flavonolglycosides had been isolated from irradiated parsley cell cultures. From such cultures, two malonyltransferases were later isolated that catalyze the transfer of the malonyl residue from malonyl-CoA to the primary hydroxyl group of the glucosyl moiety (6 position) (Matern *et al.*, 1981, 1983). One enzyme was most active with flavone and flavonol 7-*O*-glycosides; the other enzyme preferentially malonylated flavonol 3-*O*-glucosides. The malonylated flavonoid glycosides are synthesized in the cytoplasm and are deposited in the vacuoles. The latter result was obtained by ultraviolet (UV) microspectroscopy and by isolation of the vacuoles (Matern *et al.*, 1983). It was suggested that acylation of flavonoid glycosides facilitates transport into the vacuoles. Isoflavonoid glycosides have also been found in malonylated form (Köster, 1982). Malonylation of flavonoid glycosides therefore seems to be a widespread reaction.

Anthocyanins also occur in acylated form in the petals of *S. dioica*. 4-Coumaric acid is esterified with pelargonidin glycosides and caffeic acid with cyanidin glycosides. In anthocyanidin 3-glucosides, the acyl residue is esterified with the 6-hydroxyl group of glucose and in the 3-rutinosides with the 4-hydroxyl group of rhamnose. The presence of an enzyme catalyzing the transfer of the 4-coumaroyl or caffeoyl moiety of the corresponding CoA esters to the 4-hydroxyl group of the rhamnosyl moiety of anthocyanidin 3-rutinosides or 3-rutinoside 5-glucosides has been demonstrated in these petals (Kamsteeg *et al.*, 1980c). This acyltransferase is governed by the gene *Ac*.

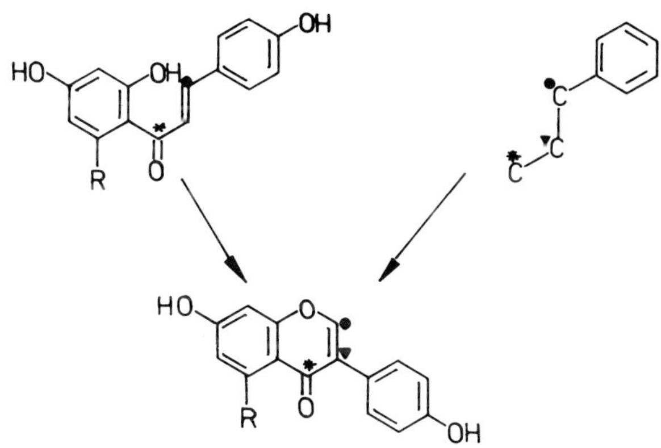

Fig. 12. Incorporation of labeled chalcone and phenylpropanoid precursor (phenylalanine, cinnamic acid) into isoflavone. Trihydroxychalcone (R = H) is incorporated only into the corresponding isoflavone daidzein (R = H). Carbons are marked for comparison.

G. *Isoflavonoids*

1. Simple Isoflavonoids

Whereas numerous enzymes involved in flavonoid biosynthesis have been identified, the pathways formulated for biosynthesis of isoflavonoids are still mostly derived from *in vivo* incorporation experiments with labeled precursors. This work has been covered in reviews by Hahlbrock and Grisebach (1975) and Dewick (1982).

Extensive tracer studies with phenylpropanoid and chalcone precursors have provided evidence for the formation of isoflavones by an 1,2-aryl migration (Grisebach and Doerr, 1960) occurring at or after the stage of the chalcone/flavanone intermediate. The labeling pattern found with phenylpropanes or chalcones is shown in Fig. 12. Oxidation of a chalcone has been proposed as one possible mechanism for aryl migration (Pelter *et al.*, 1971), but this has not yet been verified on the enzymatic level.

Isoflavones and isoflavanones are interconvertible. From incorporation studies, it is more likely that isoflavanones are formed from hydrogenation of isoflavones (Grisebach and Zilg, 1968, and unpublished work on coumestanes).

An *O*-methyltransferase catalyzing the transfer of the methyl group from *S*-adenosyl-L-methionine to the 4′ hydroxyl of isoflavones has been detected in seedlings and cell suspension cultures of *Cicer arietinum* L. (Wengenmeyer *et al.*, 1974). It was purified 360-fold from the cell culture

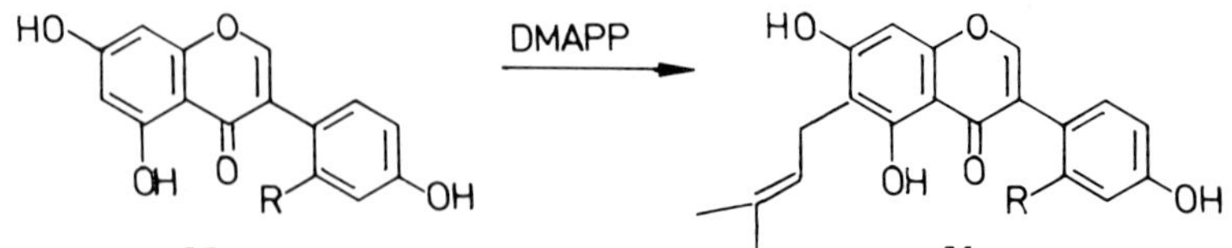

15 **16**

Fig. 13. Prenylation of genistein (**15**, R = H) or 2′-hydroxygenistein (**15**, R = OH) to wighteone (**16**, R = H) or luteone (**16**, R = OH). DMAPP = Dimethylallyl pyrophosphate.

and proved to have a low Michaelis constant of 80 μM for daidzein. 4-Hydroxycinnamic acid and 5,7,4′-trihydroxyflavanone were not methylated.

White lupine (*Lupinus albus* L.) and other *Lupinus* species contain as preinfectional agents[1] wighteone (6-isopentenylgenistein) and luteone (6-isopentenyl-2′-hydroxygenistein). A 45,000-*g* particulate fraction from hypocotyls of *Lupinus albus* catalyzes the prenylation of genistein and 2′-hydroxygenistein with dimethylallyl pyrophosphate to the respective 6-isopentenyl derivatives (Fig. 13) (Schröder *et al.*, 1979).

2. Pterocarpans

Earlier tracer studies had shown that coumestrol is biosynthetically related to isoflavones and that daidzein and dihydrodaidzein are efficient precursors for this compound (reviewed by Hahlbrock and Grisebach, 1975). Incorporation was generally low in such experiments with seedlings or cell suspension cultures. A considerable improvement was brought by the observation that biosynthesis of coumestanes and other compounds acting as phytoalexins can be stimulated considerably by infection, or, more conveniently for biosynthetic experiments, by UV light or heavy metal salts.

The pathway for medicarpin shown in Fig. 14 was deduced from extensive labeling studies in $CuCl_2$-treated seedlings of red clover (*Trifolium pratense*) (Dewick, 1982). It probably proceeds from formononetin (**17**) via 2′-hydroxylation, reduction to the isoflavanone, further reduction to the isoflavan-3-ol (**19**), followed by cyclization via the carbonium ion (**20**) to medicarpin.

A similar sequence of reactions was proposed for biosynthesis of (+)-(6a*R*,11a*S*)-maackiain and (+)-(6a*R*,11a*R*)-pisatin in $CuCl_2$-treated pea (*Pisum sativum*) tissues (Banks and Dewick, 1982) (Fig. 14).

That the reduction of isoflavone to (−)-medicarpin is stereospecific was

[1] As opposed to postinfectional agents (phytoalexins).

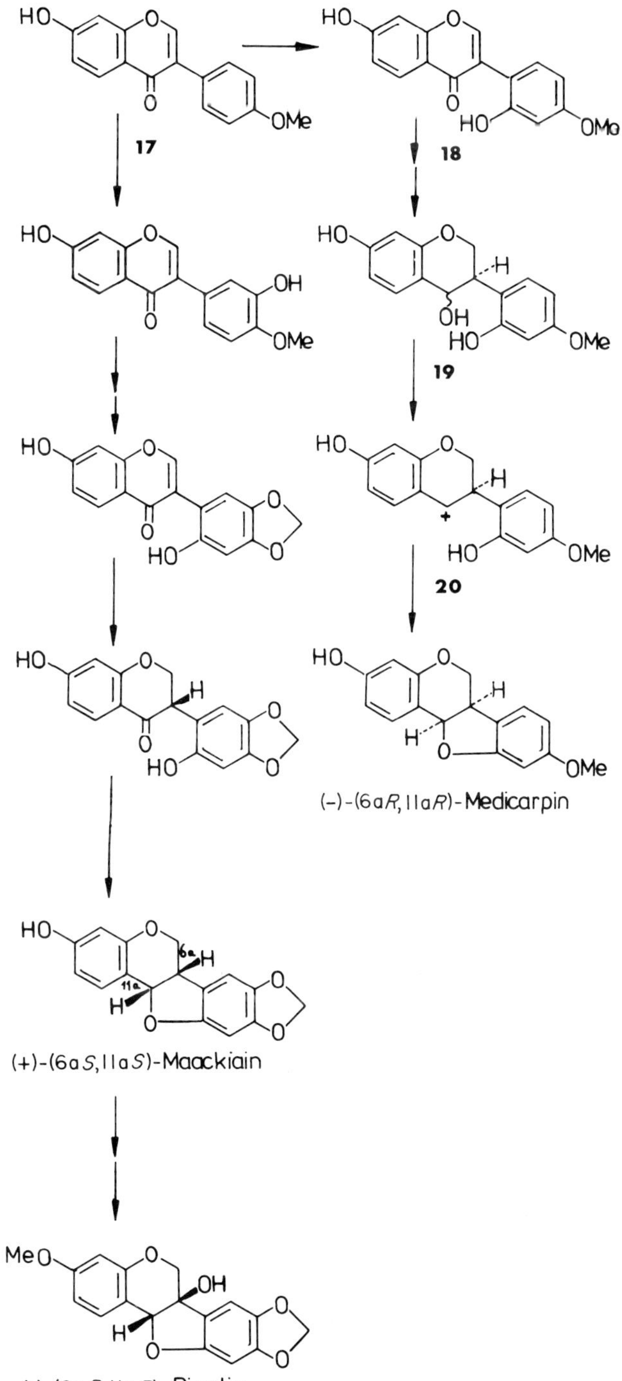

Fig. 14. Proposed reaction sequence from formononetin to medicarpin, maackiain, and pisatin.

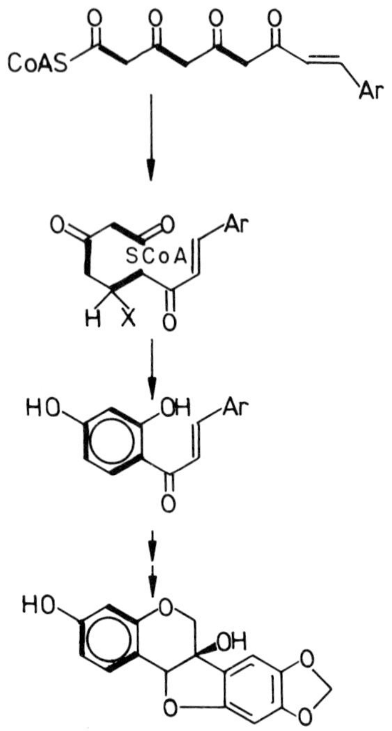

Fig. 15. Incorporation of [1,2-$^{13}C_2$]acetate into pisatin. Heavy bars indicate intact incorporation of acetate units, demonstrating a clockwise folding of the polyketide chain.

proved in fenugreek seedlings (*Trigonella foenum-graecum*), where [^{2}H]-isoflavone gave rise to (−)-medicarpin with deuterium located exclusively in the 6-pro-*R* position (Dewick and Ward, 1977). This corresponds to an overall *E*-reduction. Feeding experiments in cupric chloride-treated *Pisum sativum* pods and seedlings have demonstrated the preferential incorporation of (+)-(6a*S*,11a*S*)-[^{3}H]maackiain over (−)-(6a*R*,11a*R*)-[^{14}C]-maackiain into pisatin with a change in the ^{3}H : ^{14}C ratio from 4.7 to 30 (Banks and Dewick, 1983) (compare Fig. 14). This result establishes that the 6a-hydroxylation of pterocarpans proceeds with retention of configuration.

A specific folding of the polyketide chain and reduction of the oxygen function prior to closure of ring A was demonstrated by the intact incorporation of doubly labeled acetate units into carbons 1a-1, 2-3, and 4-4a of pisatin (Fig. 15) (Stoessl and Stothers, 1979) and phaseollin (Dewick *et al.*, 1982). In contrast, randomization of the label is found in kaempferol

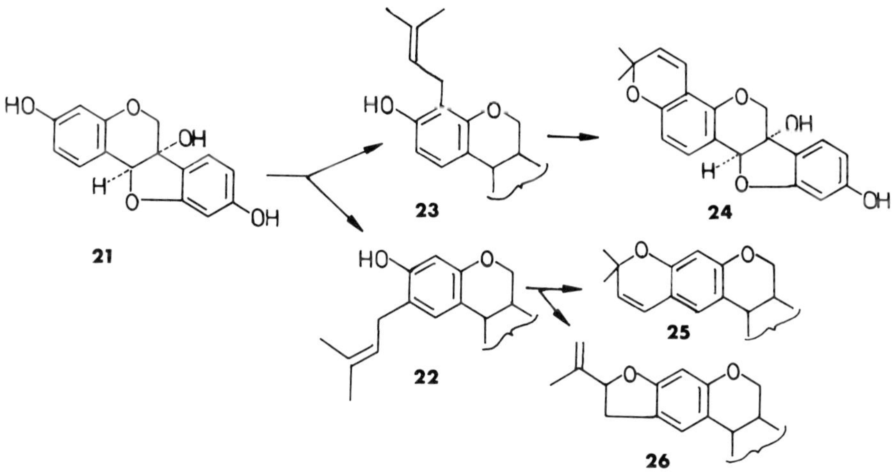

Fig. 16. Prenylation of (6a*S*,11a*S*)-3,6a,9-trihydroxypterocarpan (**21**) to the 2-prenyl (**22**) and 4-prenyl (**23**) isomers and further conversion to the glyceollin isomers **24–26**.

(Light and Hahlbrock, 1980) and kievitone (Dewick *et al.*, 1982), which is in accordance with the intermediacy of a 2′,4′,6′-trihydroxylated chalcone.

In soybean seedlings in which phytoalexin synthesis is induced by infection with *Phytophthora megasperma* f.sp. *glycinea* or by treatment with a glucan elicitor from this fungus, (6a*S*,11a*S*)-[^{14}C]3,6a,9-trihydroxypterocarpan (glycinol) was shown to be a good precursor (15% incorporation) for the glyceollin isomers **24–26** (Fig. 16) (Grisebach *et al.*, 1983). The formation of 2- and 4-dimethylallyl glycinol from glycinol and dimethylallyl pyrophosphate is catalyzed by an enzyme present only in induced soybean seedlings or cell cultures of this plant (Zähringer *et al.*, 1979, 1981).

3. Rotenoids

Investigations on the biosynthesis of rotenoids by Crombie and collaborators led to the discovery of a direct biosynthetic link between these compounds and 2′-methoxyisoflavones (Crombie *et al.*, 1971). The work has been reviewed (Hahlbrock and Grisebach, 1975). The biosynthetic route to rotenone proposed on account of the tracer experiments is shown in Fig. 17. Prenylation is probably a late step in the pathway. No enzyme studies have been reported so far in this field.

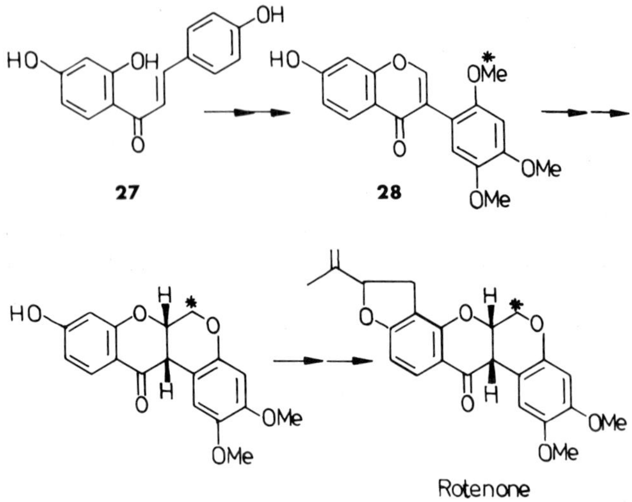

Fig. 17. Proposed biosynthesis of rotenone from trihydroxychalcone (**27**) via 2′-*O*-methylisoflavone (**28**). Asterisk indicates labeled carbon atom.

H. Regulation of Flavonoid Biosynthesis

Regulation of flavonoid biosynthesis is correlated with differentiation and development of plant tissues that have the potential to synthesize these pigments. Many external environmental factors, such as light, temperature, nutritional effects, plant hormones, mechanical damage, and pathogenic attack, influence flavonoid accumulation in plants. These factors have been reviewed by McClure (1975). A treatment of this complex subject is beyond the scope of this chapter. For recent reviews, the reader is referred to articles by Hahlbrock and Grisebach (1979) and Wiermann (1981).

Addendum

Supplementation experiments with leucoanthocyanidins (flavan-3,4-diols) with genetically defined acyanic flowers of *Matthiola incana* (blocked in the chalcone synthase gene or between dihydroflavonols and anthocyanidins) led to formation of the corresponding anthocyanidin glucosides. These results support the role of leucoanthocyanidins in anthocyanidin biosynthesis (Heller *et al., Planta,* in press). It has not yet been clarified whether the 3,4-*trans*- or the 3,4 *cis*-diols are the actual precursors. However, in view of the fact that Stafford and Lester (*Plant Physiol.*, in press) showed that a NADPH-dependent reductase from Douglas fir (Stafford and Lester, 1982) leads to the 3,4-*cis*-diol it can be assumed that the cis diol is also the precursor for anthocyanidins. Stafford and Lester also showed that the 3,4-*cis*-diol produced from (+)-dihydroquercetin was further reduced with the reductase preparation from Douglas fir to (+)-catechin (2,3-*trans* flavon-3-ol).

Important progress was achieved in the enzymology of isoflavonoid biosynthesis. A microsomal preparation from elicitor-challenged soybean cell suspension cultures or soybean seedlings catalyzes an NADPH- and dioxygen-dependent rearrangement of (2*S*)-naringenin to genistein (5,7,4′-trihydroxyisoflavone). Rearrangement to the isoflavone therefore seems to take place with the flavanone and not with the chalcone (M. Hagmann and H. Grisebach, in press).

References

Abbott, M. T., and Udenfriend, S. (1974). *In* "Molecular Mechanism of Oxygen Activation" (O. Hayaishi, ed.), pp. 167–214. Academic Press, New York.

Alston, R. E. (1964). The genetics of phenolic compounds. *In* "Biochemistry of Phenolic Compounds" (J. B. Harborne, ed.), pp. 171–204. Academic Press, London.

Amrhein, N. (1979). *Phytochemistry* **18,** 585–589.

Amrhein, N., and Gödeke, K. H. (1977). *Plant Sci. Lett.* **8,** 313–317.

Amrhein, N., Gödeke, K. H., and Kefeli, V. I. (1976). *Ber. Deutsch. Bot. Ges.* **89,** 247–259.

Arnstadt, K. I., Schindlbeck, G., and Lynen, F. (1975). *Eur. J. Biochem.* **55,** 561–571.

Banks, S. W., and Dewick, P. M. (1982). *Phytochemistry* **21,** 2235–2242.

Banks, S. W., and Dewick, P. M. (1983). *Phytochemistry* **22,** 1591–1595.

Boland, M. J., and Wong, E. (1975). *Eur. J. Biochem.* **50,** 383–389.

Britsch, L., Heller, W., and Grisebach, H. (1981). *Z. Naturforsch.* **36C,** 742–750.

Chadenson, M., Hauteville, M., Chopin, J., Wollenweber, E., Tissut, M., and Egger, K. (1971). *Compt. Rend.* **273,** 2658.

Chopin, J., Hauteville, M., Joshi, B. S., and Gawad, D. H. (1978). *Phytochemistry* **17,** 332–334.

Crombie, L., Dewick, P. M., and Whiting, D. A. (1971). *Chem. Commun.* 1182, 1183.

De Vlaming, P., and Kho, K. F. F. (1976). *Phytochemistry* **15,** 348–349.

Dewick, P. M. (1982). *In* "The Flavonoids: Advances in Research" (J. B. Harborne and T. J. Mabry, eds.), pp. 611–632. Chapman and Hall, London.

Dewick, P. M. and Ward, D. (1977). *J. Chem. Soc. Chem. Commun.* 338–339.

Dewick, P. M., Steele, M. J., Dixon, R. A., and Whitehead, I. M. (1982). *Z. Naturforsch.* **37C,** 363–368.

Dixon, R. A., Dey, P. M., and Whitehead, I. M. (1982). *Biochim. Biophys. Acta* **715,** 25–33.

Doodeman, M., Tabak, A. J. H., Schram, A. W., and Bennink, G. J. H. (1982). *Planta* **154,** 546–549.

Dooner, H. K., and Nelson, O. E. (1977). *Biochem. Genet.* **15,** 509–519.

Ebel, J., and Hahlbrock, K. (1982). *In* "The Flavonoids: Advances in Research" (J. B. Harborne and T. J. Mabry, eds.), pp. 641–679. Chapman and Hall, London.

Ebel, J., Hahlbrock, K., and Grisebach, H. (1972). *Biochim. Biophys. Acta* **269,** 313–326.

Egin-Bühler, B., Loyal, R., and Ebel, J. (1980). *Arch. Biochem. Biophys.* **203,** 90–100.
Forkmann, G. (1977). *Planta* **137,** 159–163.
Forkmann, G. (1980). *Planta* **148,** 157–161.
Forkmann, G., and Dangelmayr, B. (1980). *Biochem. Genet.* **18,** 519–527.
Forkmann, G., and Kuhn, B. (1979). *Planta* **144,** 189–192.
Forkmann, G., and Stotz, G. (1981). *Z. Naturforsch.* **36C,** 411–416.
Forkmann, G., Heller, W., and Grisebach, H. (1980). *Z. Naturforsch.* **35C,** 691–695.
Fritsch, H., and Grisebach, H. (1975). *Phytochemistry* **14,** 2437–2442.
Fritsch, H., Hahlbrock, K., and Grisebach, H. (1971). *Z. Naturforsch.* **26B,** 581–585.
Geissman, T. A., and Mehlquist, G. A. L. (1947). *Genetics* **32,** 410.
Grambow, H. J., and Grisebach, H. (1971). *Phytochemistry* **10,** 789–796.
Grisebach, H. (1979). *In* "Recent Advances in Phytochemistry" (T. Swain, J. B. Harborne, and C. F. Van Sumere, eds.), Vol. 12, pp. 221–248. Plenum, New York.
Grisebach, H. (1982). *In* "Anthocyanins as Food Colors" (P. Markakis, ed.), pp. 69–92. Academic Press, New York.
Grisebach, H., and Doerr, N. (1960). *Z. Naturforsch.* **15B,** 284–286.
Grisebach, H., and Kellner, S. (1965). *Z. Naturforsch.* **20B,** 446–450.
Grisebach, H., and Zilg, H. (1968). *Z. Naturforsch.* **23B,** 494–504.
Grisebach, H., Börner, H., and Moesta, P. (1982). *Ber. Bot. Ges.* **95,** 619–642.
Hagmann, M., Heller, W., and Grisebach, H. (1983). *Eur. J. Biochem.* **134,** 547–554.
Hahlbrock, K., and Grisebach, H. (1975). *In* "The Flavonoids" (J. B. Harborne, T. J. Mabry, and H. Mabry, eds.), pp. 866–915. Chapman and Hall, London.
Hahlbrock, K., and Grisebach, H. (1979). *Ann. Rev. Plant Physiol.* **30,** 105–130.
Hahlbrock, K., Wong, E., Schill, L., and Grisebach, H. (1970). *Phytochemistry* **9,** 949–958.
Harrison, B. J., and Stickland, R. G. (1974). *Heredity* **33,** 112–115.
Hauteville, M., Chadenson, M., and Chopin, J. (1979). *Bull. Soc. Chem. Fr.* **3–4,** Part 2, 125–131.
Heller, W., and Hahlbrock, K. (1980). *Arch. Biochem. Biophys.* **200,** 617–619.
Hrazdina, G., Kreuzaler, F., Hahlbrock, K., and Grisebach, H. (1976). *Arch. Biochem. Biophys.* **175,** 392–399.
Ibrahim, R. K., DeLuca, V., Jay, M., and Voirin, B. (1982). *Naturwissenschaften* **69,** 41–42.
Jonsson, L. M. V., Aarsman, M. E. G., Schram, A. W., and Bennink, G. J. H. (1982). *Phytochemistry* **21,** 2457–2459.
Kamsteeg, J., van Brederode, J., and van Nigtevecht, G. (1978a). *Biochem. Genet.* **16,** 1045–1058.
Kamsteeg, J., van Brederode, J., and van Nigtevecht, G. (1978b). *Biochem. Genet.* **16,** 1059–1071.
Kamsteeg, J., van Brederode, J., and van Nigtevecht, G. (1980a). *Phytochemistry* **19,** 1459–1462.
Kamsteeg, J., van Brederode, J., and van Nigtevecht, G. (1980b). *Z. Naturforsch.* **35C,** 249–257.
Kamsteeg, J., van Brederode, J., Hommels, C. H., and von Nigtevecht, G. (1980c). *Biochem. Physiol. Pflanzen* **175,** 403–411.
Kamsteeg, J., van Brederode, J., Verschuren, P. M., and van Nigtevecht, G. (1981). *Z. Pflanzenphysiol.* **102,** 435–442.
Kho, F. F., Bennink, G. J. H., and Wiering, H. (1975). *Planta* **127,** 271–279.
Kho, K. F. F., Bolsman-Louwen, A. C., Vuik, J. C., and Bennink, G. J. H. (1977). *Planta* **135,** 109–118.
Köster, J. (1982). Untersuchungen zum Stoffwechsel von Isoflavonen in Wurzeln der Kichererbse *Cicer avietinum* L. Doctoral thesis, Münster.

Koukol, J., and Conn, E. (1961). *J. Biol. Chem.* **236,** 2692–2698.
Kreuzaler, F., and Hahlbrock, K. (1975a). *Eur. J. Biochem.* **56,** 205–213.
Kreuzaler, F., and Hahlbrock, K. (1975b). *Arch. Biochem. Biophys.* **169,** 84–90.
Kreuzaler, F., Light, R. J., and Hahlbrock, K. (1978). *FEBS Lett.* **94,** 175–178.
Kreuzaler, F., Ragg, H., Heller, W., Tesch, R., Witt, I., Hammer, D., and Hahlbrock, K. (1979). *Eur. J. Biochem.* **99,** 89–96.
Kuhn, B., Forkmann, G., and Seyffert, W. (1977). *Planta* **138,** 199–203.
Kuroki, G., and Poulton, J. E. (1981). *Z. Naturforsch.* **36C,** 916–920.
Larson, R. L., and Coe, Jr., E. H. (1977). *Biochem. Genet.* **15,** 153–156.
Legrand, M., Fritig, B., and Hirth, L. (1978). *Planta* **144,** 101–108.
Light, R. J., and Hahlbrock, K. (1980). *Z. Naturforsch.* **35C,** 717–721.
Matern, U., Potts, J. R. M., and Hahlbrock, K. (1981). *Arch. Biochem. Biophys.* **208,** 233–241.
Matern, U., Heller, W., and Himmelspach, K. (1983). *Eur. J. Biochem.* **133,** 439–448.
McClure, J. W. (1975). *In* "The Flavonoids" (J. B. Harborne, T. Mabry, and H. Mabry, eds.) pp. 970–1055. Chapman and Hall, London.
McCormick, S. (1978). *Biochem. Genet.* **16,** 777–785.
Mors, W. B., Magelheas, M. T., and Gottlieb, O. R. (1961). *Prog. Chem. Org. Natural Prod.* **20,** 131–164.
Patschke, L., and Grisebach, H. (1965). *Z. Naturforsch.* **20B,** 1039–1042.
Patschke, L., Barz, W., and Grisebach, H. (1966). *Z. Naturforsch.* **21B,** 45–47.
Pelter, A., Bradshaw, J., and Warren R. F. (1971). *Phytochemistry* **10,** 835–850.
Perrin, P. W., and Towers, G. H. N. (1973). *Phytochemistry* **12,** 589–592.
Potts, J. R., Weklych, R., and Conn, E. E. (1974). *J. Biol. Chem.* **249,** 5019–5026.
Poulton, J., Hahlbrock, K., and Grisebach, H. (1976). *Arch. Biochem. Biophys.* **176,** 449–456.
Poulton, J., Hahlbrock, K., and Grisebach, H. (1977). *Arch. Biochem. Biophys.* **180,** 543–549.
Reimold, U., Kröger, M., Kreuzaler, F., and Hahlbrock, K. (1983). *EMBO J.* **2,** 1801–1805.
Saleh, N. A. M., Fritsch, H., Witkop, P., and Grisebach, H. (1976a). *Planta* **133,** 41–45.
Saleh, N. A. M., Poulton, J. E., and Grisebach, H. (1976b). *Phytochemistry* **15,** 1865–1868.
Saleh, N. A. M., Fritsch, H., Kreuzaler, F., and Grisebach, H. (1978). *Phytochemistry* **17,** 183–186.
Schill, L., and Grisebach, H. (1973). *Hoppe-Seyler's Z. Physiol. Chem.* **354,** 1555–1562.
Schröder, G., Zähringer, U., Heller, W., Ebel, J., and Grisebach, H. (1979). *Arch. Biochem. Biophys.* **194,** 635–636.
Specht, J. E., Gorz, H. J., and Haskins, F. A. (1976). *Phytochemistry* **15,** 133–134.
Spribille, R., and Forkmann, G. (1981). *Z. Naturforsch.* **36C,** 619–624.
Spribille, R., and Forkmann, G. (1982). *Phytochemistry* **21,** 2231–2234.
Stafford, H. A., and Lester, H. H. (1982). *Plant Physiol.* **70,** 695–698.
Stafford, H. A., and Lester, H. H. (1984). *Plant Physiol.* in press.
Stickland, R. G., and Harrison, B. J. (1974). *Heredity* **33,** 108–112.
Stoessl, A., and Stothers, J. B. (1979). *Z. Naturforsch.* **34C,** 87–89.
Stotz, G., and Forkmann, G. (1981). *Z. Naturforsch.* **36C,** 737–741.
Sütfeld, R., and Wiermann, R. (1978). *Biochem. Physiol. Pflanzen* **172,** 111–123.
Sütfeld, R., and Wiermann, R. (1980). *Arch. Biochem. Biophys.* **201,** 64–72.
Sütfeld, R., and Wiermann, R. (1981). *Z. Naturforsch.* **36C,** 30–34.
Sütfeld, R., Kehrel, B., and Wiermann, R. (1978). *Z. Naturforsch.* **33C,** 841–846.
Sutter, A., and Grisebach, H. (1973). *Biochim. Biophys. Acta* **309,** 289–295.
Sutter, A., Ortmann, R., and Grisebach, H. (1972). *Biochim. Biophys. Acta* **258,** 71–87.
Sutter, A., Poulton, J., and Grisebach, H. (1975). *Arch. Biochem. Biophys.* **170,** 547–556.

Tabak, A. J. H., Meyer, H., and Bennink, G. J. H. (1978). *Planta* **139,** 67–71.
Tsang, J. F., and Ibrahim, R. K. (1979). *Phytochemistry* **18,** 1131–1136.
Vaughan, P. F. T., Butt, V. S., Grisebach, H., and Schill, L. (1969). *Phytochemistry* **8,** 1373–1378.
Wengenmayer, H., Ebel, J., and Grisebach, H. (1974). *Eur. J. Biochem.* **50,** 135–143.
Wiering, H. (1974). *Gene Phaenen* **17,** 117.
Wiermann, R. (1981). *In* "The Biochemistry of Plants" (E. E. Conn and P. K. Stumpf, eds.), Vol. 7, pp. 85–116. Academic Press, New York.
Wong, E. (1976). Biosynthesis of Flavonoids. *In* "Chemistry and Biochemistry of Plant Pigments" (T. W. Goodwin, ed.), Vol. 1, pp. 464–526. Academic Press, New York.
Zähringer, U., Ebel, J., Mulheirn, L. J., Lyne, R. L., and Grisebach, H. (1979). *FEBS Lett.* **101,** 90–92.
Zähringer, U., Schaller, E., and Grisebach, H. (1981). *Z. Naturforsch.* **36C,** 234–241.

CHAPTER 13

Biosynthesis of Tannins

W. E. HILLIS

CSIRO
Division of Chemical and Wood Technology
Highett, Victoria, Australia

I. Introduction

For thousands of years, various types of polymerized polyphenols obtained from plants have been used to convert animal hides and skins to leather, by a process known as tanning. The name vegetable "tannins," used for the polyphenols that combine with or precipitate the protein of skins to increase their stability to water, microorganisms, heat, and abrasion is appropriate for that purpose. The name is also used less appropriately when some of these polyphenols are used as bases for wood adhesives, control of viscosity of clay–water mixes, water purification, iron protection, and other uses. There is widespread use of these materials, and the world's consumption of commercially produced vegetable tannins was ~267,000 tons in 1982, having declined from over 418,000 tons in 1948.

The major portion of vegetable tannin extracts contains components with molecular weights in the range of 500 to 3000. The remainder of the

Biosynthesis and Biodegradation of Wood Components

ISBN 0-12-347880-4

extracts consists of monomeric or dimeric polyphenols, carbohydrates, and organic acids, known collectively as "nontannins," which can influence several properties of the extracts.

Vegetable tannins are classified into *condensed* or *hydrolyzable* tannins based on their structures. Condensed tannins, when heated with acid, undergo progressive polymerization to yield mainly insoluble "phlobaphenes" or red-colored substances. Condensed tannins are polymers containing flavan-3-ol units and, when treated with hydrochloric acid or butanol–hydrochloric acid, yield anthocyanidins. The monomeric polyphenols present in the nontannin fraction of vegetable tannins are considered in this chapter, to provide clues to the constitution of the polymers. They and the dimers do not have tanning properties with collagen substrates (Sykes and Roux, 1957).

The polyester structure of hydrolyzable tannins is readily hydrolyzed by acids (or some enzymes) to phenol carboxylic acids and a sugar or related polyhydroxy alcohol. The *gallotannins* yield gallic acid on hydrolysis, and the *ellagitannins* yield hexahydroxydiphenic acid, which is rapidly converted to its stable dilactone ellagic acid. Some hydrolyzable tannins yield both gallic and ellagic acids.

II. Phytochemistry and Distribution

Bate-Smith (1954, 1962; Bate-Smith and Lerner, 1954) noted the relation between the presence of proanthocyanidins (formerly known as leucoanthocyanins) and a woody habit of growth in plants. He did not propose, however, that proanthocyanidins have a connection with lignification or that they play a structural role in wood. Thus, they are wood extractives. Bate-Smith based his conclusions on the red color (supported by chromatographic identification of the anthocyanidin) produced when tissues were heated with acid. Consequently, no distinction was made among the molecular sizes of the color precursors.

Subsequently, Bate-Smith (1973a,b, 1975, 1976, 1977, 1978) undertook extensive studies on proanthocyanidin polymers (and ellagitannins) in order to ascertain their taxonomic significance. This monumental work has laid the foundations and most of the knowledge of the systematic distribution of these compounds. As has been pointed out by Foo and Porter (1980), however, there are some aspects of methodology (such as the solvent used) that limit the development of some of the conclusions, particularly those concerning woody tissues. Nevertheless, there is no reason to doubt the conclusions that procyanidin polymers predominate over other proanthocyanidins in nature.

Following a comprehensive survey (Bate-Smith, 1956) in which ellagitannins (or ellagic acid) were found consistently in the leaves of species of certain plant orders, there have been different proposals for the interrelationship of gallotannins and ellagitannins and their taxonomic significance. Owing to difficulties in distinguishing between the complexities of the different components of hydrolyzable tannins, progress has been slow. Limited success has been achieved in revealing relationships of taxonomic importance at family and genus level (e.g., Haslam, 1966; Haddock *et al.*, 1982a).

The various data on vegetable tannins in the literature relate to hot aqueous extracts of various tissues. Some polymerized polyphenols, as in the heartwood of *Eucalyptus marginata* Donn ex Sm. or the bark of *Sequoia sempervirens* (D. Don) Endl., require dilute ($<0.5\%$) alkali for their removal. A partial fractionation of the tannins is achieved by using different solvents in the extraction procedure. The tannin content is determined by shaking an aqueous extract of known solids content with hide or polyamide powder to absorb the tannins, which are estimated from the reduction of solids content in the filtrate, nontannins are represented by the material not absorbed by the powders.

Condensed tannins are more widespread through the plant kingdom than are the hydrolyzable tannins (Gnamm, 1949). Bark or wood of certain species of both softwoods (Coniferae) and hardwoods (Angiospermae) can contain large amounts, and sources that have been used for major commercial extracts are listed in Table I. Lists have been published of plant families and genera containing condensed or hydrolyzable tannins, but these can be misleading. Some genera (such as *Eucalyptus*) include species with woods that contain considerable amounts of condensed tannins, whereas other species contain large amounts of hydrolyzable tannins. Also, some data do not specify whether the tannin was detected in leaves, wood, or bark, and all three tissues in some species can contain different types of tannins. Furthermore, some sources can contain both condensed and hydrolyzable tannins.

Whereas different woods and barks provide economic sources of condensed tannins, extracts of hydrolyzable tannins are mainly obtained from fruit, pods, and galls of various trees and from the decreasing supplies of chestnut wood (Table II).

III. Structure of Tannins

The chemistry of both condensed and hydrolyzable tannins has been investigated for many years (Perkin and Everest, 1918). Only recently,

Table I
Major Sources of Condensed Tannins[a]

Family	Plant material	Yield (%)
Anacardiaceae	*Schinopsis balansae* (heartwood)	20–25
	Schinopsis lorentzii (heartwood)	16–17
Fagaceae	*Quercus robur* (bark)	12–16
Leguminosae	*Acacia catechu* (heartwood)	15
	Acacia mearnsii (bark)	35–40
Myrtaceae	*Eucalyptus astringens* (bark)	40–54
	Eucalyptus wandoo (bark and heartwood)	13–15
Pinaceae	*Picea abies* (bark)	5–10
	Pinus radiata (bark)	15–25
	Pinus sylvestris (bark)	16
	Larix decidua (bark)	5–20
Rhizophoraceae	*Bruguiera gymnorrhiza* (bark)	9–40
	Rhizophora mucronata (bark)	10–50
	Rhizophora mangle (bark)	20–30

[a] From Hathway (1962), Sands (1975), and Hillis (1956).

with the application of techniques and criteria such as ^{13}C and ^{1}H NMR (e.g., Jacques *et al.*, 1977) and IR (e.g., Foo, 1981), capable of resolving the key features of complex situations, have significant advances been made in the elucidation of their structures. Slowly, one of the classic problems of natural product chemistry is yielding to intensive study.

Table II
Sources of Hydrolyzable Tannins of Commercial Importance[a]

Source	Tannin
Galls on leaves of *Rhus semialata* (Anacardiaceae)	Chinese tannin (tannic acid)
Leaves of *Rhus coriaria* (Anacardiaceae)	Sumac tannin
Fruit of *Terminalia chebula* (Combretaceae)	Myrobalans
Galls on wood of *Quercus infectoria* (Fagaceae)	Turkish tannin
Acorn cups of *Quercus valonea* or *Q. macrolepis* (Fagaceae)	Valonea extract
Wood of various *Quercus* species (Fagaceae)	Oak extract
Wood of *Castanea sativa* and *C. dentata* (Fagaceae)	Chestnut extract
Pods of *Caesalpinia spinosa* (Leguminosae)	Tara extract
Pods of *Caesalpinia coriaria* (Leguminosae)	Divi-divi
Pods of *Caesalpinia brevifolia* (Leguminosae)	Algarobillia extract

[a] From White (1958).

Much of the recent developments concerning both classes of tannins has been initiated by Bate-Smith, largely from a phytochemical viewpoint. Those aspects, as well as the structural studies of his own and other groups, have been summarized recently by Haslam (1977, 1982a,b).

A. *Condensed Tannins*

The major portion of condensed tannins consists of proanthocyanidins, or those substances isolated from plants that form anthocyanidins when heated with hydrochloric acid. The yield of the corresponding anthocyanidin when heated with acid is increased in the presence of propanol (Pigman *et al.*, 1953; Hillis, 1954) or butanol (Swain and Hillis, 1959).

1. Flavan-3-ols and Flavan-3,4-diols

a. Flavan-3-ols (Catechins). The stability of this class of compounds and their widespread occurrence with polymeric and sometimes dimeric proanthocyanidins points to their involvement as precursors. A number of flavan-3-ols have been characterized (Haslam, 1975), and the relation with proanthocyanidins has been shown with (+)-catechin [**1** where X = H; (2*R*,3*S*)-3,5,7,3′,4′-pentahydroxyflavan]; (−)-epicatechin [**2** where X = H; (2*R*,3*R*)-3,5,7,3′,4′ pentahydroxyflavan]; (+)-gallocatechin [**1** where X = OH], (−)-epigallocatechin [**2** where X = OH]; (−)-fisetinidol [(2*R*,3*S*)-3,7,3′,4′-tetrahydroxyflavan) and (−)-robinetinidol [(2*R*,3*S*)-3,7,3′,4′,5′-pentahydroxyflavan]. They do not produce a red color when heated with hydrochloric acid.

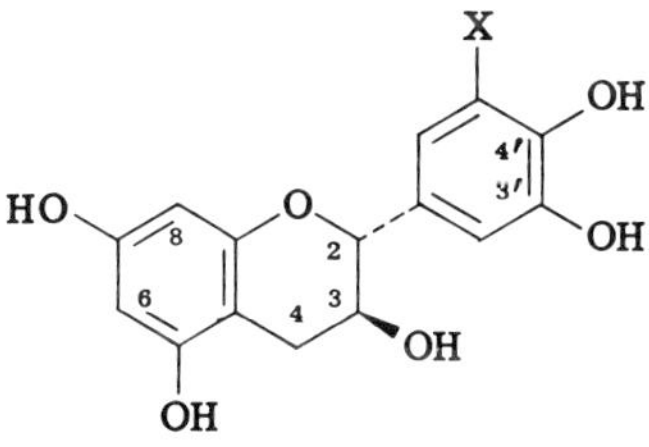

Catechin
1

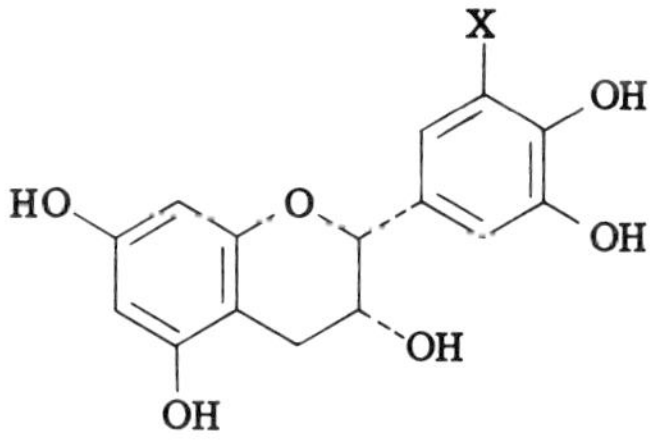

Epicatechin
2

b. Flavan-3,4-diols (Leucoanthocyanidins). The very high reactivity of the most common members of this class, namely, "leucocyanidin," "leucodelphinidin," and "leucopelargonidin," has so far prevented their characterization. The nine flavan-3,4-diols that have been fully defined, mainly by Roux and co-workers, lack the C-5 hydroxyl and have been isolated from species of *Acacia, Gleditsia, Guibourtia, Neuratanenia,* and *Robinia* (Haslam, 1982a). The flavan-3,4-diols and the oligomeric proanthocyanidins occur free and unglycosylated.

The corresponding anthocyanidin produced by flavan-3,4-diols (e.g., **3**) or proanthocyanidins when heated with acid apparently arises via an initial dehydration to a flav-3-en-3-ol, which in hot acidic media is readily solvolyzed to give the corresponding benzyl carbocation (**4**) and then the anthocyanidin (**5**) (Haslam, 1975).

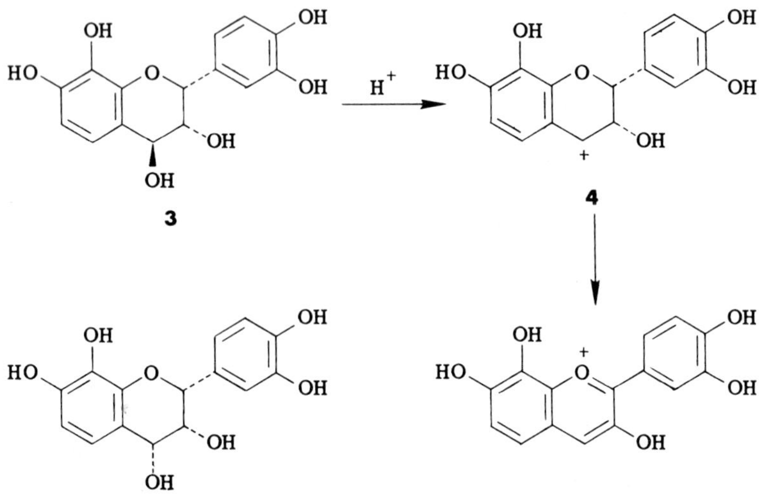

3 **4** **6** **5**

The conformation of the puckered heterocyclic ring lies somewhere between that of the sofa and half-chair forms (Roux *et al.*, 1978). The ease of ionization at the 4 position is influenced by the stereochemistry at the 3 and the 4 positions as first noticed for isomelacacidin (2*R*,3*R*,4*S*-3,4,7,8,3′,4′-hexahydroxyflavan) (**3**) (Clark-Lewis and Mortimer, 1960). This isomer was more readily solvolyzed than its epimer at the 4 position, melacacidin (**6**).

The formation of the carbocation under controlled conditions can be a key stage in further reactions, as it may be captured by nucleophilic compounds such as resorcinol, phloroglucinol, and flavans, resulting in C—C bonds, or by water, alcohols, and sulfur compounds.

2. Dimeric Proanthocyanidins with C-5 Hydroxyls

The proanthocyanidins studied contain basic units based on (+)-catechin (**1**, X = H), (−)-epicatechin (**2**, X = H), (+)-gallocatechin (**1**, X = OH), and (−)-epigallocatechin (**2**, X = OH).

Procyanidin Dimers. Four major naturally occurring procyanidin dimers (B-1, B-2, B-3, and B-4) have been isolated from a range of plant tissues and characterized. Mixtures of these dimers have not been found, but other dimers and trimers have been detected with them. The relationships among these dimers are shown in Fig. 1. The absolute stereochemistry at the C-4 positions of procyanidins B-1 and B-2 has been determined as 4*R* and of B-3 and B-4 as 4*S*. Consequently, in procyanidins B-1 and B-2, the C—C interflavan linkage at the C-4 position in the heterocyclic ring of the "upper" flavan-3-ol occupies a quasiaxial position. With B-3 and B-4, the "lower" flavan-3-ol unit occupies a quasiequatorial position at C-4 on the upper (+)-catechin unit (Haslam, 1982a).

With the nomenclature proposed by Porter *et al.* (1979), the common names of the monomeric flavan-3-ols are used, the interflavan bond and its direction is given by an arrow, and the *R* and *S* symbols signifying absolute stereochemistry are replaced by β and α, respectively. Thus, procyanidin B-1 is designated as epicatechin (4β→8)-catechin (**7**), B-2 as epicatechin-(4β→8)-epicatechin (**8**), B-3 as catechin-(4α→8)-catechin (**9**), (**9**), and B-4 as catechin(4α→8)-epicatechin (**10**).

The acid-catalyzed solvolysis of the C–C interflavan bond of the dimers B-1 through B-4 yields the benzylic carbocation (and thence cyanidin) from the "upper" part of the molecule. This part possesses a trans orientation of the hydroxyl group at C-3 and the flavan substituent at C-4. The "lower" parts of the dimers yield the appropriate flavan-3-ol, (+)-catechin (with B-1 and B-3) or (−)-epicatechin (with B-2 and B-4) (Fig. 2).

The dimer procyanidin B-5 has been found to be epicatechin-(4β→6)-epicatechin (Hemingway *et al.*, 1981a,b; Nonaka *et al.*, 1982).

Arising from the absolute configurations of 4*R* and 4*S*, restricted rotation around the interflavan linkage in the dimers results in one form of hindered rotation with B-1 and B-2 and another form with B-3 and B-4 (Fletcher *et al.*, 1977). The oligomers formed from B-1 or B-2 by the further addition of (−)-epicatechin result in left-hand helices, whereas oligomers formed from B-3 and B-4 by the further addition of (+)-catechin result in right-hand helices (Haslam, 1982a).

A prodelphinidin corresponding to the procyanidin B-4 has been characterized (Porter and Foo, 1978), as well as dimers with mixed C ring oxidation patterns of pyrogallol and catechol (Haslam, 1982a). No propelargonidin has been reported, although a polymerized form appears to exist along with (+)-afzelechin (Hillis and Carle, 1960).

3. Tri- and Polymeric Proanthocyanidins with C-5 Hydroxyls

Only trace or minor amounts of trimeric proanthocyanidins are revealed on chromatographic examination of plant extracts. Mainly, com-

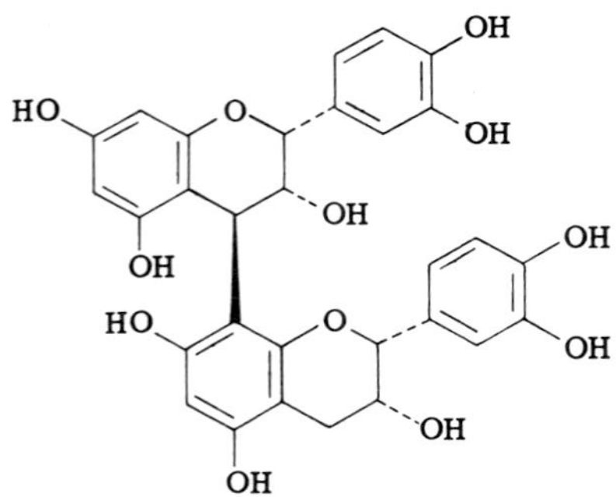

8 B-2

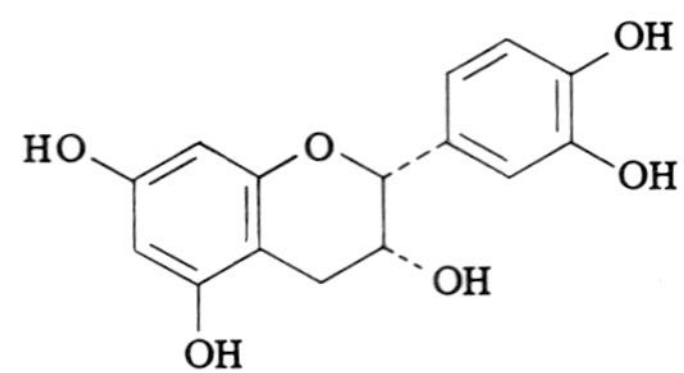

Epicatechin

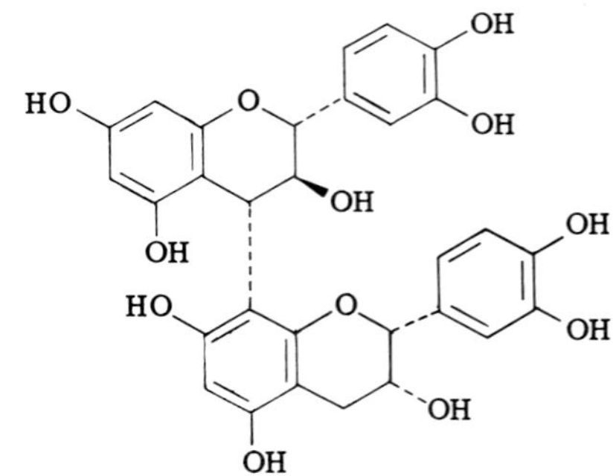

10 B-4

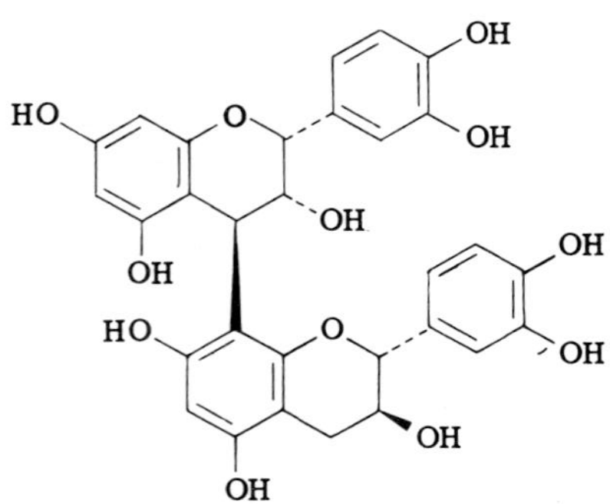

7 B-1

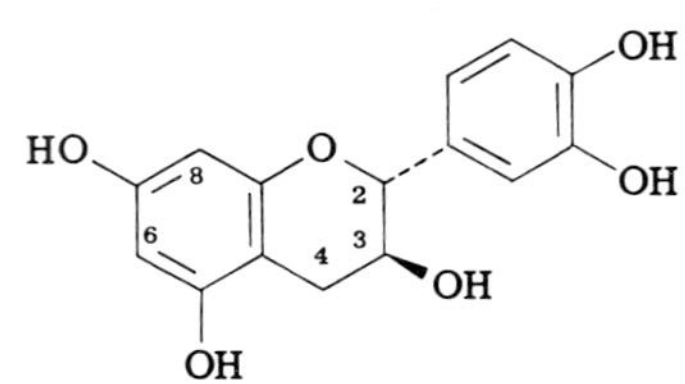

Catechin

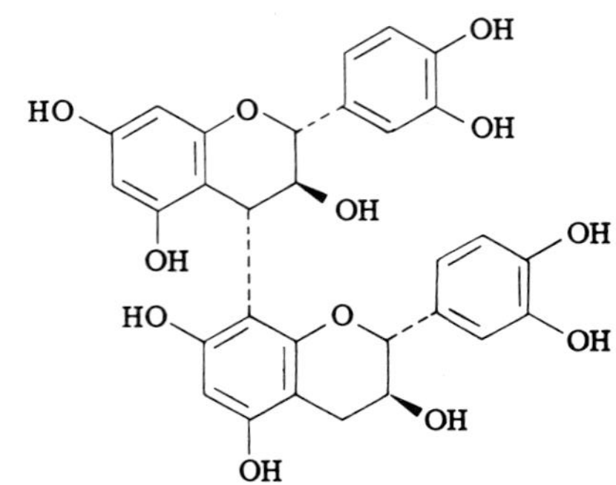

9 B-3

Fig. 1. Four dimeric procyanidins.

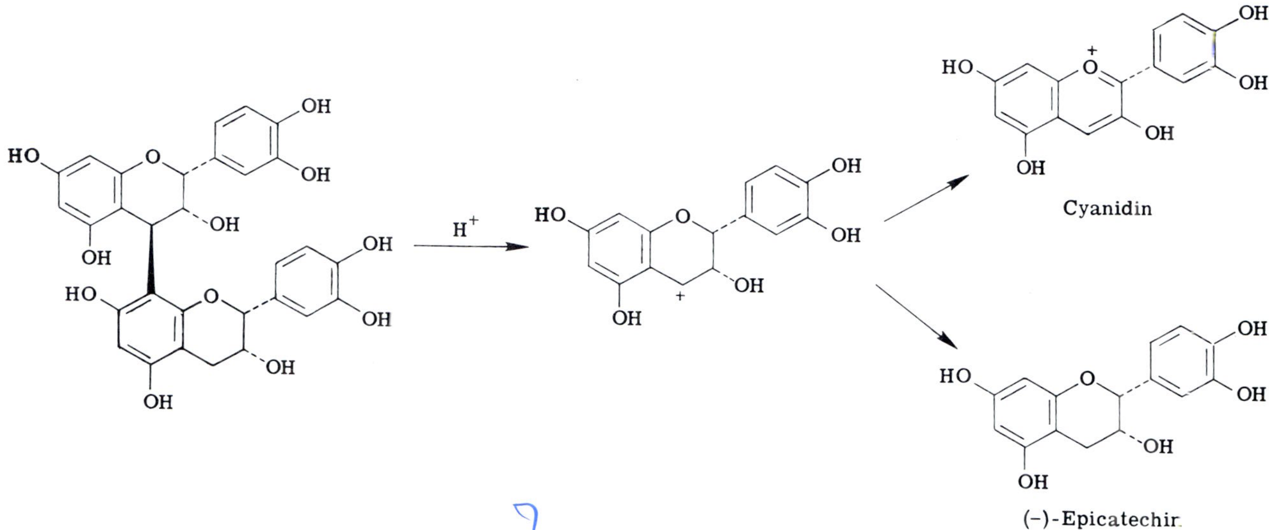

Fig. 2. Acid-catalyzed solvolysis of the C—C interflavan bond in procyanidin B-2.

plex and gradational ranges of oligomers of increasing mass and affinity for substrates exist, with resultant difficulties in their isolation and purification.

Three configurational isomers of trimeric procyanidins have been isolated, and two of these possessed a (4→6) linkage in addition to the more usual (4→8) linkage (Hemingway *et al.*, 1981a,b; Nonaka *et al.*, 1982). Two procyanidin tetramers possess (4→8) and (4→6) linkages (Nonaka *et al.*, 1981).

The structure of the procyanidin polymer from sorghum was shown to be **11** with a molecular weight of 1700 to 2000 (Gupta and Haslam, 1978). The predominantly prodelphinidin polymers from *Ribes sanguineum* with an apparent molecular weight of ~6600 were shown to be (+)-gallocatechin structures linked to a (−)-epigallocatechin terminal group (Porter and Foo, 1978; Porter *et al.*, 1979). Polymers with chain lengths of up to 20 units have also been examined (Czochanska *et al.*, 1979; Porter *et al.*, 1979). In all cases, ring C (i.e., the *ortho*-dihydroxy or *vic*-trihydroxy ring) would project laterally from the core (Haslam, 1977). In aqueous solutions some polymers appear to associate into conglomerates of large size (Yazaki and Hillis, 1977, 1980; Czochanska *et al.*, 1980).

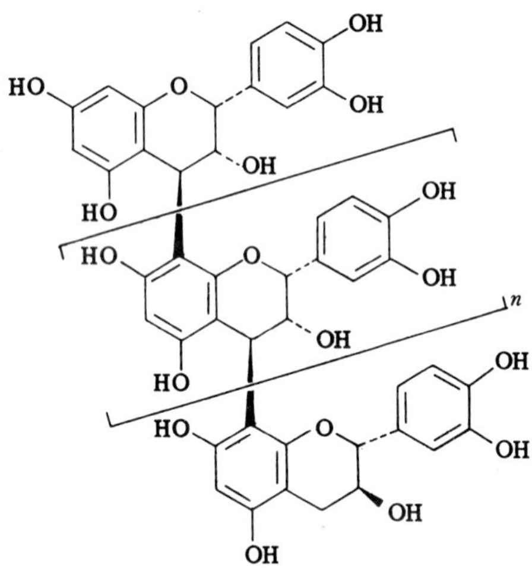

(n = 4–5)

11

McGraw and Hemingway (1982) have produced evidence in support of their view that a significant proportion of the linkages in procyanidin polymers would be (4→6) bonds in addition to the (4→8) bonds.

In contrast to the regular structures described above it has been found that there are no similarities of stereochemistry between the upper units and the additional units of oligomers from a number of coniferous barks (Samejima and Yoshimoto, 1982). The ratio of 2,3-trans units in the upper part was distinctly higher than that of the extension units.

Solely on the basis of their IR spectra, Foo (1981) grouped 26 proanthocyanidin polymers into five classes as follows.

Class A Polymers mainly of the procyanidin type with the monomers having the cis configuration (i.e., the epicatechin type).

Class B Polymers mainly of the procyanidin type with monomers having the cis and trans configurations (i.e., the units being an approximately equal mixture of the epicatechin and catechin types).

Class C Polymers of the mixed procyanidin and prodelphinidin type with the monomers having both the cis and trans configurations (i.e., the monomer units being mainly of the epicatechin and epigallocatechin types with significant proportions of catechin and gallocatechin types).

Class D Polymers mainly of the prodelphinidin type with the monomers having the cis configuration (i.e., the epigallocatechin type).

Class E Polymers mainly of the prodelphinidin type with the monomers having mainly the trans configuration (i.e., the gallocatechin type).

Of the 38 polymers studied so far (Foo and Porter, 1980), the units possess predominantly cis stereochemistry. It is common to find (+)-catechin or (−)-epicatechin (or the related gallocatechins) linked as terminal units to dimers or polymers of procyanidin or prodelphinidin with the opposite stereochemistry (Haslam, 1977; Foo and Porter, 1980; Foo, 1982).

4. Miscellaneous Proanthocyanidins

Roux *et al.* (1978) condensed (+)-mollisacacidin [(2*R*,3*S*,4*R*)-3,4,7,3′,4′-pentahydroxyflavan] with (+)-catechin at ambient temperatures with dilute hydrochloric acid to obtain two dimeric (4→8)-linked proanthocyanidins. In extensions of these studies, trimers with "angular" configurations involving (4→6) as well as (4→8) linkages have been synthesized. The structures and stereochemistry of the dimers and trimers have been found to be identical with profisetinidins isolated from the bark and wood of *Acacia mearnsii* and elsewhere (Roux and Ferreira, 1982).

5. Stilbenoids

Stilbenes have been isolated from the wood extractives of a number of species (see Chapter 14), including *Eucalyptus wandoo* Blakely (Hathway and Seakins, 1959), which formerly provided the commercial extract Myrtan. This and related eucalypts (Hillis, 1969) contain resveratrol (3,5,4′-trihydroxy stilbene, **12**, R^1, R^2 = H), its 3-β-D-glucoside, and 3,5,3′,4′-tetrahydroxy stilbene (**12**, R^1, R^2 = OH). The condensed tannins of spruce bark (*Picea abies*) contain 3,5,3′,4′-tetrahydroxy stilbene (**12**, R^1 = OH, R^2 = H) (Haslam, 1966; Gorham, 1980).

Stilbenes form polymeric materials on exposure to sunlight and under other conditions. Stilbenes may be the precursors of some condensed tannins (Grassman and Endres, 1959). but there is no information on the existence of stilbenoid polymers in wood extractives.

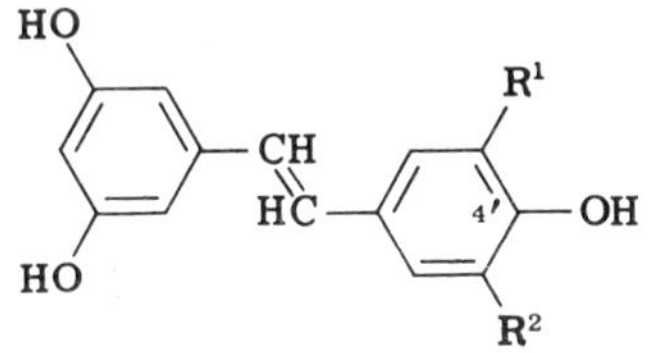

12

B. *Hydrolyzable Tannins*

1. Gallotannins

Gallic acid is present in the woody tissues of a number of eucalypt, oak, and other species, but is most commonly found in a wide range of plant tissues as esters of sugars, polyols, and other phenols. The 10 galloyl esters of D-glucose that have been characterized were originally obtained from leaves, pods, fruit, and galls, but now with β-1-*O*-galloyl-D-glucose (Fig. 3) and 2′,5-di-*O*-galloyl-D-hamamelose, also from wood and bark of various species. The most commonly reported ester is β-penta-*O*-galloyl-D-glucose, which can be further metabolized (see Fig. 4) (Haslam, 1982b). Usually the ester groups are located at positions 1, 2, and 6 on the D-glucopyranose ring. So far, no simple α-glucosides have been identified (Haddock *et al.*, 1982a), although an α-1-*O*-galloyl-2,3;4,6-dihexahydroxydiphenyl-D-glucose has been reported (Okuda *et al.*, 1982b).

Depsides. Additional gallic acid molecules may be esterified with a pre-existing galloyl ester to form meta depsides (**13**), which are found linked to the pentagalloylglucose core in "tannic acid" or "Chinese gallotannin." The depsidically linked galloyl ester residues are specifically

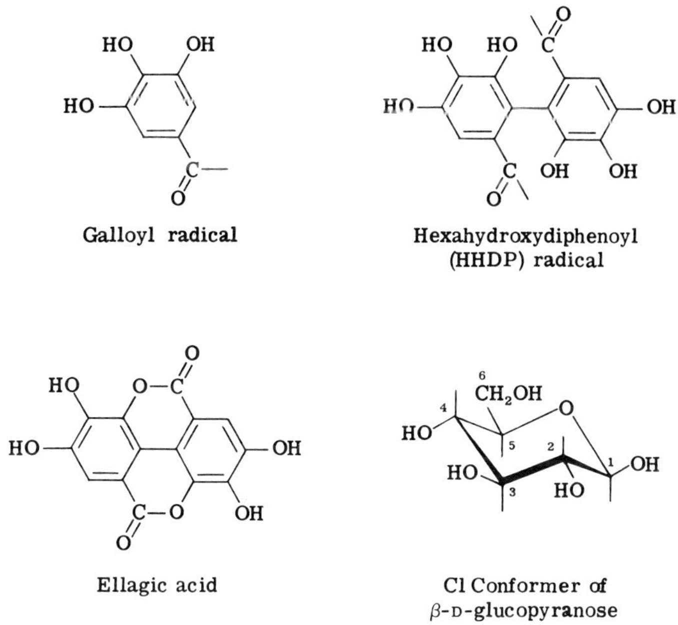

Fig. 3. Units of hydrolyzable tannins.

cleaved under very mild conditions, such as refluxing with methanol. The most common depside is the β-penta-*O*-galloyl-D-glucose carrying one or two digallate substituents on the galloyl substituent at the 2-glucose position (Haslam, 1982b; Haddock *et al.*, 1982a,b).

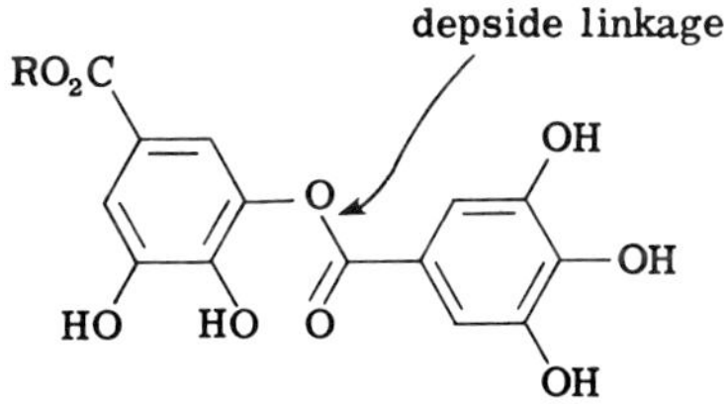

13

2. Ellagitannins

Significant quantities of these tannins are found in the wood and bark of chestnut, some oaks and eucalypts, and in the galls, acorn cups, and fruit of oaks of some *Caesalpinia* spp. and *Terminalia chebula*. The foundations leading to the elucidation of the structure of the complex compo-

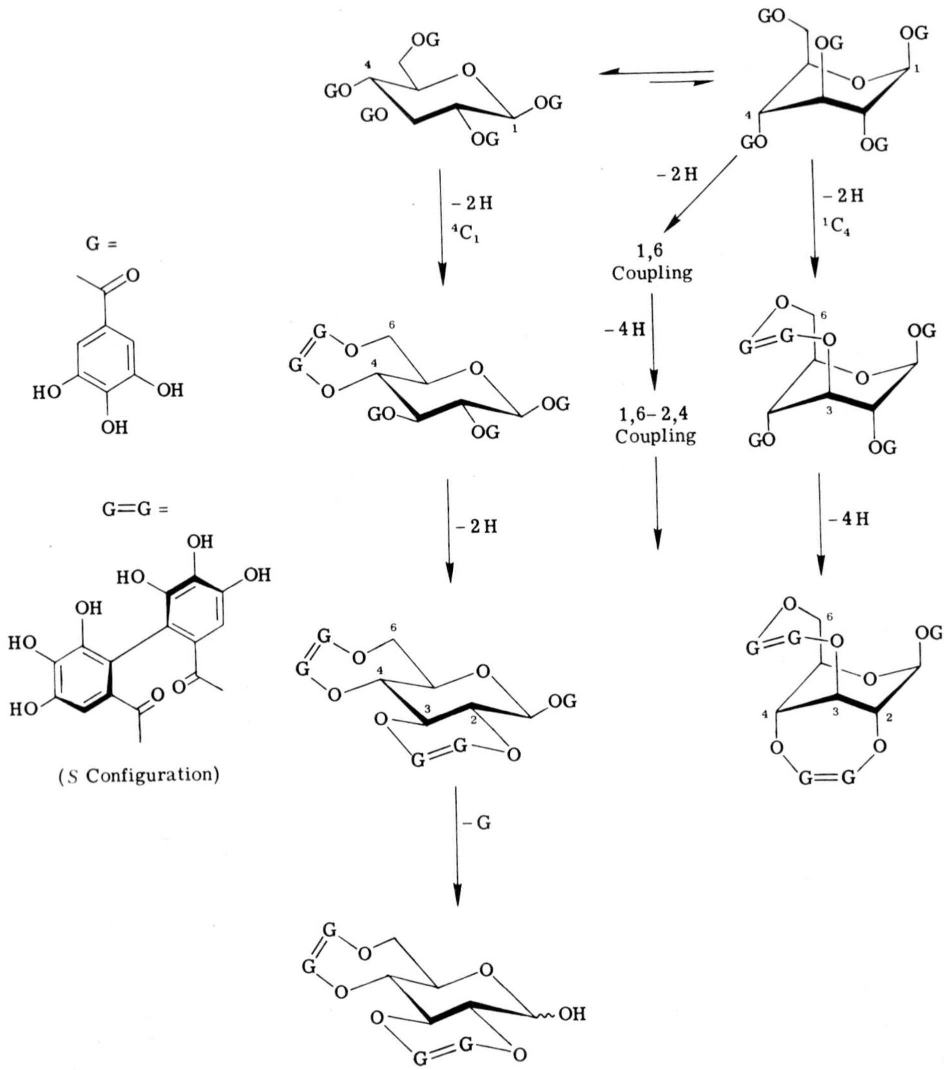

14

Fig. 4. Biogenetic relationships of pentagalloylglucose and some ellagitannins.

nents of these tannins are largely due to Schmidt and co-workers (Mayer, 1973).

Apart from the poor stability of some components (Seikel and Hillis, 1970), the various conformations of closely related compounds render their separation and purification difficult. Chromatographic examination shows a pattern of occurrence of a number of these components, and interrelationships between them have been observed (Hillis, 1969; Seikel and Hillis, 1970; Haslam, 1982b). Characterization of a new component is dependent on its isolation from a particular plant material found to be a specialized source.

a. Hexahydroxydiphenoyl Esters. It is assumed that oxidative coupling of suitably placed galloyl ester groups result in the hexahydroxydiphenoyl (HHDP) group (abbreviations for the 2,2′-dicarbonyl-4,4′,5,5′,6,6′-hexahydroxydiphenyl radical) (Fig. 3). On hydrolysis, the resultant diphenic acid rapidly lactonizes to the stable ellagic acid (Fig. 3). It has been shown that the hexahydroxydiphenoyl group that bridges the 3 and 6 positions of D-glucopyranose is the dextrorotatory form, and the bridges have the *R*-configuration. On the other hand, the HHDP that bridges the 2 and 3 or 4 and 6 positions is the levorotatory isomer, and the bridges have the *S* configuration (Haslam, 1982b). The formation of diphenoyl ester bridges in galloyl-D-glucose derivatives gives the molecule greater rigidity, and this may result in constraints on the D-glucose residue to adopt unfavorable conformations (Haslam, 1982b).

Structures of a number of isolated D-glucopyranose derivatives containing one or more —G=G— units have been reported and their interrelationships suggested by Haslam (1982b), Haddock *et al.* (1982b), Okuda *et al.* (1982a,b), and others (Fig. 4) (Haslam, 1982b). Pedunculagin (**14**) has been isolated from various sources. The ratio of these and other derivatives varies considerably within and between plant families.

Attention has been drawn to the possible effects of different conformations of D-glucose on NMR spectra and to the fact that some linkages may stabilize some unstable conformations (Seikel and Hillis, 1970; Hillis, 1972; Haslam, 1982b), The 1,6 as well as the 3,6 and 2,4 linkages require a relatively unfavorable $^1C_4$ conformation for the D-glucose (Haddock *et al.*, 1982b) and different configurations in the configurations in the HHDP ester group. An example is corilagin, the β-1-*O*-galloyl-3,6-(*R*)-HHDP-D-glucopyranose in 1B conformation. Complex ester formations have been proposed involving two glucose moieties either in a ring (Hillis, 1972; Hillis and Yazaki, 1972) or linear formation (Haslam, 1982b) or with the glucose ring ruptured (Okuda *et al.*, 1982a).

b. Hexahydroxydiphenoyl-Related Esters. The dehydrohexahydroxydiphenoyl esters based on **15** and those isohexahydroxydiphenoyl esters based on **16** provide a small group of structural variations of the more common ellagitannins (Haddock *et al.*, 1982a,b). Hydrolysis of esters of the iso-HHDP radical with hydrochloric acid yield ellagic acid, whereas esters containing the dehydro-HHDP radical (**16**) give chlorellagic acid.

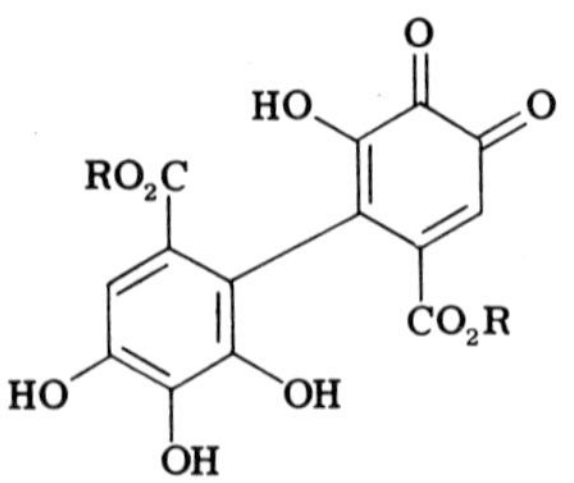

15

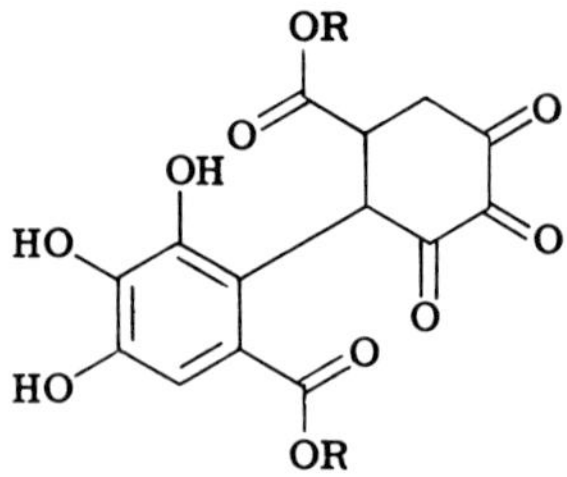

16

On base-catalyzed hydrolysis, both dehydro-HHDP and iso-HHDP give brevifolin carboxylic acid (**17**) and chebulic acid (**18**). Iso-HHDP acid is

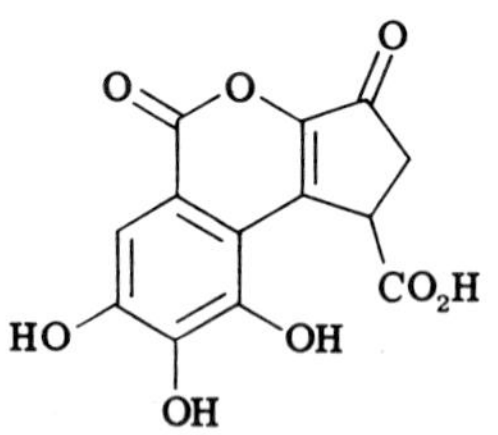

17

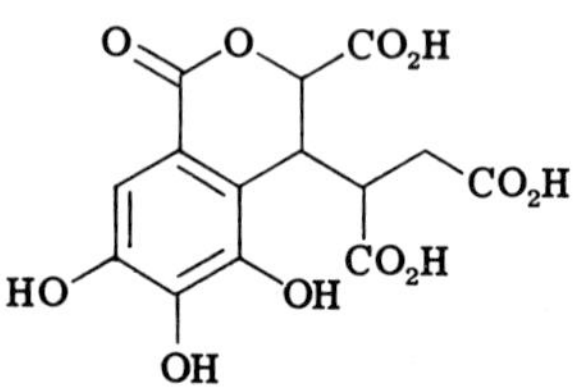

18

2,4 linked to β-1,3,6-tri-*O*-galloyl glucose in terchebin. Chebulic acid in the hydrated form is 2,4 linked to β-1,3,6-tri-*O*-galloyl glucose in chebulinic acid and to β-1-*O*-galloyl-3,6-HHDP-glucose in chebulagic acid (Haslam, 1982b). The nature of the conformation of the D-glucopyranose that has been esterified with HHDP, and of its derivatives, in the 1,6 or 2,4 or 3,5 positions is still under debate (Haddock *et al.*, 1982b).

Dehydrodigalloyl esters (**19**) can be considered as derived by a C—O coupling, as opposed to a C—C coupling, of two galloyl ester groups. A glucoside has been isolated, as well as another involving an intermolecular oxidative coupling of the same compound (Ozawa *et al.*, 1978, 1980).

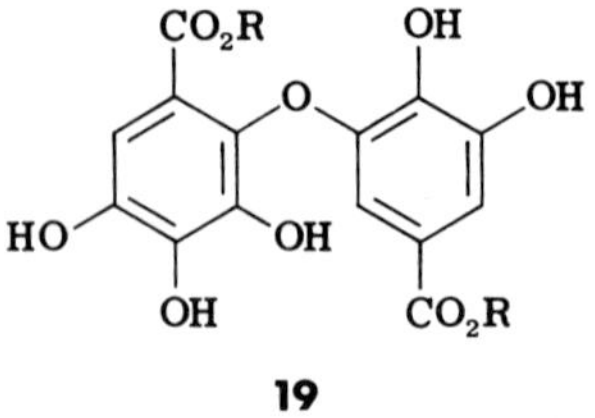

19

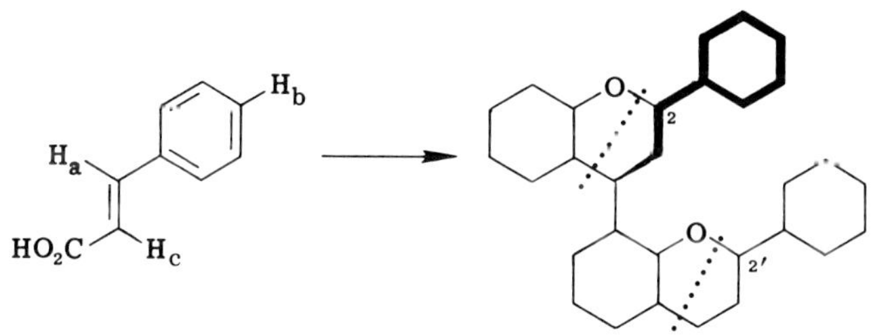

Fig. 5. Biosynthesis of procyanidins: the fate of isotopically labeled cinnamic acid is shown in the heavy outline.

IV. Biosynthesis of Tannins

A. *Proanthocyanidins*

Early proposals for the biosynthesis of proanthocyanidins arose from observations of the co-occurrence of related flavan-3-ols and flavan-3,4-diols (e.g., Grisebach, 1968). These have been supported or confirmed by administering labeled (^{14}C and ^{3}H) cinnamic acids to a number of procyanidin-forming plants (Jacques *et al.*, 1977). The major conclusions drawn from these studies included (1) the retention in both flavan units of the dimers of the proton H_a at C-2 and C-2′ (see Fig. 5), (2) the loss of the proton H_c from the cinnamic acid precursor, and (3) the differential extent of labeling obtained in the "upper" and "lower" flavan-3-ol units of the procyanidin dimers. The best explanation for the latter result is that the two "halves" of the procyanidin dimers are derived from different metabolic sources (Fig. 5) (Jacques *et al.*, 1977).

In a general scheme incorporating the results from the study of labeled compounds, Haslam and co-workers (Haslam, 1977; Jacques *et al.*, 1977; Haslam, 1982a) proposed a key intermediate for the biosynthesis of procyanidins as the flav-3-en-3-ol (**20**) or its biological equivalent. Furthermore, it was suggested a two-stage stereospecific reduction of this compound could lead to either of the two cis or trans stereoisomeric compounds (+)-catechin and (−)-epicatechin. A survey of several plant tissues (Foo and Porter, 1980) provides supporting evidence of the existence of such an intermediate. If the supply of a biological reductant such as NADPH is rate limiting, the intermediate C-4 carbocations (**21,** 3*S* or 3*R* absolute stereochemistry) might react with (+)-catechin or with (−)-epicatechin to produce dimers (**22**) or oligomers of procyanidins along the

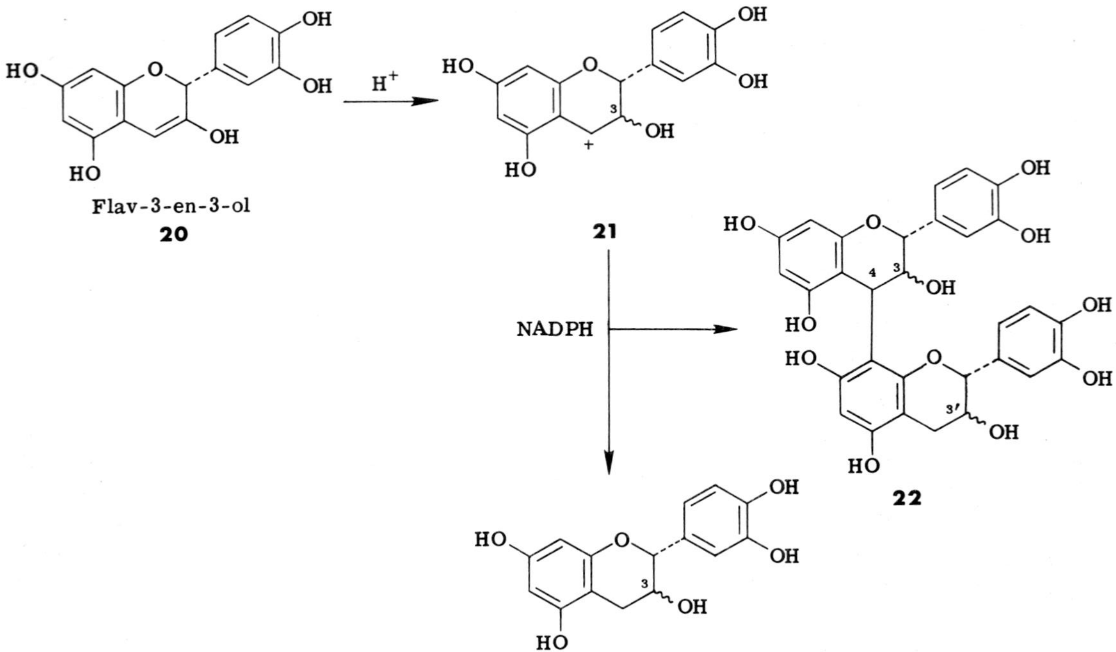

Flav-3-en-3-ol

20

21

22

(−)-3*R*-Epicatechin

(+)-3*S*-**Catechin**

lines suggested by Creasy and Swain (1965) and Geissman and Yoshimura (1966).

The nucleophilic character at C-8 or C-6 in (+)-catechin or (−)-epicatechin, or in both, could be utilized with one, or both, of the carbocations that are related in stereochemistry at C-2 and C-3. This linkage could result in the various patterns of the procyanidins. The reactions would be regulated by the stability of the carbocations arising from the flavan-3,4-diols. In addition, the electrophilic substitution reactions of the catechins would occur more readily at the C-8 than at the C-6 position. Consequently, the procyanidin polymers would be expected to have essentially linear structures with linkages via the C-4 of one flavan unit (the carbocation) and the C-8 of the next flavan. Substitution at nucleophilic centers is highly sensitive to steric factors, however, so that reactions occur preferentially at the least hindered sites, and coupled "angular" or "branched" trimers of C-5 deoxy flavans have been found in *Acacia mearnsii* and elsewhere (Roux and Ferreira, 1982). Even with the presence of a C-5 hydroxyl, linkages via the C-6 position (**23**) occur in some trimers from the phloem of *Pinus taeda* (Hemingway *et al.*, 1981a,b) and other sources (Nonaka *et al.*, 1981, 1982).

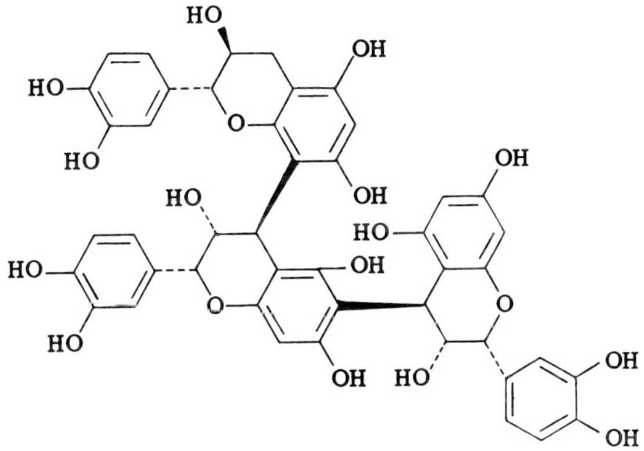

23

The synthesis of polymeric proanthocyanidins *in vitro* using dimers in media of controlled acidity (Haslam, 1982a; Roux and Ferreira, 1982) indicate mechanisms by which polymeric proanthocyanidins could be built up in nature under the control of biological reductants such as NADPH.

Chromatographic examination of plant extracts reveals a range of polymeric sizes of proanthocyanidins. The variations in the ratios of components in the extracts from similar tissues would probably be due to the changes in the metabolic conditions. If the supply of a biological reductant such as NADPH is low in relation to the rest of the metabolic processes, the more highly polymerized proanthocyanidins could result. It is assumed the higher oligomeric forms are essentially extensions of the simpler structures in which the molecular size is increased by the addition of further flavan units mainly by a (4→8) linkage.

B. *Hydrolyzable Tannins*

Gallic acid itself occurs in only very small amounts but exists in plants in a wide range of simple monoesters, such as β-D-glucogallin, to complex esters with D-glucose. As has been mentioned, various HHDP esters occur along with β-penta-*O*-gallate D-glucose (Schmidt and Mayer, 1956). Haslam (1982b; Haddock *et al.*, 1982a,b) has suggested the latter is the biogenetic precursor of the HHDP esters. This proposal is based on an oxidative coupling taking place preferentially between galloyl ester groups 2 and 3, and between 4 and 6 on the D-glucopyranose ring with both the precursor and product molecules adopting the C-1 conformation of the sugar ring.

However, not all the data fits this scheme. There are several reports of mono HHDP esters (e.g., 2,3; 4,6) that lack galloyl substituents. These two esters have been found in eucalypt wood (Seikel and Hillis, 1970) and *Eucalyptus viminalis* leaves, along with their 1-*O*-galloyl esters (Okuda *et al.*, 1982a). In other eucalypt leaves, the glucose is fully substituted by galloyl or HHDP groups (Haddock *et al.*, 1982a). Pure ellagitannins hydrolyze or change to other components on storage in water (Seikel and Hillis, 1970), and ester hydrolysis proceeds most rapidly at the anomeric center C-1 (Armitage *et al.*, 1961). It remains to be determined whether the galloyl substituents remaining from the putative pentagalloyl ester have been hydrolyzed during the long storage of the wood in the living tree or whether selective galloyl esters were formed in the first instance.

V. Relationship to the Biosynthesis and Biodegradation of Wood

Most evidence supports the view that the gallic acid (and its metabolites) of the hydrolyzable tannins arises from shikimic acid and the direct dehydrogenation of 3-dehydroshikimic acid (Conn and Swain, 1961; Cornthwaite and Haslam, 1965) and is linked to the central core of D-glucopyranose or very occasionally to a degradation product of it (Haslam, 1982b). Lignin also originates from the shikimic acid pathway and the CoA ester derivatives of substituted cinnamic acids but is not linked to a carbohydrate (see Chapter 10, Section IV). The mono- and oligomeric components of condensed tannins are derived from the shikimic acid pathway and the CoA derivatives of substituted cinnamic acids, but the additional benzene ring of this class arises from the citric acid (TCA) cycle and three molecules of malonyl-CoA (see Chapter 2); glycosides apparently do not exist (Hahlbrock and Grisebach, 1975).

There is little evidence of the mechanisms controlling the biosynthesis of one or more of these three major benzenoid classes involved in the formation of wood and the energy requirements for the different pathways. Diversion of the pathway from one class to another can take place abruptly, such as in the biosynthesis of gallotannins in the sapwood of *Rhus* spp. or the ellagitannins in the cambium of *Eucalyptus* spp., when their metabolic pathways are diverted to that of flavonoids or proanthocyanidins during the formation of heartwood or kino, respectively (Hillis, 1977). However, similar biological situations do not lead to similar biochemical changes. For example, the biosynthesis of ellagitannins in the sapwood of *E. marginata* changes abruptly to the biosynthesis of considerable amounts of proanthocyanidins at the heartwood periphery. On the other hand, with *E. regnans, E. delegatensis,* and similar eucalypts, there is a considerable increase in only the amount of the same or similar ellagitannins at the heartwood periphery.

Abrupt diversions in the formation of two subclasses of condensed tannins also occur. Just as with proanthocyanidins, stilbenes are biosynthesized from the shikimic acid pathway (Hillis and Hasegawa, 1962) and three acetate units, although the terminal carbon of the acetate chain is lost after cyclization to the benzene ring. In *Pinus radiata* sapwood wounded by *Sirex noctilio,* however, only stilbenes are formed, instead of the stilbene–flavone mixture of the heartwood (Hillis, 1977). Also, biochemical variants of a number of eucalypts possess leaves containing mostly stilbenes as well as flavonoid derivatives instead of the normal composition of flavonoids without stilbenes (Hillis, 1966), and similar diversions of pathways have been reported in the woods of other species (Hillis, 1977).

The lack of knowledge concerning the systems controlling the formation of lignins, hydrolyzable tannins, and condensed tannins extends further, to the complex and unresolved components of the tannins. These components form more than half of all extracts of wood tissues. Proposals on their possible roles in the formation of various wood tissues and on their biodegradation must await the collection of precise information of their constitution.

References

Armitage, R., Bayliss, G. S., Gramshaw, J. G., Haslam, E., Haworth, R. D., Jones, K., Rogers, H. J., and Searle, T. (1961). *J. Chem. Soc.* 1842–1853.
Bate-Smith, E. C. (1954). *Biochem. J.* **58,** 122–125.
Bate-Smith, E. C. (1956). *Chemistry & Industry. Brit. Inds. Fair. Rev.* April, R32–33.
Bate-Smith, E. C. (1962). *J. Linn. Soc. (Bot.)* **58,** 95–101.
Bate-Smith, E. C. (1973a). *Phytochemistry* **12,** 907–912.
Bate-Smith, E. C. (1973b). *Phytochemistry* **12,** 1809–1812.
Bate-Smith, E. C. (1975). *Phytochemistry* **14,** 1107–1113.
Bate-Smith, E. C. (1976). *Biochem. Syst. Ecol.* **4,** 13–23.
Bate-Smith, E. C. (1977). *Phytochemistry* **16,** 1421–1426.
Bate-Smith, E. C. (1978). *Phytochemistry* **17,** 267–270.
Bate-Smith, E. C., and Lerner, N. H. (1954). *Biochem. J.* **58,** 126–132.
Clarke-Lewis, J. L., and Mortimer, P. I. (1960). *J. Chem. Soc.* 4106–4112.
Conn, E. E., and Swain, T. (1961). *Chem. Ind.* 592–593.
Cornthwaite, D. C., and Haslam, E. (1965). *J. Chem. Soc.* 3008–3011.
Creasy, L. L., and Swain, T. (1965). *Nature (London)* **208,** 151–153.
Czochanska, Z., Foo, L. Y., Newman, R. H., Porter, L. J., Thomas, W. A., and Jones, W. T. (1979). *J. Chem. Soc., Chem. Commun.* 375–377.
Czochanska, Z., Foo, L. Y., Newman, R. H., and Porter, L. J. (1980). *J. Chem. Soc., Perkin Trans. I.* 2278–2286.
Fletcher, A. C., Porter, L. J., Haslam, E., and Gupta, R. K. (1977). *J. Chem. Soc., Perkin Trans. I* 1628–1637.
Foo, L. Y. (1981). *Phytochemistry* **20,** 1397–1402.
Foo, L. Y. (1982). *Phytochemistry* **21,** 1741–1746.
Foo, L. Y., and Porter, L. J. (1980). *J. Chem. Soc., Perkin Trans I,* 2278–2281.
Geissman, T. A., and Yoshimura, N. N. (1966). *Tetrahedron Lett.* 2669–2671.
Gorham, J. (1980). *Prog. Phytochem.* **6,** 203–252.
Gnamm, H. (1949). "Die Gerbstoffe u. Gerbmittel." Wissenschaftliche, Verlagsgesellschaft M.B.H. Stuttgart.
Grassman, W. E., and Endres, H. (1959). *Leder* **10,** 237–244.
Grisebach, H. (1968). *In* "Recent Advances in Phytochemistry" (T. Mabry, V. C. Runeckles, and R. E. Alston, eds.), Vol. 1, pp. 379–413. Appleton, New York.
Gupta, R. K., and Haslam, E. (1978). *J. Chem. Soc., Perkin Trans. I* 892–896.
Haddock, E. A., Gupta, R. K., Al-Shafi, S. M. K., Layden, K., Haslam, E., and Magnolato, D. (1982a). *Phytochemistry* **21,** 1049–1062.
Haddock, E. A., Gupta, R. K., and Haslam, E. (1982b). *J. Chem. Soc., Perkin Trans I,* 2535–2545.
Hahlbrock, K., and Grisebach, H. (1975). *In* "The Flavonoids" (J. B. Harborne, T. J. Mabry, and H. Mabry, eds.) pp. 866–915. Chapman and Hall, London.
Haslam, E. (1966). "Chemistry of Vegetable Tannins." Academic Press, London.
Haslam, E. (1975). *In* "The Flavonoids" (J. B. Harborne, T. Mabry, and H. Mabry, eds.) pp. 505–559. Chapman and Hall, London.
Haslam, E. (1977). *Phytochemistry* **16,** 1625–1640.
Haslam, E. (1982a). *In* "The Flavonoids: Advances in Research" (J. B. Harborne and T. J. Mabry, eds.) pp. 427–448. Chapman and Hall, London.
Haslam, E. (1982b). *Prog. Chem. Org. Nat. Prod.* **41,** 1–46.
Hathway, D. E. (1962). *In* "Wood Extractives" (W. E. Hillis, ed.) pp. 191–228. Academic Press, London.

Hathway, D. E., and Seakins, J. W. (1959). *Biochem. J.* **72,** 369–374.

Hemingway, R. W., Foo, L. Y., and Porter, L. J. (1981a). *J. Chem. Soc., Chem. Commun.* 320–322.

Hemingway, R. W., Foo, L. Y., and Porter, L. J. (1981b). *J. Chem. Soc., Chem. Commun.* 1209–1216.

Hillis, W. E. (1954). *J. Soc. Leather Trades Chem.* **38,** 91–102.

Hillis, W. E. (1956). *Empire For. Rev.* **35,** 420–436.

Hillis, W. E. (1966). *Phytochemistry* **5,** 541–556.

Hillis, W. E. (1969). *APPITA,* **23,** 89–101.

Hillis, W. E. (1972). *Phytochemistry* **11,** 1207–1218.

Hillis, W. E. (1977). *In* "Recent Advances in Phytochemistry" (F. A. Loewus and V. C. Runeckles, eds.), Vol. 11, pp. 247–309. Plenum, New York.

Hillis, W. E., and Carle, A. (1960). *Aust. J. Chem.* **13,** 390–395 and unpublished results.

Hillis, W. E., and Hasegawa, M. (1962). *Chem. Ind. (London)* 1300–1301.

Hillis, W. E., and Yazaki, Y. (1972). *Phytochemistry* **12**, 2963–2968.

Jacques, D., Opie, C. T., Porter, L. J., and Haslam, E. (1977). *J. Chem. Soc., Perkin I,* 1637–1643.

McGraw, G. W., and Hemingway, R. W. (1982). *J. Chem. Soc., Perkin Trans I,* 973–978.

Mayer, W. (1973). *Justus Liebigs Ann. Chem.* 1759–1776.

Nonaka, G., Hsu, F., and Nishioka, I. (1981). *J. Chem. Soc., Chem. Commun.* 781–783.

Nonaka, G., Kawahara, O., and Nishioka, I. (1982). *Chem. Pharm. Bull.* **30,** 4277–4282.

Okuda, T., Yoshida, T., Hatano, T., Yazaki, K., and Ashida, M. (1982a). *Phytochemistry* **21,** 2871–2874.

Okuda, T., Yoshida, T., Kuwahara, M., Memon, M., Shingo, T. (1982b). *J. Chem. Soc., Chem. Commun.* 163–164.

Ozawa, T., Arai, N., and Takino, Y. (1978). *Agric. Biol. Chem.* **42,** 1907.

Ozawa, T., Odaira, Y., Imagawa, H., and Takino, Y. (1980). *Agric. Biol. Chem.* **44,** 581.

Perkin, A. G., and Everest, A. E. (1918). "The Natural Organic Colouring Matters." Longmans, Green and Co., London.

Pigman, W., Anderson, E., Fischer, R., Buchanan, M. R., and Browning, B. L. (1953). *TAPPI* **36,** 4–12.

Porter, L. J., and Foo, L. Y. (1978). *J. Chem. Soc., Perkin Trans. I,* 1186–1191.

Porter, L. J., Czochanska, Z., Foo, L. Y., Newman, R. H., Thomas, W. T., and Jones, W. T. (1979). *J. Chem. Soc., Chem. Commun.* 375–377.

Roux, D. G., and Ferreira, D. (1982). *Prog. Chem. Org. Nat. Prod.* **41,** 47–76.

Roux, D. G., Engel, D. W., Hattingh, M., and Hundt, H. K. L. (1978). *J. Chem. Soc., Chem. Commun.* 695–697.

Samejima, M., and Yoshimoto, T. (1982). *Mokuzai Gakkaishi* **28,** 67–74.

Sands, R. (1975). *N. Z. J. For. Sci.* **5**(1), 74–86.

Schmidt, O. T., and Mayer, W. (1956). *Angew, Chem.* **68,** 103–115.

Seikel, M. K., and Hillis, W. E. (1970). *Phytochemistry* **9,** 1115–1128.

Swain, T., and Hillis, W. E. (1959). *J. Sci. Food Agric.* **10,** 63–68.

Sykes, R. L., and Roux, D. G. (1957). *J. Soc. Leather Trades' Chem.* **41,** 14–22.

White, T. (1958). *In* "The Chemistry and Technology of Leather" (F. O'Flaherty, W. T. Roddy, and R. M. Lollar, eds.), Vol. II, pp. 111–160. Reinhold, New York.

Yazaki, Y., and Hillis, W. E. (1977). *Holzforschung* **31,** 20–25.

Yazaki, Y., and Hillis, W. E. (1980). *Holzforschung* **34,** 125–130.

CHAPTER **14**

Biosynthesis of Stilbenes

H. KINDL

Fachbereich Chemie
Philipps-Universität Marburg
Marburg, Federal Republic of Germany

I. Introduction

Stilbenes are constituents of wood and lignified tissues. The naturally occurring stilbenes possess phenolic hydroxyl groups and exhibit typical substitution patterns, which are, in terms of biosynthesis, the distinguishing feature of different pathways. Pinosylvin, as an example of simple hydroxylation pattern, is a constituent of the heartwood of pine species

Biosynthesis and Biodegradation of Wood Components

ISBN 0-12-347880-4

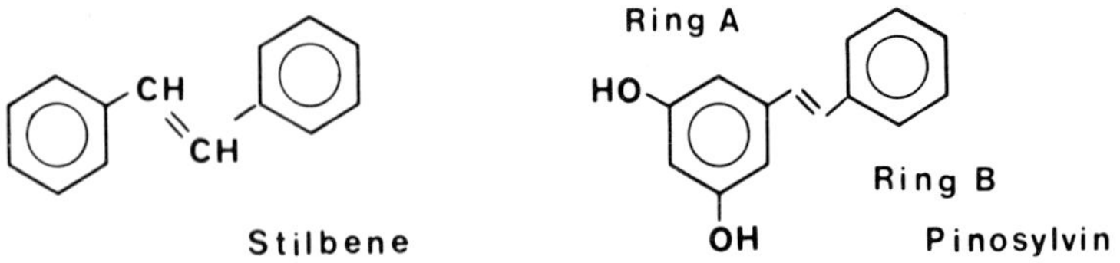

Fig. 1. Stilbene skeleton and one instance of a hydroxy derivative. The ring that bears hydroxyl groups at positions 3 and 5 is marked as ring A.

(Fig. 1). It is an almost unique compound that it can be utilized as taxonomic marker of the genus *Pinus* (see Chapter 9).

In recent years, stilbenes in plants became interesting not only as constituents. Their physiological role has attracted more and more attention, especially when stilbenes were seen in the context of the concept of phytoalexins and decay resistance. Pinosylvin is an example also in this respect: its biosynthesis as a phytoalexin is induced in young seedlings, and its role in older trees is multiple as a toxin and repellent.

Major reviews on stilbenes, also covering the aspect of biosynthesis, were written by Billek (1964) and more recently by Gorham (1980). This chapter goes beyond those in emphasizing the biochemical and enzymological aspects currently under investigation. As it has become evident that dihydrophenanthrenes, bibenzyls, and oligomers are closely related to stilbenes and occur in lignified tissues, these biosynthetic sequences will also be included.

II. The Principal Biosynthetic Sequence as Revealed by Feeding Experiments

In the early 1960s, biosynthetic studies were performed, revealing the origin of the two aromatic rings and suggesting the principles of the biosynthetic route leading to stilbenes. As an example for the biosynthesis of other stilbene derivatives, pinosylvin synthesis in wood (Billek and Ziegler, 1962; Rudloff and Jorgensen, 1963) and the biosynthesis of a stilbene carboxylic acid, hydrangeic acid (Billek and Kindl, 1962; Ibrahim and Towers, 1962), were thoroughly elucidated by feeding experiments and subsequent selective degradation of the product. After testing radioactive acetate and different aromatic compounds (e.g., L-phenylalanine) in terms of efficiency of incorporation into hydrangeic acid (Fig. 2), selectively labeled precursors—e.g., [1-^{14}C]acetate and [2-^{14}C]*p*-hydroxycinnamic acid—were administered to *Hydrangea macrophylla*. Hydrangeic acid was isolated and subjected to selective degradation (Fig. 2) in order to localize the label within single carbon atoms of the stilbene molecule.

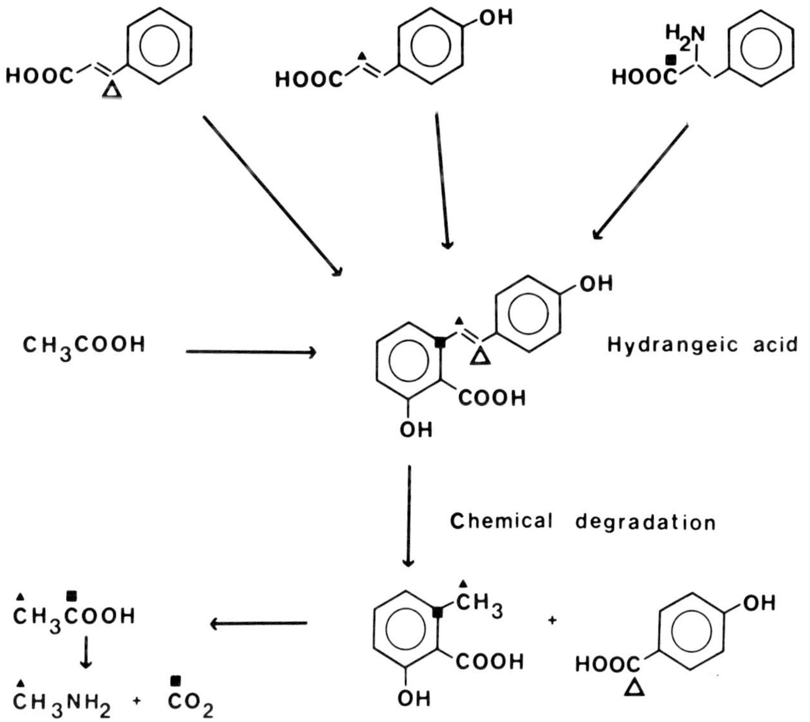

Fig. 2. Selective degradation of hydrangeic acid biosynthesized from various selectively labeled precursors. For example, [3-^{14}C]cinnamic acid gave rise to hydrangeic acid labeled in Cβ. Labeling of hydrangeic acid with acetate was restricted to ring A, and radioactivity of [*ring*-^{14}C]phenylalanine was localized in ring B of the stilbene carboxylic acid. A selective chemical degradation is required to localize the radioactivity in the single carbon atoms. A few of the degradation steps are shown here.

By these findings it became apparent that the biosynthetic pathway must proceed from phenylalanine via cinnamic acids and must then include the condensation of a highly reactive cinnamic acid derivative (Fig. 3) with acetate units. In analogy to the role of malonyl-CoA in fatty acid synthesis and in the formation of 6-methylsalicylic acid, at that time just becoming the object of enzymatic studies (Lynen and Tada, 1961), the acetate unit required for stilbene formation was expected also to be malonyl-CoA.

The scheme depicted in Fig. 3 implies the aromatization of a polyketide moiety giving rise to the formation of a 3,5-dihydroxyphenyl group. Therefore, the substitution pattern with a 3,5-dihydroxyphenyl group (if the carboxyl group is removed subsequently) or a 3-hydroxy-2-carboxylic

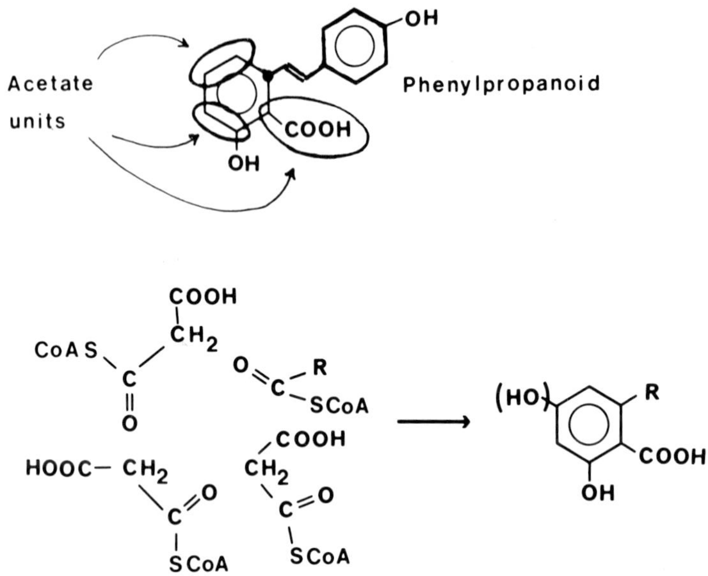

Fig. 3. Phenylpropane unit and acetate units as building blocks; condensation of a starter CoA ester with malonyl-CoA.

acid pattern (if a reducing step is included) can be utilized to assign, in a formula for a naturally occurring stilbene, the ring that is derived from the acetate units.

The aromatic ring B and the three carbon atoms adhering to the ring (including the ethene bridge) have been found to originate from phenylpropanoids (C_6C_3 bodies), either from cinnamic acid or *p*-coumaric acid (Fig. 2). According to the many investigations concerning the transformations at the level of phenylpropanoids, we know that phenylalanine must be converted to cinnamic acid by L-phenylalanine ammonia-lyase and that cinnamic acid can be hydroxylated to *p*-hydroxycinnamic acid (*p*-coumaric acid) (see Chapter 10, Section II). Hydroxylation of cinnamic acid in the para position takes place at the endoplasmic reticulum (ER), and this monooxygenase reaction is tightly coupled with the ammonia-lyase. Thus, a membrane-bound enzyme system channels phenylalanine to *p*-coumaric acid, irrespective of other phenylalanine ammonia-lyase pools in the cell (Kindl, 1979; Czichi and Kindl, 1977).

Cinnamic acid and *p*-coumaric acid, respectively, are substrates in CoA-ligase reactions. Figure 4 summarizes the enzymatic steps common to modification of C_6C_3 bodies and leading thus to the immediate precursors of stilbene formations.

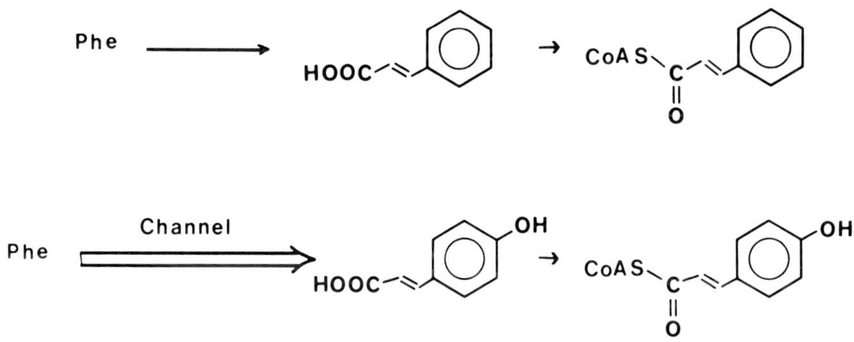

Fig. 4. Formation of cinnamic acids from phenylalanine and their conversion into cinnamoyl-CoA esters, the precursors of stilbenes.

III. Naturally Occurring Stilbenes Grouped According to Biosynthetic Aspects

Stilbenes have different patterns of hydroxyl groups, and the mechanisms underlying this phenomenon are the following. First, the hydroxylation pattern in ring B is frequently preformed at the level of phenylpropanoids. This applies to monohydroxylated derivatives. Stilbenes originate either from *o*-, *m*-, or *p*-hydroxylated or from nonhydroxylated cinnamic acids. Second, deviation from standard hydroxylation pattern in ring A occurs when reduction during the course of polyketide formation furnishes a monohydroxyl pattern in ring A of stilbenes (the part of the molecule that derives from malonyl units).

The former case, existence or lack of a 4′-hydroxyl group (at ring B) in the stilbene, is illustrated (Fig. 5) by the conversion of cinnamic acid or *p*-coumaric acid to the respective stilbenes.

Another aspect of distinguishing stilbenes according to biosynthetic mechanisms emphasizes the fact that in a few cases the carboxyl group of a postulated intermediate, a polyketo acid, remains bound to the condensation product. Stilbene-2-carboxylic acids (Fig. 6) are thus obtained. Irrespective of the primary products of stilbene synthesis, secondary hydroxylation ortho to the already existing OH group affords further variations. This holds for ring A as well as ring B.

Recent investigations on the biosynthesis of 9,10-dihydrophenanthrenes and bibenzyls revealed that we must also discriminate between pathways leading to stilbenes and the ones furnishing bibenzyls (dihydrostilbenes).

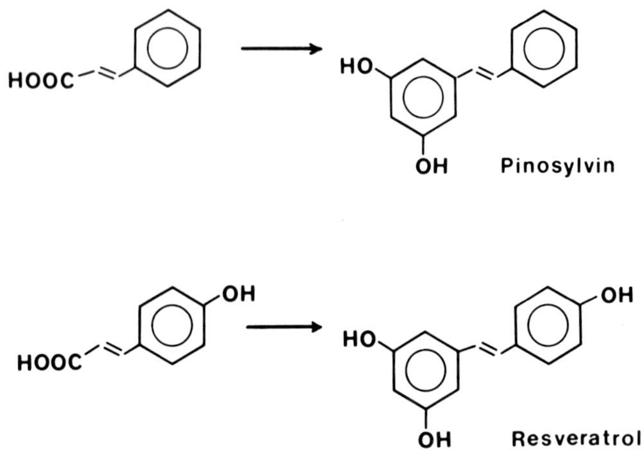

Fig. 5. Formation of pinosylvin and resveratrol.

A. *Pinosylvin*

Pinosylvin and its methyl ethers represent constituents of pine. Erdtman (1939, 1955, 1959) compared the amounts of pinosylvin and its derivatives present in various gymnosperms, of which some were very closely related. All pinosylvins were found to be confined to the genus *Pinus*.

The *p*-hydroxylated compound (at ring B), resveratrol, is much more

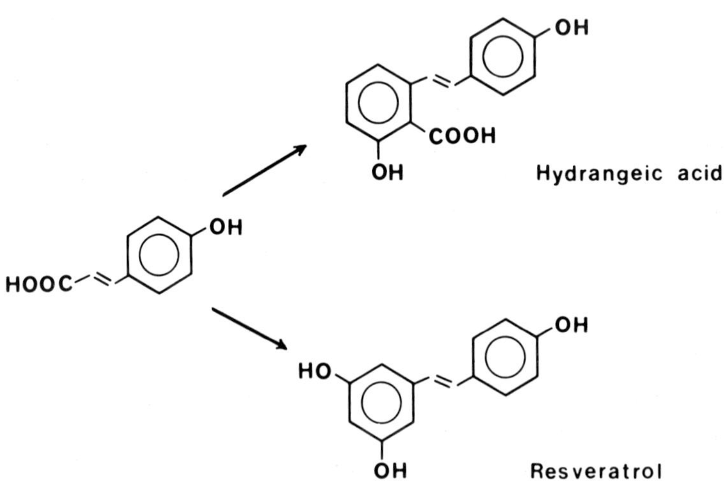

Fig. 6. Biosynthesis of hydrangeic acid compared to the formation of resveratrol.

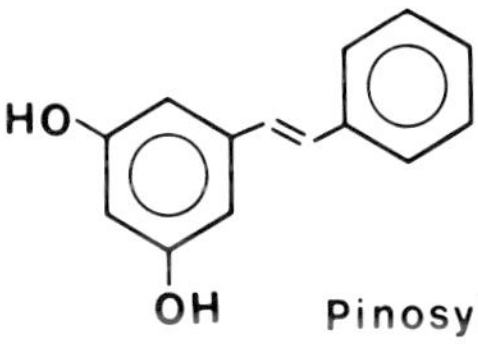

Pinosylvin

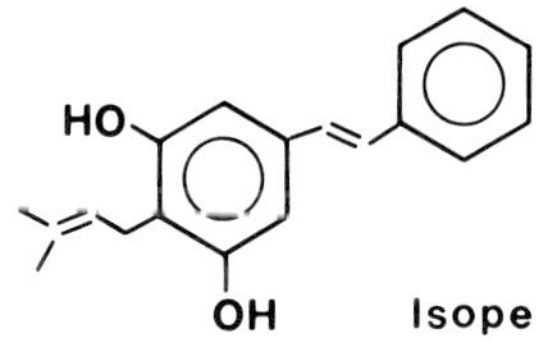

Isopentenylpinosylvin

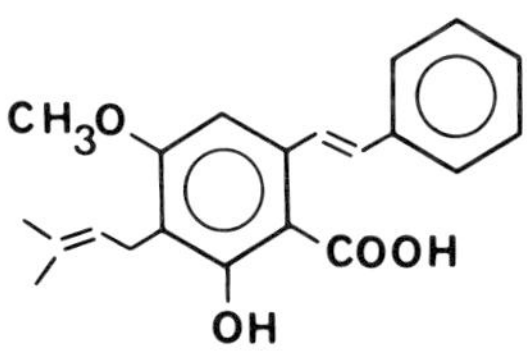

Phytoalexin from *Cajanus*

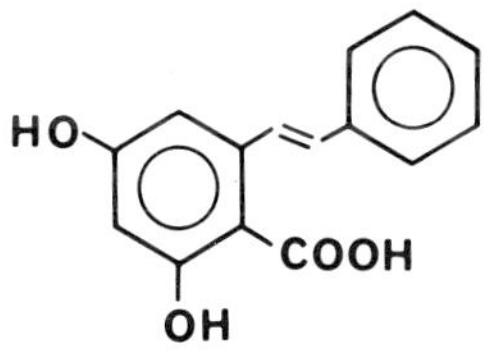

Aglycone of gaylussacin

Fig. 7. Stilbenes and stilbene carboxylic acids lacking hydroxyl groups at ring B.

widely distributed among gymnosperms and occurs, for example, in *Picea* and *Eucalyptus*. Like pine, *Picea* and *Eucalyptus* are used for the production of paper and other cellulose products. Paper mills can turn out, besides pulp, lignin wastes with high amounts of stilbenes. The subgenus *Haploxylon* contains pinosylvin ethers and 3-hydroxy-5-methoxybibenzyl. Outside the genus *Pinus*, pinosylvin derivatives are only rarely found: pinosylvin in *Alnus* spp. (Suga *et al.*, 1972) and *Nothofagus* (4 species) (Hillis and Orman, 1962; Hillis and Inoue, 1967) and stilbene-2-carboxylic acid in *Gaylussacia* spp. (Askari *et al.*, 1972). A compound that may be important in terms of fungistatic function is 4-isopentenylpinosylvin, as yet found only in *Derris* spp. (Filho *et al.*, 1975). 4-Isopentenylpinosylvin (Fig. 7) is a rather hydrophobic phenol that might be placed within the group of potent phytoalexins. Such a correlation was demonstrated for a related compound prepared from *Cajanus cajan* (Cooksey *et al.*, 1982).

B. *Resveratrol*

Resveratrol is the most widespread stilbene in nature. It is found in most species of *Picea* (Kariyone *et al.*, 1958; Kariyone *et al.*, 1959; Takahashi *et al.*, 1960) and even in species of *Pinus* that are already distinguished by the occurrence of pinosylvin. Angiosperms also frequently contain resveratrol; families such as Leguminosae (Ingham, 1976; Späth

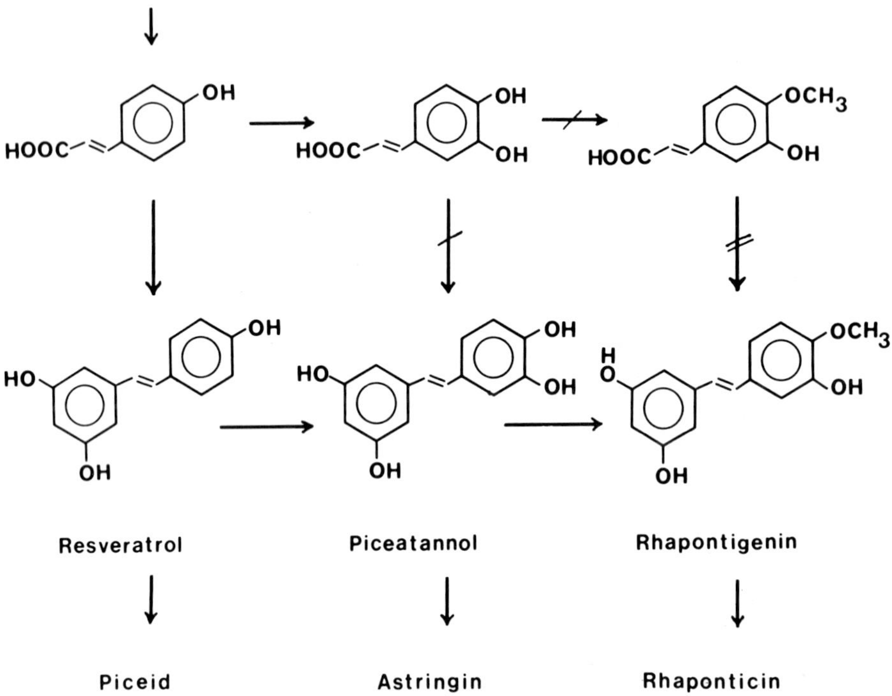

Fig. 8. Alternative pathways affording rhapontigenin and rhaponticin. Feeding experiments established that the conversion of *p*-coumaric acid into resveratrol is the step by which the stilbene skeleton really is formed. Other conversions, although imaginable in analogy to known reactions, could *not* be observed: pinosylvin → resveratrol, isoferulic acid → rhapontigenin. The last row shows the cognate glucosides.

and Schläger, 1940), Myrtaceae (Hillis *et al.*, 1974; Hillis and Carle, 1962; Hillis, 1966), and Vitaceae (Langcake and Pryce, 1976; Blaich and Bachmann, 1980; Fritzemeier and Kindl, 1981) are the thoroughly examined instances, with resveratrol being found in more than one genus. As resveratrol is the biosynthetic precursor of several polyhydroxylated stilbenes, we can propose that in other species and families, characterized by the occurrence of, e.g., piceatannol (3,3′,4′,5-tetrahydroxystilbene) (King *et al.*, 1956; Hillis and Ishikura, 1970), resveratrol must be present. The tetrahydroxystilbene is a primary phenolic constituent of the heartwood of *Vouacapoua* spp. (King *et al.*, 1956). Tetrahydroxy- and pentahydroxystilbenes originating from resveratrol seem to be constituents of many tropical trees. Competent scientists call the rain forest the world's largest reservoir of genetic traits, now largely unexploited. Breeders may some day need to fall back on wild trees to resist epidemics, and diseases

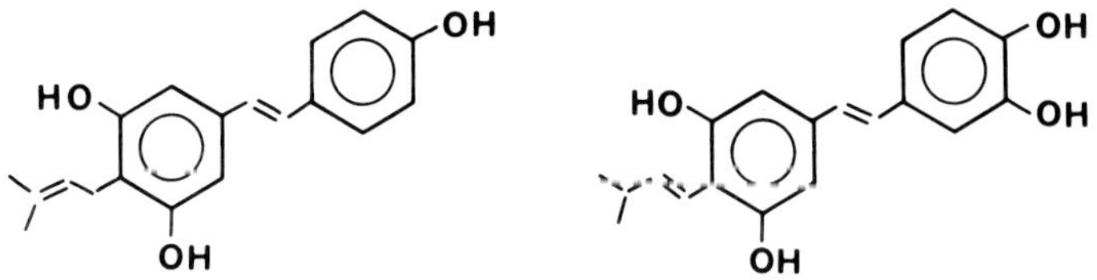

Fig. 9. Phytoalexins from *Arachis hypogaea.*

caused by fungi or insect infestations. This may also hold true for stilbenes.

Beside piceatannol, another derivative of resveratrol, namely, piceid (3-D-β-glucopyranosid of resveratrol), belongs to the widely distributed stilbenes. Its biosynthesis was studied by Hillis and Hasegawa (1962); it was one of the first examples by which the origin of ring B of stilbenes from phenylalanine was demonstrated. The conversion of resveratrol into piceatannol (this and astringenin are identical) was studied with the biosynthesis of rhapontigenin (Rupprich *et al.*, 1980) and rhaponticin (Fig. 8). Comprehensive studies corroborated that primary hydroxylation (Kindl, 1971a) had to take place at the level of cinnamic acids, while the introduction of a second hydroxyl in ring B (piceatannol, astringenin) is effected after formation of the stilbene skeleton (Rupprich *et al.*, 1980).

Resveratrol can give rise to the formation of hydrophobic polyphenols that seem to be well suited as fungistatic phytoalexins. In Leguminosae (e.g., *Arachis hypogaea*), several C-alkylated isoprene derivatives (Fig. 9) have been demonstrated to be synthesized upon fungal attack (Aguamah *et al.*, 1981; Keen and Ingham, 1976).

Derivatives of resveratrol were found (Langcake and Pryce, 1977a,b) as a class of physiologically very interesting oligomers, the viniferins. In a mechanism resembling in many aspects the dehydrogenation of coniferyl alcohol to lignin (see Chapter 7, Section III), resveratrol can form dimers or trimers by reaction between an oxygen radical at the OH group of ring

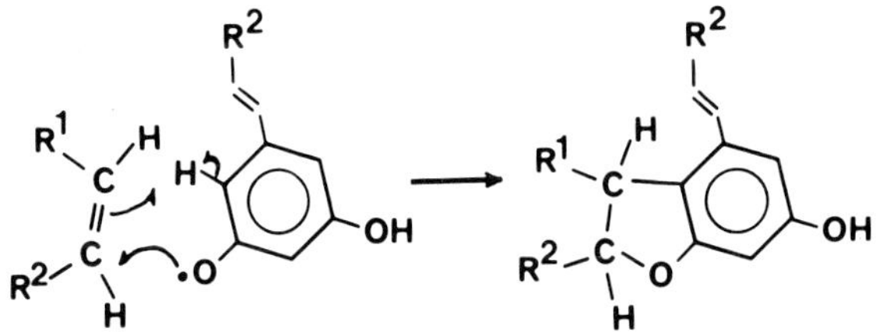

Fig. 10. Dehydrogenative oligomerization.

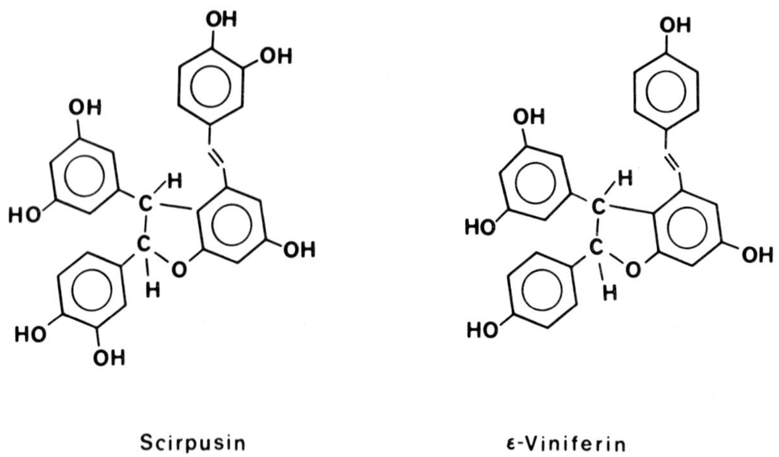

Fig. 11. Dimeric stilbenes.

A of one stilbene with the ethene double bond of another stilbene molecule. The principle is outlined in Fig. 10.

If piceatannol (3,3′,4′,5-tetrahydroxystilbene) or resveratrol (3,4′,5-trihydroxystilbene) are dimerized according to the mechanism outlined in Fig. 10, scirpusin (Nakajima *et al.*, 1978) or ε-viniferin (Langcake and Pryce, 1977a,b), respectively, are obtained (Fig. 11). Other viniferins are higher oligomers of resveratrol. Hopeaphenol (Coggon *et al.*, 1965), discovered in extracts of *Hopea odorata* and other Dipterocarpaceae, represents a tetramer of resveratrol and differs from α-viniferin (trimer) in exhibiting an additional type of C—C bond.

C. *Hydrangeic Acid*

Hydrangeic acid, formed from *p*-coumaric acid as is resveratrol, differs in the extra carboxylic group at ring A. This group is preserved in biosynthesis, during ring closure of the polyketoacid derivative and aromatization of ring A. The other difference compared to resveratrol formation is that, during formation of ring A, a reducing step is involved prior to aromatization. This leads then to a monohydroxyl pattern at ring A (Fig. 12). Hydrangeic acid binds water, and the β-hydroxy compound forms a lactone, hydrangenol.

Phyllodulcin carries the 3′-hydroxy-4′-methoxy pattern ("isovanillin structure") which is infrequent in nature. It is also carried by the stilbene rhapontigenin.

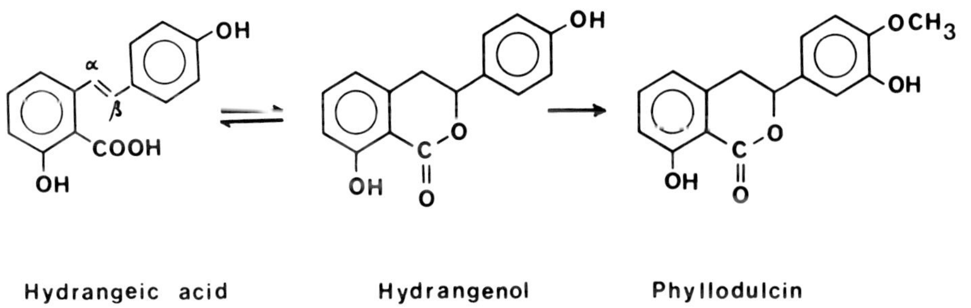

Fig. 12. Formation of a 3-aryl-3,4-dihydroisocoumarin from stilbene-2-carboxylic acid, and its further modification by hydroxylation and methylation.

D. Bibenzyls and 9,10-Dihydrophenanthrenes

In several woody plants, stilbenes are accompanied by dihydrostilbenes (bibenzyls). However, we also know of many plants that are distinguished by the exclusive content of groups of bibenzyls. They are thought to be biosynthetically related to stilbenes, but that has been investigated in very few cases only. Dihydropinosylvin is a common constituent of *Pinus* spp., and its 4-isoprenyl derivatives were characterized as antimicrobial agents (Bohlmann and Hoffmann, 1979; Mitscher *et al.,* 1983). Compretaceae (Letcher *et al.,* 1972; Letcher and Nhamo, 1972) contain bibenzyls in their heartwood that are supposed to be related to *p*-hydroxyphenylpropionic acid (dihydro-*p*-coumaric acid, Fig. 13).

While the 4′-hydroxy structure is a general feature of the *Combretum* bibenzyls, ring A is characterized by an additional hydroxyl group in position 4. Various substitution patterns are created by partial methylation of the 3 possible positions at ring A. These more hydrophobic, partially methylated phenols again seem to have physiological functions. They may be relevant in young seedlings and are implicated, in the heartwood of these trees, in the resistance of termite attack. Bibenzyls in Combretaceae are frequently accompanied by phenanthrenes, which may be products of bibenzyl metabolism.

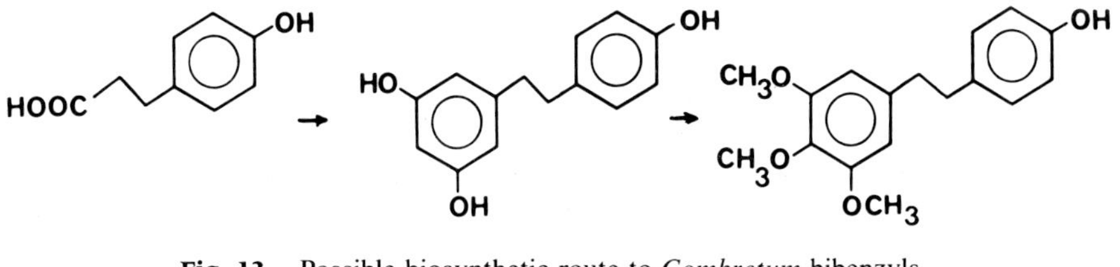

Fig. 13. Possible biosynthetic route to *Combretum* bibenzyls.

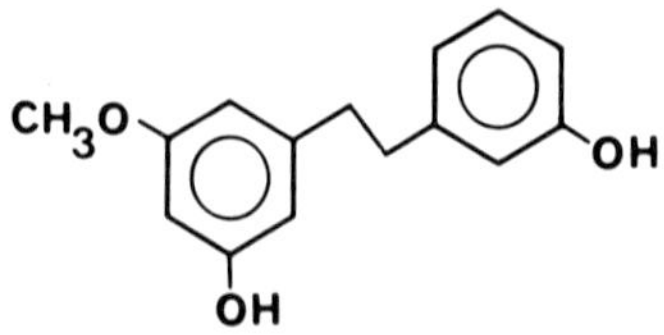

Batatasin III

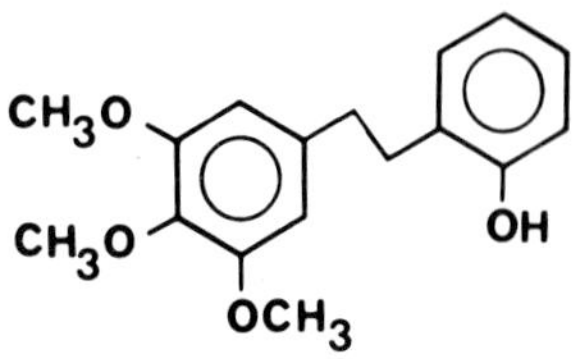

Batatasin V

Fig. 14. Examples of bibenzyls occurring in yam (*Discorea batatas*).

Bibenzyl carrying a 4′-hydroxyl group have been described to occur in *Cannabis sativa* and *Morus* species. 3,4′,5-Trihydroxybibenzyl and 3,3′-dihydroxy-5,4′-dimethoxy-2′-isopentenylbibenzyl (canniprene) were shown (L. Crombie and H. Kindl, unpublished) to be synthesized in *C. sativa* from *p*-hydroxyphenylpropionic acid (similar to the first reaction shown in Fig. 13).

A group of particularly interesting bibenzyls have been found in *Dioscorea* (Hashimoto *et al.*, 1974; Hashimoto *et al.*, 1968): batatasins (Fig. 14). They are distinguished by bearing *m*-hydroxy or *o*-hydroxy groups at ring B, which have been shown to originate from the phenyl ring of *o*- or *m*-hydroxyphenylpropionic acid, respectively (Fritzemeier *et al.*, 1984). Batatasins were described as dormancy hormones in the yam bulbs (Hasegawa and Hashimoto, 1975). The phytoalexin hircinol (2,5-dihydroxy-4-methoxy-9,10-dihydrophenanthrene), first described to be synthesized in orchid bulbs, was detected as preformed, antifungal agent in the peel of brazilian yam (*Dioscorea rotundata*) (Coxon *et al.*, 1982).

While *o*-coumaric acid and its dihydro derivative melilotic acid are well-known plant constituents whose biosyntheses have been established *in vivo* and *in vitro* (Kindl, 1971b; Gestetner and Conn, 1974; Poulton *et al.*, 1980), the conversions of cinnamic acid to *m*-hydroxycinnamic acid and *m*-hydroxyphenylpropionic acid (Fig. 15) have been newly discovered (Fritzemeier and Kindl, 1983; Fritzemeier *et al.*, 1984) and are as yet only poorly understood reactions.

Beside bibenzyls, bibenzyl carboxylic acids also occur in plants. The fern genus *Notholaena* contains 3-hydroxy-4′,5-dimethoxybibenzyl-2-carboxylic acid (Wollenweber and Favre-Bonvin, 1979).

The same 3,3′,5-trihydroxybibenzyl that is a precursor of batatasin III was proved to be an intermediate in the synthesis of several 9,10-dihydrophenanthrenes (Fritzemeier and Kindl, 1983), highly fungistatic phy-

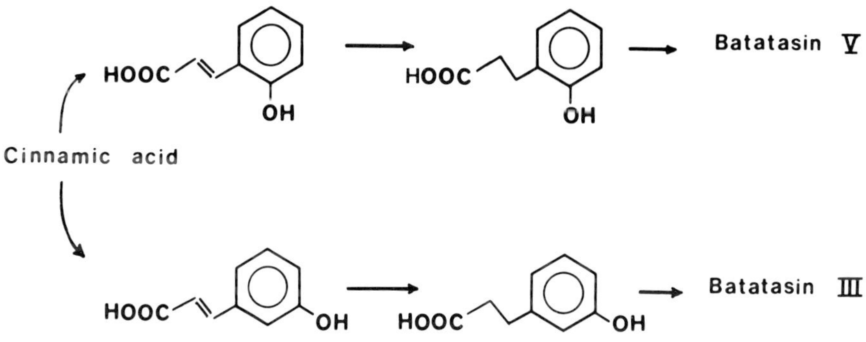

Fig. 15. Dihydrocinnamic acids as precursors of bibenzyls.

toalexins produced by orchids (Gäumann, 1963; Stoessl, 1982; Hardegger *et al.*, 1963). Heartwood, too, can contain 9,10-dihydrophenanthrenes. In *Cassia gorretiana* a 2,3,5,7-tetrahydroxy derivative was discovered (Hata *et al.*, 1979).

On the basis of other hypotheses, Stoessl (1982) proposed a pathway to dihydrophenanthrenes implicating *o*-coumaric acid or *p*-coumaric acid as precursors of bibenzyls, which are further converted to spiro compounds and 9,10-dihydrophenanthrenes (Fig. 16). Such a route was highly likely as spiro compounds, bibenzyls, and dihydrophenanthrenes can co-occur.

Recent studies with various orchids ruled out *p*-coumaric acid and *o*-coumaric acid as precursors. Instead, dihydro-*m*-coumaric acid and 3,3′,5-trihydroxybibenzyl were established (Fritzemeier and Kindl, 1983) as intermediates in the synthesis of orchid phytoalexins (Fig. 17).

Present data suggest that *m*-hydroxyphenylpropionic acid can be considered as the precursor of many naturally occurring 9,10-dihydrophenanthrenes except aristolochia acid and few alkylated phenanthrenes (juncusol, micrandol). Spiro compounds as formulated in Fig. 16 were found in *Cannabis* species (Crombie and Crombie, 1982) and may turn out as precursors of 7-hydroxylated dihydrophenanthrenes.

Phenanthrenes with a 2,4,6,7-tetrahydroxy or 2,3,4,7,8-pentahydroxy pattern were discovered in another Dioscoreaceae (*Tamus communis*, black bryony). At present it is not known whether phenanthrenes derive from 9,10-dihydrophenanthrenes by oxidation or originate from stilbenes. The close taxonomic relationship between the genera *Dioscorea* and *Tamus* suggests that bibenzyls and phenanthrenes are also related in terms of biosynthesis, in as much as also some batatas comprise bibenzyls together with phenanthrenes (Fig. 18). Evidences have been provided

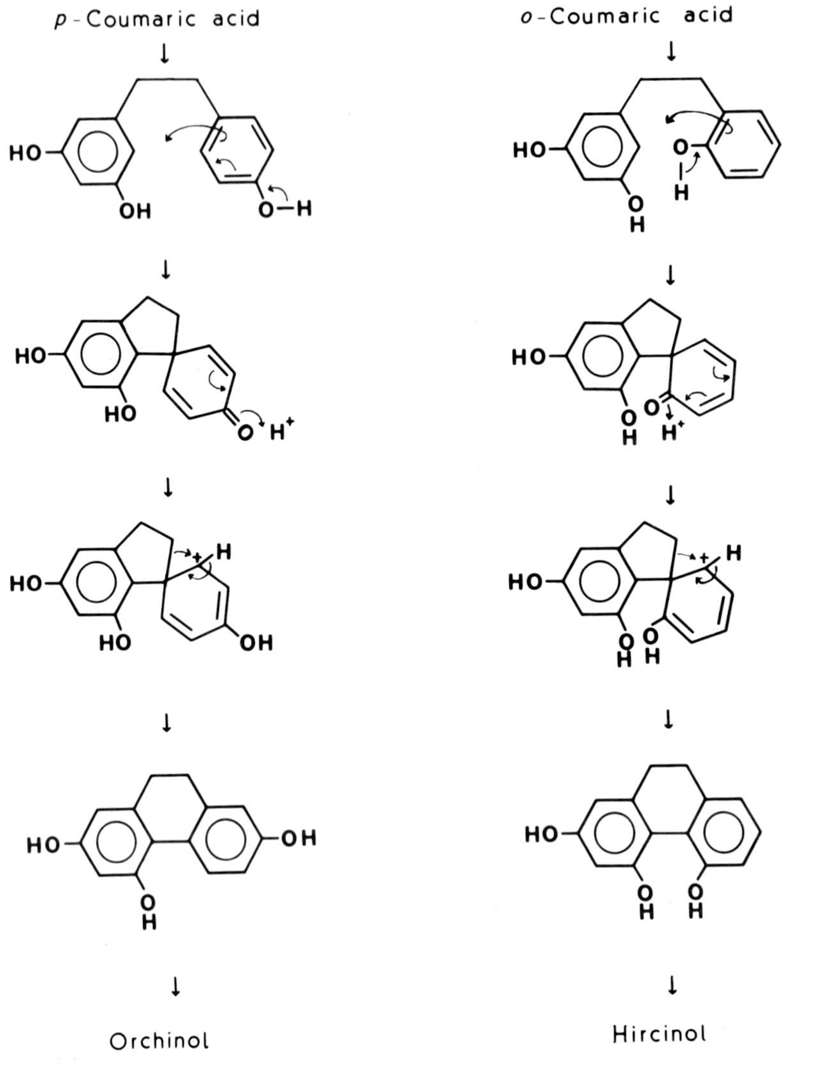

Fig. 16. Possible pathway from *o*-coumaric acid or *p*-coumaric acid to hircinol or orchinol, respectively.

Fritzemeier *et al.*, 1984) that in *Dioscorea* species two independent pathways exist, namely pathway A: *p*-coumaric acid → stilbene (resveratrol) → phenanthrene (7-hydroxy-2,4,5-trimethoxyphenanthrene); and pathway B: dihydro-*m*-coumaric acid → bibenzyl (3,3′,5-trihydroxybibenzyl) → dihydrophenanthrene (hircinol).

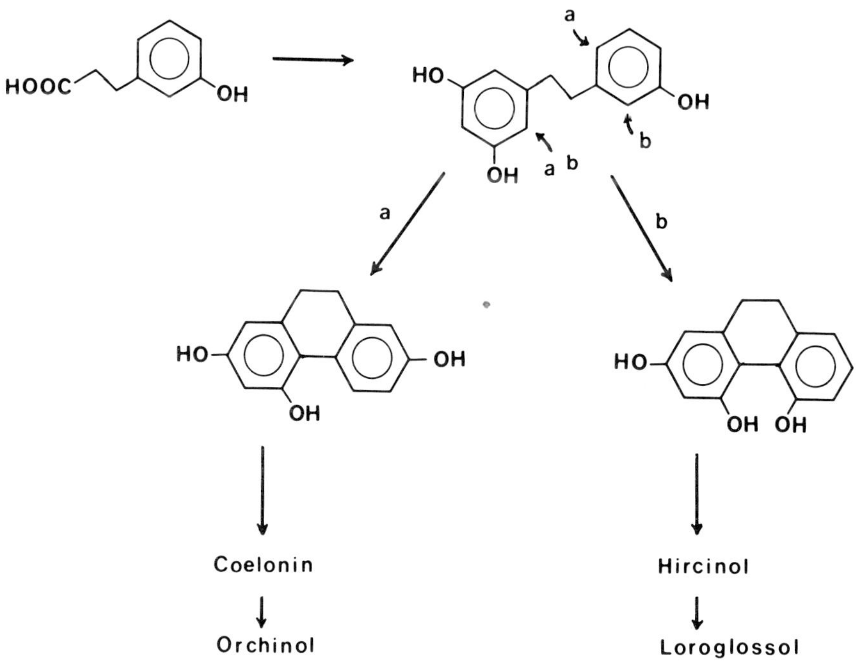

Fig. 17. Principal steps in the conversion of *m*-hydroxyphenylpropionic acid to dihydrophenanthrenes. The trihydroxydihydrophenanthrenes are modified by methylation steps and thus afford the orchid phytoalexins.

E. *Bibenzyls of Hepaticae*

Liverworts constitute a group of plants with such a particular pattern of bibenzyls that the exceedingly manifold structures are here referred to as a special section. Lunularic acid, detected in a wide variety of liverworts (Pryce, 1971; Pryce, 1972; Gorham, 1977), was originally examined as an

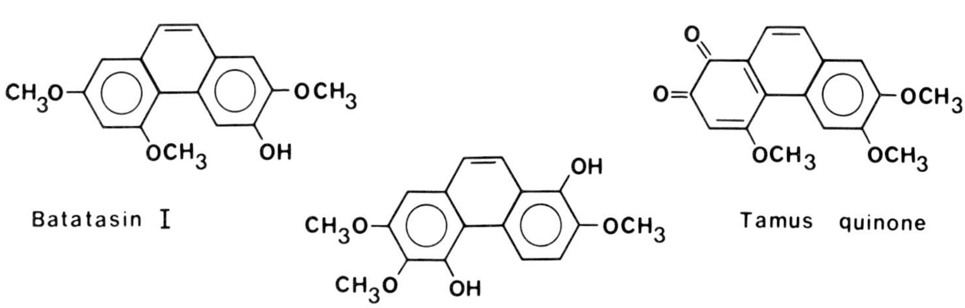

Fig. 18. Phenanthrene derivatives in Dioscoreaceae.

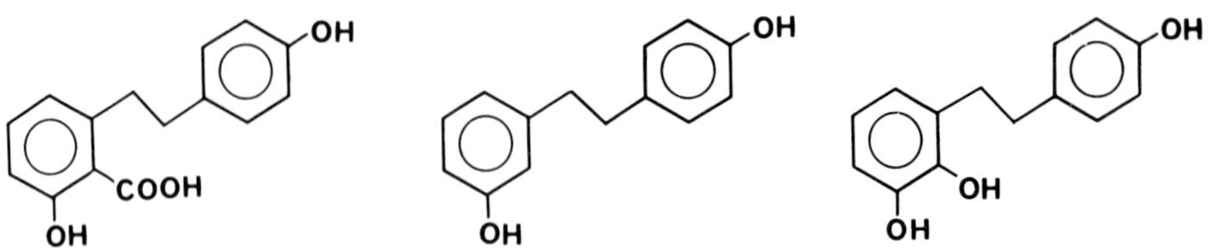

Lunularic acid — Lunularin — Basic structure of *Pellia epiphylla* bibenzyl

Fig. 19. Bibenzyls originating from lunularic acid.

endogenous growth regulator (Valio and Schwabe, 1970). Its biosynthesis was reported to proceed via hydrangeic acid (Pryce 1971b). Recent work of Ohta *et al.* (1983) and of Abe and Ohta (1984), however, strongly suggests dihydro-*p*-coumaric acid as the precursor. By demonstrating the occurrence of prelunularic acid, a prearomatic structure was implicated in the route to lunularic acid. Some of the most conspicuous chemical features of liverwort bibenzyls are indicated by the examples in Figs. 19 and 20.

Several other bibenzyls in liverworts (Asakawa *et al.*, 1982) seem to be related to dihydropinosylvin and dihydropinosylvin carboxylic acid, as exemplified in Fig. 20. But bibenzyls highly hydroxylated in ring B also occur (Asakawa *et al.*, 1981).

IV. Modification of the Stilbene Skeleton

Modifications of the stilbene skeleton can take place at the oxygen atoms or the carbon skeleton. Formation of methyl ethers or the conver-

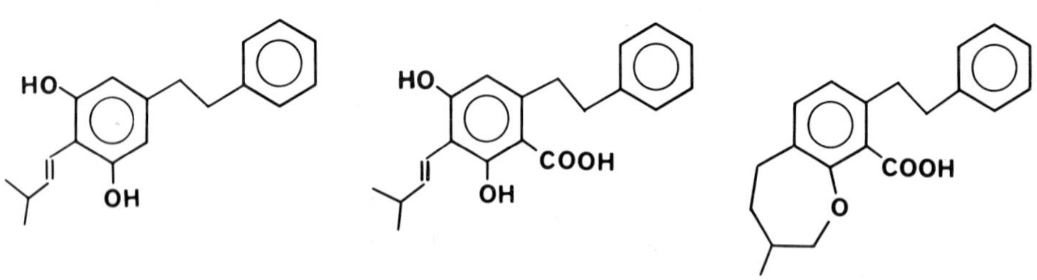

Fig. 20. Dihydropinosylvincarboxylic acid as parent substance of several 4-isopentenyl-bibenzyls (and other 4-prenyl derivatives).

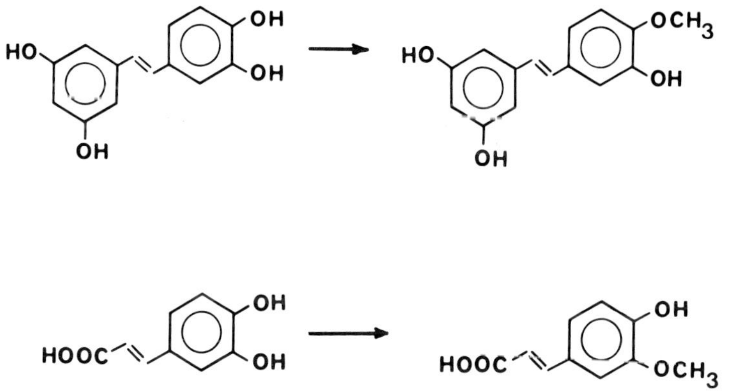

Fig. 21. Specific O-methylation of polyhydroxystilbene versus dihydroxycinnamic acid.

sion of phenols to *O*-glycosides are the most frequent derivatizations at OH groups. Stilbenes modified at the carbon structure originate by hydroxylation, preferrably ortho to an already existing OH group, and C-alkylation with isoprenyl groups.

An *O*-methyltransferase (Rupprich *et al.,* 1980) using *S*-adenosyl methionine as second substrate was detected in preparations from *Rheum* rhizomes catalyzing a very selective *O*-methylation, i.e., only at the 4′ position of a polyhydroxystilbene (3,3′,4′,5-tetrahydroxystilbene, astringenin). In the same extract another methyltransferase was characterized (Fig. 21) that exclusively methylates in the 3 position (caffeic acid to ferulic acid) Rupprich *et al.,* 1980).

In cell cultures of *Picea excelsa* an effective *O*-methyltransferase was observed (Rolfs and Kindl, 1984).

Stilbenes in equilibrium with hydroxyethane analogs can form an internal ester (lactone) if they possess a carboxylic group at position 2 of the aromatic ring A: hydrangeic acid or phyllodulcic acid being in equilibrium with 3,4-dihydroisocoumarines (hydrangenol, phyllodulcin: Fig. 12). Modification (hydroxylation followed by specific methylation as in Fig. 21) at the level of dihydroisocoumarins, too, was demonstrated, i.e., the conversion of hydrangenol to phyllodulcin (Yagi *et al.,* 1977) (Fig. 12).

Very little is known about hydroxylation reactions. The 3′-hydroxylation of a 4′-hydroxystilbene [e.g., the conversion of resveratrol into astringenin (piceatannol)] has been observed in *Rheum rhaponticum* (Rupprich *et al.,* 1980) and *Arachis hypogaea* (C.-H. Rolfs and H. Kindl, unpublished), but the enzymatic process remains obscure. As with many other instances, the analogy to the metabolism of flavanoids suggests the involvement of a microsomal NADPH-dependent hydroxylase (Britsch *et al.*, 1981). Similar comparisons of stilbenes versus flavonoids may also be drawn concerning the prenyltransferase conveying an isopentenyl group from dimethylallyl diphosphate to resveratrol, giving rise to 4-isopentenylresveratrol, the main phytoalexin of *Arachis hypogaea*. Zähringer *et al.* (1981) observed such a process during the biosynthesis of glyceollin.

A 2′,4′-dihydroxyphenyl structure, such as in oxyresveratrol and its isopentenyl derivative (chlorophorin), can either originate by introducing the 2′ OH group at the level of stilbenes (resveratrol → oxyresveratrol) or by incorporation of 2,4-dihydroxycinnamic acid (Billek and Schimpl, 1962).

V. Stilbene Synthase and Its Properties

Stilbene synthase activity in a cell-free system was first demonstrated in 1978 by Rupprich and Kindl. The enzyme preparation was obtained from rhizome of *Rheum rhaponticum* and was later ascribed to a constitutive enzyme characterized by a rather high K_m value for malonyl-CoA (Rupprich *et al.*, 1980). The partially purified enzyme (Hildebrand, 1980) exhibited resveratrol-forming activity (see Fig. 8), rather than catalyzing rhapontigenin formation from isoferuloyl-CoA. This was contrary to the specificity of the enzyme that had been searched for and that had been thought to be responsible for the direct formation of rhapontigenin (Fig. 8). Since then, resveratrol-forming stilbene synthases (type I; Fig. 22, upper row) have been found in *Picea excelsa* (Rolfs and Kindl, 1984), *Pinus griffitii* (Schöppner and Kindl, in preparation), *Arachis hypogaea* (Rolfs *et al.*, 1981; Schöppner and Kindl, 1984), *Vitis* spp. (Fritzemeier and Kindl, 1981), and other plants.

Another type of stilbene synthase (type II), selectively converting cinnamoyl-CoA to pinosylvin, was partially purified from an acetone powder of pine seedlings (*Pinus sylvestris*) (Schöppner and Kindl, 1979; Schöppner and Kindl, in preparation).

The most thoroughly investigated of the stilbene synthases has been the enzyme (type I) from cell suspension cultures of *Arachis hypogaea* (Rolfs and Kindl, 1984; Schöppner and Kindl, 1984). It was purified to apparent homogeneity and characterized by its molecular weight, subunit molecular weight, and catalytic properties.

A number of factors regulate the activity of stilbene synthase, chiefly, the state of reduction of a central SH group and the ionic strength of the medium. Modification or partial denaturation can be reversible; for example, if the enzyme has been inactivated with mercurials (*p*-hydroxymercuribenzoate), the recovery of enzyme activity depends on reactivation through an excess of dithiothreitol. Dithiothreitol is an essential reagent

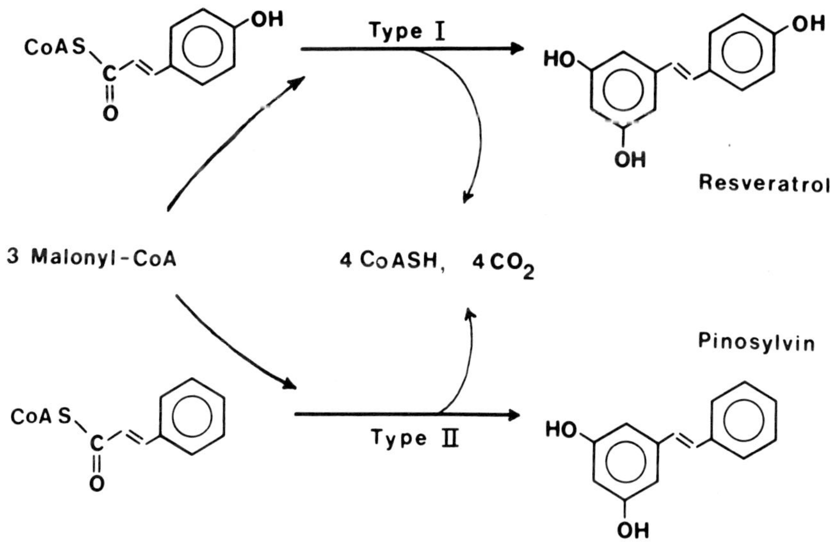

Fig. 22. Overall reaction of stilbene synthase-catalyzed reactions.

to keep the enzyme active during the course of the preparation, as is polyvinylpyrrolidone.

Another difficulty during enzyme purification was posed by the decrease in specific activity during the last step of purification and the appearance of enzyme species exhibiting somewhat lower subunit molecular weights. Any explanation of how the highest specific activities can be attained must address partial proteolytic degradation of the catalytically active enzyme.

A. *Molecular and Catalytic Properties*

When selectively induced cell suspension cultures were used as starting material, the crude extract already contained stilbene synthase of relatively high specific activity. A 100-fold purification by ion exchange chromatography and separation on hydroxylapatite sufficed to accomplish homogeneity.

The resveratrol-forming enzyme, like other stilbene synthases under investigation, is a dimer with MW 90,000. The subunit MW of 47,000 was evidenced by SDS gel electrophoresis. A rather acidic isoelectric point (pI = 5.8) was found upon isoelectric focusing.

Analyses for substrate specificity are indispensable for stilbene synthases, as several kinds of enzymes have been found that differ only by selectivity with respect to the phenylpropanoid CoA ester. The *Arachis* enzyme prefers *p*-coumaroyl-CoA to cinnamoyl-CoA by a factor of 6. The enzyme from rhizomes of *Rheum rhaponticum,* too, selectively converted *p*-coumaroyl-CoA (four times better than cinnamoyl-CoA); it did not accept benzoyl-CoA, phenylacetyl-CoA, caffeoyl-CoA, or isoferuoyl-CoA as substrates (Rupprich *et al.,* 1980; Hildebrand, 1980). The K_m values are low for *p*-coumaroyl-CoA (2 and 14 μM for *Arachis* and *Rheum* enzymes, respectively), whereas the affinity toward malonyl-CoA ranges from $K_m = 10\ \mu M$ (*Arachis* enzyme) to $K_m = 380\ \mu M$ (*Rheum* enzyme). The pH optimum of stilbene formation is between 7.6 and 8.4.

Unlike these resveratrol-forming enzymes, the stilbene synthase from *Pinus sylvestris* preferentially converts the nonhydroxylated cinnamoyl-CoA to pinosylvin. The enzyme's selectivity toward nonhydroxylated versus *p*-hydroxylated substrate is 4 : 1 (Schöppner and Kindl, 1979). The formation of pinosylvincarboxylic acid from 9-phenyl-3,5,7-triketo-non-8-enoic acid at pH 5.0 was demonstrated in a biomimetic reaction (Harris and Carney, 1967). In an analogous way, phenyltriketononanoic acid was converted to 6-phenethyl-β-resorcyclic acid and dihydropinosylvin.

B. *Comparison of Stilbene Synthase with Related Enzymes*

As both *p*-coumaroyl-CoA and malonyl-CoA are the substrates not only of stilbene synthase but also of chalcone synthase (Hahlbrock and Grisebach, 1979), enzyme-catalyzed reactions with malonyl-CoA and phenylpropanoid-CoA must be characterized by substrate specificity as well as by the compounds produced. With *p*-coumaroyl-CoA as substrate, resveratrol, naringenin (and its chalcone), and products of side reactions (Hrazdina *et al.,* 1976) (*p*-hydroxyphenylpyrone, *p*-hydroxystryrylpyrone) must be considered. It is evident, however, that purified stilbene synthases do not exhibit chalcone synthase activity (see Chapter 12, Section IV,A). Anti-stilbene synthase antibodies did not cross-react with chalcone synthase. Apart from these clear differences between the two enzymes, the molecular properties are surprisingly similar. This holds true for MW, subunit MW, and some mechanistic details (Kreuzaler *et al.,* 1978, 1979).

The principle of the stilbene synthase reaction (i.e., chain lengthening of an acyl-CoA with malonyl-CoA units and subsequent aromatization, Fig. 3), also appears in the conversion catalyzed by 6-methylsalicylic acid synthase. This enzyme is, however, distinguished by a 4′-phosphopantethein residue and an extremely high molecular weight (Dimroth *et al.,* 1970, 1976).

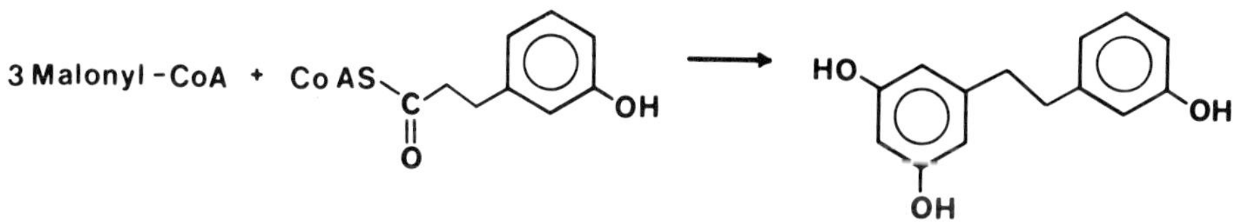

Fig. 23. Conversion of *m*-hydroxyphenylpropionyl-CoA to 3,3′,5-trihydroxybibenzyl. The enzyme has been detected in bulbs of orchid species and in various yam bulbs.

In addition to stilbene synthase converting cinnamoyl-CoA esters to stilbenes, a closely related type of enzyme has been recently discovered. Bibenzyl synthases, or dihydrostilbene synthases (Fritzemeier and Kindl, 1983), were shown to utilize dihydro derivatives of hydroxycinnamoyl-CoA esters as substrates instead of the respective cinnamoyl derivatives. The enzymes from orchids (Fritzemeier and Kindl, 1983) and Dioscoreaceae (Fritzemeier *et al.*, 1984) catalyze the condensation of *m*-hydroxyphenylpropionyl-CoA with malonyl-CoA (Fig. 23).

A dihydrostilbene (bibenzyl) of this type (3,3′,5-trihydroxybibenzyl) is the precursor of naturally occurring 9,10-dihydrophenanthrenes. It is possible that bibenzyl synthases using other phenylpropanoid esters are responsible for the synthesis of the dihydrostilbenes that occur in wood and other plant tissues—for example, for the synthesis of dihydropinosylvin in *Pinus* sp. Dihydroresveratrol (3,4′,5-trihydroxybibenzyl) is formed by a bibenzyl synthase from *Cannabis sativa* (Crombie and Kindl, in preparation).

C. *Mechanism of Stilbene Synthase-Catalyzed Reaction*

The exceedingly low K_m value for *p*-coumaroyl-CoA (2 μM) suggests a rather tight binding of this compound. As the carbonyl carbon of the thioester must be accessible for the attack of a carbanion and as the CoA moiety is removed as thiolate during the condensation, it is likely that another part of the molecule adds considerably to the binding.

Purified stilbene synthase incubated with malonyl-CoA as a sole substrate shows decarboxylase activity; the CO_2 from the unsubstituted carboxyl group of malonyl-CoA is reversibly removed. This process can best be demonstrated by the equilibrium between malonyl-CoA and enzyme-bound acetyl-CoA plus CO_2. If the latter is labeled, radioactive malonyl-CoA can be produced (Fig. 24). Thus, the first condensation step can be formulated as nucleophilic attack of an carbanion (acetyl-CoA) at the carbonyl carbon atom of the thioester, with thiolate ion as leaving group

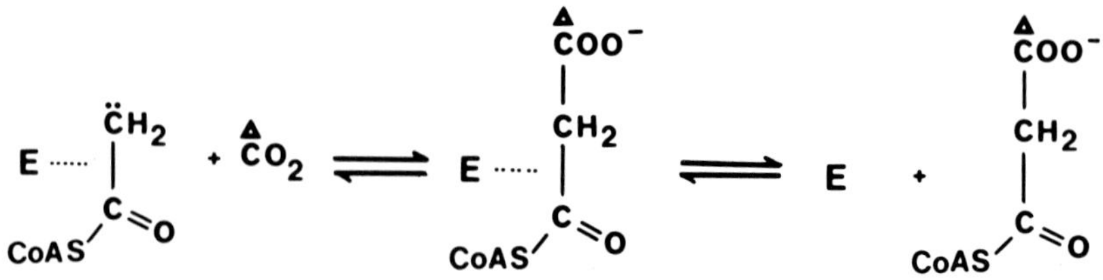

Fig. 24. Demonstration of an enzyme-bound acetyl-CoA as intermediate during stilbene synthase reaction.

(Fig. 25). In the same manner, two more malonyl-CoA units must be introduced, giving rise to a CoA ester of C_6C_9 polyketocarboxylic acid. This compound should be an appropriate substrate of an aldol reaction (Fig. 26). The resulting cyclic compound forms enols and thus becomes aromatic.

At present, it is not understood which factors cause the ester cleavage and the subsequent decarboxylation. Except for a few cases, stilbenes are formed and not the respective stilbene-2-carboxylic acid. With the purified enzyme (type I), too, no free carboxylic acid could be detected.

In keeping with all results known, we can postulate the cooperation of the substrates and the stilbene synthase during the course of the many partial reactions (Fig. 27).

VI. Induction of Stilbene Synthesis

In many hardwoods, stilbene amounts are rather high and stilbenes constitute a significant percentage within the group of phenolic compounds. The synthesis takes place, in these cases, at low rates, and is not or is only slightly inducible. Young seedlings, however, were found to

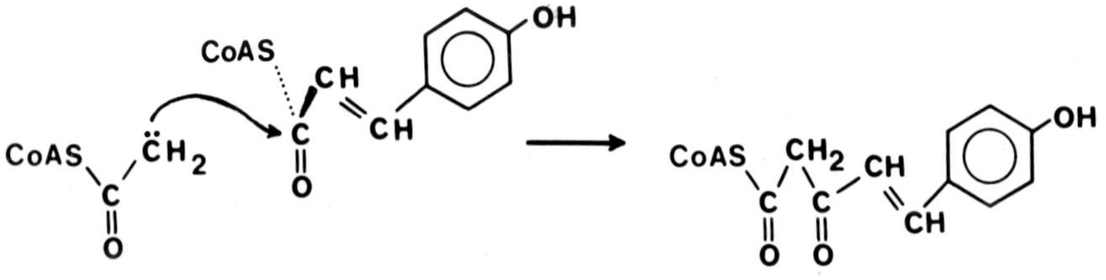

Fig. 25. Condensation of the first C_2 unit with the phenylpropanoid moiety.

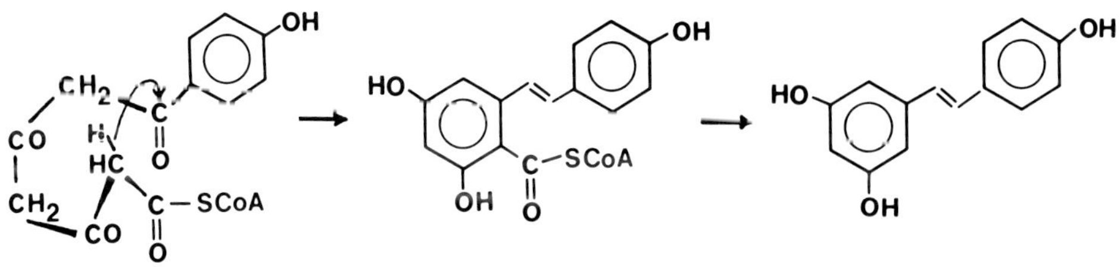

Fig. 26. Attack of the carbanion (at C-2) on the carbonyl carbon atom (C-7). Removal of H_2O and enolization of the other keto groups leads to an aromatic system.

behave differently, frequently increasing, upon induction, the amounts of stilbenes up to 100-fold. Stilbene synthesis can thus be grouped into inducible and noninducible processes. Inducible stilbene synthesis may be considered with particular interest, as physiological functions of the stilbene production can be assumed. There are some secondary reflections with regard to which kind of inducing mechanism is functioning. Undoubtedly, it is intriguing to study the mechanism responsible for the transformation of the external signal. The way that signals are passed on is mainly obscure, and the level at which the signal triggers stilbene synthesis is being elucidated.

A. *Elicitation of the Biosynthetic Branch Leading to Stilbenes*

Examples of a more or less concerted induction of several enzymes leading to stilbenes have been shown to occur. If an external stimulus is applied to intact plants, it seems to be rather common that more than one of the enzymes catalyzing consecutive steps of the pathway from phenylalanine to stilbenes is induced. In leaves of Vitaceae (e.g., grapevine), the application of elicitors—endogenous or fungal elicitors or even artificial elicitors (Fritzemeier and Kindl, 1981)—leads to an enhancement of the activities of phenylalanine ammonia-lyase, cinnamate *p*-hydroxylase, cinnamoyl-CoA ligase, and stilbene synthase. For stilbene synthase activity, an increase of more than 100-fold above the basic level was observed.

It is not yet clear whether this behavior of coordinate induction is observable only in intact tissues or may also be found in cell cultures of these organisms. In other organisms such as *Arachis hypogaea,* intact seedlings, callus cultures, and cell suspension cultures behaved differently in the way phenylalanine ammonia-lyase and stilbene synthase activities appeared (Rolfs and Kindl, in preparation; Rolfs *et al.*, 1981).

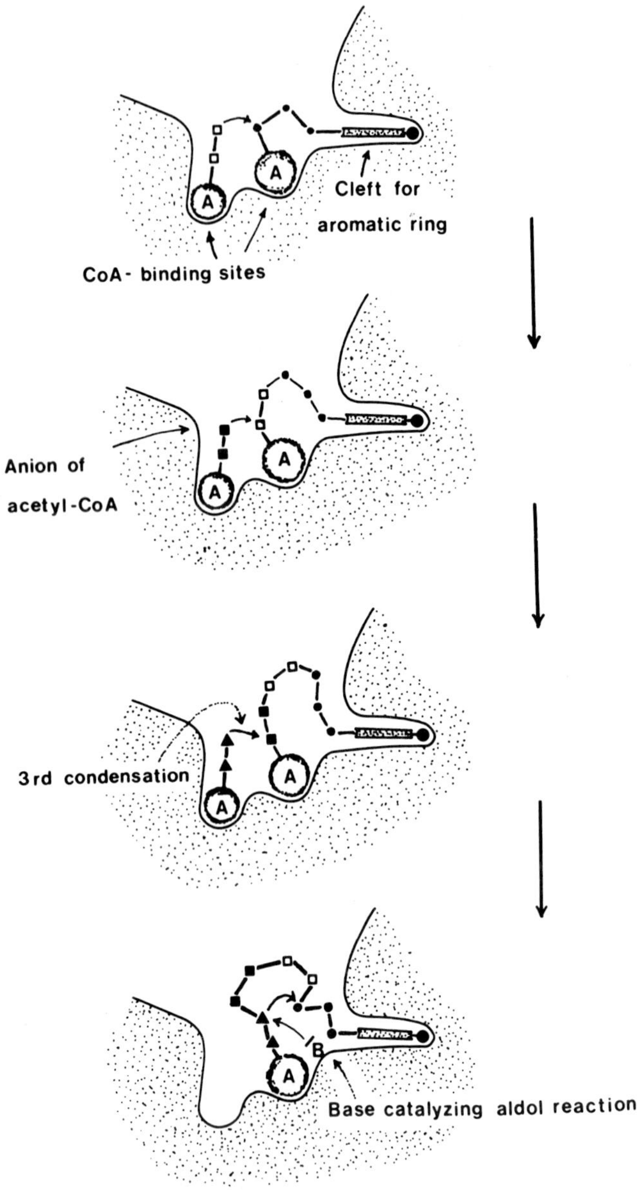

Fig. 27. Proposed mechanism of stilbene synthesis. The schematic drawing shows a phenylpropanoid bound to CoA at the right side; the *p*-hydroxy group is indicated by a solid circle. At the left side, the intermediary form of the C_2 unit (i.e., the anion of acetyl-CoA) is drawn with extra symbols for each new C_2 unit to be condensed with the growing chain of the keto ester. Following condensations, a basic group at the enzyme induces the nucleophilic attack of the anion at C-2, thus giving rise to a cyclic and finally to an aromatic structure.

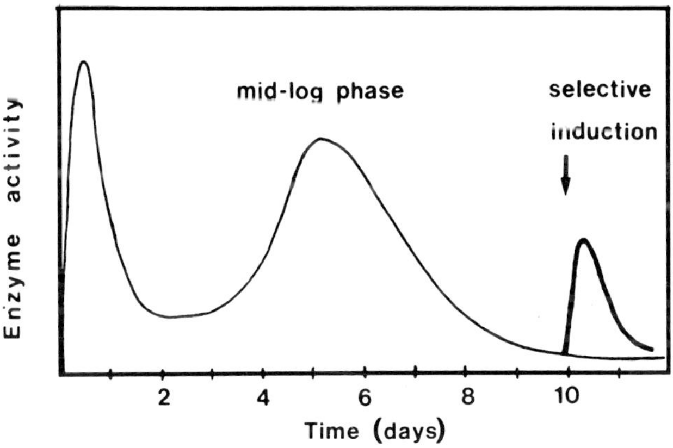

Fig. 28. Induction of stilbene synthase activity in cell suspension culture of *Arachis hypogaea*. The profile of stilbene synthase activity is shown during the growth cycle. After 10 days, stationary phase is reached, characterized by a low level of stilbene synthase activity. At this stage, selective induction of stilbene synthase can be triggered by elicitors or artificial means. It is not an activation of an already existing protein, but rather a protein *de novo* synthesis, as demonstrated by labeling of the stilbene synthase with precursor amino acids.

B. *Selective Induction of Stilbene Synthase*

Unlike intact plants, plant cell cultures may behave in a special way upon elicitor treatments, in that they increase the level of stilbene synthase more or less selectively. A rather selective induction of stilbene synthase could be demonstrated with suspension cultures of *Arachis hypogaea* (groundnut). It has been known from studies with intact seedlings that stilbenes are formed in this plant as phytoalexins in response to the attack of certain fungi. 4-Isopentenyl resveratrol and the dimethyl ether of resveratrol are the predominant stilbenes appearing in the cotyledons after infection.

In cell suspension cultures, three periods of time were established during which enhanced levels of stilbene synthase were attained. First, after dilution and transfer of cells to a new medium, a marked increase of stilbene synthase is paralleled by enhanced activities of other enzymes (e.g., phenylalanine ammonia-lyase). During mid log phase, another peak was determined, which does not, however, actually reflect a preference in stilbene synthase formation. At the end of the growth phase, a rapid, transient, and extremely selective induction of stilbene synthase activity could be elicited. The different stages having increased levels of stilbene synthase are summarized in Fig. 28.

C. *Noninducible Stilbene Synthases*

Stilbene synthase activities were first demonstrated in cell-free extracts from rhubarb rhizomes. This enzyme, being constitutive and occurring at a constant but rather low level in the rhizomes, could not be detected in leaves of this plant. This constitutive enzyme differs from other stilbene synthases in its low affinity toward malonyl-CoA ($K_m = 380\ \mu M$).

Another stilbene synthase (resveratrol-forming) whose level was found not to be affected by light or several other factors was prepared from *Picea excelsa* (Rolfs and Kindl, 1984), from young seedlings as well as from callus cultures. Neither elicitors (including UV light) nor white light influenced the level of stilbene synthase.

D. *Level of Stilbenes is Affected by Increasing the Level of PolyA mRNA Coding for Stilbene Synthase*

The underlying causes of stilbene synthase levels are primarily the level of translatable mRNA coding for this enzyme as well as the intensity of degradative steps. Regulation of the step of translation itself does not seem to occur. In cases where stilbene synthase is inducible, a rather high turnover of the messenger was observed. The decrease in enzyme activity is then the inevitable result, when mRNA formation ceases. Furthermore, it became evident that the increase in enzyme activity is paralleled by an enhanced synthesis of catalytic protein. Studies on the level of translatable mRNA, in turn, revealed that the control of stilbene synthesis must be at the level of transcription. The way elicitor signals reach the promoters of the stilbene synthase gene is not known at present.

Studies *in vitro* have demonstrated that induction at the end of the growth phase (Fig. 28) is rapid (30 min) and so selective that the messenger for stilbene synthase becomes one of the predominant polyA mRNAs under these conditions.

VII. Concluding Remarks

Any attempt to provide an overview and to understand relations of biosynthetic processes requires studies that combine surveys *in vivo* and assays *in vitro*. The situation *in vivo* can be described by stationary concentrations of intermediates and of apparent products. Rates of synthesis are then compared with those data. Investigations *in vitro* usually imply the properties of enzymes, as an isolated system as well as in relation to the many possible effectors modulating the activity. For stilbenes, both types of observations correlate well with each other and reveal pathways functioning side by side. The pathways are controlled genetically or are balanced by delicate regulation mechanisms. More than one pathway may be operative, since stilbenes with or without a 4′-hydroxy group and dihydrostilbenes can co-occur.

The biosynthesis of stilbenes is influenced by the stationary concentrations of substrates of stilbene synthase, by the affinity of stilbene synthase toward these substrates (indicated by K_m values), and, most frequently, by control of the level of translatable mRNA coding for stilbene synthase.

A number of exciting questions remain: What mechanism controls the induction of synthesis of stilbene synthase, or, for the enzyme action, which topological factors govern the kind and number of ester condensations, folding of polyketo intermediates, and consecutive processes leading specifically to stilbenes?

References

Abe, S., and Ohta, Y. (1984). *Phytochemistry* **23,** 1379–1381.
Aguamah, G. E., Langcake, P., Leworthy, D. P., Page, J. A., Pryce, R. J., and Strange, R. N. (1981). *Phytochemistry* **20,** 1381–1383.
Asakawa, Y., Matsuda, R., Toyota, M., Hattori, S., and Ourisson, G. (1981). *Phytochemistry* **20,** 2187–2194.
Asakawa, Y., Takikawa, K., Toyota, M., and Takemoto, T. (1982). *Phytochemistry* **21,** 2481–2490.
Askari, A., Worthen, L. R., and Shimizu, Y. (1972). *Lloydia* **35,** 49–54.
Billek, G. (1964). *In* "Progress in the Chemistry of Organic Natural Products" (L. Zechmeister, ed.), pp. 115–152. Springer-Verlag, Wien.
Billek, G., and Kindl, H. (1962). *Monatsh. Chem.* **93,** 814–821.
Billek, G., and Schimpl, A. (1962). *Monatsh. Chem.* **93,** 1457–1459.
Billek, G., and Ziegler, W. (1962). *Monatsh. Chem.* **93,** 1430–1440.
Blaich, R., and Bachmann, O. (1980). *Vitis* **19,** 230–240.
Bohlmann, F., and Hoffmann, E. (1979). *Phytochemistry* **18,** 1371–1374.
Britsch, L., Heller, W., and Grisebach, H. (1981). *Z. Naturforsch.* **36C,** 742–750.
Coggon, P., Janes, N. F., King, F. E., King, T. J., Molyneux, R. J., Morgan, J. W. W., and Sellars, K. (1965). *J. Chem. Soc.* 406–409.
Cooksey, C. J., Dahiya, J. S., Garratt, P. J., and Strange, R. N. (1982). *Phytochemistry* **21,** 2935–2938.
Coxon, D. T., Ogundana, S. K., and Dennis, C. (1982). *Phytochemistry* **21,** 1389–1392.
Crombie, L., and Crombie, W. M. L. (1982). *J. Chem. Soc. Perkin Trans. I,* 1455–1466.
Czichi, U., and Kindl, H. (1977). *Planta* **134,** 133–134.
Dimroth, P., Walter, H., and Lynen, F. (1970). *Eur. J. Biochem* **13,** 98–110.
Dimroth, P., Ringelmann, E., and Lynen, F. (1976). *Eur. J. Biochem* **68,** 591–596.
Erdtman, H. (1939). *Ann.* **539,** 116–127.
Erdtman, H. (1955). *Experientia, Suppl. II,* 145–180.
Erdtman, H. (1959). "Biochemistry of Wood." Pergamon, London.
Filho, R. B., Gottlieb, O. R., and Mourao, A. P. (1975). *Phytochemistry* **14,** 261–263.
Fritzemeier, K.-H., and Kindl, H. (1981). *Planta* **151,** 48–52.
Fritzemeier, K.-H., and Kindl, H. (1983). *Eur. J. Biochem* **133,** 543–550.

Fritzemeier, K.-H., Rolfs, C.-H., Pfau, J., and Kindl, H. (1983). *Planta* **159,** 25–29.
Fritzemeier, K.-H., Kindl, H., and Schlösser, E. (1984). *Z. Naturforsch.* **39c,** 217–221.
Gäumann, E. (1963). *Phytopath. Z.* **49,** 211–232.
Gorham, J. (1977). *Phytochemistry* **16,** 249–253.
Gorham, J. (1980). *Prog. Phytochem.* **6,** 203–252.
Gestetner, B., and Conn, E. E. (1974). *Arch. Biochem. Biophys.* **163,** 617–624.
Hahlbrock, K., and Grisebach, H. (1979). *Ann. Rev. Plant. Physiol.* **30,** 105–130.
Hardegger, E., Schellenbaum, M., and Corrodi, H. (1963). *Helv. Chim. Acta* **46,** 1171–1180.
Harris, T. M., and Carney, R. L. (1967). *J. Am. Chem. Soc.* **89,** 6734–6740.
Hasegawa, K., and Hashimoto, T. (1975). *J. Exp. Bot.* **26,** 757–764.
Hashimoto, T., Ikai, T., and Tamura, S. (1968). *Planta* **78,** 89–92.
Hashimoto, T., Hasegawa, K., Yamaguchi, Y., Saito, M., and Ishimoto, S. (1974). *Phytochemistry* **13,** 2849–2852.
Hata, K., Baba, K., and Kozawa, M. (1979). *Chem. Pharm. Bull.* **27,** 984–989.
Hildebrand, H. (1980). Thesis, Marburg.
Hillis, W. E. (1966). *Phytochemistry* **5,** 541–556.
Hillis, W. E., and Carle, A. (1962). *Biochem. J.* **82,** 435–439.
Hillis, W. E., and Hasegawa, M. (1962). *Chem. Ind.* 1330–1331.
Hillis, W. E., and Inoue, T. (1967). *Phytochemistry* **6,** 59–67.
Hillis, W. E., and Ishikura, N. (1970). *Phytochemistry* **9,** 1517–1528.
Hillis, W. E., and Orman, H. R. (1962). *J. Linn. Soc. (Bot.)* **58,** 175–184.
Hillis, W. E., Hart, J. H., and Yazaki, Y. (1974). *Phytochemistry* **13,** 1591–1595.
Hrazdina, G., Kreuzaler, F., Hahlbrock, K., and Grisebach, H. (1976). *Arch. Biochem. Biophys.* **175,** 392–399.
Ibrahim, R. K., and Towers, G. H. N. (1962). *Can. J. Biochem. Physiol.* **40,** 449–453.
Ingham, J. L. (1976). *Phytochemistry* **15,** 1791–1793.
Kariyone, T., Takahashi, M., Ito, T., and Masutani, K. (1958). *J. Pharm. Soc. Jpn.* **78,** 935–936.
Kariyone, T., Takahashi, M., Ito, T., and Masutani, K. (1959). *J. Pharm. Soc. Jpn.* **79,** 1326–1327.
Keen, N. T., and Ingham, J. L. (1976). *Phytochemistry* **15,** 1794–1795.
Kindl, H. (1971a). *Naturwissenschaften* **58,** 554–563.
Kindl, H. (1971b). *Hoppe-Seyler's Z. Physiol. Chem.* **352,** 78–84.
Kindl, H. (1979). *In* "Regulation of Secondary Product and Plant Hormone Metabolism" (M. Luckner and K. Schreiber, eds.), Vol. 55, pp. 49–61. Pergamon, London.
King, F. E., King, T. J., Godson, D. H., and Manning, L. C. (1956). *J. Chem. Soc.* 4477–4480.
Kreuzaler, F., Light, R. J., and Hahlbrock, K. (1978). *FEBS Lett.* **94,** 175–178.
Kreuzaler, F., Ragg, H., Heller, W., Tesch, R., Witt, I., Hammer, D., and Hahlbrock, K. (1979). *Eur. J. Biochem.* **99,** 89–96.
Langcake, P., and Pryce, R. J. (1976). *Physiol. Plant Pathol.* **9,** 77–86.
Langcake, P., and Pryce, R. J. (1977a). *Phytochemistry* **16,** 1193–1196.
Langcake, P., and Pryce, R. J. (1977b). *J. Chem. Soc. Chem. Commun.* 208–210.
Letcher, R. M., and Nhamo, L. R. M. (1972). *J. Chem. Soc., Perkin I* 2941–2946.
Letcher, R. M., Nhamo, L. R. M., and Gumiro, T. (1972). *J. Chem. Soc., Perkin I* 206–210.
Lynen, F., and Tada, M. (1961). *Angew. Chem.* **73,** 513–519.
Mitscher, L. A., Rav, G. S. R., Khanna, I., Veysoglu, T., and Drake, S. (1983). *Phytochemistry* **22,** 573–576.

Nakajima, N., Taguchi, H., Endo, T., and Yosioka, I. (1978). *Chem. Pharm. Bull.* **26,** 3050–3057.
Ohta, Y., Abe, S., Komura, H., and Kobayashi, M. (1983). *J. Am. Chem. Soc.* **105,** 4480–4481.
Poulton, J. E., McRee, D. E., and Conn, E. E. (1980). *Plant Physiol.* **65,** 171–175.
Pryce, R. J. (1971a). *Planta* **97,** 354–357.
Pryce, R. J. (1971b). *Phytochemistry* **10,** 2679–2685.
Pryce, R. J., and Langcake, P. (1977). *Phytochemistry* **16,** 1452–1454.
Rolfs, C.-H., and Kindl, H. (in preparation).
Rolfs, C.-H., and Kindl, H. (1984). *Plant Physiol.* **75,** 489–492.
Rolfs, C.-H., Fritzemeier, K.-H., and Kindl, H. (1981). *Plant Cell Rep.* **1,** 83–85.
Rudloff, E. von, and Jorgensen, E. (1963). *Phytochemistry* **2,** 297–304.
Rupprich, N., and Kindl, H. (1978). *Hoppe-Seyler's Z. Physiol. Chem.* **359,** 165–175.
Rupprich, N., Hildebrand, H., and Kindl, H. (1980). *Arch. Biochem. Biophys.* **200,** 72–78.
Schöppner, A., and Kindl, H. (1979). *FEBS Lett.* **108,** 349–352.
Schöppner, A., and Kindl, H. (1984). *J. Biol. Chem.* **259,** 6806–6811.
Schöppner, A., and Kindl, H. (in preparation).
Späth, E., and Schläger, J. (1940). *Chem. Ber.* **73,** 881–884.
Stoessl, A. (1982). *In* "Phytoalexins" (J. A. Bailey, and J. W. Mansfield, eds.), pp. 133–180. Blacki, Glasgow.
Suga, T., Iwata, N., and Asakawa, Y. (1972). *Bull. Chem. Soc. Jpn.* **45,** 2058–2061.
Takahashi, M., Ito, T., Mizutani, A., and Iso, K. (1960). *J. Pharm. Soc. Jpn.* **80,** 1488–1492.
Valio, I. F. M., and Schwabe, W. W. (1970). *J. Exp. Bot.* **21,** 122–150.
Wollenweber, E., and Favre-Bonvin, J. (1979). *Phytochemistry* **18,** 1243–1244.
Yagi, A., Ogata, Y., Yamauchi, T., and Nishioka, I. (1977). *Phytochemistry* **16,** 1098–1100.
Zähringer, U., Schaller, E., and Grisebach, H. (1981). *Z. Naturforsch.* **36C,** 234–241.

CHAPTER **15**

Biosynthesis of Terpenoid Wood Extractives

RODNEY CROTEAU
MARK A. JOHNSON

Institute of Biological Chemistry
and Biochemistry/Biophysics Program
Washington State University
Pullman, Washington

ISBN 0-12-347880-4

I. Introduction

A. *Origin and Diversity of Terpenoids*

The terpenoids that accumulate in trees and shrubs encompass the full range of possible structural types and fulfill the same biological functions as in herbaceous plants (Gottleib and Mors, 1980; Langenheim, 1981). The conifers provide an example in which the types and amounts of terpenes produced are perhaps unparalleled in nature. The role of terpenoids in defense and disease resistance is especially prevalent in woody plants (Kúc and Lisker, 1978; Dell and McComb, 1978a). Although the biochemistry of terpenoids and the mobilization of these substances in woody species are poorly understood, it is abundantly clear that these processes are of central importance in the ecological interactions of forest trees and their associated predators and parasites (Shrimpton, 1978; Raffa and Berryman, 1982; Langenheim *et al.*, 1982).

The terpenoids are derived by the formal condensation of C_5 isoprenoid units, and they range in size from the volatile "monomer" isoprene (C_5) to the polymer rubber (MW $\sim 10^6$). All plants employ the isoprenoid pathway in the synthesis of certain essential substances, including phytol (C_{20}), carotenoids (C_{40}), dolichol phosphate (C_{100}), steroids (C_{27}), prenylated quinones (plastoquinone), and plant hormones [e.g., abscisic acid (C_{15}) and gibberellins (C_{20})]. The biosynthesis of these ubiquitous terpenoids is outside the scope of this chapter and will not be discussed here. Many of these topics were discussed in proceedings of a colloquium, published in *Biochemical Society Transactions* (Vol. 11, Part 5, 1983). In addition to these common substances, there exists a diverse array of other terpenoids, classed as so-called secondary metabolites, that are often stored in large quantities in specialized compartments, the specific constituents of which are frequently characteristic of a narrow range of species. These terpenes are of long-standing interest from a chemical, biogenetic, and taxonomic point of view (Gibbs, 1974; Seigler, 1981a,b), and of more recent interest because of the ecological advantage they may impart to the species that produce and accumulate them (Langenheim, 1981). While the terpenes are structurally the most diverse group of wood extractives, they lend themselves to simple classification by size, as will be done in this chapter.

There are relatively few naturally occurring iso-branched C_5 compounds (e.g., isoprene, valine, angelic acid, and the sugar apiose), and while all these may be formally considered "hemiterpenes," only isoprene itself is likely to be of isoprenoid origin (i.e., derived from mevalonic acid). The monoterpenes (C_{10}) consist of two main groups: the lipophilic, steam-volatile constituents of essential oils and their various conjugates, and the hydrophilic, nonvolatile iridoid glycosides. The sesquiterpenes (C_{15}) encompass a wider range of skeletal types and oxidation states than are commonly found in the monoterpenes (i.e., furans, lactones, quinones) and are often found in the higher boiling fractions of essential oils. The diterpenes (C_{20}) include some volatile hydrocarbons (cembrene, rimuene) and usually occur as nonvolatile acids (resins), lactones, or polyhydroxylated compounds. The triterpenes (C_{30}), with the exception of their acyclic hydrocarbon precursor, squalene, occur as C_{30} oxygenated polycyclic compounds (3-deoxytriterpenes occur in ferns and protozoans). The triterpenes are nonvolatile and often occur as glycosylated derivatives (saponins). The polyterpenes are long-chain acyclic alcohols (polyprenols) or very high molecular weight polyisoprene polymers (e.g., rubber, gutta, chicle).

The taxonomic distribution of terpene-accumulating plants has been the subject of extensive reviews (Hegnauer, 1962–1973; Gibbs, 1974; Seigler, 1981a,b). Terpene accumulation has been observed in every plant organ and in a variety of physical forms, including oils, resins, gums, latex, and waxes (Dell and McComb, 1978a). Within the narrow range of wood extractives that are the subject of this review, they occur in oleoresins, gums, latex, and occasionally in water-soluble fractions. Bark extractives from the root and stem are often considered along with those from wood because of their similarities in chemical composition. Resins are particularly common among the conifers and among dicots of the Burseraceae, Lauraceae, Euphorbiaceae, Santalaceae, and Thymelaeaceae. Wood gums containing terpenes are prominent in the Hamamelidaceae and Styracaceae. Wood and bark lactifers are well known in the Euphorbiaceae. Members of the Rutaceae are best known for their fruit oils, although triterpenes (limonoids) have been noted in wood extracts as well. Although our botanical knowledge of woody monocots is well developed (Tomlinson, 1964; Dahlgren and Clifford, 1982), terpene extractives in this group remain obscure.

B. *Morphological Requirements for Terpene Accumulation*

Oleoresin accumulation in woody tissues may be the result of normal physiological processes or a response to wounding and infection of the cambium. In either case, the process is thought to result from the biosynthetic activities of the epithelial cells (a specialized type of parenchyma) lining the ducts or cavities in which the terpenes are eventually sequestered (Kramer and Kozlowski, 1979; Sjöström, 1981). In some genera such as *Pinus* these cell types are thought to be highly suberized (Dell and McComb, 1978a).

The vertical resin ducts in conifer wood have been studied the most extensively (Fahn, 1979). These ducts, which are a normal feature of the wood of *Pinus, Picea, Larix,* and *Pseudotsuga,* develop as a result of wounding in *Abies, Cedrus,* and *Tsuga* (traumatic resin ducts) but are absent from *Juniperus* and *Cupressus* (Fahn and Zamski, 1970). Cortical pitch blisters (ovoid cavities) occur in *Abies* as the normal site of resin storage. In Douglas fir (*Pseudotsuga menziesii*), the appearance of resin ducts is characteristic of latewood formation (Wenzl, 1970). Generally, sapwood is low in resin content compared to the heartwood, which consists of dead cells saturated with resinous material (Sjöström, 1981). However, sapwood can saturate with resin in response to woundings, as a result of the differentiation of new resin ducts (this is the physiological basis of commercial "tapping" used to obtain the resin) (Dell and McComb, 1978a; Kramer and Kozlowski, 1979).

Resin synthesis at the cellular level has been described by Benayoun and Fahn (1979) and Carde *et al.* (1980). The oleoresin is reportedly made in plastids in the epithelial cells. These plastids are in close contact with tubular endoplasmic reticulum (ER), to which they release their contents. The ER, in turn, fuses with the plasmalemma, eventually releasing the resinous contents to the duct, which may be under 3–10 bars of pressure (Dell and McComb, 1978a). It is not yet clear whether the cell membrane forms vesicles around the resin before release (Carde *et al.,* 1980) or if a form of exocytosis of the resin droplets occurs (Benayoun and Fahn, 1979). Accumulation of terpene extractives is almost always associated with the presence of extracellular storage sites. Detailed anatomical descriptions of resin ducts in all the conifer genera can be found in the works of Greguss (1972) and Phillips (1963).

C. *Biochemical Requirements for Terpene Accumulation*

Terpene biosynthesis is a metabolically expensive process, consuming three acetyl-CoA, two NADPH, and three ATP in construction of the basic isoprenoid condensing unit, isopentenyl pyrophosphate (IPP). The production of terpenoid extractives generally takes place in nonphotosynthetic tissues, and so is highly dependent on mobile photosynthate in the form of sugars for carbon and energy supply (Loomis and Croteau, 1973). Both the chain-lengthening steps (prenyl transferase) and the cyclization processes (cyclases) of isoprenoid metabolism exploit the electrophilic attack of a carbonium ion or equivalent center on a double bond as the basic carbon bond-forming process (Poulter and Rilling, 1981a). Generation of a positive center in these ligase reactions occurs either by ionization of an allylic pyrophosphate or by protonation of a double bond or epoxide, reaction types characteristic of terpenoid biochemistry (Cane, 1980). The manner in which the transferase and cyclase enzymes control the regio- and stereochemistry of bond formation remains a challenging problem, only recently beginning to yield to experimentation.

Terpene biochemistry has been the subject of extensive reviews in recent years, including a detailed two-volume set edited by Porter and Spurgeon (1981, 1983), treatises on plant terpenoids edited by Nes *et al.* (1984) and on terpenoids in glandular trichomes (Croteau and Johnson, 1984), and works of a more general nature by Loomis and Croteau (1980) and Banthorpe and Charlwood (1980). Periodic reviews covering the topic can be found in *Terpenes and Steroids, Biosynthesis, Annual Reports on the Progress of Chemistry,* and *Natural Products Reports,* which are published by the Royal Chemical Society in London.

II. Historical Perspective

A. *Classical Terpene Extractives*

The naval stores industry has been the largest processor of terpene wood extractives since ancient times (see review by Zinkel, 1981). Originally, pitch from conifer rosins was used for the waterproofing of wooden ships. Today conifer rosin (nonvolatile terpenes) and turpentine (essential oil) are utilized primarily in the manufacture of specialty chemicals and a wide range of industrial products, including solvents, adhesives, polymers, emulsifiers, coatings, and paper sizing (Weaver, 1982). The primary sources of these oleoresins are longleaf pine (*Pinus palustris*) and slash pine (*P. elliotii*) (with production centered in the state of Georgia in the United States) (Koch, 1972).

The process for obtaining resin from living trees is called turpentining and involves chipping the wood and collecting the resulting flow. Due to the depletion of virgin forests, the major sources for these raw materials will derive from the chemical stimulation of new resin flow from living stumps or young pulpwood trees (Kramer and Kozlowski, 1979; Zinkel, 1981). This process, called induced lightwood formation (lightering), is carried out by administering paraquat-type herbicides in sublethal doses (see Roberts *et al.*, 1982, for the details of this treatment). Heartwood and sapwood may become saturated with up to 40% (by weight) oleoresin, producing lightwood (so-called because it burns readily). Lightwood yields the oleoresin by extraction of the wood chips that are subsequently processed into pulp. It appears that paraquat is effective in stimulating resin production only in pines, and not in other members of the Pinaceae such as *Pseudotsuga, Abies, Picea, Tsuga,* or *Larix* (Kramer and Kozlowski, 1979). This process has been recently commercialized [*Chemical & Engineering News* **16**(3), 30, 1983].

Specialty chemicals derived from the oleoresin are obtained from the monoterpenes α- and β-pinene (see Fig. 2) and the diterpenes levopimaric and abietic acids (see Fig. 11) (Hedrick *et al.*, 1965a,b; Zinkel, 1981, Weaver, 1982). The proportion of $(-)$-β-pinene in the oleoresin may be increased over physiological levels by induced lightwood formation (Zinkel, 1981).

The natural rubber industry is a major source of polymeric terpenes. Rubber is obtained from cutting the bark of *Hevea brasiliensis* (Euphorbiaceae). Latex (which is a living plasm) containing the rubber particles pours from the wound and is collected. Since the cambium is not wounded, the bark regenerates and the tree may be tapped again (Kramer and Kozlowski, 1979). The utilization and economics of natural rubber is the subject of a book by Allen (1972) and has been discussed more recently by Stinson (1983).

The perfume and flavor industry has been the primary user over many centuries of rare and valued essential oils from the distillation of wood (see historical accounts in Gildemeister and Hoffmann, 1913). The constituents of these oils include the monoterpenes camphor from the camphor tree (*Cinnamomum camphora:* Lauraceae) and linalool from rosewood (*Aniba duckei:* Lauraceae). Sesquiterpenes used in perfumery are obtained from sandalwood (*Santalum album:* Santalaceae), agarwood (*Aquillaria:* Thymelaeaceae), and balsamics such as copaiba (*Copaifera:* Leguminosae), frankincense (*Boswellia:* Burseraceae), elemi (*Canarium:* Burseraceae), and myrrh (*Commifora*. Burseraceae). Extensive reviews on the occurrence, composition, and use of these and other extractives can be found in accounts by Dorland and Rogers (1977), Guenther (1976), Arctander (1960), and in continuing reports in *Perfumer and Flavorist* by B. Lawrence and others. The occurrence of sesquiterpene alcohols of significance to the perfume industry has been reviewed by Herout (1982).

Since many of these natural fragrance materials are obtained from unreliable cottage industries or are otherwise of limited supply, synthetic alternatives have been sought that utilize commonly available terpenes, usually those from conifers, as starting materials (Mookherjee and Mussinan, 1981). Thus, for example, ambergris-like fragrances can be obtained from the diterpenes manool (Fig. 11) (Ohloff, 1982) or podocarpic acid (Cambie and Palmer, 1982).

B. *Projected Usage of Terpene Extractives*

Terpene extractives are of interest to the pharmaceutical industry because of their inherent biological activities or their potential as starting materials for syntheses. The monoterpene iridoid catalposide from *Catalpa ovata* (Bignoniaceae) has been used as a starting material for prostanoid synthesis (Weinges *et al.*, 1980). The phorbol esters (Fig. 12), which are diterpenes from Euphorbiaceae and Thymelaeaceae, are among the most potent tumor promoters known (Cassady and Suffness, 1980). They apparently act as hormone analogs by binding an intracellular kinase (see discussions by Marx, 1983, and Kolata, 1983). Other terpenes with important biological activities include the sesquiterpene lactones (Rodriguez *et al.*, 1976; Gottleib and Mors, 1980); the adenylate cyclase activator forskolin, a labdanoid diterpene from *Coleus* (Labiatae) (Seamon and Daly, 1981); the kauranoid diterpenes, atractyloside and wedeloside, from Compositae, which inhibit the mitochondrial ADP/ATP carrier protein (Lewis *et al.*, 1981; Boulay *et al.*, 1983); and the cardiomuscular calcium transport inhibitor ryanodine (Fig. 12: $R^1 = H$, $R^2 = 1H$-pyrrole-2-carboxylate), which is a diterpene alkaloid from the wood and roots of *Ryania speciosa* (Flacourtiaceae) (Rogers *et al.*, 1948; McCann and Penefsky, 1981). The neuropharmacological activity of the monoterpene thujone (Fig. 2), which occurs in the conifer genera *Thuja*, *Juniperus*, and *Abies*, was reviewed by Vogt (1981) and Albert-Puleo (1978).

Natural terpene extractives and their derivatives are also finding use as agricultural pesticides. The monoterpene camphene (Fig. 2), when chlorinated, yields toxaphene, which is used as an insecticide. More sophisticated approaches involve using terpenes as aggregation pheromones, juvenile hormones, or feeding deterrents. The use of pheromones in pest management was reviewed by Silverstein (1981). The sesquiterpene juvabione, from *Abies balsamea,* acts as a juvenile hormone analog (see Norin, 1972). Spraying trees with turpentine has been used to deter insect attack (Nijholt *et al.*, 1981), and the drimanol-type sesquiterpene (Fig. 8) warburganal from the bark of *Warburgia ugandensis* (Canellaceae) has also been shown to be a potent insect antifeedant (Kubo and Nakanishi, 1979). Insect control methods based on naturally occurring compounds offer an alternative to the more traditional use of synthetic insecticides that can be environmentally unsuitable.

The abundant quantities of terpenes that can be obtained by tapping trees have made these compounds an attractive potential source of combustible hydrocarbon fuels. About 40 liters per year per tree of a sesquiterpene mixture can be obtained from *Copaifera multijuga* (Leguminosae) (Calvin, 1983). These compounds may also serve as a source of specialty chemicals as well (Buchanan *et al.*, 1978; Goldstein, 1981).

The types and amounts of terpenes available for commercial exploitation may be enhanced through the use of plant tissue cultures that can produce these substances directly from simple precursors or carry out various modifications of available intermediates (Schuler, 1981; Crocomo *et al.*, Kutney, 1982). The apparent requirement for tissue differentiation to obtain terpene synthesis in culture has been a major obstacle to this approach thus far (Dell and McComb, 1978a; Banthorpe and Charlwood, 1980). There have been some successes, however, including monoterpene production in *Perilla* cultures (Nabeta *et al.*, 1983) and sesquiterpenes production in liverwort cultures (*Calypogeia*) (Takeda and Katoh, 1981). The biotransformations of conjugated enones in *Mentha* cultures have been described by Aviv *et al.* (1983). Terpene metabolism in plant tissue culture has been discussed by Reinhard and Alfermann (1980), while transformations by microbes have been reviewed by Sebek and Kieslich (1977).

The future prospects of the terpene extractive industry will depend on finding new ways to stimulate production and on developing new markets. The development of new plants with desired traits of productivity and environmental adaptability should also have considerable impact in providing the full range of terpenes necessary for commercial use (Calvin, 1983; Farnum *et al.*, 1983). The propagation of sandalwood trees resistant to spike disease, through the generation of plantlets derived from callus cultures of spike-resistant plants, is a current example (Lakshmi *et al.*, 1979). New markets may arise from the more rational design of biodegradable insecticides based on terpenoids (Rogers, 1978; Prestwich *et al.*, 1983).

III. Role of Terpenes in Plant Defense

A. *Resin Mobilization*

Terpene-containing exudates constitute a major passive defense mechanism in response to wounding and fungal invasion in many coniferous and dicotyledonous trees (in particular those of the Burseraceae, Dipterocarpaceae, Euphorbiaceae, Hamamelidaceae, Leguminosae, and Thymelaeaceae). In conifers, the pathological resin exudation is called resinosis. Resinosis exhibits some aspects of an active defense (changes in composition) that are characteristic of many phytoalexins studied in herbaceous plants (Kúc and Lisker, 1978; Stoessl, 1980). The process analogous to resinosis in dicots has not been studied so well [i.e., in *Aquillaria:* Thymelaeaceae (Jalaluddin, 1977)].

Oleoresins of the Pinaceae generally consist of 50–80% nonvolatile diterpenes and 20–50% volatile mono- and sesquiterpenes (Wenzl, 1970; Kramer and Kozlowski, 1979). One obvious function of the volatile liquid fraction is to serve as a solvent for the diterpenes, allowing them to flow to the wound site (Mutton, 1962). On exposure at the wound site, the volatile fraction evaporates, leaving behind a resinous plug. The presence of diterpenes containing conjugated double bonds (especially levopimarate and palustrate, Fig. 11), which are subject to autooxidation and subsequent polymerization in the presence of light and oxygen, causes the plug to harden (Lagercrantz and Yhland, 1962; Cunningham *et al.*, 1977). The diterpene content also determines the rate of resin crystallization (Zavarin and Cobb, 1970). Thus, the amount of strobic acid (Fig. 11) in *Pinus strobus* oleoresin is directly related to the rate of crystallization, a factor important in weevil resistance (Wilkinson, 1979).

The extent of resin secretion in response to wounding and infection depends primarily on the physiological status of the tree and the morphological nature of the secretory structures. Upon wounding by a penetrating insect attack, an immediate resin flow usually occurs (Russell and Berryman, 1976; Shrimpton, 1978). This results from the severance of resin ducts and the release of their terpenoid contents, which are stored under pressure. Moisture stress may be a critical factor in determining the extent of this initial response (Kramer and Kozlowski, 1979). The primary resin flow dissipates in 1 to 2 days and a new secondary flow begins. The secondary resin is apparently generated from the nutrient stores of the tree, and flow depends to some extent on the formation of new resin ducts, a process which takes about 1–2 weeks to complete (Shrimpton, 1978). If the tree is not healthy, nutrient stores will be low and the secondary resin flow will be diminished (Wright *et al.*, 1977). The speed and duration of the secondary resin flow often determines the degree of success in resisting pathological attack (Raffa and Berryman, 1982).

Hormonal control of the mobilization of secondary resin was discussed in a review by Hoque (1982). Fahn and Zamski (1970) have shown that the auxins indole-3-acetic acid, 1-naphthylacetic acid, and 2,4-dichlorophenoxyacetic acid induced the formation of new vertical resin ducts in *Pinus halepensis* within a month after application. During this time, the cambium produced a considerable number of tracheids. This process resembles the proliferation of vertical resin ducts during latewood formation in Douglas fir (Wenzl, 1970). A wide range of other plant hormones is ineffective in promoting resin duct formation. Ethylene is commonly found at elevated levels in injured woody tissues, and this hormone may also be important in resin mobilization (Shain, 1979; Hoque, 1982; Telewski *et al.*, 1983).

B. Resin Composition

Changes in the monoterpene composition of secondary oleoresin with respect to the primary oleoresin have been correlated with an increase in antifungal and insect-repellent properties. Studies on the resistance of grand fir (*Abies grandis*) to bark beetle attack have shown that α-pinene (Fig. 2) is the first monoterpene to increase in concentration, and is also the constituent most repellent to the beetle (Bordasch and Berryman, 1977). The wound resin also contains higher levels of 3-carene and myrcene (Fig. 2), which are toxic to both the beetle and its fungal symbiont (Russell and Berryman, 1976). A similar increase in α-pinene content of the wound resin relative to the primary resin has been observed in wounded lodgepole pine (*Pinus contorta*) (Raffa and Berryman, 1982), Norway spruce (*Picea abies*) (Schuck, 1982), and eastern white pine (*Pinus strobus*) (Wilkinson, 1980), and a similar increase in myrcene content was noted in loblolly pine (*Pinus taeda*) (Gollob, 1980). A more detailed review of these changes in composition in response to pathological attack can be found in Cates and Alexander (1982). While such alterations in monoterpene content are thought to represent *de novo* synthesis, this point has not been firmly established.

The biological effects of a number of monoterpenes are highly dependent on their enantiomeric composition, thus revealing the highly specific nature of these interactions. Of particular interest in this regard is a study on the inhibition of spore germination and germ-tube growth of *Diplodia pinea* by individual monoterpenes (Chou and Zabkiewicz, 1976). The greatest difference in effect between optical antipodes was observed with (+)- and (−)-limonene. (−)-Limonene, which is common in the Pinaceae, was relatively nontoxic. Its enantiomer, (+)-limonene, which is common in *Citrus* oils and rare in the Pinaceae, strongly inhibited germination and germ-tube growth. Other monoterpenes that were inhibitory included (+)-α-pinene, (−)-β-pinene, (+)-3-carene, and terpinolene.

Another example where enantiomeric composition influences the biological effect of a monoterpene is in the conversion of (+)- and (−)-α-pinene to aggregation pheromones by *Ips typographus* bark beetles. Trees contain varying proportions of the α-pinene antipodes (see the following section on monoterpene biosynthesis), and bacteria in the gut of *Ips* beetles convert this host monoterpene to (−)-*cis*-verbenol, the key pheromone in sustaining beetle attack (Brand *et al.*, 1975). By an elegant chromatographic separation of all the verbenol stereoisomers, Klimetzek and Francke (1980) were able to show that the amount of (−)-*cis*-verbenol synthesized depended on the proportion of (−)-α-pinene present in the wood oil, providing a chemical basis for beetle–host specificity.

In conclusion, the insect-repellent and fungitoxic effects of monoterpenes, balanced against their role in aggregation pheromone biosynthesis, can determine the vulnerability of conifers to insect attack and subsequent infection by the fungal symbiont. Alteration in oleoresin composition resulting from attack may serve to shift this balance to the tree's advantage, imparting resistance. Conversely, the lack of quantitative or qualitative response may signal susceptibility.

C. Resin Elicitation

The nature of the chemical signal that triggers the response to infection has been the subject of many studies, and it is now clear that most of these substances are of fungal origin, rather than produced by the beetle vector. Thus, an early investigation of resin flow in conifers showed that the application of live pitch canker fungus (*Fusarium*) to wounds on Virginia pine (*Pinus virginiana*) stimulated an average resin flow 30 times greater than that of control wounds where dead fungus was applied (Hepting, 1947).

A wide range of fungal elicitors and their complex roles in phytoalexin production in higher plants have been described by West (1981a), Hadwiger and Loschke (1981), and Darvill and Albersheim (1984). Fatty acids from *Phytophthora* mycelia have been identified as elicitors of sesquiterpene phytoalexins (rishitin and lubimin) in potatoes (Bostock *et al.*, 1981,

1982). Fungal endopolygalacturonase and host-derived pectic polysaccharide fragments elicit the formation of casbene, a macrocyclic diterpene, in infected castor beans. (*Ricinus communis:* Euphorbiaceae) (Bruce and West, 1982). The elicitation of sesquiterpene phytoalexins in the sweet potato during weevil attack is thought to involve a weevil-derived polygalacturonase activity as well (Sato *et al.,* 1982). Soluble fragments of chitin and its deacetylation product chitosan (which are components of fungal cell walls and arthropod exoskeletons) elicit the lignification of wheat (Pearce and Ride, 1982), the production of protease inhibitors in tomatoes (Walker-Simmons *et al.*, 1983), and the synthesis of the phytoalexin pisatin in peas (Hadwiger and Loschke, 1981). The release of soluble products from the chitin of hyphal walls, by an endogenous chitinase from maple and oak species, has been demonstrated (Wargo, 1975), and an ethylene-induced chitinase has been purified from bean (*Phaseolus*) leaves (Boller *et al.*, 1983).

Although a coherent mechanism of elicitation has not yet emerged from these studies, it appears that the necessary signal for the formation of phytoalexins arises from fungal or insect attack and plant counterattack and is mediated through the release of soluble fragments by enzymatic hydrolysis of insoluble polymers. Presumably, a similar mechanism is involved in resin elicitation in conifers.

IV. Origin of Terpene Precursors

A. *Pathway to Isopentenyl Pyrophosphate*

Terpenes arise from the basic two-carbon precursor acetyl-CoA through the intermediacy of the condensation products acetoacetyl-CoA and 3-hydroxymethylglutaryl-CoA (HMG-CoA) and eventually mevalonic acid (MVA) (the committed precursor of the pathway). Mevalonic acid is, in turn, phosphorylated and converted to isopentenyl pyrophosphate, the five-carbon-containing building block of the isoprenoids (Fig. 1). Each step of the pathway has been demonstrated in higher plants, although few of the relevant enzymes have been examined in woody species (e.g., mevalonate kinase from *Pinus pinaster* seedlings by Suarez *et al.,* 1974). The general pathway and its regulation at the point of conversion of hydroxymethylglutaryl-CoA to mevalonic acid will not be considered here, since these aspects have been reviewed in detail elsewhere (Qureshi and Porter, 1981; Dugan, 1981).

The administration of exogenous precursors such as mevalonate, sugars, amino acids, and other organic acids to intact tissue is a classic

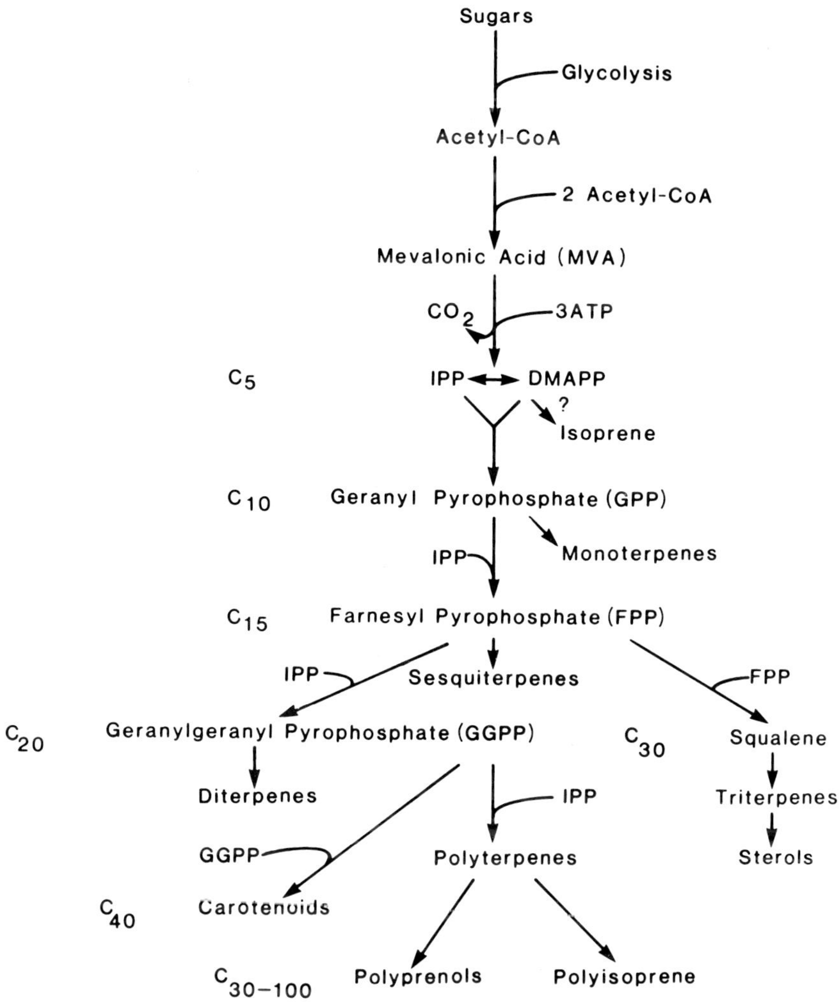

Fig. 1. Origin of the terpenoid classes.

approach to examine terpene biosynthesis (Suga *et al.*, 1980a; Tange, 1981; Groeneveld *et al.*, 1982). Compartmentation of terpene metabolism in specialized tissues such as resin duct epithelia, lactifers, or glandular trichomes often leads, however, to poor incorporation of more immediate precursors such as mevalonate, while sugars are much more efficient precursors (Croteau and Loomis, 1975: Groeneveld *et al.*, 1982). The more efficient utilization of sugars may be due in part to the ease of transport of these substances, which, additionally, can fulfill both the carbon and energy requirements for terpene biosynthesis (Croteau *et al.*, 1972). Storage carbohydrates appear to provide the major fuel for new terpene synthesis, as reflected by the utilization of xylem starch during wound resin biosynthesis resulting from bark beetle attack in conifers. Low starch reserves afford a diminished flow of secondary resin, which, in turn, diminishes resistance to insect attack (Wright *et al.*, 1977; Raffa and Berryman, 1982). Similar mobilization of starch and other carbohydrates during periods of active terpene synthesis in herbaceous plants has been observed (Vogel, 1962; DeAngelis *et al.*, 1983).

B. *Prenyl Transferase*

The enzyme prenyl transferase catalyzes the chain-lengthening process in terpene biosynthesis via the condensation of the homoallylic, five-carbon-containing substrate IPP, with an allylic (prenyl) pyrophosphate. The reaction thus yields a new allylic pyrophosphate containing five carbons (one isoprene unit) more than the original. Work on this enzyme has been reviewed in detail by Poulter and Rilling (1981a), and Cori (1983) has discussed the enzyme from higher plant sources such as *Citrus* and *Pinus*.

The initial reaction of the elongation sequence involves the condensation of IPP with dimethylallyl pyrophosphate (DMAPP), producing geranyl pyrophosphate (GPP; C_{10} trans isomer). This is the only step in which the five-carbon-containing allylic pyrophosphate is employed, and this "starter piece" is derived by the isomerization of isopentenyl pyrophosphate. The enzyme that catalyzes this reaction, isopentenyl pyrophosphate isomerase, has been discussed by Poulter and Rilling (1981a). The stereochemistry of this reaction in *Pinus* and *Citrus* has been examined by Jedlicki *et al.* (1972). Geranyl pyrophosphate-specific prenyl transferases have been isolated from a bacterium (Sagami *et al.*, 1978) and from the herbaceous plant sage (*Salvia officinalis:* Labiatae) (Croteau *et al.*, 1985). Monoterpene biosynthesis requires GPP and, by implication, a GPP-specific prenyl transferase.

The most intensely studied prenyl transferase is the *trans, trans*-farnesyl pyrophosphate (FPP) synthetase. This enzyme catalyzes the condensation of IPP and DMAPP to form GPP and, in a second step, the condensation of IPP and GPP to yield FPP (C_{15}). The second step is much faster than the first, and as a consequence GPP does not accumulate to appreciable levels (Laskovics and Poulter, 1981). Studies by Poulter and associates have provided strong evidence that FPP synthetase employs a stepwise ionization–condensation mechanism involving, as an intermediate, a rigid pyrophosphate–carbonium ion pair (Poulter *et al.*, 1981; Mash *et al.*, 1981). Farnesyl pyrophosphate serves as a precursor of sesquiterpenes (C_{15}), diterpenes (C_{20}), triterpenes (C_{30}), and sterols, and thus is a branch point metabolite (Fig. 1). As all plants produce sterols, and thus require FPP, the FPP synthetase (unlike GPP synthetase) is ubiquitous. This fact may, in part, account for the broader distribution of sesquiterpenes than monoterpenes in the plant kingdom.

The most common prenyl transferases (producing chains up to 20 carbons) catalyze condensations resulting in the formation of double bonds of the trans (*E*) configuration (Cori, 1983). Polyterpenes that contain double bonds of the cis (*Z*) configuration are derived from prenyl transferases exhibiting cis stereoselectivity. Examples of the latter are *cis*-polyisoprene rubber and malloprenols from *Mallotus* (Euphorbiaceae) (Suga *et al.*, 1980b, 1983), which contain a mixture of cis and trans double bonds, as does the undecaprenyl pyrophosphate of bacteria (Baba and Allen, 1978). The malloprenols (C_{45} to C_{55}) were shown to be derived by the action of a cis-specific prenyl transferase, which employs all *trans*-geranylgeranyl pyrophosphate (GGPP-C_{20}) as the first allylic substrate. Likewise, undecaprenyl pyrophosphate is derived by cis elongation of all *trans*-farnesyl pyrophosphate.

The C_{10} and C_{15} *cis*-prenyl pyrophosphates were once considered to be the more likely precursors of certain cyclic monoterpenes and sequiterpenes, but recent studies have shown the more common trans isomers to function as precursors of such cyclic compounds (Croteau and Karp, 1979a; Cane *et al.*, 1981a). Thus, the biological significance of prenyl transferases that synthesize short-chain (C_{10} and C_{15}) *cis*-prenyl pyrophosphates is presently uncertain (Widmaier *et al.*, 1980; Pérez *et al.*, 1983).

V. Hemiterpenes

The most prominent hemiterpene, isoprene (2-methyl-1,3-butadiene), is a volatile compound (bp 34°C) emitted from the leaves of many trees (including conifers) and herbs (Evans *et al.*, 1982). Isolated chloroplasts from cottonwood (*Populus deltoides*) were shown to produce isoprene from $^{14}CO_2$ under physiological conditions that favor the utilization of acetyl-CoA (Mgaloblishvili *et al.*, 1981). The photorespiration products glycolate and glycine were converted to isoprene in witch-hazel (*Hamamelis virginiana*) leaf disks, presumably via the intermediacy of acetyl-CoA (Jones and Rasmussen, 1975). Tingey and co-workers (1981) showed that isoprene is a normal biosynthetic product of oak leaves and that this olefin does not arise as a response to stress. This group also showed that isoprene emission decreases rapidly in the dark, but that a low level of production is maintained even under these conditions. These data support the concept that isoprene production is dependent on photorespiration and that the C_5 olefin is derived from an intermediate of this process. The immediate precursor of isoprene is presumed to be IPP or DMAPP; however, no definitive labeling experiments have been carried out to test this possibility. The association of photorespiration with isoprene synthesis has been discussed in a review by Loomis and Croteau (1980).

The survey of Evans *et al.* (1982) and a review of terpene emissions (Rogers, 1978) showed that a number of tree species not noted for their essential oil production released readily detectable levels of monoterpene hydrocarbons (in addition to isoprene in some cases). It is possible that in these instances (e.g., the sugar maple and balsam poplar) the volatile monoterpenes do not accumulate because they are released through the stomatal openings, as is isoprene. It would be of interest to determine if these monoterpenes arise from pathways related to those that afford isoprene.

The conjugated double bonds of isoprene make this compound very prone to photooxidation. The blue haze that surrounds forested areas on hot and bright days is a direct result of the volatility and photoreactivity of isoprene and related monoterpene hydrocarbons under these conditions (Rogers, 1978). Rasmussen (1972) has estimated that total annual terpene emissions exceed man-made air pollutants by a factor greater than 6; however, the impact of these natural products on the atmosphere has not been addressed.

VI. Biosynthesis of Monoterpenes

A. *Distribution of Structural Types*

Monoterpenes are fragrant principles of the wood of conifers and numerous dicots, particularly those of the Lauraceae (Gottleib, 1972). Among the Rutaceae and Myrtaceae, monoterpenes are largely restricted to fruit and leaf oils (e.g., *Citrus* and *Eucalyptus*). The distribution of monoterpenes in the Pinaceae and Cupressaceae was reviewed by von Rudloff (1975) (see also Mirov, 1961). The compounds illustrated in Figs. 2 and 3 are representative of the structural types encountered in this chapter. These monoterpenes are of the regular cyclohexanoid class, comprising a "head-to-tail" arrangement of isoprene units, or can be derived from such structures by well-known rearrangements (Ruzicka *et al.*, 1953).

Another group of monoterpenes are the iridoids, based on the cyclo-

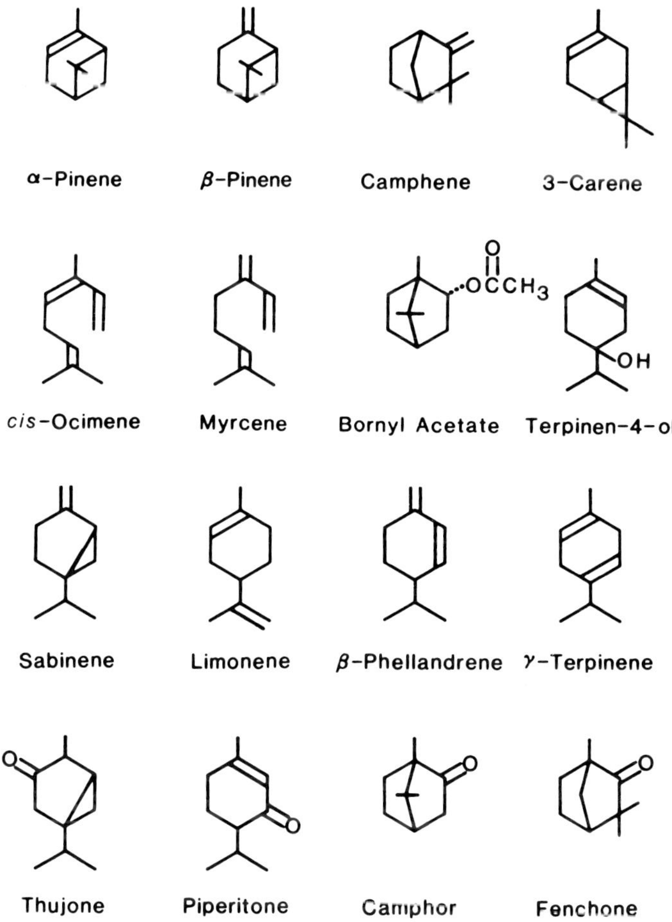

Fig. 2. Major components of the monoterpene fraction in volatile oils of the Pinaceae and Cupressaceae (>10%). After von Rudloff (1975).

pentanoid nucleus. These compounds occur as lactol glycosides in leaf and bark extracts of the Bignoniaceae (*Catalpa, Tecoma*), Fouquieriaceae, Rubiaceae, and Apocynaceae (Rowe and Conner, 1979; Gottleib and Mors, 1980; Seigler, 1981a,b). A number of alkaloids contain iridoid or *seco*-iridoid moieties (Cordell, 1977; Kapil and Brown, 1979). The structures of the iridoids have been reviewed by El-Naggar and Beal (1980).

The tropolone and related monoterpenes (Fig. 4) are novel constituents of Cupressaceae heartwood. These compounds are responsible for the

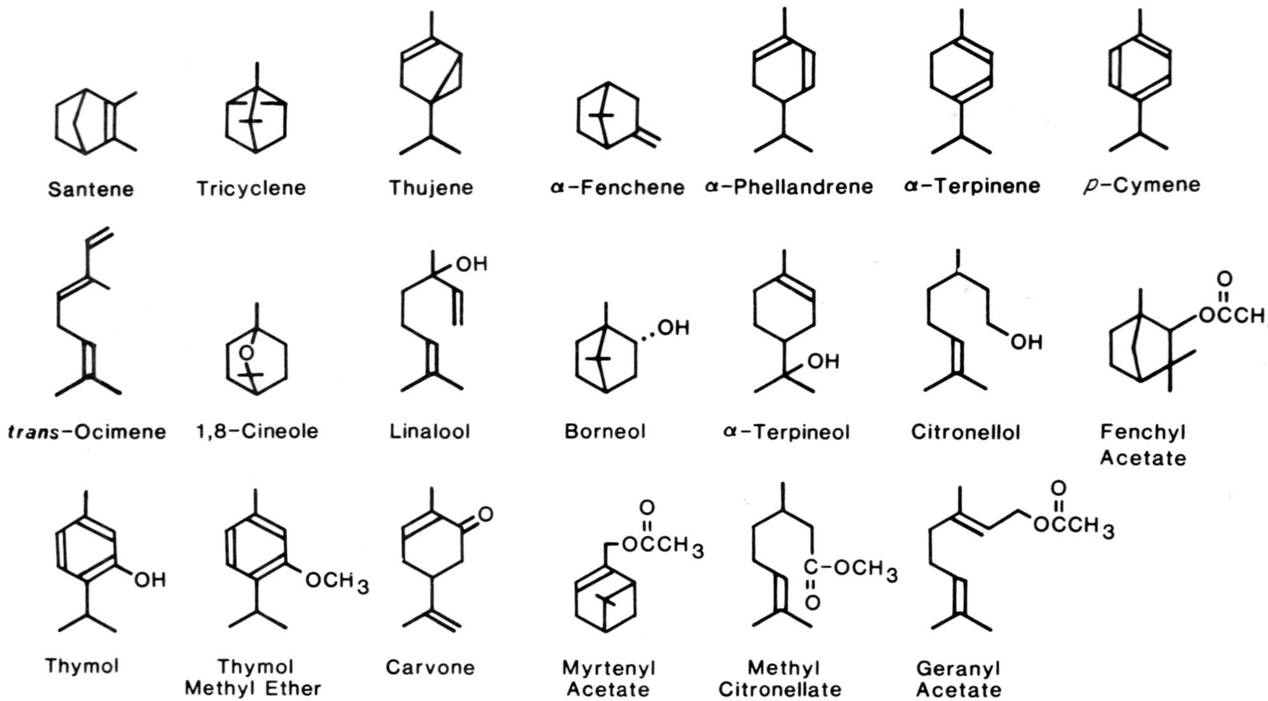

Fig. 3. Minor components of the monoterpene fraction in volatile oils of the Pinaceae and Cupressaceae (0.5–10%). After von Rudloff (1975).

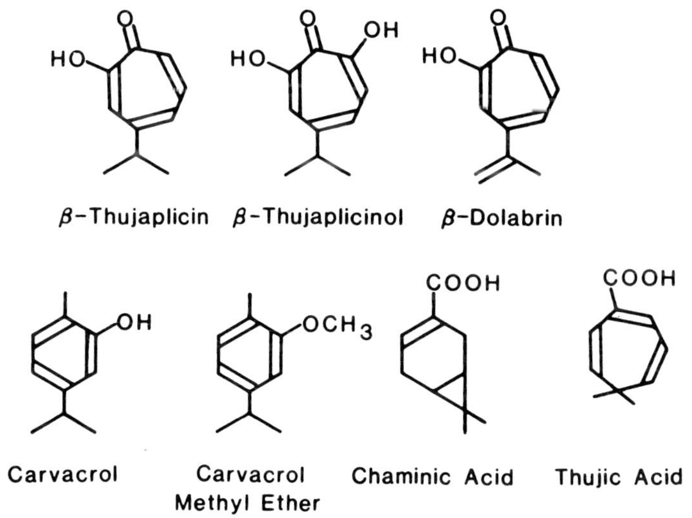

Fig. 4. Heartwood monoterpenes of the Cupressaceae.

durability and resistance to insect attack that is characteristic of cedar wood (*Thuja*) (Mabry and Gill, 1979), and their role in heartwood formation in *Thuja plicata* was discussed by Swan *et al.* (1969). Cell cultures of *Thuja occidentalis* have been shown to make several thujaplicin derivatives (Witte *et al.*, 1983), and thujaplicinol is a useful antibacterial adenyltransferase inhibitor (Allen *et al.*, 1982). Analogous sesquiterpene tropolones also occur in *Cupressus* and *Chamaecyparis* (Zavarin *et al.*, 1967). The biochemistry of the tropolones has not been explored in any detail, although these compounds are presumed to arise from the carane (to thujic acid) and thujane (to β-thujaplicin) monoterpenes, which cooccur with the tropolones in relevant species (see Charlwood and Banthorpe, 1978).

The site of monoterpene biosynthesis is the epithelial cells lining the resin ducts of *Pinus* wood (Benayoun and Fahn, 1979). The resin ducts at the base of young needles are the loci of monoterpene synthesis in *P. pinaster* (Bernard-Dagan *et al.*, 1980). Isolated leucoplasts from the fruit peel of *Citrofortunella mitis* were shown to incorporate [^{14}C]IPP into monoterpenes, supporting the plastidic origin of these compounds (Gleizes *et al.*, 1983). The highly specialized nature of monoterpene biosynthetic sites in woody species is reminiscent of the isolated glandular sites of monoterpene biosynthesis in the leaves of herbaceous species (Loomis and Croteau, 1973; Croteau, 1977; Croteau and Johnson, 1984).

Additional aspects of monoterpene biosynthesis relevant to the subject of wood extractives have been described in more general reviews by Croteau (1981, 1984), Cori (1983), Bernard-Dagan *et al.* (1982), and Charlwood and Banthorpe (1978).

B. *Biosynthesis of Specific Monoterpenes*

Early studies have long since established the mevalonoid origin of the monoterpenes (Banthorpe *et al.,* 1972), and more recent work, primarily with cell-free systems, has focused on the enzymology, stereochemistry, and mechanism of monoterpene cyclization processes. Secondary transformation of the parent cyclic compounds have also received attention over the last several years, and a few attempts have been made to answer complex questions involving the regulation of monoterpene metabolism (Croteau *et al.,* 1981; Croteau and Winters, 1982).

By the administration of [2-^{14}C]mevalonic acid to *Cinnamomum camphora* (Lauraceae), Suga *et al.* (1972) demonstrated that the label in the resulting linalool (Fig. 3) was equally distributed between the isopentenyl pyrophosphate- and dimethylallyl pyrophosphate-derived portions of the molecule. This result contrasts with those of studies on other systems in which preferential labeling of the isopentenyl pyrophosphate-derived portions of various monoterpenes were observed, presumably reflecting the participation of endogenous pools of dimethylallyl pyrophosphate in the reaction sequence (Charlwood and Banthorpe, 1978; see also Ekundayo, 1983). Subsequent studies with this system using (3*R*,5*R*)-[5-^{3}H]mevalonate as a precursor led to the conclusion that geranyl pyrophosphate was converted to linalool with overall syn stereochemistry (Godtfredsen, 1978), a result arguing for tight ion pairing between the allylic cation and pyrophosphate anion in the rearrangement (Cane *et al.,* 1981a).

Studies with a number of cell-free systems have revealed that geranyl pyrophosphate is converted to cyclic monoterpenes without the involvement of free intermediates. As the trans double bond at C-2 of geranyl pyrophosphate precludes the direct conversion to cyclohexanoid compounds, most investigators have concluded that cyclization of the trans precursor occurs through the intermediacy of a bound linalyl intermediate (see, for example, Figs. 5 and 6). Linalyl pyrophosphate (LPP) itself can serve as an effective precursor of cyclic monoterpenes in most cases, as can (in some instances) neryl pyrophosphate (NPP), the cis isomer of GPP. In no instance has the interconversion of the three acyclic substrates been observed prior to cyclization, and proposed schemes for the oxidative interconversion of such free intermediates have been largely disproven (Croteau, 1981).

The properties of cyclases responsible for the construction of *p*-methane monoterpenes (the major monocyclic class) are illustrative of this general enzyme type. Thus, studies on the partially purified cyclases responsible for the synthesis of 1,8-cineole (Croteau and Karp, 1977a), γ-terpinene (Poulose and Croteau, 1978), α-terpinene (Johnson and Croteau, 1980), and limonene (Chayet *et al.*, 1977; Portilla *et al.*, 1982; Kjonaas and Croteau, 1983) have revealed these enzymes to possess molecular weights in the 50,000–100,000 range, pH optima between 6 and 7, a requirement for a divalent cation (Mg^{2+} or Mn^{2+} often preferred), and a sensitivity to inhibition by thiol-directed reagents and inorganic pyrophosphate. As indicated previously, most cyclases can utilize GPP, NPP, or LPP as the acyclic precursor, with relative efficiencies depending on the specific enzyme (the accurate determination of relative efficiency is often complicated by the presence of competing phosphatases in the preparations) (Croteau, 1983; Croteau and Cane, 1984).

Studies on cyclizations to bicyclic monoterpenes of the pinane, bornane, and fenchane types (Figs. 2, 5, and 6) constitute the most detailed biochemical investigations in the monoterpene series thus far. While the general properties of these cyclases resemble those of the *p*-menthane series, the structural complexity of the bicyclics lends itself to a more detailed probing of the stereochemistry of cyclization processes.

The enzymes catalyzing the synthesis of α- and β-pinene have been examined in partially purified preparations from *Citrus limonum* (Chayet, *et al.*, 1977; Portilla *et al.*, 1982; Rojas *et al.*, 1983) and *Salvia officinalis* (sage) (Gambliel and Croteau, 1982). α-Pinene often occurs naturally as a mixture of enantiomers [~70% plus isomer in sage], whereas β-pinene occurs almost exclusively as the minus isomer. Fractionation of *Salvia* leaf extracts by gel filtration afforded two regions of pinene cyclase activity (cyclase I of molecular weight 96,000 and cyclase II of molecular weight 55,000) that catalyzed cyclizations of opposite enantiomeric specificity (Gambliel and Croteau, 1984). Cyclase I catalyzed the conversion of GPP to (+)-α-pinene and to smaller quantities of (+)-camphene and (+)-limonene, whereas cyclase II transformed the acyclic precursor to (−)-α-pinene and (–)-β-pinene, as well as to (–)-camphene, (–)-limonene, and the acyclic olefin myrcene. Subsequent purification of each enzyme, as well as differential inactivation studies, provided strong evidence that each set of stereochemically related olefins was synthesized by a single cyclase, probably by a mechanism like that outlined in Fig. 5. Neryl pyrophosphate and LPP could serve as alternate substrates for olefin synthesis by both cyclases, although in these cases the product distribution differed somewhat from the naturally occurring distribution in sage oil that was afforded by GPP as the precursor.

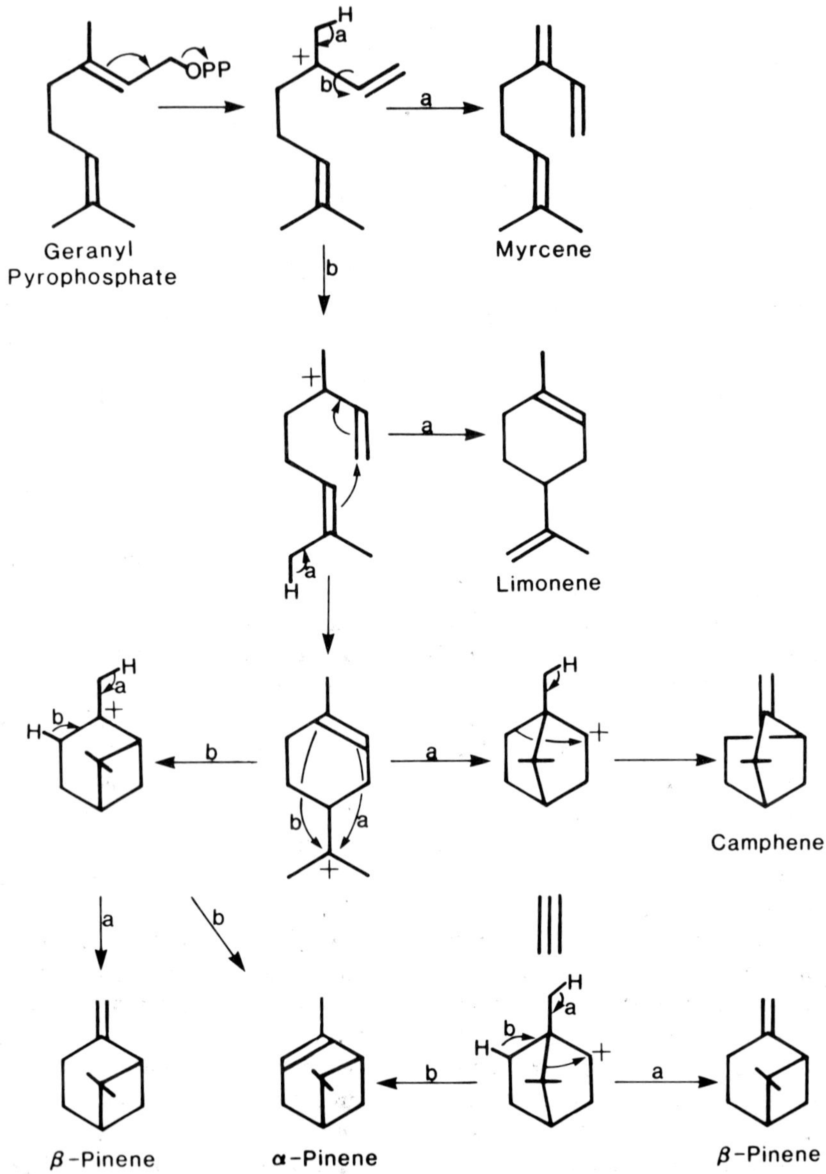

Fig. 5. Proposed ionic pathway for the formation of pinene and related monoterpenes.

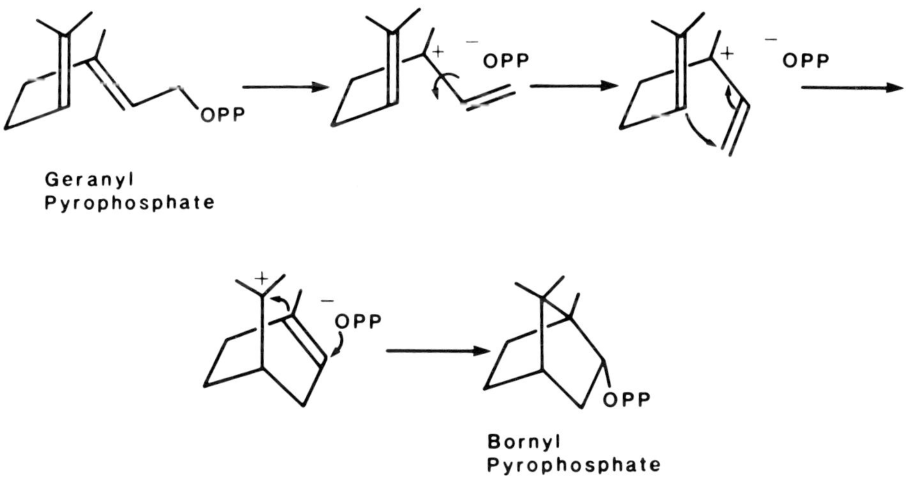

Fig. 6. Proposed mechanism for bornyl pyrophosphate biosynthesis.

Investigations of the enzymology of camphor (Fig. 2) biosynthesis have provided the most detailed information to date regarding a cyclization process. The biosynthesis of (+)-camphor in sage has been shown to involve the conversion of GPP to (+)-bornyl pyrophosphate, which is subsequently hydrolyzed by a distinct pyrophosphatase to (+)-borneol, followed by the NAD-dependent dehydrogenation of the alcohol to the ketone (Croteau and Karp, 1977b, 1979a,b; Croteau *et al.*, 1978). (−)-Camphor in tansy (*Tanacetum vulgare:* Compositae) and rosemary (*Rosemarinus officinalis:* Labiatae) is derived by a similar sequence of reactions (Croteau and Karp, 1977b; Cane *et al.*, 1982). Studies on the biosynthesis of (+)-bornyl pyrophosphate were of particular significance in establishing for the first time that GPP is the preferred substrate for cyclization and that NPP was not a mandatory intermediate, and in demonstrating that the pyrophosphate moiety of the substrate was retained in the bicyclic product bornyl pyrophosphate. This reaction type is thus far unique in the monoterpene series and has provided an unusual opportunity to examine the role of the pyrophosphate moiety in the coupled isomerization–cyclization process.

A scheme for the formation of bornyl pyrophosphate from geranyl pyrophosphate is shown in Fig. 6. The first step is the generation of the tertiary allylic isomer free to rotate about the C-2—C-3 bond. Back-side attack at C-1 of the cisoid conformer by the neighboring double bond affords the cyclohexanoid ring, which by subsequent internal addition and return of the pyrophosphate provides bornyl pyrophosphate with the correct stereochemistry. One consequence of this scheme is that cyclization to (+)-bornyl pyrophosphate should proceed with net retention of configuration at C-1 of GPP and with inversion of configuration at C-1 of NPP (i.e., without the attendant rotation at C-2—C-3); these predictions were verified by the enzymatic generation of bornyl pyrophosphate from ^{3}H-labeled 1*R* and 1*S* GPP and NPP, followed by selective exchange of the α-hydrogens of the derived camphors (Croteau, 1984).

In examining the role of the pyrophosphate moiety, it was first established, through the use of ^{32}P-labeled inorganic pyrophosphate, that exogenous pyrophosphate did not exchange with substrate-derived pyrophosphate. Experiments with ^{32}P-labeled substrates then established that the two ends of the pyrophosphate retained their identities during the cyclization to bornyl pyrophosphate. Thus, separate incubations of [1-^{3}H,α-^{32}P]GPP and [1-^{3}H,β-^{32}P]GPP afforded [^{3}H,^{32}P]bornyl pyrophosphates, which were selectively hydrolyzed to the corresponding bornyl phosphates containing, respectively, all and none of the ^{32}P. Use of ^{18}O-labeled substrates subsequently revealed an unexpectedly tight coupling of the reaction partners at the active site of the cyclase. In this instance, incubations of [1-^{18}O]geranyl pyrophosphate with the enzyme gave a product, analyzed by mass spectrometry of the derived benzoate, that indicated complete lack of positional oxygen isotope exchange during the cyclization; the product had an ^{18}O enrichment essentially identical to that of the precursor (Cane *et al.*, 1982). Each of the above experiments was carried out with the (+)-bornyl pyrophosphate cyclase (sage), and the results were confirmed with the (−)-bornyl pyrophosphate cyclase (tansy); the availability of both enantiomeric cyclases provided a particularly powerful probe of stereochemical and mechanistic questions.

The origin of the rearranged bicyclic monoterpene fenchone (Fig. 2), derived from the corresponding alcohol *endo*-fenchol, presents another interesting cyclization. Partially purified extracts from *Foeniculum vulgare* (Umbelliferae) convert [1-$^{3}H_2$]GPP to (−)-*endo*-[7-^{3}H]fenchol, establishing the course of the reaction to involve rearrangement of a pinyl intermediate (Croteau *et al.*, 1980a,b). All attempts to demonstrate the intermediacy of cyclic pyrophosphorylated intermediates in the formation of fenchol have been unsuccessful, however, and it was recently shown through the use of ^{18}O-labeled substrates that the hydroxyl oxygen of fenchol is derived from water, rather than from the ester oxygen of GPP as was the case in the bornane series (Croteau *et al.*, 1984). The manner by which the enzyme converts the pinyl intermediate (shown in Fig. 5) to the fenchyl nucleus without concomitant formation of the more favored bornane system is an example of the absolute control over Wagner–Meer-

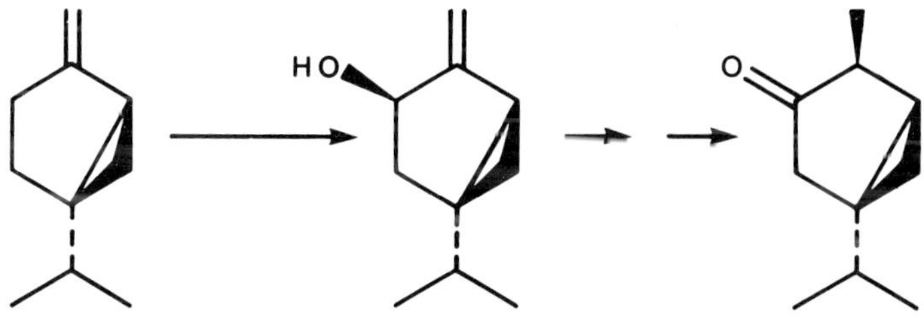

Fig. 7. Pathway of 3-thujone biosynthesis.

wein rearrangements, characteristic of terpene cyclizations but rarely observed in synthetic model reactions [e.g., see Luft (1979) and references therein].

Studies on the biosynthesis of 3-carene (Fig. 2) in *Pinus sylvestris* from specifically labeled mevalonate, nerol, and geraniol have demonstrated that the 2,3 double bond of the acyclic precursor migrates to the 3,4 position in the product (Banthorpe and Ekundayo, 1976). Additional work supports the migration of a C-1 proton of GPP to C-2 of the carane skeleton, which may be the driving force for double-bond migration and cyclopropane ring formation (Akhila and Banthorpe, 1980). Such an interesting and unusual reaction is worthy of further investigation.

The elucidation of the biosynthesis of C-3–oxygenated thujane monoterpenes (i.e., thujone, Fig. 2) has followed a circuitous path from initial hypotheses involving the photooxidation of sabinene (Fig. 2), to the suggested involvement of α-terpineol and terpinen-4-ol (Fig. 3) as intermediates, to a proposal involving a series of 1,2-hydride shifts (Charlwood and Banthorpe, 1978). Direct testing of [10-^{3}H]sabinene, as well as [10-^{3}H]α-thujene (Fig. 3), [3-^{3}H]terpinen-4-ol, and [3 ^{3}H]α-terpineol in intact sage, tansy, and wormwood (*Artemisia absinthum:* Compositae), demonstrated that sabinene, via *cis*-sabinol (Fig. 7), was the precursor of thujone (or isothujone) in these plants, and this eliminated the other cyclic compounds as possible precursors (Karp and Croteau, 1982). Furthermore, cell-free studies with these tissues established the conversion of GPP to sabinene by a soluble enzyme (Karp and Croteau, 1982) and the conversion of sabinene to *cis*-sabinol by a membranous, NADPH/O_2-dependent mixed-function oxidase (F. Karp and R. Croteau, unpublished). Subsequent steps in the synthesis of C-3–oxygenated thujanes have not yet been firmly established, but almost certainly involve, at some stage, stereoselective double-bond reduction to afford either the thujone or isothujone series (Fig. 7).

C. *Iridoid Monoterpenes*

The iridoid monoterpenes are cyclopentanoid representatives derived by a mechanism of cyclization that differs considerably from that involved in the generation of the cyclohexanoids. Although the origin of the cyclopentanoids has not been studied at the cell-free level, 2(*Z*)- or 2(*E*)-10-oxocitral, which can be derived from 10-hydroxygeraniol, serve as precursors of iridoids in *Gardenia jasminoides* (Rubiaceae) (Uesato *et al.*, 1982b). The conversion of 10-hydroxygeraniol to iridoids has been demonstrated in other species (Escher *et al.*, 1970; Battersby *et al.*, 1970), with the cyclization presumed to involve a Michael-type addition (Inouye *et al.*, 1977). The cytochrome *P*-450-mediated conversion of geraniol to 10-hydroxygeraniol in *Catharanthus roseus* (Apocynaceae) has been examined in detail (Licht and Coscia, 1978; Madyastha and Coscia, 1979). The synthesis of numerous specifically labeled acyclic precursors of the iridoids (Uesato *et al.*, 1982a,b; Balsevich, 1983) holds promise for the clarification of this aspect of monoterpene biosynthesis.

VII. Biosynthesis of Sesquiterpenes

A. *Distribution of Structural Types*

Sesquiterpenes are derived from the C_{15} prenyl pyrophosphate *trans, trans*-farnesyl pyrophosphate (FPP, Fig. 1); the relevant prenyl transferase, FPP synthetase, is essentially ubiquitous in the plant kingdom. The sesquiterpenes are a most diverse class, occurring naturally in a wide range of oxidation states, and including representatives that have undergone some of the most unusual molecular rearrangements in terpene biochemistry (see Fig. 8 for representative structures encountered in wood extractives).

Sesquiterpenes often occur as complex mixtures, but as minor constituents (<10%) in oleoresins of the Pinaceae. In some species (e.g., *Pinus armandi, P. flexis,* and *P. longifolia*), however, sesquiterpenes may exceed 10% of the oleoresin (Mirov, 1967). Wood of the genus *Cedrus* (the true cedars) is rich in himachalene sesquiterpenes (compounds related to the bicyclic intermediate in the path to longifolene, Fig. 9) (Norin and Winell, 1971). A survey of the pinyons (pines of semiarid regions) revealed a number that produce oleoresin of unusually high sesquiterpene content, with germacrene D (Fig. 10) and longifolene (Fig. 8) as principal components (Zavarin and Snajberk, 1980). Cedrol (Fig. 8) and cedrene (Fig. 9) often occur as major components of wood oils of the Cupressaceae and are characteristic of the pleasant fragrance of the wood (Piovetti and Diara, 1977; Piovetti *et al.*, 1980, 1981; Herout, 1982).

The distribution of sesquiterpenes among woody dicots (of the Thymelaeaceae, Santalaceae, Lauraceae, and Zygophyllaceae) is of interest primarily because of the value of these compounds to the fragrance industry (Guenther, 1976). The balsamics (from the Burseraceae, Leguminosae, and Hamamelidaceae) are also of importance in this regard (Boelens *et al.*, 1981). A number of sesquiterpene lactones of the wood of plants of the Compositae and Magnoliaceae are of pharmaceutical interest (Rowe and Conner, 1979; Gottleib and Mors, 1980).

Some of the calamenic sesquiterpenes from the wood of the Ulmaceae and Tiliaceae (Fig. 8) are produced in response to disease and probably function as phytoalexins (Burden and Kemp, 1983), while others may alter the commercially useful properties of the wood (Rowe and Conner, 1979). The calamenic sesquiterpenes are structurally related to the gossypol phytoalexins of cotton plants (Malvaceae) (Stipanovic *et al.*, 1975), suggesting a similar biological effect. Although the role of gossypol-type sesquiterpenes in the defensive response of cotton is well known (Mace and Bell, 1981; Mace, 1983), the role of the corresponding mansonones and calamenals of *Ulmus* in response to Dutch elm disease needs to be examined further (Overeeme and Elgersma, 1970; Jeng *et al.*, 1983). The mansonones (quinoid derivatives of calamenenes) were first discovered in wood of the Sterculiaceae (Gibbs, 1974). The distribution of these sesquiterpenes in *Ulmus* was reviewed by Rowe *et al.* (1972) and Rowe and Conner (1979).

The biosynthesis of the well-known plant hormone abscisic acid, which has an unusual sesquiterpenoid structure, has been discussed by Milborrow (1983a,b), Davies and Mansfield (1983), and Neill *et al.* (1982). Although the mevalonoid origin of abscisic acid has been established, it is not yet clear whether this hormone is derived from a C_{15} precursor such as farnesol or via cleavage of a carotenoid.

B. *Biosynthesis of Specific Sesquiterpenes*

Studies on the biosynthesis of sesquiterpenes have dealt primarily with higher plant phytoalexins or those of fungal origin. Fungal metabolites and phytoalexins have been of particular interest because of their often interesting biological properties and because of the relatively high rates of incorporation of exogenous precursors *in vivo* (Cane, 1981). Current studies in this area have focused on biogenetic information gained from NMR analysis of ^{13}C- and ^{2}H-labeled products generated from primary precursors (Cane *et al.*, 1981b; Stothers, 1981; Bradshaw *et al.*, 1982). Although

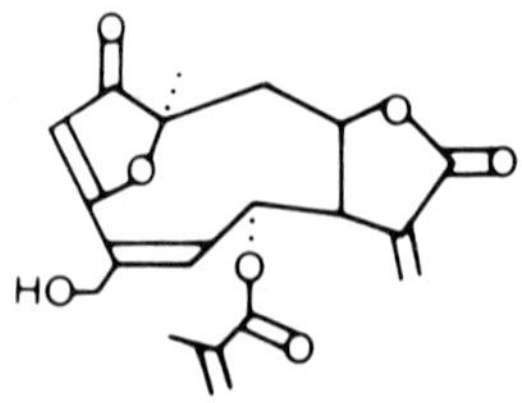

Goyazensolide
(Compositae)

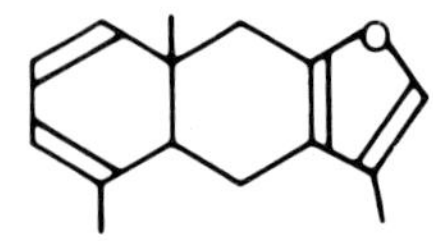

Furanoeudesma-1,3-diene
(Burseraceae)

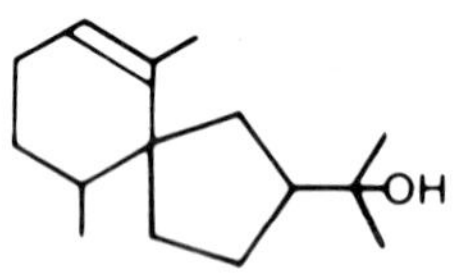

Agarospirol
(Thymelaeaceae)

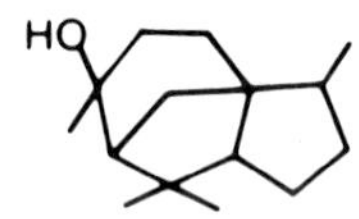

Cedrol
(Cupressaceae)

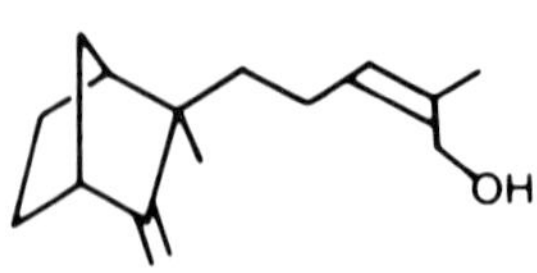

β-Santalol
(Santalaceae)

Longifolene
(Pinaceae)

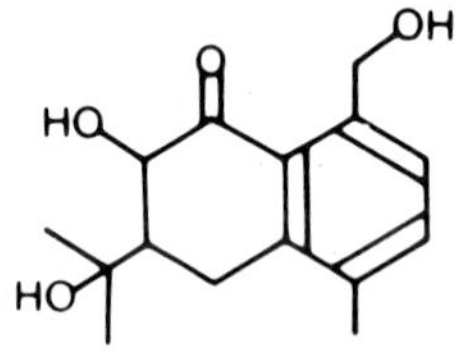

Emmotin A
(Icacinaceae)

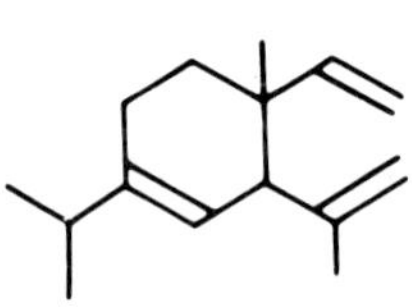

δ-Elemene
(Meliaceae)

Fig. 8. Representative sesquiterpenoid wood extractives.

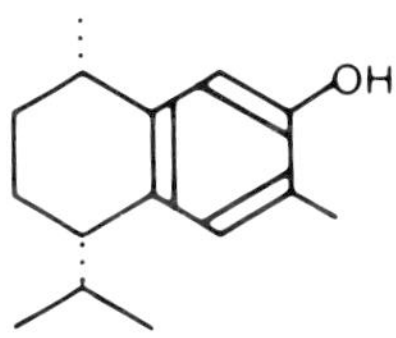

7-Hydroxycalamenene
(Ulmaceae)

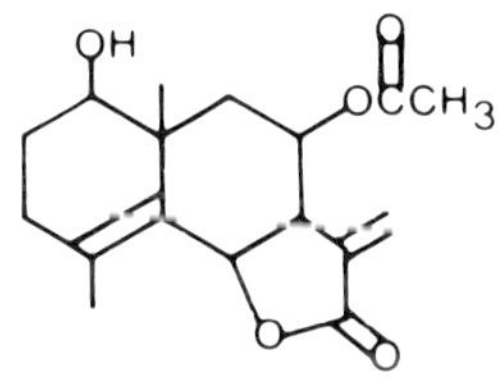

γ-Liriodenolide
(Magnoliaceae)

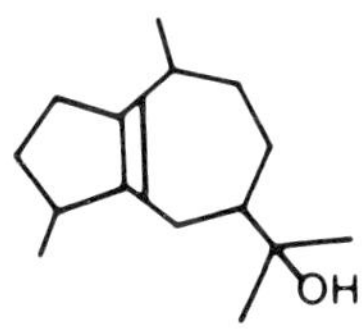

Guaiol
(Zygophyllaceae)

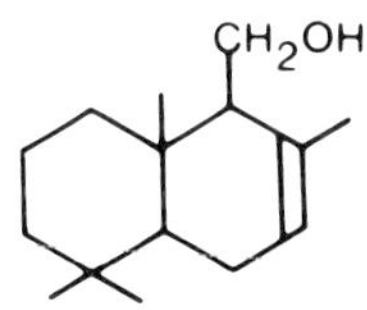

Drimenol
(Winteraceae)

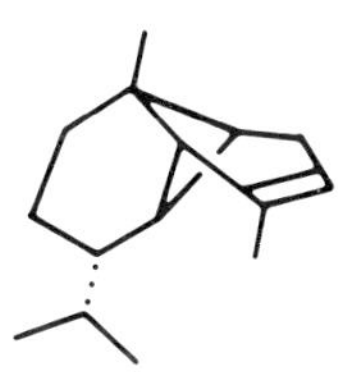

α-Copaene
(Leguminosae)

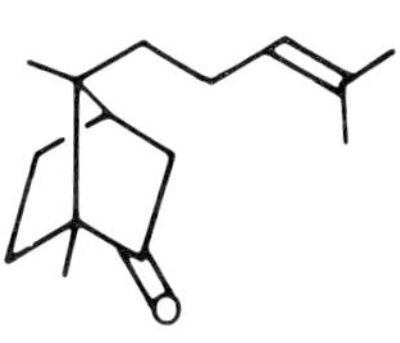

Campherenone
(Lauraceae)

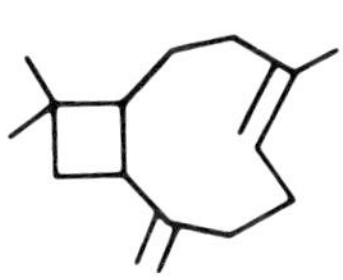

β-Caryophyllene
(Hamamelidaceae)

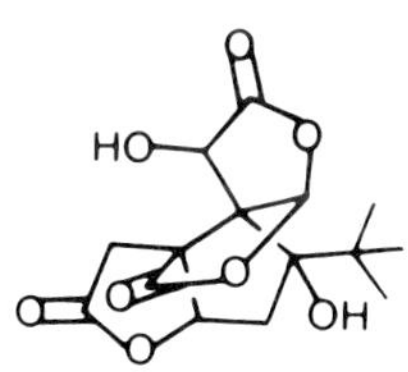

Bilobalide
(Ginkgoaceae)

Fig. 8. *(Continued)*

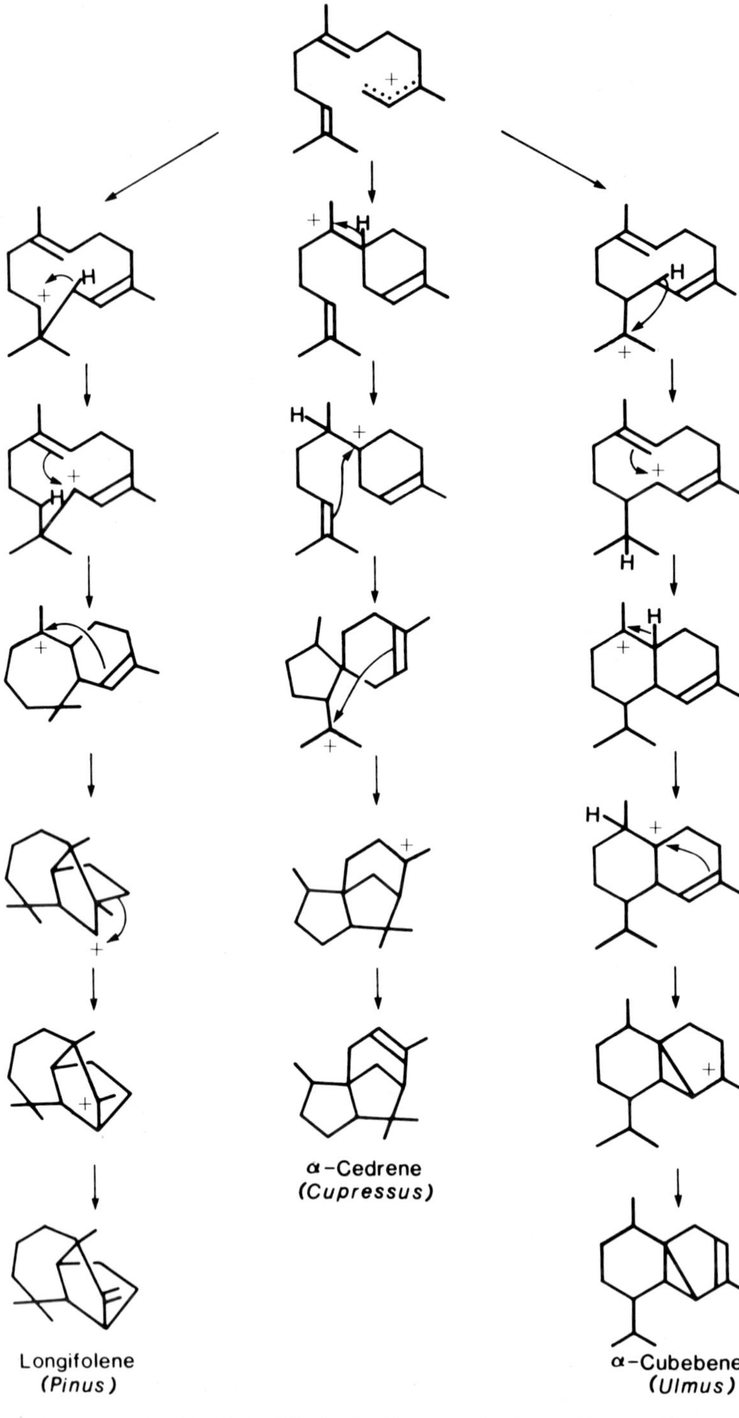

Fig. 9. Role of hydride shifts in the biosynthesis of tricyclic sesquiterpenes.

this technique has in many instances replaced the more cumbersome use of $^{3}H/^{14}C$-double-labeled substrates followed by chemical degradation, it often suffers from the necessity of feeding simple precursors that may be too remote from the analyzed end product to permit unambiguous conclusions (for example, see the discussion in Cane *et al.*, 1981b). The biosynthesis of sesquiterpenes has been reviewed in detail by Cane (1981) and by Cordell (1976). Earlier coverage by Herz (1971, 1977) and Rücker (1973) has described this area from a more theoretical perspective.

It is now accepted that all sesquiterpenes are derived from farnesyl pyrophosphate, and extensive biogenetic schemes based on this principle have been put forward. The central theme of these hypotheses is an intramolecular electrophilic attack by the carbon bearing the pyrophosphate on either the central or distal double bond of the farnesyl molecule. Further cyclizations, rearrangements, and redox interconversions lead to the array of some 200 known sesquiterpenoid skeletons. The precise nature and origin of the farnesyl pyrophosphate isomer that undergoes cyclization has been the center of some controversy; the difficulty arises from the realization that, as was the case with the cyclohexanoid monoterpenes, the trans configuration at C-2 of the precursor prevents direct conversion to numerous cyclic types. This apparent problem has lead to a number of theories, the evidence for which has been reviewed by Cane (1980). The weight of the evidence now indicates that all redox schemes for the trans to cis conversion prior to cyclization (for example, see Mackie and Overton, 1977) are untenable, supporting instead an alternate isomerization–cyclization mechanism involving (as was the case with the monoterpenes) the tertiary allylic isomer, in this case nerolidyl pyrophosphate or its enzyme-bound equivalent (Cane, 1980).

The origin of the sesquiterpenes longifolene, α-cedrene, and α-cubebene (Fig. 9) exemplify the role of hydride shifts in ring formation. The biosynthesis of longifolene in *Pinus ponderosa* was examined by Arigoni (1975), and this work has been reviewed by Cane (1980, 1981) and others. The indicated hydride shift in the humulyl cation (Fig. 9) was demonstrated by feeding [5-$^{3}H_2$,2-^{14}C]MVA followed by chemical degradation to locate all six tritium atoms. This contribution has led to the hypothesis that 1,3-hydride shifts are likely to be of central importance in sesquiterpene cyclization processes. (Thus, the hypothesis of a 1,3-hydride shift, rather than two 1,2-hydride shifts, in the case of the germacryl cation is suggested for the biosynthesis of α-cubebene in Fig. 9.) The driving force for a 1,3 shift may be that it leads to the generation of the more stable allylic cation, a not unreasonable assumption since more binding energy must be expended to achieve the conformational restriction required for orbital overlap. Thus, in contrast to the 1,2-hydride shift, the enzymatic 1,3-hydride shift may require the assistance of an adjacent double bond.

Orbital alignment is also an important consideration in sesquiterpene cyclizations initiated by double-bond protonation rather than pyrophosphate ionization. A particularly relevant example is the biosynthesis of pentalenene, the precursor of pentalenolactone in *Streptomyces* (Cane *et al.*, 1981b; Cane and Tillman, 1983). Evidence was presented that pentalenene is derived from the olefin humulene, via protonation of the conformational arrangement in which orbital overlap for the pentalene cyclization is optimal. The alternative direct route from FPP would require a 1,2-hydride shift from one secondary position to another of the humulyl cation, a step for which there is no obvious preference over humulene formation from FPP followed by protonation. Although humulene could not be established as an intermediate in cell-free trapping experiments (and was not tested directly as a substrate), the loss of the expected proton for double-bond formation was demonstrated using *trans,trans*-[9-$^3H_2$]FPP as the substrate. The observation that humulene could not be trapped is best explained by the sequential cyclizations proceeding without release of intermediates from the enzyme surface (D. E. Cane, 1983, personal communication; Cane and Tillman, 1983).

While there are uncertainties in the biosynthesis of sesquiterpenes such as pentalenene, stemming from the fact that there is no mechanistic constraint requiring the intermediacy of an olefin as opposed to a hydride shift in a cationic intermediate, the role of olefin (or epoxide) protonation instead of direct cyclization from FPP has likewise been supported in the formation of cadinenes (Arigoni, 1975), guaienols (González *et al.*, 1983), and eudesmanolides (Herz, 1971) (see Fig. 10). The role of double-bond protonation in sesquiterpene lactone biosynthesis was discussed in detail by Herz (1977). These proposals were originally based on chemical analogies that exploit the ready cyclization of germacryl and humulyl systems under acidic conditions (for a recent example, see Matthes *et al.*, 1982). Cane (1981) reviewed the data pertaining to the biosynthesis of a wide range of dimethylcyclopentane fungal sesquiterpenes for which humulene is likely to serve as a precursor. Such proton-initiated cyclizations may have biosynthetic precedent in conifers. Germacrene D is the major sesquiterpene hydrocarbon of young Douglas fir needles, whereas the disappearance of this macrocyclic sesquiterpene from mature needles coincides with the appearance of γ-cadinene (Fig. 10) (Kepner *et al.*, 1975).

The biosynthesis of sesquiterpenes in *Pinus pinaster* needles was reviewed by Bernard-Dagan *et al.* (1982). Sesquiterpene biosynthesis occurs throughout the length of the needle, in contrast to monoterpene biosynthesis, which is localized only at the base of young needles. A membranous fraction from *Pinus* needle homogenates produced cyclic sesquiterpenes (caryophyllene, humulene) from [^{14}C]IPP, whereas a solu-

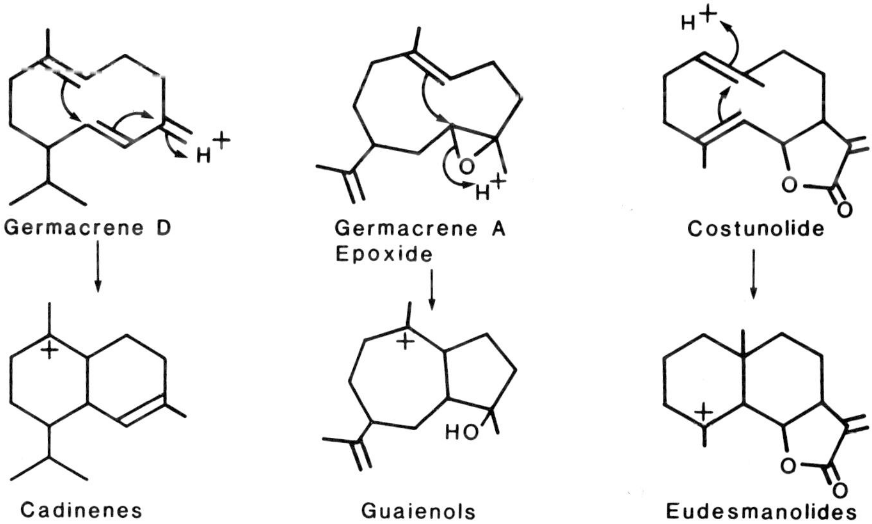

Fig. 10. Proton-initiated cyclization in sesquiterpene biosynthesis.

ble fraction afforded only farnesene (an acyclic hydrocarbon) from the same precursor (Bernard-Dagan *et al.*, 1980). The biosynthesis of mono- and sesquiterpenes in separate compartments of conifer needles is reminiscent of the phenomenon observed in herbaceous plants (Croteau and Loomis, 1972).

The biosynthesis of sesquiterpene phytoalexins of the Solanaceae has been reviewed by Stoessl (1980) and Kúc and Lisker (1978). The mode of formation of a number of these compounds was explored using $^{2}H/^{13}C$-NMR techniques (Stoessl and Stothers, 1982). Sesquiterpene phytoalexin biosynthesis in the Convolvulaceae (sweet potatoes; *Ipomoea*) was examined by Burka and Thorsen (1982) and by Ôba and Uritani (1981).

C. *Sesquiterpene Lactones*

Sesquiterpene lactones represent the highest oxidation state usually found among the sesquiterpenoid wood extractives, and they are especially prominent in the Compositae and Magnoliaceae (see Fig. 8) (Rowe and Conner, 1979; Gottlieb and Mors, 1980). Interest in the origin of sesquiterpene lactones (Herz, 1977; Fischer *et al.*, 1979) has been prompted by the potential ecological roles and pharmaceutical utility of these compounds (Rodriguez *et al.*, 1976; Mabry and Gill, 1979), but relatively little is known about their biosynthesis at either the *in vivo* or cell-free level. The origin of numerous fungal sesquiterpene lactones has been studied *in vivo,* but these compounds are of skeletal types not commonly found in higher plants (see Cane, 1981). The presence of sesquiterpene lactones has been used as a criterion in analyzing complex taxonomic relationships in guayule (*Parthenium:* Compositae) and related plants (Mears, 1980). Numerous books have emphasized this diverse class as well (Yoshioka *et al.,* 1973; Heywood *et al.,* 1977; Seaman, 1982).

VIII. Biosynthesis of Diterpenes

A. *Distribution of Structural Types*

The diterpenes most commonly encountered are nonvolatile acids in conifer and leguminous resins, phorbol irritants of the Euphorbiaceae and Thymelaeaceae, and high-boiling hydrocarbons of numerous conifers. The diterpenes are derived from the C_{20} precursor all *trans*-geranylgeranyl pyrophosphate (GGPP). Those found in wood extractives can be divided into three general biogenetic categories: (1) the acyclics, (2) the polycyclics derived from the proton-initiated cyclization of GGPP, and (3) those derived from the macrocyclics (especially cembrene). A few other structural types are known, such as ryanodol and cinnzeylanol (Fig. 12), which are water-soluble polyhydroxyditerpenes (Nakanishi *et al.,* 1974; Nohara *et al.,* 1982). The origin of the various diterpene structural classes is not well understood, with the exception of a few monocyclics and the extensively studied kaurenoid group (e.g., the gibberellins). In terms of the sheer quantity of terpenes produced in conifer tissues, the diterpenes exceed all other classes. Oleoresins characteristically contain 60% by weight or more of diterpenoid material (Squillace, 1976; Kramer and Kozlowski, 1979). The resinous nature of this material and the propensity to autooxidative polymerization and crystallization are the bases of the physical barrier to infection at wound sites. Examples of diterpenes found in oleoresins of the Pinaceae and Cupressaceae are illustrated in Fig. 11. Examples of those found in other gymnosperms and dicots are shown in Fig. 12.

The structural diversity of the naturally occurring diterpenes has been exploited to delineate taxonomic relationships among the gymnosperms and dicots (Ourisson, 1973; Zinkel, 1977). The distribution of acyclic diterpenes has been discussed by West (1981b) and Ahlquist *et al.* (1978). Prominent examples from woody plants include the fragrant compound geranylgeraniol from the wood of *Cedrela toona* (Meliaceae) (Nagasampagi *et al.,* 1967) and the 7-hydroxymethyl derivative of 6-*cis*-geranylgeraniol of *Croton* stems (Euphorbiaceae) (Sato *et al.* 1980). This latter compound forms the basis of numerous patents, since it appears to serve as a prostaglandin analog in protecting the digestive tract from the effects of ulceration and in promoting wound healing. Low levels of geranylgeranyl fatty acid esters have been identified in Norway spruce wood (*Picea abies*) (Ekman, 1980).

The macrocyclic diterpenes include cembrene and its derivatives, such as incensole (Fig. 12) and phorbol (Fig. 12). The distribution of macrocyclic diterpenes among terrestrial plants and marine organisms has been discussed by Weinheimer *et al.* (1979) and Dauben *et al.* (1975). These compounds have been isolated from numerous species of pine. The baccatins are unusual toxic macrocycles (Fig. 12) belonging to a group called the taxanes, and are found in heartwood extracts of the yews (Taxaceae) (Nakanishi *et al.,* 1974; Seigler, 1981b).

The polycyclic diterpenoids derived by proton-initiated cyclization of GGPP (see biogenetic scheme, Fig. 13) are the most common type found in wood extractives from the conifers and the Caesalpinioideae tribe of the Leguminosae (Zinkel, 1981; Langenheim, 1981). These compounds also occur in leaf-surface resins of the Ricinocarpodeae tribe of the Euphorbiaceae (Dell and McComb, 1978a; Jeffries, 1979). The bicyclic labdanoid diterpenes (Figs. 11 and 13) are characteristic of bark and needle resins in the Pinaceae, while the tricyclic group (especially the pimaranes and abietanes) dominate in the wood resin (Hannus, 1976; Hemingway, 1981; Zinkel, 1981). *Cedrus* and *Tsuga* wood, which lack resin canals, do not contain polycyclic acids, although these compounds are present in *Cedrus* cones (Norin and Winell, 1971). The wood resin of some pinyons is highly enriched in pimarane-type diterpenes (up to 70%) (Zavarin and Snajberk, 1980). The presence of strobic acid (Fig. 11) is characteristic of eastern white pine wood (*Pinus strobus*) (Zinkel, 1981).

The formation of osmiophilic droplets in the epithelial cells lining resin ducts is the only observation suggesting that diterpenes are synthesized at this site (Benayoun and Fahn, 1979). This observation is consistent with the expected isolation of diterpene biosynthesis in specialized cells, as is the case with the monoterpenes and sesquiterpenes. Dell and McComb (1978b) showed that glandular hairs isolated from the leaves of *Newcastelia viscida* (Dicrastylidaceae) were capable of incorporating [^{14}C]mevalonate into pimaradiene-type diterpenes.

B. *Biological Role of the Diterpenes*

The diterpenes appear to function in chemical defense against insect and fungal attack and in the formation of physical barriers that seal off

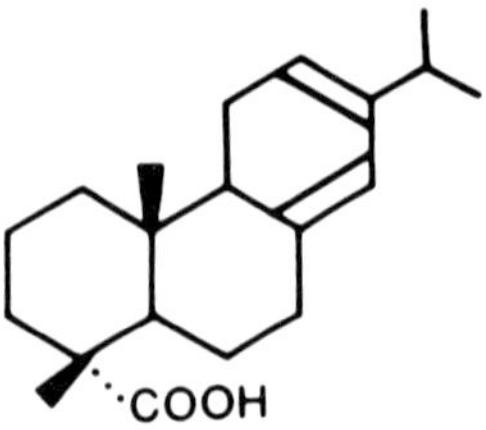

Levopimaric Acid

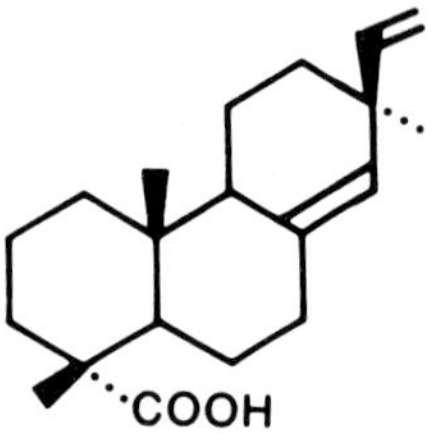

Pimaric Acid

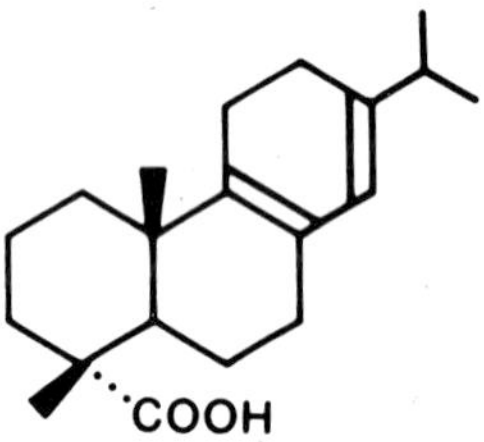

Palustric Acid

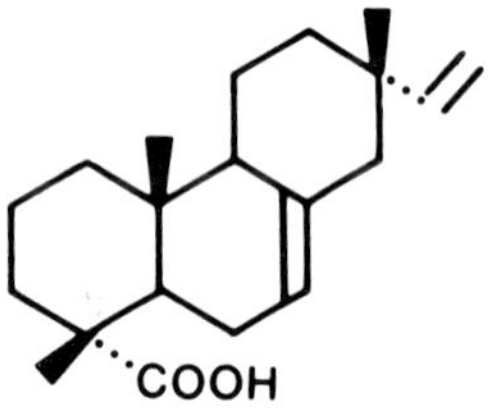

Isopimaric Acid

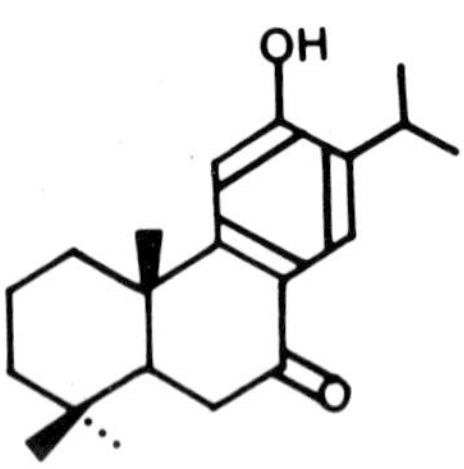

Sugiol

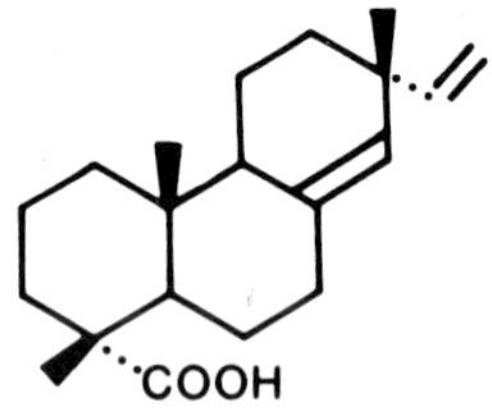

Sandaracopimaric Acid

Fig. 11. Diterpenes of the Pinaceae and Cupressaceae.

wound sites so that healing may occur. The adverse effects of individual diterpene acids on the feeding and development of the larch sawfly were examined by Wagner *et al.* (1983). Deterrence to feeding and inhibition of sawfly development were observed with abietic, neoabietic, and isopimaric acids, natural constituents of their host the tamarack (*Larix laricina*). Related resin acids were identified as the active principles in Ponderosa pine needles, which when ingested by cattle lead to reproductive failure (Kubik and Jackson, 1981). The mode of action of these diterpenes is thought to involve the interference with normal steroid metabolism, a

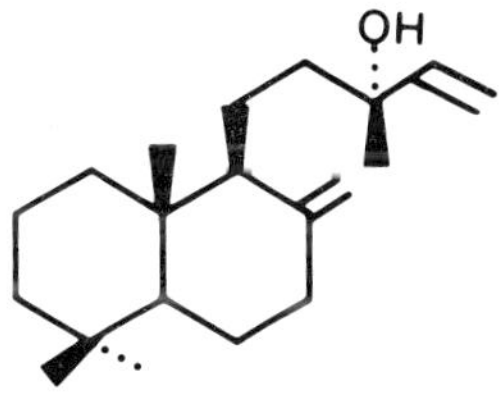

Manool

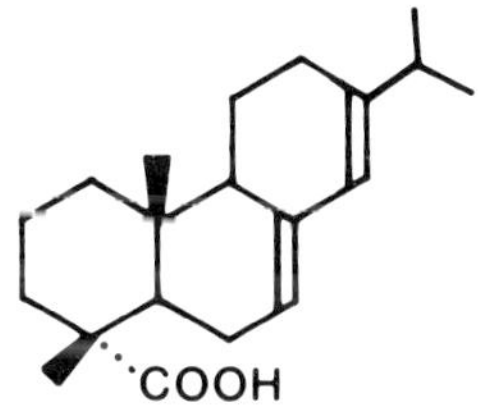

Abietic Acid

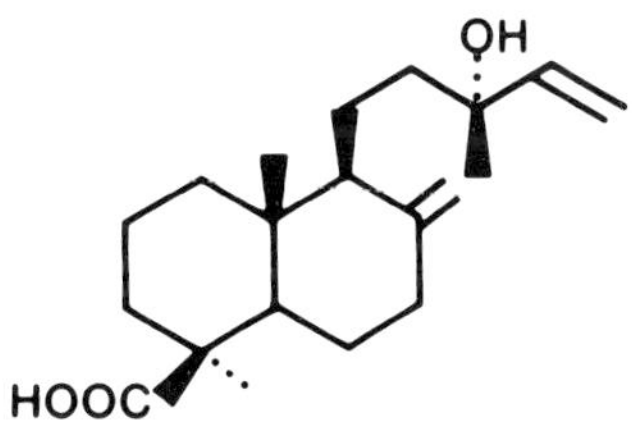

Torulosic Acid

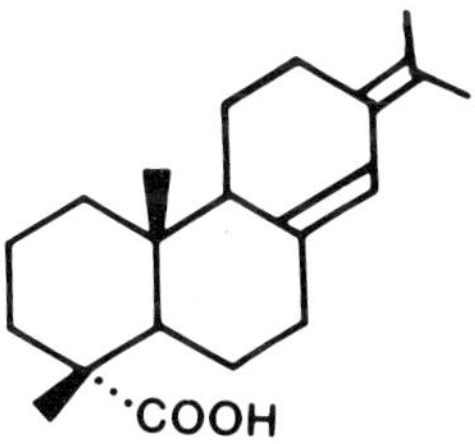

Neoabietic Acid

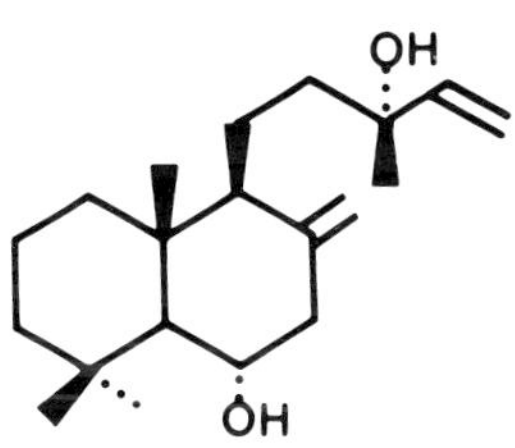

Larixol

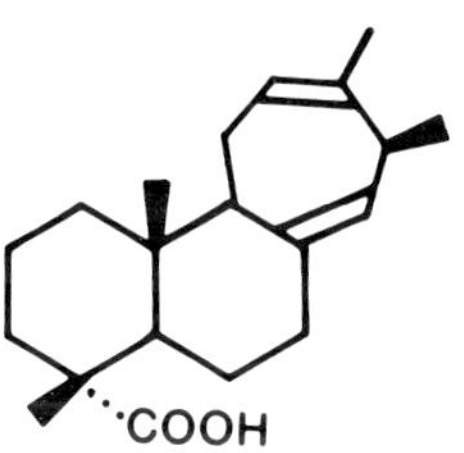

Strobic Acid

Fig. 11. (*Continued*)

likely possibility in view of the structural similarity of the tricyclic diterpenes to the C_{30} precursors of steroids in animals.

A similar mode of action is possible for the tetracyclic diterpenes of the kaurenoid and beyeranoid types (Fig. 13), which occur in the leaf and wood resin of Leguminosae and Euphorbiaceae, and which, by interfering with normal gibberellin metabolism, may serve as effective phytotoxins inhibiting the growth of competitors. The role of insect antifeedants of leguminous resins in shaping the evolution of tropical rain forest species has been described (Langenheim, 1981; Langenheim *et al.*, 1982). In no

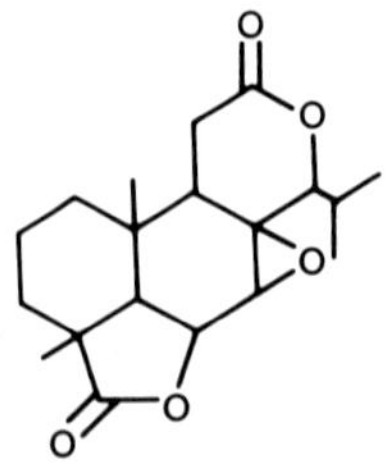

Nagilactone G
(Podocarpaceae)

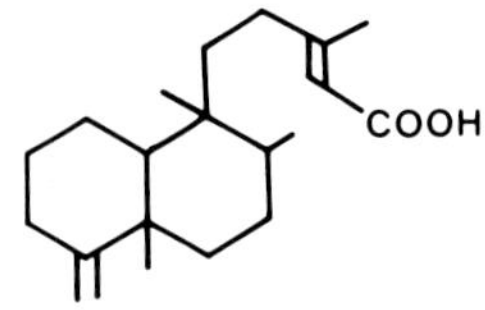

Clerodadienoic Acid
(Araucariaceae)

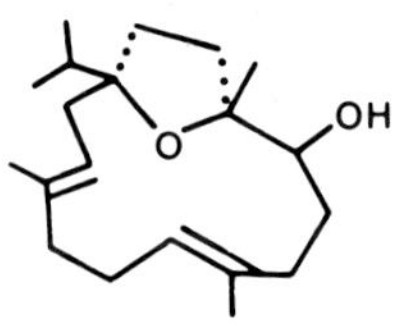

Incensole
(Burseraceae)

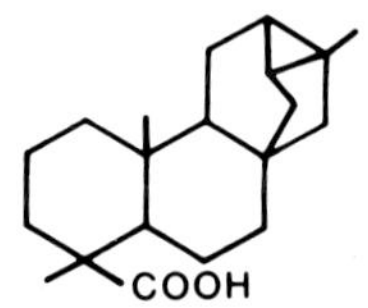

Trachylobanic Acid
(Leguminosae)

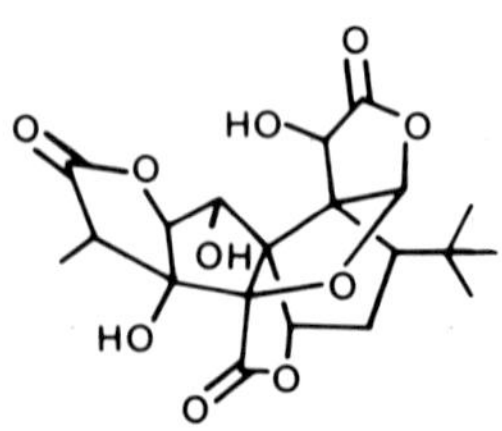

Ginkgolide B
(Ginkgoaceae)

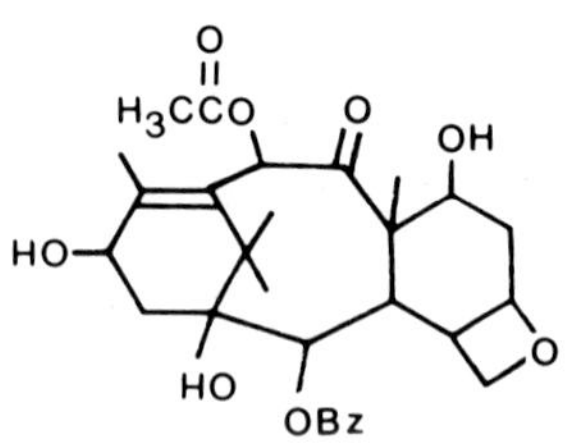

Baccatin III
(Taxaceae)

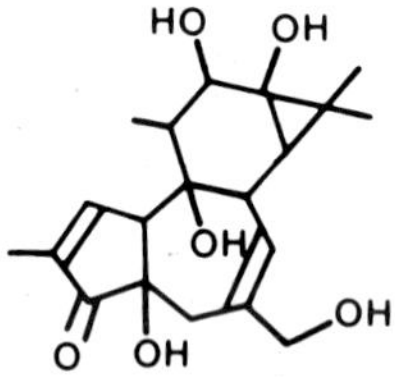

Phorbol
(Euphorbiaceae)

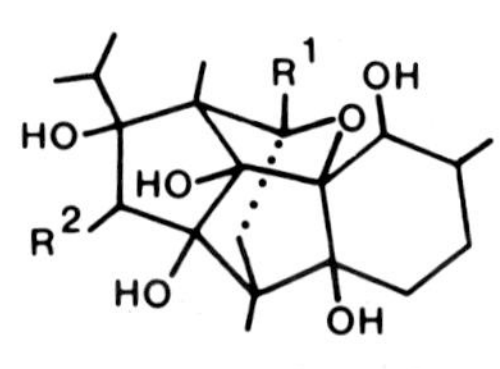

Ryanodol R^1=H, R^2=OH
(Flacourtiaceae)
Cinnzeylanol R^1=OH, R^2=H
(Lauraceae)

Fig. 12. Diterpenes of other gymnosperms and dicots.

other class of terpenoid wood extractives is an ecological role more prominent than among the diterpenes.

C. *Biosynthesis of Specific Diterpenes*

Although the critical cyclization steps and subsequent secondary transformations in construction of most of the diterpenes found in wood extractives remain largely unexplored, the biosynthesis of the structurally related gibberellin plant hormones and casbene, a macrocyclic phytoalexin from the jack bean (*Ricinus communis:* Euphorbiaceae), have been examined in detail. The biosynthesis of gibberellins and the regulation of this process were discussed by West *et al.* (1979, 1982) and by Graebe (1982). The general area of diterpene biosynthesis has been reviewed by West (1981b).

The tetracyclic diterpene *ent*-kaurene (see kaurane, Fig. 13) is the precursor of the gibberllins in the fungus *Gibberella fujikuroi* and in higher plants. The oxidation (cytochrome *P*-450 type) and ring-contraction steps in this pathway have been described (West, 1980, 1981b). The grayanoid toxins from the leaves of shrubs and trees of the Ericaceae [e.g., mountain laurel (*Kalmia*); El-Naggar *et al.,* 1980] are also presumed to be derived by ring contraction of kaurene (Nakanishi *et al.,* 1974).

The biosynthesis of kaurene in partially purified cell-free extracts from *Marah macrocarpus* endosperm (Cucurbitaceae) and several other higher plant tissues has been the object of intensive investigation by West (see Duncan and West, 1981, and references therein). Two enzymatic steps (the A and B activities) are required for the conversion of GGPP to kaurene. The first of these partially resolvable enzymes catalyzes the cyclization to copalyl pyrophosphate (CPP, Fig. 13) by protonation of the terminal double bond of GGPP. The second enzyme, by ionization of the allylic pyrophosphate, promotes the further cyclization and rearrangement of CPP to kaurene. Each enzyme has a molecular weight of 82,000 and requires Mg^{2+} for activity, and the two appear to function *in vitro* as a complex, in that the CPP produced by component A is channeled to component B without equilibration with the media. Because of the presence of endogenous inhibitors of the A activity, often only the B activity (CPP to kaurene) can be readily demonstrated in cell-free extracts. These observations have lead to the hypothesis that the gibberellin pathway is regulated via the inactivation of the A activity (West *et al.,* 1982).

The occurrence and biogenesis of the surface resins in Euphorbiaceae were reviewed by Jeffries (1979). A mechanistic scheme for the derivation of the tetracyclic diterpenes of the Araucariaceae was presented by Briggs and White (1975), and the origin of the *seco*-beyerane diterpenes

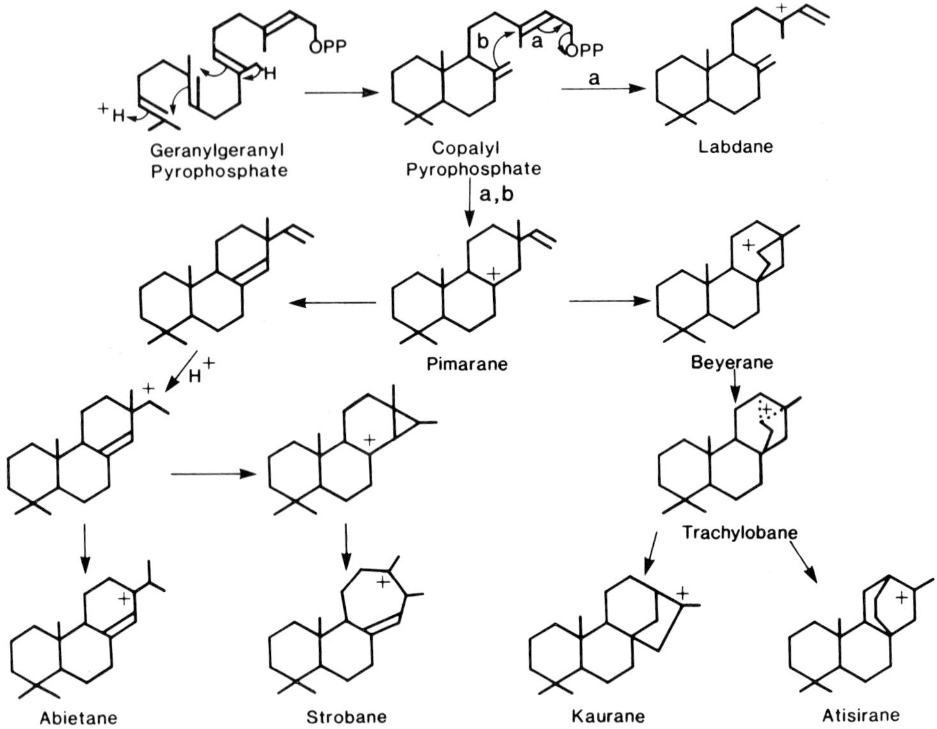

Fig. 13. Biogenesis of diterpene structural types.

was demonstrated by feeding various [^{3}H]beyerane precursors to *Beyeria calycina* (Euphorbiaceae) leaves (Bakker *et al.*, 1972; Ghisalberti *et al.*, 1978). Attempts to model the biogenesis of tetracyclic diterpenes via the synthesis of putative nonclassic ion intermediates have failed (Blunt *et al.*, 1976). The intriguing rearrangements in the genesis of the diterpene series resemble some of those occurring in lower terpene biosynthesis, and they have proved to be as difficult to mimic in model systems [see van Tamelen and Marson (1982) for pimarane series].

The biosynthesis of diterpene resin acids of the type that occur in the Pinaceae has received little attention. Gleizes (1979) has demonstrated that the tricyclic resin acids of *Pinus pinaster* (abietanes, pimaranes; Figs. 11 and 13) are synthesized at the base of needles, as is the case with the monoterpenes. The bicyclic resin acids (labdanes), on the other hand, are synthesized throughout the needles, as are the sesquiterpenes. This dichotomy is also observed in the wood, where monoterpenes and tricyclic diterpenes dominate and labdanoids and sesquiterpenes are rare. The biosynthesis of sandaracopimaradiene (see the corresponding acid, Fig. 11) from *S*-[1-^{2}H]GGPP in cell-free extracts of *Ricinus communis* has been examined by Drengler and Coates (1980), and the cyclization involving the pyrophosphate was demonstrated to proceed with overall anti stereochemistry.

Synthetic models for the biosynthesis of strobane diterpenes were examined by Herz *et al.* (1983). The origin of strobic acid from either pimarane or labdane precursors seems possible; however, the synthetic models suggest the pimarane route to be the more likely (Fig. 13). Zinkel, who discovered strobic acid (Fig. 11), lists it as a labdane type, even though no other labdanoids occur in wood extracts of *Pinus strobus* (Tobolski and Zinkel, 1982). An extensive compilation of the spectroscopic and chromatographic properties of conifer diterpenes is available (Zinkel *et al.*, 1971; Foster and Zinkel, 1982; Perry and Weavers, 1984), as are facile methods for preparing the critical precursor GGPP in labeled form (Sandmann *et al.*, 1980). These advances may lay the foundation for more extensive *in vitro* studies on the biosynthesis of resin acids and related substances.

Casbene is a rare cyclopropane-containing macrocyclic phytoalexin (14-carbon ring) produced by *Ricinus communis*. The casbene synthetase, which converts GGPP to the olefin, was partially purified from fungus-treated seedlings and characterized (Deuber *et al.*, 1978). The enzyme has a molecular weight of 53,000 and requires Mg^{2+} for catalysis. The mechanism of casbene formation was shown to involve the formal displacement of the pyrophosphate by the terminal double bond of GGPP to afford the macrocycle, followed by proton loss to generate the cyclopropane ring (Guilford and Coates, 1982). Pectic fragments, derived from the cell wall by the action of fungal endopolygalacturonase, have been identified as elicitors of casbene biosynthesis in *R. communis* (Bruce and West, 1982).

Casbene, or its cembranoid progenitor, has been proposed as the precursor of the phorbol diterpenes (Fig. 12) of Euphorbiaceae and Thymelaeaceae (Adolf and Hecker, 1977; Jeffries, 1979; Evans and Taylor, 1983). Although casbene itself has not been isolated from phorbol-producing plants, the recent isolation of a casbene diketone from twigs of *Croton nitens* supports the suggested precursor–product relationship (Burke *et al.*, 1981).

D. *Ginkgolides*

The ginkgolides (Fig. 12) are an unusual series of *seco*-diterpenes occurring in the leaves and root bark of *Ginkgo biloba*. Closely related sesquiterpenes occur in the heartwood (e.g., bilobalide, Fig. 8). The biosynthesis of ginkgolide B in 5-month-old *G. biloba* plants was examined by Nakanishi and Habaguchi (1971). The terpenoid origin of this compound was confirmed, and it was proposed that the ginkgolides are derived from a pimarane precursor via oxidative ring opening and subsequent ring contraction. It was additionally demonstrated, through the use of labeled methionine, that the unusual *tert*-butyl group of the ginkgolide was derived by methylation (presumably of an isopropylidene double bond). Double-bond methylations are common in the triterpene series [for example in the phytosterols (Goodwin, 1980, 1981)], and these reactions may also involve the generation of a *tert*-butyl group [for example, see dimethyllanostadienone (Fig. 14), which was isolated from oak leaves (Hui and Li, 1977)].

IX. Biosynthesis of Triterpenes

A. *Distribution of Structural Types*

The triterpenes constitute the waxy components of leaf surface lipids and bark extractives or occur as the aglycones of intracellular saponins having a detergent or hemolytic action. They are derived by the cyclization of the C_{30} olefin squalene, which is formed by the reductive condensation of two molecules of FPP (C_{15}) in an unusual head-to-head manner (Fig. 14). Major interest in the triterpenes has been centered on the demethylated subclass, the ubiquitous steroids.

There are three major structural classes of the nonsteroidal triterpenes: (1) the tetracyclic group, (2) the pentacyclic group, and (3) the *seco*-lactones. The three classes are biogenetically related, in that the pentacyclic and *seco*-lactone groups are derived from tetracyclic intermediates. Three basic modes for the cyclization of squalene (and its derivatives) are known. The most common involves the cyclization of squalene 2,3-epoxide [via protonation and ring opening of the epoxide (Fig. 15)]. A second, and similar, type of cyclization employs squalene 2(3),22(23)-diepoxide as the substrate. The products of the latter cyclizations (e.g., serratendiol, α-onocerin, Fig. 14) exhibit little structural diversity and are prevalent in bark extractives. The third mode of cyclization is the direct proton-initiated cyclization (nonoxidative) of squalene itself, a route found in ferns and protozoans.

Nonsteroidal triterpenes are found in a greater number of plant species than are members of any other terpene class. This is attributable, at least in part, to their very common occurrence as constituents of both leaf and fruit waxes as well as intracellular saponins. The distribution of triter-

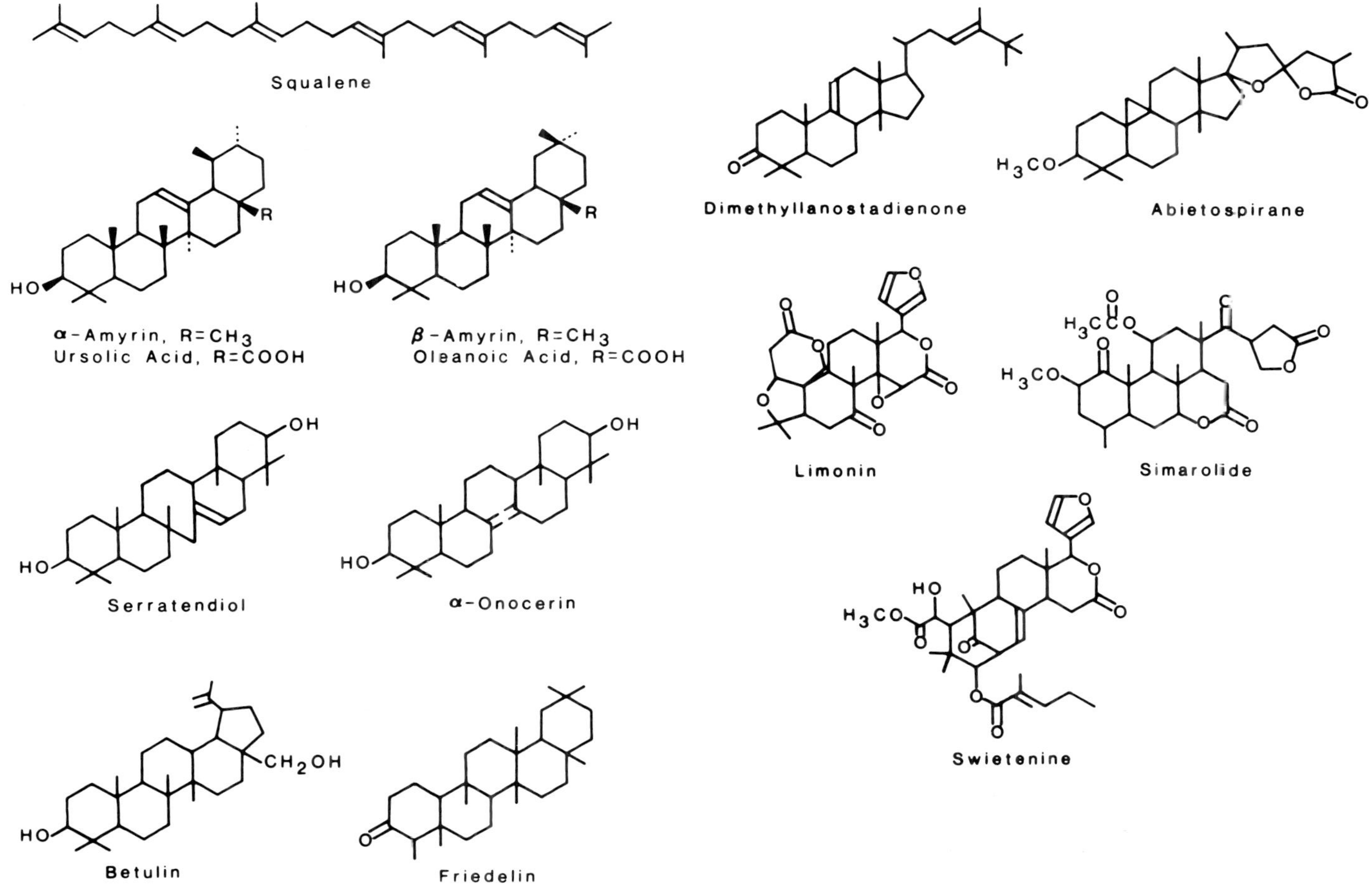

Fig. 14. Representative triterpenes from arboreal plants.

penes was reviewed by Pant and Rastogi (1979) and Das and Mahato (1983), the triterpene saponins were covered by Chandel and Rastogi (1980), and the friedelins (Fig. 14) were discussed by Chandler and Hooper (1979). The evolutionary relationship of triterpenes to steroids was examined by Ourisson *et al.* (1979) and Bouvier *et al.* (1980). The triterpenes in gum resins of the Burseraceae (Tessier *et al.*, 1982) and Styracaceae formed the focus of early investigations directed toward structural elucidation in this class (Simonsen and Ross, 1957).

The bark of trees such as *Quercus suber* (cork) and *Betula alba* (white birch) contain large amounts of triterpenes. Prominent in cork are friedelin (Fig. 14) and cerin (Rowe and Conner, 1979). The high content of betulin (up to 30%) (Fig. 14) gives birch bark its characteristic grey-white color (Gibbs, 1974; Rowe and Conner, 1979). Morolic acid is a major component of *Euclayptus* bark (Hart and Lamberton, 1965). The bark of *Abies* often contains large amounts of cycloartenol lactones (Fig. 14) (Steglich *et al.*, 1979), whereas the serratendiols (Fig. 14) are common in *Pinus* and *Picea* bark (Norin, 1972; Conner *et al.*, 1980). The great diversity of triterpene types that occur in bark may be of taxonomic significance.

The site of triterpene synthesis has been examined in mango tree saplings (*Mangifera indica:* Anacardiaceae) (Joel and Fahn, 1980). The cellular events associated with the formation and secretion of cycloartenol-type triterpenes (mangiferolic acids) in epithelial cells surrounding the resin ducts are similar to those described previously for the production of conifer resins (i.e., the triterpenes are produced in plastids called secretoplasts and are transported to the cell membrane via tubular endoplasmic reticulum).

B. *Biological Role of the Triterpenes*

Although the C_{30} triterpenes are inferior to the sterols as membrane stabilizers (Bloch, 1983), it has often been proposed that they have a role in membrane structure (Nes and Heftmann, 1981). Germinating peas synthesize β-amyrin (Fig. 14) in preference to sterols, although within 2 days postgermination, sterol synthesis dominates in new membrane formation (Baisted *et al.*, 1962). The C_{30} protosterol triterpenes may accumulate in inactive metabolic pools, later becoming a source of sterol precursors (Gibbons, 1977).

Since phytosterols serve as a source of steroidal substances in plant-eating insects (Prestwich *et al.*, 1983) and are essential for the growth of some fungi (Nes *et al.*, 1982), the presence of structurally related triterpenes in the host plants may interfere with normal steroid metabolism of these pests. Certain pentacyclic and polyhydroxylated tetracyclic triterpenes inhibit the growth of the parasitic fungi *Phytophthora cactorum* (Nes *et al.*, 1982), although the basis for this action is uncertain. The triterpene ester urs-12-en-3β-ylpalmitate, isolated from sandalwood bark, was shown to be a growth inhibitor of forest insect pests (Shankaranarayana *et al.*, 1980). Triterpene saponins, which accumulate in vacuoles and are released on cellular wounding, prevent the utilization of sterols through the formation of crystalline sterol–saponin complexes (Nes and Heftmann, 1981). A more subtle role for the triterpenes was revealed by a study of Steffens *et al.* (1983), in which haustorium development in a parasitic angiosperm was shown to be dependent on a triterpene produced by the leguminous host root.

C. *Biosynthesis of Specific Triterpenes*

The triterpenes are derived from squalene, a symmetrical C_{30} olefin formed by the head-to-head condensation of two units of FPP by way of the unusual cyclopropanyl-containing intermediate presqualene pyrophosphate. Ionization of the pyrophosphate group of this intermediate, followed by rearrangement and transfer of a hydride ion from NADPH, affords squalene. Examination of the squalene synthetase reaction using a presqualene pyrophosphate analog, in which an ammonium ion was present at the site where the carbonium ion would have formed, demonstrated inhibition of squalene synthesis only in the presence of pyrophosphate ion, thereby supporting an ion-pair mechanism for the rearrangement (Sandifer *et al.*, 1982). The squalene synthetase reaction was the subject of a thorough review by Poulter and Rilling (1981b).

The next step in the biosynthesis of most triterpenes is the epoxidation of squalene to 2,3-oxidosqualene by a mixed-function oxidase. The subsequent cyclization of oxidosqualene is thought to lead to most of the known triterpenes, although only the cyclization to cycloartenol, the essential intermediate in phytosterol biosynthesis, has been examined in any detail (Goodwin, 1980, 1981). Thus, the cyclizations leading to the lupanes and friedelins (Figs. 14 and 15) have yet to be demonstrated in cell-free systems, although the cycloartenol precedent clearly establishes 2,3-oxidosqualene as the precursor of these classes. The cyclization of oxidosqualene to β-amyrin by microsomal preparations from germinating pea cotyledons has been demonstrated, and the reaction is effectively inhibited by ammonium analogs (nitrogen placed where positive charge would develop during the normal proton-initiated cyclization), thus supporting a carbonium ion-type cyclization process (Delprino *et al.*, 1983).

The mode of formation of the ursane and oleanane skeletons (α- and β-amyrin type) was established with NMR techniques by feeding ^{13}C-labeled precursors to tissue cultures of *Isodon japonicus* (Labiatae) (Seo *et al.*, 1981); the results also confirmed the expected 1,2-methyl shift in the generation of the ursane nucleus (Fig. 15).

The precursor of the *Pinus* bark serratanes (Fig. 14) is assumed to be α-onocerin even though it has not been reported to occur in bark extracts (Banthorpe and Charlwood, 1980), and a cell-free system from *Ononis spinosa* (Leguminosae) was shown to convert 2(3),22(23)-squalene diepoxide to α-onocerin (Rowan and Dean, 1972). The seven-membered ring of serratendiol is probably derived via protonation of one methylene group of α-onocerin. Diepoxysqualene was suggested as the precursor of 24(25)-epoxylanostol (aglaiol) found in leaves of the evergreen, *Aglaia odorata* (Meliaceae) (Boar and Damps, 1973), and the relevant cyclization has since been demonstrated in yeast (Field and Holmlund, 1977).

Numerous *in vivo* studies involving the synthesis of triterpenes from labeled CO_2, acetate, and mevalonate have been reported (e.g., see Strüby *et al.*, 1972; Atallah *et al.*, 1975; and Ekundayo, 1983). As with other terpenoids, difficulties due to compartmentation and/or transport effects are often encountered. Thus, sucrose was shown to be a more effective precursor of the lactiferous triterpenes of *Euphorbia lathyrus* than were mevalonate, amino acids, or Krebs-cycle intermediates (Groeneveld *et al.*, 1982). Compartmentation and transport problems may be overcome through the use of protoplasts instead of intact tissue, as evidenced by the synthesis of β-amyrin from [^{14}C]acetate by protoplast preparations from *Calendula officinalis* (Compositae) (Auguścińska and Kasprzyk, 1982).

D. Limonoids and Quassinoids

The limonoids (e.g., limonin, swietenine, Fig. 14) are a group of highly oxygenated tetranortriterpene lactone furans occurring in Rutaceae and Meliaceae. Similar triterpenes occur in the Cneoraceae (Mondon and Epe, 1983), and quassinoids (e.g., simarolide, Fig. 14) occur in Simaroubaceae (Polonsky, 1973). The structural variations within these classes were described by Connolly and Overton (1972). The limonoids occur in *Citrus* fruits, and, as they are quite bitter tasting, many attempts have been made to minimize their presence in commercial products. Wood of the Meliaceae (mahogany) often contains limonoids such as swietenine (Fig. 14) (Adesida *et al.*, 1971).

Studies on the distribution of limonin in *Citrus paradisi* [determined by radioimmunoassay (RIA) techniques] showed that flushing leaves con-

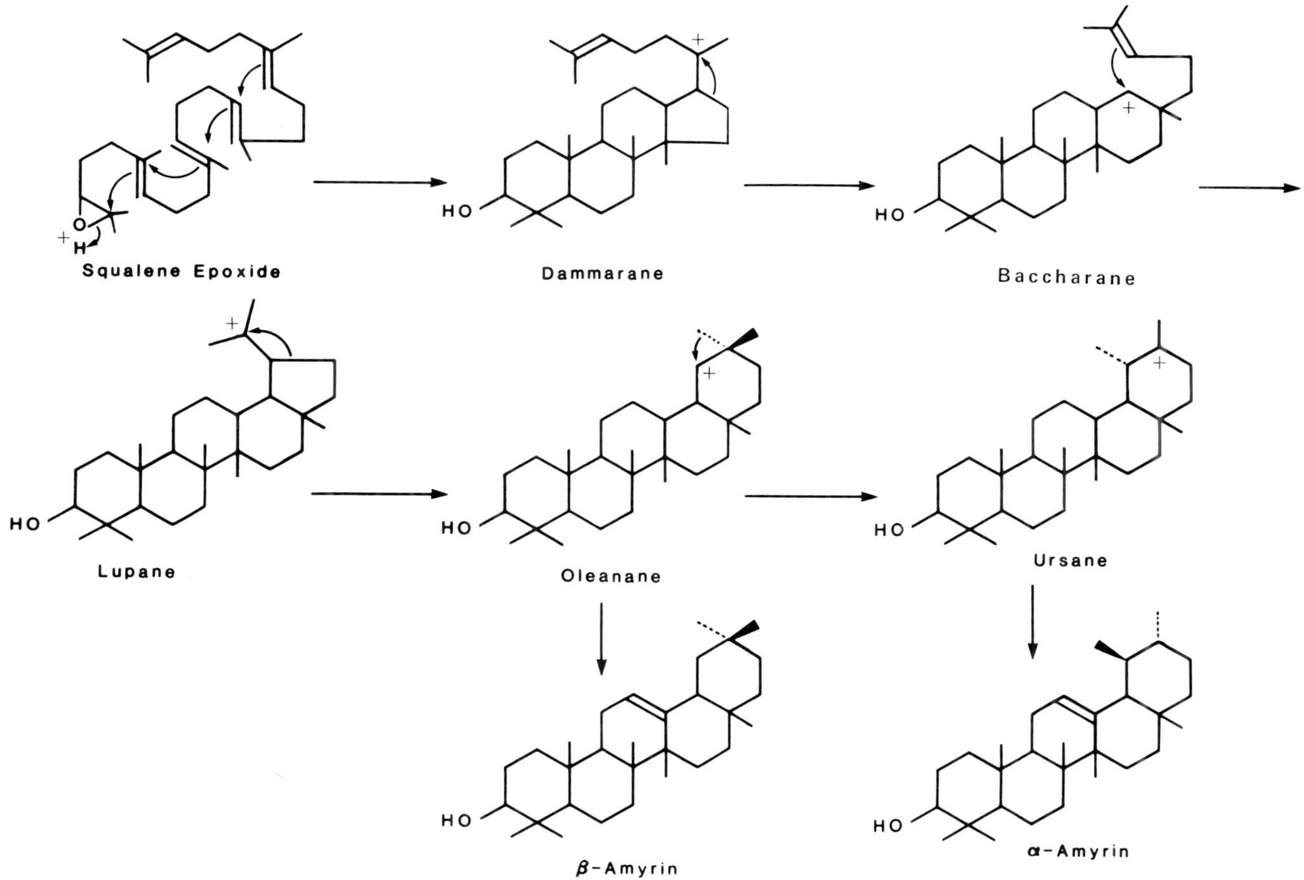

Fig. 15. Biogenesis of triterpene structural types.

tained a high concentration, while mature leaves had negligible levels (McIntosh and Mansell, 1983). Comparison between leaves, fruits, and stems showed that the highest concentration was in the seed of the fruit (Mansell and Weiler, 1980). These observations are consistent with earlier results demonstrating that the young leaves synthesize limonin, which is subsequently transported to the fruit. *Citrus* fruit itself cannot carry out limonin biosynthesis (Hasegawa and Hoagland, 1977). Foliar application of ammonium compounds (crude analogs of the previously mentioned squalene cyclase inhibitors) decrease limonin accumulation (Hasegawa *et al.*, 1977).

X. Polyterpenes

A. *Structural Types*

The naturally occurring polymeric terpenes are classified as either polyisoprenes, such as the latex rubber of the rubber tree, or as polyprenols, such as the acyclic esters that contribute to the fatty nature of birch wood. Polyisoprenes are large polymers ranging in molecular weight from 10^5 to 10^6 and are considered to be derived by "uncontrolled" prenyl transferase activity (Archer, 1980). Polyprenols range in size from 4 to 20 isoprene units, and the terminal hydroxyl is usually esterified (Hemming, 1977). Both types of polyterpenes may contain double bonds of the cis or trans geometry, depending on the specificity of the relevant prenyl transferase.

B. *Polyisoprene*

Rubber is a high molecular weight *cis*-1,4-polyisoprene polymer, of unknown functionality at the elongation terminus where a hydroxyl group would be expected (Archer, 1980). The distribution in higher plants and the biosynthesis of rubber and other polyisoprenes were reviewed by Archer (1980) and Benedict (1983). Various forms of rubber are widespread in plants, usually occurring in small amounts (Sternberg and Rodriquez, 1982). On a commercial scale, the polymer is obtained from the latex (20–60% dry weight) of *Hevea brasiliensis* (Euphorbiaceae) (Kramer and Kozlowski, 1979). An alternative source that has been studied is the woody tissue of guayule (*Parthenium argentatum:* Compositae) (Banigan *et al.*, 1982; Bhatia and Masohan, 1983).

Other forms of polyisoprene, generally of lower molecular weight, are produced in commercial quantities by trees of the Sapotaceae. These include gutta percha from *Palaquium gutta* and balata from *Mimusops balata,* both of which are forms of *trans*-1,4-polyisoprene. That gutta is also of widespread occurrence was revealed by a recent survey of the gutta-producing grasses (Buchanan *et al.,* 1979). Closely related to gutta is the material known as chicle (from *Achras sapota*), which is a mixture of all *cis*- and all *trans*-polyisoprene in an approximate 1 : 2 ratio (Gibbs, 1974; Archer, 1980).

Much of the research on the biosynthesis of rubber has been reviewed by Archer (1980) and Loomis and Croteau (1980). It has been shown that IPP, MVA, or HMG-CoA can serve as substrates for rubber biosynthesis in *Hevea* latex. Addition of NPP (C_{10} cis) to the latex as a "starter" stimulates rubber biosynthesis, while the addition of GPP (C_{10} trans) does not (Archer and Cockbain, 1969). This observation is consistent with the cis specificity of the prenyl transferase responsible for rubber formation. Although the addition of GPP has no role in rubber formation in the latex, it does give rise to FPP (C_{15}), which is utilized in the synthesis of other terpenoids. The possibility that rubber synthesis is regulated at the level of the enzyme HMG-CoA reductase (MVA synthesis) was examined in latex from mature trees by Sipat (1982). The enzyme from lutoids [vacuolar bodies analogous to lysosomes (Archer, 1980)] prefers NADPH as the reductant, and reductase activity is inhibited in the presence of MgATP. This effect is presumed to be the result of protein phosphorylation, the suggested mode of regulation. In an attempt to establish a defined system for studies on the regulation of rubber biosynthesis, Radin *et al.* (1982) demonstrated rubber accumulation in guayule tissue cultures. However, the level of rubber accumulated in such cultures was only ~7% of that observed in the intact plant. Whether the regulatory processes in callus culture will resemble those involved in rubber formation in the intact plant remains an open question. Rubber synthesis in guayule can be stimulated by a bioregulator that increases the enzyme activities of this pathway (Benedict *et al.,* 1983).

C. *Polyprenols*

The distribution and biosynthesis of the polyprenols have been described by Hemming (1977, 1983) and Threlfall (1980). The occurrence of these compounds in foliage of such trees as horsechestnut (castaprenols), fig (ficaprenols), and mulberry (moraprenols) was also surveyed by Hemming (1967). The presence of polyprenols in leaves suggests a possible biogenetic relationship to phytol, the C_{20} prenyl moiety of chlorophyll; however, it is not known if polyprenols are, like phytol, synthesized within chloroplasts. The betulaprenols, which occur as esters in the wood of the silver birch (*Betula verrucosa*), are a series of terpenols containing six to nine isoprene units, with ~60% of the constituent double bonds in the cis form (Lindgren, 1965). The pinoprenyl acetates (10–19 isoprene units) constitute up to 0.9% of the dry weight of *Pinus sylvestris* needles (Hannus and Pensar, 1974; Hannus, 1976).

The dolichoprenols (C_{80} to C_{105}; derived from cis elongation of *trans, trans*-FPP) are ubiquitous components of eukaryotic cells, where they function in glycosyl transfer reactions (Waechter and Lennarz, 1976). Similarly, bacteria require polyprenols (bactoprenols) for protein glycosylations as well as for cell wall formation. The moraprenols (from mulberry leaves) have been shown to function as effective substitutes for bactoprenol in *O*-antigen biosynthesis in *Salmonella* (Shibaev *et al.*, 1979), yet the possibility exists that other plant polyprenols could interfere with microbial glycosylations when wounding leads to the uptake of these substances by potential microbial pathogens. Infected oak leaves were shown to produce certain C_{45} to C_{60} polyprenols that were absent from healthy leaves (Monaco *et al.*, 1983).

XI. Terpene Catabolism

The metabolic turnover of terpenes in herbaceous plants, particularly those of the Labiatae, is now well established (Croteau and Johnson, 1984). By contrast, studies on the terpenoids of woody plants, while less definitive, suggest that terpenoids stored in wood may undergo various metabolic interconversions, but are unlikely to be catabolized as a source of carbon. This point of view is supported by studies on *Pinus halepensis* (Benayoun and Ikan, 1980), in which seedlings deprived of carbon nutrients were shown not to utilize endogenous resin as a carbon source.

The turnover of leaf terpenes in the leguminous tree *Copaifera* was examined by Langenheim and associates (1981), and it was shown that sesquiterpenes (labeled via $^{14}CO_2$) exhibited a significant degree of turnover in 8–10 days. It was not clear from these results whether the reported loss of labeled sesquiterpenes was due to volatilization or catabolism; however, in view of the observations with herbaceous species (Martinkus and Croteau; 1981; Croteau and Winters, 1982), catabolism in foliar tissue would not be surprising. Examples in which the metabolism of preexisting terpenes may have a biological role in arboreal plants include the conversion of limonoids to deoxylimonoids in *Citrus* leaves (Hasegawa *et al.*, 1980), and the conversion of geraniol to the corresponding glucoside in apple fruit (Wills and Scriven, 1979). In the latter instance, the glucoside is synthesized in the core and then transported to the flesh, where further metabolism affords products that may promote deterioration of the apple.

As a general conclusion, it would appear that terpenes that are stored in, or exposed to, physiologically active tissues are likely to be converted to water-soluble derivatives that can be transported and ultimately catabolized. Terpenes stored in essentially inactive tissue, such as wood resin ducts and heartwood, may be oxidized via endogenous peroxidases (Ebermann and Stich, 1982), but the isolation of these products from metabolically active tissue would seem to preclude extensive catabolism.

XII. Perspective

Terpenoid wood extractives have been used by humans since ancient times, and their continued commercial utilization well into the future is virtually assured. Our understanding of the biosynthesis and biological roles of these compounds is still primitive, however, and has yet to emerge on a solid experimental foundation. Thus, how the constituent terpenes are made and how their formation is regulated are, in many cases, still open questions, as is the ultimate fate of these compounds in the biosphere. The relationship between the type of terpenes produced, the anatomy of the associated secretory structures, and the overall development of the plant is uncertain, as are the biochemical bases of the terpenoid-mediated interactions with predators, pathogens, and beneficial organisms. Indeed, it is only relatively recently that we have realized that terpenoids can serve for both the defense of the tree and communication among its insect enemies, establishing their importance in forest ecosystems. The importance of forest products and their vulnerable position with respect to the deteriorating environment can provide no stronger stimulus to explore the origin and function of terpenoids in arboreal species.

Acknowledgments

The work by the authors was supported in part by research grants from the National Science Foundation (PCM 8204391), the National Institutes of Health (GM 31354), the Department of Energy (DE-AT06-82ER12027), the Washington Mint Commission and Mint Industry Research Council, and the Haarmann and Reimer Corporation.

References

Adesida, G. A., Adesogan, E. K., Okorie, D. A. and, Taylor, D. A. H. (1971). *Phytochemistry* **10,** 1845–1853.

Adolf, W., and Hecker, E. (1977). *Isr. J. Chem.* **16,** 75–83.

Ahlquist, L., Bergstrom, G., and Liljenberg, C. (1978). *Prog. Chem. Fats Other Lipids* **16,** 231–255.

Akhila, A., and Banthorpe, D. V. (1980). *Phytochemistry* **19,** 1691–1694.
Albert-Puleo, M. (1978). *Econ. Bot.* **32,** 65–74.
Allen, P. W. (1972). "Natural Rubber and the Synthetics." Wiley, New York.
Allen, N. E., Alborn, W. E., Hobbs, J. N., and Kirst, H. A. (1982). *Antimicrob. Agents Chemother.* **22,** 824–831.
Archer, B. L. (1980). *In* "Encyclopedia of Plant Physiology, New Series" (E. A. Bell and B. V. Charlwood, eds.), Vol. 8, pp. 309–327. Springer-Verlag, Berlin.
Archer, B. L., and Cockbain, E. G. (1969). *Methods Enzymol.* **15,** 476–480.
Arctander, S. (1960). "Perfume and Flavor Materials of Natural Origin." Arctander, Elizabeth, New Jersey.
Arigoni, D. (1975). *Pure Appl. Chem.* **41,** 219–245.
Atallah, A. M., Aexel, R. T., Ramsey, R. B., and Nichols, H. J. (1975). *Phytochemistry* **14,** 1529–1533.
Auguścińska, E., and Kasprzyk, Z. (1982). *Acta Biochem. Pol.* **29,** 7–14.
Aviv, D., Dantes, A., Krochmal, E., and Galun, E. (1983). *Planta Med.* **47,** 7–10.
Baba, T., and Allen, Jr., C. M. (1978). *Biochemistry* **17,** 5598–5604.
Baisted, D. J., Capstack, E., and Nes, W. R. (1962). *Biochemistry* **1,** 537–541.
Bakker, H. J., Ghisalberti, E. L., and Jeffries, P. R. (1972). *Phytochemistry* **11,** 2221–2231.
Balsevich, J. (1983). *Can. J. Chem.* **61,** 1053–1059.
Banigan, T. F., Verbiscar, A. J., and Weber, C. W. (1982). *J. Agric. Food Chem.* **30,** 427–431.
Banthorpe, D. V., and Charlwood, B. V. (1980). *In* "Encyclopedia of Plant Physiology, New Series" (E. A. Bell and B. V. Charlwood, eds.), Vol. 8, pp. 185–220. Springer-Verlag, Berlin.
Banthorpe, D. V., and Ekundayo, O. (1976). *Phytochemistry* **15,** 109–112.
Banthorpe, D. V., Charlwood, B. V., and Francis, M. J. O. (1972). *Chem. Rev.* **72,** 115–155.
Battersby, A. R., Brown, S. H., and Payne, T. G. (1970). *J. Chem. Soc., Chem. Commun.* 827–828.
Benayoun, J., and Fahn, A. (1979). *Ann. Bot.* **43,** 179–181.
Benayoun, J., and Ikan, R. (1980). *Ann. Bot.* **45,** 645–648.
Benedict, C. R. (1983). *In* "Biosynthesis of Isoprenoid Compounds." (J. W. Porter and S. L. Spurgeon, eds.), Vol. 2, pp. 335–369. Wiley, New York.
Benedict, C. R., Reibach, P. H., Madhavan, S., Stipanovic, R. V., Keithly, J. H., and Yokoyama, H. (1983). *Plant Physiol.* **72,** 897–899.
Bernard-Dagan, C., Gleizes, M., Pauly, G., and Marpeau, A. (1980). *In* "Biogenesis and Function of Plant Lipids" (P. Mazliak, P. Benveniste, C. Costes, and R. Douce, eds.), pp. 437–440. Elsevier, New York.
Bernard-Dagan, C., Pauly, C., Marpeau, A., Gleizes, M., Carde, J., and Baradat, P. (1982). *Physiol. Veg.* **20,** 775–795.
Bhatia, V. K., and Masohan, A. (1983). *J. Sci. Ind. Res.* **42,** 273–280.
Bloch, K. (1983). *CRC Cr. Rev. Biochem.* **14,** 47–92.
Blunt, J. W., Boyd, G. S., Hartshorn, M. P., and Munro, M. H. (1976). *Aust. J. Chem.* **29,** 987–993.
Boar, R. B., and Damps, K. (1973). *J. Chem. Soc., Chem. Commun.* 115–116.
Boelens, H. M., de Rijke, D., and Haring, H. G. (1981). *Perfum. Flavor.* **6,** 7–14.
Boller, T., Gehri, A., Mauch, F., and Vögeli, V. (1983). *Planta* **157,** 22–31.
Bordasch, R. P., and Berryman, A. A. (1977). *Can. Entomol.* **109,** 95–100.
Bostock, R. M., Kúc, J. A., and Laine, R. A. (1981). *Science* **212,** 67–69.

Bostock, R. M., Laine, R. A., and Kúc, J. A. (1982). *Plant Physiol.* **70,** 1417–1424.
Boulay, F., Lauquin, G. J. M., Tsugita, A., and Vignais, P. V. (1983). *Biochemistry* **22,** 477–484.
Bouvier, P., Berger, Y., Rohmer, M., and Ourisson, G. (1980). *Eur. J. Biochem.* **112,** 549–556.
Bradshaw, A. P. W., Hanson, J. R., Nyfeler, R., and Sadler, I. H. (1982). *J. Chem. Soc., Perkin Trans I* 2187–2192.
Brand, J. M., Bracke, J. W., Markovetz, A. J., Wood, D. L., and Browne, L. E. (1975). *Nature (London)* **254,** 136–137.
Briggs, L. H., and White, G. W. (1975). *Tetrahedron* **31,** 1311–1314.
Bruce, R. J., and West, C. A. (1982). *Plant Physiol.* **69,** 1181–1188.
Buchanan, R. A., Otey, F. H., Russell, C. R., and Cull, I. M. (1978). *J. Am. Oil Chem. Soc.* **55,** 657–662.
Buchanan, R. A., Swanson, C. L., Weisleder, D., and Cull, I. M. (1979). *Phytochemistry* **18,** 1069–1071.
Burden, R. S., and Kemp, M. S. (1983). *Phytochemistry* **22,** 1039–1040.
Burka, L. T., and Thorsen, A. (1982). *Phytochemistry* **21,** 869–870.
Burke, B. A., Chan, W. R., Pascoe, K. O., Blount, J. F., and Manchard, P. S. (1981). *J. Chem. Soc., Perkin Trans. I* 2666–2669.
Calvin, M. (1983). *Science* **219,** 24–26.
Cambie, R. C., and Palmer, B. D. (1982). *Aust. J. Chem.* **35,** 601–612.
Cane, D. E. (1980). *Tetrahedron* **36,** 1109–1159.
Cane, D. E. (1981). *In* "Biosynthesis of Isoprenoid Compounds" (J. W. Porter and S. L. Spurgeon, eds.), Vol. 1, pp. 283–374. Wiley, New York.
Cane, D. E., and Tillman, A. M. (1983). *J. Am. Chem. Soc.* **105,** 122–124.
Cane, D. E., Iyengar, R., and Shiao, M. (1981a). *J. Am. Chem. Soc.* **103,** 914–931.
Cane, D. E., Rossi, T., Tillman, A. M., and Pachlatko, J. P. (1981b). *J. Am. Chem. Soc.* **103,** 1838–1843.
Cane, D. E., Saito, A., Croteau, R., Shaskus, J., and Felton, M. (1982). *J. Am. Chem. Soc.* **104,** 5831–5833.
Carde, J., Bernard-Dagan, C., and Gleizes, M. (1980). *In* "Biogenesis and Function of Plant Lipids" (P. Mazliak, P. Benveniste, C. Costes, and R. Douces, eds.), pp. 441–444. Elsevier, New York.
Cassady, J. M., and Suffness, M. (1980). *In* "Anticancer Agents Based on Natural Product Models" (J. M. Cassady and J. D. Douros, eds.), pp. 201–270. Academic Press, New York.
Cates, R. G., and Alexander, H. (1982). *In* "Bark Beetles in North American Conifers" (J. B. Mitton and K. B. Sturgeon, eds.), pp. 212–263. Univ. Texas Press, Austin, Texas.
Chandel, R. S., and Rastogi, R. P. (1980). *Phytochemistry* **19,** 1889–1908.
Chandler, R. F., and Hooper, S. N. (1979). *Phytochemistry* **18,** 711–724.
Charlwood, B. V., and Banthorpe, D. V. (1978). *In* "Progress in Phytochemistry" (L. Reinhold, J. B. Harborne, and T. Swain, eds.), Vol. 5, pp. 65–125. Pergamon, Oxford.
Chayet, L., Rojas, C., Cardemil, E., Jabalquinto, A. M., Vicuna, R., and Cori, O. (1977). *Arch. Biochem. Biophys.* **180,** 318–327.
Chou, C. K. S., and Zabkiewicz, J. A. (1976). *Eur. J. For. Path.* **6,** 354–359.
Conner, A. H., Nagasampagi, B. A., and Rowe, J. W. (1980). *Phytochemistry* **19,** 1121–1131.
Connolly, J. D., and Overton, K. H. (1972). *In* "Chemistry of Terpenes and Terpenoids" (A. A. Newman, ed.), pp. 207–287. Academic Press, New York.
Cordell, G. A. (1976). *Chem. Rev.* **76,** 425–460.

Cordell, G. A. (1977). *In* "The Alkaloids" (R. H. F. Manske, ed.), Vol. 16, pp. 432–510. Academic Press, New York.

Cori, O. (1983). *Phytochemistry* **22,** 331–341.

Crocomo, O. J., Aquarone, E., and Gottleib, O. R. (1981). *In* "Plant Tissue Culture" (T. A. Thorpe, ed.), pp. 359–372. Academic Press, New York.

Croteau, R. (1977). *Plant Physiol.* **59,** 519–520.

Croteau, R. (1981). *In* "Biosynthesis of Isoprenoid Compounds" (J. W. Porter and S. L. Spurgeon, eds.), Vol. 1, pp. 225–282. Wiley, New York.

Croteau, R. (1984). *In* "Isopentenoids in Plants" (W. D. Nes, G. Fuller, and L. S. Tsai, eds.), pp. 31–64. Dekker, New York.

Croteau, R., and Cane, D. E. (1984). *Methods Enzymol.* **100,** in press.

Croteau, R., and Johnson, M. A. (1984). *In* "Biology and Chemistry of Plant Trichomes" (E. Rodriguez, P. Healey, and I. Mehta, eds.), pp. 133–186. Plenum, New York.

Croteau, R., and Karp, F. (1977a). *Arch. Biochem. Biophys.* **179,** 257–265.

Croteau, R., and Karp, F. (1977b). *Arch. Biochem. Biophys.* **184,** 77–86.

Croteau, R., and Karp, F. (1979a). *Arch. Biochem. Biophys.* **198,** 512–522.

Croteau, R., and Karp, F. (1979b). *Arch. Biochem. Biophys.* **198,** 523–532.

Croteau, R., and Loomis, W. (1972). *Phytochemistry* **11,** 1055–1066.

Croteau, R., and Loomis, W. (1975). *Int. Flavours Food Addit.* **6,** 292–296.

Croteau, R., and Winters, J. N. (1982). *Plant Physiol.* **69,** 975–977.

Croteau, R., Burbott, A. J., and Loomis, W. (1972). *Phytochemistry* **11,** 2937–2948.

Croteau, R., Hooper, C. L., and Felton, M. (1978). *Arch. Biochem. Biophys.* **188,** 182–193.

Croteau, R., Felton, M., and Ronald, R. C. (1980a). *Arch. Biochem. Biophys.* **200,** 524–533.

Croteau, R., Felton, M., and Ronald, R. C. (1980b). *Arch. Biochem. Biophys.* **200,** 534–546.

Croteau, R., Felton, M., Karp, F., and Kjonaas, R. (1981). *Plant Physiol.* **67,** 820–824.

Croteau, R., Purkett, P., and Felton, M. (1985). Submitted.

Croteau, R., Shaskus, J., Cane, D., Saito, A., and Chang, C. (1984). *J. Am. Chem. Soc.* **106,** 1142–1143.

Cunningham, A., West, P. R., Hammond, G. S., and Langenheim, J. H. (1977). *Phytochemistry* **16,** 1441–1443.

Dahlgren, R. M. T., and Clifford, H. T. (1982). "The Monocotyledons: A Comparative Study," pp. 48–51. Academic Press, New York.

Darvill, A. G., and Albersheim, P. (1984). *Ann. Rev. Plant Physiol.* **35,** 243–275.

Das, M. C., and Mahato, S. B. (1983). *Phytochemistry* **22,** 1071–1095.

Dauben, W. G., Beasley, G. H., Broadhurst, M. D., Muller, B., Peppard, D. J., Pesnelle, P., and Suter, C. (1975). *J. Am. Chem. Soc.* **97,** 4973–4980.

Davies, W. J., and Mansfield, T. A. (1983). *In* "Abscisic Acid" (F. T. Addicott, ed.), pp. 237–268. Praeger, New York.

DeAngelis, J. D., Marin, A. B., Berry, R. E., and Krantz, G. W. (1983). *Environ. Entomol.* **12,** 522–527.

Dell, B., and McComb, A. J. (1978a). *In* "Advances in Botanical Research" (H. W. Woolhouse, ed.), Vol. 6, pp. 227–316. Academic Press, New York.

Dell, B., and McComb, A. J. (1978b). *J. Exp. Bot.* **29,** 89–95.

Delprino, L., Balliano, G., Cattel, L., Benveniste, P., and Bouvier, P. (1983). *J. Chem. Soc., Chem. Commun.* 381–382.

Dorland, W. E., and Rogers, Jr. J. A. (1977). "The Fragrance and Flavor Industry." Dorland, Mendham, New Jersey.

Drengler, K. A., and Coates, R. M. (1980). *J. Chem. Soc., Chem. Commun.* 856–857.

Dueber, M. T., Adolf, W., and West, C. A. (1978). *Plant Physiol.* **62,** 598–603.

Dugan, R. E. (1981). *In* "Biosynthesis of Isoprenoid Compounds" (J. W. Porter and S. L. Spurgeon, eds.), Vol. 1, pp. 95–160. Wiley, New York.

Duncan, J. D., and West, C. A. (1981). *Plant Physiol.* **68,** 1128–1134.
Ebermann, R., and Stich, K. (1982). *Phytochemistry* **21,** 2401–2402.
Ekman, R. (1980). *Phytochemistry* **19,** 321–322.
Ekundayo, O. (1983). *Z. Pflazenphysiol.* **112,** 139–146.
El-Naggar, L. J., and Beal, J. L. (1980). *J. Nat. Prod.* **43,** 694–707.
El-Naggar, S. F., Doskotch, R. W., O'Dell, T. M., and Girard, L. (1980). *J. Nat. Prod.* **43,** 617–631.
Escher, S., Loew, P., and Arigoni, D. (1970). *J. Chem. Soc., Chem. Commun.* 823–825.
Evans, F. J., and Taylor, S. E. (1983). *Prog. Chem. Org. Nat. Prod.* **44,** 1–99.
Evans, R. C., Tingey, D. T., Gumpertz, M. L., and Burns, W. F. (1982). *Bot. Gaz.* **143,** 304–310.
Fahn, A. (1979). "Secretory Tissues in Plants." Academic Press, New York.
Fahn, A., and Zamski, E. (1970). *Isr. J. Bot.* **19,** 429–446.
Farnum, P., Timmis, R., and Kulp, L. (1983). *Science* **219,** 694–702.
Field, R. B., and Holmlund, C. E. (1977). *Arch. Biochem. Biophys.* **180,** 465–471.
Fischer, N. H., Oliver, E. J., and Fischer, H. D. (1979). *Prog. Chem. Org. Nat. Prod.* **38,** 47–390.
Foster, D. O., and Zinkel, D. F. (1982). *J. Chromatrogr.* **248,** 89–98.
Gambliel, H., and Croteau, R. (1982). *J. Biol. Chem.* **257,** 2335–2342.
Gambliel, H., and Croteau, R. (1984). *J. Biol. Chem.* **259,** 740–748.
Ghisalberti, E. L., Jeffries, P. R., and Sefton, M. A. (1978). *Phytochemistry* **17,** 1961–1965.
Gibbons, G. F. (1977). *Biochem. Biophys. Res. Commun.* **75,** 995–1003.
Gibbs, R. D. (1974). "Chemotaxonomy of Flowering Plants" Vol. 2, pp. 770–872. McGill-Queen's University Press, Montreal.
Gildemeister, E., and Hoffmann, F. (1913). "The Volatile Oils" (translated by E. Kremers), 2nd Ed., Vol. 1, pp. 3–244. Wiley, New York.
Gleizes, M. (1979). Thése Doct. Etat., Université de Bordeaux-I, Bordeaux. *In* Bernard-Dagan *et al.* (1982). *Physiol. Veg.* **20,** 775–795.
Gleizes, M., Pauly, G., Carde, J. P., Marpeau, A., and Bernard-Dagan, C. (1983). *Planta* **159,** 373–381.
Godtfredsen, S. E. (1978). *Diss. ETH (Zurich),* 6243. *In* Cane *et al.* (1981a). *J. Am. Chem. Soc.* **103,** 914–931.
Goldstein, I. S. (1981). *In* "Organic Chemicals from Biomass" (I. S. Goldstein, ed.), pp. 9–18. CRC Press, Boca Raton, Florida.
Gollob, L. (1980). *Naturwissenschaften* **67,** 409–410.
González, A. G., Galindo, A., Mansilla, H., and Palenzuela, J. A. (1983). *Tetrahedron Lett.* **24,** 969–972.
Goodwin, T. W. (1980). *In* "The Biochemistry of Plants" (P. K. Stumpf, ed.), Vol. 4, pp. 485–508. Academic Press, New York.
Goodwin, T. W. (1981). *In* "Biosynthesis of Isoprenoid Compounds" (J. W. Porter and S. L. Spurgeon, eds.), Vol. 1, pp. 443–480. Wiley, New York.
Gottleib, O. R. (1972). *Phytochemistry* **11,** 1537–1570.
Gottlieb, O. R., and Mors, W. B. (1980). *J. Agric. Food Chem.* **28,** 196–215.
Graebe, J. E. (1982). *In* "Plant Growth Substances 1982" (P. F. Wareing, ed.), pp. 71–80. Academic Press, New York.
Greguss, P. (1972). "Xylotomy of the Living Conifers." Akademiae Kiado, Budapest.
Groeneveld, H. W., Hageman, J., and Vellenga, A. T. N. (1982). *Phytochemistry* **21,** 1589–1597.
Guenther, E. (1976). "The Essential Oils." Krieger, Huntington, New York. (Originally published 1948–1952 by Van Nostrand-Reinhold, Princeton, New Jersey.)

Guilford, W. J., and Coates, R. M. (1982). *J. Am. Chem. Soc.* **104,** 3506–3508.
Hadwiger, L. A., and Loschke, D. C. (1981). *Phytopathology* **71,** 756–762.
Hannus, K. (1976). *Appl. Polym. Symp.* **28,** 485–501.
Hannus, K., and Pensar, G. (1974). *Phytochemistry* **13,** 2563–2566.
Hasegawa, S., and Hoagland, J. E. (1977). *Phytochemistry* **16,** 469–471.
Hasegawa, S., Yokoyama, H., and Hoagland, J. E. (1977). *Phytochemistry* **16,** 1083–1085.
Hasegawa, S., Bennett, R. D., and Verdon, C. P. (1980). *Phytochemistry* **19,** 1445–1447.
Hart, N. K., and Lamberton, J. A. (1965). *Aust. J. Chem.* **18,** 115–116.
Hedrick, G. W., Fisher, G. S., Lawrence, R. V., and Mazzeno, Jr. L. W. (1965a). *Agric. Res. Service/USDA* 72–34.
Hedrick, G. W., Fisher, G. S., Lawrence, R. V., and Mazzeno, Jr. L. W. (1965b). *Agric. Res. Service/USDA* 72–35.
Hegnauer, R. (1962–1973). "Chemotaxonomie der Pflanzen," Vols. 1-6. Birkhauser Verlag, Basel.
Hemming, F. W. (1967). *In* "Terpenoids in Plants" (J. B. Pridham, ed.), pp. 223–240. Academic Press, New York.
Hemming, F. W. (1977). *In* "Lipids and Lipid Polymers in Higher Plants" (M. Tevini and H. K. Lichtenthaler, eds.), pp. 183–198. Springer-Verlag, New York.
Hemming, F. W. (1983). *In* "Biosynthesis of Isoprenoid Compounds" (J. W. Porter and S. L. Spurgeon, eds.), Vol. 2, pp. 305–354. Wiley, New York.
Hemmingway, R. W. (1981). *In* "Organic Chemicals from Biomass" (I. S. Goldstein, ed.), pp. 189–248. CRC Press, Boca Raton, Florida.
Hepting, G. H. (1947). *Science* **105,** 209.
Herout, V. (1982). *In* "Frangrance Chemistry" (E. T. Theimer, ed.), pp. 221-265. Academic Press, New York.
Herz, W. (1971). *In* "Pharmacognosy and Phytochemistry" (H. Wagner and L. Hörhammer, eds.), pp. 64–92. Springer-Verlag, Berlin.
Herz, W. (1977). *Isr. J. Chem.* **16,** 32–44.
Herz, W., Prasad, J. S., and Mohanraj, S. (1983). *J. Org. Chem.* **48,** 81–90.
Heywood, V. H., Harborne, J. B., and Turner, B. L. (eds.) (1977). "The Biology and Chemistry of the Composite" Vols. 1, 2. Academic Press, New York.
Hoque, E. (1982). *Eur. J. For. Path.* **12,** 280–296.
Hui, W., and Li, M. (1977). *J. Chem. Soc., Perkin Trans. I,* 897–904.
Inouye, H., Ueda, S., and Uesato, S. (1977). *Phytochemistry* **16,** 1669–1675.
Jalaluddin, M. (1977). *Econ. Bot.* **31,** 222–224.
Jedlicki, E., Jacob, G., Faini, F., and Cori, O. (1972). *Arch. Biochem. Biophys.* **152,** 590–596.
Jeffries, P. R. (1979). *J. R. Soc. West. Aust.* **62,** 95–107.
Jeng, R. S., Alfenas, A. C., Hubbes, M., and Dumas, M. (1983). *Eur. J. For. Pathol.* **13,** 239–244.
Joel, D. M., and Fahn, A. (1980). *Ann. Bot.* **46,** 779–783.
Johnson, M., and Croteau, R. (1980). *Plant Physiol.* **65,** (suppl.), 96.
Jones, C. A., and Rasmussen, R. A. (1975). *Plant Physiol.* **55,** 982–987.
Kapil, R. S., and Brown, R. T. (1979). *In* "The Alkaloids" (R. H. F. Manske and R. G. A. Rodrigo, eds.), Vol. 17, pp. 543–588. Academic Press, New York.
Karp, F., and Croteau, R. (1982). *Arch. Biochem. Biophys.* **216,** 616–624.
Kepner, R. E., Ellison, B. O., and Muller, C. J. (1975). *Phytochemistry* **14,** 808.
Kjonaas, R., and Croteau, R. (1983). *Arch. Biochem. Biophys.* **220,** 79–89.
Klimetzek, D., and Francke, W. (1980). *Experientia* **36,** 1343–1345.

Koch, P. (1972). "Utilization of the Southern Pines" Agriculture Handbook 420, Vols. I, II. U.S. Dept. Agric., Forest Service, Washington, D.C.
Kolata, G. (1983). *Science* **220,** 291–292.
Kramer, P. J., and Kozlowski, T. T. (1979). "Physiology of Woody Plants." Academic Press, New York.
Kubik, Y. M., and Jackson, L. L. (1981). *Cornell Vet.* **71,** 34–42.
Kubo, I., and Nakanishi, K. (1979). *In* "Advances in Pesticide Science, Part 2" (H. Geissbühler, G. T. Brooks, and P. C. Kearney, eds.), pp. 284–294. Pergamon, Oxford.
Kúc, J., and Lisker, N. (1978). *In* "Biochemistry of Wounded Plant Tissues" (G. Kahl, ed.), pp. 203–242. Walter de Gruyter, Berlin.
Kutney, J. P. (1982). *Pure Appl. Chem.* **54,** 2523–2536.
Lagercrantz, C., and Yhland, M. (1962). *Acta Chem. Scand.* **16,** 505–506.
Lakshmi, S. G., Raghava, R. N. V., and Vaidyanathan, C. S. (1979). *Plant Sci. Lett.* **15,** 265–270.
Langenheim, J. H. (1981). *In* "Advances in Legume Systematics" (R. M. Polhill and P. H. Raven, eds.), Part 2, pp. 627–655. Royal Botanic Gardens, Kew, Richmond, Surrey.
Langenheim, J. H., Arrhenius, S. P., and Nascimento, J. C. (1981). *Biochem. Syst. Ecol.* **9,** 27–37.
Langenheim, J. H., Lincoln, D. E., Stubblebine, W. H., and Gabrielli, A. C. (1982). *Am. J. Bot.* **69,** 595–607.
Laskovics, F. M., and Poulter, C. D. (1981). *Biochemistry* **20,** 1893–1901.
Lewis, I. A. S., MacLeod, J. K., and Delrichs, P. B. (1981). *Tetrahedron* **37,** 4305–4311.
Licht, H. J., and Coscia, C. J. (1978). *Biochemistry* **17,** 5638–5646.
Lindgren, B. O. (1965). *Acta Chem. Scand.* **19,** 1317–1326.
Loomis, W. D., and Croteau, R. (1973). *In* "Recent Advances in Phytochemistry" (V. C. Runeckles and T. J. Mabry, eds.), Vol. 6, pp. 147–186. Academic Press, New York.
Loomis, W. D., and Croteau, R. (1980). *In* "The Biochemistry of Plants" (P. K. Stumpf, ed.), Vol. 4, pp. 364–419. Academic Press, New York.
Luft, R. (1979). *J. Org. Chem.* **44,** 523–524.
Mabry, T. J., and Gill, J. E. (1979). *In* "Herbivores: Their Interaction with Secondary Metabolites" (G. A. Rosenthal and D. H. Janzen, eds.), pp. 502–537. Academic Press, New York.
McCann, F. V., and Penefsky, Z. J. (1981). *Comp. Biochem. Physiol.* **70C,** 185–193.
Mace, M. E. (1983). *New Phytol.* **95,** 115–119.
Mace, M. E., and Bell, A. A. (1981). *Can. J. Bot.* **59,** 951 955.
McIntosh, C. A., and Mansell, R. L. (1983). *J. Agric. Food. Chem.* **31,** 319–325.
Mackie, H., and Overton, K. H. (1977). *Eur. J. Biochem.* **77,** 101–106.
Madyastha, K. M., and Coscia, C. J. (1979). *J. Biol. Chem.* **254,** 2419–2427.
Mansell, R. L., and Weiler, E. W. (1980). *Phytochemistry* **19,** 1403–1407.
Martinkus, C., and Croteau, R. (1981). *Plant Physiol.* **68,** 99–106.
Marx, J. L. (1983). *Science* **219,** 158–159.
Mash, E. A., Gurria, G. M., and Poulter, C. D. (1981). *J. Am. Chem. Soc.* **103,** 2927–2929.
Matthes, H. W. D., Luu, B., and Ourisson, G. (1982). *Tetrahedron* **38,** 3129–3135.
Mears, J. A. (1980). *Proc. Acad. Natur. Sci. (Philadelphia)* **132,** 156–172.
Mgaloblishvili, M. P., Litvinov, A. I., and Sanadze, G. A. (1981). *Doklady Botanical Sciences* (English translation of *Doklady Akademii Nauk SSSR)* **259,** 72–75.
Milborrow, B. V. (1983a). *In* "Biosynthesis of Isoprenoid Compounds" (J. W. Porter and S. L. Spurgeon, eds.), Vol. 2, pp. 413–436. Wiley, New York.
Milborrow, B. V. (1983b). *In* "Abscisic Acid" (F. T. Addicott, ed.), pp. 79–111. Praeger, New York.

Mirov, N. (1961). "Composition of Gum Turpentines of Pines" (Tech. Bull. 1239). U.S. Dept. Agric., Forest Service, Washington, D.C.
Mirov, N. T. (1967). "The Genus *Pinus*." Ronald Press Company, New York.
Monaco, P., Previtera, L., and Belardini, M. (1983). *J. Nat. Prod.* **46,** 174–177.
Mondon, A., and Epe, B. (1983). *Prog. Chem. Org. Nat. Prod.* **44,** 101–187.
Mookherjee, B. D., and Mussinan, C. J. (eds.) (1981). "Essential Oils" Allured, Wheaton, Illinois.
Mutton, D. B. (1962). *In* "Wood Extractives and Their Significance to the Pulp and Paper Industries" (W. E. Hillis, ed.), pp. 331–363. Academic Press, New York.
Nabeta, K., Ohnishi, Y., Hirose, T., and Sugisawa, H. (1983). *Phytochemistry* **22,** 423–425.
Nagasampagi, B. A., Yankov, L., and Dev, S. (1967). *Tetrahedron Lett.* 189–192.
Nakanishi, K., and Habaguchi, K. (1971). *J. Am. Chem. Soc.* **93,** 3546–3547.
Nakanishi, K., Goto, T., Itô, S., Natori, S., and Nozoe, S. (1974). "Natural Products Chemistry" Vol. 1. Academic Press, New York.
Neill, S. J., Horgan, R., Walton, D. C., and Griffin, D. (1982). *In* "Plant Growth Substances, 1982" (P. F. Wareing, ed.), pp. 315–324. Academic Press, New York.
Nes, W. D., and Heftmann, E. (1981). *J. Nat. Prod.* **44,** 337–400.
Nes, W. D., Fuller, G., and Tsai, L. S. (eds). (1983). "Biochemistry of Isopentenoids in Plants." Dekker, New York.
Nes, W. D., Saunders, G. A., and Heftmann, E. (1982). *Lipids* **17,** 178–183.
Nijholt, W. W., McMullen, L. H., and Safranyik, L. (1981). *Can. Entomol.* **113,** 337–340.
Nohara, T., Kashiwada, Y., Tomimatsu, T., and Nishioka, I. (1982). *Phytochemistry* **21,** 2130–2132.
Norin, T. (1972). *Phytochemistry* **11,** 1231–1242.
Norin, T., and Winell, B. (1971). *Phytochemistry* **10,** 2818–2821.
Ôba, K., and Uritani, I. (1981). *Agric. Biol. Chem.* **45,** 1635–1639.
Ohloff, G. (1982). *In* "Fragrance Chemistry" (E. T. Theimer, ed.), pp. 535–574. Academic Press, New York.
Ourisson, G. (1973). *In* "Chemistry in Botanical Classification" (G. Bendz, J. Santesson, and V. Runnström-Reio, eds.), pp. 129–134. Nobel Foundation, Stockholm. Academic Press, New York.
Ourisson, G., Rohmer, M., and Anton, R. (1979). *In* "Recent Advances in Phytochemistry" (T. Swain and G. R. Waller, eds.), Vol. 13, pp. 131–162. Plenum, New York.
Overeem, J. C., and Elgersma, D. M. (1970). *Phytochemistry* **9,** 1949–1952.
Pant, P., and Rastogi, R. P. (1979). *Phytochemistry* **18,** 1095–1108.
Pearce, R. B., and Ride, J. P. (1982). *Physiol. Plant Path.* **20,** 119–123.
Pérez, L. M., Lozada, R., and Cori, O. (1983). *Phytochemistry* **22,** 431–433.
Perry, N. B., and Weavers, R. T. (1984). *J. Chromatogr.* **284,** 478–481.
Phillips, E. W. J. (1963). "Identification of Softwoods by Their Microscopic Structure" (Forest Products Research Bull. No. 22). Her Majesty's Stationery Office, London. (Reprinted from 1948 edition.)
Piovetti, L., and Diara, A. (1977). *Phytochemistry* **16,** 103–106.
Piovetti, L., Combant, G., and Diara, A. (1980). *Phytochemistry* **19,** 2117–2120.
Piovetti, L., Francisco, C., Pauly, G., Benchabane, O., Bernard-Dagan, C., and Diara, A. (1981). *Phytochemistry* **20,** 1299–1302.
Polonsky, J. (1973). *In* "Recent Advances in Phytochemistry" (V. C. Runeckles and T. J. Mabry, eds.), Vol. 6, pp. 31–64. Academic Press, New York.
Porter, J. W., and Spurgeon, S. L. (eds). (1981). "Biosynthesis of Isoprenoid Compounds," Vol. 1. Wiley, New York.
Porter, J. W., and Spurgeon, S. L. (eds). (1983). "Biosynthesis of Isoprenoid Compounds," Vol. 2. Wiley, New York.

Portilla, G., Rojas, M. C., Chayet, L., and Cori, O. (1982). *Arch. Biochem. Biophys.* **218,** 614–618.
Poulose, A. J., and Croteau, R. (1978). *Arch. Biochem. Biophys.* **191,** 400–411.
Poulter, C. D., and Rilling, H. C. (1981a). *In* "Biosynthesis of Isoprenoid Compounds" (J. W. Porter and S. L. Spurgeon, eds.), Vol. 1, pp. 161–224. Wiley, New York.
Poulter, C. D., and Rilling, H. C. (1981b). *In* "Biosynthesis of Isoprenoid Compounds" (J. W. Porter and S. L. Spurgeon, eds.), Vol. 1, pp. 413–442. Wiley, New York.
Poulter, C. D., Wiggins, P. L., and Le, A. T. (1981). *J. Am. Chem. Soc.* **103,** 3926–3927.
Prestwich, G. D., Gayen, A. K., Phirwa, S., and Kline, T. B. (1983). *Bio/Technology* **1,** 62–65.
Qureshi, N., and Porter, J. W. (1981). *In* "Biosynthesis of Isoprenoid Compounds" (J. W. Porter and S. L. Spurgeon, eds.), Vol. 1, pp. 47–94. Wiley, New York.
Radin, D. N., Behl, H. M., Proksch, P., and Rodriguez, E. (1982). *Plant Sci. Lett.* **26,** 301–310.
Raffa, K. F., and Berryman, A. A. (1982). *Environ. Ent.* **11,** 486–492.
Rasmussen, R. A. (1972). *J. Air Pollut. Cont. Assn.* **22,** 537–543.
Reinhard, E., and Alfermann, A. W. (1980). *Adv. Biochem. Eng.* **16,** 49–84.
Roberts, D. R., Williams, I. L., and Kossuth, S. V. (1982). *For. Prod. J.* **32,** 74–78.
Rodriguez, E., Towers, G. H. N., and Mitchell, J. C. (1976). *Phytochemistry* **15,** 1573–1580.
Rogers, I. H. (1978). *J. Am. Oil Chem. Soc.* **55,** 113A–118A.
Rogers, E. F., Konuiszy, F. R., Shavel, Jr., J., and Folkers, K. (1948). *J. Am. Chem. Soc.* **70,** 3086–3088.
Rojas, M. C., Chayet, L., Portilla, G., and Cori, O. (1983). *Arch. Biochem. Biophys.* **222,** 389–396.
Rowan, M. G., and Dean, P. D. G. (1972). Phytochemistry **11,** 3111–3118.
Rowe, J. W., and Conner, A. H. (1979). "Extractives in Eastern Hardwoods" (General Technical Report, FPL-18) U.S. Dept. Agric., Forest Service, Forest Products Laboratory, Madison, Wisconsin.
Rowe, J. W., Seikel, M. K., Roy, D. N., and Jorgensen, E. (1972). *Phytochemistry* **11,** 2513–2517.
Rücker, G. (1973). *Angew. Chem. Int. Ed. Engl.* **12,** 793–806.
Russell, C. E., and Berryman, A. A. (1976). *Can. J. Bot.* **54,** 14–18.
Ruzicka, L., Eschenmoser, A., and Heusser, H. (1953). *Experientia* **9,** 357–396.
Sagami, H., Ogura, K., Seto, S., and Kurokawa, T. (1978). *Biochem. Biophys. Res. Commun.* **85,** 572–578.
Sandifer, R. M., Thompson, M. D., Gaughan, R. G., and Poulter, C. D. (1982). *J. Am. Chem. Soc.* **104,** 7376–7378.
Sandmann, G., Hilgenberg, W., and Böger, P. (1980). *Z. Naturforsch.* **35C,** 927–930.
Sato, A., Ogiso, A., and Kuwano, H. (1980). *Phytochemistry* **19,** 2207–2209.
Sato, K., Uritani, I., and Saito, T. (1982). *Appl. Entomol. Zool.* **17,** 368–374.
Schuck, H. J. (1982). *Eur. J. For. Path.* **12,** 175–181.
Schuler, M. L. (1981). *Ann. N.Y. Acad. Sci.* **369,** 65–79.
Seaman, F. C. (1982). "Sesquiterpene Lactones in Taxonomic Characterization in Asteraceae." New York Botanical Garden, Bronx, New York.
Seamon, K. B., and Daly, J. W. (1981). *J. Cyclic Nucleotide Res.* **7,** 201–224.
Sebek, O. K., and Kieslich, K. (1977). *In* "Annual Reports on Fermentation Processes" (D. Perlman and G. T. Tsao, eds.), Vol. 1, pp. 267–297. Academic Press, New York.
Seigler, D. S. (1981a). *In* "The Biochemistry of Plants" (E. E. Conn, ed.), Vol. 7, pp. 139–176. Academic Press, New York.
Seigler, D. S. (1981b). *In* "Phytochemistry and Angiosperm Phylogeny" (D. A. Young and D. S. Seigler, eds.), pp. 117–148. Praeger, New York.

Seo, S., Tomita, Y., and Tori, K. (1981). *J. Am. Chem. Soc.* **103,** 2075–2080.
Shain, L. (1979). *Phytopathology* **69,** 1143–1147.
Shankaranarayana, K. H., Ayyar, K. S., and Rao, G. S. K. (1980). *Phytochemistry* **19,** 1239–1240.
Shibaev, V. N., Druzhinina, T. N., Popova, A. N., Rozhnova, S., and Kilesso, V. A. (1979). *Eur. J. Biochem.* **101,** 309–316.
Shrimpton, D. M. (1978). *In* "Theory and Practice of Mountain Pine Beetle Management in Lodgepole Forests" (A. A. Berryman, G. D. Amman, R. W. Stark, and D. L. Kibbee, eds.), pp. 64–76. Forest, Wildlife and Range Experiment Station, University of Idaho, Moscow.
Silverstein, R. M. (1981). *Science* **213,** 1326–1331.
Simonsen, J., and Ross, W. C. J. (1957). "The Terpenes," Vol. 5. Cambridge University Press, Cambridge.
Sipat, A. B. (1982). *Phytochemistry* **21,** 2613–2618.
Sjöström, E. (1981). "Wood Chemistry." Academic Press, New York.
Squillace, A. E. (1976). *In* "Modern Methods of Forest Genetics" (J. P. Miksche, ed.), pp. 120–157. Springer-Verlag, Berlin.
Steffens, J. C., Roark, J. L., Lynn, D. G., and Riopel, J. L. (1983). *J. Am. Chem. Soc.* **105,** 1669–1671.
Steglich, W., Klaar, M., Zechlin, L., and Hecht, H. J. (1979). *Angew. Chem. Int. Ed. Engl.* **18,** 698.
Sternburg, C., and Rodriguez, E. (1982). *Am. J. Bot.* **69,** 214–218.
Stinson, S. C. (1983). *Chem. Eng. News* **61**(17), 23–40.
Stipanovic, R. D., Bell, A. A., Mace, M. E., and Howell, C. R. (1975). *Phytochemistry* **14,** 1077–1081.
Stoessl, A. (1980). *Phytopath. Z.* **99,** 251–272.
Stoessl, A., and Stothers, J. B. (1982). *J. Chem. Soc., Chem. Commun.* 880–881.
Stothers, J. B. (1981). *Pure Appl. Chem.* **53,** 1241–1258.
Strüby, K., Janiszowska, W., and Kaspryzk, Z. (1972). *Phytochemistry* **11,** 1733–1736.
Suarez, D., Garcia-Peregrín, E., and Mayor, F. (1974). *Phytochemistry* **13,** 1059–1063.
Suga, T., Shishibori, T., and Bukeo, M. (1972). *Bull. Chem. Soc. Jpn.* **45,** 1480–1482.
Suga, T., Shishibori, T., and Morinaka, H. (1980a). *J. Chem. Soc., Chem. Commun.* 167–168.
Suga, T., Shishibori, T., and Nakaya, K. (1980b). *Phytochemistry* **19,** 2337–2330.
Suga, T., Hirata, T., Aoki, T., and Shishibori, T. (1983). *J. Am. Chem. Soc.* **105,** 6178–6179.
Swan, E. P., Jiang, K. S., and Gardner, J. A. F. (1969). *Phytochemistry* **8,** 345–351.
Takeda, R., and Katoh, K. (1981). *Planta* **151,** 525–530.
Tange, T. (1981). *Bull. Chem. Soc. Jpn.* **54,** 2763–2769.
Telewski, F. W., Wakefield, A. H., and Jaffe, M. J. (1983). *Plant Physiol.* **72,** 177–181.
Tessier, A. M., Delaveau, P., and Piffault, N. (1982). *Planta Med.* **44,** 215–217.
Tingey, D. T., Evans, R., and Gumpertz, M. (1981). *Planta* **152,** 565–570.
Tobolski, J. J., and Zinkel, D. F. (1982). *For. Sci.* **28,** 785–796.
Tomlinson, P. B. (1964). *In* "Formation of Wood in Forest Trees" (M. H. Zimmerman, ed.), pp. 65–86. Academic Press, New York.
Threlfall, D. R. (1980). *In* "Encyclopedia of Plant Physiology, New Series" (E. A. Bell and B. V. Charlwood, eds.), Vol. 8, pp. 288–308. Springer-Verlag, Berlin.
Uesato, S., Kobayashi, K., and Inouye, H. (1982a). *Chem. Pharm. Bull.* **30,** 927–940.
Uesato, S., Kobayashi, K., and Inouye, H. (1982b). *Chem. Pharm. Bull.* **30,** 3942–3950.
van Tamelen, E. E., and Marson, S. A. (1982). *Bioorg. Chem.* **11,** 219–249.
Vogel, S. (1962). *Akad. Wiss. Lit. Mainz. Abh. Math.-Naturwiss. Kl.* **10,** 599–763.
Vogt, D. D. (1981). *J. Ethnopharmacol.* **4,** 337–342.

von Rudloff, E. (1975). *Biochem. Syst. Ecol.* **2,** 131–167.

Waechter, C. J., and Lennarz, W. J. (1976). *Ann. Rev. Biochem.* **45,** 95–112.

Wagner, M. R., Benjamin, D. M., Clancy, K. M., and Schuh, B. A. (1983). *J. Chem. Ecol.* **9,** 119–127.

Walker-Simmons, M., Hadwiger, L., and Ryan, C. A. (1983). *Biochem. Biophys. Res. Commun.* **110,** 194–199.

Wargo, P. M. (1975). *Physiol. Plant Path.* **5,** 99–105.

Weaver, J. C. (1982). *In* "Kirk-Othmer Encyclopedia of Chemical Technology" 3rd Ed. (M. Grayson and D. Eckroth, eds.), Vol. 20, pp. 197–206. Wiley, New York.

Weinges, K., von der Eltz, H., and Tran-Viet, D. (1980). *Angew. Chem. Int. Ed. Engl.* **19,** 268.

Weinheimer, A. J., Chang, C. W. J., and Matson, J. A. (1979). *Prog. Chem. Org. Nat. Prod.* **36,** 285–387.

Wenzl, H. F. J. (1970). "The Chemical Technology of Wood" (translated by F. E. Brauns and D. A. Brauns). Academic Press, New York.

West, C. A. (1980). *In* "The Biochemistry of Plants" (D. D. Davies, ed.), Vol. 2, pp. 317–364. Academic Press, New York.

West, C. A. (1981a). *Naturwissenschaften* **68,** 447–457.

West, C. A. (1981b). *In* "Biosynthesis of Isoprenoid Compounds" (J. W. Porter and S. L. Spurgeon, eds.), Vol. 1, pp. 375–412. Wiley, New York.

West, C. A., Dudley, M. W., and Dueber, M. T. (1979). *In* "Recent Advances in Phytochemistry" (T. Swain and G. R. Waller, eds.), Vol. 13, pp. 163–198. Plenum, New York.

West, C. A., Shen-Miller, J., and Railton, I. D. (1982). *In* "Plant Growth Substances, 1982" (P. F. Wareing, ed.), pp. 81–90. Academic Press, New York.

Widmaier, R., Howe, J., and Heinstein, P. (1980). *Arch. Biochem. Biophys.* **200,** 609–616.

Wilkinson, R. C. (1979). "Oleoresin Crystallization in Eastern White Pine: Relationship with Chemical Components of Cortical Oleoresin and Resistance to the White Pine Weevil" (Research Paper NE–438). U.S. Dept. Agric., Forest Service, Northeast Forest Experiment Station, Broomall, Pennsylvania.

Wilkinson, R. C. (1980). *For. Sci.* **26,** 581–589.

Wills, R. B. H., and Scriven, F. M. (1979). *Phytochemistry* **18,** 785–786.

Witte, L., Berlin, J., Wray, V., Schubert, W., Kohl, W., Höfle, G., and Hammer, J. (1983). *Planta Med.* **49,** 216–221.

Wright, L. C., Berryman, A. A., and Gurusiddaiah, S. (1977). *Can. Entomol.* **111,** 1255–1262.

Yoshioka, H., Mabry, T. J., and Timmermann, B. N. (1973). "Sesquiterpene Lactones." University of Tokyo Press, Tokyo.

Zavarin, E., and Cobb, Jr., F. W. (1970). *Phytochemistry* **9,** 2509–2515.

Zavarin, E., and Snajberk, K. (1980). *J. Agric. Food Chem.* **28,** 829–834.

Zavarin, E., Smith, L. V., and Bicho, J. G. (1967). *Phytochemistry* **6,** 1387–1394.

Zinkel, D. F. (1977). *In* "TAPPI Conference Papers Forest Biology/Wood Chemistry Conference," pp. 53–56. Technical Association of the Pulp and Paper Industry, Atlanta, Georgia.

Zinkel, D. F. (1981). *In* "Organic Chemicals from Biomass" (I. S. Goldstein, ed.), pp. 163–187. CRC Press, Boca Raton, Florida.

Zinkel, D. F., Zank, L. C., and Wesolowski, M. F. (1971). "Diterpene Resin Acids—A Compilation of Infrared, Mass, Nuclear Magnetic Resonance, Ultraviolet Spectra and Gas Chromatographic Retention Data." USDA Forest Service, Forest Products Laboratory, Madison, Wisconsin.

CHAPTER **16**

Degradation of Wood by Microorganisms

KATIA RUEL
FERNAND BARNOUD

Centre de Recherches sur les Macromolécules Végétales
CNRS et Université de Grenoble I
St. Martin d'Heres, France

I. Introduction

In unfavorable environmental conditions, wood is severely degraded by microorganisms. The most efficient wood destroyers are the white-rot and brown-rot fungi in the class Basidiomycetes. Ligninolytic activity has been found in more than 600 species of white rot. Other fungi, belonging to the Ascomycetes and Fungi imperfecti which cause typical soft rot, are also able to damage wood. Some bacteria are agents of wood deterioration, but the degradation process is usually limited and slow and is found in aquatic habitats.

Because of the considerable importance that wood has gained in the life of humans, these biological causes of wood damage have been the subject of intensive investigations. Timber, woodpiles, timber for mines, and pulpwood (logs and chips) must be protected against degradation by microorganisms. An improved knowledge of the decaying process will en-

Biosynthesis and Biodegradation of Wood Components

ISBN 0-12-347880-4

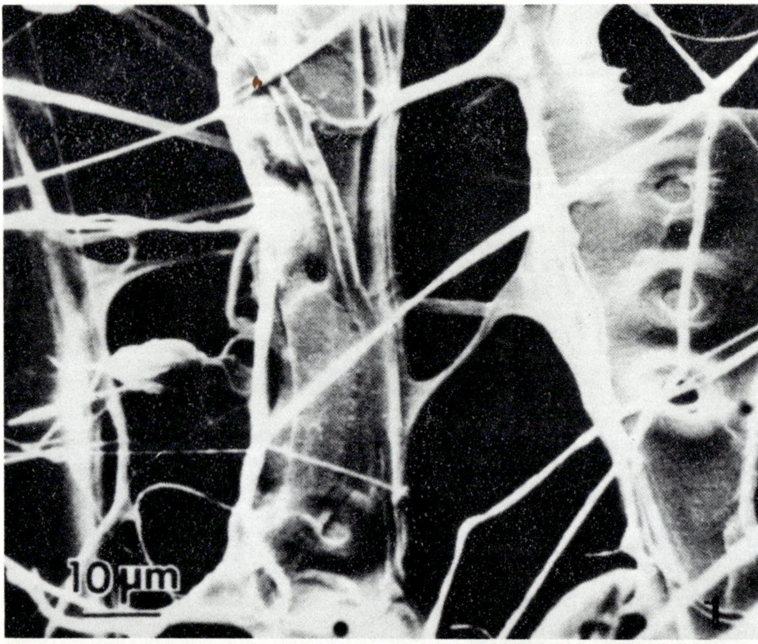
10 μm
1

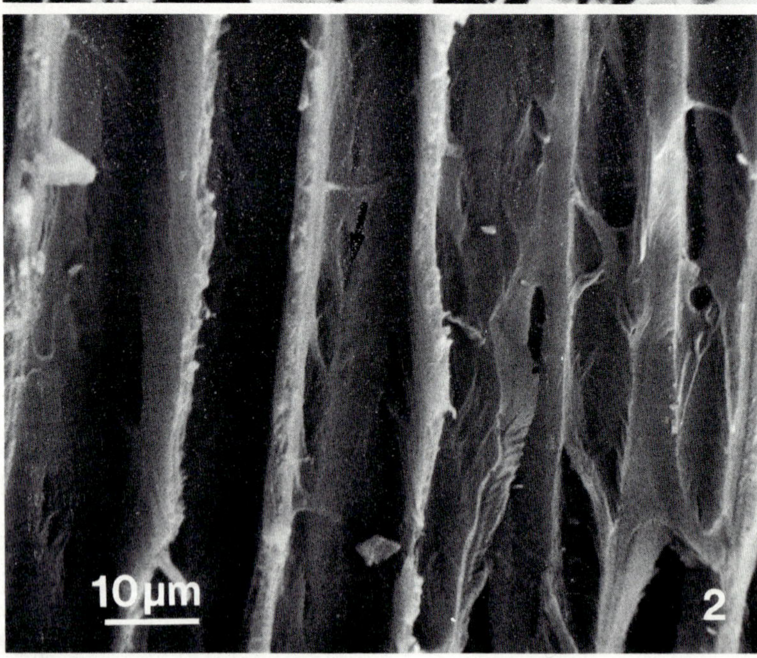
10 μm
2

sure progress in the present techniques of wood preservation, which is of great economic importance.

In the past, excellent reviews have covered the many publications dealing with the microscopic study of wood degradation by microorganisms. Areas of research that have been previously reviewed include the various types of decay caused by bacteria and fungi, the morphological changes provoked by the digestion of woody tissues (Meier, 1955; Juraček, 1964; Wilcox, 1968, 1970, 1973; Liese, 1970) and the various patterns of propagation of hyphae (Cowling, 1961). The aim of this chapter is to present the major results obtained since 1970 that describe the types of decay at the ultrastructural level.

II. Micromorphological Aspects of Wood Decay

A. *Scanning Electron Microscopy*

Since 1970, scanning electron microscopy (SEM) has been extensively used in the field of wood pathology (Blanchette *et al.*, 1978). The possibility of three-dimensional observations makes SEM particularly attractive for the exploration of infected wood. The most commonly investigated subjects have been the size and shape of hyphae, the mode of microcolonization of tissues, the growth of the hyphae into tracheids, fibers, vessels, and ray cells, and the spreading of hyphae from cell to cell in woody tissues (Fukuda and Haraguchi, 1975; Blanchette *et al.*, 1981; Eriksson *et al.*, 1980a; Dirol, 1976, 1978; Ravilly and Dirol, 1977; Blanchette, 1980b; Otjen and Blanchette, 1982).

The degradation of wood by white-rot, white-pocket-rot, brown-rot, and canker-rot fungi has been described in detail by several authors (Bravery, 1971; Jutte and Zabel, 1974; Dirol, 1976, 1978; Blanchette, 1980a,b 1982a, 1983; Otjen and Blanchette, 1982). For example, erosion troughs have been visualized beneath the hyphae and are characteristic of the early stages of decay by the white-rot fungi (Blanchette, 1980b), whereas numerous holes within the tracheids are visible at more advanced stages of decay (Jutte and Zabel, 1974; Blanchette, 1979; Ravilly and Dirol, 1977;

Fig. 1. Scanning electron micrograph of white fir wood decayed by *P. pini* var. *cancriformans*. Erosion troughs or holes, typical of a white-rot pattern of decay in tracheids, are shown. From Blanchette (1982b) with permission.

Fig. 2. Scanning electron micrograph of Scots pine sapwood cell walls degraded by the brown-rot fungus *Serpula lacrymans*. Buttonhole-shaped cavities are visible (arrows). From Dirol (1978) with permission.

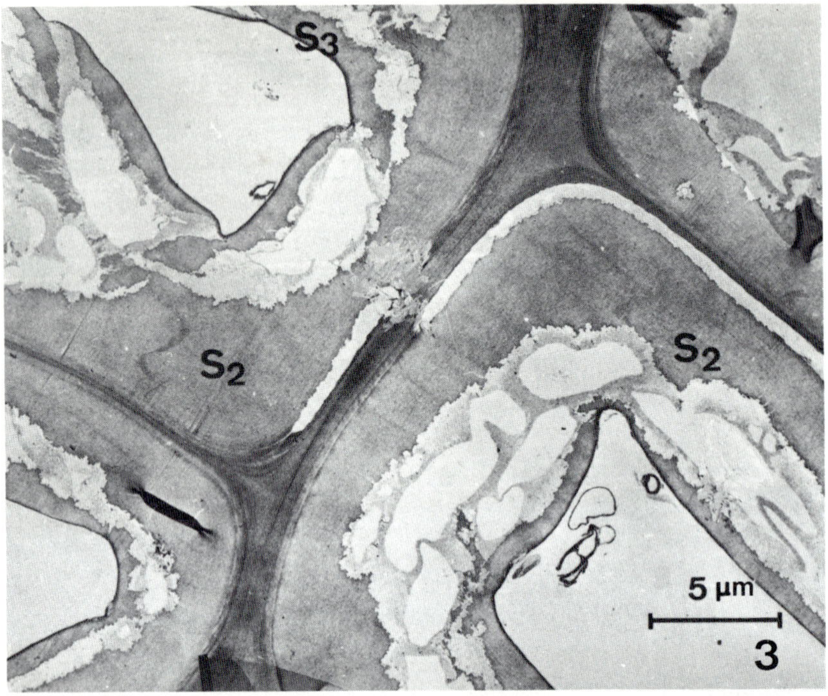

S3
S2
S2
5 µm
3

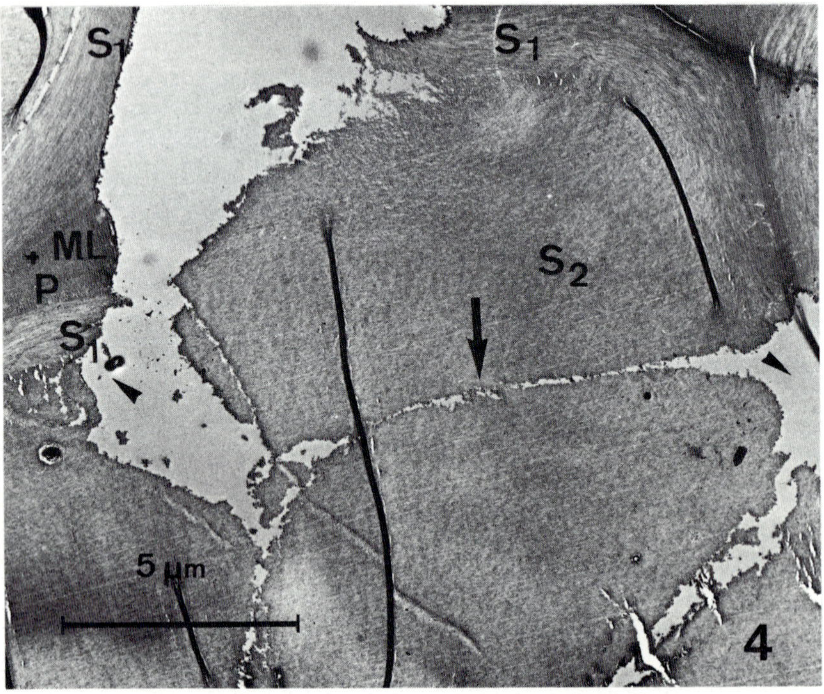

S1
S1
ML
P
S1
S2
5 µm
4

Dirol and Ravilly, 1979; Eriksson *et al.*, 1980b). Figure 1 shows the typical pattern of decay by white-rots in troughs or holes in tracheids (Blanchette, 1982b). Spindle-shaped zones of cellulosic microfibrils surrounded by apparently sound wood are characteristic of the white-pocket type of decay (Blanchette, 1980a,b, 1982b, 1983; Dirol, 1980; Otjen and Blanchette, 1982).

Brown-rot fungi remove cellulose and leave lignin relatively unaltered. In advanced stages of degradation, buttonhole-shaped cavities are visible (see Fig. 2) (Dirol, 1978). Upon drying, the wood cracks, breaks into cubical blocks, and finally crumbles (Blanchette *et al.*, 1978; Dirol, 1978; Blanchette, 1980b).

Scanning electron microscopy has also been used in the study of the associations among bacteria, yeasts, ascomycetes, and basidiomycetes during natural wood decay processes (Kilbertus *et al.*, 1976; Blanchette and Shaw, 1978; Blanchette *et al.*, 1978; Blanchette, 1979). Resolution in SEM studies is affected by the necessity of metallizing samples before observation of wood. Although transmission electron microscopy (TEM) is limited to two-dimensional observation, it allows high-resolution images to be produced and specific stain to be used to identify the wood components.

B. *Transmission Electron Microscopy*

The micromorphological aspects of wood decay can be conveniently visualized by low magnification observation of ultrathin sections of wood (Chou and Levi, 1971; Wilcox *et al.*, 1974; Ouellette, 1978; Peterson and Thomas, 1978) or other lignified tissues like barks (Parameswaran *et al.*, 1976). The pattern of degradation of the different layers of tracheids or fibers can easily be followed. The propagation of fungal hyphae across the cell walls of woody tissues can also be conveniently studied.

Ruel *et al.* (1981) have used TEM to study ultrathin sections of spruce and birch wood degraded by the white-rot fungus *Sporotrichum pulverulentum*. After a 6-week incubation period, the fungal hyphae and microhyphae were observed in both tracheids and parenchyma ray cells, the

---

Fig. 3. An example of the degradation of tracheid walls of spruce by the wild-type strain of the white-rot fungus *Sporotrichum pulverulentum*. S_2 and S_3, middle and inner layers of the secondary cell wall. Bore holes tend to join and coalesce to form large galleries. From K. Ruel, unpublished.

Fig. 4. Brown-rot degraded pine wood tracheids (*Coniophora puteana*). S_1 and S_2, outer and middle layers of secondary cell wall; ML, middle lamella; P, primary wall. The organization of the wall is maintained, but radial dislocations (arrow) giving rise to holes near the S_1 layer and near the lumen are visible (arrowheads). From K. Ruel, unpublished.

latter being usually the first infected. In spruce (Fig. 3), the tracheids appear to be the most altered structures. A quite characteristic feature of hyphal propagation is the disorganization of the wood structure, suggesting a possible diffusion of the enzymes through the transition layers S_2/S_3 and S_1/S_2 of the tracheids walls. The easier fungal progression into these transition layers could be due to a looser fibrillar texture, a change in direction of cellulose fibrils or a lesser degree of lignification.

The three layers of the tracheids walls are not degraded at the same rate. In spruce tracheids, the S_3 layer appears particularly resistant. This layer is progressively detached along the S_2/S_3 transition layer and finally torn off, but it is not lysed.

The S_2 layer is the most degraded, mainly in its inner part. Several holes are bored concentrically to the lumen and tend to coalesce to form large galleries, which result in the complete disorganization of the inner part of S_2. It can be understood why wood suffers dramatic modifications of its physical properties after such a complete disorganization of the internal portion of the S_2 layer. The S_1 layer is degraded only in the later stages of decay.

In the case of a brown-rot fungus, such as *Coniophora puteana,* the aspects of wood degradation are more discrete (K. Ruel, unpublished data). It can be established by the observation of ultrathin sections of *Pinus sylvestris* wood that the organization of the tissues is maintained after a 4-week period of incubation. Nevertheless, radial dislocations turning into holes oriented toward the lumen and outward, toward the S_1 layer, are visible (Fig. 4). The S_2 layer appears to be less affected by the enzymes of the brown-rot fungi than by those of the white-rot fungi. In more advanced stages of degradation the wall disorganization progresses from the holes, concentrically modifying the lignin texture and provoking separation of a ring consisting of the S_3 and the internal part of the S_2. This pattern was previously noted by Wilcox *et al.* (1974) in Douglas fir tracheids.

In summary, as has been stated by Liese (1970), the differences in susceptibility of the three main layers of the tracheid (or fiber) walls to various degrading organisms could be ascribed to the different chemical composition and architecture of these layers.

III. Ultrastructural Aspects of the Degradation of Wood Components

A major difficulty at this level of observation lies in the identification without ambiguity of the three main wall components, cellulose, hemicelluloses, and lignin. Until relatively recently there was no specific stain for localizing wood cell wall components, except for lignin (Hepler *et al.*, 1970). Lignin is very susceptible to oxidizing agents and can be localized by manganese oxide deposition after oxidation by potassium permanganate. The conditions of the use of potassium permanganate have been defined by Kerr and Goring (1975). Bland *et al.* (1971) suggested that the more reactive group of lignin are the free phenolic hydroxyls belonging to syringic and guaiacyl monomers. Potassium permanganate can also react with acidic groups, but since acidic polysaccharides are very weakly represented in the wood cell wall, it can be considered that lignin alone is contrasted in the secondary wall of wood cells.

The polysaccharide part of the wood cell wall can be visualized by the periodic acid–thiocarbohydrazide–silver proteinate method (PATAg) (Thiery, 1969), provided that a carefully controlled time of oxidation with periodate is applied (Ruel *et al.*, 1977).

A clear distinction between hemicelluloses and cellulose is difficult to observe because of the lack of specificity of the method, which cannot make a distinction between $\beta(1\rightarrow4)$ polysaccharides. The enzymatic modifications undergone by lignin, cellulose, and hemicelluloses at the site of hyphal progression, however, can provide an indirect approach for visualizing the ultrastructural organization of the cell wall.

In order to determine the effect of the different degrading activities of the fungi on the respective cell wall constituents, we used a white-rot and a brown-rot strain. The white-rot, *Sporotrichum pulverulentum* Novabrovna, was selected for its ability to excrete different enzymes according to the conditions of growth (Eriksson *et al.*, 1980a) and to give a cellulase-less mutant Cel 44 (Eriksson and Goodell, 1974). The wild-type strain of *S. pulverulentum* can degrade lignin and all other wood components (Ander and Eriksson, 1978). When growing on glucose-impregnated samples, the wild type shows a repression of its cellulase and hemicellulase activities (Eriksson and Hamp, 1978).

The cellulase-less mutant Cel 44, which was found to be totally unable to degrade cellulose, could attack both lignin and hemicelluloses (xylans and glucomannans) (Ander and Eriksson, 1975). On the other hand, the brown-rot strain used in our experiments was *Coniophora puteana*. The activity of this fungus is characterized by a two-step degradation. The first step is an intensive depolymerization of the cellulose. In the second step, there is a further degradation of the depolymerized cellulose and an attack of the hemicelluloses (Kirk, 1975). Owing to the particular enzymatic equipment of each of these fungi, information could be obtained both on the degradation and the localization of the wall components.

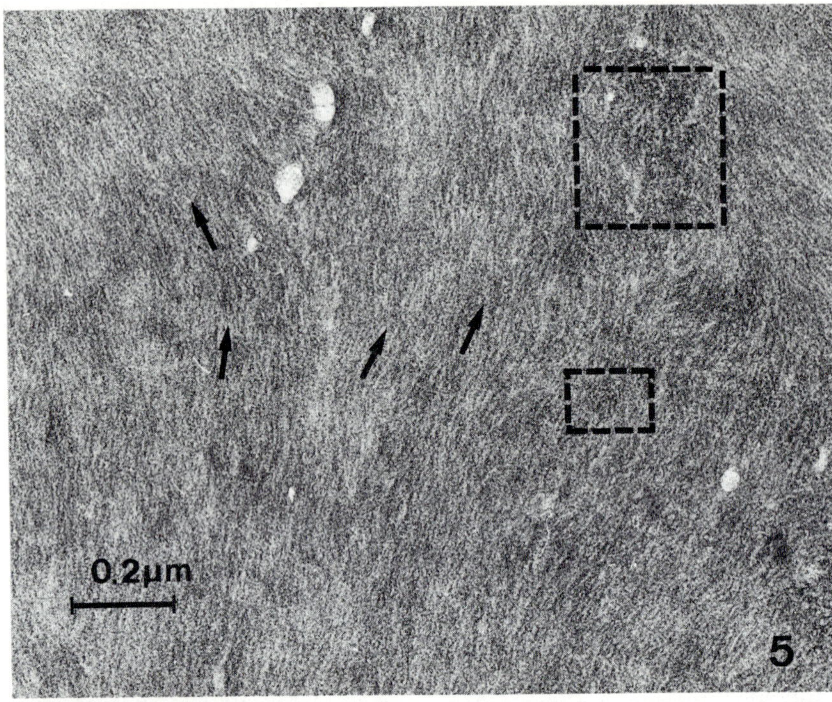
0.2µm
5

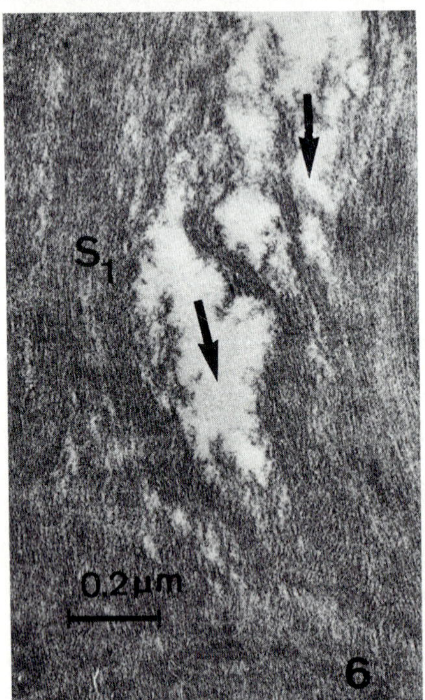
S1
0.2µm
6

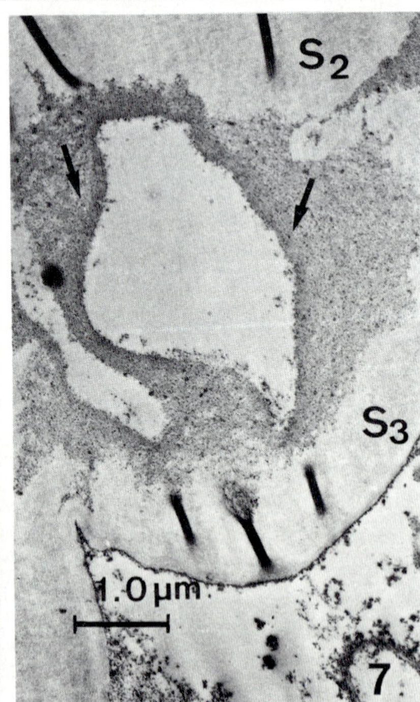
S2
S3
1.0µm
7

A. *Aspects of Lignin Degradation*

The actions of wild-type *S. pulverulentum* and Cel 44 on ultrathin sections of spruce and birch wood were studied (Ruel *et al.*, 1981). The changes brought about by the ligninolytic enzymes were established by comparison of the lignin ultrastructural aspects in decayed and sound areas.

In transverse sections of sound wood (Fig. 5) lignin can be seen as random spots or as a network of short and interrupted lamellae that vary slightly in orientation. This heterogeneity in the ultrastructural distribution of lignin (Ruel and Barnoud, 1983) must be related to variations in the microfibril orientation inside the S_2 layer; the microfibrils are probably not associated in two or three units along their tangential side but are isolated from one another as described by Revol *et al.* (1982).

In spruce wood decayed by Cel 44, the walls of the tracheids become swollen and dislocated. This is particularly visible in the S_1 layer where cellulose microfibrils are lying in the plane of the section (Fig. 6). Bundles of fibrillar material are separated by more or less distinct open spaces. Whether the rupture of the cohesion between the wall components was due to ligninolytic or hemicellulolytic enzymes is still an open question, although we have evidence that the breakdown of the hemicelluloses, or at least a part of them, is the first manifestation of the enzymes diffusing in the tracheid walls (Ruel *et al.*, 1981). These observations are in agreement with the chemical data of Kirk and Highley (1973) and those of Ander and Eriksson (1975).

Initial swelling of the lignin network is observed only in wood decayed by the cellulase-less mutant Cel 44. It must be recalled that the lignin-degrading activity of this strain is approximately five times lower than that of the wild type (Ander and Eriksson, 1975). This low degrading activity could be explained by the loss of the enzyme cellobiose : quinone oxidore-

---

Fig. 5. Sound wall of a spruce tracheid. Cross section through the S_2 layer after $KMnO_4$ staining. Lignin appears as short and interrupted lamellae varying in orientation under the region observed (arrows) and sometimes as random spots (framed in dotted lines). From Ruel and Barnoud (1983) with permission.

Fig. 6. Spruce wall tracheid degraded by Cel 44 showing the swelling and the rupture of the lignin network in the S_1 (outer) layer. Bundles of fibrillar material are separated by more or less distinct open spaces (arrow). From Ruel *et al.* (1981) with permission.

Fig. 7. Tracheid wall degraded by the wild-type *S. pulverulentum*. S_2 and S_3, middle and inner layers of the secondary cell wall. After $KMnO_4$ staining, bore holes are surrounded by a strongly contrasted crown of disorganized lignin (arrows). From Ruel and Barnoud (1981).

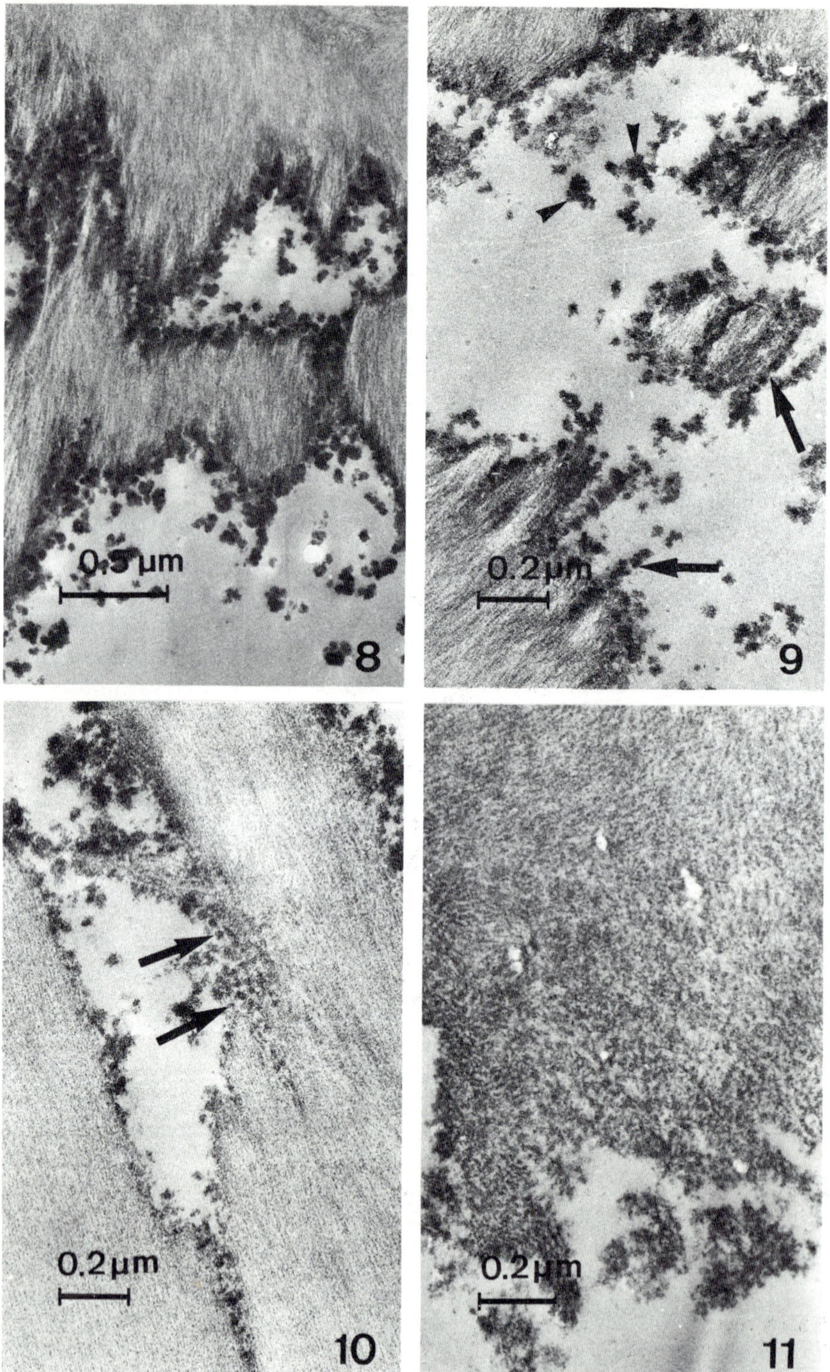
0.5 µm
8
0.2 µm
9
0.2 µm
10
0.2 µm
11

ductase, which has an important role in lignin degradation (Westermark and Eriksson, 1974a,b, 1975). The requirement that polysaccharides in wood must be degraded simultaneously with lignin (Ander and Eriksson, 1975, 1977, 1978; Kirk and Moore, 1972; Kirk *et al.*, 1976, Hiroi and Eriksson, 1976) could also explain the reduced activity of this mutant.

When spruce wood is decayed by the wild-type strain, lignin degradation appears different. Observations of cross sections show an area of granular structure around the bore holes extending 2–3 μm in diameter. This area is more strongly contrasted by MnO_2 deposits than is the surrounding sound part of the wall (Fig. 7). These ultrastructural images of lignin degradation can be related to the chemical data obtained by Ander *et al.* (1980), Kirk (1981), Chen *et al.* (1981), Higuchi (1982), Chen *et al.* (1983), and Tai *et al.* (1983). The white-rot degradative processes of lignin give in part monomeric units bearing α-carboxyl groups as a result of C_α—C_β cleavage in the side chains and C_α-oxidation (Kirk and Chang, 1974, 1975; Kirk, 1981a,b; Chen *et al.*, 1981a, 1982a,b; Chua *et al.*, 1982; Tai *et al.*, 1983). Lignin is also partly degraded by cleavage of the aromatic rings when still bound in the polymer (Buswell and Eriksson, 1979; Ellwardt *et al.*, 1981; Chua et al., 1982). Ring cleavage results in an enrichment of degraded lignin with carboxyl groups (Higuchi *et al.*, 1955; Kirk and Chang, 1975; Kirk *et al.*, 1978; Ellwardt *et al.*, 1981; Chua *et al.*, 1982; Chen *et al.*, 1983). Furthermore, an increase in phenolic hydroxyl groups has been established at an early stage of lignin degradation (Eriksson, 1981b).

All of these chemical changes would cause the degraded lignin to be more reactive to $KMnO_4$. We have effectively observed a more intense staining of the lignin around the decayed zones as the degradation progresses.

Fig. 8. Longitudinal section through a tracheid wall decayed by the wild-type strain *S. pulverulentum*. Many characteristic deposits of electron-dense granules border the decayed areas. From K. Ruel, unpublished.

Fig. 9. Longitudinal tangential section through the S_1 layer (wild-type decayed). Dark granules of modified lignin can be seen still aligned (arrows). When detached, they tend to form large clusters (arrowheads). From Ruel *et al.* (1981) with permission.

Fig. 10. Birch wood fibers decayed by the Cel 44 mutant of *S. pulverulentum*. The images of lignin degradation are identical to those observed in softwoods. Lined-up dark granules are formed before being detached from the wall (arrows). From Ruel and Barnoud (1981).

Fig. 11. Brown-rotted tracheid cell wall of pine. Lignin network takes on a foamy aspect, and lignin distribution becomes both scattered and loosened ($KMnO_4$ fixed section). From K. Ruel, unpublished.

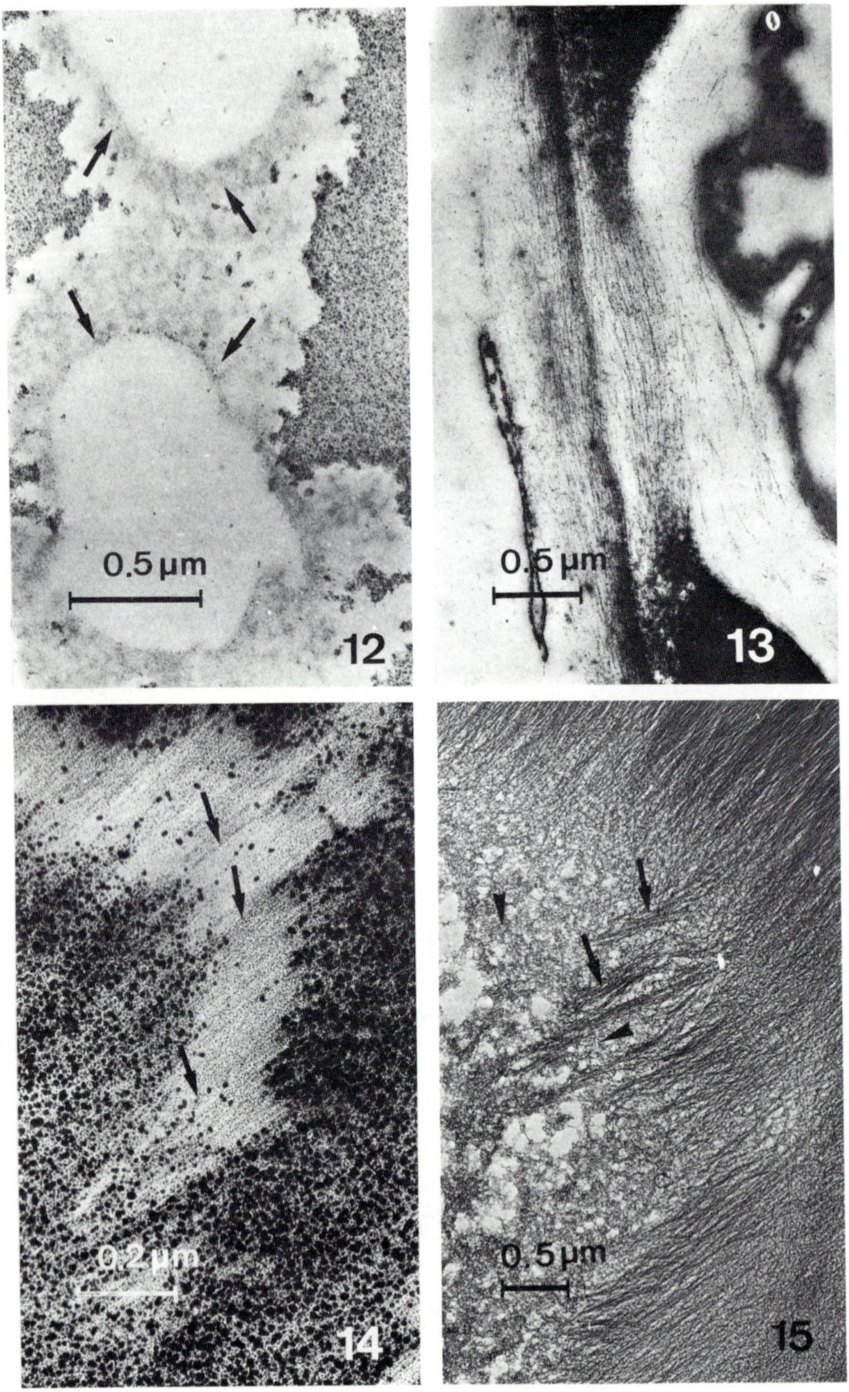
0.5 µm
12
0.5 µm
13
0.2 µm
14
0.5 µm
15

The same phenomenon was also observed on longitudinal sections of spruce wood, and characteristic deposits of electron-dense granules are abundant in strongly degraded areas (Fig. 8). These granules, which are very reactive to potassium permanganate, fail to stain with reagents for polysaccharides (neutral or acidic) or proteins. Accordingly, they can be considered as degraded forms of lignin not yet metabolized by the fungus (Chen *et al.*, 1983). This interpretation seems in agreement with Kirk's assumption that lignin degradation is a slow process relative to the secondary stage of growth of the fungus (Kirk, 1981b; Chua *et al.*, 1983).

In sections cut tangentially through the S_1 layer during the early stages of degradation (Fig. 9), the dark granules of modified lignin are still aligned and about 5–10 nm in diameter. At a more advanced stage of degradation, they become isolated and then tend to aggregate to form large clusters of 30–50 nm in size. These dark clusters observed could be interpreted as images of the repolymerization of low molecular weight degradation products (Iwahara, 1980; Chua *et al.*, 1983).

When birch wood is degraded by the Cel 44 strain, which is more active on hardwoods than on softwoods, aligned dark granules are also formed before being detached from the wall (Fig. 10). These observations indicate that at least during the first steps of the degradation, birch and spruce lignins could be degraded by a similar biochemical process, a process that seems independent of the guaiacyl or syringyl nature of the lignin.

It has long been known that brown-rot fungi can degrade both cellulose and hemicelluloses but do not significantly lyse the lignin part of the wall (Cowling, 1961; Wilcox, 1968, 1970; Liese, 1970). In early stages of degradation by *Coniophora puteana,* tracheid cell walls take on a foamy appearance (Fig. 11) (K. Ruel, unpublished data). The usual distribution of the lignin in short and interrupted lamellae in transverse sections of sound wood is modified. Lignin becomes both scattered and loosened as decay

---

Fig. 12. Hemicelluloses degradation in a tracheid spruce wall degraded by the wild-type strain. A crown failing to stain with the PATAg reagent is clearly visible around the bore holes (arrows). From Ruel *et al.* (1981) with permission.

Fig. 13. A pit area in spruce wood degraded by Cel 44 after PATAg staining. The solubilization of the hemicelluloses leave fibrillar structures. From K. Ruel, unpublished.

Fig. 14. Degradation of the S_2 layer by Cel 44, PATAg staining. In the clear area corresponding to hemicelluloses solubilization, weakly stained fibrillar structures (arrows) can be regarded as cellulose microfibrils. From K. Ruel, unpublished.

Fig. 15. Birch ray parenchyma cell degraded by Cel 44 observed in minimum beam exposure conditions (without any staining). Long bundles of cellulose microfibrils (arrows) still surrounded with matrix components (arrowheads) are visible. From Ruel and Barnoud (1981) with permission.

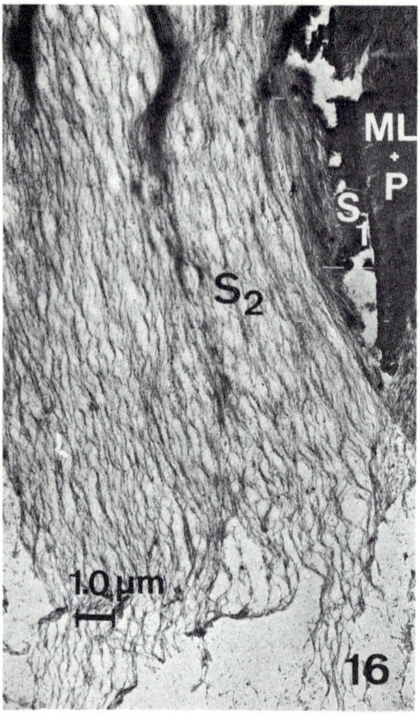
ML
+
P
S1
S2
1.0 µm
16

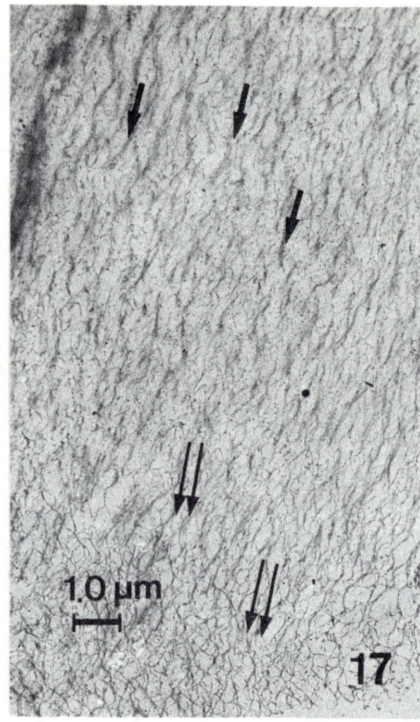
1.0 µm
17

18

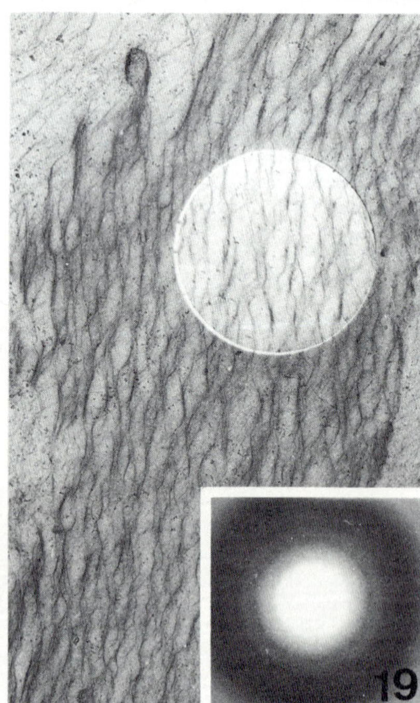
19

progresses, taking on a granular aspect. Although modified in its organization and chemically changed by a decrease in its content of methoxyl groups (Kirk, 1975) lignin keeps a coherent structure. There is no apparent enhancement of the reactivity of the lignin to potassium permanganate oxidation as was seen after white-rotting of the cell walls. The demethylation described by Kirk (1975) seems to have no consequence on lignin susceptibility to permanganate.

B. *Aspects of the Degradation of the Hemicelluloses and Cellulose*

The polysaccharide and lignin constituents of the walls are known to be degraded simultaneously by certain white-rot fungi (Cowling, 1961; Liese, 1970; Kirk and Moore, 1972; Kirk, 1973; Ander and Eriksson, 1975; Eriksson, 1978, 1981a). The hemicellulose part of the polysaccharides is the first to be degraded, as shown by both chemical analysis (Kirk and Highley, 1973; Ander and Eriksson, 1975, 1978; Eriksson, 1978, 1981a,b; Kirk, 1981a) and ultrastructural observations (Ruel and Barnoud, 1981). This is illustrated in Fig. 12, which shows a section of spruce wood degraded by the wild-type strain and in which glucomannans and galactoglucomannans have been strongly solubilized, partly demasking the fibrillar cellulosic framework (Figs. 13–15). This degradation took place principally around the bore holes. A crown-shaped area failing to stain by the PATAg technique appears as a clear zone, whereas the sound part is heavily deposited with silver grains. In the case of the walls degraded by Cel 44, which degrades slower than the wild-type, glucomannan-depleted areas are essentially limited to the pits (Fig. 13).

Fig. 16. Longitudinal section through a spruce tracheid cell wall degraded by the wild-type strain of *S. pulverulentum* (glucose-impregnated wood). S_1 and S_2, outer and middle layers of the secondary cell wall; ML, middle lamella; P, primary wall. Lignin solubilization leads to a trellis-like structure, clearly seen in minimum beam exposure conditions of observations. From Ruel *et al.* (1984) with permission.

Fig. 17. Detail of the honeycomb structure of the S_2 layer. This structure shows wrapped fibrils with ill-defined edges in the upper part (arrow) and thinner fibrils with well defined edges in the lower part (double arrow). From Ruel *et al.* (1984) with permission.

Fig. 18. Transverse section as described in Fig. 16. The diffraction diagram from the outer part of the S_2 layer (clear circle) is typical of cellulose I (×5000). From Ruel *et al.* (1984) with permission.

Fig. 19. Same type of degradation as in Fig. 18 but in a more degraded area, near the lumen. The diffraction diagram shows an absence of crystallinity (×5000). From Ruel *et al.* (1984) with permission.

In Fig. 14, the removal of hemicelluloses in the S_2 layer creates an opening in the compact structure of the wall and exposes the cellulose framework. Owing to the weak reactivity of cellulose microfibrils to the PATAg staining, the framework appears as a very discrete structure. This resistance of cellulose to staining with PATAg in the secondary wall was overcome by the use of the minimum beam exposure technique [''low dose'' (Beer, 1980)]. This is a convenient method to prevent cellulose degradation under the electron beam. When applied to ray parenchyma cells of birch wood degraded with Cel 44 (Fig. 15) long bundles of cellulose microfibrils still surrounded with matrix components can be observed.

When the wild-type strain is grown on glucose-impregnated wood cellulose and hemicelluloses are not degraded, and lignin solubilization leads to a trellis-like pattern (Fig. 16), which tends to give individualized microfibrillar structures (Fig. 17), of which the finest observed are 35–40 nm in diameter.

It is noteworthy that whereas the typical electron diffraction diagram of cellulose I was obtained on the parts of the wall where the matrix components are still present (Fig. 18), this typical diagram disappeared in the strongly depleted areas although a fibrillar structure was still present (Fig. 19). The loss of the characteristic pattern of diffraction of cellulose raises questions about the nature of these fibrils:

- Are they only hemicelluloses, and if so, should cellulose have been digested by cellulolytic enzymes, glucose not being a cellulase inhibitor during all the period of the attack of the wood?
- Are they cellulose microfibrils having lost their crystallinity? The cellulose I characteristic diagram of electron diffraction would have been lost due to an attack of cellulases extensive enough to destroy cellulose I crystallinity but not extensive enough to destroy the morphology of the cellulose microfibrils.

More recently, electron diffraction observations have been done again in collaboration with H. Chanzy (1983), and new information has been obtained:

1. Cellulose is really present in these trellis-like structures.
2. Cellulose crystallinity is strongly reduced.
3. The poor resolution of the diagram obtained could be due to the presence of noncrystalline polysaccharides.

The second proposition suggests that the repression of the cellulasic activity was not complete. This observation seems in agreement with

Eriksson's results showing that the exclusively ligninolytic activity of this strain occurred only during a limited period of time.

The third proposition is in agreement with the observations of Ruel *et al.* (1984), who found, using a double staining method (PATAg on $KMnO_4$ fixed sections), that both lignin and hemicelluloses were still associated with cellulose microfibrils.

At present, the finding of two specific new methods for staining of xylans and glucomannans (Joseleau and Ruel, 1984; Ruel and Joseleau, 1984) permits the characterization of the presence of both glucomannans and xylans on these trellis-like structures demasked by the strong ligninolytic activity of *S. pulverulentum* growing on glucose-impregnated spruce wood samples (Figs. 20 and 21).

IV. Diffusion of Wood-Degrading Enzymes

The lignolytic action of microorganisms is usually considered as the consequence of the action of intra- and extracellular enzymes.

Microscopic observations show that during fungal decay the penetration and the progression of the fungus correspond to a lytic action localized in the vicinity of the hyphae. However, decayed areas can also be observed completely disconnected from the presence of the hyphae. This action that we shall call "at a distance" has been most frequently interpreted as a proof of the capability for the enzymes to diffuse through the walls (Cowling, 1961; Juraček, 1964; Liese, 1970; Wilcox, 1970).

Because they are extracellular, glycanases (Ander and Eriksson, 1975; Eriksson, 1981a,b) are considered as the agents capable of diffusion through the wood cell wall, in opposition to ligninolytic enzymes, which have been described as intracellular (Chua *et al.*, 1983). Rosenberg (1980) showed that polysaccharides, but not lignin, can effectively be degraded at a distance. The more general hypothesis is that ligninolytic enzymes cannot degrade lignin without being in true contact or narrow binding with the lignin polymer (Chua *et al.*, 1983).

Authors who considered degradation areas without visible hyphae as resulting from the diffusion of enzymes have in fact never localized the enzymes along the degradation areas (Cowling, 1961; Blanchette *et al.*, 1978; Blanchette, 1980b). It is indeed difficult to visualize enzymes inside the walls, and at the present time, neither ligninolytic enzymes nor polyosidases have ever been characterized *in situ*.

If we consider the case of lignin degradation, it has been chemically proved that the degradation is not due solely to the enzymatic action

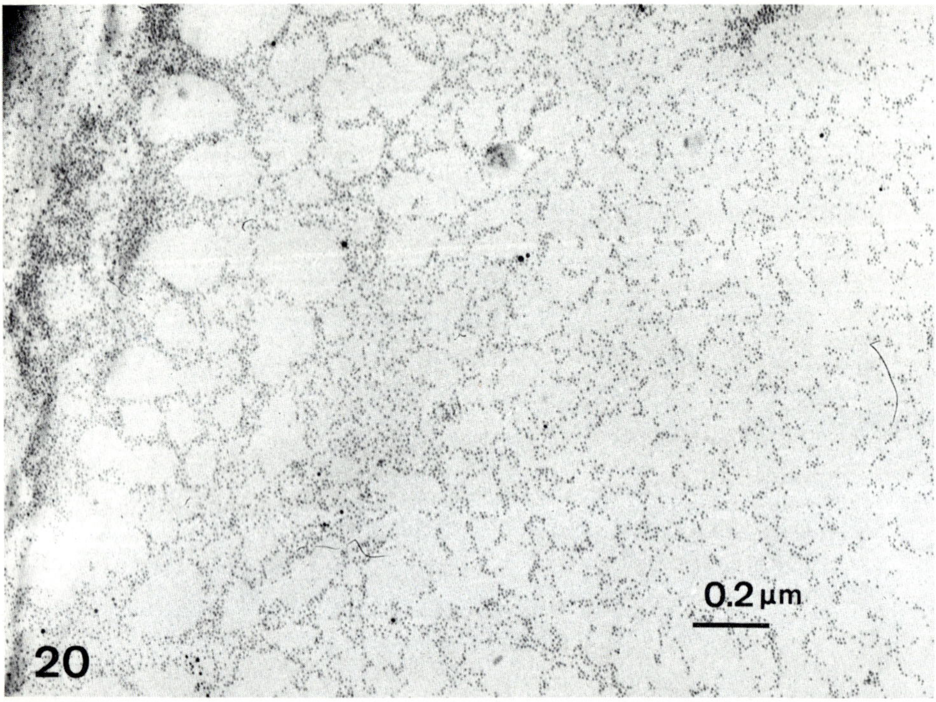
0.2 µm
20

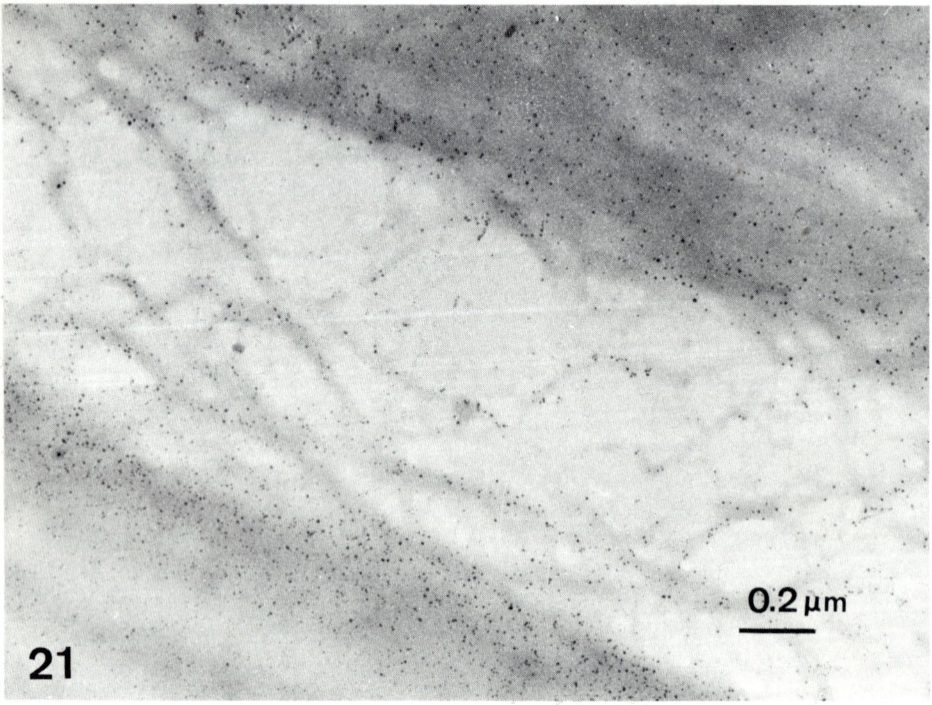
0.2 µm
21

(Higuchi, 1981). Lignin degradation has also been demonstrated as resulting from, among other modifications, an oxidative cleavage of aromatic rings inside the macromolecules (Kirk and Chang 1974, 1975; Ellwardt *et al.*, 1981; Haider *et al.*, 1981). This oxidative degradation implies the action of activated forms of molecular oxygen ($^1O_2$) or radical reactions due to the hydroxyl radical OH· (Hall, 1980; Nakatsubo *et al.*, 1981). Bes *et al.* (1983) have shown that these activated forms of oxygen are not specific for a functional group and can act indifferently on the aromatic rings or on the side chains. These chemically reactive agents can be produced by fungi. It has been demonstrated (Bes *et al.*, 1983) that *Phanerochaete chrysosporium* [cospecific species of *S. pulverulentum* (Burdsall and Eslyn, 1974)] is capable of releasing OH· that degrades lignin. Furthermore, the obligatory presence of phenoloxidase activity in lignin degradation and the impossibility for a mutant of *Ph. chrysosporium* (104-2) to elaborate the hydroxyl radical and consequently to degrade lignin (Gold *et al.*, 1982) are also arguments in favor of a radical mechanism of degradation of lignin by *S. pulverulentum*. Moreover, an enhancement of lignolytic activity was shown when oxygen concentration increases (Kirk *et al.*, 1978; Bar-Lev and Kirk, 1981). It has been also observed that ligninolytic activity decreases when atmospheric nitrogen concentration increases (Keyser *et al.*, 1978; Buswell *et al.*, 1981; Leatham and Kirk, 1983).

Electron microscopic observations of $KMnO_4$-stained sections degraded by *S. pulverulentum* that show a stronger reactivity to permanganate oxidation due to the presence of carboxyl groups (Chen *et al.*, 1982a, 1983) prove that lignin has effectively been oxidized (Figs. 8–10). No hyphae are visible in contact with lignin-degraded areas. The progressive modifications of the lignin network observed in Fig. 23 could be interpreted as ultrastructural modifications due to both a radical mechanism of degradation of lignin and the diffusion of ligninolytic enzymes.

In the case of the polysaccharide part of the wall, it is well established that hemicellulases and cellulases can migrate far from the tips of the hyphae, which are known to be the active site of secretion. Electron microscopy shows PATAg-negative channels clearly visible between

Fig. 20. Trellis-like structure demasked by the action of *S. pulverulentum* growing on glucose-impregnated spruce wood stained by a mannanase–gold complex. Gold grains corresponding to glucomannan deposits followed the trellis like organization of the S_2 layer. From Ruel and Joseleau (1984) with permission.

Fig. 21. Same type of sample as in Fig. 20 contrasted by the ETAg method. The reducing ends created by a xylanase are stained with silver grains and deposited on the fibrillar structure. From Joseleau and Ruel (1984) with permission.

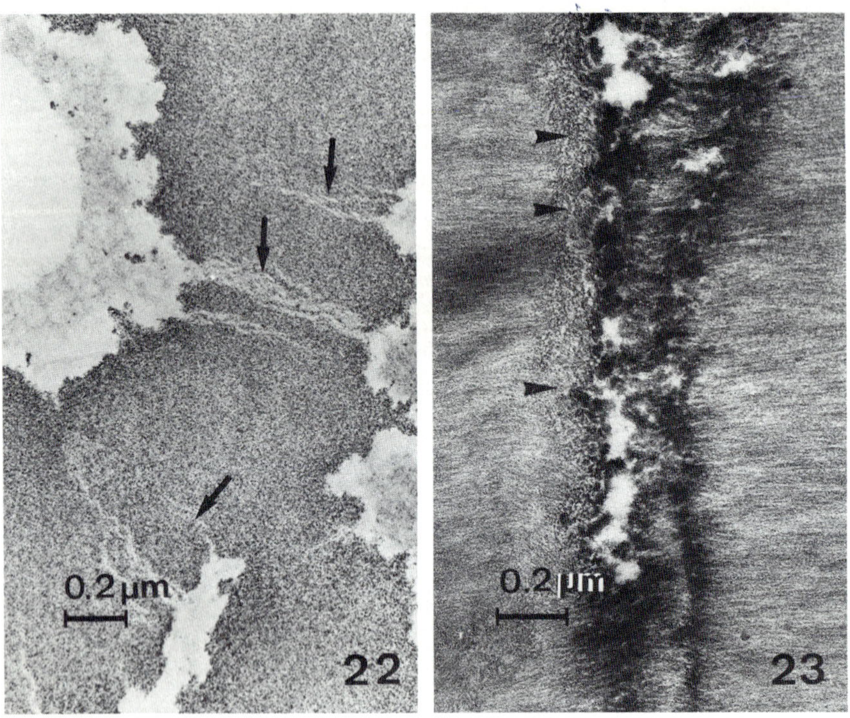

Fig. 22. Cross section of a PATAg-stained tracheid of spruce degraded by the wild-type *S. pulverulentum* showing the diffusion of hemicellulolytic enzymes from one hole to another. Traces of diffusion can be followed by thin PATAg-negative channels (arrows). From Ruel and Barnoud (1981) with permission.

Fig. 23. Longitudinal tangential section through the S_1 layer of a tracheid wall showing lignin modifications due to ligninolytic enzymes after $KMnO_4$ staining. The modified area is about 0.2 to 0.3 μm in size (arrowhead). From Ruel *et al.* (1981) with permission.

holes created in the S_2 layer of spruce tracheids by the wild-type strain of *S. pulverulentum* (Fig. 22), which indicates that polysaccharides, and particularly hemicelluloses [the first polysaccharides lysed by *S. pulverulentum* (Ander *et al.*, 1975; Ruel and Barnoud, 1981)], have been depleted. The diameter of these channels could be sufficient to ensure the progression of hemicellulases. From a chemical point of view, it must be noted that the possibility of degradation of polysaccharides at a distance has been proposed by Halliwell (1965) and Koenigs (1974).

Presently, it is impossible to establish if wall components are degraded by a radical mechanism or by the propagation of the enzymes "at a distance" from the hyphae. It seems probable that fungi are capable of producing active radicals (Ander *et al.*, 1984; Bes *et al.*, 1983), and it could be supposed that both diffusion of the enzymes acting directly on the molecules and "at a distance" action by the production of radical forms could be two complementary mechanisms of the mode of attack on wall components by the fungi.

V. Degradation of Wood by Bacteria

Degradation of woody tissues by mixed bacterial flora or even bacterial monocultures has been the object of microbiological, chemical, and microscopic investigations for many years. Several reviews on these subjects have been presented since 1970 (Liese, 1970; Wilcox, 1970, 1973; Liese and Greaves, 1975). It is generally recognized that bacterial degradation of wood has a technical importance only in special environmental conditions, especially in aquatic habitats, such as submerged pilings or cooling towers, for example (Fazzani *et al.*, 1975; Kohlmeyer, 1980; Holt and Jones, 1983). The bacterial factor in wood decay and its influence on physical properties and microscopic features of wood has been studied by Greaves (1970, 1971, 1973).

The description of bacterial degradation is mainly derived from observations of timber decayed under natural conditions. The complete degradation of wood is evidently the result of an association between several groups of microorganisms (Blanchette *et al.*, 1978; Blanchette, 1979). Accordingly, the sole role of bacteria is difficult to establish clearly since other microorganisms may be the agents of initial attack.

As stated by Holt and Jones (1983), one of the major problems in studying wood degradation by bacteria is the difficulty of research workers to isolate the bacterial strains and to monitor the different stages of the decay process in the laboratory. *Streptomyces* strains were shown to be able to degrade softwood, hardwood, and grass lignocelluloses (Antai and Crawford, 1981). Holt (1982) has shown the effect of preconditioning wood with fungi on its subsequent degradation by cellulolytic bacteria. The most important fact that remained doubtful for years was the ability of some strains of bacteria to degrade lignin effectively, breaching this efficient barrier against lysis of wood polysaccharides. Until the 1970s, no rigorous investigations were made to establish clearly the ability of pure strains of bacteria to metabolize lignin when associated with other wood components in lignified secondary walls.

Haider *et al.* (1978), Haider and Trojanowski (1980), and Crawford and Crawford (1980) have established, by the use of ^{14}C-labeled lignins, the capability of several species of bacteria to degrade lignin in its isolated

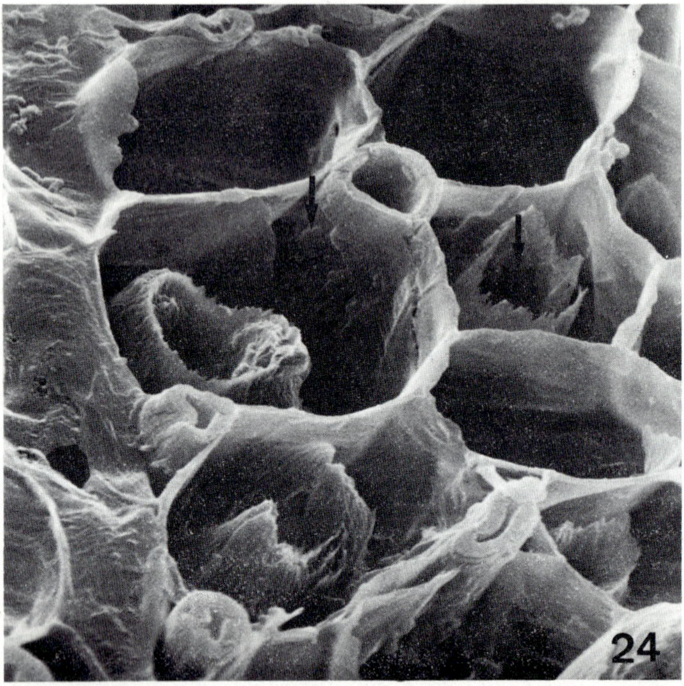

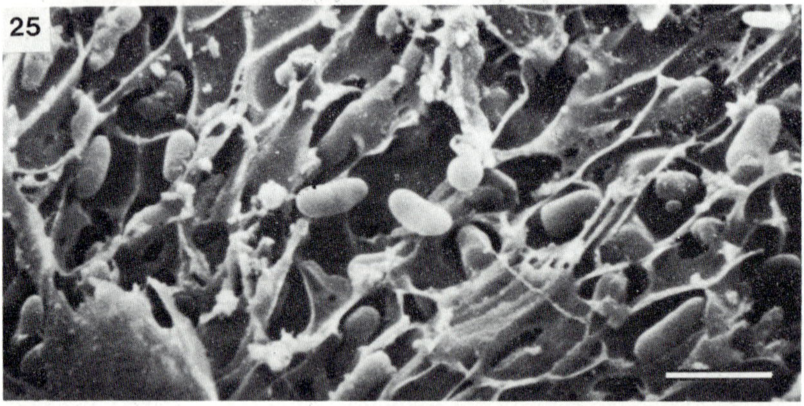

Fig. 24. Bacterial decomposition of the inner secondary wall layer of a pretreated beech wood section with adhering bacteria (arrows). From Schmidt and Liese (1982) with permission.

Fig. 25. Scots pine tracheid with bacterial erosion in the S_2 layer (testblocks were submerged in the sea for 9 weeks). Late stages of degradation. Deep channels are clearly visible. Bar = 2 μm. From Holt (1983) with permission.

polymeric form or in lignified tissues. *Pseudomonas, Bacillus,* and *Nocardia* strains were active in lignin degradation. Bacterial delignification of sound and untreated poplar wood (20-μm sections) has been investigated by Monties *et al.* (1981). The authors claimed that high rates of delignification (52%) were obtained in a 1-month incubation period with a *Pseudomonas* species. Studying the decomposition of lignified cell wall of *Fagus sylvatica* and *Pinus sylvestris* by 66 selected bacterial strains, Schmidt and Liese (1982) have shown that none of them could deteriorate sapwood cross sections; after mild delignification with chlorite, several strains were able to degrade the cell wall heavily, but without lignin metabolization (Fig. 24).

Recent studies (Nilsson and Holt, 1983) have focused attention on special strains of single-celled bacteria called tunneling bacteria. They are responsible for typical decay patterns, mainly in the S_2 layer of the fibers of softwoods and hardwoods (Holt, 1983; Holt and Jones, 1983). General background on tunneling bacteria has been presented by Nilsson and Daniel (1983). Several different species are involved, but up to now, attempts to isolate pure cultures have had only limited success (D. M. Holt, personal communication). Scanning and transmission electron microscopy have clearly confirmed that the attack is caused by these peculiar bacteria (Fig. 25). Chamber-like structures in tunnels are very characteristic patterns of decay. Although mainly concentrated in the S_2 layer of the fibers, tunneling bacteria are able to pierce the highly lignified middle lamella of adjacent spruce tracheids (Nilsson and Daniel, 1983).

Laboratory studies indicate that tunneling bacteria are able to attack a wide variety of wood species that are more or less resistant against other forms of decay. The importance of tunneling bacteria for timbers in the field has not been well investigated, but the decay of wood in marine or fresh water environments is now clearly established.

VI. Conclusion

In this chapter some important recent developments obtained in the field of wood degradation by microorganisms have been presented. Most of the reported results are due to electron microscopic investigations, with either the scanning technique or the transmission technique. In addition to improvements in making ultrathin section of wood, more specific stains for wood components are now available. Also, the considerable chemical and biochemical data presented by a great number of research groups are an aid to a better interpretation of the electron micrographs. No doubt, progresses in the different fields of research will have positive consequences for better protection and better use of wood.

References

Ander, P., and Eriksson, K.-E. (1975). *Sven. Papperstidn.* **78,** 643–652.

Ander, P., and Eriksson, K.-E. (1977). *Physiol. Plant.* **41,** 239–248.

Ander, P., and Eriksson, K.-E. (1978). *Prog. Ind. Microbiol.* **14,** 1–58.

Ander, P., Hatakka, A., and Eriksson, K.-E. (1980). *Arch. Microbiol.* **125,** 189–202.

Ander, P., Eriksson, K.-E., and Romanov, S. (1984). To be published.

Antai, S. P., and Crawford, D. L. (1981). *Appl. Environ. Microbiol.* **42,** 378–380.

Bar-Lev, S. S., and Kirk, T. K. (1981). *Biochem. Biophys. Res. Commun.* **99,** 373–378.

Beer, M. (1980). *Anal. Chem. Rev.* **52,** 40R–42R.

Bes, B., Ranjeva, R., and Boudet, A. M. (1983). *Biochimie* **65,** 283–289.

Blanchette, R. A. (1979). *Can. J. For. Res.* **9**(4), 464–469.

Blanchette, R. A. (1980a). *Can. J. Bot.* **58**(13), 1496–1503.

Blanchette, R. A. (1980b). *J. For.* **78**(12), 734–737.

Blanchette, R. A. (1982a). *Phytopathology* **72**(10), 1272–1277.

Blanchette, R. A. (1982b). *Can. J. For. Res.* **12**(2), 304–310.

Blanchette, R. A. (1983). *Mycologia* **75**(3), 552–556.

Blanchette, R. A., and Shaw, C. G. (1978). *Phytopathology* **68,** 631–637.

Blanchette, R. A., Shaw, C. G., and Cohen, A. L. (1978). *In* "Scanning Electron Microscopy," Vol. 2, pp. 61–67. SEM, O'Hare, Illinois.

Blanchette, R. A., Sutherland, J. B., and Crawford, D. L. (1981). *Can. J. Bot.* **59**(1), 1–7.

Bland, D. E., Foster, R. C., and Logan, A. F. (1971). *Holzforschung* **25**(5), 137–143.

Bravery, A. F. (1971). *J. Inst. Wood Technol.* **30,** 13–17.

Burdsall, H. H., and Eslyn, W. E. (1974). *Mycotaxon* **1**(2), 123–133.

Buswell, J. A., and Eriksson, K.-E. (1979). *FEBS Lett.* **104,** 258–261.

Buswell, J. A., Ander, P., and Eriksson, K.-E. (1981). *Int. Symp. Wood Pulping Chem. Stockholm* **3,** 88–92.

Chen, C.-L., Chang, H.-M., and Kirk, T. K. (1981a). *Holzforschung* **35**(3), 99–103.

Chen, C.-L., Chua, M. G. S., Evans, J. E., Chang, H.-M., and Kirk, T. K. (1981b). *Int. Symp. Wood Pulping Chem. Stockholm* **3,** 75–87.

Chen, C.-L., Chang, H.-M., and Kirk, T. K. (1982a). *Holzforschung* **36,** 3–9.

Chen, C.-L., Chua, M. G. S., Evans, J., and Chang, H.-M. (1982b). *Holzforschung* **36,** 239–247.

Chen, C.-L., Chang, H.-M., and Kirk, T. K. (1983). *J. Wood Chem. Technol.* **3**(1), 35–57.

Chou, C. K., and Levi, M. P. (1971). *Holzforschung* **4,** 107–112.

Chua, M. G. S., Chen, C. L., Chang, H.-M., and Kirk, T. K. (1982). *Holzforschung* **36,** 165–172.

Chua, M. G. S., Choi, S., and Kirk, T. K. (1983). *Holzforschung* **37,** 55–61.

Cowling, E. B. (1961). Technical Bulletin No. 1258. U.S. Department of Agriculture, Forests Products Laboratory, Forest Service, Washington, D.C.

Crawford, D. L., and Crawford, R. L. (1980). *Enzyme Microb. Technol.* **2,** 11–22.

Dirol, D. (1976). *Revue Mycol.* **40,** 295–315.

Dirol, D. (1978). *Revue Mycol.* **42,** 277–292.
Dirol, D. (1980). *Cryptogam. Mycol.* **1,** 187–199.
Dirol, D., and Ravilly, F. (1979). *Ann. Biol.* **18,** 11–12, 478–492.
Ellwardt, P. C., Haider, R., and Ernst, L. (1981). *Holzforschung* **35**(3), 99–103.
Eriksson, K.-E. (1978). *Biotechnol. Bioeng.* **20,** pp. 317–332.
Eriksson, K.-E. (1981a). *Pure Appl. Chem.* **53,** 33–43.
Eriksson, K.-E. (1981b). *Int. Symp. Wood Pulping Chem. Stockholm.* **3,** 60–65.
Eriksson, K.-E., and Goodell, E. W. (1974). *Can. J. Microbiol.* **20,** 371–378.
Eriksson, K.-E., and Hamp, S. G. (1978). *Papier (Darmstadt),* **32**(12), 545–550.
Eriksson, K.-E., and Pettersson, B. (1975). *Eur. J. Biochem.* **51,** 193–206.
Eriksson, K.-E., Grünewald, A., and Vallander, L. (1980a). *Biotechnol. Bioeng.* **22,** 363–376.
Eriksson, K.-E., Grünewald, A., Nilsson, T., and Vallander, L. (1980b). *Holzforschung* **34,** 207–213.
Fazzani, K., Furtado, S. E. J., Eaton, R. A., and Jones, E. B. G. (1975). *In* "Microbial Aspects of the Deterioration of Materials" (D. W. Lovelock and R. J. Gilbert, eds.), pp. 39–58. Academic Press, London.
Fukuda, K., and Haraguchi, T. (1975). *J. Jpn. Wood. Res. Soc.* **21**(1), 43–46.
Gold, M. H., Mayfield, M. B., Cheng, T. M., Krisnangkura, K., Shimada, M., Enoki, A., and Glenn, J. K. (1982). *Arch. Microbiol.* **132,** 115–122.
Greaves, H. (1970). *Holzforschung,* **24,** 6–14.
Greaves, H. (1971). *Wood Sci. Technol.* **5,** 6–16.
Greaves, H. (1973). *Holzforschung,* **27,** 80–88.
Haider, K., and Trojanowski, J. (1980). *In* "Lignin biodegradation: Microbiology, Chemistry and Potential Applications" (T. K. Kirk, T. Higuchi, and H.-M. Chang, eds.), pp. 111–134. CRC Press, Inc., Boca Raton, Florida.
Haider, K., Ellwardt, P. C., and Ernst, L. (1981). *Int. Symp. Wood Pulping Chem. Stockholm.* **3,** 93–98.
Haider, D., Trojanowski, J., and Sundman, U. (1978). *Arch. Microbiol.* **119,** 103–106.
Hall, P. L. (1980). *Enzyme Microbiol. Technol.* **2,** 170–176.
Halliwell, G. (1965). *Biochem. J.* **95,** 35–40.
Hepler, P. K., Fosket, D. E., and Newcomb, E. H. (1970). *Am. J. Bot.* **57,** 85–96.
Higuchi, T. (1981). *Int. Symp. Wood Pulping Chem. Stockholm* **3,** 16–24.
Higuchi, T. (1982). *Experientia* **38,** 159–166.
Higuchi, T., Kawamura, I., and Kawamura, H. (1955). *J. Jpn. For. Soc.* **37,** 298–302.
Hiroi, T., and Eriksson, K.-E. (1976). *Sven. Papperstidn.* **79,** 157–161.
Holt, D. M. (1982). *Beih. Mater. Org.* **17,** 1–20.
Holt, D. M. (1983). *J. Inst. Wood Sci.* **9**(5), 212–223.
Holt, D. M., and Jones, E. B. G. (1983). *Appl. Envir. Microbiol.* **46**(3), 722–727.
Iwahara, S. (1980). *In* "Lignin Biodegradation: Microbiology, Chemistry and Potential Applications, (T. K. Kirk, T. Higuchi, and H.-M. Chang, eds.), Vol. 1, CRC Press, Inc., Boca Raton, Florida.
Joseleau, J. P., and Ruel, K. (1984). *Biol. Cell.* in press.
Juraček, L. (1964). *Drev. Vyskum* **3**(2), 126–141.
Jutte, S. M., and Zabel, R. A. (1974). *IITRI/SEM* 445–452.
Kerr, A. J., and Goring, D. A. I. (1975). *Cellul. Chem. Technol.* **9,** 563–573.
Keyser, P., Kirk, T. K., and Zeikus, J. G. (1978). *J. Bacteriol.* **135,** 790.
Kilbertus, G., Rohr, R., and Mangenot, F. (1976). *Holzforschung* **30**(5), 156–164.

Kirk, T. K. (1973). *In* "Wood Deterioration and its Prevention by Preservative Treatment" (D. D. Nicholas, ed.), Vol. 1, pp. 149–181. Syracuse Univ. Press, New York.
Kirk, T. K. (1975). *Holzforschung* **29,** 99–107.
Kirk, T. K. (1981a). *Int. Symp. Wood Pulping Chem. Stockholm* **3,** 66–70.
Kirk, T. K. (1981b). *In* "Trends in the Biology of Fermentations for Fuels and Chemicals" (A. Hollaender, ed.), pp. 131–148. Plenum, New York.
Kirk, T. K., and Chang, H.-M. (1974). *Holzforschung* **28,** 217–222.
Kirk, T. K., and Chang, H.-M. (1975). *Holzforschung* **29,** 56–64.
Kirk, T. K., and Highley, T. L. (1973). *Phytopathology* **63**(11), 1338–1342.
Kirk, T. K., and Moore, W. E. (1972). *Wood Fiber* **4,** 72–79.
Kirk, T. K., Connors, W. J., and Zeikus, I. G. (1976). *Appl. Environ. Microbiol.* **32,** 192–194.
Kirk, T. K., Yang, H. H., and Keyser, H. H. (1978). *Prod. Dev. Ind. Microbiol.* **19,** 51–61.
Kohlmeyer, J. (1980). *In* "Biodeterioration" *Proc. 4th Int. Symp., Berlin* (T. A. Oxley, G. Becker, and D. Allsopp, eds.), pp. 187–192. Pitman, London.
Koenigs, J. W. (1974). *Arch. Microbiol.* **99,** 129–145.
Leatham, G., and Kirk, T. K. (1983). *FEMS Microbiol. Lett.* **16,** 65–67.
Liese, W. (1970). *Ann. Rev. Phytopathol.* **8,** 231–258.
Liese, W., and Greaves, H. (1975). *In* "Biological Transformation of Wood by Microorganisms" (W. Liese, ed.), pp. 74–88. Springer-Verlag, Berlin.
Meier, H. (1955). *Holz Roh Werkst.* **13,** 323–338.
Monties, B., Odier, E., Janin, G., and Czaninski, Y. (1981). *Holzforschung* **35,** 217–222.
Nakatsubo, F., Reid, D., and Kirk, T. K. (1981). *Biochem. Biophys. Res. Commun.* **102,** 484–491.
Nilsson, T., and Daniel, G. (1983). Document No. IRG/WP/1186. Int. Res. Group on Wood Preserv., pp. 1–19. IRG Secretariat, Stockholm.
Nilsson, T., and Holt, D. M. (1983). *Holzforschung* **37**(2), 107–108.
Otjen, L., and Blanchette, R. A. (1982). *Can. J. Bot.* **60**(12), 2770–2794.
Ouellette, G. B. (1978). *Can. J. Bot.* **56**(21), 2666–2693.
Parameswaran, N., Wilhelm, G. E., and Liese, W. (1976). *J. For. Pathol.* **5,** 274–286.
Peterson, M. D. and Thomas, R. J. (1978). *Wood Fiber* **10**(3), 150–163.
Ravilly, F., and Dirol, D. (1977). *Mater. Org.* **12**(1), 37–48.
Revol, J. F., Gancet, C., and Goring, D. A. I. (1982). *Wood Sci.* **14**(3), 120–126.
Rosenberg, S. L. (1980). *Mycologia* **72,** 798–812.
Ruel, K., and Barnoud, F. (1981). *Int. Symp. Wood Pulping Chem. Stockholm* **1,** 11–16.
Ruel, K., Comtat, J., and Barnoud, F. (1977). *C. R. Hebd. Seances Acad. Sci. Sér. D* **284,** 1421–1424.
Ruel, K., and Barnoud, F. (1983). *C. R. Hebd. Seances Acad. Sci.* **296,** 231–234.
Ruel, K., and Joseleau, J. P. (1984). *Histochemistry* (in press).
Ruel, K., Barnoud, F., and Eriksson, K.-E. (1981). *Holzforschung* **35,** 157–171.
Ruel, K., Barnoud, F., and Eriksson, K.-E. (1984). *Holzforschung* **38,** 61–68.
Schmidt, O., and Liese, W. (1982). *Int. J. Wood Preserv.* **2,** 13–19.
Tai, D., Terazawa, M., Chen, C.-L., Chang, H.-M., and Kirk, T. K. (1983). *In* "Recent Advances in Lignin Biodegradation Research: Fundamentals and Biotechnology" (T. Higuchi, H.-M. Chang, and T. K. Kirk, eds.), pp. 44–63. Uni, Tokyo.
Thiery, J. P. (1969). *J. Microsc.* **6,** 987–1018.
Westermark, U., and Eriksson, K.-E. (1974a). *Acta Chem. Scand.* **B28,** 204–208.
Westermark, U., and Eriksson, K.-E. (1974b). *Acta Chem. Scand.* **B28,** 209–214.
Westermark, U., and Eriksson, K.-E. (1975). *Acta Chem. Scand.* **B29,** 419–424.

Wilcox, W. W. (1968). Forest Products Laboratory Report No. 70. U.S. For. Serv. Res., Washington, D.C.

Wilcox, W. W. (1970). *Bot. Rev.* **36**(1), 1–27.

Wilcox, W. W. (1973). *In* "Wood Deterioration and Its Prevention by Preservative Treatments" (D. D. Nicholas, ed.), Vol. 1, pp. 107–148. Syracuse Univ. Press, Syracuse, New York.

Wilcox, W. W., Parameswaran, N., and Liese, W. (1974). *Holzforschung* **6,** 211–217.

CHAPTER **17**

Biodegradation of Cellulose

K.-E. ERIKSSON

Swedish Forest Products Research Laboratory
Stockholm, Sweden

T. M. WOOD

Rowett Research Institute
Aberdeen, United Kingdom

I. Introduction

Living organisms produce a wide variety of polymeric carbohydrates, of which cellulose is the most abundant. It was recognized as early as 1883 by Hoppe-Seyler and in 1886 by de Bary that cellulose is decomposed in nature by animals and plants. The main deterioration of cellulose is caused by microorganisms, and the microbiological degradation of lignocellulosic materials is one of the most important processes in nature.

The fermentation of cellulose by microorganisms was first studied by Omelianski (1902). Since then, hundreds of microorganisms have been reported to be cellulolytic. Lists of cellulolytic organisms are given by Siu (1951), Gascoigne and Gascoigne (1960), and Mandels and Weber (1969). The first observation of *in vitro* action of cellulolytic enzymes was proba-

Biosynthesis and Biodegradation of Wood Components

ISBN 0-12-347880-4

bly that of Seilliere in 1906. He reported that digestive juice of the snail *Helix pomatia* caused a slow breakdown of native cellulose.

The cellulolytic enzymes have been defined as the enzymes that hydrolyze cellulose, thereby ultimately yielding soluble sugars small enough to pass through cell walls (Eriksson, 1967). Today a modification of this definition seems necessary, since, in addition to hydrolytic enzymes, oxidative enzymes (Ayers *et al.*, 1978; Vaheri, 1982a) and an oxidoreductive enzyme (Westermark and Eriksson, 1974a,b) are known to participate in cellulose degradation.

During the past 20 years, extensive research has been carried out on cellulolytic enzymes of fungal and bacterial origin. Reviews of these works are given by Reese (1963), Norkrans (1967), Ghose (1977), Goksøyr and Eriksen (1980), Eriksson (1981b), and Wood (1970).

Anyone who has read the early literature on cellulolytic enzymes attentively must have noticed that conflicting reports flourished concerning the composition of the cellulase system and the multiplicity and mechanism of action of the cellulolytic enzymes. Some cellulolytic microorganisms can degrade cellulosic materials only after some kind of modification of the cellulose has occurred. Other organisms also degrade cellulose in native forms, such as cotton fibers. Only the latter organisms have been considered as truly cellulolytic.

In 1950, Reese and co-workers introduced their "C_1, C_x" hypothesis. It was postulated that "truly cellulolytic microorganisms" are equipped with both C_1 and C_x enzymes, while microorganisms able to hydrolyze only modified cellulose lack the C_1 enzyme. In a later publication, Mandels and Reese (1964) gave their C_1, C_x concept the following schematic description:

cellulose → reactive cellulose → cellobiose → glucose

reactive cellulose	cellobiose	glucose
C_1	C_x	β-glucosidase
?	hydrolytic	hydrolytic

The important conclusion of the work by Reese *et al.* (1950), which is still valid, is that crystalline cellulose is hydrolyzed, at least by *Trichoderma* fungi, by a two- or three-step reaction sequence. However, the scheme proposed by Reese and co-workers no longer holds true. Eriksson (1969) and co-workers suggested that the C_1 enzyme was an endwise-acting enzyme that splits off cellobiose units from the cellulose. They envisaged (Eriksson, 1969) the C_x-type enzymes, the endo-1,4-β-glucanases, to initiate the attack by hydrolyzing a few β(1→4)-glucosidic linkages and the new nonreducing chain ends generated to be attacked by the exo-1,4-β-glucanase (formerly called C_1). Exoglucanases of this type have now been isolated from a number of fungi (Eriksson and Pettersson, 1975b; Halliwell *et al.*, 1972; Wood and McCrae, 1972; Berghem and Pettersson, 1973; Emert *et al.*, 1974; Eriksson, 1978), and there is now a consensus of opinion that sequential action of endo- and exo-1,4-β-glucanases effects the hydrolysis of cellulose. Finally, a 1,4-β-glucosidase converts water-soluble cellodextrins to glucose. Thus, the terms C_x and C_1 should be avoided, since they are incorrect and will cause confusion.

II. Enzyme Mechanisms Involved in Cellulose Degradation by Fungi

A. *White-Rot Fungi*

The white-rot group of fungi have in common the capacity to degrade lignin as well as the wood polysaccharides. The relative amounts of lignin and cellulose degraded and utilized by these fungi vary, as does the order of preferential attack (Eriksson, 1981a). However, lignin metabolism seems to occur only during secondary metabolism (Kirk *et al.*, 1978; Keyser *et al.*, 1978), while the cellulose and hemicelluloses are degraded during primary growth. In the filamentous fungi, the primary growth phase continues until the growth medium becomes limited for an essential nutrient such as nitrogen, carbon, sulfate, or phosphate.

The most studied white rot fungus concerning the enzyme mechanisms involved in cellulose degradation is *Sporotrichum pulverulentum* (Fig. 1). This fungus is thermotolerant and has optimal growth around 38–39°C. Details of cultivation and concentration of the culture solutions for production of cellulolytic enzymes are given by Eriksson and Pettersson (1975a).

Sporotrichum pulverulentum degrades cellulose with the usual pattern of hydrolytic enzymes and produces these enzymes:

a. Five endo-1,4-β-glucanases (Eriksson and Pettersson, 1975a) that attack the cellulose chain at random, splitting β(1 → 4)glucosidic linkages.
b. One exo-1,4-β-glucanase (Eriksson and Pettersson, 1975b) that splits off either cellobiose or glucose from the nonreducing end of the cellulose.

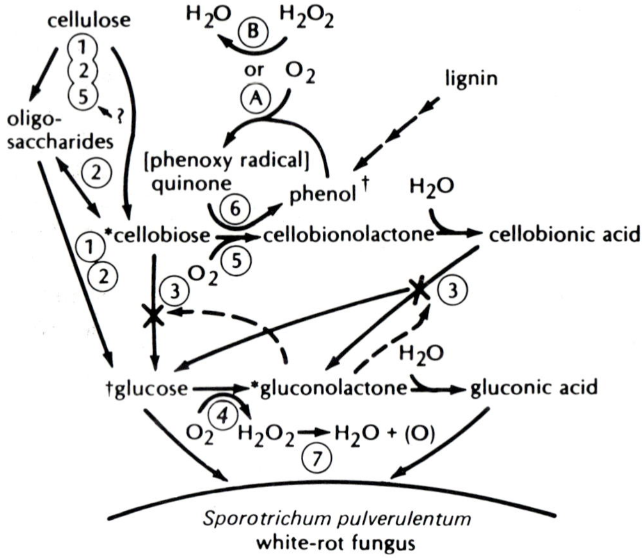

Fig. 1. Enzyme mechanisms for cellulose degradation and extracellular regulation of the process in *Sporotrichum pulverulentum*. Enzymes involved in cellulose degradation: 1, endo-1,4-β-glucanases; 2, exo-1,4-β-glucanase; 3, 1,4-β-glucosidases; 4, glucose oxidase; 5, cellobiose oxidase; 6, cellobiose : quinone oxidoreductase; 7, catalase. Enzymes involved in lignin degradation: A, laccase; B, peroxidase. An asterisk denotes products regulating enzyme activity: gluconolactone inhibits 3; cellobiose increases transglycosylations. A dagger denotes products regulating enzyme synthesis: glucose, gluconic acid → catabolite repression; phenols → repression of glucanases. From Eriksson (1978). Copyright 1978; reprinted by permission of John Wiley & Sons, Inc.

c. Two 1,4-β-glucosidases (Deshpande *et al.*, 1978) that hydrolyze cellobiose and other water-soluble cellodextrins to glucose.

All these enzymes have been purified and characterized both physicochemically (Keyser *et al.*, 1978; Eriksson and Pettersson, 1975a,b) and functionally (Deshpande *et al.*, 1978). The synergistic action between the endoglucanases and the exoglucanases has been investigated in some detail (Streamer *et al.*, 1975). A strong synergistic response between the five endoglucanases and the exoglucanase was observed when these enzymes were allowed to degrade dewaxed cotton and Avicel. No synergism was observed when Walseth cellulose—an acid-swollen cellulose—was used as substrate.

The endoglucanases were found to vary in their substrate preferences. The most striking difference between them was, however, their transglycosylation activities (Streamer *et al.,* 1975). The exoglucanase releases its products, glucose and cellobiose, in the α-configuration (Streamer *et al.,* 1975). In contrast to the endoglucanases, four of which were glycoproteins, the exoglucanase contains no carbohydrate. The molecular weight of the exoglucanase is higher than the molecular weights of the endoglucanases, around 50,000 compared to 30,000–35,000.

The two β-glucosidases have much higher molecular weights (165,000 and 182,000) (Deshpande *et al.,* 1978) than the endo- and exoglucanases. It has been suggested that fungal extracellular enzymes, cellulases, mannanases, and xylanases—i.e., those capable of hydrolyzing insoluble high molecular weight polysaccharides—are smaller molecules than the corresponding β-glucosidases hydrolyzing soluble oligosaccharides (Ahlgren and Eriksson, 1967). This, however, does not apply to *Trichoderma koningii* (see below).

In *S. pulverulentum,* an oxidative enzyme of importance for cellulose degradation has been discovered in addition to the hydrolytic enzymes already described (Ayers *et al.,* 1978). The enzyme has been purified and characterized and found to be a cellobiose oxidase that oxidizes cellobiose and higher cellodextrins to their corresponding -onic acids using molecular oxygen. The enzyme is a hemoprotein and also contains an FAD group. It is not yet known whether this enzyme also oxidizes the reducing end group formed when a $\beta(1\rightarrow4)$-glycosidic bond is split in insoluble cellulose through the action of the endoglucanases. It was recently reported by Vaheri (1982a) that culture solutions of *T. reesei* grown on cellulose also contain gluconic and cellobionic acids. These findings may indicate that *T. reesei* also produces an oxidative enzyme involved in cellulose degradation (see Section II,C,4).

The fungus *S. pulverulentum* has two pathways to oxidize cellobiose. The first is through the enzyme cellobiose oxidase, while the other involves the enzyme cellobiose : quinone oxidoreductase (cellobiose dehydrogenase) (Westermark and Eriksson, 1974a,b). The latter enzyme reduces quinones and phenoxy radicals in the presence of cellobiose and is of importance for the degradation of both cellulose and lignin. Although the enzyme seems to be involved in both lignin and cellulose degradation, the highest yields of the enzyme were obtained when cellulose powder was used as a carbon source. In *S. pulverulentum,* development of cellobiose : quinone oxidoreductase activity and cellulolytic activity occurred simultaneously. The enzyme is relatively specific for its disaccharide substrate, while the requirements of the quinone structure are less specific, and the enzyme is able to reduce both *ortho-* and *para-*quinones. A reaction scheme for the enzyme is presented in Fig. 2 (Westermark and Eriksson, 1974b).

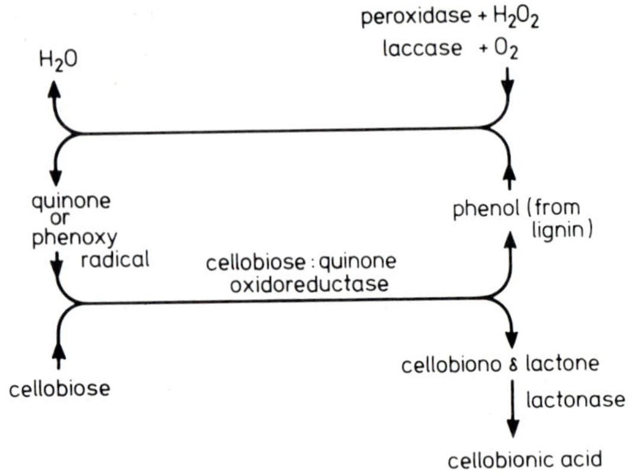

Fig. 2. Reaction mechanism of the enzyme cellobiose : quinone oxidoreductase. The enzyme is of importance in both cellulose and lignin degradation. From Westermark and Eriksson (1974a), with permission of *Acta Chemica Scandinavica*.

The enzyme cellobiose dehydrogenase has now been studied in many laboratories and has been demonstrated to exist in all white-rot fungi so far investigated (Ander and Eriksson, 1977; Kelleher, 1981; Hüttermann and Noelle, 1982), in the imperfect fungus *Monilia* (Dekker, 1980), and in the fungus *Sporotrichum thermophile* (Coudray *et al.*, 1982), phylogenetically belonging to the Ascomycetes. The properties of all these enzymes seem to be very similar to the well-characterized cellobiose dehydrogenase from *S. pulverulentum* (Westermark and Eriksson, 1974a,b).

Two acidic proteases, protease I and protease II, from culture solutions of *S. pulverulentum* have been purified and partly characterized (Eriksson and Pettersson, 1982). These proteases have been demonstrated to be of importance for the degradation of cellulose by the fungus, since the proteases increase the activity of the endoglucanases. If a culture solution of *S. pulverulentum* grown on cellulose is treated with the individual proteases or a mixture of both proteases, the endoglucanase activity in the solution increases. Treatment with both proteases gives a 10-fold increase of the endoglucanase activity (Fig. 3). A possible explanation is that the endoglucanases are activated through the proteases or that the proteases destroy an endoglucanase inhibitor of protein origin. The proteases from *S. pulverulentum* can also increase the activity of endoglucanases other than its own. It has thus been demonstrated that the proteases from *S. pulverulentum* can double the endoglucanase activity of certain commercial cellulases (Eriksson and Pettersson, 1982).

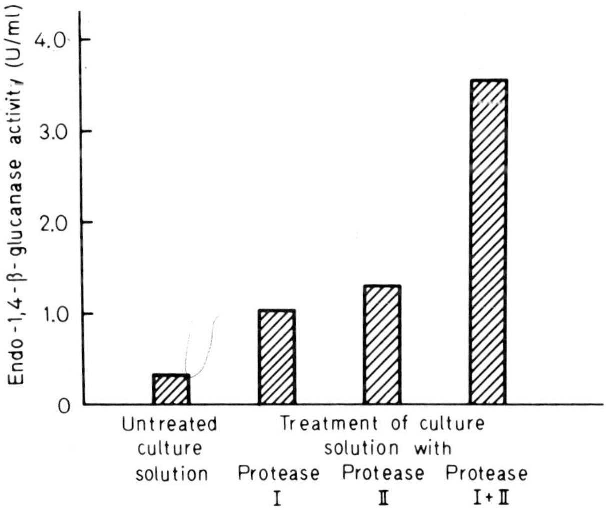

Fig. 3. Activation of endo-1,4-β-glucanases from *Sporotrichum pulverulentum* after treatment with protease I and protease II. From Eriksson and Pettersson (1982).

B. *Brown-Rot Fungi*

Analyses of wood decayed by brown-rot fungi show that the hemicellulose fraction is virtually absent and that the degree of polymerization (DP) of the cellulose has been drastically decreased (Cowling, 1961; Rypacek, 1975). The lignin seems to be modified mainly by the removal of methoxyl groups (Ander and Eriksson, 1978; Crawford, 1981). Brown-rot attack on wood is characterized by a gradual thinning of the wood fiber cell wall as carbohydrates are removed, leaving a framework of lignified tissues (Jutte and Sachs, 1976; see Chapter 16). While scanning electron micrographs of wood degraded by white-rot fungi show erosion troughs intimately associated with the fungal hyphae, similar techniques used for the study of wood attack by brown-rot fungi show attack at some distance from the fungal cell wall (Blanchette *et al.*, 1978; Eriksson *et al.*, 1980).

Brown-rot fungi, which can degrade cellulose extensively, seem to utilize mechanisms different from those operating in *S. pulverulentum* and *T. reesei*. Brown-rot fungi produce endo-1,4-β-glucanases but seem to lack the exo-1,4-β-glucanases (Highley, 1975b). In brown-rot fungi, crystalline cellulose is thus not degraded by a synergistic action between endo- and exoglucanases as is the case in *T. reesei* and in *S. pulverulentum*. Brown-rot fungi depolymerize cellulose rapidly during early stages of wood decay, in contrast to white-rot fungi, which depolymerize cellulose in wood more slowly and utilize the degradation products simultaneously. Koenigs (1974) has found that brown-rot fungi are more powerful producers of H_2O_2 than are the white-rot fungi (Koenigs, 1972). He therefore suggests (Koenigs, 1974) that the initial attack on crystalline cellulose by brown-rot fungi is via an H_2O_2/Fe^{2+} system.

It has been reported by several authors that brown-rot fungi oxidize cellulose (Koenigs, 1974). It is therefore an attractive hypothesis that their initial attack on cellulose takes place via a chemical agent of low molecular weight, which can easily diffuse through the wood fiber walls, rather than via an enzyme with limited diffusibility.

It was found by Highley (1975a) that the brown-rot fungus *Poria placenta* was unable to utilize cellulose unless placed in contact with wood. Nilsson (1974) found no morphological change on cotton fibers placed between wood pieces inoculated with brown-rot fungus but reported that after this exposure no tensile strength was left in the cotton fibers, indicating attack by the brown-rot fungus. These studies suggest that when cultured on wood, brown-rot fungi produce a factor able to diffuse into the pure cellulose and depolymerize it. In a more recent study, Highley (1977) found that addition of other carbohydrate supplements such as cellobiose, glucose, and mannan results in depolymerization of cellulose by *P. placenta*.

In experiments at the Swedish Forest Products Research Laboratory, it was found that H_2O_2/Fe^{2+} pretreatment of cotton cellulose did not enhance the action of an endo-1-4-β-glucanase attack on cellulose. If a mixture of H_2O_2/Fe^{2+} and oxalic acid was used as the pretreating agent, however, an increased release of reducing sugars from a pretreated cotton cellulose was obtained when endoglucanases were added to the pretreated cotton fiber suspension.

In thesis work, Cobb (1982) was able to demonstrate clearly that a nonenzymatic cellulose decay mechanism exists in brown-rot fungi. An ultrafiltration membrane that did not allow passage of enzymes was placed between a fungal culture and the cellulose substrate. It could be demonstrated that a fungal culture grown on hemicellulose monosaccharides decomposed the cellulose substrate more than a fungal culture grown on ball-milled pine, which is a poor substrate for H_2O_2 production.

More work, however, needs to be done before the mechanisms for cellulose degradation by brown-rot fungi are understood.

C. *Soft-Rot Fungi*

Wood polysaccharides are the main targets for soft-rot fungi. Lignin is largely undigested, attack being confined to a few methoxyl groups (Levi and Preston, 1965). Soft-rot attack is typified by boreholes in the S_2 layer that have wide cavities with characteristic conic ends (Liese, 1964; Levi and Preston, 1965). The regularity of the shape of these cavities has been the subject of much discussion. It would appear to be the result of restricted diffusion of enzymes that are secreted by regions of hypae growing along the microfibrils (Levi and Preston, 1965; Liese and Schmid, 1966). The S_3 layer and the middle lamella are resistant to attack (Meier, 1955), probably because of their high lignin content. Even if cavity formation is the most typical type of soft-rot attack on wood, however, it is not the only type of attack. Other patterns of attack on wood can be a combination of cavity formation and erosion, or erosion only (Nilsson, 1973). Fungi like *Trichoderma, Aspergillus,* and *Penicillium* can only form the erosion pattern in their attack on wood, but nevertheless they do to some extent degrade wood (Nilsson, 1973).

Although soft-rot fungi are capable of degrading crystalline, highly ordered cellulose under laboratory conditions, only a few of the species studied seem to release an extracellular enzyme system with this capability. Notable in this respect are the cellulases liberated by *Trichoderma viride* (Nisizawa *et al.,* 1972; Ogawa and Toyama, 1972; Selby and Maitland, 1972), *T. reesei* (Mandels and Reese, 1964, Gritzali and Brown, 1979) (formerly *T. viride* QM6a and mutants of this strain), *T. koningii* (Wood, 1968; Halliwell, 1965), *Penicillium funiculosum* (Wood *et al.,* 1980; Selby, 1968), and *Fusarium solani* (Wood and Phillips, 1969), and these have been used extensively for studies on the mechanism of cellulase action (Wood and McCrae, 1972; Nisizawa *et al.,* 1972; Ogawa and Toyama, 1972; Selby and Maitland, 1972; Mandels and Reese, 1964; Gritzali and Brown, 1979; Wood, 1968; Halliwell, 1965; Wood *et al.,* 1980; Selby, 1968; Wood and Phillips, 1969; Wood, 1969, 1971; Berghem *et al.,* 1975, 1976; Berghem and Pettersson, 1974; Wood and McCrae, 1977a,b, 1978; Halliwell and Griffin, 1973) in this group of fungi. Fractionation studies (Wood and McCrae, 1972; Nisizawa *et al.,* 1972; Ogawa and Toyama, 1972; Selby and Maitland, 1972; Mandels and Reese, 1964; Gritzali and Brown, 1979; Wood, 1968; Halliwell, 1965; Wood *et al.,* 1980; Selby, 1968; Wood and Phillips, 1969; Wood, 1969, 1971; Berghem *et al.,* 1975, 1976; Berghem and Pettersson, 1974; Wood and McCrae, 1977, 1978; Halliwell and Griffin, 1973) performed on these cellulases have shown that they are similar in many respects to the cellulases of the white-rot fungi. They contain endo-1,4-β-glucanases, exo-1,4-β-glucanases (normally cellobiohydrolases), and 1,4-β-glucosidases or cellobiases. Cell-free culture filtrates of most other soft-rot fungi contain appreciable amounts of endoglucanase, variable amounts of β-glucosidases, but little exoglucanase. These fungi are capable of degrading only the more disordered amorphous celluloses. The reason why some soft-rot fungi liberate a "complete" enzyme complex while others do not is not clear, but this may be of crucial importance to our understanding of the mechanism of breakdown of cellulose by these fungi. The possibility that more than one mechanism of action operates in the soft-rot fungi cannot be excluded (Wood, 1983).

There is a consensus that cellulase enzymes synthesized by the soft-rot fungi overcome, by their sequential and concerted action, problems created by the degree of crystallinity of the cellulose and the inhibition caused by accumulation of end products (Wood and McCrae, 1979a). There are, however, significant differences in opinion regarding the specificities of the various enzymes and the roles they play in the dissolution of cellulose in its various forms (Wood and McCrae, 1979a; Reese, 1976). Some of these differences may arise directly from the use by the various investigators of a variety of poorly characterized substrates to study the cellulase components, but the incomplete separation of enzymes with overlapping substrate specificities cannot be excluded. This does not mean to say that fractionation of the individual enzymes has been badly performed, for it could merely indicate the existence of enzyme–enzyme complexes that are difficult to disrupt. Such complexes have recently been reported to exist in *T. reesei* (Sprey and Lambert, 1983). Notwithstanding these difficulties, some generalizations regarding the modes of action and properties of the individual enzymes comprising the cellulase complex can be made, and these are summarized next.

1. Endo-1,4-β-Glucanases

Endo-1,4-β-glucanases are found in varying concentrations in cell-free culture filtrates of all soft-rot fungi. They have properties and modes of action similar to the endo-1,4-β-glucanases of the white-rot fungi. They hydrolyze cellulose chains at random to produce a rapid decrease in chain length, together with a slow increase in reducing power (Wood and McCrae, 1978; Shoemaker and Brown, 1978; Wood and McCrae, 1977a; Gilligan and Reese, 1954). Substrates include carboxymethylcellulose and H_3PO_4- or alkali-swollen cellulose, but cotton fiber (Wood and McCrae, 1978) or Avicel (Shoemaker and Brown, 1978) is not attacked to a significant extent. The β-glucose configuration is retained on hydrolysis, yielding a mixture of glucose, cellobiose, or other soluble cello–oligosaccharides (Wood and McCrae, 1978; Shoemaker and Brown, 1978). The rate of hydrolysis of the longer chain cello–oligosaccharides is high, and the rate increases with degree of polymerization (Whitaker, 1954): glucose and cellobiose are the principal products of the reaction. Cellobiose is a powerful inhibitor of the action on carboxymethylcellulose (Wood and McCrae, 1975; Mandels and Reese, 1965; Reese *et al.,* 1952). Transglycosylation has been reported to a feature of *T. viride* endoglucanase (Gritzali and Brown, 1979), but apparently it is not a property common to all endoglucanases. Molecular weights are normally in the range 37,000–52,000 (Gritzali and Brown, 1979; Wood, 1971; Berghem *et al.,* 1976; Pettersson *et al.,* 1981), but many culture filtrates of *Trichoderma* species have endoglucanases of low molecular weight (12,000–20,000) (Selby and Maitland, 1972; Wood, 1968; Berghem *et al.,* 1976). This low molecular weight enzyme may be absent in cultures in which the pH is not allowed to fall below 3.5. (Pettersson *et al.,* 1981). Most endoglucanases are associated with varying amounts (<20%) (Gritzali and Brown, 1979) of carbohydrate; an endoglucanase from *T. viride,* however, has been found to be devoid of carbohydrate (Håkansson *et al.,* 1979).

2. Exo-1,4-β-Glucanases

Exo-1,4-β-glucanases of the soft-rot fungi are cellulases that remove glucose or cellobiose from the nonreducing end of the chain. Cellobiohydrolase is the most common enzyme (Wood and McCrae, 1972; Gritzali and Brown, 1979; Wood *et al.,* 1980; Wood and McCrae, 1977b; Halliwell and Griffin, 1973; Pettersson *et al.,* 1981). Most cellobiohydrolases are reported to release traces of glucose from the cellulose. However, it is not clear in all cases if this is a feature of the enzyme or the result of the presence of some associated hydrolase. Cotton fiber is not attacked to a significant extent (Wood and McCrae, 1972), but H_3PO_4-swollen cellulose is hydrolyzed with a characteristic slow fall in degree of polymerization (Wood and McCrae, 1977b; Berghem and Pettersson, 1973; Wood and McCrae, 1979b). Carboxymethylcellulose and cellobiose are not substrates, but cellotriose and longer chain cello–oligosaccharides are hydrolyzed, with the rate increasing with degree of polymerization (Wood and McCrae, 1977b; McHall and Coughlan, 1981). Avicel (Gritzali and Brown, 1979; Pettersson *et al.,* 1981; Emert *et al.,* 1984) is a substrate, and Avicelase is now commonly regarded as being synonymous with exo-1,4-β-glucanase. Avicel has certainly proved to be a useful substrate for isolating exoglucanases, but there is a regrettable tendency to regard even the smallest release of reducing sugars from Avicel as being indicative of the presence of an exoglucanase enzyme. A specific assay procedure for exoglucanases in a mixture of cellulases has been presented by Eriksson and Johnsrud (1982), utilizing amorphous cellulose as substrate.

A characteristic property of the exoglucanases (Nisizawa *et al.,* 1978) is that the β-configuration is inverted on hydrolysis; this property can be used to distinguish these enzymes from β-glucosidases (Reese *et al.*, 1968). Cellobiose is a powerful inhibitor (Halliwell and Griffin, 1973; Nisizawa *et al.*, 1978) of exoglucanases, but gluconolactone is not (Wood, 1972). Molecular weights of the cellobiohydrolases range from 45,000 to 62,000 (Gritzali and Brown, 1979; Wood *et al.*, 1980; Pettersson *et al.*, 1981; Wood and McCrae, 1979a, Li *et al.*, 1965), and isoelectric points are usually acidic (3.8–5.6) (Gritzali and Brown, 1979; Wood *et al.*, 1980; Wood and McCrae, 1977a; Li *et al.*, 1965). All are glycoproteins with proportions of carbohydrates ranging from 50 (Selby and Maitland, 1972) to 1.4% (Gritzali and Brown, 1979), most being in the region 5–15% (Wood *et al.*, 1980; Wood and McCrae, 1977b; Halliwell and Griffin, 1973; Pettersson *et al.*, 1981; Wood, 1972). Glucohydrolases have been found in cultures of soft-rot fungi, but they are not common.

3. β-Glucosidases

β-Glucosidases of soft-rot fungi are normally active on both aryl β-glucosides and cellobiose, but there are exceptions (Jermyn, 1975; Hash and King, 1958; Montenecourt *et al.*, 1981). Cello–oligosaccharides are hydrolyzed at a rate that decreases with increasing degree of polymerization (Reese *et al.*, 1968; Hash and King, 1958; Sternberg *et al.*, 1977). Other characteristics are that (a) the configuration is retained on hydrolysis (Reese *et al.*, 1968), (b) β(1→2)-, β(1→3)-, and β(1→6)-linked dimers of glucose are hydrolyzed as easily as β(1→4)-linkages (Reese *et al.*, 1968; Wood and McCrae, 1982b), (c) transferase activity that acts on glucose units to form alcohols or other sugar molecules such as dimers, trimers, and higher oligosaccharides (Gritzali and Brown, 1979; Wood, 1968; Halliwell, 1965; Wood *et al.*, 1980; Selby, 1968; Wood and Phillips, 1969; Wood, 1969; Emert *et al.*, 1974), and (d) the action on nitrophenyl β-glucosides is competitively inhibited by glucose [K_i 1.05 m*M*; 0.66 m*M*, *T. konigii* (Wood and McCrae, 1982b); 1.01 m*M*, *T. viride* (Emert *et al.*, 1974); 0.71 m*M*, *Talaromyces emersonii* (McHall and Coughlan, 1981)], gluconolactone (K_i 1.8 μ*M*, *T. koningii* (Wood and McCrae, 1982b); 32 μ*M*, *T. viride* (Emert *et al.*, 1974)], or nojirimycin [e.g., 28 μ*M* to give 50% inhibition, *Aspergillus niger* (Reese and Parrish, 1971)]. Aryl β-glucoside values of K_m of 1.8 m*M* and 1.5 m*M* are quoted for *T. viride* (Berghem and Pettersson, 1974; Emert *et al.*, 1974), and these are typical for *Trichoderma* species [0.14 m*M*, *T. emersonii* (McHall and Coughlan, 1981)]. Molecular weight values cover a very wide range: 39,800 (*T. koningii*) (Wood and McCrae, 1982b); 76,000 (Li *et al.*, 1965) and 47,000 (Berghem and Pettersson, 1974) (*T. viride*); 400,000 (*F. solani*) (Wood, 1969); and 170,000 (*Aspergillus wentii*) (Legler, 1967). Trichoderma is a poor source of extracellular β-glucosidase, most of this activity being cell wall bound (Montenecourt *et al.*, 1981; Inglin *et al.*, 1980).

These generalizations (Sections II,C,1–II,C,3), while acceptable for the purposes of review, are oversimplifications, and can therefore be misleading. Certain qualifications are therefore necessary. One of the most important relates to the specificity of the cellobiohydrolase component and its ability to attack highly ordered cellulose when acting alone.

When cotton fiber is used as a model substrate for highly ordered cellulose, it would appear that highly purified cellobiohydrolases have, in the main, little capacity for hydrolysis when acting in isolation. When Avicel, the microcrystalline α-cellulose from wood pulp, is used, however, extensive degradation can be effected. Indeed, 45% hydrolysis has been recorded for a cellobiohydrolase from *T. viride*, and this could be increased to 79% when incubation was done in an ultrafiltration cell under conditions that made it possible to remove the inhibitory products of the reaction (Berghem *et al.*, 1976). These observations are consistent with the hypothesis that only one enzyme is necessary for the hydrolysis of crystalline cellulose, and this is how Li, Flora, and King (1965) interpreted their data obtained with a hydrocellulase isolated from a commercial preparation of *T. viride*. Using a dialysis arrangement similar to that used with *T. viride*, the cellobiohydrolase of *T. koningii* was found to produce only limited hydrolysis (7%) of Avicel (Wood and McCrae, 1979a).

4. Oxidative Enzymes

In addition to the hydrolytic enzymes, there is some evidence that *T. reesei* (Vaheri, 1982a,b) and *T. koningii* (Wood and McCrae, 1977a) may also produce an oxidative enzyme similar to that already purified and characterized in *S. pulverulentum* (see Section II,A). The oxidative reaction has been reported to influence the generation of "short fibers," in addition to the formation of onic acids (Vaheri, 1982a). It has been suggested by Vaheri (1982b) that in the initial stage of breakdown of cotton, oxidative reactions increase the accessibility of the fibers to the endoglucanases, which confirms an earlier observation made by Marsh (1966).

5. Synergistic Action between the Hydrolytic Enzymes

Synergism between enzymes of the cellulase complex of soft-rot fungi was first demonstrated by Reese and colleagues using H_3PO_4-swollen cellulose (Walseth cellulose) (Gilligan and Reese, 1954). Cotton fiber, however, was subsequently shown to be a better substrate to demonstrate

Table I

Cellulase Activity of the Components of *Trichoderma koningii* and *Fusarium solani* Cellulase When Acting Alone and in Combination[a]

Enzyme	Cellulase activity (solubilization, %)	
	T. koningii	*F. solani*
Cellobiohydrolase (CBH)	1	2
Endo-1,4-β-glucanase (Endo)	1	1
β-Glucosidase	Nil	Nil
Endo + CBH	55	58
Endo + β-glucosidase	6	4
CBH + β-glucosidase	20	18
Endo + CBH + β-glucosidase	72	71
Original unfractionated enzyme	71	71

[a] All enzymes were present in the same proportions in which they were present in the original unfractionated enzyme. From Wood and McCrae (1979b).

this effect with highly active cellulase preparations (Mandels and Reese, 1964). Indeed, when the enzymes are highly purified and the cotton fiber has received the minimum of pretreatment, it would appear that solubilization is almost completely dependent on the synergistic action of the enzymes (Table I) (Wood and McCrae, 1979b). Cooperation is evident between cellobiohydrolase and endoglucanase and to a limited extent between cellobiohydrolase and β-glucosidase, but for maximum rate and extent of hydrolysis it seems that all three hydrolytic enzymes must be present at the same time.

On the basis of this and other evidence (Eriksson, 1978; Wood and McCrae, 1975; Wood, 1972), two types of synergism appear to operate between the three enzymes. In the first stage of the reaction, an enzyme A (endoglucanase) is producing new chain ends for enzyme B (cellobiohydrolase) to attack (Wood, 1972), while in a second stage, an enzyme C (β-glucosidase) is removing an inhibitory product (cellobiose) of the action of enzyme B (Halliwell and Griffin, 1973; Wood and McCrae, 1975).

A discovery of Fägerstam and Petterson (1980) and Brown and colleagues (1981) provides evidence for yet another type of synergism. Working with *T. reesei* cellulase the two groups have isolated a cellobiohydrolase (so-called cellobiohydrolase II) that is quite different from that previously investigated (cellobiohydrolase I) in terms of composition, sequence of amino acids, isoelectric pH, immunology, and in the pattern of

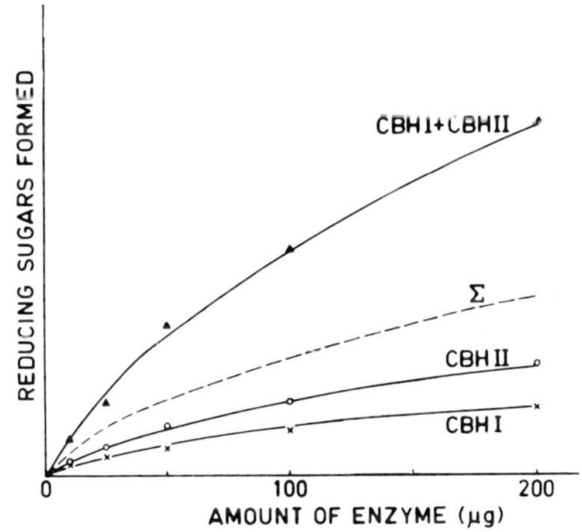

Fig. 4. Synergism between the two cellobiohydrolases of *Trichoderma reesei* QM 9414. The nonsynergistic sum of the two activities of the two enzymes is indicated by Σ. From Fägerstam and Pettersson (1980).

bonds cleaved in cello–oligosaccharides. Remarkably, cellobiohydrolase II had a marked capacity for cooperating with cellobiohydrolase I in solubilizing both Avicel and cotton cellulose (Fig. 4). This exo–exo cooperation is clearly a new type of synergism, and one that is difficult to explain in terms of sequential action on the cellulose chain. Fägerstam and Pettersson speculate (1980) that it may indicate some structural anomaly in the arrangement of the cellulose chains. Whatever the explanation, this new cellobiohydrolase (II) obviously has a crucial role to play in the dissolution of crystalline cellulose. Brown and colleagues (1981) have reported that for optimal activity on "swollen" reactive cellulose, only one type of cellobiohydrolase is found on the cellulose, but when crystalline cellulose is the substrate there is a ratio of the two cellobiohydrolases that produces maximum catalytic activity. No reports have appeared describing the existence of two unrelated cellobiohydrolases in an extracellular cellulase from any other soft-rot fungus.

Wood and McCrae (1979b, 1982a) have found two exo-1,4-β-glucanases in the cellulase of *P. funiculosum,* but these did not cooperate at all to solubilize highly ordered cellulose. One of the enzymes was a cellobiohydrolase, and the other was a glucohydrolase. Thus, if exo–exo synergism is a feature of soft-rot fungi, it would appear that it is restricted to the interaction of two cellobiohydrolases (Wood, 1983). More unexpectedly, the glucohydrolase, unlike the cellobiohydrolase, did not act in synergism with the separated endoglucanase (Wood and McCrae, 1982a). Wood and

McCrae (1979a) point out that these results raise the interesting question of why an enzyme that removes glucose units two at a time from the end of a cellulose chain (cellobiohydrolase) can act synergistically with endoglucanase (and indeed another type of cellobiohydrolase), whereas an enzyme that removes only one residue cannot. They conclude that the explanation is likely to be connected in some way with the fact that cellobiose is the repeating unit in the cellulose crystallite.

Synergism between endo- and exoglucanases is reported to be a feature of many other studies (Gritzali and Brown, 1979; Wood and McCrae, 1979a; Ogawa *et al.*, 1982; Gum and Brown, 1977; Gong *et al.*, 1979; McHall and Coughlan, 1980; Eriksen and Goksøyr, 1977; Halliwell and Riaz, 1971).

If the hydrolysis of crystalline cellulose can be explained in terms of sequential attack on the substrate by randomly acting endo-1,4-β-glucanases and endwise-acting cellobiohydrolase, it is reasonable to expect that the cellobiohydrolase from one fungus should be able to act in synergism with the endo-1,4-β-glucanase from another.

Observations made with several cellulases of the soft-rot fungi indicate, however, that cooperation is very variable (Wood and McCrae, 1979a; Wood, 1980, 1982). Some rationalization is possible, but an explanation of the phenomenon appears difficult. It has been pointed out by Wood (1980) that synergism is best between enzymes (cellobiohydrolase and endoglucanase) from different fungi when the fungal sources produce a "complete" cellulase complex, i.e., one capable of extensive hydrolysis of crystalline cellulose. Thus, synergism that is high between any combination of the cellobiohydrolases and endoglucanases of *T. koningii, F. solani,* and *P. funiculosum* (Table II) is low when the cellobiohydrolase of *T. koningii, F. solani,* or *P. funiculosum* are mixed with the endoglucanases isolated from culture filtrates of fungi that do not apparently liberate exoglucanase, namely, *Myrothecium verrucaria, Stachybotrys atra* (Table II), *Gliocladium roseum* (Wood, 1980), and *Memnoniella echinata* (Wood, 1980). In keeping with this speculation is the observation that synergism between the cellobiohydrolases of the soft-rot fungi *T. koningii, F. solani,* and *P. funiculosum* and an endo-1,4-β-glucanase of the white-rot fungus *S. pulverulentum* is high (Wood, 1983), but it is nonexistent when the cellobiohydrolases are acting in admixture with the endoglucanases of the rumen anaerobic bacteria *Ruminococcus albus, R. flavefaciens,* and *Bacteroides succinogenes* (Wood, 1983). None of the rumen bacteria produce a cellulase that contains an exo-1,4-β-glucanase, but, of course, *S. pulverulentum* does.

It is of interest in the context of the mechanism of cellulase action that not all endo-1,4-β-glucanases and cellobiohydrolases are completely in-

Table II

Cellulase (Cotton Solubilizing) Activity Shown by Cellobiohydrolase and Endo-1,4-β-glucanase from Various Fungi Alone and in Combination[a]

Source of cellobiohydrolase (CBH)	Source of endo-1,4-β-glucanase (Endo)	Solubilization (%)		
		CBH alone	Endo alone	CBH + Endo
Trichoderma koningii	*T. koningii*	2	1	51
T. koningii	*Fusarium solani*	1	1	79
F. solani	*F. solani*	1	1	59
F. solani	*T. koningii*	1	1	51
Penicillium funiculosum	*T. koningii*	1	1	51
P. funiculosum	*F. solani*	1	1	51
P. funiculosum	*M. verrucaria*	1	6	20
P. funiculosum	*Stach. atra*	1	4	21
T. koningii	*M. verrucaria*	1	6	20
T. koningii[b]	*S. pulverulentum*	1	5	61
P. funiculosum[b]	*S. pulverulentum*	2	5	52
F. solani	*S. pulverulentum*	8	5	43

[a] All assays contained the same amount of endo-1,4-β-glucanase, cellobiohydrolase, and β-glucosidase. β-Glucosidase from *F. solani* was added where necessary to make all assays the same in terms of this activity. From Wood (1980), in part.

[b] Recent preparations of enzyme.

terchangeable in these studies. Wood (1980) discusses several possible explanations, but suggests that the problem can be a stereochemical one, namely, that the configuration of the nonreducing end of cellulose chain generated by one endo-1,4-β-glucanase may not fit the active site of an exo-1,4-β-glucanase from another source (Fig. 5). This situation arises because of the fact that the nonreducing end group generated by the endoglucanase can have two different configurations, depending on which of the two glycosidic linkages is cleaved.

An alternative speculation suggested by Wood (Wood and McCrae, 1977a; Wood, 1980) for the variation in cooperation between endoglucanase and the cellobiohydrolase is that only those enzymes that can interact in some way to form an enzyme–enzyme complex on the face of the cellulose crystallite can cooperate effectively. Rapid sequential action could be effected easily in these circumstances, and this would prevent re-formation of the glycosidic linkage between two glucose residues held rigidly in position of hydrogen bonds.

Whatever the true explanation, it is clear that highly ordered crystalline

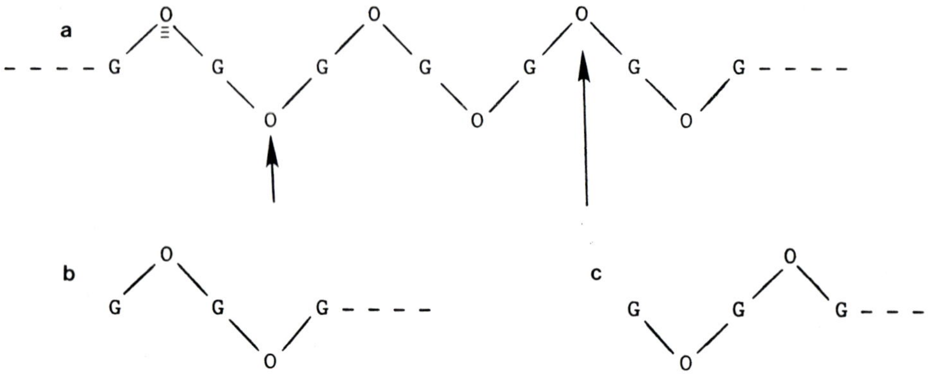

Fig. 5. Diagram of cellulose chain showing different points of cleavage by endo-1,4-β-glucanases. Cleavage of a cellulose chain (**a**) by endo-1,4-β-glucanases can result in the production of two sterically different end groups: nonreducing end group type I (**b**) or nonreducing end group type II (**c**). Attack on these end groups will require enzymes with quite different active sites. From Wood (1980).

cellulose can only be degraded by enzymes with very special properties, and these properties cannot be defined only in terms of the parameters normally evaluated.

6. Multiple Enzyme Components and Their Significance

So far, the mechanism of cellulase action in the soft-rot fungi has been discussed in terms of the type of enzyme rather than in terms of the number of enzymes required. In fact, multiple forms of all the enzyme types exist in most cellulases. Just how many of these are actually elaborated by the fungus and how many are artefacts resulting from post-translational modification (Nakayama *et al.*, 1976) are the subject of considerable debate. There seems to be a consensus emerging, however, that the enzyme systems required for hydrolysis of crystalline cellulose are less complex than those normally reported to have been fractionated from many extracellular cellulases (Gritzali and Brown, 1979; Gong *et al.*, 1979).

Cultures of *T. reesei* (Gritzali and Brown, 1979) in which the fungus was first grown in a medium containing glucose as the sole carbon source and then the enzymes induced by the addition of sophorose (1,2-β-diglucose) yielded a very simple enzyme complex. This complex contained cellobiohydrolase I and II (Fägerstam and Pettersson, 1980; Brown *et al.*, 1981) in the proportion (by weight) 60 and 25%, respectively, 15% endoglucanase, and some β-glucosidase; the complex reportedly could solubilize Avicel at the same initial rate as a cellulase complex induced on cellulose. Unfortunately, it is not clear from the published data if the enzyme system was capable of degrading Avicel to an extent that demonstrates unequivocally that crystalline cellulose is being hydrolyzed (Wood, 1983).

When cellulose is the carbon source, the cellulase pattern in the culture filtrate is much more complex. *Trichoderma koningii* cellulase obtained by stationary culture, for example, has been found to contain two cellobiohydrolase components (both cellobiohydrolase type I), two β-glucosidase components, and four major endoglucanase components (Koenigs, 1974). *Trichoderma viride* (Gritzali and Brown, 1979; Emert *et al.*, 1974; Nisizawa, 1973) and *T. reesei* (Gritzali and Brown, 1979; Ogawa *et al.*, 1982; Nakayama *et al.*, 1976; Ogawa and Toyama, 1982) cellulases induced in a similar fashion have been found to be equally heterogeneous, but, as already discussed, they have the added complication of containing two unrelated cellobiohydrolases.

Heterogeneity of both types of cellobiohydrolase in *T. reesei* (Gritzali and Brown, 1979; Berghem and Pettersson, 1973), *T. viride* (Emert *et al.*, 1974; Gum and Brown, 1977), and *F. solani* (Wood and McCrae, 1977b) has been ascribed to the variation in the amount of carbohydrate associated with the protein; however, some of the multiplicity of the β-glucosidase observed may result from the appearance in solution of some cell-bound β-glucosidase (Montenecourt *et al.*, 1981; Inglin *et al.*, 1980). Carbohydrase action and cell lysis may therefore account for much of the heterogeneity in these enzymes.

Multiplicity in terms of endoglucanases is certainly the rule, and these have been shown to have different modes of action on "swollen" reactive cellulose and soluble derivatives of cellulose. Nakayama *et al.* (1976) have reported that partial proteolysis of an endoglucanase from *T. viride* yielded enzymes with altered substrate specificities, and this would suggest that some secondary modification of a single endoglucanase might account for the multiplicity of endoglucanases found in some cellulases. In another publication (Gong *et al.*, 1979), young cultures of *T. reesei* have been reported to contain only a small amount of protease activity, but this increased with the age of the culture. In the same study, the number of extracellular endoglucanase components increased from one to four as the fermentation progressed. It is interesting that only two of the four endoglucanases of *T. koningii* have been reported (Wood and McCrae, 1978) to be able to act synergistically with the cellobiohydrolase to solubilize highly ordered cellulose. This might indicate that the tertiary structure of the two of the enzymes has been altered in some way that makes it impossible for them to bind to the cellulose and generate end groups for the endwise-acting cellobiohydrolase to attack (Wood, 1983).

III. Regulation of Enzymes of the Cellulase Complex

There seem to be considerable differences between *S. pulverulentum* and *T. reesei* in regulating the production of cellulases. While the endoglucanase activity in *S. pulverulentum* is induced by cellobiose in as low a concentration as 1 mg/liter, cellobiose at the same or even higher concentrations does not induce endoglucanase activity at all in *T. reesei* (Eriksson and Hamp, 1978). However, sophorose causes induction of endoglucanases in this organism at a concentration of 1 mg/ml. Comparison of the regulation of endoglucanase production in the two fungi also demonstrates several other important differences. Thus, a solution of carboxymethylcellulose (CMC) alone gives rise to endoglucanase formation in *S. pulverulentum* but not in the *T. reesei* strain. Under the experimental conditions used (Eriksson and Hamp, 1978), no endoglucanases were actively excreted into the solution of *T. reesei,* while this was the case with *S. pulverulentum*.

In a specific study of induction of cellulolytic enzymes in *Trichoderma reesei* QM6a, Sternberg and Mandels (1979) found that sophorose induces endoglucanases in the fungus within 1.5 to 2 hr. The induction response to sophorose concentration shows saturation kinetics. Most of the cellulase appears after most of the sophorose has been taken up, but the presence of an inducer is required to maintain enzyme synthesis. Endoglucanase production ceases after separation of the mycelium from the induction medium. As indicated by enzymatic and electrophoretic studies, sophorose appears to induce an incomplete array of cellulases under the cultural conditions used.

Catabolite repression of endoglucanase production is obtained in *S. pulverulentum* at glucose concentrations of around 50 mg/liter. The glucose level causing catabolite repression of endoglucanases in *T. reesei* has as far as is known not been determined.

The production of cellulases in fungi can also be hampered by factors other than catabolite repression. Thus, it was demonstrated by Varadi (1972) that a wide variety of phenols repress the production of both cellulases and xylanases in the fungi *Schizophyllum commune* and *Chaetomium globosum*. At concentrations of less than 1 m*M*, vanillic acid, vanillyl alcohol, and vanillin considerably repress the production of these enzymes. Ander and Eriksson (1976) showed that in a phenol oxidase-less mutant (Phe 3) of *S. pulverulentum,* the production of endoglucanases was drastically repressed in the presence of kraft lignin and phenols at a concentration of 10^{-3} *M*. However, both the wild type (WT) and a phenol-oxidase-positive revertant (Rev 9), obtained by ultraviolet (UV) irradiation of Phe 3, produced the enzymes without significant repression by phenols. Furthermore, if a highly purified laccase preparation was added to the growth medium of Phe 3 in the presence of phenols, the endoglucanase production increased to normal. These results indicate that kraft lignin and phenols decrease endoglucanase synthesis in Phe 3 due to the absence of phenol-oxidizing enzymes. Phenol oxidases may thus function in regulating the production of cellulases by polymerizing lignin-related phenols, which act as repressors of enzyme production when *S. pulverulentum* is growing on wood.

The activity of cellulose hydrolyzing enzymes in culture filtrates of *S. pulverulentum* is dependent not only on mechanisms regulating their biosynthesis but also on the presence of specific regulatory inhibitors of the enzymes themselves. One such inhibitor is gluconolactone, which is produced by oxidation of glucose by the enzyme glucose oxidase or by hydrolytic cleavage of cellobionolactone (Eriksson, 1978). The importance of gluconolactone for the regulation of 1,4-β-glucosidases from *S. pulverulentum* has been studied by Deshpande *et al.* (1978). Both extracellular 1,4-β-glucosidase and cell-bound enzymes appear in *S. pulverulentum*. The extracellular β-glucosidase activity contains two main peaks separable by chromatography. The K_i values for gluconolactone as a competitive inhibitor for the two free β-glucosidases were 3.5×10^{-7} and 15×10^{-7} *M*, respectively. A K_i value of 3.2×10^{-5} *M* was obtained for the *T. reesei* QM 9414 β-glucosidase for the same inhibitor. The corresponding K_i value for inhibition of this enzyme by glucose was 1×10^{-3} *M* (Eriksson and Pettersson, 1975).

IV. Mutant Fungal Strains Hyperproducing Cellulases

The last 5–10 years has seen a great increase in interest in the utilization of cellulose as a source of energy and chemicals. Glucose can be generated from cellulose either chemically using acids or biochemically using enzymes. Processes based on acid hydrolysis already exist, but they suffer from a number of disadvantages. These include degradation and contamination of the product, corrosion of equipment, and high capital cost. Enzymatic hydrolysis is clearly preferable, but it is not, as yet, attractive to the industrialist because of the high cost of the cellulase enzymes.

Considerable effort has therefore been focused on obtaining high cellulase-producing mutants from *T. reesei* (Montenecourt and Eveleigh, 1979, 1977a, 1977b, Shoemaker *et al.*, 1981; Nevalainen *et al.*, 1980; Nevalainen and Palva, 1978; Ghosh *et al.*, 1982; Mishra *et al.*, 1982; Beja Da Costa and Van Uden, 1980; Mandels *et al.*, 1971; Mandels, 1975), and other fungi (Sadana *et al.*, 1979; Selby, 1969), a thermophilic actinomycete (Fennington *et al.*, 1982), and thermophilic anaerobes (Lee and Blackburn, 1975). The most significant strain improvement program has been performed on *T. reesei,* since it has been shown to produce the most active and well-balanced cellulase for the saccharification of cellulose (Ryu and Mandels, 1980).

The mutant strains of *T. reesei* currently being used include QM 9123 (Mandels *et al.*, 1971; Mandels, 1975), QM 9414 (Mandels *et al.*, 1971; Mandels, 1975), MCG 77 (Gallo *et al.*, 1978), Rut C30 and NG 14 (Montenecourt and Eveleigh, 1977, 1979, 1978), L-series (Shoemaker *et al.*, 1981) (Cetus Corp.), VTT series (Finland) (Nevalainen *et al.*, 1980; Nevalainen and Palva, 1978), and D series (Ghosh *et al.*, 1982; Mishra *et al.*, 1982) (India). All have been derived from QM 6a, which was originally named *T. viride* but is now recognized to be a new species, *T. reesei* (Simmons, 1977).

Strains QM 9123 and 9414 were isolated in the Natick Laboratories after linear accelerator irradiation of QM 6a (Mandels *et al.*, 1971; Mandels, 1975); further treatment of 9414 with UV irradiation and Kabicidin treatment yielded strain MCG 77 (Gallo *et al.*, 1978). Strains NG 14 and C30 (Montenecourt and Eveleigh, 1979a, 1978) were produced at Rutgers after UV irradiation and nitrosoguanidine treatment of QM 9414; other mutant strains mentioned were produced in a similar fashion. All of these mutants were reported to be resistant to catabolite repression to varying degrees. Rut C30 is considered to be completely catabolite repression resistant in the presence of the repressor glycerol (Montenecourt and Eveleigh, 1979b). Good levels of cellulase are reported to be produced constitutively with the Cetus (Shoemaker *et al.*, 1981) and D-series mutants (Mishra *et al.*, 1982).

The mutants have been selected on the basis of their abilities to effect clearing of H_3PO_4-swollen cellulose incorporated into agar petri plates in the presence of the catabolite repressors glycerol or glucose (Montenecourt and Eveleigh, 1979b). Rose bengal (Montenecourt and Eveleigh, 1977a, 1979b), oxgall (Montenecourt and Eveleigh, 1977a, 1979b), Saponin (Ghosh *et al.*, 1982; Mishra *et al.*, 1982; Eriksson and Johnsrud, 1983), or Phosphon (Montenecourt and Eveleigh, 1977a, 1979b) have been used to restrict the colony size. A combination of oxgall and Phosfon D (Montenecourt and Eveleigh, 1979b) has proved most effective, in that it

Table III
Cellulase Production by Wild-Type and Mutant Strains of *Trichoderma reesei*

Strain	CMCase (units/ml)	Activity to filter paper (units/ml)	β-Glucosidase (units/ml)	Productivity (FPU/liter · hr)	Soluble protein (mg/ml)
QM 6A	88	5	0.3	15	7
QM 9414	109	10	0.6	30	14
MCG 77	104	11	0.9	33	16
C 30	150	14	0.3	42	19
NG 14	133	15	0.6	45	21
L 27	—	18	—	94	22
D 1-6	94	11	—	57	19

[a] Data from Ryu and Mandels (1980), Shoemaker *et al.* (1981), and Ghosh *et al.* (1982).

appears to retard hyphal spreading and enhances visualization of cellulase activity. Rut NG 14 was selected using this medium, but it was only partially resistant to catabolite repression and was further mutated (Montenecourt and Eveleigh, 1979b). Rut C30 was selected from the parent NG 14 using agar medium containing 1% cellobiose and 0.25% of the antimetabolite 2-deoxyglucose (Montenecourt and Eveleigh, 1979b). Screening methods for the selection of various types of mutant are discussed by Montenecourt and Eveleigh (1979b).

In general, the strain development program has concentrated on increasing the yield of cellulase. Just how successful this has been is shown in Table III, which provides a comparison of cellulase production by various mutant strains. Progress, clearly, has been quite striking, for some of the mutants isolated to date are capable of synthesizing as much as 22 mg extracellular protein/ml (Shoemaker *et al.*, 1981), much of which is cellulase. This may be the maximum yield obtainable from this organism (Montenecourt *et al.*, 1981). Yields of this order have now been produced from several sources (Shoemaker *et al.*, 1981; Ghosh *et al.*, 1982; Ryu and Mandels, 1980).

Notwithstanding these developments in the production of enzyme, the specific activity of the cellulase appears to be unchanged. These are of the order of 25 units/mg protein for carboxymethyl cellulase (endoglucanase) and 0.6 units/mg protein for filter-paper cellulase (measures cooperation of endo- and exoglucanases (Selby, 1969; Mandels, 1982). Specific activity of the cellobiase is high, but the yield is low. Indeed, cellobiase accounts for only about 0.2% of the total extracellular protein (Ryu and Mandels, 1980). Some strain improvement programs have concentrated on increasing the yield of extracellular β-glucosidase (Shoemaker *et al.*,

1981; Beja Da Costa and Van Uden, 1980; Montenecourt and Eveleigh, 1979b). Success has been limited. This has resulted in attempts to improve saccharification efficiency by the addition of supplemental β-glucosidase from *Aspergillus phoenicis* (Sternberg *et al.*, 1977; Bisset and Sternberg, 1978).

The efficiency of saccharification depends on the proportion of the various enzymes in the complex (Wood, 1975). In *T. reesei* strains, cellobiohydrolase is the largest single component. Indeed, the cellulase produced by the NG 14 mutant contains 10 times as much cellobiohydrolase as endoglucanase (Montenecourt *et al.*, 1981). This has been shown to represent about 70% of the dry weight of the filtrate when determined by quantitative precipitation with antibodies specific to the enzyme (Montenecourt *et al.*, 1980). This compares with 48% for QM 9414 and MCG 77 and 52% for Rut C30. As these figures do not include the other immunologically unrelated cellobiohydrolase (i.e., cellobiohydrolase II), which is also a major protein component of *T. reesei* cellulase (Pettersson *et al.*, 1981), it would appear that most of the extracellular protein is composed of cellobiohydrolase.

The implication of these results is that as the composition of the cellulase complex alters in favor of the cellobiohydrolase, so does the ability to accomplish the breakdown of highly ordered cellulose. That this is indeed the case has been demonstrated with *T. koningii* cellulase, when it was established that a ratio of purified cellobiohydrolase : endo-1,4-β-glucanase of at least 2 : 1 is required for maximum filter paper degrading activity, providing β-glucosidase/cellobiase is present in excess (Wood, 1983). Excess for *T. koningii* cellulase was 0.5 IU cellobiase. For maximum effect on *T. reesei* cellulase, a ratio of approximately 1.5 cellobiase units per filter paper cellulase unit of activity was required (Ryu and Mandels, 1980; Montenecourt *et al.*, 1980).

The importance of cellobiase in the cellulase–glucose conversion is clearly established. Indeed, it may be regarded as the limiting enzyme in *T. reesei* cellulase (Montenecourt *et al.*, 1981). It is limiting, however, not only because of the small amount present, but also because it is strongly inhibited by glucose. Montenecourt and colleagues (1980) have recently isolated mutants that produced cellulases containing β-glucosidases with K_i for glucose considerably higher (1.02–1.11 m*M*) than that produced by Rut NG 14 (0.56 m*M*). The screening procedures used to select these mutants with improved resistance to end-product inhibition were based on the use of gluconolactone in a cellobiose–agar plate, or 4-methylumbelliferyl β-glucoside as an overlay on pregrown colonies on cellobiose agar plates (Montenecourt *et al.*, 1981). Gluconolactone is a powerful inhibitor of β-glucosidase (see Section II,C).

Empirical methods of inducing mutations coupled with suitable screening procedures are obviously producing the desired results. Clearly, however, the incomplete knowledge of the enzymatic interactions that result in the hydrolysis of crystalline cellulose and the lack of understanding of the mechanisms controlling the synthesis and secretion of the cellulases make it difficult to select mutants on a truly scientific basis. It seems unlikely in these circumstances that the problem of enzyme synthesis at the genetic level can be tackled confidently in the immediate future. Little is known about the genetics of protein synthesis in *Trichoderma* species. There is debate even on the question of whether or not the various enzymes have the same regulatory mechanism. Gong and Tsao (1979) suggest that in *Trichoderma* species, exo- and endoglucanase synthesis is coordinately controlled. Montenecourt and colleagues (1980) on the basis of their work with mutants of *Trichoderma,* however, suggest that the two types of enzyme are regulated by different control mechanisms. All are agreed, however, that β-glucosidase synthesis is subject to independent control.

V. Enzyme Mechanisms Involved in Cellulose Degradation by Bacteria

Yields of extracellular cellulases from bacteria are generally much lower than those from the fungi. Perhaps it is for this reason that the action patterns used by the bacteria to degrade cellulose have received less attention in the past than have the fungi. With the interest now being generated worldwide in the utilization of cellulose as a chemical and energy resource, however, there has been a renewed activity in the study of bacterial cellulases.

Most enzyme preparations from cellulolytic bacteria are found to be unable to solubilize highly ordered cellulose to an extent that ensures that the crystalline component is being attacked. This is true of cellulolytic bacteria such as *Cellvibrio gilvus* (Storvick and King, 1960), *Thermomonospora* sp. (Hägerdal *et al.*, 1978, 1979), *Acetivibrio cellulyticus* (Saddler and Khan, 1980), *Bacteroides succinogenes* (Groleau and Forsberg, 1981), and *Cellvibrio fulvus* (Berg *et al.*, 1972); it is even true of *Sporocytophaga myxococcoides* (Osmundsvag and Goksøyr, 1975; Berg *et al.*, 1972), which has been described as the most cellulolytic bacterium known. Mandels and Andreotti (1978) have pointed out that degradation of cellulose by a microorganism does not necessarily mean that the cellulase liberated or isolated from the microorganism will be highly active. Clearly, this applies as much to the bacteria as it does to the fungi.

According to Hofsten (1975), the fact that some bacterial cellulases are very effective when bound to the cell may be explained by their localized high concentration and favorable alignment on the surface of the cell. He envisages that the formation of a series of enzyme–substrate complexes along the cellulose chains would cause sufficient weakening of the hydrogen bonding to allow hydrolysis to take place. The inability of cell-free enzyme preparations to attack the cellulose may be because the necessary topological conformation is not attained (Groleau and Forsberg, 1981).

Some bacteria certainly adhere very closely to the substrate. Indeed, so close is the attachment in the case of the rumen anaerobe *B. succinogenes* that the bacterium seems to modify its shape to conform with the contours of the plant cell wall (Stewart *et al.*, 1981). Electron micrographs show that there is no obvious interface between bacterial and plant cell walls, and it is, indeed, possible that enzymes mediate attachment in this and other cases (Wood, 1981, 1983).

While topological considerations are obviously important in any discussion of the mechanism of cellulase action in the bacteria, there is nevertheless a tendency to discuss the action patterns of these cellulases in terms of the interactions between different types of enzyme. Most bacterial cellulase preparations contain endo-1,4-β-glucanases but little or no exo-1,4-β-glucanase activity (Mandels, 1982), and it is assumed that by analogy with the fungal cellulases that the deficiency in the endwise-acting enzyme explains the lack of activity on highly ordered substrates. This assumption may not always be valid, however, for it has already been demonstrated that the cellulase of *C. gilvus* (Storvick, 1960) does not hydrolyze highly ordered cellulose, despite containing both a cellobiohydrolase and an endoglucanase. Another observation that might be relevant in this context is that the addition of purified cellobiohydrolases from the fungi *T. koningii* or *P. funiculosum* does not show any capacity for cooperating with the endoglucanase of *Ruminococcus albus, Ruminococcus flavefaciens,* or *B. succinogenes* in solubilizing cotton fiber (Wood, 1983).

As already stated, claims that bacteria such as *A. cellulolyticus* (Saddler and Khan, 1980), a *Thermoactinomyces* (Hägerdal *et al.*, 1978), and *Pseudomonas fluorescens* var. *cellulosa* (Suzuki *et al.*, 1969) have yielded enzymes capable of acting on highly ordered crystalline cellulose have appeared, and it is implied or stated that they must therefore contain exoglucanase activity. Most, however, show only limited hydrolysis of crystalline substrates, and this does not necessarily demonstrate that crystalline cellulose is being solubilized or that exoglucanase is present. Selective hydrolysis of the amorphous easily hydrated cellulose chains of some native celluloses can be effected by most endoglucanases acting alone (Wood and McCrae, 1978). "True" cellulase action can only be considered proven when highly ordered cellulose is hydrolyzed extensively (Mandels, 1975).

Of the bacterial cellulases, that elaborated by the thermophilic anaerobe *Clostridium thermocellum* has been studied most. Several isolates have been obtained: some are reported to grow on glucose, and some apparently do not. Because the enzymes are difficult to purify, crude extracellular enzymes have been used in several studies.

Crude extracellular protein preparations of the strain LQRI have been demonstrated to contain endoglucanase but not cellobiase or xylosidase (Ng and Zeikus, 1981).

The final products after incubation of cellulose were glucose and cellobiose, and it was suggested that an exoglucanase was present. However, no direct evidence for this type of enzyme was reported. The enzymes in the crude cellulase preparation were not sensitive to air or oxygen and were stable and active at 60°C. A high thermostability is a general characteristic of enzymes from thermophilic bacteria (Ljungdahl, 1979).

It has been reported from studies using two *C. thermocellum* strains, ATCC 27/405 (Shinmyo *et al.*, 1979) and LQRI (Ng and Zeikus, 1981), that their endo- and exoglucanases are not inhibited by glucose or cellobiose, while for a third strain, JW 20, glucose, cellobiose, and also xylose may be inhibitory (Ljungdahl *et al.*, 1981). It is difficult to understand how the degradation products of cellulose, cellobiose, and glucose would fail to exhibit some degree of inhibition on the cellulase component. A purified endoglucanase from *C. thermocellum* has been shown to be inhibited by cellobiose (Petre *et al.*, 1981).

Johnson and colleagues (1982) showed that a cell-free preparation from *C. thermocellum* that was highly active on highly ordered celluloses could be obtained provided Ca^{2+} and a thiol reducing agent such as dithiothreitol were present. This cellulase degraded crystalline cellulose extensively and at rates comparable with the cellulase from *T. reesei*. Clearly, it is now of great interest to ascertain whether or not the cooperative action of endo- and exoglucanase can account for the unique activity of this bacterial cellulase. As yet, synergistic action of the type observed in the fungal cellulases has not been demonstrated in the bacteria.

As in the fungal cellulases, the endo-1,4-β-glucanases found in bacterial cellulase systems appear to exist in multiple forms. Thus, cell-free supernatants of *Clostridium thermocellum* have been reported to contain four endoglucanases, while those from a *Cellulomonas* sp. and a species of *Thermomonospora* were found to have three. Cultures of *Pseudomonas fluorescens* var. *cellulosa* (Suzuki *et al.*, 1969; Yamane *et al.*, 1970) and an unidentified species of *Pseudomonas* isolated from sewage sludge (Hägerdal *et al.*, 1979) contained two extracellular enzymes and one cell-bound enzyme when cellulose was the carbon source. Only cell-bound enzymes, however, were detected after growth on cellobiose, and this is similar to the observation made with the cells of *B. succinogenes* (Groleau and Forsberg, 1981).

Despite the recent interest in *Clostridium thermocellum* cellulase, the enzyme system is not well characterized. Ait *et al.* (1979) have partially purified an enzyme complex from cellulose-grown cultures of *C. thermocellum,* strain NCIB 1068 (Bary, 1886). The molecular weight of the endoglucanase-type enzyme was 125,000 ± 10,000. The enzyme contained ~10% carbohydrate and consisted of at least five protein components. More recently, endoglucanases were purified to apparent homogeneity from *C. thermocellum* NCIB 10682 and LQRI by Petre *et al.* (1981) and by Ng and Zeikus (1981), respectively. The reported properties for the enzymes purified by the two groups were very different with respect to the molecular weight: the enzyme isolated by Petre *et al.* (1981) was monomeric with a molecular weight of about 56,000, while the endoglucanase isolated by Ng and Zeikus (1981) had a molecular weight of around 90,000.

Clostridium thermocellum produces a yellow substance when grown on cellulose. It was discovered by Ljungdahl *et al.* (1983) that this low molecular weight water-insoluble substance is attached to the cellulose and that cellulose complexed with the yellow substance effectively binds the endoglucanase activity produced by the bacterium. This yellow cellulose can be used as an affinity column for purification of the enzyme (Fig. 6). The endoglucanase that is bound to the yellow cellulose can be extracted by distilled water. The authors propose that the yellow substance is involved in the cellulolytic system of *C. thermocellum*.

That the cellulase system of *C. thermocellum* is different from and probably more complex than fungal systems is also supported by a finding by Lamed *et al.* (1983). These authors have isolated a high molecular weight factor that binds both the cellulases and the bacterium to the cellulose. The factor has been termed "cellusome," and the data indicate that a defined arrangement of endo- and exocellulases are organized in a factor-bound complex.

The apparent lack of extracellular β-glucosidase in *C. thermocellum* has led to speculation that cellodextrins up to cellohexaose are transported into the cells (Avgerinos and Wang, 1980). In the cell, they may be phosphorolytically cleaved to form glucose 1-phosphate and a nonphosphorylated cellodextrin with one less glucose unit. Sheth and Alexander (1967, 1968, 1969) have thus isolated from *C. thermocellum* both a cello-

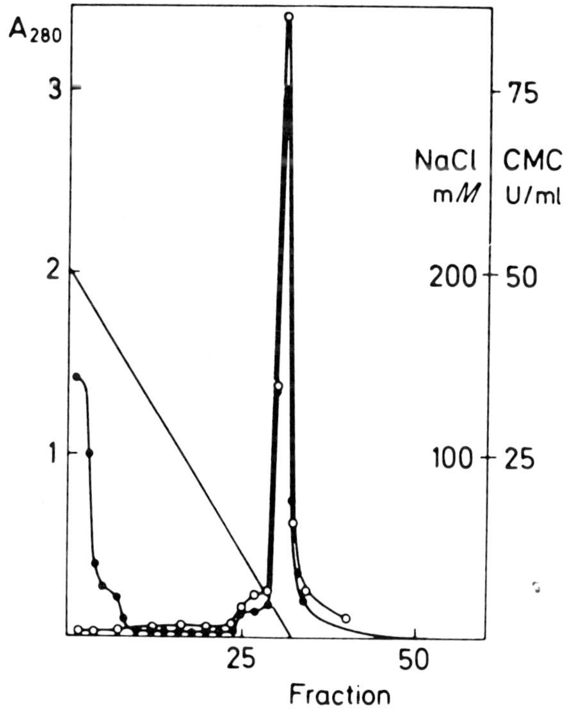

Fig. 6. Affinity chromatography of endo-1,4-β-glucanase from *Clostridium thermocellum* on cellulose with attached yellow substance. The enzyme was eluted with a gradient of 100 ml salt solution and 100 ml H_2O. ●, A_{280}; ○, CMC units; the straight line is the concentration of salt gradient. From Ljungdahl *et al.* (1983).

dextrin phosphorylase and a cellobiose phosphorylase, which are apparently two distinctly different enzymes.

Phosphorylases rather than β-glucosidase also seem to operate in *Cellvibrio gilvus* (Sheth and Alexander, 1967) and *R. flavefaciens* (Ayers, 1959) for the breakdown of cellobiose. *Cellvibrio gilvus* (Schafer and King, 1965), like *Clostridium thermocellum* (Sheth and Alexander, 1967), possesses a cellodextrin phosphorylase. No phosphorylase activity seems to be present in *B. succinogenes* (Selby and Maitland, 1972).

The aryl β-glucosidases of *B. succinogenes* (Groleau and Forsberg, 1981), *Thermomonospora* sp. (Hägerdal *et al.*, 1979), *A. cellulolyticus* (Saddler and Khan, 1980), *P. fluorescens* var. *cellulosa* (Suzuki *et al.*, 1969; Johnson *et al.*, 1980), and another species of *Pseudomonas* (Ramasamy and Verachtert, 1980) are all reported to be cell bound. In *Pseudo-*

monas (Yamane *et al.*, 1970; Ramasamy and Verachtert, 1980), they may be located in the periplasmic space or the cytoplasm in forms that are electrophoretically distinguishable. According to Ramasamy and Verachtert (1980), the relative amounts of the various forms in *Pseudomonas* sp. depend on whether cellulose or cellobiose is the energy source. The form with the highest activity towards cellobiose is located most externally in the periplasm. Although at least some of the β-glucosidase in *Pseudomonas* is inducible, β-glucosidase in *B. succinogenes* (Groleau and Forsberg, 1981) is constitutive. Groleau and Forsberg (1981) came to this conclusion when they found that similar activities were found in extracts prepared from glucose, cellobiose, and cellulose-grown cells. Unlike some β-glucosidase isolated from fungal sources, the β-glucosidase of the cellulolytic bacteria seem to be able to hydrolyze cellobiose in addition to aryl β-glucosides.

References

Ahlgren, E., and Eriksson, K.-E. (1967). *Acta Chem. Scand.* **21,** 1193.

Ait, N., Creuzet, N., and Forget, P. (1979). *J. Gen Microbiol.* **113,** 399.

Alexander, J. K. (1968). *J. Biol. Chem.* **243,** 2899.

Ander, P., and Eriksson, K.-E. (1976). *Arch. Microbiol.* **109,** 1.

Ander, P., and Eriksson, K.-E. (1977). *Physiol. Plant.* **41,** 239.

Ander, P., and Eriksson, K.-E. (1978). *Prog. Ind. Microbiol.* **14,** 1.

Avgerinos, G. C., and Wang, D. I. C. (1980). *Am. Rep. Ferm. Proc.* **4,** 165.

Ayers, W. A. (1959). *J. Biol. Chem.* **234,** 2819.

Ayers, A. R., Ayers, S. B., and Eriksson, K.-E. (1978). *Eur. J. Biochem.* **90,** 1.

Bary, A. de. (1886). *Bot. Z.* **44,** 377.

Beja Da Costa, M., and Van Uden, N. (1980). *Biotechnol. Bioeng.* **24,** 2429.

Berg, B., Hofsten, B., and Pettersson, G. J. (1972). *Appl. Bacteriol.* **35,** 201.

Berghem, L. E. R., and Pettersson, L. G. (1973). *Eur. J. Biochem.* **37,** 21.

Berghem, L. E. R., and Pettersson, L. G. (1974). *Eur. J. Biochem.* **46,** 295.

Berghem, L. E. R., Pettersson, L. G., and Fredriksson, U.-B. (1975). *Eur. J. Biochem.* **53,** 55.

Berghem, L. E. R., Pettersson, L. G., and Fredriksson, U.-B. (1976). *Eur. J. Biochem.* **61,** 621.

Bisset, F., and Sternberg, D. (1978). *Appl. Environ. Microbiol.* **35,** 750.

Blanchette, R. A., Shaw, C. G., and Cohen, A. L. (1978). *Scan. Electron Micros.* **2,** 61.

Brown, R. D., Gritzali, M., and Chirico, W. J. (1981). *Symp. Wood Pulping Chem. Stockholm* **3,** 28.

Cobb, T. L. (1982). "The Nonenzymatic Decomposition of Cellulose by the Brown-rot Fungus *Gloeophyllum trabeum.*" Thesis, Michigan Technical University.

Coudray, M.-R., Canevascini, G., and Meier, H. (1982). *Biochem. J.* **203,** 277.

Cowling, E. B. (1961). *USDA Tech. Bull.* **1258,** 1.

Crawford, R. L. (1981). "Lignin Biodegradation and Transformation." Wiley, New York.

Dekker, R. F. H. (1980). *J. Gen. Microbiol.* **120,** 309.

Deshpande, V., Eriksson, K.-E., and Pettersson, B. (1978). *Eur. J. Biochem.* **90,** 191.
Emert, G. H., Gum, E. K., Lang, J. A., Liu, T. H., and Brown, R. D. (1974). *Adv. Chem. Ser.* **136,** 79.
Eriksen, J., and Goksøyr, J. (1977). *Eur. J. Biochem.* **77,** 445.
Eriksson, K.-E. (1967). *Sven. Kem. Tidskr. (Eng)* **79,** 660.
Eriksson, K.-E. (1969). *Adv. Chem. Ser.* **95,** 59.
Eriksson, K.-E. (1978). *Biotechnol. Bioeng.* **20,** 317.
Eriksson, K.-E. (1981a). *Pure Appl. Chem.* **53,** 33.
Eriksson, K.-E. (1981b). *In* "Trends in the Biology of Fermentation for Fuels and Chemicals" (E. A. Hollaender, ed.), p. 19. Plenum, New York.
Eriksson, K.-E., and Hamp, S. G. (1978). *Eur. J. Biochem.* **90,** 183.
Eriksson, K.-E., and Johnsrud, S. C. (1982). *In* "Experimental Microbial Ecology" (R. G. Burns and J. H. Slater, eds.), pp. 134–153. Blackwell, London.
Eriksson, K.-E., and Johnsrud, S. C. (1983). *Enzyme Microb. Technol.* **5,** 425.
Eriksson, K.-E., and Pettersson, B. (1975a). *Eur. J. Biochem.* **51,** 193.
Eriksson, K.-E., and Pettersson, B. (1975b). *Eur. J. Biochem.* **51,** 213.
Eriksson, K.-E., and Pettersson, B. (1982). *Eur. J. Biochem.* **124,** 635.
Eriksson, K.-E., and Pettersson, B. *In* "Biodeterioration of Materials" (H. H. Walters and E. H. Hueck-van der Plas, eds.), Vol. 2, p. 116. Applied Sciences, London.
Eriksson, K.-E., Grünewald, A., Nilsson, T., and Vallander, L. (1980). *Holzforschung* **34,** 207.
Fägerstam, L. G., and Pettersson, L. G. (1980). *FEBS Lett.* **119,** 97.
Fennington, G., Lupo, D., and Stutzenberger, F. (1982). *Biotechnol. Bioeng.* **24,** 2487.
Gallo, B. J., Andreotti, R., Roche, C., Ryu, D., and Mandels, M. (1978). *Biotechnol. Bioeng. Symp.* **8,** 89.
Garcia-Martinez, D. V., Shinmyo, A., Madia, A., and Demain, A. L. (1980). *Eur. J. Appl. Microbiol. Biotechnol.* **9,** 187.
Gascoigne, J. A., and Gascoigne, M. M. (1960). "Biological Degradation of Cellulose." Butterworths, London.
Ghose, T. K. (1977). *Adv. Biochem. Eng.* **6,** 39.
Ghosh, V. K., Ghose, T. K., and Gopalkrishman, K. S. (1982). *Biotechnol. Bioeng.* **24,** 241.
Gilligan, W., and Reese, E. T. (1954). *Can. J. Microbiol.* **1,** 90.
Goksøyr, J., and Eriksen, J. (1980). Cellulases. *In* "Economic Microbiology: Microbial Enzymes and Bioconversions" (A. H. Rose, ed.), Vol. 5, p. 283. Academic Press, New York.
Gong, C. S., and Tsao, C. T. (1979). *Ann. Rep. Ferment. Proc.* **3,** 111.
Gong, C-S., Ladisch, M. R., and Tsao, G. T. (1979). *Adv. Chem. Ser.* **181,** 261.
Gritzali, M., and Brown, R. D. (1979). *Adv. Chem. Ser.* **181,** 237.
Groleau, D., and Forsberg, C. W. (1981). *Can. J. Microbiol.* **27,** 517.
Gum, E. K., and Brown, R. D. (1977). *Biochim. Biophys. Acta* **492,** 225.
Hägerdal, B. G. R., Ferchak, J. D., and Pye, E. K. (1978). *Appl. Environ. Microbiol.* **36,** 606.
Hägerdal, B., Ferchak, J., Pye, E. K., and Forro, J. R. (1979a). *Adv. Chem. Ser.* **181,** 331.
Hägerdal, B., Harris, H., and Pye, E. (1979b). *Biotechnol. Bioeng.* **21,** 345.
Häkansson, U., Fägerstam, L. G., Pettersson, G., and Andersson, L. (1979). *Biochem. J.* **179,** 141.
Halliwell, G. (1965). *Biochem. J.* **95,** 270.
Halliwell, G., and Griffin, M. (1973). *Biochem. J.* **135,** 587.
Halliwell, G., and Riaz, M. (1971). *Asch. Mikrobiol.* **78,** 295.

Halliwell, G., Griffin, M., and Vincent, R. (1972). *Biochem. J.* **127,** 43.
Han, Y. W., and Smith, S. K. (1978). Report ARSW-53. U.S. Dept. Agriculture, Washington, D.C.
Hash, J. H., and King, K. (1958). *J. Biol. Chem.* **232,** 395.
Highley, T. L. (1975a). *For. Prod. J.* **25,** 38.
Highley, T. L. (1975b). *Wood Fiber* **6,** 275.
Highley, T. L. (1977). *Mater. Org.* **12,** 25.
Hofsten, B. (1975). *In* "Enzymatic Hydrolysis of Cellulose" (M. Bailey, T. M. Enari, and M. Links, eds.), p. 281. SITRA, Aulanko.
Hoppe-Seyler, F. (1883). *Chem. Ber.* **16,** 122.
Humphrey, A. E., Moreira, A., Armiger, W., and Zabriskie, D. (1977). *Biotechnol. Bioeng. Symp.* **7,** 45.
Hüttermann, A., and Noelle, A. (1982). *Holzforschung* **36,** 283.
Inglin, M., Feinberg, B. A., and Loewenberg, J. R. (1980). *Biochem. J.* **185,** 515.
Jermyn, M. A. (1975). *Aust. J. Biol. Sci.* **8,** 541.
Johnson, E. A., Sakajoh, M., Halliwell, G., Madia, A., and Demain, A. L. (1982). *Appl. and Environ. Microbiol.* **43,** 1125.
Jutte, S. M., and Sachs, I. B. (1976). *IITRI/SEM* **2,** 535
Kelleher, Jr., T. J. (1981). The ligninolytic activity of *Phanerochaete chrysosporium* Burds: Regulation and application. Dissertation, Rutgers University, New Brunswick, New Jersey.
Keyser, P., Kirk, T. K., and Zeikus, J. G. (1978). *J. Bacteriol.* **135,** 790.
Kirk, T. K., Schultz, E., Connors, W. J., Lorenz, L. F., and Zeikus, J. G. (1978). *Arch. Microbiol.* **117,** 277.
Koenigs, J. W. (1972). *Phytopathology* **62,** 100.
Koenigs, J. W. (1974). *Wood Fiber* **6,** 66.
Lamed, R., Setter, E., and Kenig, R. (1983). "The Cellulosome—A Discrete Cell Surface Organelle of *Clostridium thermocellum* which Exhibits Separate Antigenic, Cellulose-binding and Various Cellulolytic Activities." Presented at the Biotechnology for Fuels and Chemicals Symposium, Gatlinburg, Tennessee.
Lee, B. H., and Blackburn, T. H. (1975). *Appl. Microbiol.* **30,** 346.
Legler, G. (1967). *Hoppe-Seyler's Z. Physiol. Chem.* **348,** 1359.
Levi, M. P., and Preston, R. D. (1965). *Holzforschung* **19,** 183.
Li, L. H., Flora, M., and King, K. W. (1965). *Arch. Biochem. Biophys.* **111,** 439.
Liese, W. (1964). *Holz Roh-Werkst.* **22,** 289.
Liese, W., and Schmid (1966). *Holz Roh-Werkst.* **24,** 454.
Ljungdahl, L. G. (1979). *Adv. Microbiol. Physiol.* **19,** 149.
Ljungdahl, L. G., Carreira, L., and Wiegel, J. (1981). *Int. Symp. Wood Pulping Chem., Stockholm* **4,** 23.
Ljungdahl, L. G., Pettersson, B., Eriksson, K.-E., and Wiegel, J. (1983). *Curr. Microbiol.* **9,** 195.
Mandels, M. (1975). *Biotechnol. Bioeng. Symp.* **5,** 81.
Mandels, M. (1982). *Annu. Rep. Ferm. Proc.* **5,** 35.
Mandels, M., and Andreotti, R. E. (1978). *Proc. Biochem.* **13,** 6.
Mandels, M., and Reese, E. T. (1964). *Dev. Ind. Microbiol.* **5,** 5.
Mandels, M., and Reese, E. T. (1965). *Ann. Rev. Phytopathol.* **3,** 85.
Mandels, M., and Weber, J. (1969). *Adv. Chem. Ser.* **95,** 391.
Mandels, M., Kostick, J., and Parizek, R. (1971). *Appl. Microbiol.* **21,** 152.
Marsh, C. (1966). *Biochim. Biophys. Acta* **122,** 367.
McHall, A., and Coughlan, M. P. (1980). *FEBS Lett.* **117,** 319.

McHall, A., and Coughlan, M. P. (1981). *Biochim. Biophys. Acta* **662,** 152.
Meier, H. (1955). *Holz Roh-Werkst.* **13,** 323.
Mishra, S., Gopalkrishnan, K. S., and Ghose, T. K. (1982). *Biotechnol. Bioeng.* **24,** 251.
Montenecourt, B. S., and Eveleigh, D. E. (1977a). *Appl. Environ. Microbiol.* **33,** 178.
Montenecourt, B. S., and Eveleigh, D. E. (1977b). *Appl. Environ. Microbiol.* **34,** 777.
Montenecourt, B., and Eveleigh, D. E. (1978). *Proc. 2nd Annual Symp. Fuels Biomass.* **2,** 613.
Montenecourt, B., and Eveleigh, D. E. (1979a). *Tappi Proc. Annu. Meeting New York* **101.**
Montenecourt, B. S., and Eveleigh, D. E. (1979b). *Adv. Chem. Ser.* **181,** 289.
Montenecourt, B. S., Kelleher, T. J., and Eveleigh, D. E. (1980). *Biotechnol. Bioeng. Symp.* **10,** 15.
Montenecourt, B. S., Cuskey, S. M., Nhlapo, S. D., Trimino-Vazquez, H., and Eveleigh, D. E. (1981). *Proc. Int. Symp. Wood Pulping Chem. Stockholm* **3,** 43.
Nakayama, M., Tomita, Y., Suzuki, H., and Nisizawa, K. (1976). *J. Biochem. (Tokyo),* 955.
Nevalainen, K. M. H., and Palva, E. T. *Appl. Environ. Microbiol.* **35,** 11.
Nevalainen, K. M. H., Palva, E. T., and Bailey, N. J. (1980). *Enzyme Microb. Technol.,* 59.
Ng, T. K., and Zeikus, J. G. (1981a). *Biochem. J.* **199,** 341.
Ng, T. K., and Zeikus, J. G. (1981b). *Appl. Environ. Microbiol.* **42,** 231.
Nilsson, T. (1973). "Studia Forestalia Suecica," No. 104. Royal College of Forestry, Stockholm.
Nilsson, T. (1974). *Mater. Org.* **9,** 173.
Nisizawa, K. (1973). *J. Ferment. Technol.* **51,** 267.
Nisizawa, K., Kanda, T., Shikata, S., and Wakabayashi, K. (1978). *J. Biochem.* (*Tokyo*) **83,** 1625.
Nisizawa, K., Tomita, Y., Kanda, T., Suzuki, H., and Wakabayashi, K. (1972). *Ferment. Technol. Today* 719.
Norkrans, B. (1967). *Adv. Appl. Microbiol.* **9,** 91.
Ogawa, K., and Toyama, N. (1982). *J. Ferment. Technol.* **60,** 349.
Ogawa, K., and Toyama, N. (1972). *J. Ferment. Technol.* **50,** 236.
Ogawa, K., Toyama, H., and Toyama, N. (1982). *Bull. Fac. Agric. Miyazaki Univ.* **29,** 191.
Omelianski, W. (1902). *Cent. Bakt. Parasitenk. Abt. 2* **8,** 193.
Osmundsvag, K., and Goksøyr (1975). *J. Eur. J. Biochem.* **57,** 405.
Petre, J., Longin, R., and Millet, J. (1981a). *Biochimie* **63,** 629.
Pettersson, G., Fägerstam, L., Bhikhabhai, R., and Leandor, K. (1981). *Proc. Int. Symp. Wood Pulping Chem. Stockholm* **3,** 39.
Pollock, M. R. (1980). *J. Gen. Microbiol.* **117,** 181.
Reese, E. T. (1963). "Advances in Enzymic Hydrolysis of Cellulose and Related Materials." Pergamon, New York.
Reese, E. T. (1976). *In* "Biotechnol. and Bioeng. Symp.," No. 6. (E. L. Gaden, M. H. Mandels, E. T. Reese, and L. A. Spano, eds.), p. 9. Interscience, New York.
Reese, E. T., and Parrish, F. W. (1971). *Carbohyd. Res.* **18,** 381.
Reese, E. T., Gilligan, W., and Norksans, B. (1952). *Physiol. Plant.* **5,** 345.
Reese, E. T., Maguire, A. H., and Parrish, F. W. (1968). *Can. J. Biochem.* **46,** 25.
Reese, E. T., Siu, R. G. H., and Levinson, H. S. (1950). *J. Bacteriol.* **59,** 485.
Rypacek, V. (1975). *Drevarsky Vyskum* **20,** 1.
Ryu, D. E., and Mandels, M. (1980). *Enzyme Microb. Technol.* **2,** 91.
Sadana, J. C., Shewale, J. G., and Deshpande, M. V. *Appl. Environ. Microbiol.* **38,** 730.
Saddler, J. N., and Khan, A. W. (1980). *Can. J. Microbiol.* **26,** 760.
Schafer, M. L., and King, K. W. (1965). *J. Bacteriol.* **89,** 113.
Seilliere, G. (1906). *Compt. Rend. Soc. Biol.* **61,** 205.

Selby, K. (1968). *Proc. Int. Biodeterioration Symp. 1st Elsevier, Barking,* p. 62. Applied Science Publishers, London.
Selby, K. (1969). *Adv. Chem. Ser.* **95,** 34.
Selby, K., and Maitland, C. C. (1972). *Biochem. J.* **128,** 1183.
Sheth, K., and Alexander, J. K. (1967). *Biochim. Biophys. Acta* **148,** 808.
Sheth, K., and Alexander, J. K. (1969). *J. Biol. Chem.* **244,** 457.
Shinmyo, A., Garcia-Martinez, D. V., and Demain, A. L. (1979). *J. Appl. Biochem.* **1,** 202.
Shoemaker, S. P., and Brown, R. D. (1978). *Biochim. Biophys. Acta* **523,** 133.
Shoemaker, S. P., Raymond, J. C., and Brewer, R. (1981). *In* "Trends in the Biology of Fermentations for Fuels and Chemicals" (A. Hollaender, ed.), p. 89. Plenum, New York.
Simmons, E. G. (1977). *Abst. 2nd Int. Mycol. Cong.,* 618.
Siu, R. G. H. (1951). "Microbial Demomposition of Cellulose." Reinhold, New York.
Sprey, B., and Lambert, C. (1983). Paper given at Workshop on Biochemistry of Cellulose Degrading Enzymes, KFA Jülich, West Germany.
Sternberg, D., and Mandels, G. R. (1979). *J. Bacteriol.* **139,** 761.
Sternberg, D., Vijayakumur, P., and Reese, E. T. (1977). *Can. J. Microbiol.* **23,** 139.
Stewart, C. S., Paniagna, M., Dinsdale, D., Cheng, K.-J., and Garrow, S. H. (1981). *Appl. Environ. Microbiol.* **41,** 504.
Storvick, W. O., and King, K. W. (1960). *J. Biol. Chem.* **235,** 303.
Streamer, M., Eriksson, K.-E., and Pettersson, B. (1975). *Eur. J. Biochem.* **59,** 607.
Suzuki, H., Yamane, K., and Nisizawa, K. (1969). *In* "Cellulases and Their Applications" (Adv. in Chem. Ser. 95), (G. J. Hajuy and E. T. Reese, eds.). American Chemical Society, Washington, D.C.
Varadi, J. (1972). *Biodeterior. Mater.* **2,** 129.
Vaheri, M. P. (1982a). *J. Appl. Biochem.* **4,** 153.
Vaheri, M. P. (1982b). *J. Appl. Biochem.* **4,** 356.
Westermark, U., and Eriksson, K.-E. (1974a). *Acta Chem. Scand. Ser. B* **28,** 209.
Westermark, U., and Eriksson, K.-E. (1974b). *Acta Chem. Scand. Ser. B* **28,** 204.
Whitaker, D. R. (1954). *Arch. Biochem. Biophys.* **53,** 439.
Wood, T. M. (1968). *Biochem. J.* **109,** 217.
Wood, T. M. (1969). *Biochem. J.* **115,** 457.
Wood, T. M. (1970). *In* "World Review of Nutrition and Dietetics" (G. H. Bourne, ed.), p. 227. Karger, Basel.
Wood, T. M. (1971). *Biochem. J.* **121,** 353.
Wood, T. M. (1972). *Proc. IV IFS Ferment. Technol. Today,* 711.
Wood, T. M. (1975). *Biotechnol. Bioeng. Symp.* **5,** 111.
Wood, T. M. (1980). "Colloque Cellulolyse Microbienne," p. 167. CNRS, Marseille.
Wood, T. M. (1981). *Proc. Seminar, Degradation of Plant Cell-wall Material,* p. 24. A. R. C., London.
Wood, T. M. (1982). *In Proc. of the Technical Association of the Pulp and Paper Industry Research Division Conference,* Asheville, North Carolina.
Wood, T. M. (1983). *Proc. Seminar 'La Biomasse, Source D'Industriels,' A.D.E.P.R.I.N.A.,* Paris.
Wood, T. M., and McCrae, S. I. (1972). *Biochem. J.* **128,** 1183.
Wood, T. M., and McCrae, S. I. (1975). *In* "Symposium on Enzymatic Hydrolysis of Cellulose." (M. Barley, T. M. Enari, and M. Linko, eds.), p. 231. Aulanko, Finland.
Wood, T. M., and McCrae, S. I. (1977a). *Proc. of Bioconversion Symp. IIT,* Delhi, p. 111.
Wood, T. M., and McCrae, S. I. (1977b). *Carbohyd. Res.* **57,** 117.
Wood, T. M., and McCrae, S. I. (1978). *Biochem. J.* **171,** 61.

Wood, T. M., and McCrae, S. I. (1979a). *Adv. Chem. Ser.* **181,** 181.
Wood, T. M., and McCrae, S. I. (1979b). *In* "Developments in Food Science" (H. Chiba, M. Fujmaki, K. Iwai, H. Mitsuda, and Y. Morita, eds.), Vol. 2, p. 257. Elsevier, Amsterdam.
Wood, T. M., and McCrae, S. I. (1982a). *Carbohyd. Res.* **110,** 291.
Wood, T. M., and McCrae, S. I. (1982b). *J. Gen. Microbiol.* **128,** 2973.
Wood, T. M., and Phillips, D. R. (1969). *Nature (London)* **222,** 986.
Wood, T. M., McCrae, S. I., and McFarlane, C. M. (1980). *Biochem. J.* **189,** 51.
Yamane, K., Suzuki, H., and Nisizawa, K. (1970). *J. Biochem.* **67,** 19.

CHAPTER **18**

Biodegradation of the Hemicelluloses

ROBERT F. H. DEKKER

CSIRO
Division of Chemical and Wood Technology
Clayton, Victoria, Australia

I. Introduction

The collective term "hemicellulose" is ambiguously defined as "those polysaccharides soluble in alkali that are associated with cellulose of the plant cell wall." Hemicelluloses defined in this way would include the noncellulosic β-D-glucans, the pectic substances (polygalacturonans), and several complex heteropolysaccharides such as those enriched in galactose (e.g., arabinogalactans), mannose (galactogluco- and glucomannans), and xylose (arabinoglucurono- and glucuronoxylans). It is this last group of heteropolysaccharides that this chapter will deal with. This group of polysaccharides is normally classified as those constituting the "hemicelluloses," and this terminology will be retained throughout this chapter. The hemicelluloses as such represent a distinct group of polysaccharides and are classified according to their chemical composition and structure (Aspinall, 1970). Three predominant types are recognized within wood species. They include the 1,3- and 1,4-β-D-galactans, 1,4-β-D-mannans, and 1,4-β-D-xylans, and usually occur as heteroglycans containing differ-

ISBN 0-12-347880-4

ent types of sugar residues, often as short mono- or oligosaccharide appendages linked to the main glycan backbone chain. Their structures have been discussed in detail by Aspinall (1959, 1970), Timell (1964, 1965, 1967), and Wilkie (1979).

The hemicelluloses rank next to cellulose in abundance as naturally occurring compounds, and some hardwood plant species have been reported to contain up to 35% hemicellulose (Timell, 1967). The hemicelluloses are present in both primary and secondary layers of the plant cell wall, where they are closely associated with lignin and cellulose. Certain hemicelluloses (e.g., arabinogalactan, galactan, and galactoglucomannan) are also found in compression wood, in addition to being present in normal wood (Timell, 1982).

The structures of some commonly occurring hemicelluloses found in woods are shown in Fig. 1.

II. Enzymes Degrading the Hemicelluloses

Enzymes attacking the hemicelluloses are hydrolytic by nature and are referred to as hemicellulolytic enzymes or hemicellulases (Dekker, 1979; Dekker and Lindner, 1979; Dekker and Richards, 1976). Hemicellulases (glycan hydrolases, EC 3.2.1) specifically degrade those glycans that make up the backbone chain of the hemicelluloses. Typical hemicellulases are therefore β-D-galactanases, β-D-mannanases, and β-D-xylanases. Excluded from this group of enzymes are the specific exoglycosidases such as α- and β-D-galactosidases, β-D-mannosidases, α-L-arabinosidases, and β-D-xylosidases. These enzymes are capable of hydrolyzing not only low molecular weight (MW) glycosides but also the short-chain or monosaccharide appendages from the main hemicellulosic backbone chain. The action of the glycosidases, however, is necessary to achieve total hydrolysis of the hemicelluloses into monosaccharides, because these enzymes act in synergism with the hemicellulases. This aspect will be discussed with biodegradative pathways in Section V.

Hemicellulases, like most other polysaccharide-degrading enzymes (glycan hydrolases), attack their substrates in two ways, exo and endo hydrolytic attack. An exo enzyme degrades the polysaccharide by successive removal of terminal glycose or oligosaccharide units and proceeds in a stepwise manner, usually from the nonreducing end of the polysaccharide chain. An inverted configuration of the glycose residue released usually denotes exo enzyme action. Examples of exoglycanases that attack hemicellulose are two relatively recently discovered exo-β-D-mannanases (McCleary, 1982; and Araki and Kitamikado, 1982).

(a) *O*-Acetyl-4-*O*-methylglucuronoxylan (hardwood)

→4)-β-D-Xyl*p*-(1→4)-β-D-Xyl*p*-(1→4)-β-D-Xyl*p*-(1→4)-β-D-Xyl*p*-(1→
3 2
↑ ↑
Acetyl 1
4-*O*-Me-α-D-Glc*p*A

(b) Glucomannan (hardwood)

→4)-β-D-Glc*p*-(1→4)-β-D-Man*p*(1→4)-β-D-Man*p*-(1→4)-β-D-Man*p*(1→

(c) Arabino-4-*O*-methylglucuronoxylan (softwood)

→4)-β-D-Xyl*p*(1→4)-β-D-Xyl*p*-(1→4)-β-D-Xyl*p*-(1→4)-β-D-Xyl*p*-(1→
2 3
↑ ↑
1 1
4-*O*-Me-α-D-α-D-Glc*p*A α-L-Ara*f*

(d) *O*-Acetylgalactoglucomannan (softwood)

→4)-β-D-Glc*p*-(1→4)-β-D-Man*p*-(1→4)-β-D-Man*p*-(1→4)-β-D-Man*p*-(1→
6 2(3)
↑ ↑
1 Acetyl
α-D-Gal*p*

(e) Arabinogalactan (larch)

→3)-β-D-Gal*p*-(1→3)-β-D-Gal*p*-(1→3)-β-D-Gal*p*-(1→3)-β-D-Gal*p*-(1→3)-β-D-Gal*p*-(1→
6 6 6 6 6
↑ ↑ ↑ ↑ ↑
1 1 1 1 1
β-D-Gal*p* β-D-Gal*p* β-D-Gal*p* R L-Ara*f*
6 6 6 3
↑ ↑ ↑ ↑
1 1 1 1
β-D-Gal*p* β-D-Gal*p* β-D Gal*p* β-L-Ara*f*

(f) Galactan (compression wood)

→4)-β-D-Gal*p*-(1→4)-β-D-Gal*p*-(1→4)-β-D-Gal*p*-(1→4)-β-D-Gal*p*-(1→
6
↑
1
β-D-Gal*p*A

Fig. 1. Structures of hemicelluloses commonly found in wood species.

Endo enzymes attack polysaccharides in a random manner, causing multiple scission that is accompanied by a marked decrease in the degree of polymerization (DP) of the substrate. The polymer is progressively degraded into shorter fragments until nondegradable products (usually mono- and disaccharides) are formed. Hemicellulases of the endo type constitute the most common group of the hemicellulases.

III. Occurrence of Hemicellulases

The occurrence of the hemicellulases and especially the xylanases is widespread. They are reported (Dekker and Richards, 1976) to occur in bacteria from marine and terrestrial environments, fungi (the phytopathogens and the rot fungi being the most active producers of hemicellulases), rumen microorganisms (bacteria and protozoa), cecal bacteria of ruminants and nonruminants, yeasts, xylophagous insects, certain molluscs (e.g., snails), crustaceans, marine algae, and in germinating seeds of land plants. Hemicellulases have not been detected in the mucosal cells lining the alimentary canal of vertebrates, including humans. These animals rely upon the hemicellulases of microbial origin in their digestive tracts, e.g., bacteria and protozoa in the rumen (Dekker, 1976), bacteria in the cecum (Bailey and MacRae, 1970), and bacteria in the human colon (Salyers *et al.*, 1977).

Most of the bacteria, fungi, and yeasts producing hemicellulases secrete the enzyme extracellularly (Dekker and Richards, 1976), although there are some reports of intracellular localization [e.g., a xylanase isolated from the fungus *Tyromyces palustris* (Ishihara *et al.*, 1975)]. Cell-bound hemicellulases have also been described (Dekker and Candy, 1979; Nisizawa *et al.*, 1971; Notario *et al.*, 1979; Pettipher and Latham, 1979).

Hemicellulases of microbial origin have been reported (Dekker and Richards, 1976) to be produced both constitutively (produced irrespective of growth substrate) and inductively (produced when grown on hemicellulose or similar inducers). Cellulases, by analogy, are inductive enzymes and are produced only when the microorganism is grown on cellulose or cellobiose (Ryu and Mandels, 1980).

IV. Mode of Action of the Hemicellulases

Studies concerned with the mode of action of the hemicellulases are meaningful only if the enzyme preparations used are homogeneous—i.e., they consist of a single protein component. This criterion is used throughout this section, in which will be reviewed some of the more recent developments in the field of hemicellulose-degrading enzymes. For earlier reviews on this subject, the reader is referred to Dekker (1979) and Dekker and Richards (1976).

A. *1,4-β-D-Xylanases*

1,4-β-D-Xylanases (EC 3.2.1.8, 1,4-β-D-xylan xylanohydrolases) of the endo enzyme type are the only xylanases that have been unequivocally characterized (Dekker, 1979). This does not, however, preclude the possible existence of exoxylanases [1,4-β-D-xylan xylo(bio)hydrolases]. The xylanases hydrolyze the 1,4-β-D-xylopyranosyl linkages of D-xylans such as L-arabino-D-xylans, L-arabino-D-glucurono-D-xylans, and D-glucurono-D-xylans. Certain highly purified endoxylanases have also been reported (Dekker and Richards, 1976) that are capable of hydrolyzing cellulose in addition to xylans. Furthermore, cellulases have also been described (Shikata and Nisizawa, 1975; Toda *et al.*, 1971) that can attack xylan. A case in point is a novel exo-1,4-β-D-glucanase (a carboxymethylcellulase) produced by *Trichoderma viride* (Shikata and Nisizawa, 1975). This highly purified enzyme preparation degraded carboxymethylcellulose (CMC) to cellobiose and glucose and degraded D-xylan to xylobiose (Xyl_2), without the concomittant production of D-xylose. Although this enzyme possessed exoxylanase activity (a 1,4-β-D-xylan xylobiohydrolase), it showed a higher affinity for cellulose than for xylan and is better classified as an exocellulase with multisubstrate specificity.

1. Bacterial Xylanases

There are relatively few studies dealing with xylanases of bacterial origin. This is because eukaryotic microorganisms such as fungi are rather better producers of xylanases. Most more recent studies on bacterial xylanases are confined to the genera *Bacillus* and *Streptomyces*.

a. Bacillus *Xylanases.* Species of *Bacillus* from which xylanases have been isolated and characterized include an alkalophile (Horikoshi and Atsukawa, 1973) and a thermophilic acidophile (Uchino and Nakane, 1981). Other species studied include *B. circulans* (Esteban *et al.*, 1982), *B. pumilus* (Panbangred *et al.*, 1983), and *B. subtilis* (Takahashi and Hashimoto, 1963).

A xylanase preparation [isoelectric point (pI) 6.3] from the alkalophilic *Bacillus* C-59-2 degraded rice straw arabinoxylan to xylobiose and xylotriose (Xyl_3) as major end products, with smaller amounts of higher xylooligosaccharides, some of which contained L-arabinose residues (Horikoshi and Atsukawa, 1973). D-Xylose was not produced in short digestion times (1 hr). The acidophilic *Bacillus* 11-1S (Uchino and Nakane, 1981) also produced a single xylanase (MW 56,000), which degraded xylans of larchwood and rice straw in an endo manner, liberating Xyl_2, Xyl_3, and xylose as chief end products of hydrolysis. An arabinose-containing xyloside ($AraXyl_3$) was also produced in these hydrolyzates. The purified xylanase preparation showed a higher affinity for CMC than for xylan and also attacked CMC at a faster rate than xylan. On this basis, this enzyme is better classified as a cellulase. *Bacillus circulans* WL-12 (Esteban *et al.*, 1982) produced two xylanases that attacked xylan differently, as judged by the array of products appearing in the hydrolyzate. Xylanase (MW 85,000, pI 4.5) degraded xylan in short durations to mainly xylotetraose (Xyl_4), and, upon prolonged hydrolysis times, Xyl_2, Xyl_3, and xylose appeared in the hydrolysate. The other xylanase (xylanase B, MW 15,000, pI 9.1) produced by this bacillus did not liberate xylose from xylan. The chief hydrolysis products were identified as Xyl_4, Xyl_3, and Xyl_2. The minimum substrate binding requirement for xylanase B appeared to be filled by a chain of at least four 1,4-linked-β-D-xylopyranoside units with bond cleavage occurring at the second β-D-xylosidic linkage. Thus, Xyl_4 yielded two molecules of Xyl_2.

b. Streptomyces *Xylanases.* A highly purified xylanase preparation (MW 40,500, pI 7.3) from a *Streptomyces* isolate (Kusakabe *et al.*, 1977a) degraded several different xylans to mainly Xyl_3 and D-xylose. Enzymatic hydrolyzates of corn hull and cob arabinoxylans also yielded several major oligosaccharides of mixed constitution containing both D-xylose and L-arabinose (Kusakabe *et al.*, 1977b). The hydrolysis of arabinoxylan was strongly influenced by the degree of substitution (DS) by arabinose. When the DS was high (i.e., Xyl/Ara of 2.6) the extent of hydrolysis was low, and increased (up to 65%) as the DS decreased (e.g., Xyl/Ara of 17.5). An unsubstituted xylan (cotton seed) was degraded to the same extent as the xylans of low DS (i.e., 65–68% hydrolysis). The *Streptomyces* xylanase preferentially attacked the xylan chain where the degree of branching by arabinose was less. *Streptomyces* xylanase attack on xylan yielded mainly Xyl_3 in the early stages of hydrolysis, with lesser amounts of Xyl_4 and Xyl_2 also being present. As hydrolysis proceeded, xylose oligosaccharides of DP 3–5 were further degraded into Xyl_2 and some xylose. Long-term hydrolysis (120 hr) showed that Xyl_2 was also degraded, with xylose being the major end product of hydrolysis of the xylan. Attack on xylose oligosaccharides of DP 3–6 showed that these substrates were likewise degraded to Xyl_2. Xyl_2 was slowly hydrolyzed to xylose.

Another *Streptomyces, S. exfoliatus,* has been found to produce five isoenzymes, three of which were purified to homogeneity (Sreenath and

Joseph, 1982). Two of the xylanases liberated an array of xylose oligosaccharides from larchwood xylan, which was typical of endo-acting enzymes. In both cases, no xylose could be detected in the enzymic hydrolysates. A third xylanase produced by this microorganism (xylanase Ib) degraded xylan to xylose and some xylose oligosaccharides. This enzyme was demonstrated to possess "exo-like" enzyme activity. The authors claimed that the higher oligosaccharides may have been due to contamination of the enzyme with some endoxylanase activity (polyacrylamide gel electrophoresis at pH 8.6, nevertheless, revealed only one protein band), or that the enzyme (if exo) possessed xylosyltransferase activity.

2. Yeast Xylanases

Yeasts have only recently been recognized to produce xylanases, and these enzymes appear to be confined to the genera *Aureobasidium, Cryptococcus,* and *Trichosporon* (Biely *et al.,* 1978).

a. Trichosporon *Xylanases.* *Trichosporon cutaneum* when grown on xylan produced a single xylanase component (extracellular, MW 45,000), which degraded several xylans but not cellulose or Xyl_2 (Stuttgen and Sahm, 1982). Enzymatic degradation of oat husk arabinoxylan yielded an array of xylose oligosaccharides, some of mixed constitution containing L-arabinose after short hydrolysis times (15 min). The major end products after prolonged hydrolysis were xylose, Xyl_2, and Xyl_3. Arabinose was not released during hydrolysis.

b. Cryptococcus *Xylanases.* *Cryptococcus albidus* also produced one type of extracellular xylanase (MW 26,000), an endo type (Biely *et al.,* 1980). A cell wall-associated xylanase has also been demonstrated in another *C. albidus* strain (Notario *et al.,* 1979). The purified extracellular enzyme exhibited a low affinity for *p*-nitrophenyl-β-D-xyloside, which was ascribed to the ability of the endoxylanase to hydrolyze the aryl-β-D-xyloside. Xyl_2 was also attacked, but at a very slow rate (Biely *et al.,* 1981b), and cleavage was detected only when radioactively labeled [1-^{3}H]Xyl_2 was used. Bond-cleavage frequencies of xylose oligosaccharides of DP 3–5 by *C. albidus* xylanase were dependent on the substrate concentration. The substrate binding site of *C. albidus* xylanase was demonstrated to be composed of four subsites, two of which (subsites −II and II) showed a strong affinity for 1,4-β-D-xylopyranosyl residues (Biely *et al.,* 1981a) (Fig. 2). The catalytic site of the *C. albidus* xylanase substrate binding site is located between subsites −I and I. The negative affinities of subsites −I and I explain why Xyl_2 was a very poor substrate for the enzyme. Subsites to the left of subsite −II, and to the right of subsite II

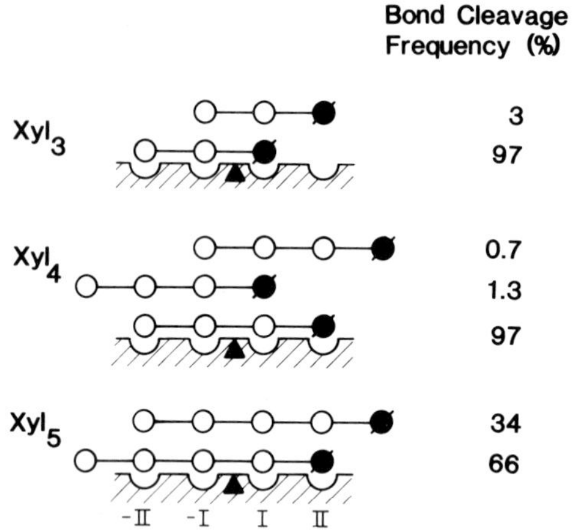

Fig. 2. Mechanisms for the binding and hydrolysis of xylose oligosaccharides by endo-1,4-β-D-xylanase from *Cryptococcus albidus* (Biely *et al.*, 1981a,b). Symbols: ○, xylosyl residue; ●, reducing-end xylosyl residue; □, 1,3-linked α-L-arabinosyl unit; ——, 1,4-β-D-xylosidic bond; ▲, catalytic cleavage site; and −I, I, etc., D-xylosyl binding subsites.

showed almost no affinity for binding xylose oligosaccharides of DP 3–8. Bond-cleavage frequencies (determined at low substrate concentration with conditions favoring unimolecular hydrolysis) of xylose oligosaccharides of DP 3 to >8 were shown, with the exception of Xyl_3, to favor specific cleavage of the second β-D-xylosidic bond from the reducing end of this oligosaccharide. Xylotriose was hydrolyzed 100 times more slowly than Xyl_4, while xylose oligosaccharides of DP 4–7 did not show appreciable differences in the rate of hydrolysis. Xylooctaose (Xyl_8) and larger oligosaccharides were hydrolyzed at a slower rate than Xyl_4–Xyl_7. Xyl_4 was the smallest substrate that could form an enzyme–substrate complex amenable to rapid hydrolysis.

Cryptococcus albidus xylanase formed two productive complexes with Xyl_3, one of which predominated (Biely *et al.*, 1981b). At low concentrations (unimolecular hydrolysis conditions), Xyl_3 was cleaved almost exclusively at the first β-D-xylosidic linkage, with xylose being released as the chief product of hydrolysis. Bond-cleavage frequencies showed that 97% of Xyl_3 was cleaved when bound to subsites −II, −I, and I of the xylanase (Fig. 2). As the substrate concentration increased, the mechanism of hydrolysis of Xyl_3 by *C. albidus* xylanase changed to a bimolecular one.

At high substrate concentrations (>5 mM Xyl_3), a termolecular 2,1-shifted enzyme–Xyl_3 complex was formed (i.e., a second Xyl_3 molecule "pushes" the first Xyl_3 into a position at the enzyme binding site so as to fill the binding site and promote scission of the second xylosidic bond), yielding Xyl_2 as the major hydrolysis product. Evidence from studies using Xyl_3 labeled with both tritium (1-^3H) and carbon-14 (uniformly) also indicated that xylosyl transfer occurred during Xyl_3 degradation at high substrate concentrations (presence of Xyl_4 among the reaction products). Xylotetraose on binding formed mostly one type of productive complex with the enzyme, occupying all subsites of the binding site (Fig. 2). Ninety-seven percent of bond-cleavages yielded Xyl_2 as end product, which was consistent with a unimolecular mechanism. Xylopentaose was hydrolyzed to Xyl_2 and Xyl_3 in the molar ratio of 2 : 1 via two asymmetric, productive, enzyme–substrate complexes. Bond-cleavage frequencies of Xyl_5 indicated that 66% of bond scission occurred at the second β-D-xylosidic bond, while 34% occurred at the third β-D-xylosidic linkage (Biely *et al.*, 1981a). The possible productive enzyme-substrate complexes for *C. albidus* xylanase are summarized in Fig. 2.

Cryptococcus albidus xylanase also manifested transfer reactions from hydrolysis products during degradation (Biely *et al.*, 1981b). Both xylose and Xyl_2 could be utilized as acceptors by the xylanase for glycosyl transfer during the hydrolysis of higher oligosaccharides of DP 3–5. Xylosyl transfer was demonstrated using [1-^3H]xylose and Xyl_2, with Xyl_2 being a better glycosyl acceptor than xylose.

Finally, the mode of action of *C. albidus* xylanase was found to be very similar to that of some other polysaccharidases, such as lyzozyme and α-amylase (Biely *et al.*, 1981b; Paice and Jurasek, 1979).

3. Fungal Xylanases

Xylanases of fungal origin are well characterized and have been described in much detail (Dekker and Richards, 1976; Reilly, 1981). Among the endoxylanases produced by fungi, two types are at present recognizable: debranching xylanases (L-arabinose liberated from arabinoxylans and arabinoglucuronoxylans) and nondebranching xylanases (L-arabinose not released during hydrolysis). Both types are also capable of attacking glucuronoxylans and unsubstituted 1,4-β-D-xylans. The release of 4-*O*-methyl-D-glucuronic acid from glucuronoxylans has also been reported (Ishihara *et al.*, 1978). Whether this release is the action of a debranching activity or a contaminant exo-1,2-α-D-glucuronidase is not known, as the enzyme preparation used also showed activity toward other substrates.

a. Debranching Xylanases. Debranching endoxylanases of the L-arabinose-liberating type from several different fungal sources are listed in Table I. These xylanases are capable of cleaving the 1,3-α-L-arabino-

Table I
Debranching Endo-1,4-β-D-Xylanases of the L-Arabinose-Liberating Type Produced by Fungi

Fungal origin	Enzyme	Molecular weight	Isoelectric point	Reference
Agaricus bisporus	Xylanase A	NS[a]	NS	Hashimoto and Takahashi (1970)
Aspergillus niger	Xylanase I	NS	NS	Iwamoto *et al.* (1973)
A. niger str. 14	Xylanase	24,700	4.2	Gorbacheva and Rodinova (1977)
A. niger van tieghem	Xylanase II	NS	NS	Takeniski and Tsujisaka (1975)
	Xylanase III	NS	NS	
A. niger str. 11	Xylanases I (A and B)	50,000	NS	John *et al.* (1979)
	Xylanase II (A–D)	31,000	NS	
Cephalosporium sacchari	Xylanase I	10,700	9.4	Dekker *et al.* (1975)
Ceratocystis paradoxa	Xylanase I	NS	9.17	Dekker and Richards (1975a)
Oxiporus sp.	Xylanase A_2	26,900	8.6	Sinner and Dietrichs (1975, 1976)
Talaromyces byssochlamydoides	Xylanase X-a	76,000	4.3	Yoshioka *et al.* (1981)
Trichoderma reesei QM 9414	Xylanase 1	14,400	6.6	G. P. Candy and R. F. H. Dekker (unpublished)
	Xylanase 2	40,700	8.5	
	Xylanase 3	8,500	10.3	
	Xylanase 4	11,700	8.6	
Trichoderma viride	Xylanase	17,800	9.2	Sinner and Dietrichs (1975, 1976)

[a] Not specified.

furanosyl residues from L-arabinofuranosyl-containing 1,4-β-D-xylans, in addition to their ability to hydrolyze the 1,4-β-D-xylopyranosyl linkages of the xylan backbone chain. Whether both types of linkages (1,4-β-D-xylopyranosyl and 1,3-α-L-arabinofuranosyl) are hydrolyzed at the same active site is not known, although evidence arising from studies on *Ceratocystis paradoxa* xylanase I (pI 9.17) indicated the involvement of two separate catalytic sites (Dekker and Richards, 1975a). In these studies, Xyl_3 (not degraded by xylanase I) suppressed "arabinosidase" activity (i.e., the liberation of L-arabinose) without affecting the hydrolysis of the 1,4-β-D-xylosidic linkages. The possible involvement of two catalytic sites in this enzyme and by other fungal enzymes in the degradation of arabinoxylans is feasible, since the two sugar residues involved are structurally dissimilar, and sterically are not expected to be accommodated by the same binding site. The D xylose residues of hemicellulosic xylans exist in the pyranoid form, while L-arabinose is generally present in the furanoid form (Aspinall, 1970).

A strain of *Aspergillus niger* produced six endoxylanases, which were all of the L-arabinose-debranching type (John *et al.*, 1979). With the exception of two of the six xylanases (xylanase IIB and IIC), each enzyme degraded larchwood arabinoglucuronoxylan differently, yielding a different array of end products. Major end products from prolonged hydrolysis (48 hr) of xylan by xylanase IA (MW 50,000) included Xyl_2, Xyl_3, xylose, arabinose, and $AraXyl_3$, whereas xylanase IB (MW 50,000) yielded Xyl_2, Xyl_3, and $AraXyl_3$ as chief products. Both enzymes also attacked microcrystalline cellulose. Under similar conditions, xylanase IIA (MW 31,000) liberated xylose, Xyl_2, and arabinose, but no xylose oligosaccharides of mixed constitution. This enzyme removed L-arabinose directly from the xylan backbone chain. Xylanases IIB and IIC (MW 31,000) yielded mainly Xyl_2, xylose, and $AraXyl_4$. Arabinose was slowly released from $AraXyl_4$ and $AraXyl_5$, and Xyl_3 was degraded to Xyl_2 and xylose. Finally, xylanase IID (MW 31,000) degraded xylan to xylose and arabinose with small amounts of Xyl_2 and Xyl_3 also being present, which were eventually hydrolyzed to xylose. This enzyme preparation degraded xylan extensively to mainly monosaccharides. An array of xylose oligosaccharides of DP 2–7 were produced during short hydrolysis times, indicating that the mode of action of this enzyme was endo rather than exo.

Another species of *A. niger* (*A. niger* van tieghem) was also reported to produce several endoxylanases, two of which were of the L-arabinose-debranching type (Takenishi and Tsujisaka, 1975). One of these xylanases (II) degraded rice straw arabinoxylan to mainly xylose, arabinose, and Xyl_2. Xylose oligosaccharides containing L-arabinose were not produced, indicating that arabinose was cleaved from the xylan backbone. Xylanase II degraded xylose oligosaccharides of DP 3–5 to mainly Xyl_2 and xylose ($Xyl_2 \simeq$ xylose). Xyl_2 was not attacked. Attack on the arabinose–xylose oligosaccharides $AraXyl_2$ and $AraXyl_3$ yielded L-arabinose and Xyl_2, and L-arabinose, xylose, Xyl_2, and Xyl_3, respectively. In the case of $AraXyl_3$, L-arabinose was first removed, yielding Xyl_3, which was subsequently hydrolyzed to Xyl_2 and xylose. At high substrate concentrations, xylanase II also exhibited transglycosylation activity. Xylanase III from *A. niger* van tieghem degraded rice straw arabinoxylan to xylose, L-arabinose, and a mixture of xylose oligosaccharides of DP > 2, some of which contained L-arabinose. $AraXyl_2$ was not released. Attack on xylose oligosaccharides of DP 3–5 yielded mainly xylose and Xyl_2, while $AraXyl_2$ and $AraXyl_3$ were not attacked (Takenishi and Tsujisaka, 1975). Transglycosylation also occurred at high substrate concentrations.

Another strain of *A. niger* (str. 14) produced a xylanase (MW 25,000, pI 4.2) that exhibited a different mode of attack than the aforementioned *A. niger* xylanases (Gorbacheva and Rodionova, 1977). Arabinoglucuronoxylans were degraded to L-arabinose, xylose, and a mixture of xylose oligosaccharides of mixed constitution. The extent of hydrolysis was strongly influenced by the degree of branching of both L-arabinose and D-glucuronic acid residues. Xylotriose and Xyl_4 were degraded to Xyl_2 and xylose in about equal amounts, while Xyl_5 and Xyl_6 yielded Xyl_4 and Xyl_5 as major products. Xyl_2 was not attacked. This enzyme preparation also exhibited transglycosylation activity.

A xylanase component of *Oxiporus* sp. (xylanase A_2, MW 27,000; pI 8.6) degraded *Picea abies* (softwood) arabinoglucuronoxylan to mainly L-arabinose, xylose, Xyl_2, and an acidic xylose oligosaccharide (4-*O*-methyl glucuronoxylotriose, $\dot{G}lcAXyl_3$), with smaller amounts of higher oligosaccharides also being present (Sinner and Dietrichs, 1976). The same polysaccharide substrate was also degraded by a xylanase (MW 17,800, pI 9.2) from *Trichoderma viride* but in this case yielded L-arabinose, xylose, Xyl_2, and $GlcAXyl_2$ as major products of hydrolysis (Sinner and Dietrichs, 1976).

The mode of action of a xylanase isolated from *Ceratocystis paradoxa* (xylanase I, pI 9.17) has been described in detail (Dekker and Richards, 1976; Dekker, 1979). The substrate binding site of this enzyme occupied five subsites for the binding of 1,4-β-D-xylopyranosyl residues with the catalytic site located between subsites −I and I (see Fig. 3). A noteworthy point about this enzyme was its mode of attack on Xyl_4, which was degraded to Xyl_2 and Xyl_3 without the concomitant release of xylose (Dekker and Richards, 1975a). A mechanism can be proposed for the production of Xyl_2 and Xyl_3 from Xyl_4 that involves the formation of a termolecular complex (Fig. 3). This would involve a shifted 2,1 binding of Xyl_4 with subsequent xylosyl transfer and formation of Xyl_5, which is hydrolyzed more rapidly to Xyl_2 and Xyl_3. Rapid attack on $AraXyl_5$ and the presence of $AraXyl_3$ in these hydrolysates indicated that the L-arabinose substituent did not interfere with the binding of the β-D-xylosyl residues at the binding site.

b. Nondebranching Xylanases. The nondebranching group of xylanases are by far the most common and degrade heteroxylans randomly. Noteworthy enzymes among this group are the xylanases produced by *A. niger,* which have been studied in rather more detail than those from other fungi.

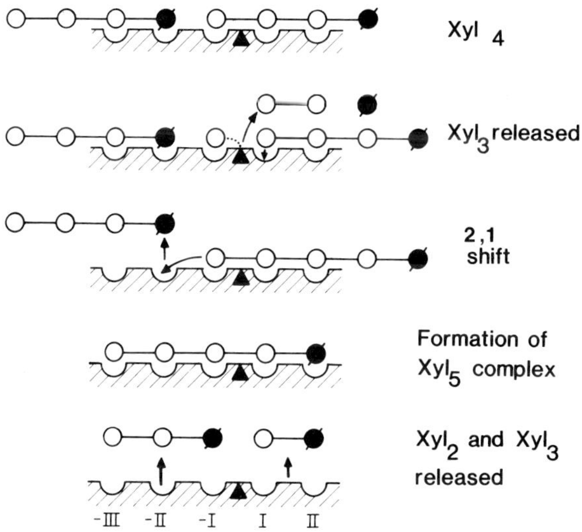

Fig. 3. Mechanisms for the binding and hydrolysis of xylose oligosaccharides by endo-1,4-β-D-xylanase from *Ceratocystis paradoxa* (Dekker and Richards, 1975a) via a termolecular enzyme–Xyl_4 complex. For symbols, see the legend to Fig. 2.

Reilly and co-workers (Reilly, 1981; Frederick *et al.*, 1981) isolated five xylanases of different specificities from a commercial enzyme preparation (Rhozyme HP-150) that did not debranch arabinoglucuronoxylans. One of these xylanases (MW 20,800, pI 6.7) degraded heteroxylans and xylose oligosaccharides of DP 3–9 to mainly Xyl_2 and xylose. Although the configuration of the products of hydrolysis were inverted (typical of exohydrolases), the enzyme liberated xylose oligosaccharides of DP 2–6 during short hydrolysis times and is thus of the endo type. Four other xylanases isolated from this enzyme preparation degraded heteroxylan to large xylose oligosaccharide fragments, but no xylose was liberated (Reilly, 1981). Furthermore, each of the four xylanases (MW ranging from 12,000 to 28,000) did not attack xylose oligosaccharides of DP <4. This is an uncommon feature of most xylanases and suggests that the substrate binding sites of these enzymes are relatively large (i.e., greater than five or six subsites).

More details on the quantitative binding and hydrolysis of xylose oligosaccharides by an *A. niger* xylanase has been described by Vrsanska *et al.* (1982). The xylanase (MW 16,000, pI 4.0) produced by *A. niger* str. 15 did not degrade Xyl_2 or aryl-β-D-xylopyranosides. The substrate binding site is composed of seven subsites (Fig. 4). Bond-cleavage frequencies of oligosaccharides by this enzyme, like that of a xylanase from *C. albidus*

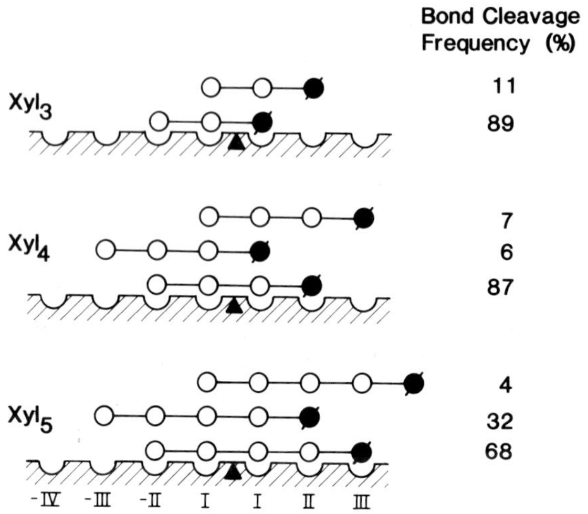

Fig. 4. Mechanisms for the binding and hydrolysis of xylose oligosaccharides by endo-1,4-β-D-xylanase from *Aspergillus niger* str. 15 (Vrsanska *et al.*, 1982). For symbols, see the legend to Fig. 2.

(see Section IV,2,b), were also dependent on the concentration of substrates. At low substrate concentrations, hydrolysis of Xyl_3–Xyl_5 series occurred via a unimolecular mechanism with the formation of productive enzyme complexes as shown in Fig. 4. Thus, Xyl_3 yielded xylose, Xyl_4 yielded Xyl_2, and Xyl_5 yielded Xyl_2 and Xyl_3 as major end products of hydrolysis. At higher concentrations of substrate (e.g., Xyl_3), there was a deviation from a unimolecular mechanism to one approaching a termolecular shifted hydrolysis with subsequent change in product distribution. Thus, Xyl_3 formed Xyl_2, while hydrolysis of Xyl_4 and Xyl_5 was more complex, and these substrates tended to be utilized during degradation for resynthesis into xylose oligosaccharides of higher DP, which were subsequently degraded to mainly Xyl_2 and Xyl_3. Xyl_2 and xylose could be utilized as glycosyl acceptors in transfer reactions.

Another interesting xylanase was that produced by *Trichoderma lignorum* (John *et al.*, 1981). Two xylanases were isolated (xylanase A, MW 21,000; xylanase B, MW 20,000) that showed different specificities toward arabinoxylan. Xylanase A degraded arabinoxylan to Xyl_2, $AraXyl_4$, $AraXyl_5$, and xylose as chief end products. Also released but in lesser amounts were Xyl_3, $AraXyl_3$, and mixed oligosaccharides substituted with two L-arabinose residues, Ara_2Xyl_3 to Ara_2Xyl_7. A branched-xylose trisaccharide was also identified but was not present in hydrolysates from another xylan, which was degraded mainly to Xyl_2 and xylose.

Xylobiose was not attacked, and Xyl_3 was hydrolyzed to Xyl_2 and xylose. Xylanase B yielded a similar array of xylose oligosaccharides. Branched xylose oligosaccharides of DP 2–4 were also produced and may have resulted from the transferase activity exhibited by this enzyme. This enzyme degraded xylan to mainly Xyl_2 and some unidentified oligosaccharides of DP 3–6. The chemical structures of the branched oligosaccharides were not reported.

There are relatively fewer studies on the nature of acidic xylose oligosaccharides released by xylanases acting on glucuronoxylans. The mode of action of a xylanase from an *Aspergillus* sp. and an *Oxiporus* sp. showed that these enzymes degraded several glucuronoxylans in a similar manner, yielding xylose, Xyl_2, and $GlcAXyl_3$ as major end products of hydrolysis (Sinner and Dietrichs, 1976). A xylanase produced by *Trichoderma viride* degraded the same glucuronoxylans to mainly xylose, Xyl_2, and $GlcAXyl_2$. When the latter enzyme was immobilized (covalently bound to polymaleic acid anhydride), its mode of action on *Fagus sylvatica* 4-*O*-methylglucuronoxylan changed and a larger acidic xylose oligosaccharide ($GlcAXyl_4$) was released in addition to xylose and Xyl_2 as major end products (Puls *et al.*, 1974). No free D-glucuronic acid or GlcAXyl were detected in these hydrolysates.

A xylanase (xylanase II) produced by *Sporotrichum dimorphosporum* degraded redwood arabinoglucuronoxylan to mainly Xyl_2, xylose, $AraXyl_2$, $AraXyl_3$, $GlcAXyl_3$, and $GlcAXyl_4$ (Comtat and Joseleau, 1981). The branch point (L-arabinose and D-glucuronic acid) was identified as being present on the nonreducing end of the D-xylosyl chain of the oligosaccharides. L-Arabinose, D-glucuronic acid and its 4-*O*-methyl derivative, AraXyl, GlcAXyl, and $GlcAXyl_2$ were not released as a result of hydrolysis. The binding site of this xylanase can accommodate at least three 1,4-β-D-xylosyl residues, and the presence of a substituent (L-arabinose) at O-3 on the nonreducing xylosyl residue did not appear to hinder binding or catalysis. The presence of a D-glucopyranosyluronic group at O-2 on the nonreducing xylosyl residue, however, appeared to require a fourth D-xylosyl residue as a minimum requirement for binding and formation of a complex amenable to hydrolysis.

A xylanase (MW 23,000) produced by the wood-rot fungus *Trametes hirsuta* degraded willow 4-*O*-methyl glucuronoxylan to mainly Xyl_4, Xyl_5, 4-*O*-methyl Glc*p*$AXyl_3$, and 4-*O*-methyl Glc*p*$AXyl_5$ (Kubackova *et al.*, 1979). Small amounts of xylose oligosaccharides of DP >6, with and without 4-*O*-methyl Glc*p*A substituents, were also present in these digests. Xylose, 4-*O*-methyl Glc*p*A, and aldobio- and aldotriouronic acids were not produced during hydrolysis. The absence of acidic xylose oligosaccharides of DP <4 indicated that the 4-*O*-methyl Glc*p*A residue

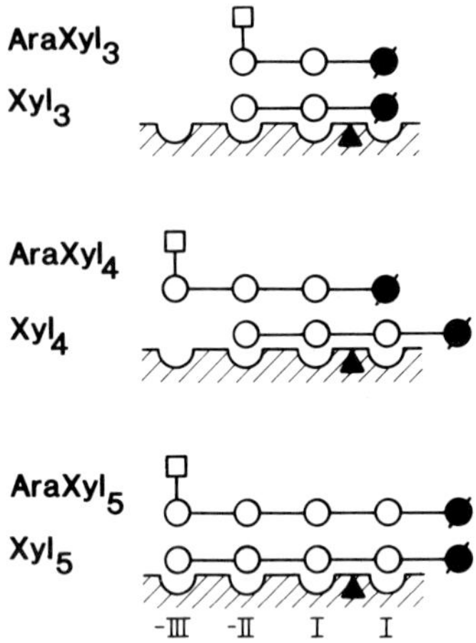

Fig. 5. Mechanisms for the binding and hydrolysis of xylose oligosaccharides by endo-1,4-β-D-xylanase from *Ceratocystis paradoxa* (Dekker and Richards, 1982b). For symbols, see the legend to Fig. 2.

(linked to C-2 of the terminal D-xylosyl group) sterically hindered the hydrolysis of the two 1,4-β-D-xylosyl linkages to the right of the branch point (Fig. 6). Enzymatic degradation of a debranched xylan containing 2% uronic acid likewise liberated Xyl_4 and Xyl_6 as the chief products of hydrolysis. In this study, the hydrolysis products were continuously removed by dialysis. The xylanase did not attack Xyl_2 but degraded Xyl_3 slowly to Xyl_2 and xylose. Xylotetraose was cleaved to Xyl_2, while Xyl_5 was rapidly hydrolyzed to Xyl_2 and Xyl_3. Attack on reduced xylose oligosaccharides of DP 3–6 indicated that these substrates were degraded from the nonreducing end. The binding site of this enzyme appeared to require a chain of at least five D-xylosyl residues for the formation of a productive complex. The proposed mode of action by which this enzyme degraded glucuronoxylan into xylose oligosaccharides of rather large DP will be shown in Fig. 6.

A second xylanase (II, pI 4.5) produced by *Ceratocystis paradoxa* has a substrate binding site that is effectively filled by a chain of four 1,4-β-D-xylosyl residues with the catalytic cleavage site asymmetrically located between the first two subsites (Dekker and Richards, 1975b). The preferred modes of binding of xylose oligosaccharides by xylanase II is shown in Fig. 5. An arabinose substituent on the nonreducing end of the xylose oligosaccharide did not impede binding of these substrates. The β-D-xylosidic linkage to the right of the substituent was resistant to cleavage. The novelty of both xylanases isolated from *C. paradoxa* was such that, unlike most other xylanases, they showed a preference to attack the xylan chain at those regions highly substituted with arabinose or uronic acid, leaving a less substituted, although somewhat shorter, fragment that retrograded.

4. Specificities of Xylanases Acting on Heteroxylans

Several xylose oligosaccharides of mixed constitution containing 1,3-linked α-L-arabinofuranosyl or 1,2-linked α-D-glucopyranosyl uronic acid residues have been isolated from enzymatic hydrolysates of different heteroxylans. Their structures have been characterized by specific enzyme cleavage, methylation, periodate oxidation, NMR, and mass spectrometry. The chemical nature of the oligosaccharides released during hydrolysis is not so much a function of the complex structure of the heteroxylan but rather a property (i.e., specificity) of the xylanase itself. The nature of the products released during hydrolysis allows one to predict the mode of attack on the polymeric substrate. This aspect is reviewed in Fig. 6.

Several interesting acidic xylose oligosaccharides have been isolated from enzymatic digests of *Betula platyphylla* 4-*O*-methylglucuronoxylan (Ishihara *et al.*, 1978). Of these, one was shown (methylation analysis and mass spectrometry) to contain interposed L-rhamnose and D-galacturonic acid residues within the 1,4-β-D-xylosyl chain (Fig. 7a). This finding suggests that the xylan chain is linked to the rhamnogalacturonans (pectic substances). A model depicting how a xylan may be linked to a rhamnogalacturonan is shown in Fig. 7.

In another finding, an arabinoxylose oligosaccharide (AraXyl$_3$) was isolated from enzymatic hydrolysates of corn cob arabinoxylan (Kusakabe *et al.*, 1977b). This oligosaccharide [*O*-β-D-Xyl*p*-(1→2)-*O*-α-L-Ara*f* (1→3)-*O*-β-D-Xyl*p*-(1→4)-D-Xyl*p*] contained an interposed L-arabinose residue attached to a terminal D-xylose unit via a 1,2-glycosidic bond. The isolation of this branched oligosaccharide provided evidence that not all of the substituents attached to the xylan backbone exist as single monosaccharide stubs. Also isolated from these hydrolysates was a disaccharide (AraXyl) that was characterized as *O*-L-Ara*f*-(1→3)-*O*-β-D-Xyl*p*. The liberation of this oligosaccharide from arabinoxylan indicated that the *Streptomyces* xylanase was also capable of cleaving the 1,4-β-xylosidic linkage to the immediate right of the substituent group.

B. *1,4-β-D-Mannanases*

β-D-Mannanases of both the exo and endo types have been characterized and hydrolyze the 1,4-β-D-mannopyranosyl linkages of branched

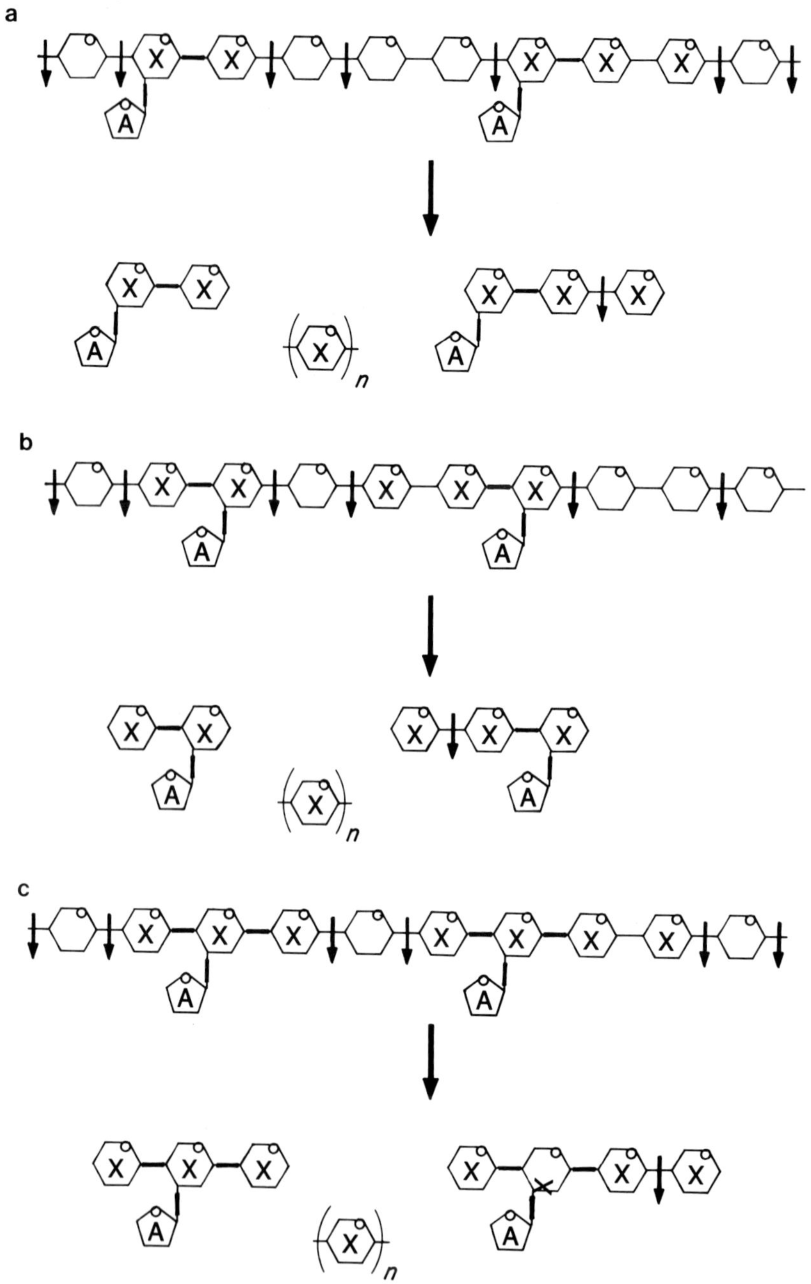

Fig. 6. The structures of some xylose oligosaccharides released during enzymolysis of heteroxylans and the mode of action of endo-1,4-β-D-xylanases from (**a**) *Ceratocystis paradoxa* (Dekker and Richards, 1975a,b,c), *Sporotrichum dimorphosphorum* (Comtat and Joseleau, 1981), and *Streptomyces* sp. (Kusakabe *et al.*, 1977b); (**b**) *Aspergillus niger* (Takenishi and Tsujisaka, 1973); (**c**) *Cephalosporium sacchari* (Dekker *et al.*, 1975); (**d**) *Trichoderma viride* (Sinner and Dietrichs, 1976); (**e**) *Aspergillus niger* and *Oxiporus* sp. (Sinner and

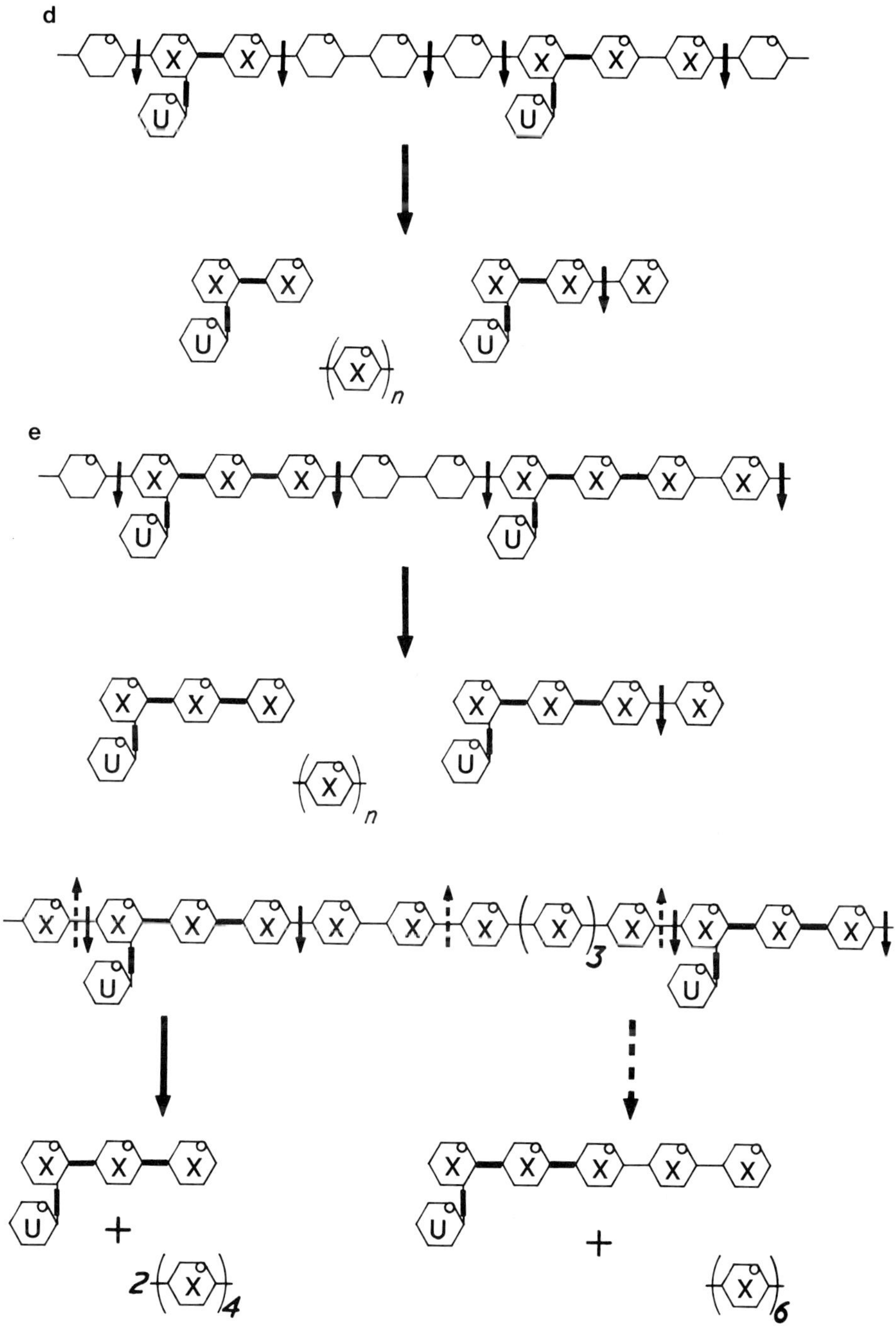

Dietrichs, 1976), and *Sporotrichum dimorphosphorus* (Comtat and Joseleau, 1981); and (**f**) *Trametes hirsuta* (Kubackova *et al.*, 1979). X, 1,4-linked β-D-xylosyl residue; A, 1,3-linked α-L-arabinofuranosyl residue; and U, 1,2-linked 4-*O*-methyl-α-D-glucopyranosyl uronic acid residue; ↓, site of cleavage; heavy bonds, enzyme-resistant linkages.

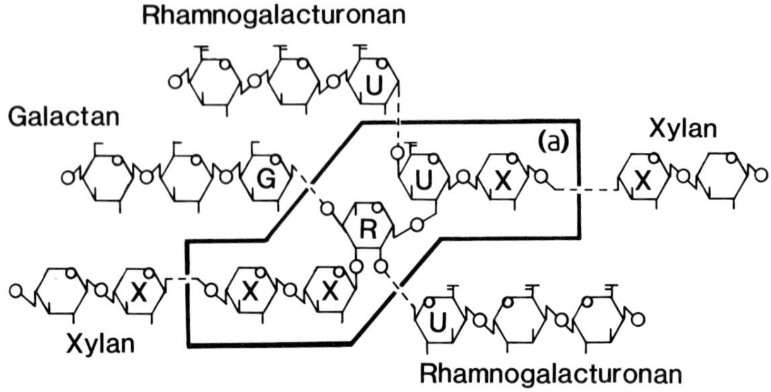

Fig. 7. Model depicting the linkage of a xylan to a rhamnogalacturonan (pectic substance) in the cell wall of *Betula platyphylla*. **Inset a,** structure of an acidic xylose oligosaccharide isolated from an enzymic hydrolyzate of *B. platyphylla* 4-*O*-methylglucuronoxylan showing interposed L-rhamnose and D-galacturonic acid residues. X, β-D-Xylosyl unit; U, α-D-galactopyranosyl uronic acid; R, α-L-rhamnosyl residue; and G, β-D-galactosyl residue.

mannans (e.g., D-galacto-D-mannans), copolymer mannans (e.g., D-gluco- and D-galacto-D-gluco-D-mannans), and linear D-mannans (Dekker, 1979; Dekker and Richards, 1976).

1. Exo-1,4-β-D-Mannanases

Two exo-acting 1,4-β-D-mannanases have been described. One of these was produced by a species of *Aeromonas* (str. F.25) isolated from the intestines of rainbow trout (Araki and Kitamikado, 1982). The other was isolated from germinated guar seeds (*Cyamopsis tetragonolobus*) (McCleary, 1982).

The exo-β-mannanase (MW 64,000, pI 5.9) produced by *Aeromonas* sp. degraded 1,4-β-D-mannans from *Codium fragile* (a green alga) and coffee bean to only mannobiose (Man_2), and mannose oligosaccharides of DP 3–5 to mainly Man_2 and some mannose. Mannobiose was not attacked, nor were aryl-β-D-mannopyranosides and β-mannans of mixed constitution (e.g., galacto- and glucomannans). Exo-β-mannanase activity was detected by its action on both *Codium* β-mannan and mannotriose (Man_3). An endo-β-mannanase produced by this bacterium did not attack Man_3. The relative rates of hydrolysis of mannose oligosaccharides by the exo-β-mannanase increased as the DP of the oligosaccharides increased. This observation and that of Man_2 not being attacked indicated that the enzyme was a polysaccharide hydrolase with a preference for substrates of DP >5, rather than a glycosidase. Furthermore, only linear β-D-mannans were degraded (to Man_2); branched and copolymer β-D-mannans were not attacked. This observation is consistent with that of exo-type polymer hydrolases, which terminate hydrolysis when they encounter branch points. Since Man_2 was the only product resulting from enzyme attack on β-mannans, the enzyme was an exo-β-mannanase of the mannobiohydrolase type, i.e., 1,4-β-D-mannan mannobiohydrolase. Attack on reduced mannose oligosaccharides of DP 3–5 showed that the exo-β-mannanase degraded its substrate from the nonreducing end of the molecule. The appearance of some oligosaccharides of higher DP amongst the products arising from attack on mannose oligosaccharides indicated transglycosylation activity.

The exo-β-mannanase (MW 59,000, pI 9.4) isolated from germinating guar seeds exhibited a different specificity toward its substrate than the exo-β-mannanase produced by *Aeromonas* sp. Guar exo-β-mannanase degraded carob galactomannan and mannose oligosaccharides of DP 2–6 to mainly D-mannose (McCleary, 1982). Aryl-β-D-mannopyranosides were also attacked. The initial rate of hydrolysis of mannose oligosaccharides increased as the DP increased. The enzyme thus showed a preference for mannose oligosaccharides of high DP and in this respect is different from exoglycosidases, which show a high affinity for low MW glycosides. The configuration of the mannose released during hydrolysis was inverted (i.e., the α anomer predominated), which is typical of exo-type enzymes (Reese *et al.*, 1968). Mannopentaose was hydrolyzed at twice the rate of Man_4 and five times faster than Man_3. The enzyme did not attack reduced Man_2, and it degraded reduced mannose oligosaccharides of DP 3–5 more slowly than the natural mannose oligosaccharides. Reduced Man_5 was degraded to mainly mannose and reduced Man_3, which was further slowly hydrolyzed to mannose and reduced Man_2. Since this exo-β-mannanase preparation liberated D-mannose as major hydrolysis product from mannose oligosaccharides and carob galactomannan, it can be classified as a 1,4-β-D-mannan, or 1,4-β-D-oligomannoside, mannohydrolase.

2. Endo-1,4-β-D-Mannanases

Endo-β-mannanases from bacterial, fungal, and plant sources and their actions on different 1,4-β-D-mannans have been described in detail by Dekker and Richards (1976) and Dekker (1979). The application of highly purified endo-β-mannanases in the analysis of the fine structure of several non-cell wall galactomannans has also been described (McCleary, 1979; McCleary *et al.*, 1982).

Studies on the actions of endo-β-mannanases on the plant cell wall β-mannans (galactogluco- and glucomannans) are rather limited and have usually employed enzymes of fungal origin. Most studies on the mode of action of endo-β-mannanases, however, have been confined to the galactomannans of guar and carob, substrates that are readily available commercially. Endo-β-mannanases (EC 3.2.1.78, 1,4-β-D-mannanohydrolases), in general, degrade β-D-mannans to D-mannose and a series of mannose oligosaccharides of DP 2–6. Their action on larch glucomannan (mannose : glucose ratio, 3 : 1) has also yielded D-glucose, in addition to mannose and mannose oligosaccharides, some of which contained glucose (Ishihara and Shimizu, 1980). Cellobiose was not produced. The purified endo-β-mannanase (MW 61,000, pI 3.45) used in this study was capable of cleaving the Man*p*-1,4-Glc*p* and Glc*p*-1,4-Man*p* bonds in larch glucomannan, in addition to the 1,4-β-D-mannopyranosyl linkages.

In a similar manner, galactoglucomannans from spruce and Canadian hemlock were degraded by an *Aspergillus niger* endo-β-mannanase (pI 4.1) to Man_2, mannose, glucose, and mannose oligosaccharides (Yamazaki, 1975; Sinner *et al.*, 1979). Cellobiose (Glc_2) was not detected in these hydrolysates, indicating that the Man*p*-1,4-Glc_2 and Glc_2-1,4-Man*p* linkages were not cleaved by this enzyme. Both polysaccharides were slowly attacked, resulting in limited hydrolysis (9 and 25% after 80 hr for the hemlock and spruce heteromannans, respectively). Several mannose oligosaccharides containing glucose and galactose were identified in enzymatic hydrolysates of Canadian hemlock and included Glc*p*-1,4-Man*p*, several trisaccharides (glucose : mannose ratio of 1 : 1, 1 : 2, and 2 : 1), and several tetrasaccharides (glucose and mannose in the ratio 2 : 2), including one containing galactose, glucose, and mannose in the ratio of 1 : 1 : 2. The chemical structures of these oligosaccharides were not characterized. The presence of *O*-β-D-Glc*p*-(1→4)-D-Man*p* in these hydolysates indicated that the *A. niger* endo-β-mannanase could also cleave the Man*p*-1,4-Glc*p* bond. This enzyme did not attack Man_2, and it degraded Man_3 and Man_4 to mainly Man_2, with some mannose also being present (Yamazaki, 1975).

C. β-D-Galactanases

Galactanases are hydrolytic enzymes that degrade the D-galactans and L-arabino-D-galactans. Two types of endo-β-D-galactanases are recognized (Dekker and Richards, 1976; Dekker, 1979). Although D-galactanases of the exo type have not been unequivocally characterized, an unusual 1,4-β-D-galactanase (MW 40,000) isolated from *Bacillus subtilis* liberating mainly Gal_4 from 1,4-β-D-galactan, but also other galactose oligosaccharides, was reported to possess both exo and endo activity

(Labavitch *et al.*, 1976). The mixed enzyme activity was claimed to be a function of the one enzyme component.

Endogalactanases degrading the 1,4-β-D-galactosyl linkages of D-galactans (such as from citrus, lupin seed, soybean, and those present in the cell walls of sycamore and potato cells) are known as endo-1,4-β-galactanases, and they degrade these galactans randomly to produce D-galactose and galactose oligosaccharides, some of which contain L-arabinose residues (Dekker, 1979; Bauer *et al.*, 1977). Endogalactanases of the other type—i.e., those that attack the 1,3-β-D-linked galactosyl bonds of arabinogalactans—liberated D-galactose and several 1,3- and 1,6-linked β-D-galactose oligosaccharides, some of which contained L-arabinose (Hashimoto, 1971). These enzymes are known as endo-1,3-β-D-galactanases.

V. Biodegradative Pathways of the Hemicelluloses

The biodegradative pathways by which microbial enzymes can attack the hemicelluloses are shown in Fig. 8. The modes of attack on the hemicellulose are envisaged as occurring in two ways:

1. The attack by the exoglycosidases precedes that of the hemicellulases (i.e., xylanase, galactanase, and β-mannanase), removing the side chain substituents, thereby "opening up" or exposing the backbone glycan chain. In this manner, the glycan chain can be more easily attacked by the hemicellulases, as steric hindrance by the side chain residues is reduced.
2. Alternatively, the endohemicellulases attack the regions of the glycan chain that are unbranched or relatively moderately branched by substituents. The actions of the endohemicellulases yield an array of oligosaccharides of mixed constitution, which are further degraded by both exoglycosidases and endohemicellulases.

The actions of the debranching endoxylanases and exohemicellulases are, however, two exceptions to the envisaged modes of attack. The debranching endoxylanases have been shown to be capable of also cleaving L-arabinose units from arabinoxylans and arabinoglucuronoxylans and therefore do not require the assistance of the α-L-arabinosidases to open up the xylan chain. With respect to the exohemicellulases, one would expect only a limited hydrolysis from such enzymes acting on hemicelluloses, as their actions are terminated when branch-point substituents are encountered. Such limitations, however, would depend on whether or not

a

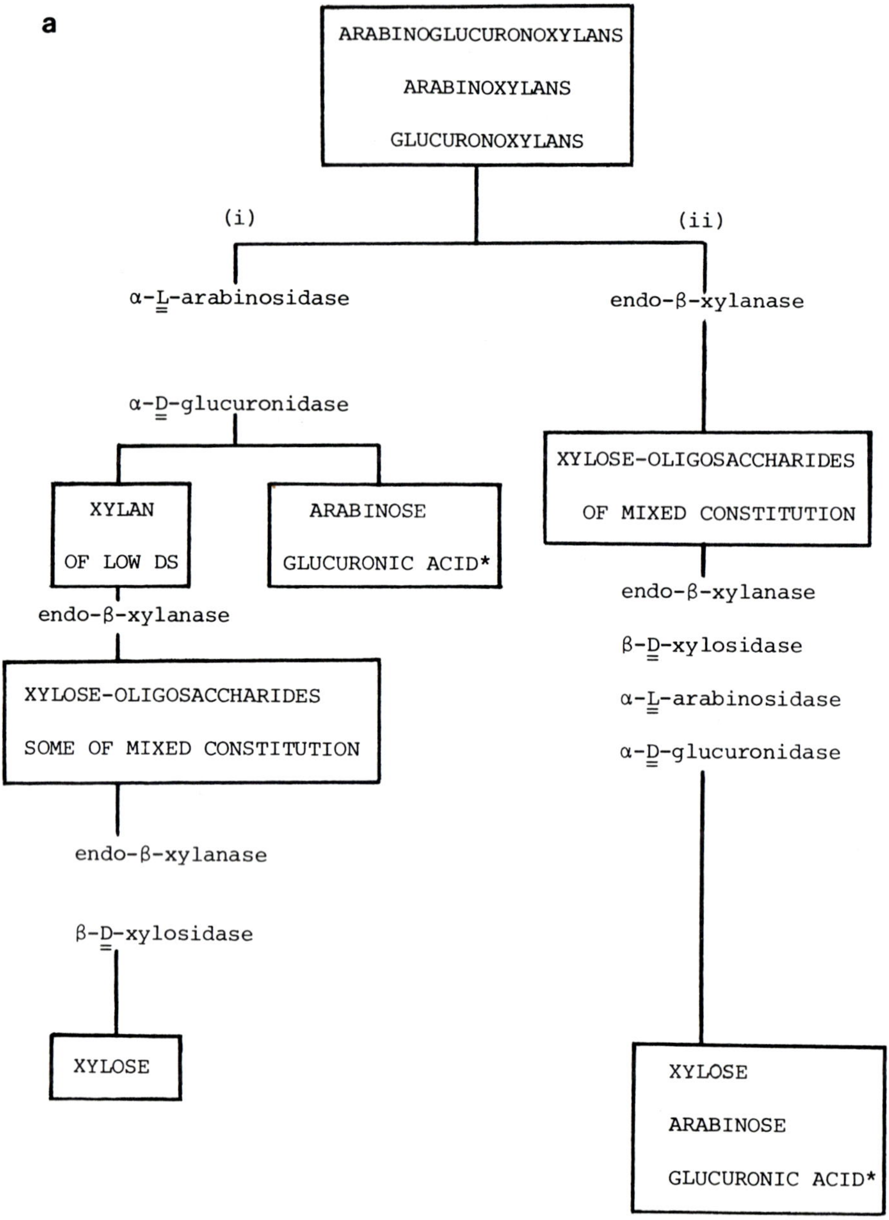

Fig. 8. Biodegradative pathways of the hemicelluloses: (**a**) heteroxylans, (**b**) hetero-β-mannans, and (**c**) heterogalactans. *And its 4-*O*-methyl derivative.

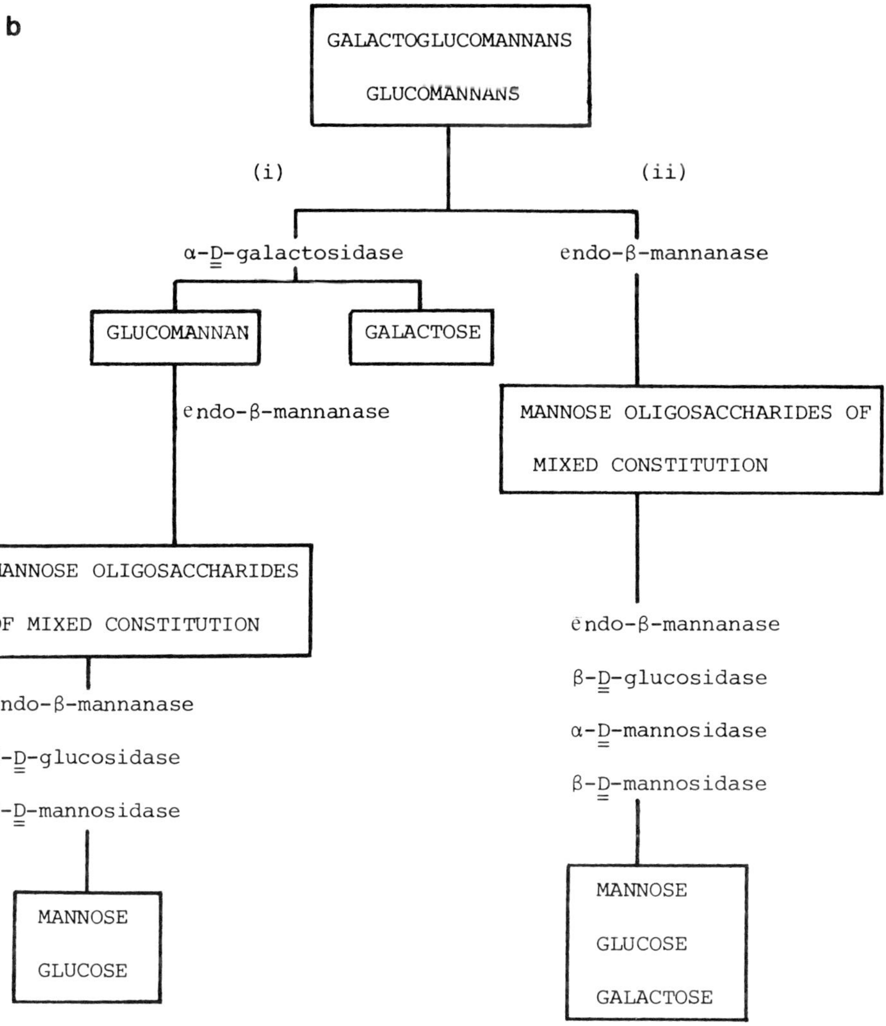

Fig. 8. (*Continued*)

c

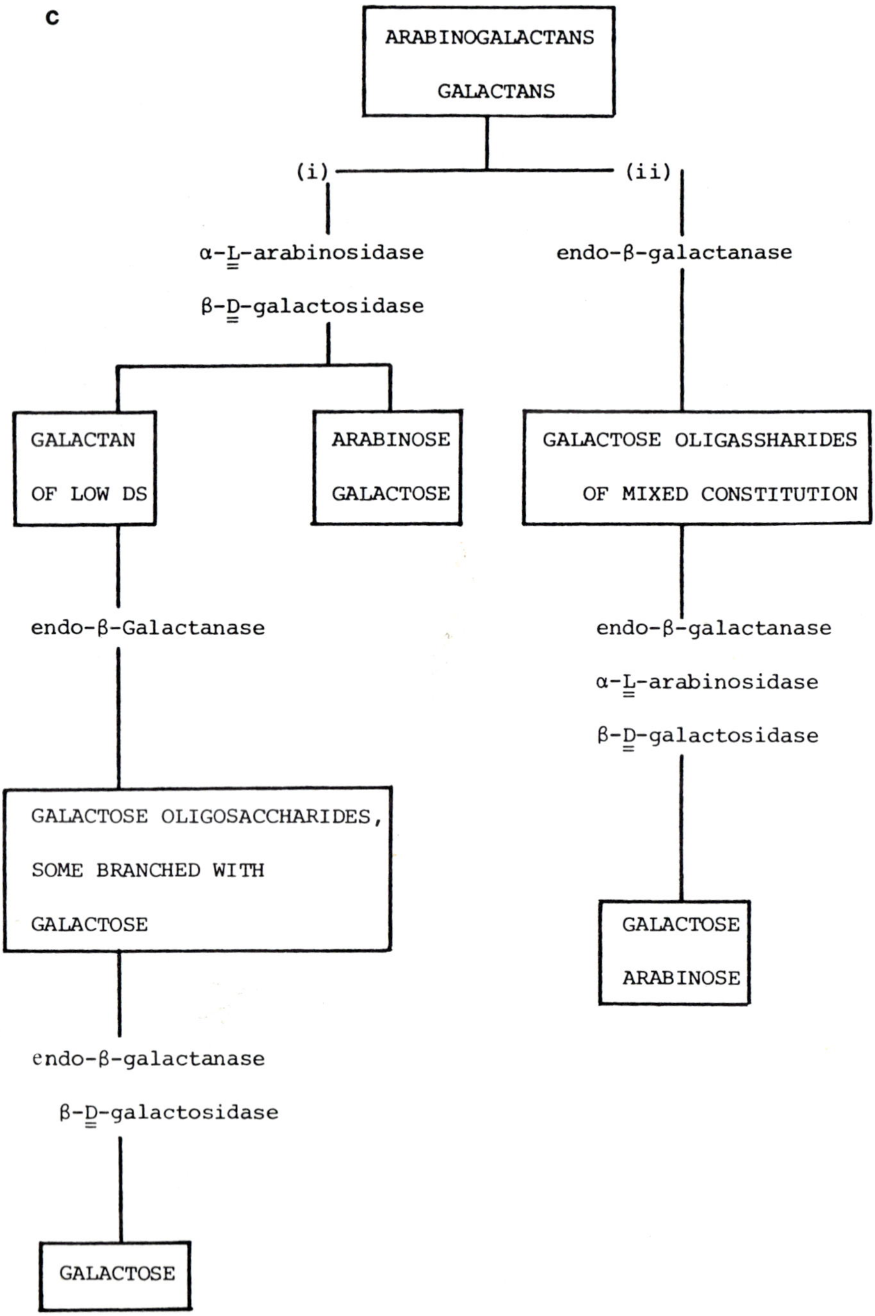

Fig. 8. (*Continued*)

the attack by the hemicellulases was preceded by the action of exoglycosidases.

The exoglycosidases attack the low molecular weight oligosaccharides produced by the action of endohemicellulases and are also capable of hydrolyzing the branch points on the hemicellulose backbone chain. Thus, the α-L-arabinosidases and α-D-glucuronidases remove the L-arabinose and D-glucuronic acid substituents from arabinoglucuronoxylans and glucuronoxylans; α-D-galactosidases cleave the α-D-galactose residues from galactoglucomannans; and the L-arabinosidases and β-D-galactosidases are responsible for the hydrolysis of the L-arabinose and D-galactose appendages from arabinogalactans (see Fig. 8). The oligosaccharides arising from attack on hemicellulose by the hemicellulases and exoglycosidases are further degraded through the actions of another group of exoglycosidases. These include the β-D-xylosidases, β-D-mannosidases, and β-D-galactosidases, which attack low MW 1,4-β-D-xylosides, 1,4-β-D-mannosides, and 1,4-β-D- and 1,3-β-D-galactosides, respectively.

References

Araki, T., and Kitamikado, M. (1982). *J. Biochem.* **91,** 1181–1186.

Aspinall, G. O. (1959). *Adv. Carbohyd. Chem.* **14,** 429–468.

Aspinall, G. O. (1970). "Polysaccharides." Pergamon Press, Oxford.

Bailey, R. W., and MacRae, J. C. (1970). *J. Agric. Sci.* **75,** 321–326.

Bauer, W. D., Bateman, D. F., and Whalen, C. H. (1977). *Phytopathology.* **67,** 862–868.

Biely, P., Kratky, Z., Kockova-Kratochvilova, A., and Bauer, S. (1978). *Folia Microbiol.* **23,** 366–371.

Biely, P., Vrsanska, M., and Kratky, Z. (1980). *Eur. J. Biochem.* **108,** 313–321.

Biely, P., Kratky, Z., and Vrsanska, M. (1981a). *Eur. J. Biochem.* **119,** 559–564.

Biely, P., Vrsanska, M., and Kratky, Z. (1981b). *Eur. J. Biochem.* **119,** 565–571.

Comtat, J., and Joseleau, J. P. (1981). *Carbohyd. Res.* **95,** 101–112.

Dekker, R. F. H. (1976). *In* "Carbohydrate Research in Plants and Animals," (Miscellaneous Papers), Vol. 12, pp. 43–54. Landbouwhogeschool, Wageningen, The Netherlands.

Dekker, R. F. H. (1979). *In* "Polysaccharides in Food" (J. M. V. Blanshard and J. R. Mitchell, eds.), pp. 93–108. Butterworths, London.

Dekker, R. F. H., and Candy, G. P. (1979). *Arch. Microbiol.* **122,** 297–299.

Dekker, R. F. H., and Lindner, W. A. (1979). *S. Afr. J. Sci.* **75,** 65–71.

Dekker, R. F. H., and Richards, G. N. (1975a). *Carbohyd. Res.* **39,** 97–114.

Dekker, R. F. H., and Richards, G. N. (1975b). *Carbohyd. Res.* **42,** 107–123.

Dekker, R. F. H., and Richards, G. N. (1975c). *Carbohyd. Res.* **43,** 335–344.

Dekker, R. F. H., and Richards, G. N. (1976). *Adv. Carbohyd. Chem. Biochem.* **32,** 277–352.

Dekker, R. F. H., Richards, G. N., and Shambe, T. (1975). *Biochem. Soc. Trans.* **3,** 1081–1082.

Esteban, R., Villanueva, J. R., and Villa, T. G. (1982). *Can. J. Microbiol.* **28,** 733–739.

Frederick, M. M., Frederick, J. R., Fratzke, A. R., and Reilly, P. J. (1981). *Carbohyd. Res.* **97,** 87–103.

Gorbacheva, I. V., and Rodinova, N. A. (1977). *Biochim. Biophys. Acta* **484,** 94–102.

Hashimoto, Y. (1971). *Nippon Nogei Kagaku Kaishi* **45,** 147–150.

Hashimoto, K., and Takahashi, Z. (1970). *Toyo Shokuhin Kogyo Tanki Daigaku, Toyo Shokuhin Kenkyusho Kenkyu Hokokusho.* **9,** 320–326.

Horikoshi, K., and Atsukawa, Y. (1973). *Agric. Biol. Chem.* **37,** 2097–2013.

Ishihara, M., and Shimizu, K. (1980). *Mokuzai Gakkaishi* **26,** 811–818.

Ishihara, M., Shimizu, K., and Ishihara, T. (1975). *Mokuzai Gakkaishi* **21,** 680–685.

Ishihara, M., Shimizu, K., and Ishihara, T. (1978). *Mokuzai Gakkaishi* **24,** 108–115.

Iwamoto, T., Sasaki, T., and Inaoka, M. (1973). *Mem. Ehime Univ.* **17,** 13–25; 185–187.

John, M., Schmidt, J., Sahm, H., and Wandrey, C. (1981). *Biochem. Soc. Trans.* **9,** 116P.

John, M., Schmidt, B., and Schmidt, J. (1979). *Can J. Biochem.* **57,** 125–134.

Kubackova, M., Karacsonyi, S., Bilisics, L., and Toman, R. (1979). *Carbohyd. Res.* **76,** 177–188.

Kusakabe, I., Yasui, T., and Kobayashi, T. (1977a). *Nippon Nogei Kagaku Kaishi.* **51,** 439–448.

Kusakabe, I., Yasui, T., and Kobayashi, T. (1977b). *Nippon Nogei Kagaku Kaishi.* **51,** 669–679.

Labavitch, J. M., Freeman, L. E., and Albersheim, P. (1976). *J. Biol. Chem.* **251,** 5904–5910.

McCleary, B. V. (1979). *Carbohyd. Res.* **71,** 205–230.

McCleary, B. V. (1982). *Carbohyd. Res.* **101,** 75–92.

McCleary, B. V., Taravel, F. R., and Cheetham, N. W. H. (1982). *Carbohyd. Res.* **104,** 285–297.

Nisizawa, T., Suzuki, H., and Nisizawa, K. (1971). *J. Biochem.* **70,** 387–393.

Notario, V., Villa, T. G., and Villanueva, J. R. (1979). *J. Gen. Microbiol.* **114,** 415–422.

Paice, M. G., and Jurasek, L. (1979). *In* "Hydrolysis of Cellulose, Mechanisms of Enzymatic and Acid Catalysis" (R. D. Brown and L. Jurasek, eds.) (*Adv. Chem. Ser.* 181), pp. 361–374. American Chemical Society, Washington, D.C.

Panbangred, W., Shinmyo, A., Kinoshita, S., and Okada, H. (1983). *Agric. Biol. Chem.* **47,** 957–963.

Pettipher, G. L., and Latham, M. J. (1979). *J. Gen. Microbiol.* **110,** 29–38.

Puls, J., Sinner, M., and Dietrichs, H. H. (1974). *Holzforschung* **28,** 106–113.

Reese, E. T., Maguire, A. H., and Parrish, F. W. (1968). *Can. J. Biochem.* **46,** 25–34.

Reilly, P. J. (1981). *In* "Trends in the Biology of Fermentations for Fuel and Chemicals" (A. Hollaender, ed.), pp. 111–129. Plenum, New York.

Ryu, D. D. Y., and Mandels, M. (1980). *Enzyme Microb. Technol.* **2,** 91–102.

Salyers, A. A., West, S. E. H., Vercellotti, J. R., and Wilkins, T. D. (1977). *Appl. Environ. Microbiol.* **34,** 529–533.

Shikata, S., and Nisizawa, K. (1975). *J. Biochem.* **78,** 499–512.

Sinner, M., and Dietrichs, H. H. (1975). *Holzforschung* **29,** 207–214.

Sinner, M., and Dietrichs, H. H. (1976). *Holzforschung* **30,** 50–59.

Sinner, M., Parameswaran, N., and Dietrichs, H. H. (1979). *In* "Hydrolysis of Cellulose: Mechanisms of Enzymatic and Acid Catalysis" (*Adv. Chem. Ser.* No. 181) (R. D. Brown and L. Jurasek, eds.) pp. 303–329. American Chemical Society, Washington, D.C.

Sreenath, H. K., and Joseph, R. (1982). *Folia Microbiol.* **27,** 107–115.

Stuttgen, E., and Sahm, H. (1982). *Eur. J. Appl. Microbiol. Biotechnol.* **15,** 93–99.

Takahashi, M., and Hashimoto, Y. (1963). *J. Ferment. Technol.* **41,** 181–186.

Takenishi, S., and Tsujisaka, Y. (1973). *Agric. Biol. Chem.* **37,** 1385–1391.

Takenishi, S., and Tsujisaka, Y. (1975). *Agric. Biol. Chem.* **39,** 2315–2323.
Timell, T. E. (1964). *Adv. Carbohyd. Chem.* **19,** 247–295.
Timell, T. E. (1965). *Adv. Carbohyd. Chem.* **20,** 409–483.
Timell, T. E. (1967). *Wood Sci. Technol.* **1,** 45–70.
Timell, T. E. (1982). *Wood Sci. Technol.* **16,** 83–122.
Toda, S., Suzuki, H., and Nisizawa, K. (1971). *J. Ferment. Technol.* **49,** 499–521.
Uchino, F., and Nakane, T. (1981). *Agric. Biol. Chem.* **45,** 1121–1127.
Vrsanska, M., Gorbacheva, I. V., Kratky, Z., and Biely, P. (1982). *Biochim. Biophys. Acta* **704,** 114–122.
Wilkie, K. C. B. (1979). *Adv. Carbohyd. Chem. Biochem.* **36,** 215–264.
Yamazaki, N. (1975). Ph.D. Thesis, University of Hamburg, West Germany.
Yoshioka, H., Nagato, N., Chavanich, S., Nilubol, N., and Hayashida, S. (1981). *Agric. Biol. Chem.* **45,** 2425–2432.

CHAPTER **19**

Chemistry of Lignin Biodegradation

CHEN-LOUNG CHEN
HOU-MIN CHANG

Department of Wood and Paper Science
North Carolina State University at Raleigh
Raleigh, North Carolina

I. Introduction

Lignin is a complex three-dimensional polymer produced *in vivo* by an enzyme-initiated dehydrogenative polymerization of three phenylpropane monomers, *p*-hydroxycinnamyl, coniferyl, and sinapyl alcohols. The monomeric phenylpropane units in lignin, unlike those of other naturally occurring polymers, are linked to each other not by a single intermonomeric linkage but by several different carbon-to-carbon and ether linkages, most of which are not readily hydrolyzable (see Chapter 7). Thus, it is not surprising that, in contrast to cellulose and other polysaccharides in plants, lignin is resistant to degradation by most microorganisms. Indeed, a main function of lignin probably is to protect plant cell walls from microbial degradation, in addition to serving as a binding agent and rein-

Biosynthesis and Biodegradation of Wood Components

ISBN 0-12-347880-4

forcing material for cell walls (Wardrop, 1971; Eriksson, 1980). Only certain fungi and probably certain strains of bacteria are capable of decomposing lignin efficiently in nature.

The best known and clearly the most important lignin-degrading microorganisms are the white-rot fungi, the wood-decaying fungi that are known to decompose lignin macromolecules rapidly and completely. Thus, most of our current knowledge of lignin biodegradation is derived from studies with these fungi, particularly from studies on the chemistry of lignin biodegradation by a single fungus, *Phanerochaete chrysosporium* (= *Sporotrichum pulverulentum*). Although the results of these studies on *P. chrysosporium* form the basis of our current knowledge on the chemistry of lignin biodegradation discussed in this chapter, there is little evidence to suggest that lignin degradation by other white-rot fungi is substantially different from what is discussed here. In fact, the limited available data seem to suggest that lignin biodegradation by white-rot fungi is similar.

Significant advances toward elucidation of the chemistry of lignin biodegradation by white-rot fungi have been made during the past few years. In addition to the contribution from the physiological and biochemical studies discussed in Chapter 21, the rapid progress in this field can be attributed to three distinct but complementary approaches:

1. Characterization of polymeric fungus-degraded lignin isolated from decayed wood.
2. Identification of low molecular weight fungal degradation products from fungus-decayed wood.
3. Fungal degradation of dimeric lignin model compounds.

In this chapter, the results of the first two approaches are reviewed and their implication on lignin biodegradation pathways discussed. The results of the third approach are discussed in Chapter 20. Characterizing the degraded lignin is especially important to understanding the chemistry of lignin biodegradation because any structural changes in this polymeric lignin would point to the pathways by which the lignin macromolecule is degraded by fungi.

II. Characterization of Polymeric Fungus-Degraded Lignin

White-rot fungi inhabit the wood cell lumens, from which the hyphae penetrate from one cell to another through bore holes or pits. The fungal degradation of cell wall components proceeds from the cell lumens to-

Table I

Elemental Analyses and Methoxyl Content of Fungus-Degraded Lignins and Milled Wood Lignins from Spruce and Birch

Source	C (%)	H (%)	O (%)	OCH_3 (%)	Reference
Spruce					
MWL	62.85	6.08	31.07	15.11	Kirk *et al.*, 1974
MWL	62.81	5.88	31.31	15.24	Hata, 1966
Lignin decayed by					
P. anceps	58.71	4.99	36.30	11.21	Kirk *et al.*, 1974
Coriolus versicolor	57.97	4.70	37.23	11.33	Kirk *et al.*, 1974
Poria subacida	58.54	5.23	36.22	11.75	Hata, 1966
Birch					
MWL	57.47	6.22	36.31	20.52	Tai *et al.*, 1982
Lignin decayed by					
Phanerochaete chrysosporium					
M-HA-1[a]	55.29	5.87	38.84	12.50	Tai *et al.*, 1982
M-HA-2	49.09	4.81	46.10	7.52	Tai *et al.*, 1982
M-HA-3	63.74	6.67	29.59	11.31	Tai *et al.*, 1982

[a] M-HA- indicates the high molecular weight acidic fraction of the methanol extract of decayed wood.

ward the compound middle lamella, with gradual thinning of the cell wall (Cowling, 1961; Schmid and Liese, 1964; Wilcox, 1970; Kirk *et al.*, 1975; Eriksson, 1980). Thus, changes in lignin caused by fungal degradation at any given time during decay are limited to only a small portion of the total lignin, and the bulk of the lignin remains structurally unchanged (Cowling, 1961; Kirk and Lundquist, 1970). Fortunately, fungal decay renders this degraded portion of lignin more soluble, and thus it can be separated from the bulk of the unchanged lignin by extraction with neutral organic solvents. Thus, for studies of the degradation products, the decayed wood is first extracted successively with petroleum ether and chloroform to remove wax-like and resinous materials. Most of the polymeric degraded lignin and the low molecular weight degradation products are isolated by subsequent extraction with methanol. The methanol extract is further fractionated to separate the low molecular weight degradation products from the polymeric fungus-degraded lignin (Chua *et al.*, 1982). Degraded lignins from both spruce (*Picea glauca*) and birch (*Betula papyrifera*) decayed by several white-rot fungi have been isolated and characterized (Hata, 1966; Kirk and Chang, 1974, 1975; Chang *et al.*, 1980; Chua *et al.*, 1982; Tai *et al.*, 1982, 1983a,b).

Table I shows the results of elemental and methoxyl analyses, compared with those of milled wood lignin (MWL). The most obvious changes are the decrease in methoxyl content and, with the exception of one fraction of birch decayed lignin, an increase in oxygen content. The increase in oxygen content is not surprising, since it has been established that microbial degradation of lignin is aerobic, requiring the participation of oxygen. On the other hand, the fraction of decayed birch lignin (M-HA-3) that has lower oxygen and higher hydrogen content than MWL adds to accumulating evidence that the ligninolytic system in *Phanerochaete chrysosporium* possesses a strong reducing activity and is capable of reducing some of the oxidized degradation products (Tai *et al.*, 1983b). As will be discussed later, the nature and the role of this reducing activity is not clear at the present. However, it is now evident that biodegradation of lignin by white-rot fungi is a combination of an array of oxidative and reductive reactions, rather than oxidative reactions alone as indicated in our earlier review (Chang *et al.*, 1980).

In addition to elemental and methoxyl analyses, the fungus-degraded lignins from both spruce and birch have been characterized by functional group analyses, spectroscopic analyses, and degradative studies, including nitrobenzene oxidation, permanganate oxidation, and acidolysis (Hata, 1966; Kirk and Chang, 1974, 1975; Chang *et al.*, 1980). These early studies led to the conclusion that at least three modes of degradative reactions are operative in the decomposition of the lignin macromolecule by white-rot fungi (Kirk and Chang, 1975; Chang *et al.*, 1980):

1. Oxidative cleavage of side chains between α- and β-carbons leading to the formation of aromatic acids.
2. Cleavage of β-aryl ether bonds and modification of side chain structures.
3. Degradation of aromatic nuclei through oxidative ring opening.

While the presence of covalently bonded aromatic acid groups was clearly indicated by the spectroscopy and by the acidolysis data, no direct evidence was obtained for cleavage of either β-aryl ether bonds or aromatic rings. Cleavage of β-aryl ether bonds was implicated by the extremely low yield of ketonic degradation products on acidolysis of polymeric fungus-degraded lignins. Aromatic ring cleavage was supported by several instances of indirect evidence, such as the presence of large aliphatic carboxylic acid groups in addition to aromatic structures, and the large content of oxygen that cannot be accounted for by side chain oxidation alone.

Relatively recent studies not only provided direct evidence in support of the aforementioned modes of lignin biodegradation reactions, but also gave considerable insight into the probable pathways by which the lignin macromolecule is degraded. In addition, these results also suggest that

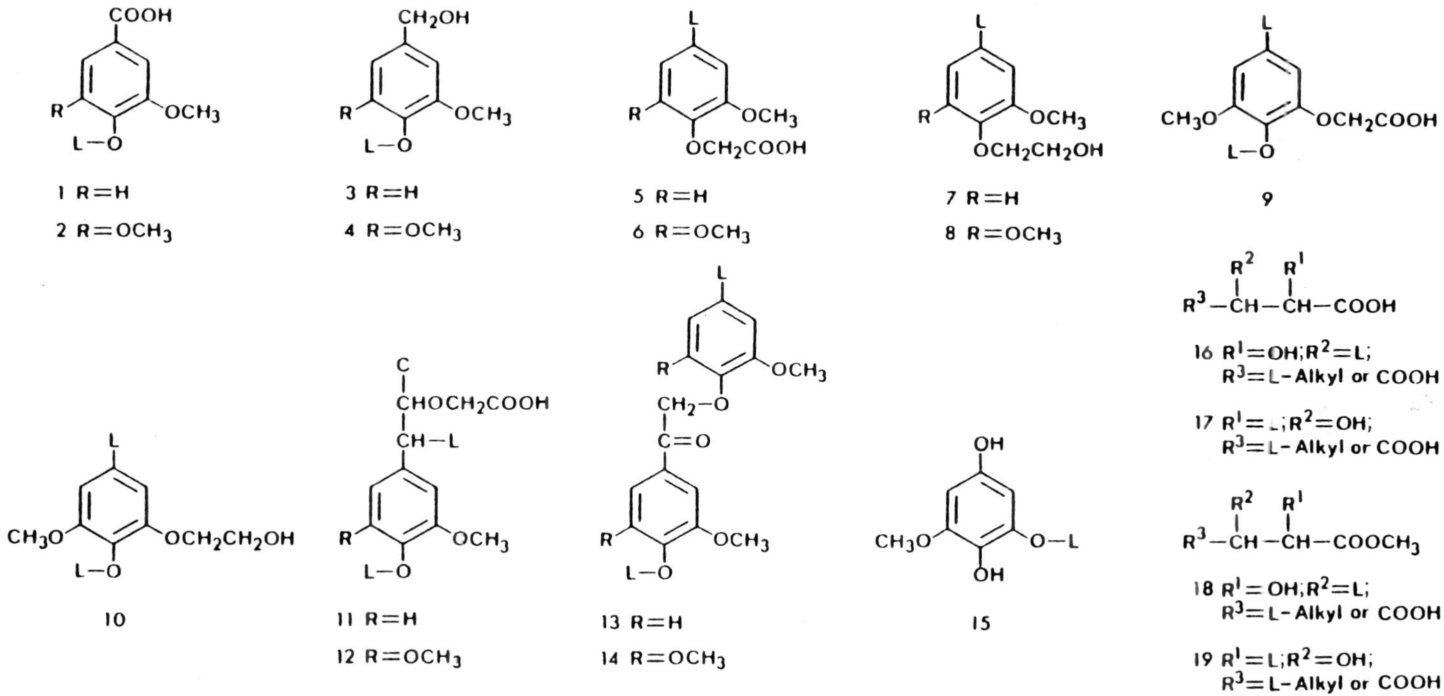

Fig. 1. New substructures in biodegraded spruce and birch lignins (M-HAs) elucidated by ^{13}C-NMR spectroscopy; L = lignin moiety.

other modes of degradative reactions may also be involved. These studies include ^{13}C-NMR characterization of fungus-degraded lignin (Chua *et al.*, 1982; Chen *et al.*, 1982b; Tai *et al.*, 1982, 1983a,b), and isolation and structural studies of the low molecular weight degradation products, discussed below (Chen *et al.*, 1982a, 1983; Tai *et al.*, 1982, 1983b; Terazawa *et al.*, 1983).

Figure 1 shows various structural features of biodegraded lignin as elucidated by detailed analyses of ^{13}C-NMR spectra of both spruce and birch biodegraded lignins, using the spectra of MWLs and model compounds as references (Chua *et al.*, 1982; Chen *et al.*, 1982b; Tai *et al.*, 1982, 1983a,b). The modifications in the side chain structures and the depletion of β-*O*-4 structures are clearly seen. Structures **9**, **10**, **11**, and **12**, as well as others, give direct support to the ring-opening reactions. In addition, the reducing activity of the ligninolytic system is clearly indicated by the presence of alcohol structures. Furthermore, the presence of signals in the lower field in the spectra of the polymeric biodegraded lignins also indicates the presence of saturated aliphatic carbons. These signals are not found in the spectra of MWL, and their origin is unclear.

III. Identification of Low Molecular Weight Lignin Biodegradation Products

In addition to characterization of the polymeric biodegraded lignins, the identification of the low molecular weight components in woods decayed by white-rot fungi has been of importance in elucidating possible lignin degradation pathways. Until relatively recently, however, analysis of the low molecular weight fractions received little attention, except for the identification of vanillin, syringaldehyde, coniferyl aldehyde, vanillic acid, and syringic acid (Henderson, 1955; Higuchi *et al.*, 1955, 1956). Constraints causing the slow progress in this field are traceable to the rather low yield and complex nature of the low molecular weight components and, most importantly, to the absence of effective analytical methods.

Progress in the methods of microanalysis, such as high-performance liquid chromatography (HPLC), gas chromatography (GC) with capillary columns, and gas chromatography–mass spectrometry (GC–MS), has led to renewed interest in the analysis of the low molecular weight fractions. These analytical methods were applied in comprehensive investigations of the constituents of low molecular weight fractions obtained from methanol extracts of spruce and birch woods decayed by the white-rot fungus

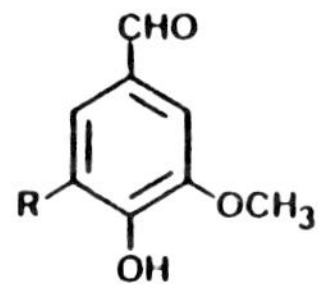

20 R = H

21 R = OCH_3

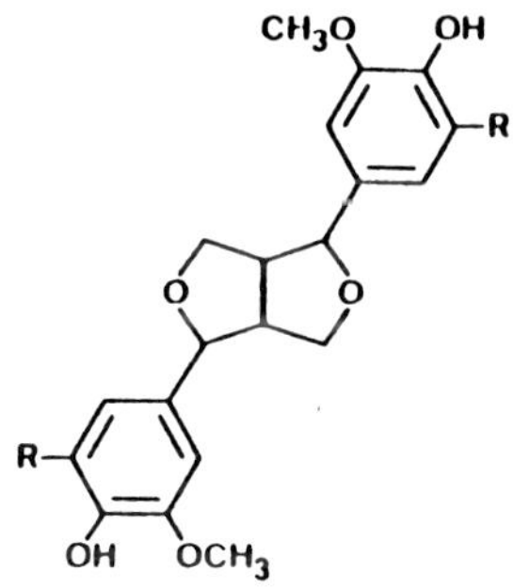

22 R = H

23 R = OCH_3

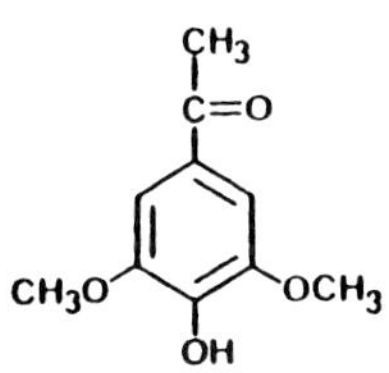

24

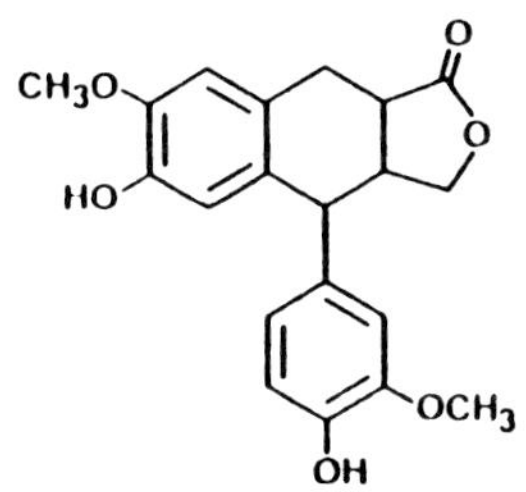

25

Fig. 2. Compounds identified in spruce and birch methanol extracts low molecular weight phenolic fractions.

P. chrysosporium (Chen *et al.*, 1982a, 1983; Tai *et al.*, 1982, 1983b; Terazawa *et al.*, 1983).

The low molecular weight components in the methanol extract were separated from the polymeric biograded lignin by extraction with chloroform. The chloroform-soluble components were further fractionated into low molecular weight acidic (M-LA), phenolic (M-LP), and neutral (M-LN) fractions according to their different solubilities in dilute sodium bicarbonate and sodium hydroxide solutions. The amount of the neutral fractions from both spruce and birch was negligible, and hence they were not studied.

From the phenolic fractions, vanillin (**20**), D,L-pinoresinol (**22**), and α-conidendrin (**25**) were isolated from spruce (Chen *et al.*, 1981, unpublished results), and vanillin, syringaldehyde (**21**), acetosyringone (**24**), and D,L-syringaresinol (**23**) from birch (Terazawa *et al.*, 1983, unpublished

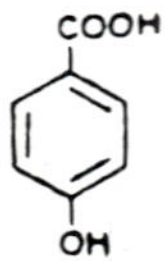

26

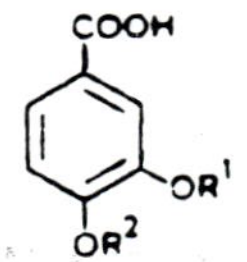

27 $R^1=R^2=H$
28 $R^1=H$; $R^2=CH_3$
29 $R^1=CH_3$; $R^2=H$
30 $R^1=R^2=CH_3$

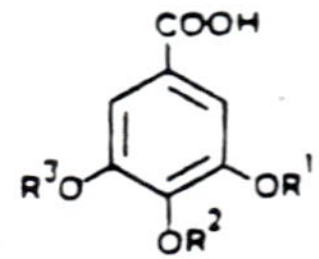

31 $R^1=R^2=H$; $R^3=CH_3$
32 $R^1=H$; $R^2=R^3=CH_3$
33 $R^1=R^3=CH_3$; $R^2=H$
34 $R^1=R^2=R^3=CH_3$

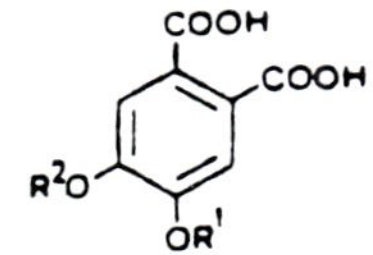

35 $R^1=R^2=H$
36 $R^1=H$; $R^2=CH_3$
37 $R^1=R^2=CH_3$

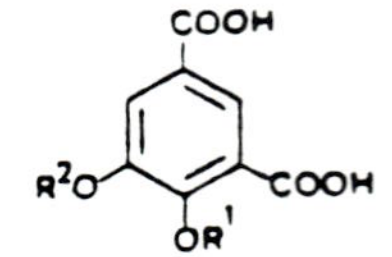

38 $R^1=H$; $R^2=CH_3$
39 $R^1=R^2=CH_3$

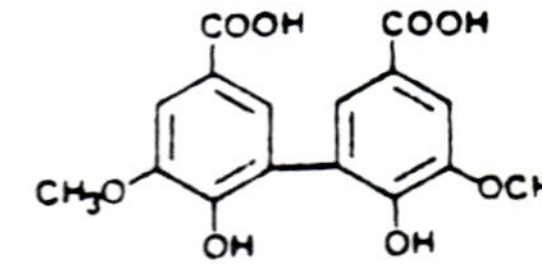

40

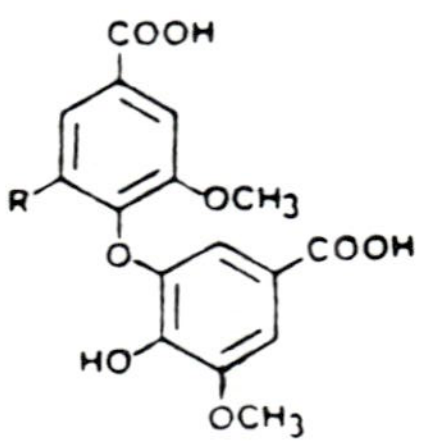

41 R=H
42 $R=OCH_3$

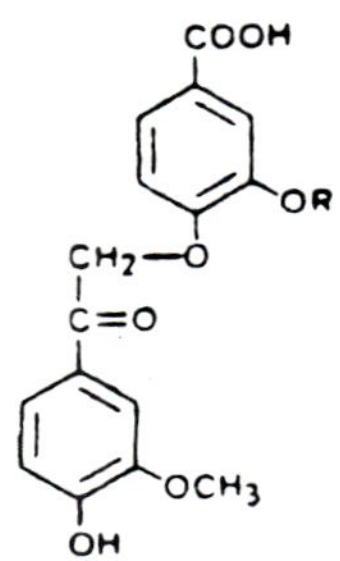

43 R=H
44 $R=OCH_3$

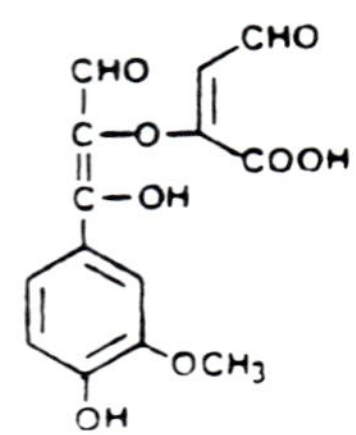

45

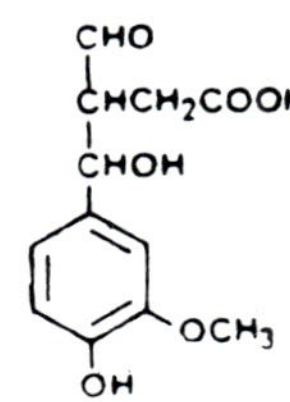

46

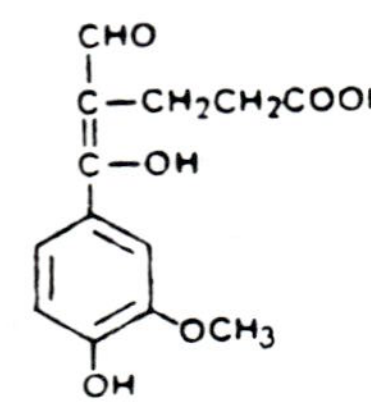

47

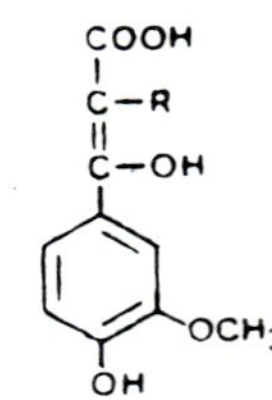

48 $R=CH_2OH$
49 $R=CH_2COCOOH$
50 $R=CH_2CH_2OH$

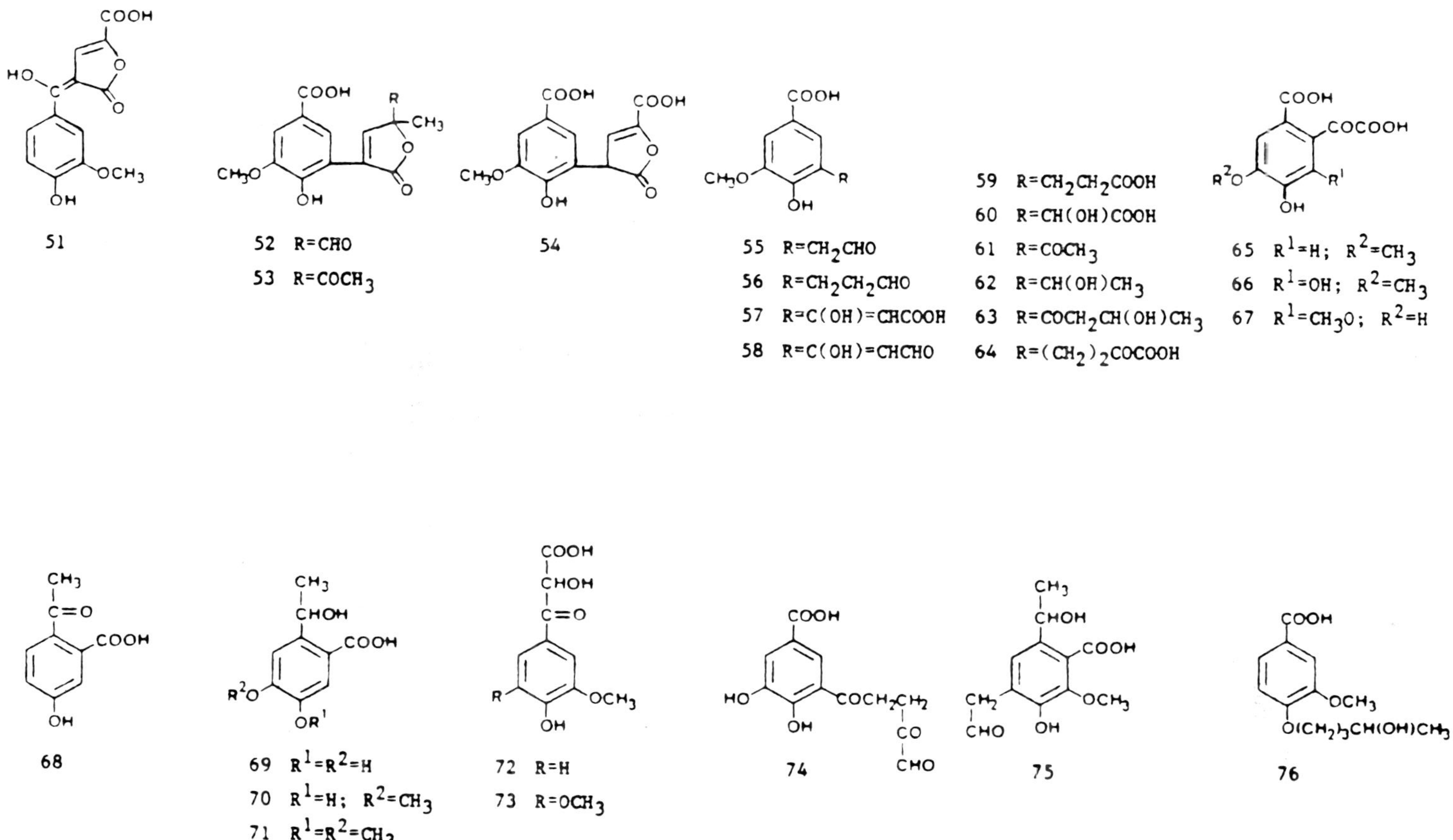

Fig. 3. Compounds identified in spruce and birch methanol extracts low molecular weight acidic fractions (M-LAs) by gas chromatography/mass spectrometry.

results), as shown in Fig. 2. No attempt was made to identify numerous minor constituents in the phenolic fractions of both species. Among the compounds isolated, α-conidendrin is probably not a lignin biodegradation product but is a part of the extractives of spruce wood and was not removed completely during extraction of wood prior to fungal decay. Spruce wood is known to contain a considerable amount of this compound (Freudenberg and Knof, 1957).

A total of 28 carboxylic acids were identified in the M-LA fraction of spruce (Chen *et al.*, 1982a, 1983). These include compounds **26**, **28–30**, **36–41**, and **43–58** in Fig. 3, as well as ferulic acid and 2-(2-methylpropyl)-3-phenylpropanoic acid. Vanillic acid (**29**) was the major component, accounting for ~20% of the acidic fraction. Isovanillic acid (**28**), veratric acid (**30**), *meta*-hemipinic acid (**36**), isohemipinic acid (**39**), 2-(2-hydroxy-3-methoxy-5-caroxyphenyl)-4-carboxy-β-butenolide (**54**), and 5-(2-oxoethyl)vanillic acid (**55**) were moderately abundant components, each forming more than ~3% but less than 10% of the fraction. The other compounds were minor components.

The M-LA fraction from birch was even more complicated. About 100 compounds were identified, including aromatic acids, aliphatic acids, phenols, and even some hydrocarbons (Tai *et al.*, 1982, 1983b; Terazawa *et al.*, 1983). Among the carboxylic acids identified in the fraction, 37 acids contained either a guaiacyl or syringyl moiety and must have been produced by degradation of the lignin. These carboxylic acids were compounds **26–42**, **55**, **57**, and **59–76** in Fig. 3. Syringic acid (**33**) and vanillic acid (**29**) were the major components, accounting for about 20 and 10% of the acidic fraction, respectively. Isovanillic acid (**28**), 5-*O*-methylgallic acid (**31**), 4,5-*O*-dimethylgallic acid (**32**), 3,4,5-*O*-trimethylgallic acid (**34**), 2-carboxy-5-hydroxy-4-methoxybenzoylformic acid (**65**), 2-carboxy-4,5-dihydroxy-6-methoxybenzoylformic acid (**67**), and 2-hydroxy-3-(4-hydroxy-3,5-dimethoxyphenyl)-3-oxopropanoic acid (**73**) were moderately abundant components, each forming ~5% of the fraction. The other compounds were minor components.

It should be noted that the identities of compounds **20** through **44** were established by comparison of their retention times in HPLC and GC, including spiking, and by comparison of their mass spectral data with those of the authentic samples. The structures of the other compounds were elucidated solely on the basis of GC–MS analyses and therefore should be regarded as tentative. In spite of this uncertainty, the identification of these numerous compounds structurally related to lignin provides collectively a valuable basis for understanding the pathways by which lignin is degraded in nature.

IV. Reaction Pathways Involved in Lignin Biodegradation by White-Rot Fungi

While characterization of polymeric degraded lignins isolated from decayed woods provides valuable information concerning the type of reactions and pathways involved in the fungal degradation of lignin at the macromolecular level, a major limitation of this approach is the difficulty of detailed structural characterization of the complicated lignin macromolecule. On the other hand, structural identification of the low molecular weight compounds, as discussed in the previous section, is relatively easier, but again there are inherent limitations associated with this approach. These include several uncertainties: (1) whether an identified compound is a fungal degradation product or a fungal metabolite, (2) whether it is a primary degradation product or a secondary degradation product, (3) whether a major identified product is a true major lignin degradation product or an accumulated product that the fungus is not capable of assimilating, and (4) whether there are major degradation products not identified because of rapid assimilation by the fungus after their formation.

In spite of these limitations, the combination of these two approaches provides a good basis for deducing the common reaction pathways involved in the degradation of lignin by *P. chrysosporium*. These pathways, discussed in this section, are further supported by the results of the physiological and biochemical studies discussed in Chapter 21 and by the results of the biodegradation of lignin model compounds discussed in Chapter 20.

A. *Cleavage of C_α—C_β Bonds in Side Chains*

It has been shown that cleavage of the C_α—C_β bond occurs at the macromolecular level during the fungal degradation of lignin. The identification of many benzoic acid, benzyl alcohol, and benzaldehyde derivatives in both the polymeric fungus-degraded lignins and among the low molecular weight degradation products clearly demonstrates that C_α—C_β cleavage is a major degradative reaction pathway. Recently, a purified enzyme catalyzing this reaction has been isolated from ligninolytic cultures of *P. chrysosporium* by Tien and Kirk (1983; see also Chapter 21). The reaction pathway has also been demonstrated in the degradation of dilignols (dimeric model compounds of lignin) in ligninolytic cultures of *P. chrysosporium* (see Chapter 20).

Cleavage of the C_α—C_β bond in lignin by the fungus is probably the most important reaction involved in lignin biodegradation. The mechanism of the reaction has been elucidated by model compound studies, using isotopic techniques, and by enzymatic degradation. The enzyme catalyzing this reaction is an oxygenase, incorporating an oxygen atom into the β-carbon and requiring an α-hydroxyl for the cleavage reaction to proceed, as shown in Scheme 1. The cleavage results in the formation of benzaldehyde derivatives that are either further oxidized to benzoic acid derivatives or reduced to benzyl alcohol derivatives (see Chapters 20 and 21 for details).

The enzyme catalyzing the cleavage of the C_α—C_β bond appears to be rather nonstereospecific. The C_α—C_β cleavage probably also occurs in many different lignin substructures as long as there is a hydroxyl group at the α-carbon. It is known to operate in β-*O*-4, and β-1 substructures, and in both phenolic or etherified units. Judging from the many aromatic dicarboxylic acids identified (Fig. 3), the C_α—C_β cleavage probably also occurs in condensed structures such as β-5 and β-6 substructures.

B. *Cleavage of C_β—C_γ Bonds in Side Chains*

In addition to cleaving the C_α—C_β bonds, the fungus appears to be able to cleave the C_β—C_γ bonds in side chains also. An example of the C_β—C_γ cleavage is shown in Scheme 2 for a β-*O*-4 substructure. The side chain of the A ring is subsequently cleaved via C_α—C_β cleavage to form a benzoic acid derivative, whereas the α-hydroxyl group of the side chain of the B ring is oxidized to an α-carbonyl group. The γ-hydroxyl group may or may not be oxidized to a carboxylic acid. The resulting intermediate presumably is converted to a substructure of the type **13** via a reverse aldol-type reaction, and subsequent release of this substructure from the lignin macromolecule would give compound **44.** Whereas structure **13** and its syringyl counterpart **14** are detected by ^{13}C NMR in polymeric degraded lignins (Fig. 1), compound **44** and acetosyringone (**24**) are identified among the low molecular weight degradation products of lignin (Figs. 2 and 3). Furthermore, oxidation of the α-hydroxyl group to an α-carbonyl group is known to occur in the fungal degradation of lignin (Hata, 1966; Kirk and Chang, 1975), and indeed, the very enzyme catalyzing the C_α—C_β cleavage was also found to catalyze C_α-oxidation (Tien and Kirk, 1983; cf. Chapter 21). Interestingly, when the α-hydroxyl group is oxidized to an α-carbonyl group, C_α—C_β cleavage is no longer feasible, and instead the C_β—C_γ cleavage evidently results. To what extent the C_α-oxidation and the subsequent C_β—C_γ cleavage occur and how this is regulated by the fungus are not understood at the present.

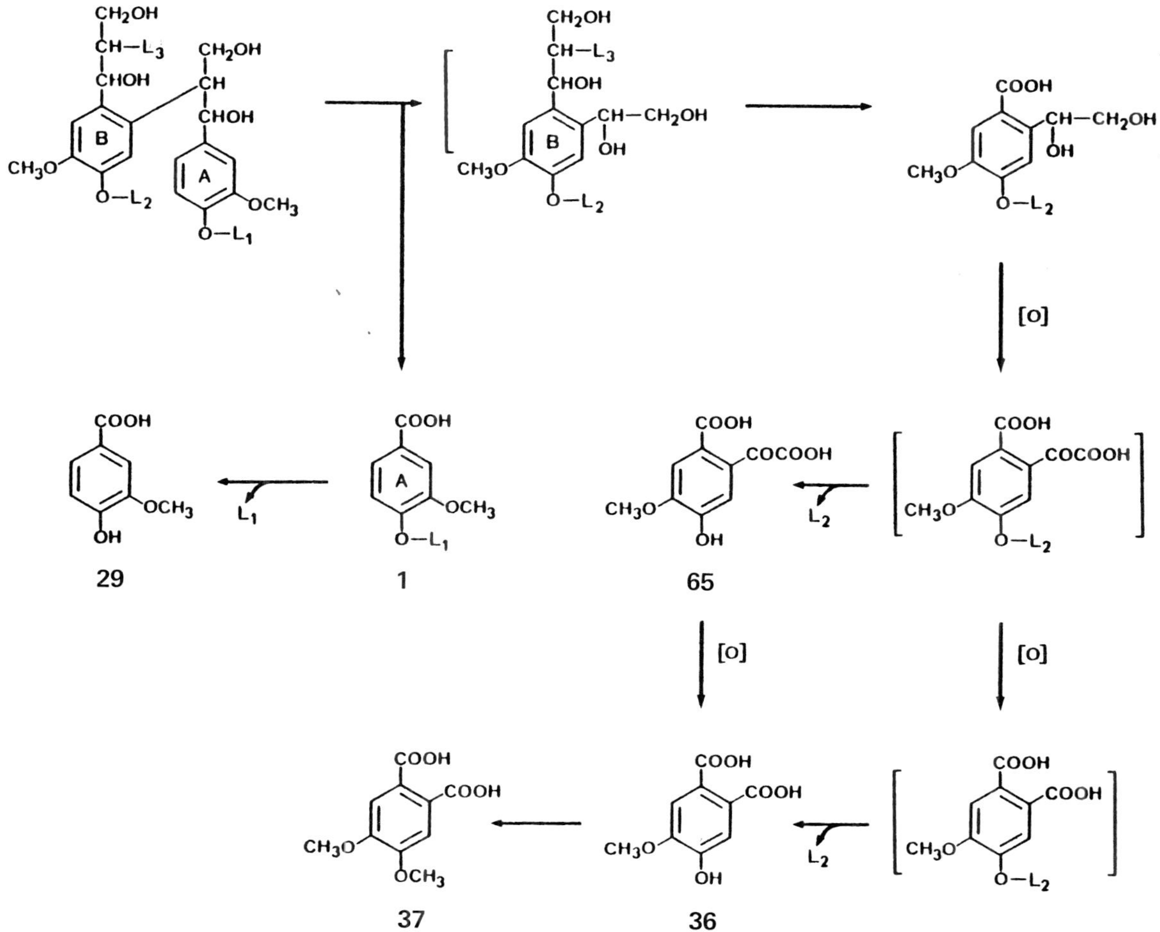

Scheme 1. Possible pathways for fungal degradation of nonphenolic β-O-4 and β-6 substructures involving cleavage of C_α—C_β bonds in side chains; L_1, L_2, L_3 = lignin moiety.

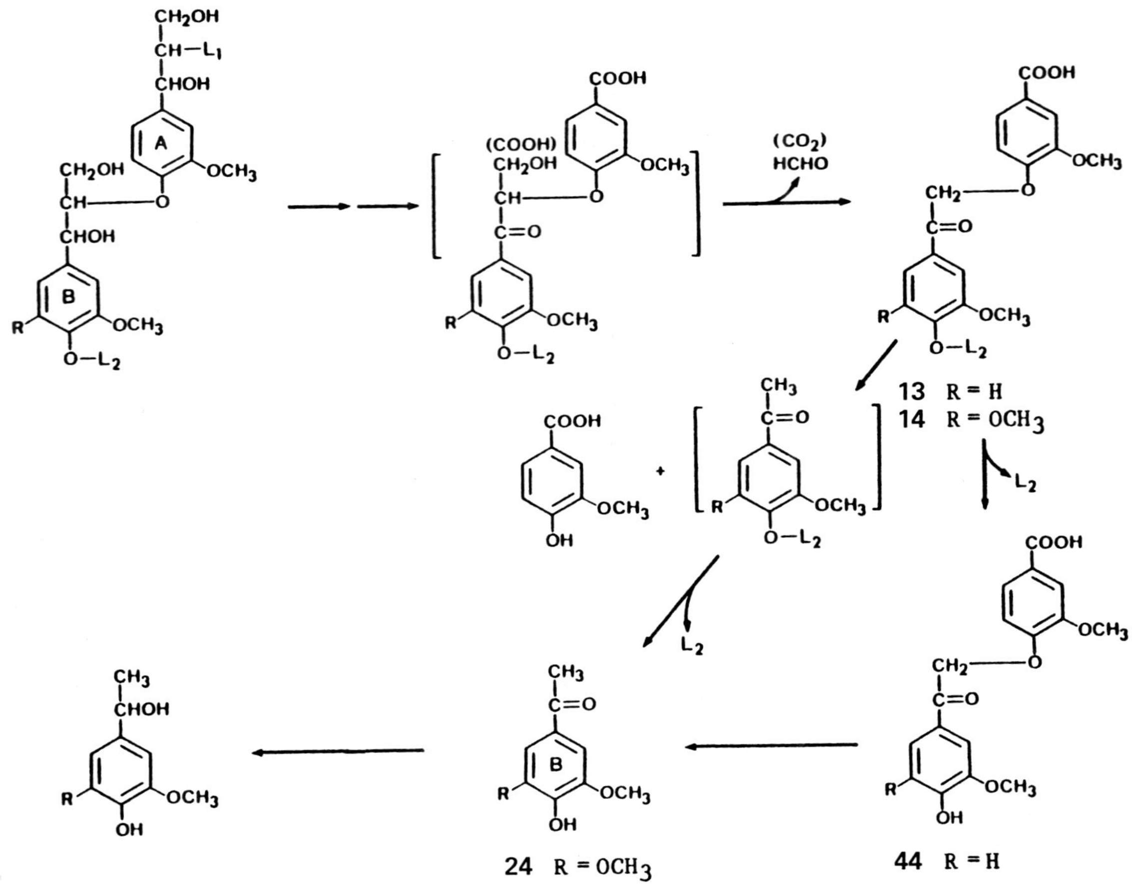

Scheme 2. Possible pathways for fungal degradation of nonphenolic β-O-4 substructures involving cleavages of

C. *Reductive Cleavage of β-O-4 Bonds*

The formation of acetosyringone (**24**) and compounds **68–71** and **75** (from β-*O*-4 substructures) can only occur via the cleavage of the C_β—C_γ bond followed by reductive cleavage of the β-*O*-4 bond, as shown in Scheme 2. Although there is no other evidence in support of this reductive cleavage of the β-*O*-4 structure in the ligninolytic system of *P. chrysosporium,* the reductive cleavage is by no means an uncommon reaction. It occurs in degradation of lignin under sod-AQ pulping conditions (Schroeder, 1980).

D. *3-O- and 5-O-Demethylation and 4-O-Methylation*

Formation of compounds **27**, **28**, **31**, **32**, **35**, **43**, **66**, **67**, **69**, **72**, and **74** involves 3-O-demethylation of the guaiacyl moiety or 3-O- or 5-O-demethylation of the syringyl moiety in corresponding *O*-methyl derivatives of the aforementioned compounds. It is not certain, however, whether these reactions occurred in polymeric fragments or after the *O*-methyl derivatives were released from lignin fragments. Lignin substructures derived from oxidative cleavage of aromatic moieties were detected in polymeric biodegraded spruce and birch lignins (Chua *et al.,* 1982; Tai *et al.,* 1983a,b). This fact suggests that 3-O- and 5-O-demethylations could occur in part at the macromolecular level, since oxidative cleavage of aromatic rings in phenols related to lignin by wood-rotting fungi involves intradiol cleavage of a catechol structure (Buswell and Eriksson, 1979; Buswell *et al.,* 1979) produced presumably via 3-O-demethylation of guaiacyl moieties and 3-O- or 5-O-demethylation of syringyl moieties according to the nature of lignins. Thus, it is conceivable that some of the O-demethylated substructures could be released from lignin fragments prior to further fungal degradation of the substructures via oxidative cleavage of aromatic moieties.

Indeed, Frick and Crawford (1983) reported an enzyme in ligninolytic cultures of *P. chrysosporium* that, in the presence of H_2O_2, catalyzes the demethylation of polyguaiacol. They also reported that the same reaction is also catalyzed by horseradish peroxidase in the presence of H_2O_2. Higuchi *et al.* (1983) demonstrated that syringic acid (**33**) is demethylated in ligninolytic cultures of *P. chrysosporium.*

In addition to 3-O-demethylation and 5-O-demethylation, the formation of compounds **28**, **30**, **32**, **36**, **37**, **39**, and **71** implies methylation of the phenolic hydroxyl group at C-4 of both guaiacyl and syringyl groups. Methylation of the 4-hydroxyl group in compounds related to lignin was reported earlier to occur in cultures of *Lentinus lepideus* (a basidiomycete) (Shimazano, 1959; Wat and Tower, 1975). Lundquist and Kirk (1978) found that *P. chrysosporium* synthesizes *de novo* 3,4-dimethoxybenzyl alcohol. This finding indicates occurrence of 4-O-methylation, and indeed, Shimada *et al.* (1981) have shown that ferulic acid undergoes 4-O-methylation in cultures of the fungus to give 3,4-dimethoxycinnamic acid, which is, in turn, degraded to 3,4-dimethoxybenzyl alcohol by way of 3,4-dimethoxycinnamyl alcohol (see Chapter 21). It has also been found that chlorinated guaiacols and vanillin are methylated in ligninolytic cultures of *P. chrysosporium* (Chang *et al.*, 1983). Thus, 4-O-methylation activity in ligninolytic cultures of this fungus has been demonstrated, but its role in lignin biodegradation is not clear.

E. *Oxidative Cleavage of Aromatic Rings*

The involvement of oxidative cleavage of aromatic rings in lignin biodegradation has long been suspected from earlier studies (Kirk and Chang, 1975; Chang *et al.*, 1980; Ellwardt *et al.*, 1981). Characterization of fungus-degraded lignin with ^{13}C NMR (Chua *et al.*, 1982; Tai *et al.*, 1983a,b) and identification of compounds **45–64** and **72–76** among the fungal degradation products (Chen *et al.*, 1983; Tai *et al.*, 1982, 1983b; Terazawa *et al.*, 1983) give strong support to the involvement of aromatic ring cleavage. Compounds **39**, **66**, and **67** are produced from β-*O*-4 substructures, compounds **46–51** from β-1 substructures, compounds **52–64** and **75** from 5-5′ substructures, and compound **76** from 4-*O*-5′ substructures. Thus, it appears that this oxidative cleavage of aromatic rings is operative indiscriminately in most major substructures.

Closer analysis of the structure of compounds resulting from aromatic ring cleavage suggests that two separate pathways are involved in the oxidative cleavage of aromatic rings. For cleavage of aromatic rings with a free phenolic hydroxyl group, cleavage occurs between C-3 and C-4 after demethylation at C-3. For cleavage of aromatic rings with an etherified phenolic hydroxyl group, the cleavage occurs between C-2 and C-3 after demethylation at C-3 and hydroxylation at C-2. Examples of these two pathways are shown in Schemes 3 and 4, respectively. Once the rings are cleaved by either of the two pathways, fragmentation of the resulting muconic acid derivatives appears to proceed with a surprisingly simple pattern: the addition of a water molecule to either one of the two double bonds, followed by reverse aldol addition of the resulting β-hydroxy carboxylic acid. This results in the formation of many of the compounds and structures identified as fungal degradation products and shown in Figs. 2 and 3.

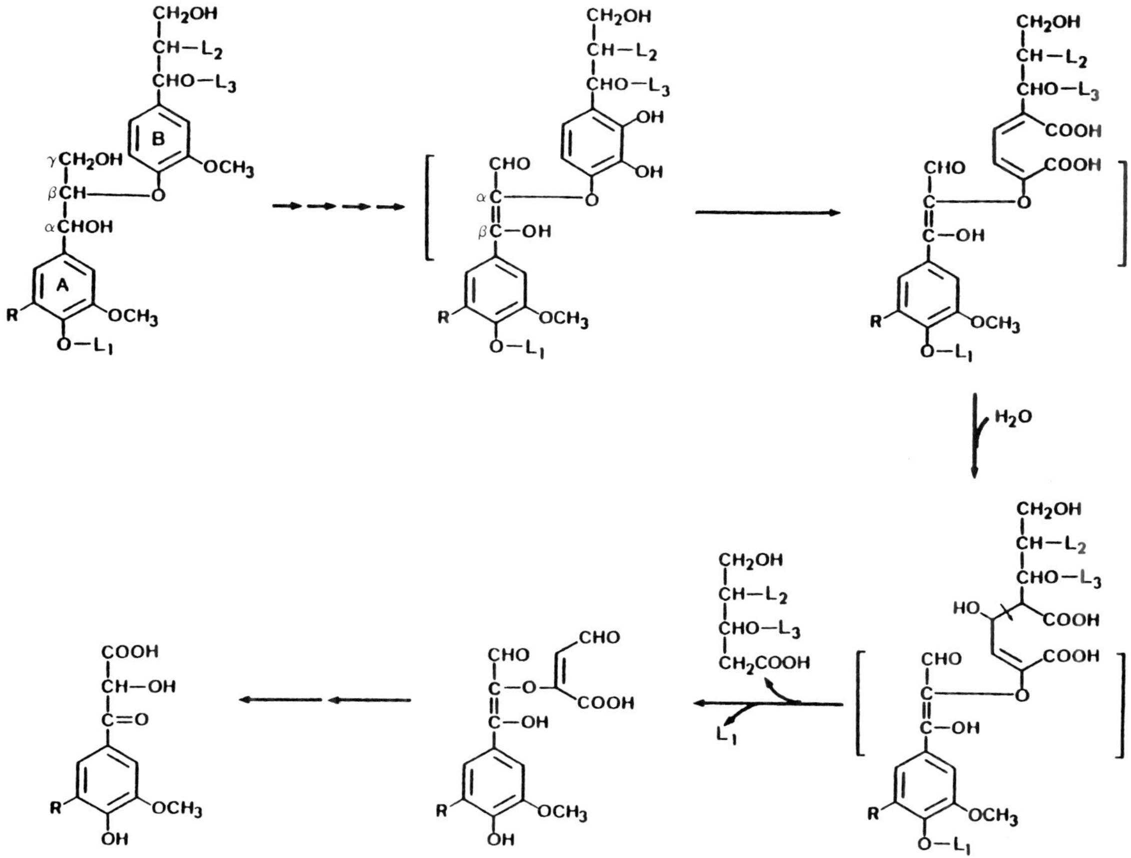

72 R = H
73 R = OCH3

45 R = H

Scheme 3. Possible pathways for fungal degradation of nonphenolic *β-O*-4 substructures involving oxidative cleavage of aromatic ring; L_1, L_2, L_3 = lignin moiety.

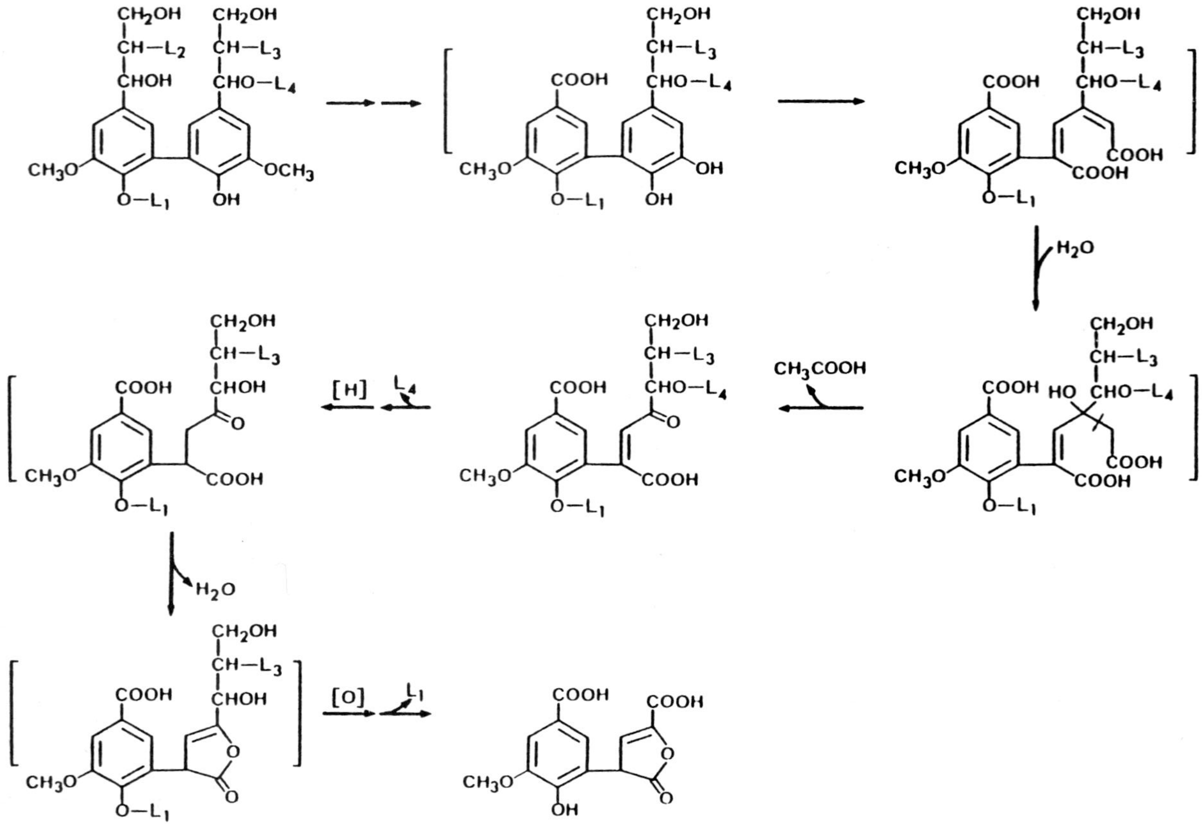

54

Scheme 4. Possible pathways for fungal degradation of phenolic 5—5′ substructures involving cleavage of C_α—C_β bonds in side chains and oxidative cleavage of aromatic rings; L_1, L_2, L_3, L_4 = lignin moiety.

It is interesting to note that formation of all three compounds resulting from ring cleavage of β-*O*-4 structures (**45**, **72**, and **73** in Scheme 3) involves oxidation of hydroxyl groups at C_α and C_γ, resulting in the formation of an enolic cinnamaldehyde structure prior to oxidative ring cleavage (Scheme 3). Formation of the cinnamaldehyde structure apparently results in decreased susceptibility of the side chain to oxidative cleavage of the C_α—C_β bond and thereby facilitates oxidative ring cleavage at ring B.

F. *Other Reactions*

In addition to the reaction pathways discussed above, there are other reactions that are implicated by the results of studies discussed in Sections II and III. Especially noteworthy is the formation of compounds **56–64** as fungal degradation products of lignin. These compounds derive mainly from oxidative ring cleavage and have been reduced to various extents. The reduced structures are also detected by ^{13}C NMR in fungus-degraded lignins. The nature and role of these reducing reactions in lignin biodegradation is not clear, but it is well established that the ligninolytic system of *P. chrysosporium* is capable of reducing vanillic acid and vanillin to vanillyl alcohol (Nakatsubo *et al.*, 1980).

Other reactions that may be involved are hydroxylation reactions. Whereas hydroxylation of the aromatic ring is implicated in the ring cleavage reaction of nonphenolic substructures, hydroxylation of the vinylene group and even the saturated benzyl carbon has been reported to be catalyzed by the oxygenase enzyme that catalyzes C_α—C_β bond cleavage (see Chapter 21). The hydroxylation of carbon–carbon double bonds also occurs in degradation of lignin model compounds in ligninolytic cultures of *P. chrysosporium* (Nakatsubo *et al.*, 1980).

V. Concluding Remarks

The progress in the chemistry of lignin biodegradation has been very rapid during the past few years. This progress is the result of contributions by several different but complementary approaches, summarized in this chapter and Chapters 20 and 21. This chapter summarizes the contribution from a pure chemical approach.

There appear to be some discrepancies between the results from studies on the fungal degradation of lignin model compounds (Chapter 20) and those from studies on the fungal degradation of lignins occurring in woods as summarized in this chapter. Some pathways for fungal degradation of lignins suggested by the latter studies were not found in the former studies, and vice versa. Three major factors may have contributed to these discrepancies. First, the culture conditions were quite different. The degradation of model compounds was carried out in liquid ligninolytic cultures in which the model compounds were dissolved. In contrast, the degradation of lignins was conducted by growing the fungus on preextracted wood chips without submerging them in liquid. Second, molecular weights of the model compounds are substantially less than those of lignins occurring in woods (300–600, versus >20,000). Finally, lignin is one of the integral components of wood tissues, in addition to cellulose and hemicelluloses. These components constitute the ultrastructure of wood cells in which lignin occurs between cell walls (Chapters 2 and 3). The composite nature of wood and the structure of wood tissues are not taken into consideration in studies on the fungal degradation of lignin model compounds. Thus, the process of fungal degradation of lignins occurring in woods must have been somewhat different from that of fungal degradation of model compounds. An important point is that in the model compound studies, degradation products could be rapidly assimilated by the fungus under the liquid culture conditions, whereas similar products could diffuse into inaccessible areas of wood cell walls in the solid wood studies.

In spite of this limitation, the model compound studies have provided valuable basic knowledge to elucidate possible pathways for the formation of fungal degradation products from naturally occurring lignins. Moreover, once a degradation pathway has been elucidated, the model compound approach makes it possible to study in detail the nature of reactions involved. Models are also valuable in accessing the various reactions that a ligninolytic enzyme or enzyme system can catalyze.

Lignin is a complex macromolecular natural product in which the monomers are linked to each other by many different linkages. Considering the complex nature of lignin in terms of chemical structure, the degradation pathways discussed in this chapter are rather simple and straightforward. It is interesting to note that *P. chrysosporum* appears to degrade preferentially the side chains and aromatic rings rather than intermonomeric linkages. The enzyme isolated from the ligninolytic cultures of the fungus also exhibits a low specificity with respect to substrate as well as to configuration of its substrates. Further research is required to isolate and characterize other enzymes involved in the degradation of lignins in order to examine the veracity of the pathways proposed here. Conversely, the pathways provide the basis to search for the enzymes involved.

References

Buswell, J. A., and Eriksson, K.-E. (1979). *FEBS Lett.*, **104,** 258–260.

Buswell, J. A., Ander, P., Petterson, B., and Eriksson, K. E. (1979). *FEBS Lett.*, **103,** 98–101.

Chang, H.-M., Chen, C.-L., and Kirk, T. K. (1980). *In* "Lignin Biodegradation: Microbiology, Chemistry and Potential Applications" (T. K. Kirk, T. Higuchi, and H.-M. Chang, eds.), Vol. 1, pp. 215–230. CRC Press, Boca Raton, Florida.

Chang, H.-M., Joyce, T. W., Campbell, A. G., Gerrard, E. D., Huynh, V.-B., and Kirk, T. K. (1983). *In* "Recent Advances in Lignin Biodegradation Research" (T. Higuchi, H.-M. Chang, and T. K. Kirk, eds.), pp. 257–268. Uni Publishers, Tokyo, Japan.

Chen, C.-L., Chang, H.-M., and Kirk, T. K. (1982a). *Holzforschung* **36,** 3–9.

Chen, C.-L., Chua, M. G. S., Evans, J., Chang, H.-M., and Kirk, T. K. (1982b). *Holzforschung* **36,** 239–247.

Chen, C.-L., Chang, H.-M., and Kirk, T. K. (1983). *J. Wood Chem. Technol.* **3,** 35–59.

Chen, C.-L., and Chang, H.-M. (1981). Unpublished results.

Chua, M. G. S., Chen, C.-L., Chang, H.-M., and Kirk, T. K. (1982). *Holzforschung* **36,** 165–172.

Cowling, E. B. (1961). *U.S. Dept. Agric. Tech. Bull.* No. 1258, 1–79.

Ellwardt, P.-Ch., Haider, K., and Ernst, L. (1981). *Holzforschung* **35,** 103–109.

Eriksson, K.-E. (1980). *Pure Appl. Chem.* **53,** 33–43.

Freudenberg, K., and Knof, L. (1957). *Chem. Ber.* **90,** 2857–2869.

Frick, T. D., and Crawford, R. L. (1983). *In* "Recent Advances in Lignin Biodegradation Research" (T. Higuchi, H.-M. Chang, and T. K. Kirk, eds.), pp. 143–152. Uni Publishers, Tokyo, Japan.

Hata, K. (1966). *Holzforschung* **20,** 142–147.

Henderson, M. E. K. (1955). *Nature (London)* **175,** 634–635.

Higuchi, T., Kawamura, I., and Kawamura, H. (1955). *J. Jpn. For.* **37,** 298–302.

Higuchi, T., Kawamura, I., and Hayashi, I. (1956). *Mokuzai Gakkaishi* **2,** 31–35.

Higuchi, T., Nakatsubo, F., Kamaya, Y., and Umezawa, T. (1983). *Proc. Int. Symp. Wood Pulping Chem.* May 23, 27. **3,** 32–37.

Kirk, T. K., and Lundquist, K. (1970). *Seven. Papperstidn.* **73,** 294–306.

Kirk, T. K., and Chang, H.-M. (1974). *Holzforschung* **28,** 218–222.

Kirk, T. K., and Chang, H.-M. (1975). *Holzforschung* **29,** 56–64.

Kirk, T. K., Chang, H.-M., and Lorenz, L. F. (1975). *Wood Sci. Technol.* **9,** 81–86.

Lundquist, K., and Kirk, T. K. (1978). *Phytochemistry* **17,** 1676.

Nakatsubo, F., Kirk, T. K., Shimada, M., and Higuchi, T. (1980). *Arch. Microbiol.* **128,** 416–420.

Schmid, R., and Liese, W. (1964). *Fr. Arch. Microbiol.* **47,** 260–276.

Shimada, M., Nakatsubo, F., Kirk, T. K., and Higuchi, T. (1981). *Arch. Microbiol.* **129,** 321–324.

Shimazano, H. (1959). *Arch. Biochem. Biophys.* **83,** 206–215.

Schroeter, M. (1980). "Hydrolysis of models featuring β-ether structure of lignin in alkali and sodium anthraquinone monosulfonate," Ph.D. Thesis, North Carolina State University, Raleigh, North Carolina.

Tai, D., Terazawa, M., Chen, C.-L., Chang, H.-M., and Kirk, T. K. (1982). *Proc. TAPPI Research and Development Division Conference,* Aug. 29–Sept. 1, 1982, Asheville, NC.

Tai, D., Terazawa, M., Chen, C.-L., Chang, H.-M., and Kirk, T. K. (1983a). *Proc. Int. Symp. Wood Pulping Chem.* **4,** 144–149.

Tai, D., Terazawa, M., Chen, C.-L., Chang, H.-M., and Kirk, T. K. (1983b). *In* "Recent Advances in Lignin Biodegradation Research" (T. Higuchi, H.-M. Chang, and T. K. Kirk, eds.), pp. 44–63. Uni Publishers, Tokyo, Japan.

Terazawa, M., Tai, D., Chen, C.-L., Chang, H.-M., and Kirk, T. K. (1983). *Proc. Int. Symp. Wood Pulping Chem.* **4,** 150–155.

Terazawa, M., Chen, C.-L., and Chang, H.-M. (1983). Unpublished results.

Tien, M., and Kirk, T. K. (1983). *Science* **221,** 661–663.

Wardrop, A. B. (1971). *In* "Lignins" (K. V. Sarkanen and C. H. Ludwig, eds.), pp. 19–41. Wiley(Interscience), New York.

Wat, C.-K., and Tower, G. H. N. (1975). *Phytochemistry* **14,** 663–666.

Wilcox, W. W. (1970). *Bot. Rev.* **36,** 1–28.

CHAPTER **20**

Degradative Pathways of Lignin Model Compounds

TAKAYOSHI HIGUCHI

Wood Research Institute
Kyoto University
Uji, Kyoto, Japan

I. Introduction

Lignin is a three-dimensional anomalous polymer containing many different stable carbon-to-carbon and ether linkages between monomeric phenylpropane units. Lignin is racemic and amorphous. It has no easily hydrolyzable repeating units along a linear backbone as do polysaccharides, proteins, and nucleic acids (Chapter 7, Section II). Hence, investigations on the biodegradation of the model compounds containing major substructures (Chapter 7, Section I) of lignin macromolecules have largely contributed to elucidating the degradation of lignin.

Higuchi and associates (Higuchi and Nakatsubo, 1980) synthesized sev-

Biosynthesis and Biodegradation of Wood Components

ISBN 0-12-347880-4

eral oligolignols and used them for the elucidation of the mechanism of lignin biodegradation with the soil isolate *Fusarium solani* M-13-1 using coniferyl alcohol dehydrogenation polymer (DHP, artificial lignin) as the sole carbon source (Iwahara *et al.*, 1977), and with *Phanerochaete chrysosporium* Burds, a white-rot fungus. Oligolignols were incubated with shaking cultures of *F. solani* M-13-1 and with ligninolytic cultures (Kirk *et al.*, 1978) of *P. chrysosporium*. The degradation products were isolated and identified by NMR and mass spectrometry (MS) to establish the biodegradation pathways for the oligolignols.

The results show that the principal reactions involved in initial degradation of the lignols are oxidation and oxygenation reactions. Oxidation occurs at the α- and γ-hydroxyl groups of the side chains and result in α-carbonyl and α- and γ-carboxyl groups. Oxygenation occurs with C_α—C_β cleavage of the side chains and possibly with aromatic ring cleavage. The results further suggest that various substructure models are degraded by the mediation of extracellular and/or membrane-bound enzymes that attack both low molecular weight and polymeric substrates (Chapter 21, Section IV) via several different pathways involving similar principal reactions.

Lignin macromolecules are composed mainly of various etherified substructures connected by α- and β-*O*-4 linkages, and some of the substructures at the end of macromolecules (10–20%) occur as nonetherified phenolic units. Hence, both phenolic and nonphenolic model compounds have been used for the degradation studies of lignin.

II. Degradation of Phenolic Dilignols

White-rot wood decay fungi, which are efficient lignin-degraders, typically produce phenol-oxidizing enzymes (POs), including laccase and peroxidase (see Chapter 21). Phenolic model compounds added to cultures of white-rot fungi invariably undergo PO-catalyzed oxidations in addition to the other degradative reactions, including oxygenative C_α—C_β cleavage, aromatic ring cleavage, etc., discussed in Section III. Phenolic models are oxidized by POs to their phenoxy radicals, which polymerize via radical coupling. Oxidation of benzyl hydroxyl groups to carbonyl groups, alkyl–phenyl cleavage, and demethylation are, in part, mediated by POs (Kirk, 1983).

A. *Arylglycerol-β-aryl Ether (β-O-4′) Models*

The arylglycerol-β-aryl ether substructure is the most abundant linkage between phenylpropanes in lignin (40–60%), and therefore most of the earlier investigations with lignin substructure model compounds were made using β-O-4′ models (Russell *et al.*, 1961; Ishikawa *et al.*, 1963a,b; Fukuzumi and Shibamoto, 1965).

Ishikawa *et al.* (1963a) reported that guaiacylglycerol-β-guaiacyl ether is converted via β-ether cleavage to guaiacylglycerol, β-hydroxyconiferyl alcohol, guaiacylpyruvic acid, vanillin, and vanillic acid by mycelial pellets and culture filtrates of the white-rot fungi *Fomes fomentarius* and *Collybia vellutipes*. In contrast, Kirk *et al.* (1968a) found that guaiacylglycerol-β-guaiacyl ether is converted to dehydrodiguaiacylglycerol-β-guaiacyl ether by radical coupling mediated by fungal laccase and that the β-ether linkage is not cleaved by the white-rot fungi *Coriolus versicolor* and *Stereum frustulatum*. Syringylglycol-β-guaiacyl ether, which cannot undergo biphenyl coupling, is converted to guaiacoxyacetaldehyde, guaiaicoxyacetic acid, and 2,6-dimethoxy-p-benzoquinone by alkyl–phenyl cleavage (Kirk *et al.*, 1968b).

On the other hand, β-O-4 cleavage of guaiacylglycerol-β-aryl ethers by the bacteria *Pseudomonas acidovorans* and *P. putida* FK-2 has been reported (Crawford *et al.*, 1973; Fukuzumi and Katayama, 1977).

In a series of studies of the degradation of phenolic dilignols by *Fusarium solani* M-13-1, Higuchi and associates (Higuchi, 1982, 1983; Katayama *et al.*, 1980) found that guaiacylglycerol-β-coniferyl ether (**1**) is oxidized to the coniferaldehyde ether (**2**) and to the ferulate ether (**3**), successively, and then the side chain of the ferulate is cleaved between $C_{\alpha'}$ and $C_{\beta'}$ to produce the vanillin ether (**4**), which is oxidized to the vanillic acid ether (**5**). The vanillic acid ether undergoes alkyl–phenyl cleavage to produce glycerol-2-vanillic acid ether (**6**), which is converted to vanillic acid (**8**) with several other compounds. Syringylglycerol-β-vanillic acid ether (**5′**) is cleaved to 2,6-dimethoxy-p-benzoquinone (**7′**) and glycerol-2-vanillic acid ether (**6**), confirming the alkyl–phenyl cleavage of the β-vanillic acid ether. The cleavage was possibly mediated by POs. The degradative pathway of **1** is shown in Fig. 1 (Katayama *et al.*, 1981).

Veratrylglycerol-β-vanillin ether is oxidized to the vanillic acid ether, which is not degraded further by *F. solani*. This indicates that the phenolic hydroxyl group is necessary for degradation of the arylglycerol moiety by this fungus and adds credence to the contention that POs are involved in degradation.

Phanerochaete chrysosporium degrades both phenolic and nonphenolic β-O-4′ compounds. The phenolic β-O-4′ model guaiacylglycerol-β-coniferyl ether is converted to guaiacylglycerol-β-guaiacylglycerol ether (**2′**) possibly via an α'–β' epoxide (Nakatsubo *et al.*, 1981; Higuchi, 1981), which is converted to the vanillic acid ether via the vanillin ether (Fig. 1).

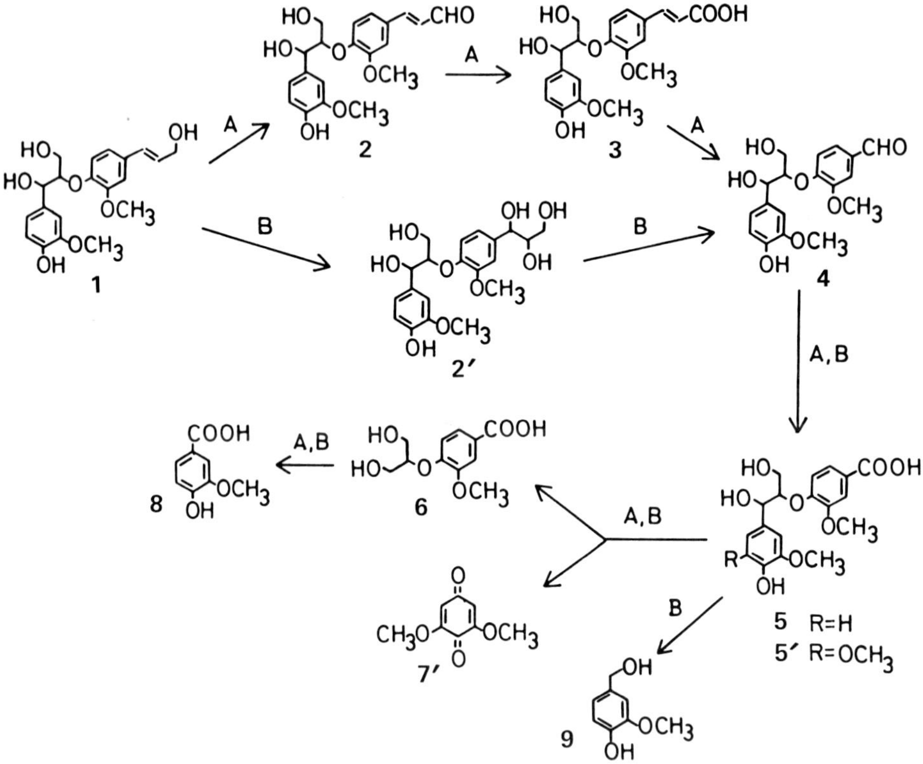

Fig. 1. Degradation pathways of guaiacylglycerol-β-coniferyl ether by *Fusarium solani* M-13-1 and *Phanerochaete chrysosporium*. A, Pathway by *Fusarium;* B, pathway by *Phanerochaete*.

The vanillic acid ether is then degraded either to glycerol-2-guaiacyl ether by alkyl–phenyl cleavage, as in the case of *F. solani* M-13-1, or to vanillyl alcohol (**9**), probably via vanillin as primary product by oxygenative C_α—C_β cleavage, as found for nonphenolic β-*O*-4 models discussed in Section III (Fig. 1). As with *F. solani,* the alkyl–phenyl cleavage is probably PO mediated.

The results showed that the degradation of phenolic arylglycerol-β-aryl ether by *F. solani* proceeds mainly via oxidative cleavage of the carbon-to-carbon linkage between the aromatic ring and the propyl side chain, whereas *P. chrysosporium* degrades phenolic β-*O*-4′ models both via alkyl–phenyl cleavage and C_α—C_β cleavage.

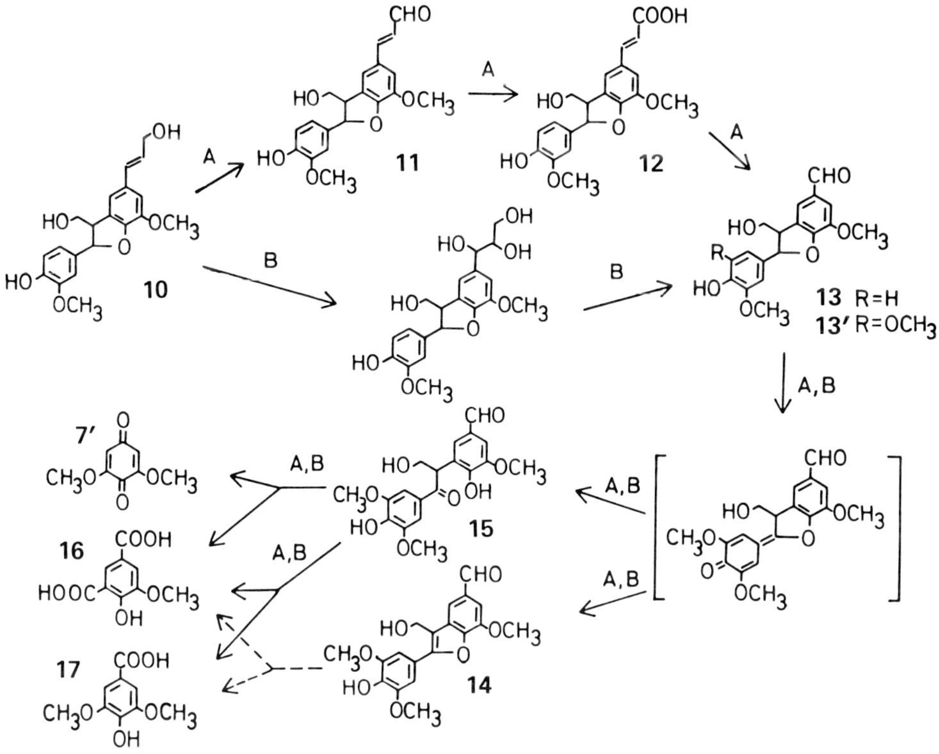

Fig. 2. Degradation pathways of dehydrodiconiferyl alcohol by *Fusarium solani* M-13-1 and *Phanerochaete chrysosporium*. A, Pathway by *Fusarium;* B, pathway by *Phanerochaete*.

B. Phenylcoumaran (β-5′) Models

Dehydrodiconiferyl alcohol (**10**), which is a major substructure (10–15% in lignin), is degraded by *F. solani* M-13-1 to phenylcoumaran α'-aldehyde (**13**) via phenylcoumaran γ'-aldehyde (**11**) and phenylcoumaran γ'-carboxylic acid (**12**), successively. Then phenylcoumaran α'-aldehyde is converted to 5-acetylvanillyl alcohol and 5-carboxyvanillyl alcohol (Ohta *et al.*, 1979). Phenylcoumaran α'-aldehyde with a syringyl group (**13′**) is partly converted either to phenylcoumarone (**14**) or to a propiosyringone derivative (**15**). The latter is converted to 5-carboxyvanillic acid (**16**) and syringic acid (**17**), or to 5-carboxyvanillic acid and 2,6-dimethoxy-*p*-benzoquinone (**7′**), respectively, by both *Fusarium* and *Phanerochaete* (Fig. 2) (Umezawa *et al.*, 1982a; Katayama and Higuchi,

1980). The results indicate that degradation of phenolic phenylcoumaran by these fungi proceeds either via alkyl–phenyl cleavage mediated by POs or via C_α—C_β cleavage mediated by another enzyme, probably an oxygenase, involved in the inherent catabolic system.

C. *Pinoresinol (β-β′) Models*

With *F. solani* M-13-1, syringaresinol (**18**), a major substructure of hardwood lignin (5–10% in lignin), is converted to the α-oxidized compound **20**, γ-lactone (**22**), and 2,6-dimethoxy-*p*-benzoquinone (**7′**). The formation of these degradation compounds indicates oxidative cleavage of the alkyl–phenyl linkage, as shown in Fig. 3 (Kamaya *et al.*, 1981, 1983). Dibenzyl and dimethyl ethers of **18** are not degraded, but the monomethyl ether **19** is converted to **21**, which is not metabolized further by this fungus. This suggests that both phenolic hydroxyl groups are necessary for the complete degradation of syringaresinol by this fungus.

d,l-Syringaresinol monomethyl ether (**19**) and *d,l*-pinoresinol monomethyl ether are converted to analogous degradation products **21**, **23**, and **7′**, as in the degradation of **18**, with *P. chrysosporium* (Kamaya and Higuchi, 1983) (Fig. 3). Unlike the result with *F. solani* M-13-1, a γ-lactone derivative with 3,4,5-trimethoxyphenyl group (**23**) that is not degraded further by *F. solani* M-13-1 is degraded via C_α—C_β cleavage to yield 3,4,5-trimethoxybenzoic acid (**24**) with *P. chrysosporium*. In addition, part of **19** is converted to a lactol (**25**) and 3-O-demethylated to a catechol derivative (**26**) by *P. chrysosporium*. Compound **26** may be degraded via ring cleavage, but dealkylation of the nonphenolic moiety is not observed (Kamaya and Higuchi, 1983) (Fig. 3). The α-oxidized compounds **20**, **21**, and **26** can occur in two chemical structures (hemiketal–keto alcohol) due to the ring–chain tautomerism (Fig. 3).

A trilignol, arylglycerol-β-syringaresinol ether (**27**, R = H or OCH_3), was found to be degraded via the combined pathway of those for phenolic β-*O*-4′ and β-β′ dilignols by *F. solani* M-13-1. Oxidation of the benzylic position, subsequent ring opening of the syringaresinol side chain (see Fig. 3), and cleavage of alkyl–phenyl C—C bonds of both arylglycerol and syringaresinol portions occurs (Kamaya *et al.*, 1983), as shown in Fig. 4. It was also found that syringylglycerol-β-syringaresinol ether (**27**, R = OCH_3) is preferentially oxidized at the benzyl position (α) of the arylglycerol moiety. Isolation of syringaresinol suggests alkyl–phenyl cleavage followed by cleavage of the β-aryl ether linkage in the degradation of the trilignol.

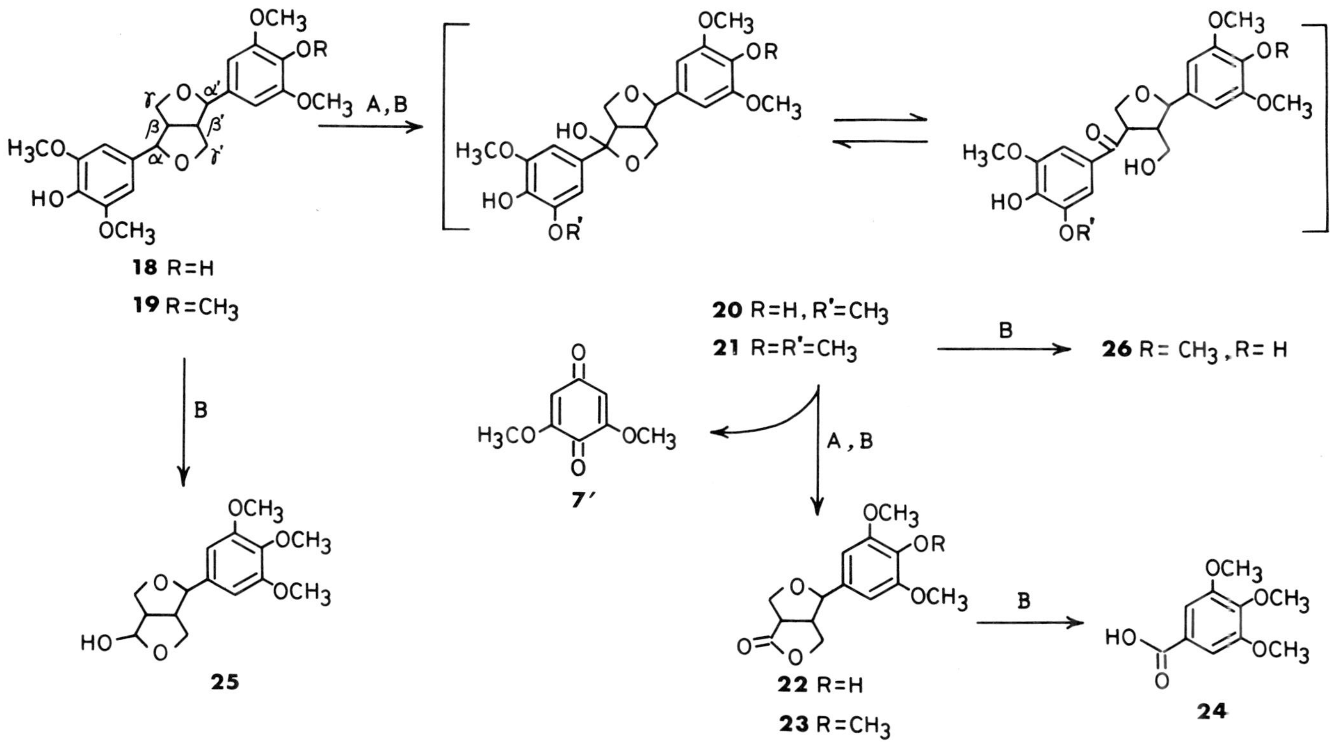

Fig. 3. Degradation pathways of syringaresinol derivatives by *Fusarium solani* M-13-1 and *Phanerochaete chrysosporium*. A, Pathway by *Fusarium;* B, pathway by *Phanerochaete*.

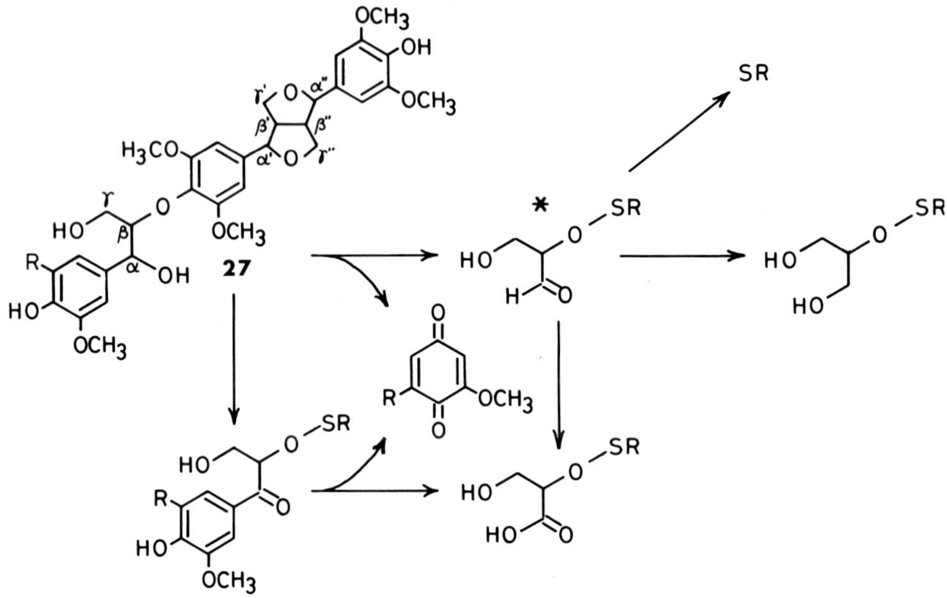

Fig. 4. Degradation pathways of arylglycerol-β-syringaresinol ether by *Fusarium solani* M-13-1. SR, Syringaresinol; asterisk, assumed compound.

D. Diarylpropane-1,3-diol (β-1′) Models

Diarylpropane-1,3-diol is one of the major substructures of lignin (5–15% in lignin). With *F. solani* M-13-1, 1,2-diguaiacylpropane-1,3-diol (**28**) is degraded via alkyl–phenyl cleavage to 2-guaiacylpropane-1,3-diol (**29**) and to its biphenyl dimer (**30**), formed by radical coupling mediated by POs. 1,2-Disyringylpropane-1,3-diol (**28′**), on the other hand, is degraded to 2,6-dimethoxy-*p*-benzoquinone (**7′**), **29′**, syringylglycol (**31**), and syringaldehyde (**32**). The latter two compounds could be formed in part via oxygenative C_α—C_β cleavage (Namba *et al.*, 1983).

With *P. chrysosporium,* 1,2-disyringylpropane-1,3-diol (**28′**) is converted to syringylglycol (**31**), α-hydroxyacetosyringone (**33**), syringaldehyde (**32**), and syringyl alcohol (**34**), but 2,6-dimethoxy-*p*-benzoquinone (**7′**) was scarcely detected (Kamaya and Higuchi, 1984a). This suggests that the degradation of phenolic β-1 models with *F. solani* M-13-1 proceeds via either alkyl–phenyl cleavage or C_α—C_β cleavage, but with *P. chrysosporium* the latter pathway dominates (Fig. 5).

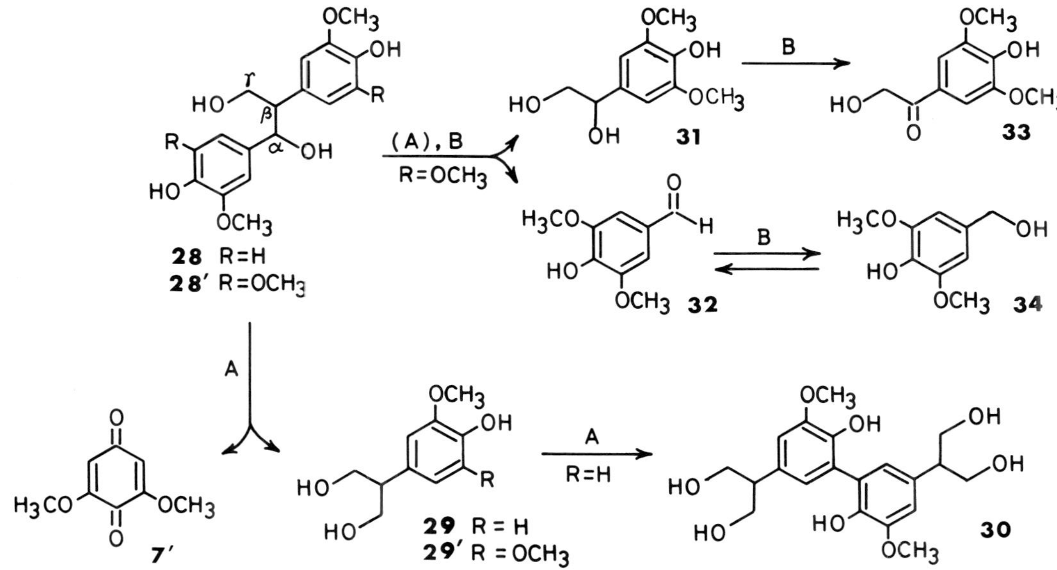

Fig. 5. Degradation pathways of 1,2-diarylpropane-1,3-diols by *Fusarium solani* M-13-1 and *Phanerochaete chrysosporium*. A, Pathway by *Fusarium;* B, pathway by *Phanerochaete*.

III. Degradation of Nonphenolic Dilignols

A. *Arylglycerol-β-aryl Ether (β-O-4′) Models*

Ishikawa *et al.* (1963b) reported that veratrylglycerol-β-guaiacyl ether is first 4-O-demethylated to produce phenolic guaiacylglycerol-β-guaiacyl ether by cultures of the white-rot fungi *Fomes fomentarius* and *Poria subacida*. The latter compound is degraded to guaiacylglycerol via β-ether cleavage, as described in Section II,A. Ishikawa and Oki (1964) later reported that enzyme preparations from *Collybia velutipes* and *Coriolus versicolor* cleave guaiacylglycerol-β-guaiacyl ether, yielding guaiacylglycerol and guaiacol. More recently Oki *et al.* (1982) have reported that an enzyme preparation from *Lentinus edodes* cleaves the β-ether linkage to give guaiacylglycerol, guaiacol, and other products, but the enzyme preparation was not characterized.

In a similar way, Fukuzumi *et al.* (1969) reported 4-O-demethylation and guaiacol liberation from veratrylglycerol-β-guaiacyl ether by an enzyme preparation from *Poria subacida*. In the degradation, O_2 is consumed, and guaiacol liberation is enhanced by adding NADH. They postulated a monooxygenase-catalyzed C_β-hydroxylation of guaiacylglycerol-β-guaiacyl ether to produce a hemiketal, which could be hydrolyzed to guaiacol and aryldihydroxyacetone; the latter is suspected to be reduced to guaiacylglycerol (see Chapter 21).

On the degradation of 4-ethoxy-3-methoxyphenylglycerol-β-guaiacyl ether (**35**) with ligninolytic cultures of *Phanerochaete chrysosporium*, Enoki *et al.* (1980, 1981a,b) identified 4-ethoxy-3-methoxyphenylglycerol (**36**), 4-ethoxy-3-methoxybenzyl alcohol (**37**), guaiacol (**38**), glycerol-2-guaiacyl ether (**39**), and guaiacoxyethanol (**40**) by GC–MS, and concluded that nonphenolic β-*O*-4′ models are converted via three pathways (Fig. 6): alkyl–phenyl cleavage (pathway A), presumably preceeded by 4-O-dealkylation to give guaiacylglycerol-β-guaiacyl ether and catalyzed by POs (see Fig. 1); C_α—C_β cleavage (pathway B), presumed to involve C_γ-oxidation, which has not been established, followed by a reversed aldol reaction; and β-ether cleavage (pathway C), followed by C_α—C_β cleavage in the phenylglycerol product (pathway D). Pathway C–D was presumed to be the major one and is analogous to that reported earlier by Ishikawa and Fukuzumi.

On the other hand, Nakatsubo (1981) (see Higuchi, 1981) found that on incubation of veratrylglycerol-β-coniferyl ether (**41**) with ligninolytic cultures of *P. chrysosporium*, the substrate is converted to the glycerol ether (**42**), which is then converted to the vanillic acid ether (**44**) via the vanillin ether (**43**) (Fig. 7). The vanillic acid ether was then degraded to 4-ethoxy-

3-methoxybenzyl alcohol. Later, Umezawa *et al.* (1983a) found that 4-ethoxy-3-methoxyphenylglycerol-β-guaiacyl-γ-benzyl diether is cleaved to 2-benzyloxyethanol, a C_β—C_γ fragment of the glycerol side chain, 4-ethoxy-3-methoxybenzyl alcohol, and vanillyl alcohol. They further found that α,β-dideuterated 4-ethoxy-3-methoxyphenylglycerol-β-vanillin-γ-benzyl diether (**45**) was degraded to deuterated 4-ethoxy-3-methoxyphenylglycerol-γ-benzyl ether (**46**), vanillyl alcohol (**47**), deuterated 2-benzyloxyethanol (**48**), and deuterated 4-ethoxy-3-methoxybenzyl alcohol (**37**) (Fig. 8). Deuterium at the β position of 4-ethoxy-3-methoxyphenylglycerol-γ-benzyl ether and in 2-benzyloxyethanol is retained almost 100%, indicating that the monooxygenase-catalyzed C_β-oxidation postulated by Fukuzumi *et al.* (1969) is not involved in ligninolytic cultures of *P. chrysosporium*. Moreover, on the degradation of 4-ethoxy-3-methoxyphenylglycerol-β-(5-deuterated vanillin)-γ-benzyl diether (**45′**) under $^{18}O_2$, ^{18}O is not incorporated into 4-ethoxy-3-methoxyphenylglycerol-γ-benzyl ether (**46′**), 2-benzyloxyethanol (**48′**), or into 5-deuterated vanillyl alcohol (**47′**) derived from the β-ether, but it is incorporated into the benzyl position of 4-ethoxy-3-methoxybenzyl alcohol (**37′**), and into the methoxyl groups of veratryl alcohol synthesized *de novo*. 4-Ethoxy-3-methoxybenzyl alcohol (**37**) derived from 4-ethoxy-3-methoxyphenylglycerol-γ-benzyl ether (**46′**) (pathway D, Fig. 8) contains no ^{18}O (Umezawa *et al.*, 1983b). Umezawa and Higuchi (1984) further found that ^{18}O is incorporated into the α-hydroxyl group but not the β-hydroxyl group of 4-ethoxy-3-methoxyphenylglycerol on the degradation of 4-ethoxy-3-methoxyphenylglycerol-β-vanillin ether with ligninolytic cultures of *P. chrysosporium* under $H_2{}^{18}O$ (Fig. 8). The absence of incorporation of ^{18}O into the β-hydroxyl group shows that direct hydrolysis of β-*O*-4′ models is not involved in the formation of 4-ethoxy-3-methoxyphenylglycerol.

Umezawa *et al.*, (1983a,b, 1984) concluded that nonphenolic β-*O*-4′ models are mainly degraded via two pathways, C–D and E, in ligninolytic cultures of *P. chrysosporium* (Fig. 8). In pathway C, 4-ethoxy-3-methoxyphenylglycerol-γ-benzyl ether could not be produced via direct hydrolysis, and is possibly produced by oxygenative degradation of the ether-linked vanillin moiety. 4-Ethoxy-3-methoxyphenylglycerol-γ-benzyl ether is cleaved to 4-ethoxy-3-methoxybenzyl alcohol and 2-benzyloxyethanol (pathway D). Pathway E is a direct oxygenative C_α—C_β cleavage of the substrate to produce [*benzyl*-^{18}O]4-ethoxy-3-methoxybenzyl alcohol and benzyloxyethanol without involvement of 4-ethoxy-3-methoxyphenylglycerol-γ-benzyl ether as intermediate.

In accord with pathway E, Tien and Kirk (1983, 1984) have isolated and characterized an enzyme that directly cleaves the C_α–C_β linkage of aryl-

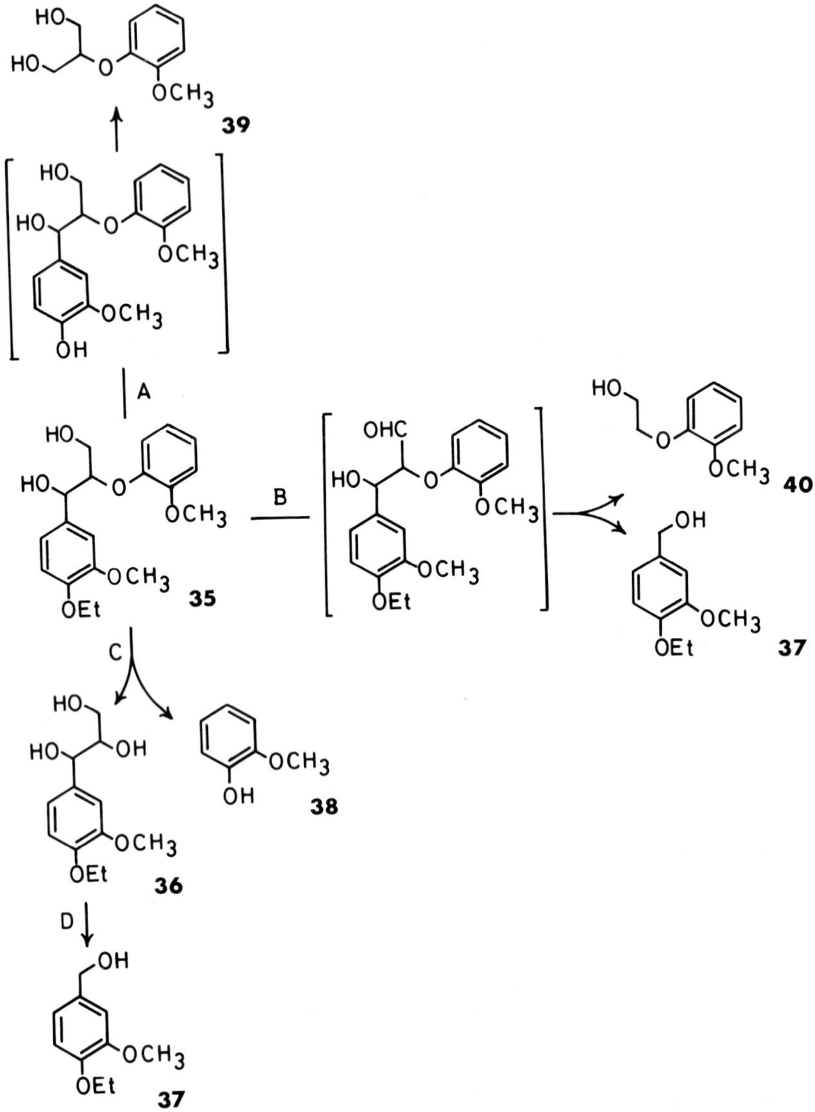

Fig. 6. Degradation pathways of nonphenolic arylglycerol-β-aryl ether models by *Phanerochaete chrysosporium*.

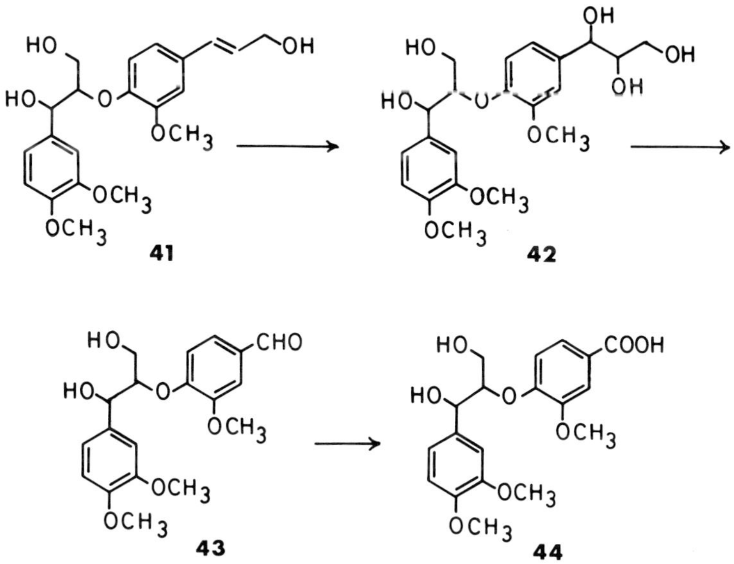

Fig. 7. Oxidative degradation of allyl alcohol side chain of veratrylglycerol-β-coniferyl ether by *Phanerochaete chrysosporium*.

glycerol-β-aryl ether. The enzyme has been purified from the culture filtrate of *P. chrysosporium* (Chapter 21, Section IV).

It is conceivable that β-aryl ether linkages of lignins are mainly cleaved via the two pathways (C–D and E, Fig. 8) to result in low molecular weight fragments that are further degraded and mineralized via C_α—C_β cleavage, aromatic ring cleavage, etc. Tien and Kirk (1983) found that the purified enzyme mentioned above actually depolymerizes spruce and birch lignins via C_α—C_β cleavage.

B. 4-O-Methyldehydrodiconiferyl Alcohol (β-5′) Model

In ligninolytic cultures of *P. chrysosporium*, 4-*C*-methyldehydrodiconiferyl alcohol (**49**) is converted to the corresponding glycerol (**50**), which is then degraded to a vanillic acid derivative (**51**). The coumaran ring is partly converted to coumarone (**52**), which is degraded slowly to a C_6—C_1 acid, as shown in Fig. 9 (Nakatsubo *et al.*, 1981). However, the mechanism of C_α—C_β cleavage of nonphenolic phenylcoumaran has not yet been established.

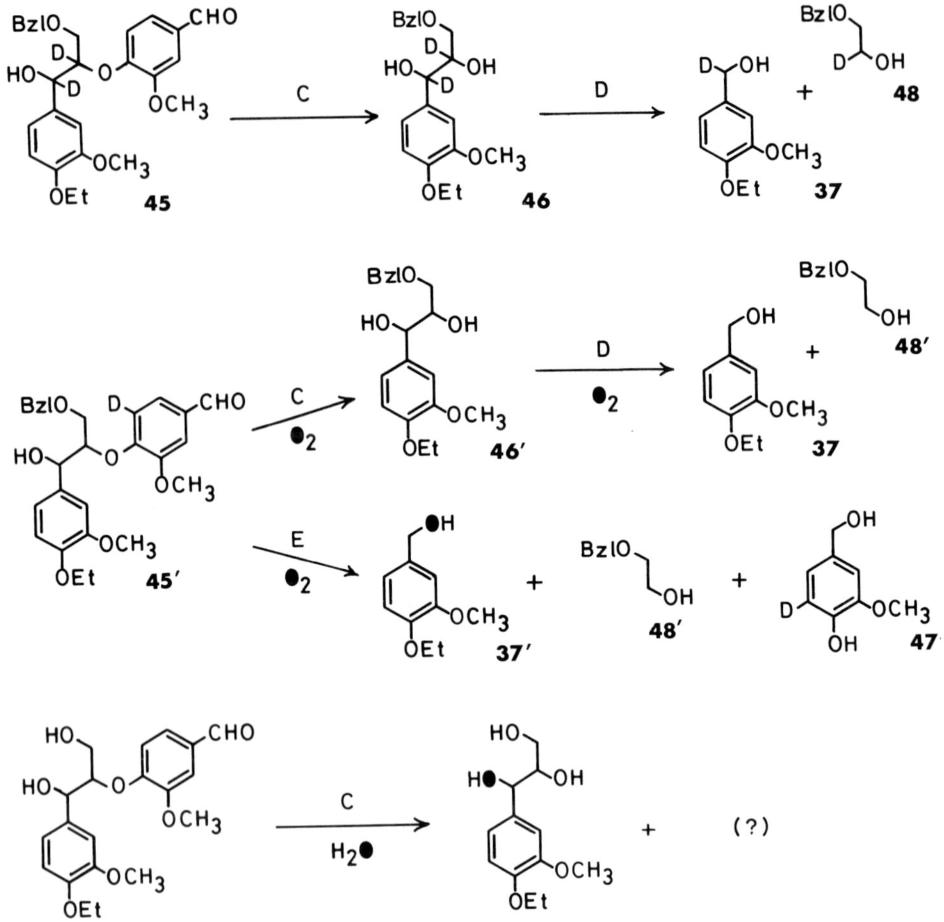

Fig. 8. Degradation pathways (based on isotopic experiments) of 4-ethoxy-3-methoxyphenylglycerol-β-vanillin-γ-benzyl diether by *Phanerochaete chrysosporium*. D, Deuterium; ●, ^{18}O; Bzl, benzyl.

C. *Syringaresinol Dimethyl Ether (β-β′) Model*

d,l-Syringaresinol dimethyl ether is hardly degraded by either *F. solani* M-13-1 or by ligninolytic cultures of *P. chrysosporium*. No C_α—C_β cleavage products such as 3,4,5-trimethoxybenzoic acid and the corresponding aldehyde or alcohol could be detected.

These results suggest that the nonphenolic resinol substructure is comparatively stable to degradation by *P. chrysosporium*, and that in lignin its

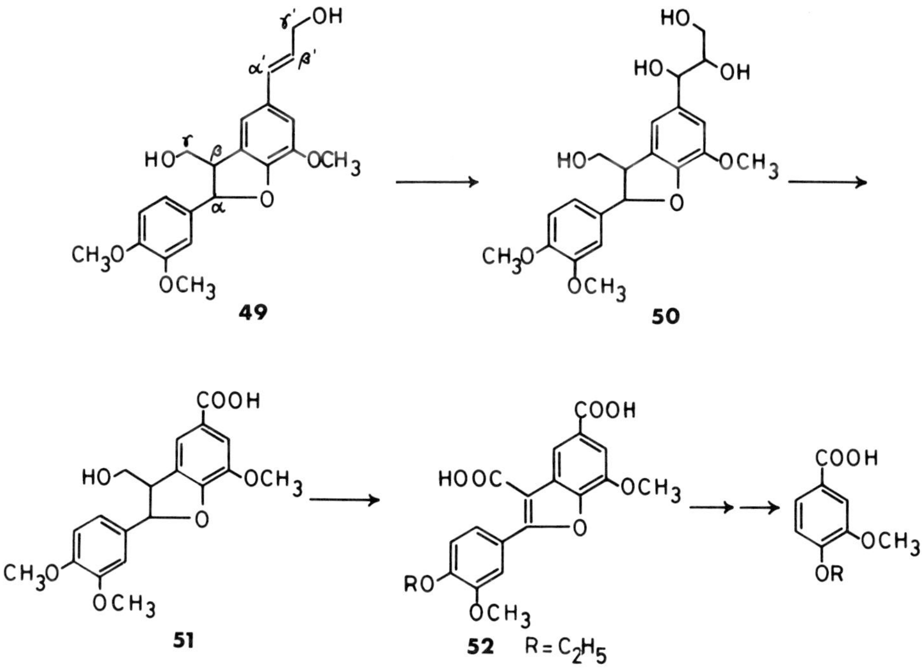

Fig. 9. Degradation pathway of 4-*O*-methyldehydrodiconiferyl alcohol by *Phanerochaete chrysosporium*.

degradation occurs in the phenolic substructure, liberated by cleavage of an ether linkage. The resistance of the resinol substructure to microbial attack was suggested by ^{13}C-NMR spectroscopy using ^{13}C-labeled DHP (Ellwardt *et al.*, 1981).

D. Diarylpropane-1,3-diol (β-1′) Models

The nonphenolic diarylpropane model, 1,2-bis(3-methoxy-4-alkoxyphenyl)propane-1,3-diol (**53**), was found to be degraded to three major products, **54**, **55**, and **56** (Nakatsubo *et al.*, 1982). An analogous result was obtained by Enoki and Gold (1982). The result showed that C_α—C_β cleavage is the dominant reaction. Kirk and Nakatsubo (1983) further showed, by mass spectrometric analysis of the phenylglycol (**54**) and the aromatic alcohol (**57**) that was derived from **56** formed from deuterated β-1 substrate under an ^{18}O atmosphere, that hydrogen atoms at the α and β positions are not lost, and that the benzyl hydroxyl oxygen atom of the phenylglycol (**54**) is derived from molecular oxygen (Kirk and Nakatsubo,

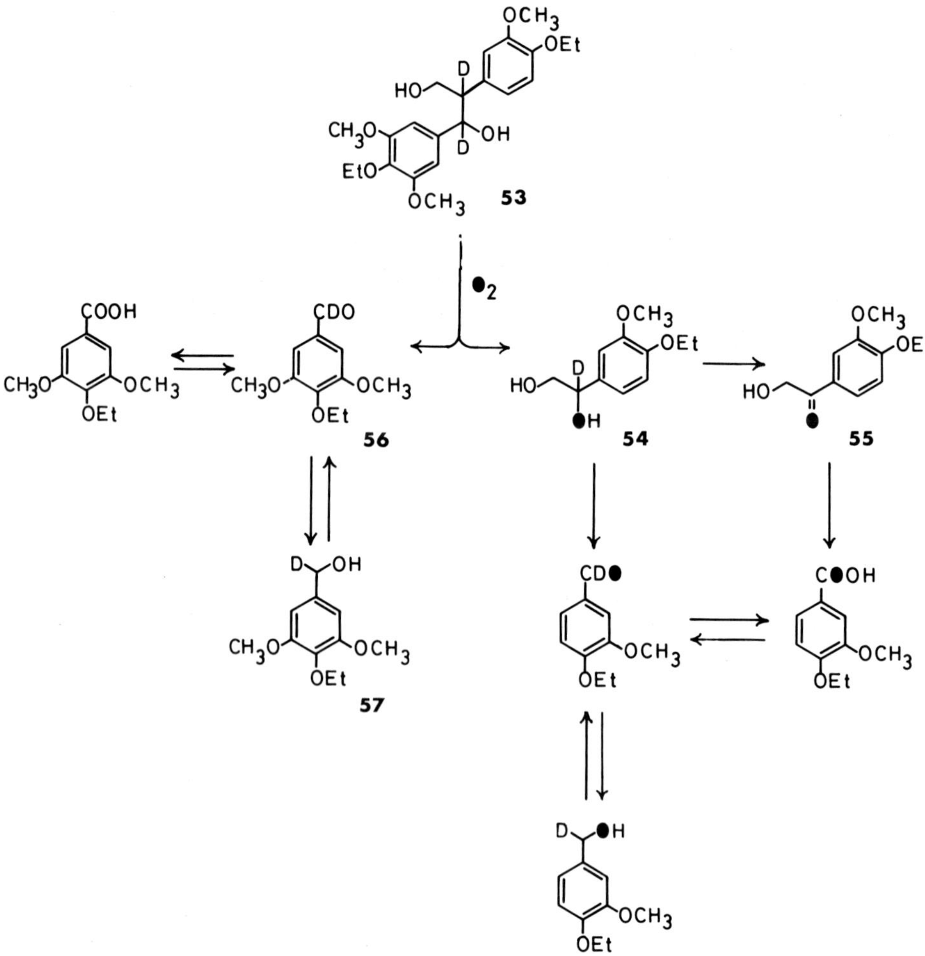

Fig. 10. Degradation pathways of nonphenolic 1,2-diarylpropane-1,3-diols by *Phanerochaete chrysosporium*. D, Deuterium; ●, ^{18}O.

1983). These results established that neither C_α-oxidation, C_β-hydroxylation, nor formation of a C_α—C_β double bond occurs prior to C_α—C_β cleavage (Fig. 10). It was further shown that C_γ is not required for the cleavage (Enoki and Gold, 1982; Kirk and Nakatsubo, 1983), and that molecular oxygen is not incorporated into the C_α position (Umezawa *et al.*, 1983b). The cleavage products were further converted as shown in Fig. 10. Kamaya and Higuchi (1984b) found that *C. versicolor* in low-nitrogen (ligninolytic) cultures degrades nonphenolic diarylpropane-1,3-diol via the same pathway.

Tien and Kirk (1983) found that β-1 models are cleaved to benzaldehyde and phenylglycol products by the extracellular H_2O_2-dependent enzyme preparation from *P. chrysosporium* that also cleaves the β-*O*-4′ model, as described in Section III,A. A similar result was reported recently by Glenn *et al.* (1983): arylglycol (as well as arylglycerol) derived from β-1′ and β-*O*-4′ models are further metabolized via diol C_α—C_β cleavage. Cleavage of the vicinal diol of dihydroanisoin to form *p*-anisaldehyde in ligninolytic cultures of *P. chrysosporium* was shown by Shimada and Gold (1983).

IV. Degradation of Monomeric Aromatic Compounds

It was shown that noncondensed guaiacyl model compounds were converted to vanillic acid and methoxyhydroquinone by *Sporotrichum pulverulentum* (= *P. chrysosporium*) (Ander *et al.*, 1980), whereas condensed guaiacyl substructure models and syringyl models were degraded to 5-carboxyvanillic acid, and to syringic acid and 2,6-dimethoxy-*p*-benzoquinone, respectively, by *P. chrysosporium* (Umezawa *et al.*, 1982b). Buswell *et al.* (1979) demonstrated that with *S. pulverulentum,* vanillic acid (**58**) is decarboxylated to methoxyhydroquinone (**59**), which is metabolized further to CO_2 (Fig. 11). Yajima *et al.* (1979) also studied the vanillate decarboxylase (vanillate hydroxylase) of *P. chrysosporium* and confirmed Buswell's results. On the other hand, Kuwahara *et al.* (1981) found that vanillic acid is metabolized in part via guaiacol with *F. solani* M-13-1. Nonoxidative decarboxylation of vanillic acid by bacteria to yield guaiacol and CO_2 had been reported by Crawford and Olson (1978). Protocatechuic acid, which is a common intermediate in aromatic metabolism in bacteria, has not rigorously been identified as a fungal product.

Umezawa *et al.* (1982b) found that dehydrodivanillic acid (**60**) is converted to 5-carboxyvanillic acid (**61**), which is then converted to 5-hydroxyvanillic acid (**62**) and/or 5-carboxyprotocatechuic acid (**63**) by ligninolytic cultures of *P. chrysosporium*. The latter two compounds are converted to gallic acid (**64**); part of the 5-hydroxyvanillic acid is methylated to 3,4-dimethoxy-5-hydroxybenzoic acid (**65**). On the other hand, syringic acid (**17**) is 3-O-demethylated to 5-hydroxyvanillic acid (**62**), which is metabolized via gallic acid (**64**) by *P. chrysosporium* (Fig. 12).

These results strongly suggest that phenylcoumaran and biphenyl sub-

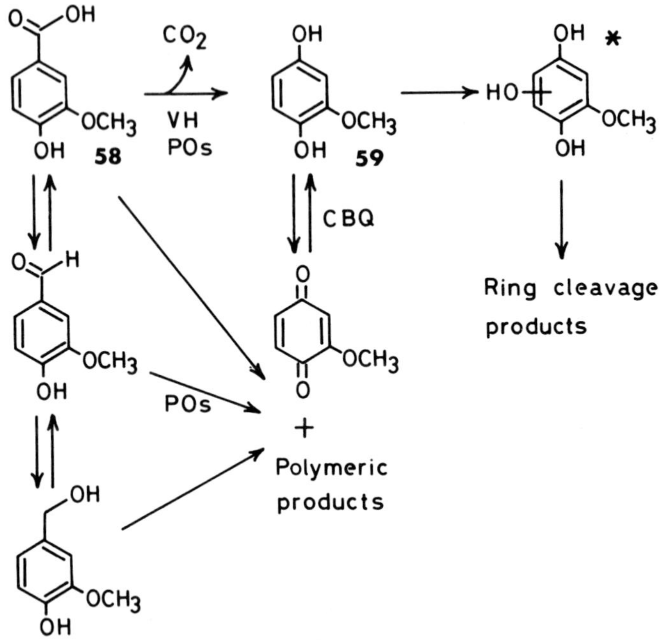

Fig. 11. Degradation pathways of vanillic acid by *Sporotrichum pulverlentum* (*Phanerochaete chrysosporium*). VH, Vanillate hydroxylase; CBQ, cellobiose : quinone oxidoreductase; POs, phenol-oxidizing enzymes; asterisk, assumed compound.

structures of lignins are mainly degraded to 5-carboxyvanillic acid, and syringyl moieties of lignins to syringic acid. Both are converted to gallic acid as the final aromatic monomer.

Oxidative vanillate decarboxylase has been shown to be relatively inactive toward syringic acid but active toward gallic acid (Buswell *et al.*, 1979). Hence, it is conceivable that gallic acid is metabolized further via ring cleavage before and/or after decarboxylation. However, decarboxylation of 5-carboxylvanillic acid to 5-hydroxyvanillic acid and decarboxylation of 5-carboxyprotocatechuic acid to gallic acid could be catalyzed by enzymes other than vanillate decarboxylase. The decarboxylation seems to be specific to 5-carboxy groups, since the isomeric products of C-1 decarboxylation could not be identified.

V. Concluding Remarks

Studies with lignin substructure models have shown that the aromatic moieties with free phenolic hydroxyl groups are initially attacked either

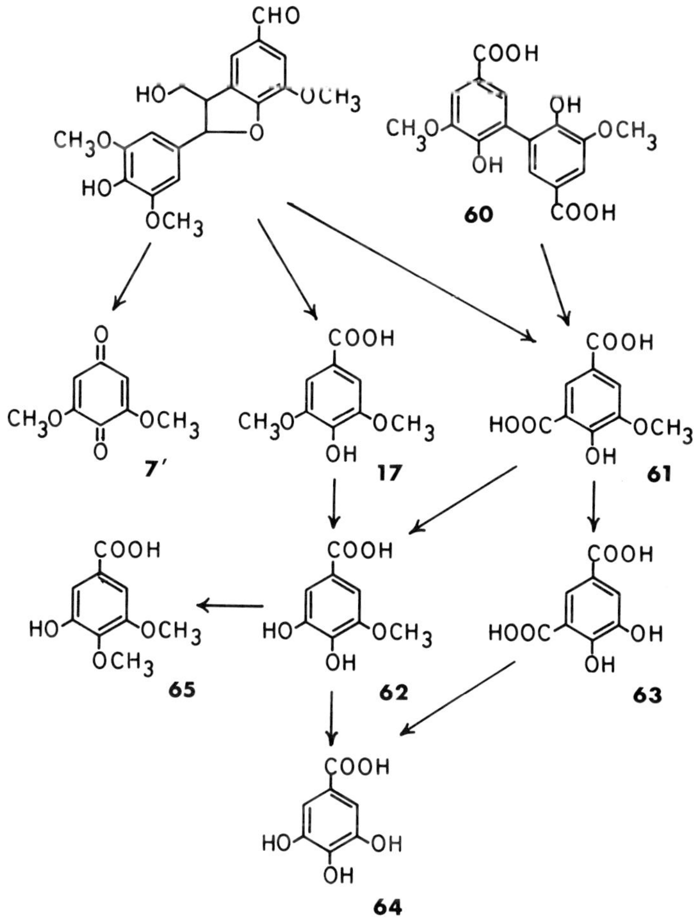

Fig. 12. Degradative transformation of syringic acid and 5-carboxyvanillic acid by *Phanerochaete chrysosporium*.

via alkyl–phenyl cleavage or C_α—C_β cleavage. The preference of the respective cleavage reactions involved seems to depend on the reactivity of the substrate and the physiological condition of the fungi. It seems that the oxygenative C_α—C_β cleavage reaction is generally masked with a dominant alkyl–phenyl cleavage mediated by POs, whose role may not be specific for the inherent catabolism of lignin but rather accidental and complementary for lignin degradation. With *P. chrysosporium*, both phenolic and nonphenolic β-1′ models are degraded rapidly to yield mainly the C_α—C_β cleavage products. Even with *F. solani* M-13-1, a phenolic β-1′ model is degraded via C_α—C_β cleavage as well as via alkyl–phenyl cleavage.

Nonphenolic substructures, which comprise the major portion (80%) of lignin macromolecules, could mainly be degraded via C_α—C_β cleavage mediated by enzymes, probably oxygenases that are involved in the inherent catabolism of lignin.

The formation of alkoxyacetic acid, C_γ-elimination with retention of C_α and C_β, and aromatic ring-cleavage products found in the degradation of spruce and birch protolignins with *P. chrysosporium* (Chapter 19) have not been reported in the investigation with substructure models. However, it is evident that the investigations with substructure models have provided essential information on the mechanism of lignin biodegradation, especially on cleavage of the linkages between monomeric units. Degradation products from substructure models have made it possible to detect specific degrdative reactions, such as oxygenative C_α—C_β cleavage, and the enzymes involved (see Chapter 21).

Several investigations have shown that the ability of microorganisms to degrade simple methoxylated aromatic acids such as vanillic acid, syringic acid, veratric acid, etc. does not necessarily correlate with an ability to degrade lignin. It was shown, on the other hand, that lignin-degrading microorganisms are also degraders of vanillic acid (Haider and Grabbe, 1967; Gradziel *et al.*, 1978). Thus, it is plausible that the correlation between utilization of a model (substructure) and lignin degradation is not that model-compound degraders decompose lignin, but that lignin degraders are able to decompose lignin models (Crawford, 1981).

Additional treatment of the microbial catabolism of lignin model compounds is provided by the reviews of Kirk (1983) and Crawford (1981).

References

Ander, P., Hatakka, A., and Eriksson, K.-E. (1980). *In* "Lignin Biodegradation: Microbiology, Chemistry and Potential Applications" (T. K. Kirk, T. Higuchi, and H.-M. Chang, eds.), Vol. 2, pp. 1–16. CRC Press.

Buswell, J. A., Ander, P., Pettersson, B., and Eriksson, K.-E. (1979). *FEBS Lett.* **103,** 98–101.

Chen, C.-L., Chang, H.-M., and Kirk, T. K. (1983). *J. Wood Chem. Technol.* **3,** 35–57.

Crawford, R. L. (1981). "Lignin Biodegradation and Transformation." Wiley(Interscience) Wiley, New York.

Crawford, R. L., and Olson, P. E. P. (1978). *Appl. Environ. Microbiol.* **36,** 539–543.

Crawford, R. L., McCoy, E., Kirk, T. K., and Harkin, J. M. (1973). *App. Microbiol.* **25,** 322–324.

Ellwardt, P.-C., Haider, K., and Ernst, L. (1981). *Holzforschung* **35,** 103–109.

Enoki, A., and Gold, M. H. (1982). *Arch. Microbiol.* **132,** 123–130.
Enoki, A., Goldsby, G. P., and Gold, M. H. (1980). *Arch. Microbiol.* **125,** 227–232.
Enoki, A., Goldsby, G. P., Krisnangkura, K., and Gold, M. H. (1981a). *FEMS Microbiol. Lett.* **10,** 373–377.
Enoki, A., Goldsby, G. P., and Gold, M. H. (1981b). *Arch. Microbiol.* **139,** 141–145.
Fukuzumi, T., and Katayama, Y. (1977). *Mokuzai Gakkaishi* **23,** 214–215.
Fukuzumi, T., and Shibamoto, T. (1965). *Mokuzai Gakkaishi* **11,** 248–252.
Fukuzumi, T., Takatsuka, H., and Minami, K. (1969). *Arch. Biochem. Biophys.* **129,** 396–409.
Glenn, J. K., Morgan, M. A., Mayfield, M. B., Kuwahara, M., and Gold, M. H. (1983). *Biochem. Biophys. Res. Commun.* **114,** 1077–1083.
Gradziel, K., Haider, K. Kochmanska, J., Malarczyk, E., and Trojanowski, J. (1978). *Acta Microbiol. Pol.* **27,** 103–109.
Haider, K., and Grabbe, K. (1967). *Zentralbl. Bakteriol. Parasitenkd. Infectionskr. Hyg.* **205,** 91–96.
Higuchi, T. (1981). *Wood Res. (Kyoto)* **67,** 47–58.
Higuchi, T. (1982). *Experientia* **38,** 159–166.
Higuchi, T. (1983). *In* "The New Frontiers in Plant Biochemistry" (T. Akazawa, T. Asahi, and H. Imaseki, eds.), pp. 23–46. Japan Sci. Soc. Press, Tokyo.
Higuchi, T., and Nakatsubo, F. (1980). *Kemia-Kemi* **9,** 481–488.
Ishikawa, H., and Oki, T. (1964). *Mokuzai Gakkaishi* **10,** 207–213.
Ishikawa, H., Schubert, W. J., and Nord, F. F. (1963a). *Arch. Biochem. Biophys.* **100,** 131–139.
Ishikawa, H., Schubert, W. J., and Nord, F. F. (1963b). *Arch. Biochem. Biophys.* **100,** 140–149.
Iwahara, S., Kuwahara, M., and Higuchi, T. (1977). *Hakkokogaku* **55,** 325–329.
Kamaya, Y., and Higuchi, T. (1983). *Mokuzai Gakkaishi* **29,** 789–794.
Kamaya, Y., and Higuchi, T. (1984a). *Mokuzai Gakkaishi.* **30,** 237–239.
Kamaya, Y., and Higuchi, T. (1984b). *FEMS Microbiol. Lett.* **22,** 89–92.
Kamaya, Y., Nakatsubo, F., Higuchi, T., and Iwahara, S. (1981). *Arch. Microbiol.* **129,** 305–309.
Kamaya, Y., Nakatsubo, F., and Higuchi, T. (1983). *Agric. Biol. Chem.* **47,** 299–308.
Katayama, T., and Higuchi, T. (1980). *Abst. 25th Symp. Lignin Chem.*, pp. 5–8. Fukuoka, Japan.
Katayama, T., Nakatsubo, F., and Higuchi, T. (1980). *Arch. Microbiol.* **126,** 127–132.
Katayama, T., Nakatsubo, F., and Higuchi, T. (1981). *Arch. Microbiol.* **130,** 198–203.
Kirk, T. K. (1983). *Proc. Int. Symp. Wood Pulping Chem.* **3,** 7–12.
Kirk, T. K., and Nakatsubo, F. (1983). *Biochim. Biophys. Acta* **756,** 376–384.
Kirk, T. K., Harkin, J. M., and Cowling, E. B. (1968a). *Biochim. Biophys. Acta* **165,** 134–144.
Kirk, T. K., Harkin, J. M., and Cowling, E. B. (1968b). *Biochim. Biophys. Acta* **165,** 145–163.
Kirk, T. K., Schultz, E., Connors, W. J., Lorenz, L. F., and Zeikus, J. G. (1978). *Arch. Microbiol.* **117,** 277–285.
Kuwahara, M., Takagami, H., Yonehara, M., Sato, T., and Iwahara, S. (1981). *Mokuzai Gakkaishi* **27,** 885–892.
Nakatsubo, F., Kirk, T. K., Shimada, M., and Higuchi, T. (1981). *Arch. Microbiol.* **128,** 416–420.
Nakatsubo, F., Reid, I. D., and Kirk, T. K. (1982). *Biochim. Biophys. Acta* **719,** 284–290.
Namba, H., Nakatsubo, F., and Higuchi, T. (1983). *Wood Res. (Kyoto)* **69,** 52–60.

Ohta, M., Higuchi, T., and Iwahara, S. (1979). *Arch. Microbiol.* **121,** 23–28.
Oki, T., Senba, Y., and Ishikawa, H. (1982). *Mokuzai Gakkaishi* **28,** 800–808.
Russel, J. D., Henderson, M. E. K., and Farmer, V. C. (1961). *Biochim. Biophys. Acta* **52,** 565–570.
Shimada, M., and Gold, M. (1983). *Arch. Microbiol.* **134,** 299–302.
Tien, M., and Kirk, T. K. (1983). *Science* **221,** 661–663.
Tien, M., and Kirk, T. K. (1984). *Proc. Natl. Acad. Sci. U.S.A.* **81,** 2280–2284.
Umezawa, T., and Higuchi, T. (1984). *Agric. Biol. Chem.* **48,** 1917–1921.
Umezawa, T., Nakatsubo, F., and Higuchi, T. (1982a). *Arch. Microbiol.* **131,** 124–128.
Umezawa, T., Higuchi, T., and Nakatsubo, F. (1982b). *Abst. 32th Ann. Meeting Wood Res. Soc.,* p. 234. Fukuoka, Japan.
Umezawa, T., Nakatsubo, F., and Higuchi, T. (1983a). *Agric. Biol. Chem.* **47,** 2677–2681.
Umezawa, T., Higuchi, T., and Nakatsubo, F. (1983b). *Agric. Biol. Chem.* **47,** 2945–2948.
Yajima, Y., Enoki, A., Mayfield, M. B., and Gold, M. H. (1979). *Arch. Microbiol.* **123,** 319–321.

CHAPTER **21**

Lignin Biodegradation: The Microorganisms Involved and the Physiology and Biochemistry of Degradation by White-Rot Fungi

T. KENT KIRK

Forest Products Laboratory
Forest Service
U.S. Department of Agriculture
Madison, Wisconsin

MIKIO SHIMADA

Wood Research Institute
Kyoto University
Uji, Kyoto, Japan

I. Introduction

Chapters 19 and 20 provide a summary of the chemistry of lignin degradation by white-rot fungi. The major purpose of this chapter is to build on that background to discuss the biochemistry of this degradation. As a complement to the biochemistry, we have also summarized the physiological features of lignin metabolism by the white-rot fungi. For completeness, we begin the chapter with a brief discussion of the microbiology of lignin biodegradation, pointing to the prominent—if not preeminent—role of the white-rot fungi.

Lignin biodegradation has been a popular topic for research in recent years, and has been reviewed many times. The most recent reviews (Higuchi, 1982; Janshekar and Fiechter, 1983; Kirk, 1984) and two books (Crawford, 1981; Higuchi *et al.* 1983a) provide detailed coverage of certain aspects.

II. The Microbiology of Lignin Degradation

Older published reports indicated that lignin is degraded by many different bacteria and fungi, including the microbes in both aerobic and anaerobic environments (see Kirk, 1971). Newer studies have been conducted with sensitive assays for degradation based on ^{14}C-labeled lignins. Those investigations have shown that only aerobes, primarily wood-decomposing members of the basidiomycetes, metabolize lignin efficiently. Members of other taxa cause partial degradation (Table I). Lignin may also be degraded by consortia of microbes, none of which alone can completely metabolize it. This section treats the various taxa reported to degrade wood and/or lignin.

The microbial groups that can decompose lignin in axenic culture are essentially the same ones that can decay intact wood. This correlation is not surprising, because lignin in wood physically protects the cellulose and hemicelluloses from enzymatic attack. Even though a wide range of bacteria and fungi degrade isolated cellulose and hemicelluloses, only those with ligninolytic capability gain access to these same polymers in intact wood. Fine grinding of wood disrupts the lignin barrier, permitting the wood to be degraded by a much wider range of microorganisms (Kirk and Cowling, 1984). Various wood-inhabiting insects grind wood fine enough to render the cellulose and hemicelluloses accessible to their intestinal microflora, which is often, if not always, anaerobic.

Anaerobic decay of intact wood, if it occurs at all, is very slow, and searches for anaerobic degradation of lignin, using assays based on ^{14}C-labeled lignins, have not revealed any (Hackett *et al.*, 1977; Odier and Monties, 1983). On the other hand, anaerobic metabolism of isolated cellulose and hemicelluloses (Ng *et al.*, 1981) and of finely ground wood is

Table I

Microorganisms That Degrade Lignin[a]

Microbe classification (representative) species)	Experimental substrate	Experimental method[b]	Relative rate[c]
Actinomycetes			
Nocardia[d]	^{14}C-Labeled synthetic lignin	I	Low
Streptomyces[d]	^{14}C-Labeled lignocelluloses	I, III, IV	Low
Fungi imperfecti, Ascomycetes			
Fusarium[d]	^{14}C-Labeled synthetic lignin	I, II	Low
Chaetomium[e]	^{14}C-Labeled synthetic lignin, solid wood	I, III	Moderate
Xylaria[f]	Solid wood	III	Low
Basidiomycetes			
Collybia[g]	Forest litter	III	Moderate
Gloeophyllum[h]	Solid wood, ^{14}C-labeled synthetic lignin	I, III, IV	Low
Phanerochaete[i]	Solid wood, ^{14}C-labeled synthetic lignin	I, III, IV	High

[a] Adapted from Kirk (1984).
[b] Methods: I, [^{14}C]lignins → $^{14}CO_2$; II, growth on lignin; III, mass balance study based on gravimetric determination of lignin; IV, chemical structural studies.
[c] Crude estimate based on literature reports; lignin degradation by most microbes has not been optimized.
[d] Soil organisms.
[e] Soft-rot wood decay fungus.
[f] Pseudo-white-rot wood decay fungus.
[g] Litter-decomposing basidiomycete.
[h] Brown-rot wood decay fungus.
[i] White-rot wood decay fungus.

well established (see Kirk and Cowling, 1984). The apparent absence of lignin metabolism in the absence of molecular oxygen may be related to the molecular size of the lignin. Thus, low molecular weight aromatics ($<\sim 300$ daltons) related to lignin, including those derived from lignin by chemical degradation, are metabolized anaerobically, whereas high molecular weight ones ($M > \sim 850$) are not (Zeikus *et al.*, 1982). This result suggests that the initial fragmentation of the lignin polymer apparently requires oxygen (Zeikus *et al.*, 1982; see also Colberg and Young, 1982). Resistance of lignified tissues to anaerobic microbial degradation might account for the Earth's coal deposits; indeed, evidence for a lignin origin of coal has been presented (e.g., Hayatsu *et al.*, 1979).

Aerobic decay of intact wood by bacteria (other than by actinomycetes) is also unknown, and lignin degradation by these organisms has not been demonstrated. As with anaerobic bacteria, many aerobic bacteria degrade isolated cellulose and hemicelluloses. They also readily degrade low molecular weight aromatics related to lignin (Cain, 1980; Kawakami, 1980). The basis for their apparent failure to metabolize lignin itself is unclear.

Limited degradation of intact wood by actinomycetes has been observed (e.g., Baecker and King, 1980), and these organisms can cause slow and partial degradation of isolated lignin (Haider and Trojanowski, 1980; Crawford and Sutherland, 1980). Based on the published work, it seems that grass lignin and the lignin in phloem tissues may be more susceptible to actinomycete degradation than are the lignins in xylem tissues of woody plants.

The lower fungi (Phycomycetes) are not known to degrade wood, and lignin degradation by these organisms is also unknown.

Various soft rots of wood are caused by members of the fungi imperfecti and Ascomycetes. These decays involve lignin degradation, although the polysaccharides are preferentially attacked (Levi and Preston, 1965; Eslyn *et al.*, 1975). Rather rapid degradation of isolated lignin by one soft-rotting ascomycete, however, has been demonstrated (Haider and Trojanowski, 1980). Degradation of lignin by soft-rot fungi has received very little research attention.

The major wood decays—white and brown rots—are caused by Basidiomycetes. Evidence indicates that brown-rot fungi degrade cellulose and hemicelluloses in intact wood initially with a small nonenzymatic oxidative agent that can penetrate the woody matrix (see Kirk and Cowling, 1984). Because lignin apparently presents no real barrier to the agent, the brown-rot fungi would seem to be an exception to the generalization that only lignin-degraders can degrade intact wood. They do degrade the lignin to a limited extent, however, even oxidizing it in part to CO_2 (Kirk *et al.*, 1975; Haider and Trojanowski, 1980). Their major effect is demethylation of methoxyl groups (Kirk and Adler, 1970; Kirk, 1975).

The white-rot wood decay fungi and closely related litter-decomposing Basidiomycetes are probably the major lignin-degraders. Chapters 19 and 20 deal primarily with degradation by white-rot fungi, and the remaining sections of this chapter, too, deal with that group. The most studied species from the standpoint of lignin degradation is *Phanerochaete chrysosporium* Burdsall [previously named *Penophora* "G," *Sporotrichum pruinosum* Gilman et Abbott, *S. pulverulentum* Novobranova, and *Chrysosporium lignorum* Bergman et Nilsson (Burdsall and Eslyn, 1974)]. *Coriolus versicolor* (L.: Fries) Quél [previously named *Polystictus versicolor* (L.: Fries) Sacc. and *Polyporus versicolor* L.: Fries] has also been a favorite white-rot fungus for study. Perhaps 25–30 additional species—of the several thousand that are known—have been studied, but only to a limited extent (Crawford, 1981).

Kirk and Cowling (1984) provide a more comprehensive treatment of wood biodecomposition. Several reviews of lignin biodegradation have included sections on microbiology and provide additional detail and a more complete introduction to the literature (Crawford, 1981; Janshekar and Fiechter 1983; Kirk, 1984).

III. Physiology of Lignin Degradation by White-Rot Fungi

Little is yet known about the physiological aspects of lignin degradation by microorganisms other than white-rot fungi. Even though much remains to be learned about the white-rot fungi as well, several of the features that have been investigated are unusual and deserve discussion as an introduction to the biochemistry of lignin biodegradation. Almost all investigations have been with *P. chrysosporium*.

A. *Lignin as a Source of Carbon and Energy*

Although there is no obvious reason why the oxidative degradation of lignin by white-rot fungi should not provide sufficient energy for growth and lignin degradation, it apparently does not, as suspected by Cowling 24 years ago (Cowling, 1961). Reports of growth of white-rot fungi on lignin (Pelczar *et al.*, 1950) have not been substantiated (Kirk *et al.*, 1976; Leisola *et al.*, 1983b; Ulmer *et al.*, 1983). Furthermore, degradation of only the lignin in wood during decay has not been observed—except in one study with *P. chrysosporium*, in which glucose was provided as a cosubstrate (Yang *et al.*, 1980). In another study, a cellulase-negative mutant of *S. pulverulentum* (*P. chrysosporium*) still degraded lignin in wood without attacking the cellulose, but it simultaneously degraded the hemicelluloses (Ander and Eriksson, 1976a). The basis for a cosubstrate requirement is not yet known but evidently lies in the balance between energy-producing and energy-consuming reactions.

B. *Lignin Degradation as a Secondary Metabolic Event*

In some white-rot fungi, lignin is degraded only during secondary metabolism, that is, when primary growth ceases because of depletion of some nutrient. Several ligninolytic fungi apparently exhibit this feature

(Freer and Detroy, 1982; Hatakka *et al.*, 1983; Leatham and Kirk, 1983). However, only *P. chrysosporium* has been studied in any detail in this regard. Secondary metabolism in *P. chrysosporium* is triggered by nitrogen, carbon, or sulfur limitation but not by phosphorus limitation (Jeffries *et al.*, 1981). Lignin degradation stops when primary growth is restored by adding excess carbon or nitrogen to cultures limited for that nutrient (Jeffries *et al.*, 1981). In nitrogen-limited cultures, ligninolytic activity is associated with formation of new, "secondary" mycelia (Leisola *et al.*, 1983b). Cultures can obviously cope well with nitrogen limitation, so that sustained degradation of lignin occurs. Indeed, nitrogen is limiting in wood, the natural substrate for the white-rot fungi (Cowling and Merrill, 1966). Cultures limited for nutrient carbon, on the other hand, degrade lignin only transiently, until they deplete their own reserves (Jeffries *et al.*, 1981).

Studies with nitrogen-limited cultures of *P. chrysosporium* have shown that the concentration of intracellular glutamate is correlated with the primary/secondary transition; thus, a low concentration of intracellular glutamate is observed in ligninolytic cultures. Addition of glutamate or ammonium to such cultures raises the concentration of intracellular glutamate, restores primary growth, and stops lignin degradation (Fenn *et al.*, 1981). This nitrogen-regulated transition from primary to secondary metabolism, and vice versa, has now been shown to be associated with the level of cyclic adenosine monophosphate (cAMP) in the cells; addition of glutamate causes a sharp decrease in the cAMP level in nitrogen-starved cultures, whereas nitrogen starvation causes a sharp increase in the cAMP level (MacDonald *et al.*, 1984).

Lignin degradation by some white-rot fungi—e.g., *Lentinus edodes*—is not affected by nutrient nitrogen (Freer and Detroy, 1982; Leatham and Kirk, 1983). It is not yet known whether such fungi simply exhibit no clear separation of primary and secondary metabolism, or whether lignin degradation is regulated differently in them. Interestingly, *P. chrysosporium* can be maintained in a mixed primary/secondary state by slow feeding of nutrient nitrogen to nitrogen-limited cultures; under these conditions, the cultures grow and degrade lignin simultaneously (Fenn *et al.*, 1981), as does *L. edodes*.

One intriguing aspect of the physiology of lignin degradation by *P. chrysosporium* is the association of degradation with *de novo* synthesis of the secondary metabolite 3,4-dimethoxybenzyl (veratryl) alcohol (Lundquist and Kirk, 1978). Veratryl alcohol is synthesized from phenylalanine (Shimada *et al.*, 1981a,b) (Fig. 1A). Lignin degradation and veratryl alcohol biosynthesis are closely intertwined and are regulated together by nutrient nitrogen. Other culture parameters that affect lignin

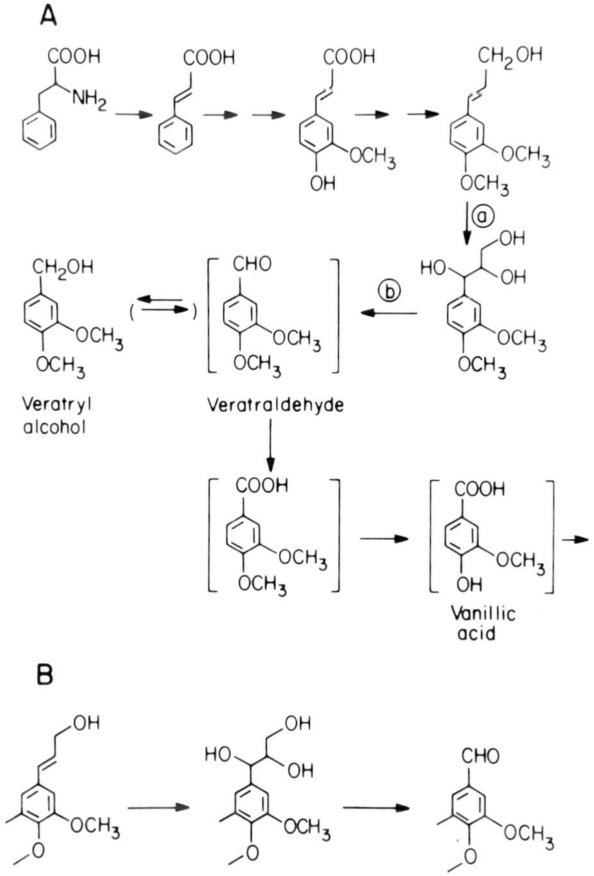

Fig. 1. (A) Veratryl alcohol is synthesized and degraded by *Phanerochaete chrysosporium*. Compounds in brackets are suspected intermediates. (B) Oxidative reactions in the cinnamyl alcohol side chains of lignin models parallel reactions (a) and (b) in veratryl alcohol synthesis.

degradation also affect veratryl alcohol synthesis. Thus, culture agitation inhibits both, and elevated concentrations of molecular oxygen in cultures increase ligninolytic activity and veratryl alcohol biosynthesis (Shimada *et al.*, 1981a,b).

The interrelation between ligninolytic activity and veratryl alcohol biosynthesis in *P. chrysosporium* might have a straightforward physiological basis. Research has shown that veratryl alcohol induces a lignin-degrading enzyme (discussed in Section IV,C,1) during secondary metabolism (Faison and Kirk, 1985). Thus it is possible that induction of the ligninase, and hence lignin degradation, occurs only following veratryl alcohol synthesis. Interestingly, however, addition of veratryl alcohol to nonligninolytic primary growth-stage cultures does not induce the enzyme in the primary-stage cultures (Faison and Kirk, 1985), and does not trigger ligninolytic activity (Kirk and Fenn, 1982). This indicates that induction by veratryl alcohol is *permitted* only after the cultures enter the secondary metabolic state.

Further overlap between veratryl alcohol metabolism and lignin metabolism probably occurs in the degradation of both by ligninolytic cultures of *P. chrysosporium*. Veratryl alcohol is metabolized to CO_2 (Lundquist and Kirk, 1978), presumably via oxidation to veratric acid and subsequent demethylation to vanillic acid (Shimada *et al.*, 1982; Fig. 1A). Vanillic acid is a major intermediate of lignin degradation by *P. chrysosporium* (Chen *et al.*, 1982).

C. Molecular Oxygen as a Regulatory Agent

Another notable feature of the physiology of lignin degradation in *P. chrysosporium* is the powerful influence of molecular oxygen (Kirk *et al.* 1978; Reid and Seifert, 1980; Bar-Lev and Kirk, 1981); the effect of oxygen on some other white-rot fungi is less dramatic or even negligible (Reid and Seifert, 1980).

Increasing the concentration of oxygen in the culture atmosphere has a dual effect: it increases the titer of the total ligninolytic system (lignin → CO_2), and it increases the rate of action of the existing system (Bar-Lev and Kirk, 1981). The former effect is probably related to a marked stimulation of an H_2O_2-generating system (Faison and Kirk, 1983); as discussed in Section IV, H_2O_2 is required for lignin degradation. The effect of oxygen on the titer of the ligninolytic system may also be related in part to veratryl alcohol concentration. As mentioned, elevating the oxygen concentration increases the veratryl alcohol level in cultures. In turn, this increase could increase the titer of the ligninolytic enzyme. Several steps in veratryl alcohol biosynthesis probably require molecular oxygen, suggesting a basis for the influence of oxygen in its biosynthesis. The second influence of molecular oxygen—on the activity of the existing ligninolytic system—could reflect the fact that molecular oxygen is itself a reactant in lignin degradation, being involved directly in substrate oxygenation (Nakatsubo *et al.*, 1982; Tien and Kirk, 1984). It is also the precursor of H_2O_2.

Leisola *et al.* (1983a), using *P. chrysosporium,* showed that reducing the depth of culture medium below 1.5 mm has the same enhancing effect on lignin degradation as increasing the oxygen concentration. Reid and

Seifert (1980) showed that increasing the air pressure above cultures has the same effect.

D. *Induction of the Ligninolytic System*

Keyser *et al.* (1978) showed that the ligninolytic system (lignin → CO_2) is synthesized in the absence of lignin as a part of secondary metabolism (again with *P. chrysosporium* as the experimental organism). Preincubation of the cultures with synthetic lignin (50 mg/liter) did not further increase their ligninolytic activity ([^{14}C]lignin → $^{14}CO_2$), leading to the conclusion that the rate-limiting component is not lignin induced. However, Ulmer *et al.* (1984) have now shown that preincubation of *P. chrysosporium* cultures with much higher concentrations of a lignin preparation (1000–2000 mg/liter of a dioxane/HCl lignin from straw) does increase their ligninolytic activity. Approximately 24 hr is needed for the increase. This time requirement, together with the requirement for large quantities of lignin, suggests that a lignin degradation product is responsible. Ulmer *et al.* (1984) did not determine whether the activity increase involves protein synthesis, and it is not yet known whether the increase is actually due to induction. Other work with *P. chrysosporium* (Faison and Kirk, 1985) shows that veratryl alcohol apparently induces the ligninase discussed in Section IV,C,1. That same work shows that lignin (38 mg/liter), too, induces the ligninase within 12 hr. The increased enzyme activity, however, is not reflected in a significant increase in the total ligninolytic system (lignin → CO_2), indicating that the enzyme is not the rate-limiting component.

IV. Biochemistry of Lignin Degradation by White-Rot Fungi

A. *Background*

The biochemical system that degrades the lignin polymer is extracellular, oxidative, and relatively nonspecific. That the ligninolytic system (lignin → CO_2) is in part extracellular is apparent in the fact that it degrades a polymeric substrate. Lignin is polydisperse, but, except for a very small fraction, it is certainly too large to be taken up by microbial cells. That the major reactions of its degradation are oxidative has been determined by characterizations of fungal degradation products of lignin, both polymeric and low molecular weight, and of lignin model compounds, as discussed in the preceding chapters. That the fungal attack is nonspecific is apparent (a) from a consideration of the heterogeneous nature of lignin, including the fact that the several types of intermonomer linkages in lignin lack stereoregularity (Harkin, 1967); (b) from the fact that industrially modified lignins, including kraft lignins, are still readily degraded by the fungi (Lundquist *et al.*, 1977); and (c) from the recently described absence of stereospecificity in fungal metabolism of lignin model compounds (Nakatsubo *et al.*, 1982).

Hall (1980) speculated that "reactive diffusible" agents, such as activated or reduced oxygen species, rather than oxidative enzymes, are directly responsible for the extracellular degradation of lignin. This suggestion was attractive for two reasons: (a) extracellular nonspecific oxidizing *enzymes* that catalyze the kinds of reactions comprising lignin degradation (Chapter 19) had not been described for any system, i.e., there was no precedent for the existence of such enzymes; and (b) nonenzymatic oxidations have been implicated in a number of biochemical systems. Subsequent research at first supported the hypothesis that nonenzymatic oxidations are important in lignin biodegradation. The most recent findings, however, do not. Section IV,B provides a brief summary of those various investigations. The summary is followed by a discussion of a newly discovered, extracellular, ligninolytic oxygenase (Section IV,C,1). Those sections are followed by a discussion of several enzymes or reports of enzymes suspected in earlier research to be involved in lignin breakdown (Section IV,C,2).

B. *"Active Oxygen Species" as Ligninolytic Agents*

The term "active oxygen species" refers to those partially reduced or excited molecules derived from ground-state molecular oxygen, including hydrogen peroxide (H_2O_2), superoxide radical ($O_2^{\cdot -}$), hydroxyl radical ($\cdot OH$), and singlet oxygen ($^1O_2$). Of these species, only $\cdot OH$ and $^1O_2$ are reactive enough to be considered as lignin oxidants in living systems. Both reportedly have roles in other biological systems (Auclair *et al.*, 1978; Krinsky, 1979). H_2O_2 and $O_2^{\cdot -}$ are thought to be biochemical precursors of the reactive species.

A suggestion that $^1O_2$ is involved in lignin oxidation in the white-rot fungus *P. chrysosporium* was based on the following evidence (Nakatsubo *et al.*, 1981): (a) the $^1O_2$-trapping agent anthracene-9,10-bisethane sulfonate (AES) strongly inhibits the oxidation of lignin, but not of glucose, in ligninolytic cultures; (b) the cultures rapidly bleach AES, in accord with the production of $^1O_2$; and (c) a photochemical $^1O_2$-generating system and cultures produced identical degradation products from a lignin model compound. Subsequent study has discredited the possible role of $^1O_2$ by showing that a chemical system for generating $^1O_2$ ($HOCl + H_2O_2$) does not mimic cultures in its effect on the model compound (Kirk *et al.*, 1983) and that $^1O_2$ is, in fact, not the only active species in the photochemical system (Kirk *et al.*, 1983; Kutsuki *et al.*, 1983). The kinds of reactions usually attributed to $^1O_2$ (Wasserman and Murray, 1979) are probably not of importance in lignin degradation. It is not yet clear whether $^1O_2$ is actually produced by the cultures. Unpublished results (B. D. Faison and T. K. Kirk) have shown that AES strongly inhibits the ligninolytic oxygenase discussed in Section IV,C,1, perhaps explaining the powerful and selective effect that AES has on lignin degradation.

Three independent groups have concluded that the hydroxyl radical ($\cdot OH$) is involved in lignin degradation by *P. chrysosporium* (Forney *et al.*, 1982b; Kutsuki and Gold, 1982; Bes *et al.*, 1983). All three groups showed that $\cdot OH$ is produced by ligninolytic cultures and that $\cdot OH$ scavengers partially suppress lignin degradation. Two of the investigations showed, in addition, that chemically generated $\cdot OH$ degrades lignin (Kutsuki and Gold, 1982; Bes *et al.*, 1983). These three studies, however, provide only circumstantial and indirect evidence for $\cdot OH$ involvement. Scavengers of $\cdot OH$ cannot be considered to be specific for that species. A fourth study showed that although $\cdot OH$ is produced by *P. chrysosporium*, it is not correlated temporally or quantitatively with ligninolytic activity (Faison and Kirk, 1983).

Another study compared degradation of model compounds by chemically generated $\cdot OH$ with degradation by ligninolytic cultures of *P. chrysosporium* (Kirk *et al.*, 1985). The Fenton reaction was used to generate $\cdot OH$: $Fe^{2+} + H_2O_2 \rightarrow Fe^{3+} + \cdot OH + OH^-$ (Walling, 1975). Results showed that the cultures do not degrade the models to the same products as the Fenton system, thus providing evidence that $\cdot OH$ is not a component of the ligninolytic system.

Even though $^1O_2$ and $\cdot OH$ do not seem to us to be involved in lignin degradation, the investigations of active oxygen species in *P. chrysosporium* have provided good evidence that H_2O_2 is involved. Forney *et al.* (1982b) cited a correlation between H_2O_2 production by broken-cell extracts and lignin degradation by intact cultures. Their later work suggested that peroxisome-like structures just inside the cytoplasmic membrane are the site of H_2O_2 synthesis (Forney *et al.*, 1982a). Research by Faison and Kirk (1983) established the presence of extracellular H_2O_2 in ligninolytic cultures, showed that specific scavengers of H_2O_2, including catalase, strongly suppress ligninolytic activity, and showed that H_2O_2 production is temporally and quantitatively correlated with ligninolytic activity.

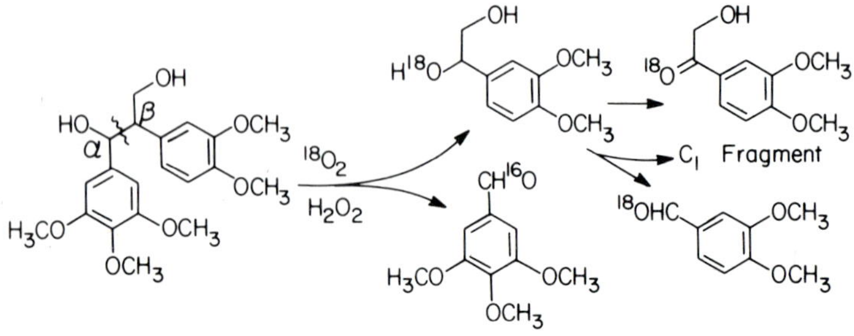

Fig. 2. Cleavage of a β-1 model compound by a ligninolytic oxygenase from *Phanerochaete chrysosporium* involves C_α—C_β cleavage.

Faison and Kirk (1983) similarly showed that superoxide dismutase partially inhibits ligninolytic activity in *P. chrysosporium,* suggesting that $O_2^{\cdot-}$ might also be involved. However, a satisfactory assay for $O_2^{\cdot-}$ in intact cultures could not be developed in that study. Amer and Drew (1980) reported detection of $O_2^{\cdot-}$ in cultures of *Coriolus versicolor,* but they did not determine whether their cultures were ligninolytic, nor did they present other evidence for the existence of $O_2^{\cdot-}$, such as that superoxide dismutase destroyed the agent they were measuring.

C. *Lignin-Degrading Enzymes*

1. Ligninolytic Oxygenase

Knowledge that H_2O_2 is required for lignin degradation by *P. chrysosporium* has facilitated the discovery of a lignin-degrading enzyme. The discovery followed investigations into the nature of the agent in ligninolytic cultures that is responsible for an oxidative cleavage reaction that had been well characterized in intact cultures using dimeric model compounds (Enoki and Gold, 1982; Nakatsubo *et al.,* 1982; Kirk *et al.,* 1983). Cleavage activity was detected in concentrated extracellular fluid and was dependent on added H_2O_2. Subsequent work showed that the responsible agent is an enzyme (Tien and Kirk, 1983). It is macromolecular, heat-labile, resolvable by polyacrylamide gel electrophoresis, and stainable with protein stains, and the C_α—C_β cleavage reaction (Fig. 2) exhibits saturation kinetics. It has been detected only in the extracellular fluid, where it is the major protein. The enzyme is strongly inhibited by azide and, surprisingly, by the $^1O_2$ scavenger AES (Section IV,B).

The enzyme has now been purified to homogeneity by DEAE-Biogel A chromatography and further characterized (Tien and Kirk, 1984). Sodium

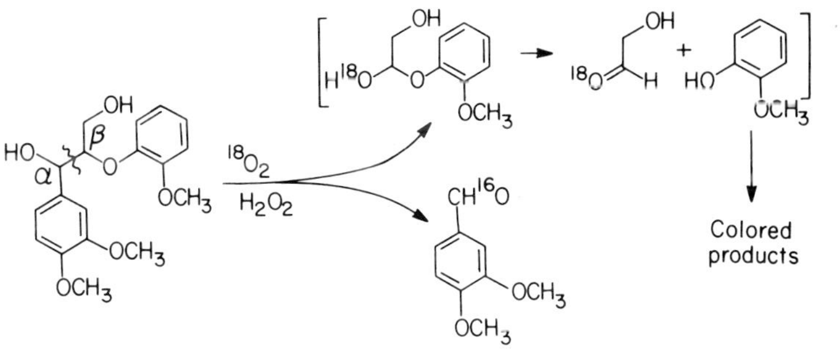

Fig. 3. Cleavage of a β-*O*-4 model compound by a ligninolytic oxygenase from *Phanerochaete chrysosporium* is analogous to cleavage of β-1 models (Fig. 2). Compounds in brackets are suspected products, based on work with related models. From Tien and Kirk (1984).

dodecyl sulfate–polyacrylamide gel electrophoresis (SDS–PAGE) and isoelectric focusing showed that it has a molecular weight of 42,000 and an isoelectric point of 3.5. The enzyme is a hemoprotein containing a single heme moiety derived from protophorphyrin IX. Further study (Tien and Kirk, 1984) has shown that it is a glycoprotein containing ~15% carbohydrate, and that it binds to concanavalin A, indicating α-D-linked terminal mannose, glucose, and/or *N*-acetylglucosamine in the carbohydrate.

The C_α—C_β cleavage reaction in β-1 models with which the enzyme was first detected (Fig. 2) also occurs in arylglycerol-β-aryl ether (β-*O*-4) models (Fig. 3); such models represent the most frequent substructure in lignins (Adler, 1977). As a consequence, lignin itself is partially depolymerized by the purified enzyme (Fig. 4). The aromatic aldehydes resulting from the C_α—C_β cleavages of models (Figs. 2 and 3) are also produced from (methylated) spruce and birch lignins (Tien and Kirk, 1983). Model studies have shown that the C_α—C_β cleavages occur without stereospecificity, which is important for degradation of the nonstereoregular lignin polymer (Tien and Kirk, 1984).

In addition to the C_α—C_β cleavage, the purified enzyme catalyzes the oxidation of benzyl alcohols to aldehydes and ketones, intradiol cleavage in phenylglycol structures, hydroxylation of benzylic methylene groups, oxidation of phenols, and hydroxylation of the olefinic bond in styryl structures (Fig. 5).

Experiments with $^{18}O_2$ have established that the enzyme is an oxygenase (Tien and Kirk, 1984). When the enzymatic cleavage of a β-1 model is done under $^{18}O_2$ and with $H_2{}^{16}O_2$, ^{18}O is incorporated into the phenylglycol product (Fig. 2). Oxygen is also incorporated from $^{18}O_2$ into the benzyl hydroxyl groups during hydroxylation of aryl methylene and

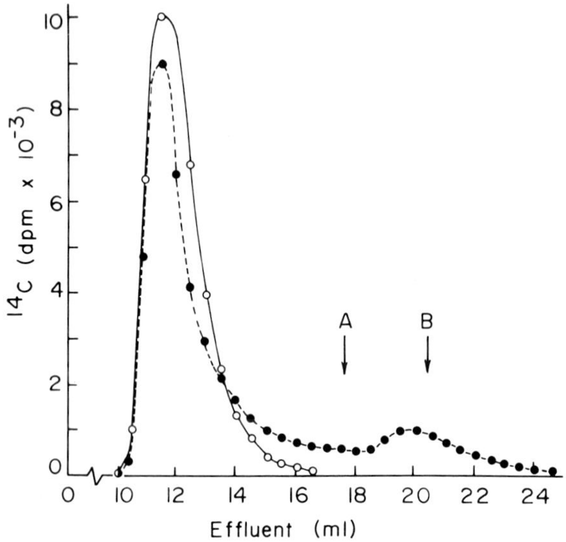

Fig. 4. Spruce lignin is partially depolymerized by a ligninolytic oxygenase form *Phanerochaete chrysosporium*. Molecular weight markers: A, 348; B, 166. From Tien and Kirk (1983).

styryl structures (Fig. 5, reactions C and E). The enzyme is an oxygenase, therefore, and is apparently unique in its requirement for H_2O_2. The role of the H_2O_2 is under investigation.

Reactions B and E in Fig. 5 have also been reported by Glenn *et al.* (1983), who observed them by using crude culture filtrates of ligninolytic cultures of *P. chrysosporium* with added H_2O_2. These workers also reported β-ether cleavage in a β-*O*-4 model. That report was based on detection of traces of the phenylglycerol product of cleavage.

2. Other Enzymes in Lignin (Polymer) Degradation

The fact that the ligninolytic oxygenase described above does not completely depolymerize lignin (Kirk and Tien, 1983) suggests that other enzymes are involved. Characterization of partially white-rotted but still polymeric lignins has also established that more is involved than cleavage between C_α and C_β in side chains, although C_α—C_β cleavage is clearly a prominent reaction (Chapter 19). The characterization studies, summarized in Chapter 19, point to several additional reactions in the polymer,

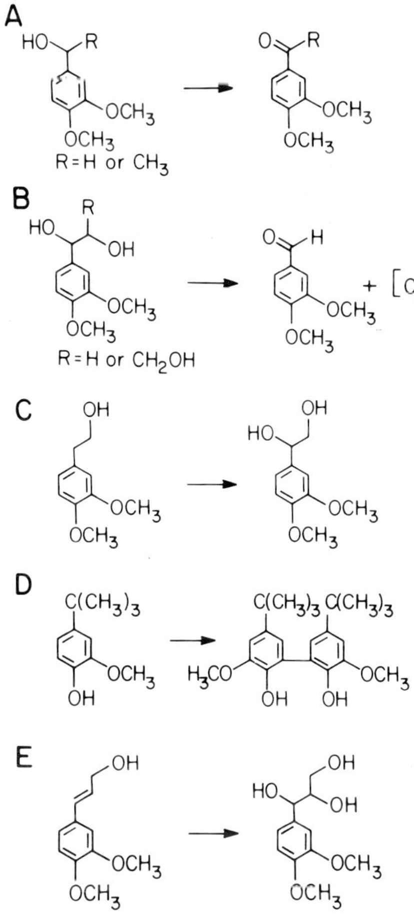

Fig. 5. A ligninolytic oxygenase from *Phanerochaete chrysosporium* catalyzes several reactions in addition to those in Figs. 2 and 3.

suggesting other enzymes: (a) demethylation (or demethyoxylation) of aryl methoxyl groups, (b) hydroxylation of aromatic rings ortho to methoxyl or hydroxyl groups, (c) cleavage of aromatic rings, (d) degradation of ring cleavage fragments, and, possibly, (e) various side-chain reductions. Enzymatic activity catalyzing these reactions has been reported only for demethylation or demethoxylation of aryl methoxyl groups, and is discussed in Section IV,C,2,a.

Enzymes other than those indicated by the chemical studies have also been implicated. These include the enzymes that oxidize phenols, the enzyme cellobiose : quinone oxidoreductase, and putative enzymes that we refer to collectively as "β-etherases." These other enzyme activities are discussed in the following sections.

a. Aryl Methoxyl Demethylase (or Demethoxylase). Frick and Crawford (1983) reported that the extracellular fluid of ligninolytic cultures of *P. chrysosporium* contains a heat-labile agent that, in the presence of H_2O_2, releases a volatile ^{14}C-labeled product (probably methanol) from a [^{14}C]methoxyl-labeled polymer of guaiacol ("polyguaiacol"). They reported, too, that a similar reaction is catalyzed by horseradish peroxidase/H_2O_2. As Frick and Crawford (1983) recognized, the reaction in both cases might well involve intermediate formation of phenoxy radicals. Figure 6 includes a possible mechanism (path D) for release of methanol from lignin and related *o*-methoxyphenols via phenol oxidation. Because the ligninolytic oxygenase discussed above (Section IV,C,1) has phenol-oxidizing activity, it may be responsible for the partial demethoxylation of polyguaiacol. It seems rather unlikely, however, that demethoxylation via phenol oxidation (as in Fig. 6, reaction D) is actually involved in lignin degradation by white-rot fungi, because *in vitro* only a small percentage of substrate can be demethoxylated by phenol-oxidizing enzymes; the reaction is simply too limited. For example, in the work of Frick and Crawford (1983), the maximum demethoxylation of polyguaiacol by culture fluid of *P. chrysosporium* was 3.6%, and by horseradish peroxidase, 5.6%; Ishihara and Miyazaki (1974) found that the maximum demethoxylation of spruce milled wood lignin by purified fungal laccase/O_2 was 1.4%.

b. Phenol Oxidases. Phenol-oxidizing enzymes of both the peroxidase and laccase types catalyze removal of one electron from phenolic hydroxyl groups in lignin-related compounds and in lignin, producing phenoxy radicals that undergo a variety of reactions in addition to demethoxylation (Fig. 6). Extracellular phenol-oxidizing enzyme activity has been detected in virtually all white-rot fungi examined; because of this association with ligninolytic capacity, extracellular phenol-oxidizing enzyme activity has long been presumed to play a role in degrading lignin (see Kirk, 1971; Ander and Eriksson, 1978).

Early research by Kirk and co-workers (1968) and more recent investigations by Higuchi (Chapter 20) with model compounds led to the conclusion that reactions B and C in Fig. 6 are important in the degradation of phenolic units in lignin and in phenolic fragments released during lignin degradation. Ishihara and Miyazaki (1972), however, were unable to degrade milled wood lignin significantly with purified fungal laccase; they and others (e.g., Hüttermann *et al.*, 1980) showed instead that the major effect of isolated laccase is further polymerization (Fig. 6, reaction A).

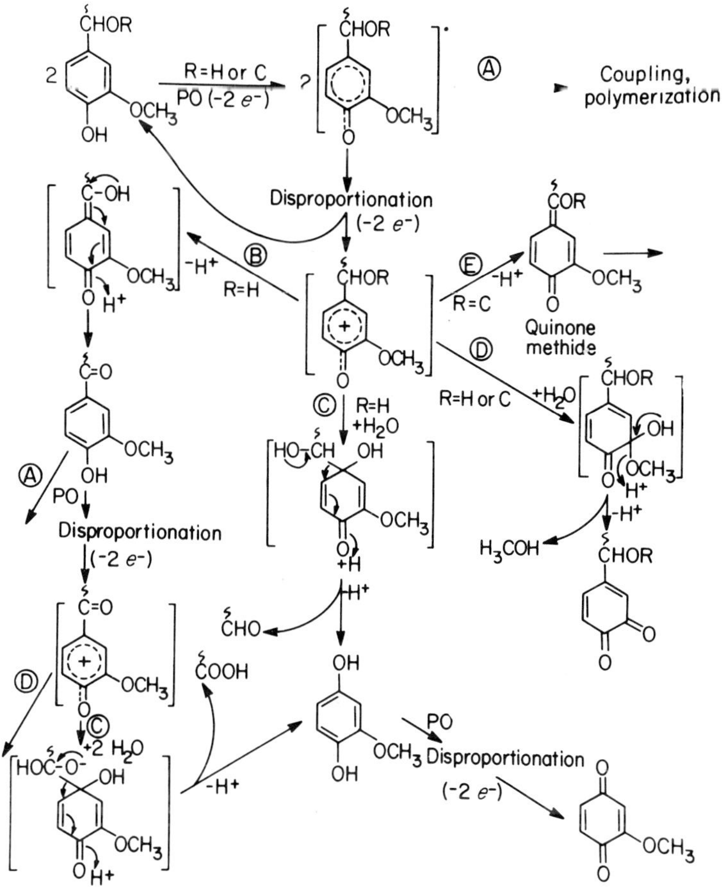

Fig. 6. A variety of reactions follow single-electron oxidation of lignin-related phenols, catalyzed by phenol-oxidizing enzymes.

Such polymerization, however, may be prevented *in vivo* by the other enzymes (see below).

Some genetic evidence suggests that phenol-oxidizing enzymes are involved in lignin degradation by *P. chrysosporium*. Ander and Eriksson (1976a) produced a mutant of *Sporotrichum pulverulentum* (= *P. chrysosporium*) that lacked phenol-oxidizing enzyme activity; this PO^- mutant no longer degraded lignin, whereas a PO^+ revertant did. Addition of purified fungal laccase to kraft lignin-containing agar cultures of the mutant restored its ability to degrade the kraft lignin. They concluded that phenol-oxidizing enzymes are required for lignin degradation and suggested that the role might be regulatory. A PO^- mutant was also isolated by Gold and co-workers (1982). It, too, did not degrade lignin, whereas a PO^+ revertant did. In this case, the addition of the phenol-oxidizing enzyme peroxidase (with H_2O_2) did not restore ligninolytic activity. The mutant was pleiotropic: it did not synthesize veratryl alcohol, and it did not form sexual fruiting bodies, whereas the revertant had regained these functions. Because PO activity in *P. chrysosporium,* as well as ligninolytic activity and veratryl alcohol synthesis, only occurs during secondary metabolism (Keyser *et al.,* 1978; Kirk and Fenn, 1982), as does fruit-body formation (Gold and Cheng, 1979), Gold and co-workers (1982) suggested that the primary mutational lesion that produced their PO^- mutant interfered with the onset of secondary metabolic functions in general.

The discovery that a true ligninase also possesses intrinsic phenol-oxidizing activity (Section IV,C,2) helps explain the connection between lignin degradation and phenol oxidation. It does not, however, explain the restoration of ligninolytic activity in the PO^- mutant by exogenous laccase reported by Ander and Eriksson (1976b).

Shimada (1980) suggested that reaction B in Fig. 6, catalyzed by phenol-oxidizing enzymes, is important in lignin degradation. Formation of "α-carbonyl" groups as in that reaction is prominent during lignin biodegradation (Chapter 19), and during metabolism of a dimeric model compound of the β-*O*-4 type (Fenn and Kirk, 1984). The importance of C_α-oxidation was suggested (Shimada, 1980) to be in reducing the steric complexity of the polymer by destroying chiral centers not only at C_α but also, via enolization, at C_β. Experiments verified that such reactions are catalyzed in lignin by phenol-oxidizing enzymes (Shimada, 1980). The study of Shimada (1980) was the first to focus on the significance that the nonstereoregularity of lignin has for its enzymatic degradation. Another showed that whereas C_α-oxidation actually retards degradation of a β-*O*-4 model, C_α-oxidation of lignin enhances its depolymerization (Fenn and Kirk, 1984). A possible explanation is that the decreased steric complexity following C_α-oxidation opens up the polymer, making it more accessible to lignin-degrading enzymes. The result with the model shows nevertheless that C_α-oxidation decreases the intrinsic degradability of the β-*O*-4 structure. That study (Fenn and Kirk, 1984) showed, too, that C_α-oxidation is a prominent reaction in degradation of the model compound, just as in lignin itself (Kirk and Chang, 1975). In the model study, however, α-carbonyl formation did not involve phenol oxidation, because the model was nonphenolic. Thus, even though laccase and peroxidase can cause C_α-oxidation in phenolic units (Shimada, 1980), C_α-oxidation takes place also in nonphenolic units, as had been indicated by studying the partially white-rotted lignin polymer (Kirk and Chang, 1975; Chapter 19). As mentioned, the ligninolytic oxygenase discussed in section IV,C,1 catalyzes C_α-oxidation, requiring no phenolic hydroxyl to do so (Fig. 5, reaction A).

c. Cellobiose : Quinone Oxidoreductase. Westermark and Eriksson (1974) discovered the enzyme cellobiose : quinone oxidoreductase (CBQase), which catalyzes the oxidation of cellobiose-δ-lactone, with transfer of electrons to quinones or phenoxy radicals, reducing them to phenols. CBQase is a flavoprotein, and its synthesis is regulated together with the cellulolytic enzymes. Thus a mutant of *P. chrysosporium* lacking the cellulase system lacked CBQase as well. That mutant still degraded lignin in wood. Similarly, repression of the activity by glucose did not interfere with the oxidation of synthetic ^{14}C-labeled lignin to $^{14}CO_2$ by *P. chrysosporium* (T. J. Kellerher, T. W. Jeffries, and T. K. Kirk, unpublished). Thus, CBQase is not required for lignin degradation. Nevertheless, in the presence of cellobiose, CBQase does prevent the polymerization of phenols and the accumulation of quinones caused by phenol-oxidizing enzymes, and in so doing it might facilitate lignin degradation indirectly by thwarting side reactions and by reducing potentially inhibitory quinones (Westermark and Eriksson, 1974). An intracellular NAD(P)H : quinone oxidoreductase discovered in *P. chrysosporium* by Buswell *et al.* (1979a) might have a similar function.

d. "β-Etherase." The β-*O*-4 linkage makes up approximately half of the interunit linkages in conifer lignins and a higher proportion in hardwood lignins (Adler, 1977). Its cleavage, therefore, would extensively depolymerize lignin. Early studies of the chemistry of the degradation of lignin and of model compounds in cultures of white-rot fungi were interpreted as indicating that the cleavage occurs directly (Ishikawa *et al.*, 1963). Such "β-etherase" activity from white-rot fungi was subsequently reported by Fukuzumi and Shibamoto (1965; Fukuzumi *et al.*, 1969) and by Ishikawa and Oki (1966) in studies with model compounds. Both research groups reported that crude enzyme preparations liberated phenylglycerol cleavage products from the model compounds, and Fukuzumi *et al.* (1969) reported that the phenolic product (guaiacol) was also formed (Fig. 7).

In more recent studies, both the phenolic and phenylglycerol products of β-ether cleavage (Fig. 7) have been detected in whole cultures of *P. chrysosporium* (Enoki *et al.*, 1981; Higuchi *et al.*, 1983b; Umezawa *et al.*, 1983; Chapter 20). Furthermore, as mentioned above, Glenn *et al.* (1983) reported formation of traces of the phenylglycerol product on incubation of a β-*O*-4 model with *P. chrysosporium* culture filtrates in the presence of H_2O_2. Notably, however, only in the work of Fukuzumi *et al.*

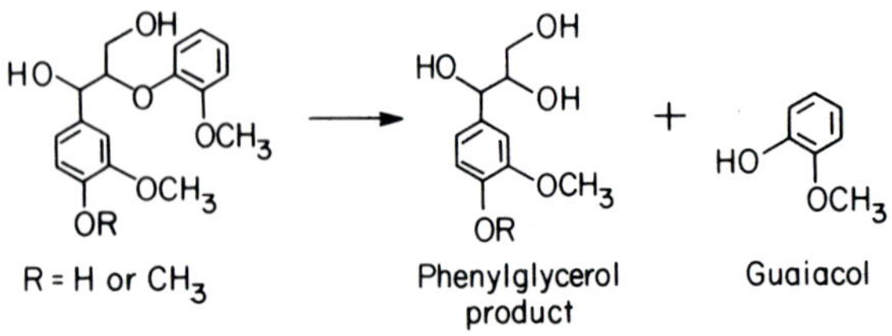

Fig. 7. Direct β-ether cleavage by white-rot fungi has been reported but has not been proved.

(1969) were both the phenolic and phenylglycerol products reported together with a cell-free enzyme preparation.

Direct enzymatic β-ether cleavage to produce phenylglycerol and phenolic products (Fig. 7) must involve direct hydrolysis, as proposed by Ishikawa and Oki (1966). However, Umezawa and Higuchi (1984) have ruled out hydrolysis in experiments with intact cultures of *P. chrysosporium*. They showed that the phenylglycerol product does not contain ^{18}O when degradation occurs in the presence of $H_2{}^{18}O$. In fact, their work seems to rule out formation of the phenylglycerol product via any reaction in the alkyl (glycerol) moiety of the model; protons at C_α and C_β are retained in the phenylglycerol product. Thus, even oxygenative cleavage via C_β-hydroxylation is ruled out.

We conclude that the existence of a β-etherase has not been established. The β-ether-linked phenolic moiety is released indirectly by C_α—C_β cleavage by the ligninolytic oxygenase isolated from *P. chrysosporium* (Fig. 3), which can explain formation of the phenolic product (guaiacol in Fig. 7) in the investigations cited above. Furthermore, direct degradative cleavage of the β-ether-linked aromatic ring, without prior cleavage of the β-ether linkage, occurs during the degradation of spruce lignin by *P. chrysosporium* (Chen *et al.*, 1983; Chapter 19). The eventual cleavage of an aliphatic ester or other labile linkage in an intermediate product, rather than direct β-ether cleavage, could be responsible for the release of phenylglycerol products in cultures.

The involvement of at least one extracellular oxygenase in lignin degradation by *P. chrysosporium* has been established, as discussed in Section IV,C,1. Indirect evidence for the participation of monooxygenases of the *P*-450 type—which the above ligninolytic oxygenase is not (Tien and Kirk, 1984)—has been obtained recently by Shimada and Higuchi (1983): the *P*-450 inhibitors metyrapone, menadione, and tyrosine–Cu^{2+} complex strongly suppressed degradation of synthetic ^{14}C-labeled lignin by *P. chrysosporium*. Oxygenases such as *P*-450 are multifunctional hemopro-

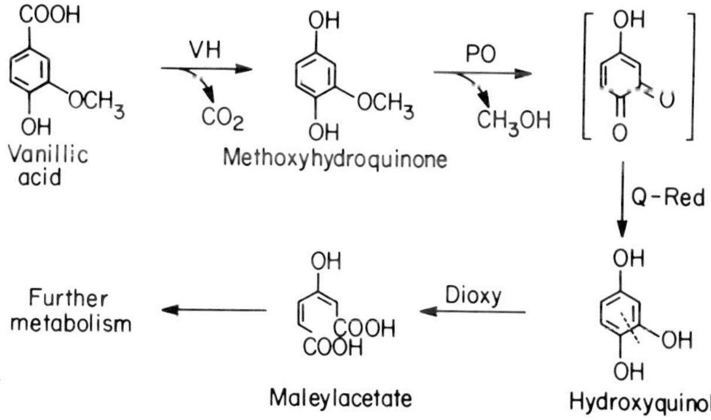

Fig. 8. Vanillic acid is degraded by *Phanerochaete chrysosporium* via oxidative decarboxylation, demethoxylation, probably quinone reduction, and ring cleavage. Enzymes: VH, vanillate hydroxylase (decarboxylase); PO, phenol-oxidizing enzyme; Q-Red, quinone reducing enzyme; Dioxy, ring-cleaving dioxygenase. Compound in brackets is suspected only.

teins that are relatively nonspecific, and some are able to utilize H_2O_2, as well as O_2 (plus a reducing cofactor), as oxidant.

e. Intracellular Enzymes. Further degradation of the low molecular weight aliphatic and aromatic fragments of lignin probably occurs inside the hyphae. Although metabolism of low molecular weight aromatics in white-rot fungi has received relatively little attention (Kirk, 1984), vanillic acid, which is the single most abundant lignin degradation product detected in spruce wood during degradation by *P. chrysosporium* (Chen *et al.*, 1982), has been studied in that fungus (Ander *et al.*, 1980). Its metabolism probably proceeds as shown in Fig. 8. In addition to the degradative reactions shown in Fig. 8, vanillic acid is also 4-O-methylated to veratric acid (Chen *et al.*, 1982) and reduced to vanillyl alcohol (Ander *et al.*, 1980). The enzyme activities responsible for decarboxylation and ring cleavage are enhanced by growing cultures with vanillic acid, and they apparently occur during both the primary and secondary metabolic phases.

Oxidative decarboxylation of vanillate to methoxyhydroquinone was first reported in the white-rot fungus *Polyporus dichrous* by Kirk and Lorenz (1973). The responsible cell-free enzyme activity was later reported simultaneously by Buswell *et al.* (1979b) and Yajima *et al.* (1979) in *Phanerochaete chrysosporium*. The purified enzyme (Buswell *et al.*, 1981) is intracellular, requires NADPH or NADH for activity, and is further stimulated by FAD. It has a molecular weight of approximately 65,000, and is strongly inhibited by heavy metals, by the sulfhydryl blocker *p*-chloromercuribenzoate, and by the chelator catechol 3,5-disulfonate.

The intracellular ring-cleaving oxygenase involved in vanillate metabolism (reaction B, Fig. 8) has not yet been purified. It was first detected with 1,2,4-trihydroxybenzene (hydroxyquinol) as substrate (Buswell and Eriksson, 1979), but formation of that compound by demethylation of methoxyhydroquinone has not been demonstrated. Because intact cultures released labeled $^{14}CO_2$ from ring carbons of methoxyhydroquinone ahead of methoxyl carbons, Ander *et al.* (1980) and later Buswell *et al.* (1982) concluded that demethylation does not precede ring cleavage, and that hydroxyquinol is not an intermediate of vanillic acid metabolism. Their more recent work indicates that demethoxylation does in fact precede ring cleavage (Ander *et al.*, 1983). They propose that demethoxylation yields methanol and an *o*-quinone. The latter is presumed then to be reduced to hydroxyquinol (Fig. 8).

Syringic acid, as well as vanillic acid, is a major intermediate of hardwood lignin degradation (Tai *et al.*, 1983). Its metabolism by *P. chrysosporium* has been studied by Umezawa *et al.* (see Chapter 20) and by Eriksson *et al.* (1984). In contrast to vanillic acid degradation, its degradation proceeds via an initial demethylation to 3-*O*-methylgallic acid (= 5-hydroxyvanillic acid). This product is also formed on degradation of dimeric models of the biphenyl type (Chapter 20) and of lignin itself (Chapter 19). In the case of the models, the immediate precursor is 5-carboxyvanillic acid, which is oxidatively decarboxylated selectively to 3-*O*-methylgallic acid. This decarboxylation might involve a different enzyme from the vanillate decarboxylase discussed above (see Chapter 20). Nevertheless, the 3-*O*-methylgallic acid is a good substrate for vanillate decarboxylase (Buswell *et al.*, 1981), and its metabolism might proceed via such decarboxylation and subsequent ring cleavage (Eriksson *et al.*, 1984). Further demethylation of 3-*O*-methylgallic acid to gallic acid also occurs, however, leading Umezawa and Higuchi (see Chapter 20) to suggest that gallic acid is the actual intermediate. Because gallic acid, too, is a good substrate for the vanillate decarboxylase (Buswell *et al.*, 1979b, 1981), it may be converted to a tetrahydroxybenzene ring cleavage substrate. In any event, demethoxylating and demethylating enzyme systems have not been isolated or characterized. The pathways for syringic acid metabolism are summarized in Fig. 9. In addition to the degradative reactions shown, syringic acid, like vanillic acid, is 4-*O*-methylated and is also reduced to syringyl alcohol by *P. chrysosporium* (Eriksson *et al.*, 1984).

Although three hydroxyl groups seem to be required for ring cleavage in the degradation of vanillic and syringic acids, catechol ring-cleavage activity has been demonstrated in intact cultures of *P. chrysosporium*

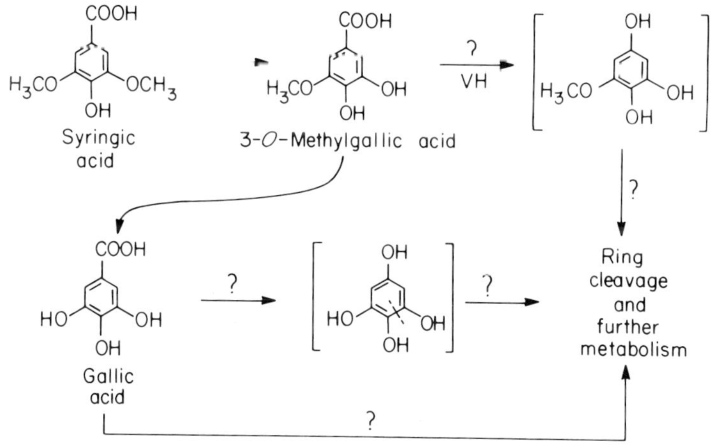

Fig. 9. Syringic acid metabolism by *Phanerochaete chrysosporium* proceeds via an initial demethylation, or via dual demethylations, probably followed by oxidative decarboxylation and ring cleavage. VH, Vanillate hydroxylase (decarboxylase). Compounds in brackets are suspected only.

(Leatham *et al.*, 1983). Its possible significance in lignin metabolism is not yet known. Clearly, much remains to be learned about the intracellular metabolism by white-rot fungi of the low molecular weight products of lignin breakdown.

V. Concluding Comments

Much remains to be learned, in fact, about each of the three main topics treated here. The *microbiology* of lignin degradation is not an easy research area, but with the techniques now available—especially the ^{14}C-based assays—many questions can be answered. The soft-rot wood decay fungi and related "soil fungi" and the litter-decomposing Basidiomycetes in particular deserve attention; they all degrade lignin, but none has been studied in any detail. The *physiology* of lignin degradation has been studied only with *P. chrysosporium* to any appreciable extent. Studies with other white-rot fungi, some of which are already known to differ from *P. chrysosporium,* and studies with other ligninolytic fungi are needed. Finally, as is apparent from this brief review, the *biochemistry* of ligninolysis is only now beginning to yield to investigation. A fuller understanding in that rapidly developing area probably has the greatest potential for payoff in terms of both basic science and technical application.

References

Adler, E. (1977). *Wood Sci. Technol.* **11,** 169–218.

Amer, G. I., and Drew, S. W. (1980). *Dev. Ind. Microbiol.* **22,** 479–484.

Ander, P., and Eriksson, K.-E. (1976a). *Mater. Org.* **3,** 129–140.

Ander, P., and Eriksson, K.-E. (1976b). *Arch. Microbiol.* **109,** 1–8.

Ander, P., and Eriksson, K.-E. (1978). *In* "Progress in Industrial Microbiology" (M. T. Bull, ed.) pp. 1–58. Elsevier, Amsterdam.

Ander, P., Hatakka, A., and Eriksson, K.-E. (1980). *Arch. Microbiol.* **125,** 189–202.

Ander, P., Eriksson, K.-E., and Yu, H.-S. (1983). *Arch. Microbiol.* **136,** 1–6.

Auclair, C., Torres, M., and Hakim, J. (1978). *Biochem. Biophys. Res. Commun.* **81,** 1067–1072.

Baecker, A. A. W., and King, B. (1980). In "Biodeterioration" (T. A. Oxley, G. Becker, and D. Allsop, eds.), pp. 53–58. Pitman, London.

Bar-Lev, S. S., and Kirk, T. K. (1981). *Biochem. Biophys. Res. Commun.* **99,** 373–378.

Bes, B., Ranjeva R., and Bondet, A. M. (1983). *Biochimie* **65,** 283–289.

Buswell, J. A., and Eriksson, K.-E. (1979). *FEBS Lett.* **104,** 258–260.

Buswell, J. A., Hamp, S., and Eriksson, K.-E. (1979a). *FEBS Lett.* **108,** 229–232.

Buswell, J. A., Ander, P., Pettersson, B., and Eriksson, K.-E. (1979b). *FEBS. Lett.* **103,** 98–101.

Buswell, J. A., Eriksson, K.-E., and Pettersson, B. (1981). *J. Chromatogr.* **215,** 99–108.

Buswell, J. A., Eriksson, K.-E., Gupta, J. K., Hamp, S. G., and Nordh, I. (1982). *Arch. Microbiol.* **131,** 366–374.

Cain, R. B. (1980). *In* "Lignin Biodegradation: Microbiology, Chemistry and Potential Applications" (T. K. Kirk, T. Higuchi, and H.-M. Chang, eds.), Vol. I, pp. 21–60. CRC Press, Boca Raton, Florida.

Chen, C.-L., Chang, H.-M., and Kirk, T. K. (1982). *Holzforschung* **36,** 3–9.

Chen, C.-L., Chang, H.-M., and Kirk, T. K. (1983). *J. Wood Chem. Technol.* **3,** 35–57.

Colberg, P. J., and Young, L. Y. (1982). *Can. J. Microbiol.* **28,** 886–889.

Cowling, E. B. (1961). USDA Tech. Bull. 1258. Superintendent of Documents, U.S. Govt. Printing Office, Washington, D.C.

Cowling, E. B., and Merrill, W. (1966). *Can. J. Bot.* **44,** 1539–1554.

Crawford, R. L. (1981). "Lignin Biodegradation and Transformation." Wiley, New York.

Crawford, D. L., and Sutherland, J. B. (1980). *In* Lignin Biodegradation: Microbiology, Chemistry and Potential Applications" (T. K. Kirk, T. Higuchi, and H.-M. Chang, eds.), Vol. II, pp. 95–101. CRC Press, Boca Raton, Florida.

Enoki, A., and Gold, M. H. (1982). *Arch. Microbiol.* **132,** 123–130.

Enoki, A., Goldsby, G. P., and Gold, M. H. (1981). *Arch. Microbiol.* **129,** 141–145.

Eriksson, K.-E., Gupta, J. K., and Rao, M. (1984). *Arch. Microbiol.* in press.

Eslyn, W. E., Kirk, T. K., and Effland, M. J. (1975). *Phytopathology* **65,** 473–476.

Faison, B. D., and Kirk, T. K. (1983). *Appl. Environ. Microbiol.* **46,** 1140–1145.

Faison, B. D., and Kirk, T. K. (1985). *Appl. Environ. Microbiol.* in press.

Fenn, P., and Kirk, T. K. (1984). *J. Wood Chem. Technol.* **4,** 131–148.

Fenn, P., Choi, S., and Kirk, T. K. (1981). *Arch. Microbiol.* **130,** 66–71.

Forney, L. J., Reddy, C. A., and Pankratz, H. S. (1982a). *Appl. Environ. Microbiol.* **44,** 732–736.

Forney, L. J., Reddy, C. A., Tien, M., and Aust, S. D. (1982b). *J. Biol. Chem.* **257,** 11455–11462.

Freer, S. N., and Detroy, R. W. (1982). *Mycologia* **74,** 943–952.

Frick, T., and Crawford, R. L. (1983). *In* "Recent Advances in Lignin Biodegradation Research" (T. Higuchi, H.-M. Chang, and T. K. Kirk, eds.), pp. 143–152. Uni, Tokyo.

Fukuzumi, T., and Shibamoto, T. (1965). *J. Jpn. Wood Res. Soc.* **11,** 248–252.
Fukuzumi, T., Takatuka, H., and Minami, K. (1969). *Arch. Biochem. Biophys.* **129,** 396–409.
Glenn, J. K., Morgan, M. A., Mayfield, M. B., Kawahara, M., and Gold, M. H. (1983). *Biochem. Biophys. Res. Commun.* **114,** 1077–1083.
Gold, M. H., and Cheng, T. M. (1979). *Arch. Microbiol.* **121,** 37–41.
Gold, M. H., Mayfield, M. B., Cheng, T. M., Krisnangkura, K., Shimada, M., and Enoki, A. (1982). *Arch. Microbiol.* **132,** 115–122.
Hackett, W. F., Connors, W. J., Kirk, T. K., and Zeikus, J. G. (1977). *Appl. Environ. Microbiol.* **33,** 43–51.
Haider, K., and Trojanowski, J. (1980). *In* "Lignin Biodegradation: Microbiology, Chemistry and Potential Applications" (T. K. Kirk, T. Higuchi, and H.-M. Chang, eds.), Vol. I, pp. 111–134. CRC Press, Boca Raton, Florida.
Hall, P. (1980). *Enz. Microbial Technol.* **2,** 170–176.
Harkin, J. M. (1967). *In* "Oxidative Coupling of Phenols" (W. I. Taylor and A. R. Battersby, eds.), pp. 243–321. Dekker, New York.
Hatakka, A. I., Buswell, J. A., Pirhonen, T. I., and Uusi-Rauva, A. K. (1983). *In* "Recent Advances in Lignin Biodegradation Research" (T. Higuchi, H.-M. Chang, and T. K. Kirk, eds.), pp. 176–187. Uni, Tokyo.
Hayatsu, R., Winans, R. E., McBett, R. L., Scott, R. G. Moore, L. P., and Studier, M. H. (1979). *Nature (London)* **278,** 41–43.
Higuchi, T. (1982). *Experientia* **38,** 159–166.
Higuchi, T., Chang, H.-M., and Kirk, T. K. (eds.) (1983a). "Recent Advances in Lignin Biodegradation Research." Uni, Tokyo.
Higuchi, T., Nakatsubo, F., Kamaya, Y., and Umezawa, T. (1983b). *In* "Recent Advances in Lignin Biodegradation Research" (T. Higuchi, H.-M. Chang, and T. K. Kirk, eds.), pp. 209–218. Uni, Tokyo.
Hüttermann, A., Herche, C., and Haars, A. (1980). *Holzforschung* **34,** 64–66.
Ishihara, T., and Miyazaki, M. (1972). *Mokuzai Gakkaishi* **18,** 415–419.
Ishihara, T., and Miyazaki, M. (1974). *Mokuzai Gakkaishi* **20,** 39–41.
Ishikawa, H., and Oki, T. (1966). *J. Jpn. Wood Res. Soc.* **12,** 101–107.
Ishikawa, H., Schubert, W. J., and Nord, F. F. (1963). *Arch. Biochem. Biophys.* **100,** 140–149.
Janshekar, H., and Fiechter, A. (1983). Pentoses and Lignin. *In "Advances in Biochemical Engineering/Biotechnology"* (A. Fiechter and T. W. Jeffries, eds.), Vol. 27, pp. 119–178. Springer, Berlin.
Jeffries, T. W., Choi, S., and Kirk, T. K. (1981). *Appl. Environ. Microbiol.* **42,** 290–296.
Kawakami, H. (1980). *In* "Lignin Biodegradation: Microbiology, Chemistry and Potential Applications" (T. K. Kirk, T. Higuchi, and H.-M. Chang, eds.), Vol. II, pp. 103–125. CRC Press, Boca Raton, Florida.
Keyser, P., Kirk, T. K., and Zeikus, J. G. (1978). *J. Bacteriol.* **135,** 790–797.
Kirk, T. K. (1971). *Ann. Rev. Phytopathol.* **9,** 185–210.
Kirk, T. K. (1975). *Holzforschung* **29,** 99–107.
Kirk, T. K. (1984). *In* "Biochemistry of Microbial Degradation" (D. T. Gibson, ed.), pp. 399–437. Dekker, New York.
Kirk, T. K., and Adler, E. (1970). *Acta Chem. Scand.* **24,** 3379–3390.
Kirk, T. K., and Chang, H.-M. (1975). *Holzforschung* **29,** 56–64.
Kirk, T. K., and Cowling, E. B. (1984). The Chemistry of Solid Wood. In "Advances in Chemistry Series" (R. M. Rowell, ed.), Vol. 207, pp. 435–487. American Chemical Society, Washington, D.C.

Kirk, T. K., and Fenn, P. (1982). *In* "Decomposer Basidiomycetes" (J. C. Frankland, J. N. Hedger, and M. J. Swift, eds.), pp. 67–90. Cambridge Univ. Press, Cambridge.
Kirk, T. K., and Lorenz, L. F. (1973). *Appl. Environ. Microbiol.* **26,** 173–175.
Kirk, T. K., Harkin, J. M., and Cowling, E. B. (1969). *Biochim. Biophys. Acta* **165,** 145–163.
Kirk, T. K., Connors, W. J., and Zeikus, J. G. (1976). *Appl. Environ. Microbiol.* **32,** 192–194.
Kirk, T. K., Schultz, E., Connors, W. J., Lorenz, L. F., and Zeikus, J. G. (1978). *Arch. Microbiol.* **117,** 277–285.
Kirk, T. K., Tien, M., and Faison, B. D. (1975). *In* "Biotechnology in the Pulp and Paper Industry" (D. Attwood, ed.). Pergamon, Oxford.
Kirk, T. K., Nakatsubo, F., and Reid, I. D. (1983). *Biochem. Biophys. Res. Commun.* **111,** 200–204.
Kirk, T. K., Mozuch, M. D., and Tien, M. (1985). *Biochem J.* in press.
Krinsky, N. I. (1979). *In* "Singlet Oxygen" (H. H. Wasserman and R. W. Murray, eds.), pp. 597–641. Academic Press, New York.
Kutsuki, H., and Gold, M. H. (1982). *Biochem. Biophys. Res. Commun.* **109,** 320–327.
Kutsuki, H., Enoki, A., and Gold, M. H. (1983). *Photochem. Photobiol.* **37,** 1–7.
Leatham, G. F., and Kirk, T. K. (1983). *FEMS Microbiol. Lett.* **16,** 65–67.
Leatham, G. F., Crawford, R. L., and Kirk, T. K. (1983). *Appl. Environ. Microbiol.* **46,** 191–197.
Leisola, M., Ulmer, D., and Fiechter, A. (1983a). *Eur. J. Appl. Microbiol. Biotechol.* **17,** 113–116.
Leisola, M., Ulmer, D., Haltmeier, T., and Fiechter, A. (1983b). *Eur. J. Appl. Microbiol. Biotechnol.* **17,** 117–120.
Levi, M. P., and Preston, R. D. (1965). *Holzforschung* **19,** 183–190.
Lundquist, K., and Kirk, T. K. (1978). *Phytochemistry* **17,** 1676.
Lundquist, K., Kirk, T. K., and Connors, W. J. (1977). *Arch. Microbiol.* **112,** 291–296.
MacDonald, M. J., Paterson, A., and Broda, P. (1984). *J. Bacteriol.* in press.
Nakatsubo, F., Reid, I. D., and Kirk, T. K. (1981). *Biochem. Biophys. Res. Commun.* **102,** 484–491.
Nakatsubo, F., Reid, I. D., and Kirk, T. K. (1982). *Biochim. Biophys. Acta* **719,** 284–291.
Ng, T. K., Ben-Bassat, A., and Zeikus, J. G. (1981). *Appl. Environ. Microbiol.* **41,** 1337–1343.
Odier, E., and Monties, B. (1983). *Appl. Environ. Microbiol.* **46,** 661–665.
Pelczar, Jr., M. J., Gottlieb, S., and Day, W. C. (1950). *Arch. Biochem.* **25,** 449–451.
Reid, I. D., and Seifert, K. A. (1980). *Can. J. Microbiol.* **26,** 1168–1171.
Shimada, M. (1980). *In* "Lignin Biodegradation: Microbiology, Chemistry and Potential Applications" (T. K. Kirk, T. Higuchi, and H.-M. Chang, eds.), pp. 195–213. CRC, Boca Raton.
Shimada, M., Higuchi, T., and Katayama, T. (1982). *Proc. Tappi Res. Dev. Conf.,* pp. 241–248, Aug. 29-Sept. 1, 1982, Asheville, North Carolina.
Shimada, M., Nakatsubo, F., Higuchi, T., and Kirk, T. K. (1981a). *Proc. Int. Symp. Wood Pulping Chem. Stockholm,* **3,** 99–104.
Shimada, M., Nakatsubo, F., Kirk, T. K., and Higuchi, T. (1981b). *Arch. Microbiol.* **129,** 321–324.
Shimada, M., and Higuchi, T. (1983). *In* "Recent Advances in Lignin Biodegradation Research" (T. Higuchi, H.-M. Chang, and T. K. Kirk eds.), pp. 195–208. Uni, Tokyo.

Tai, D., Terasawa, M., Chen, C.-L., Chang, H.-M. and Kirk, T. K. (1983). *In* "Recent Advances in Lignin Biodegradation Research" (T. Higuchi, H.-M. Chang, and T. K. Kirk, eds.), pp. 44–63. Uni, Tokyo

Tien, M., and Kirk, T. K. (1983). *Science* **221,** 661–663.

Tien, M., and Kirk, T. K. (1984). *Proc. Natl. Acad. Sci. U.S.A.* **81,** 2280–2284.

Ulmer, D. C., Leisola, M. S. A., Schmidt, B. H., and Fiechter, A. (1983). *Appl. Environ. Microbiol.* **45,** 1795–1801.

Ulmer, D. C., Leisola, M. S. A., and Fiechter, A. (1984). *J. Biotechnol.* **1,** 13–24.

Umezawa, T., Nakatsubo, F., and Higuchi, T. (1983). *Agric. Biol. Chem.* **47,** 2677–2681.

Umezawa, T., and Higuchi, T. (1984). *Agric. Biol. Chem.*

Walling, C. (1975). *Accts. Chem. Res.* **8,** 125–131.

Wasserman, H. H., and Murray, R. W. (eds.) (1979). "Singlet Oxygen." Academic Press, New York.

Westermark, U., and Eriksson, K.-E. (1974). *Acta Chem. Scand.* **B28,** 209–214.

Yajima, Y., Enoki, A., Mayfield, M. B., and Gold, M. H. (1979). *Arch. Microbiol.* **123,** 319–321.

Yang, H. H., Effland, M. J., and Kirk, T. K. (1980). *Biotech. Bioeng.* **22,** 65–77.

Zeikus, J. G., Wellstein, A. L., and Kirk, T. K. (1982). *FEMS Microbiol. Lett.* **15,** 193–197.

CHAPTER **22**

Biodegradation of Aromatic Extractives of Wood

WOLFGANG BARZ
KLAUS-MICHAEL WELTRING

Lehrstuhl für Biochemie der Pflanzen
Westfälische Wilhelms-Universität
Münster, Federal Republic of Germany

I. Introduction

Degradation of endogenous or exogenous carbon compounds by living organisms is an integral part of life. The carbon cycle in Nature requires that all organic compounds synthesized by organisms must again be metabolized and finally degraded by other organisms in order to keep the equilibrium between anabolic and catabolic routes constant. Highly developed catabolic pathways are, therefore, essential to avoid the longtime accumulation of inhibitory concentrations of organic chemicals in Nature.

For exogenous compounds—i.e., foreign substrates synthesized by organisms other than the degrader—sophisticated mechanisms for the metabolism and dissimilation have been detected. If the foreign compounds

ISBN 0-12-347880-4

can be used as growth substrates, such mechanisms comprise both processes of cellular uptake into the digesting cells and routes of energy-yielding degradation. In cases where foreign compounds do not represent degradable nutrients, well-established mechanisms for inactivation and detoxification followed by excretion must be available. In the latter case, foreign compounds are mostly modified by simple introductory reactions and excreted in the form of water-soluble conjugates.

This chapter focuses on *aromatic extractives* of wood, because the pathways of metabolism and degradation of aromatic substrates have been studied very extensively. Because of the comparatively high chemical stability of aromatic compounds and the contrasting ease of their biological disintegration, this group of natural compounds has long fascinated the biochemists. Furthermore, wood as obtained from many different plants represents a very rich source of numerous classes of aromatic compounds. Emphasis here is placed on the biological fate of monomeric, low molecular weight compounds; therefore, the hydrolytic and oxidative processes for the disintegration of oligomeric or high molecular weight, polymeric aromatic compounds is not reviewed. For the biological disintegration of other classes of wood components such as tannins, cellulose, or other polyglucans or terpenoids, the reader is referred to other reviews (Chapter 17, Chapter 18; Kieslich, 1976; Loewus and Runeckles, 1977).

As far as the organisms are concerned, the main principles of aromatic metabolism and degradation in animals and microorganisms are discussed. Subsequently, a short section on aromatic metabolism in plants demonstrates how aromatics and secondary plant products are handled and disintegrated in these organisms.

II. Metabolism and Degradation of Aromatic Compounds in Microorganisms

Since the early days of biochemistry and microbiology, numerous bacteria and fungi have been demonstrated as potent degraders of aromatic compounds. This potential for the disintegration of a great variety of exogenous carbon compounds assigns to the microflora an important biological function as part of the carbon cycle in Nature.

Over the years, a multitude of investigations has elucidated numerous pathways for the disintegration of many different plant compounds. For example, the catabolism of aromatic acids (benzoic, phenylacetic, and cinnamic acids), various phenols, and flavonoids has intensively been investigated (Barz and Hösel, 1975). The related aspects of enzymology and metabolic regulation have also been studied in great detail. In general, the field of microbial catabolism is a well understood area of biochemistry. In order to cover the various aspects of this field, the reader is referred to comprehensive reviews (Dagley, 1975, 1978; Leisinger *et al.*, 1981; Rosazza and Smith, 1979; Kieslich, 1976).

Despite all efforts, however, the microbial degradation of several other well-known classes of plant constituents, such as quinones, alkaloids, or higher terpenoids, has been only poorly investigated, so that the exact routes of microbial disintegration of numerous carbon skeletons are still unknown. Furthermore, catabolic routes are mainly elucidated by isolation of excreted catabolites from the culture medium of either growing or resting cells (Goodhue, 1982). If intermediates are not excreted or not accumulated in sufficient quantities, the resulting catabolic reaction sequences must necessarily be incomplete. Therefore, in the absence of enzyme studies on the conversion of intermediates, many catabolic routes only show the conversion of those compounds that accumulated transiently.

Only a limited number of bacteria and fungi have so far been characterized for their potential to metabolize aromatic compounds. Among the eubacteria, members of the families Coccaceae, Mycobacteriaceae, Pseudomonaceae, Spirillaceae, Bacteriaceae, and Bacillaceae are able to disintegrate aromatic structures. Several yeasts, such as *Oospora, Candida, Pichia,* and *Saccharomyces,* and certain fungi (i.e., *Aspergillus, Penicillium, Neurospora,* and *Fusarium*) catabolize aromatics.

The majority of investigations have dealt with aerobic catabolic pathways. Therefore, this chapter concentrates on oxygen-requiring catabolic routes, because anaerobic transformation of aromatic compounds is insufficiently understood. However, for future studies the great economic potential of anaerobic digestion of plant biomass should be mentioned here (Sahm, 1984). Some aspects of anaerobic microbial metabolism of aromatic compounds are described in Section III,A,3 with contributions of the gut microflora to mammalian aromatic metabolism.

A. *The Rationale of Aromatic Metabolism in Microorganisms*

Disintegration of aromatic products by microorganisms is governed by the principle that these compounds are used as a source of carbon and energy. The exogenous aromatic substrates must therefore be transformed in such a way that aliphatic products are obtained that can be further metabolized in the high energy release reactions of primary metabolism. These terminal cycles of metabolism (i.e., glycolysis, Krebs cycle, and some closely related routes) can accept only a small number of suitable intermediates, such as formate, acetate, propionate, pyruvate, succinate, malate, oxaloacetate, and some other dicarboxylic acids. Therefore, all degradative pathways of aromatic catabolism must operate in such a way that the aromatic substrates, even if highly substituted or branched, are transformed by various introductory reactions (e.g., N-demethylation, O-demethylation, hydroxylation, deamination) and intermediate steps (ring fission, chain shortening, hydrolytic or oxidative removal of substituents) to yield the substrates of primary metabolism. To fulfill this task, surprising biochemical reactions have been developed especially in case of polycyclic aromatic compounds, elegant solutions for this "unfolding" of cyclic compounds to aliphatic linear acids have been found. The economy and efficiency of microbial catabolic pathways is documented by the evolution of branched, interrelated pathways where the basic carbon skeleton of different classes of substrates can eventually be dissimilated along a common single pathway with a minimum of enzymatic machinery (see Dagley, 1971).

In view of the great number of microorganisms found to metabolize aromatic substrates and because of the extraordinary ability of the microflora to metabolize a wide range of compounds, it is not surprising that a bewildering number of catabolic pathways has been described. Therefore, different solutions for the same biochemical problem (i.e., disintegration of a given aromatic compound) have evolved. On the other hand, the ability to degrade aromatics along different pathways has played an important role in taxonomic studies on *Pseudomonas* spp. and *Acinetobacter* spp. (Stanier *et al.,* 1966; Baumann *et al.,* 1968).

1. Classification of Aromatic Metabolism

The basic types of biochemical reactions involved in the transformation of aromatic substrates by microorganisms are summarized in Table I. It is evident that a comparatively small number of reactions are sufficient for the degradation of an enormous number of substrates. In most cases these introductory and intermediate steps in aromatic degradation are catalyzed by induced enzymes that are only formed in the presence of the substrate. The various aspects of enzyme formation as caused by substrate induction have been reviewed by Dagley (1971).

2. Oxygen-Introducing Enzymes

The most notable reactions in Table I are the oxygen-introducing steps, which are catalyzed by monooxygenases and dioxygenases.

Monooxygenases are involved in aromatic and aliphatic hydroxyl-

Table I

Basic Types of Reactions Involved in Introductory and Intermediate Steps of Aromatic Degradation by Microorganisms

Oxidation

$$-CH_3 \longrightarrow -CH_2OH \longrightarrow -CHO \longrightarrow -COOH$$

$$>CHOH \longrightarrow >C{=}O$$

$$-CH_2CH_2- \longrightarrow -CH{=}CH-$$

$$>CHCO_2H \longrightarrow >CHOH + CO_2$$

$$-CH_2CH_2NH_2 \longrightarrow -CH_2COOH$$

$$>S \longrightarrow >S{=}O$$

Reduction

$$-CH{=}CH- \longrightarrow -CH_2CH_2-$$

$$-CH{=}CH-C{-}O \longrightarrow -CH_2CH_2\underset{|}{C}O \longrightarrow CH_2CH_2CH_2OH$$

$$>CO \longrightarrow >CHOH$$

$$-CHO \longrightarrow -CH_2OH \longrightarrow -CH_3$$

$$-NO_2 \longrightarrow -NH_2$$

Hydroxylation

$$-\!>CH \longrightarrow -\!>COH$$

$$>CH_2 \longrightarrow >CHOH$$

$$-CH_3 \longrightarrow -CH_2OH$$

Demethylation

$$-OCH_3 \longrightarrow -OH + HCHO$$

$$-NHCH_3 \longrightarrow -NH_2 + HCHO$$

$$-N{<}^{CH_3}_{CH_3} \longrightarrow -N{<}^{H}_{CH_3} + HCHO$$

Elimination

$$>CHCH(OH)- \longrightarrow >C{=}CH- + H_2$$

$$>CHCH(OR)- \longrightarrow >C{=}CH- + HOR$$

$$>CHCH(NH_2)- \longrightarrow >C{=}CH- + H_3N$$

Table I (*Continued*)

Isomerization	Cleavage of C—C Bonds
Trans $\rightleftharpoons$ Cis	
D $\rightarrow$ L	$-CH_2CO_2H \rightarrow -CH_3 + CO_2$
β-OH $\rightleftharpoons$ α-OH	$>C=C< \rightarrow >C=O + O=C<$
Hydrolysis	Insertion of Heteroatoms
$-COOR \rightarrow -COOH + HOR$	
$-CONHR \rightarrow -COOH + H_2NR$	$\geq C{-}C{=}O \rightarrow \geq C{-}O{-}C{=}O$
Addition	
$-CH{=}CH- \rightarrow -CH_2CH(OH)-$	

ations, epoxidations, N-, O-, and S-dealkylations, and N- and O-oxidations. They incorporate one oxygen atom of an oxygen molecule into the substrate while the other oxygen atom is reduced to water. Such enzymes, which have been isolated from bacteria and fungi (for review, see Smith and Rosazza, 1975) as well as mammalian systems (see Section III,A,1), all catalyze the same basic reaction, as shown in Eq. 1.

$$RH + O=O + NAD(P)H + H^+ \longrightarrow ROH + H_2O + NAD(P). \qquad (1)$$

Basic reaction of monooxygenase. Bacterial monooxygenases often use NADH, while fungal and mammalian enzymes require NADPH.

Monooxygenases consist of a cytochrome *P*-450 hemoprotein linked to an electron transport system operative in providing the reducing equivalents. A general scheme of such a monooxygenase is shown in Fig. 1. The monooxygenases bind substrate and molecular oxygen and interact with reductases in a two-step sequence to activate oxygen. The systems found in bacteria are considerably different from those characterized in yeasts and fungi. Prokaryotic organisms appear to possess monooxygenase systems (cytochrome *P*-450, flavoprotein reductases, nonheme iron–sulfur proteins) that are nearly identical to those found in adrenal mitochondria. In contrast, eukaryotic organisms, such as yeasts and fungi, possess monooxygenases more similar to those found in mammalian hepatic microsomes. The various differences of monooxygenases isolated from pro-

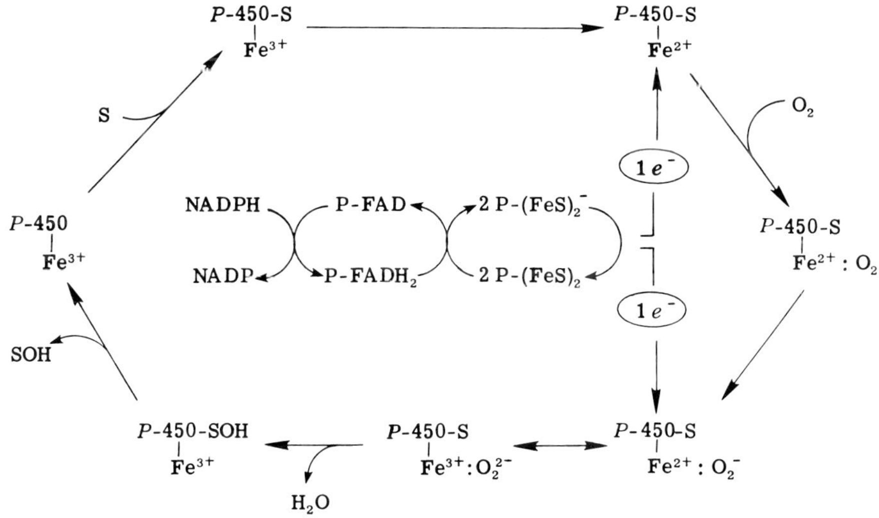

Fig. 1. Oxidation of a substrate S via a cytochrome *P*-450-linked monooxygenase.

karyotic and eukaryotic sources have been reviewed (Rosazza and Smith, 1979). Such differences are also of special interest with regard to either narrow or broad substrate specificity and to inducibility by substrate (microbial enzymes) and/or general cytochrome *P*-450 inducers (mammalian systems).

Regardless of the various differences observed so far, a great number of monooxygenase-catalyzed oxidations of a wide variety of substrates have been observed. These monooxygenases are especially involved in introducing a second hydroxyl group into a monophenol; such second hydroxylations prepare aromatic compounds for subsequent ring fission reactions.

Dioxygenases introduce both oxygen atoms of an oxygen molecule into one substrate. Reactions of this type are pertinent for ring fission reactions. Universal substrates for dioxygenase reactions are ortho- or paradihydric phenols that are cleaved alternatively to (a) dicarboxylic acids, (b) monocarboxy aldehydes, or (c) oxohexadienoic acids (Fig. 2). According to these findings, the ring cleavage reactions can thus be arranged in three groups.

1. The ortho cleavage (intradiol fission) opens the ring at the bond between the two carbon atoms that bear the hydroxyl groups. This reaction (enzyme: catechol 1,2-oxygenase or pyrocatechase) yields *cis,cis*-muconic acid and leads to the so-called β-ketoadipic acid pathway.

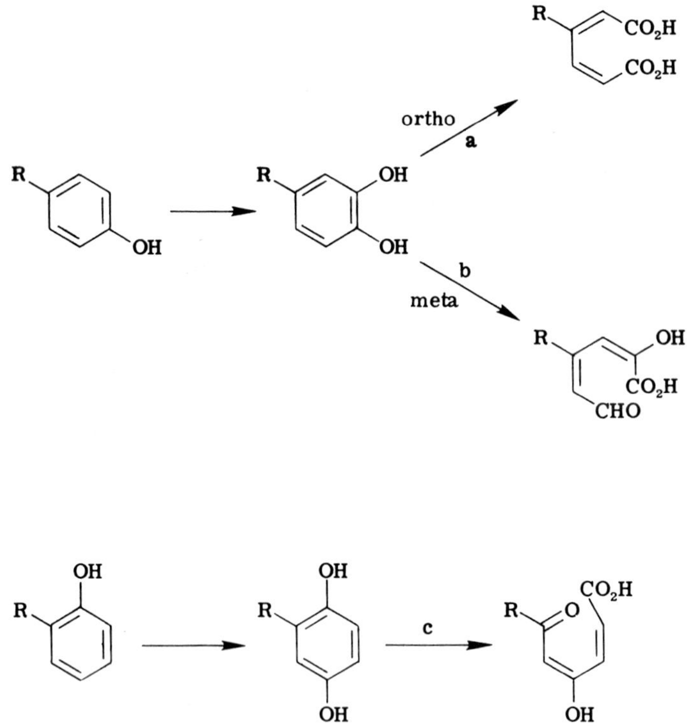

Fig. 2. Fundamental mechanisms of ring-fission reactions in microorganisms showing (a) ortho or intradiol cleavage, (b) meta or extradiol cleavage, and (c) *p*-dihydroxyphenol cleavage.

2. The meta cleavage (extradiol fission) attacks the ring between the carbon atoms bearing a hydroxyl group and an adjacent unsubstituted carbon. This mode of cleavage (enzyme: catechol 2,3-oxygenase or meta-pyrocatechase) leads to α-hydroxy-*cis,cis*-muconic acid semialdehyde and opens the α-ketocarboxylic acid pathway.
3. The *p*-dihydroxyphenol cleavage breaks the ring between the carbon atom bearing a hydroxyl group and an adjacent carbon atom bearing a hydrogen, a carboxyl group, or a side chain. This mode of ring cleavage ("gentisic acid pathway," enzyme: gentisate oxygenase) results in the formation of, for instance, 4-hydroxy-2-oxo-*cis,cis*-3,5-heptadienedioic acid from 2,5-dihydroxybenzoic acid (gentisic acid).

Several ring-cleavage dioxygenases have been purified and studied with regard to enzyme mechanism, substrate specificity, and inducibility by

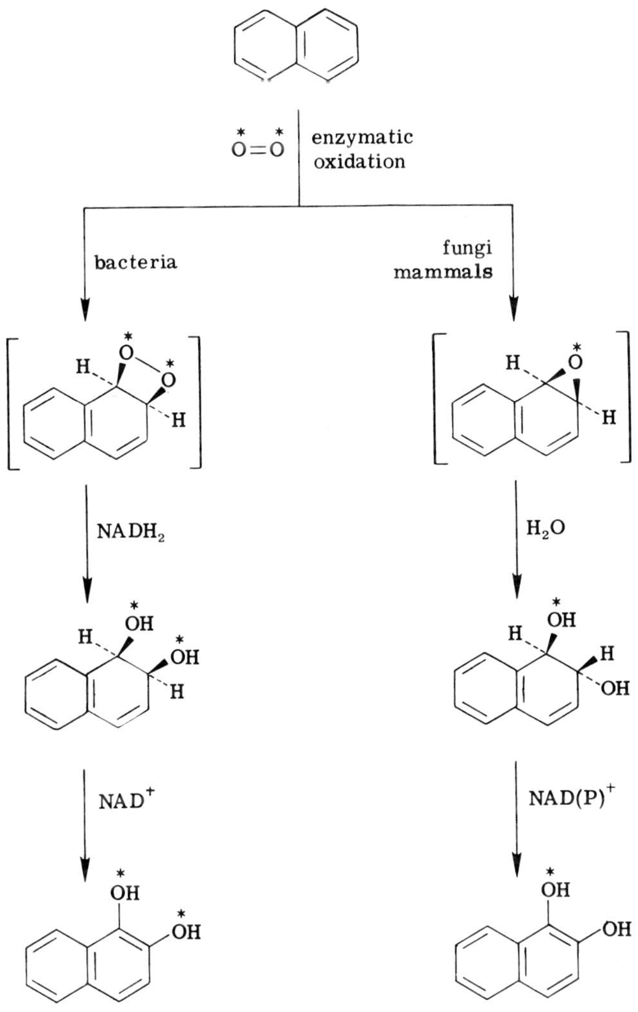

Fig. 3. Monooxygenase (fungi, mammals) versus dioxygenase (bacteria) transformations of naphthalene, depicting the origins of the oxygen atoms (*) in the *cis*-dihydrodiol formed by bacteria and the *trans*-dihydrodiol formed by mammals and fungi.

their substrates (Dagley, 1975, 1978). The intradiol class of catechol dioxygenases contain Fe^{3+}, whereas the extradiol dioxygenase appear to contain Fe^{2+}. No redox change of the ferric ion has been detected in various enzyme–substrate complexes so far reported (Walsh and Ballou, 1983).

A second important class of bacterial dioxygenases catalyze double hydroxylation reactions of aromatic compounds. This type of enzyme is suitably exemplified by anthranilate dioxygenase (Hayaishi, 1966), which forms catechol from anthranilic acid, benzoate 1,2-dioxygenase (conversion of benzoic acid to catechol) (Yamaguchi and Fujisawa, 1980), and several other dioxygenases catalyzing double hydroxylations of benzene (Crutcher and Geary, 1979), toluene (Yeh *et al.*, 1977), or naphthalene (Laborde and Gibson, 1976). Such oxygenases consists of two proteins, an NADH–cytochrome *c* reductase, and a terminal dioxygenase (Yamaguchi and Fujisawa, 1980).

As shown in Fig. 3, using naphthalene as model, a *cis*-1,2-dihydro-1,2-dihydroxyl compound has been found as an intermediate. Labeling studies with $^{18}O_2$ demonstrated that these enzymes are true dioxygenases, and NMR studies revealed that the *cis*-dihydrodiols are characteristic intermediates (Rosazza and Smith, 1979).

Much work on fungi and mammals implies that during the aerobic conversion of aromatic structures to 1,2-dihydroxy compounds by eukaryotic organisms, alternative pathways are functioning (Smith and Rosazza, 1982). First, arene oxides are formed by action of monooxygenases; epoxide hydrases subsequently yield *trans*-1,2-dihydro-1,2-dihydroxy structures which are further converted by dehydrogenases to the 1,2-dihydroxy-aromatic compounds (Fig. 3).

These enzymes allow bacteria and fungi to hydroxylate a great number of aromatic and heterocyclic structures, so that the 1,2- or 1,4-dihydroxyl structures may be obtained. These are the key compounds necessary for ring fission to occur.

3. Aspects of Catabolic Routes

The three basic types of aromatic ring fission reactions allow for the conversion of mono- and polycyclic as well as numerous substituted aromatic structures to aliphatic compounds. Depending on the chemical structure of the compound under investigation, the type of microorganism used, and the cleavage enzyme involved, several possibilities for the breakdown of a particular aromatic compound may exist. Generalization is to be established with caution.

The various possible substituents of aromatic compounds are either shortened or oxidatively removed prior to ring cleavage, or they remain on the ring and are eliminated after ring fission at the stage of aliphatics.

The ortho cleavage pathway, as shown in Fig. 4 for protocatechuate and catechol, leads to β-ketoadipic acid, which is further split in a coenzyme A-dependent reaction to yield succinic acid and acetyl coenzyme A.

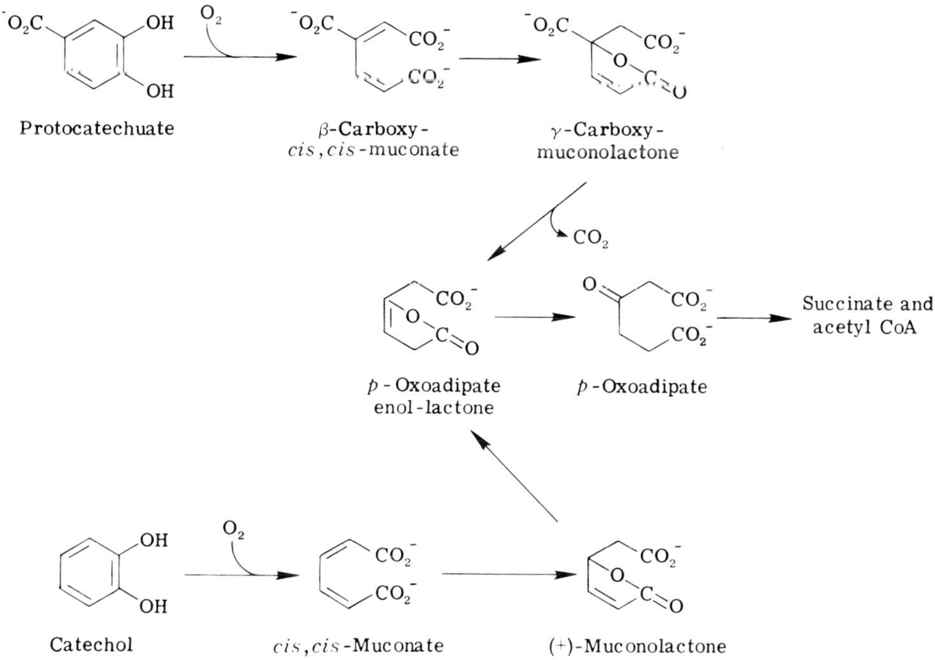

Fig. 4. The β-ketoadipic acid pathway for the disintegration of protocatechuic acid and catechol is opened by an ortho cleavage mechanism.

The utilization of this pathway absolutely requires that any substituents (for example hydrogen, carboxyl, chlorine) at the tertiary carbon atom (γ-position) of the intermediate muconolactone can be removed as a neutral molecule (carbon dioxide) or anion (chloride) prior to isomerization to yield β-oxoadipate enol-lactone. Since many natural products and wood extractives are converted to either protocatechuic acid or catechol (see the following sections), the β-ketoadipate pathway is of great importance for aromatic metabolism.

The basic pattern of the meta cleavage pathway, as shown for catechol in Fig. 5, documents that this degradative scheme must be especially suited to the disintegration of substituted aromatics. Important information relating to this function of meta fission has emerged from studies of the microbial degradation of steroids (Schoemer and Martin, 1980).

It is evident from Fig. 5 that substituents (alkyl, aryl, heterocyclic structures) in position 3 lead, after meta cleavage and further metabolism of the intermediate α-hydroxymuconic acid semialdehyde, to a carboxylic acid instead of simply to formate. Substituents in positions 4 and/or 5 will

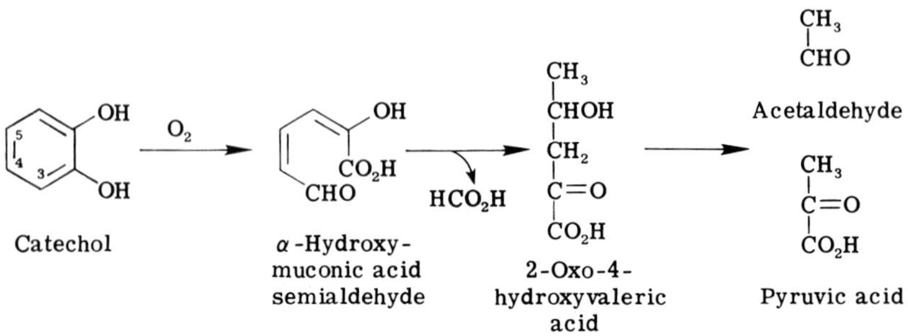

Fig. 5. The α-ketocarboxylic acid pathway of catechol resulting from a meta cleavage mechanism.

end up in the acetaldehyde moiety. In general, this type of ring cleavage constitutes a highly versatile mechanism for the disintegration of a wide spectrum of aromatic structures.

The *p*-dihydroxyphenol cleavage pathway (Fig. 6) is employed for the disintegration of numerous benzoic and phenylacetic acids, substituted phenols, and polycyclic systems (see Section II,B,6). In the case of gentisic acid (2,5-dihydroxybenzoic acid), the intermediate maleyl pyruvate is isomerized to fumaryl pyruvate, which is further split to fumaric acid and pyruvate. During the metabolism of homogentisic acid (2,5-dihydroxyphenylacetic acid derived from tyrosine), the isomerization process converts maleylacetoacetate (see Section IV,C), which yields fumarate and acetoacetate.

Application of these three ring-cleavage mechanisms on suitably substituted substrates will allow the disintegration of the wide array of natural aromatic compounds.

B. *Microbial Degradation of Different Classes of Aromatic Compounds*

This section outlines characteristic principles and important pathways for the disintegration of prominent classes of aromatic wood extractives.

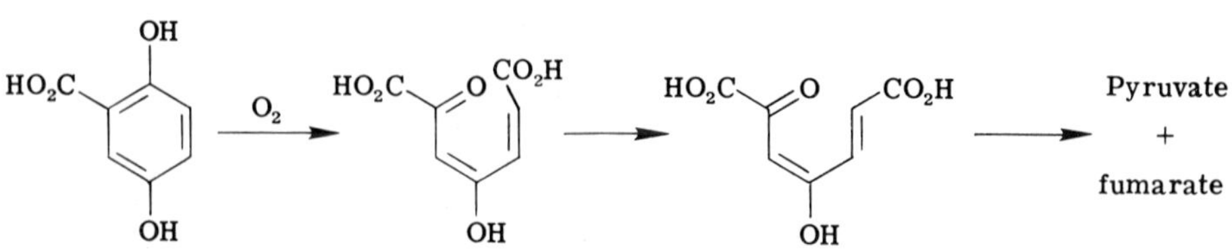

Fig. 6. The *p*-dihydroxyphenol cleavage pathway shown for gentisic acid.

In view of the many reports dealing with aromatic microbial catabolism, only representative examples can be given. Emphasis is placed on the chemistry and biochemistry of catabolic routes, whereas the many different bacteria and fungi involved are not dealt with from a taxonomic point of view.

1. Phenols

Figures 4 and 5 demonstrate that phenols such as catechol are key intermediates in aromatic catabolism. A multitude of bacteria, fungi, and yeasts have been shown to catabolize phenol, catechol, and simple substituted derivatives of these compounds. A characteristic feature of this field of catabolism is the fact that the various groups of organisms choose the alternative pathways of ortho or meta cleavage for the dissimilation of phenolic key intermediates. Details of monophenol and catechol degradation need not be given here. The relevant aspects of hydroxylation, dioxygenase ring cleavage mechanisms, enzyme induction by substrates, and regulation of catabolic routes have been summarized (Dagley, 1971, 1975, 1977; Kieslich, 1976; Stanier and Ornston, 1973).

With regard to the catabolism of aromatic wood extractives, the dissimilation of compounds with other substitution patterns (Fig. 7) such as hydroquinone (**1**), resorcinol (**2**), phloroglucinol (**3**), and pyrogallol (**4**) are of interest. Figure 7 summarizes findings on polyphenol degradation that show the various alternative pathways. Phloroglucinol (**3**) may be degraded either via resorcinol (**2**) or through pyrogallol (**4**) (Blackwood *et al.*, 1970; Walker and Taylor, 1983). The remarkable conversion of phloroglucinol to pyrogallol requires an intermolecular rearrangement of its oxygen, possibly via a transient resorcinol epoxide. Alternative pathways for resorcinol catabolism include pyrogallol (**4**) (Groseclose and Ribbons, 1981) or 1,2,4-trihydroxybenzene (**5**) (Chapman and Ribbons, 1976a,b). The latter compound is also formed by hydroxylation of hydroquinone (**1**). The 1,2,4-trihydroxybenzene (**5**) may be catabolized by meta or ortho cleavage (Chapman and Ribbons, 1976a,b; Gaal and Neujahr, 1979; Buswell and Eriksson, 1979). Figure 7 also demonstrates that alkyl substituted polyphenols with these relevant substitution patterns can be dissimilated through the same basic routes.

2. C_6–C_1 Compounds

Investigations on the microbial degradation of C_6–C_1 compounds such as benzoic acids, benzaldehydes, or benzylalcohols have greatly contributed to our understanding of aromatic catabolism. Basic aspects of ring-cleavage reactions, hydroxylations at aromatic or aliphatic positions, and

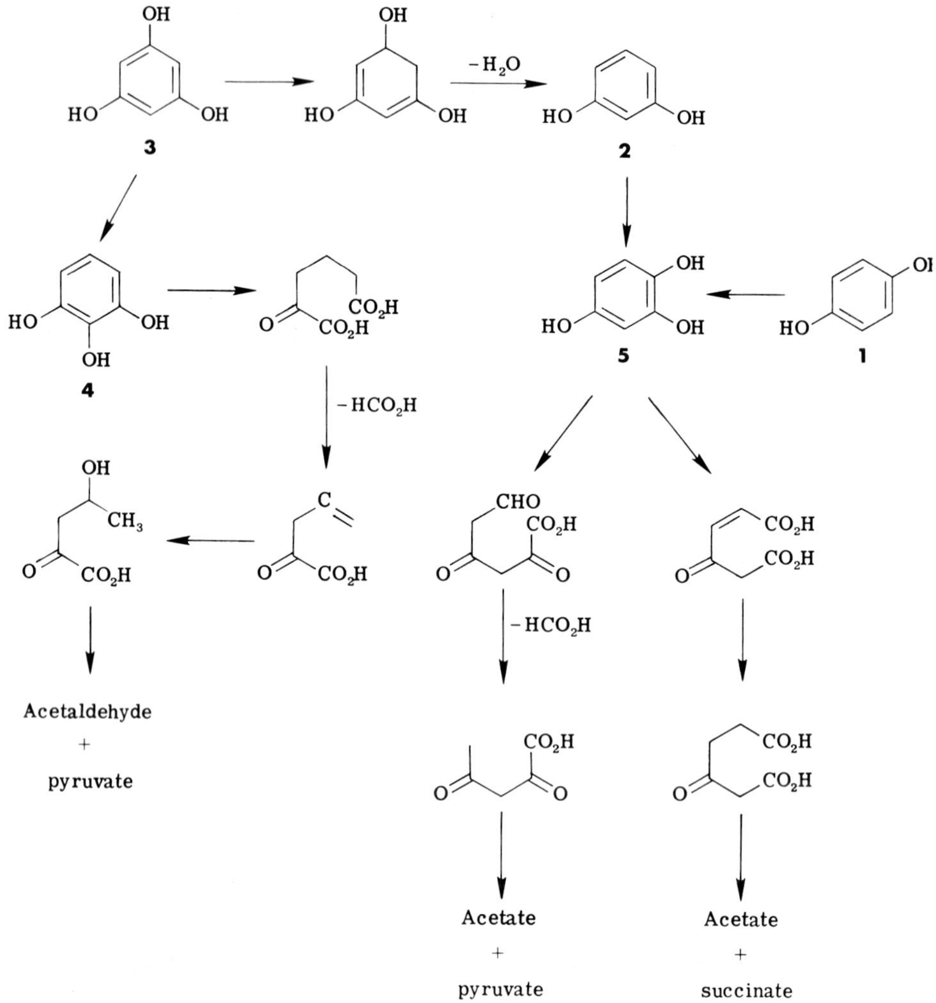

Fig. 7. Alternative degradative routes for hydroquinone (**1**), resorcinol (**2**), phloroglucinol (**3**), and pyrogallol (**4**) in microorganisms. Hydroxyhydroquinone (**5**) is an important intermediate.

regulation of catabolic pathways have been elucidated using these simple aromatic compounds. It must also be remembered that several classes of natural products are catabolized through C_6–C_1 compounds (see Section II,B,6). Thus, benzoic acids such as 3,4-dihydroxybenzoic acid (protocatechuate) or 2,5-dihydroxybenzoic acid (gentisic acid) function as key ring-fission substrates (see Figs. 4 and 6).

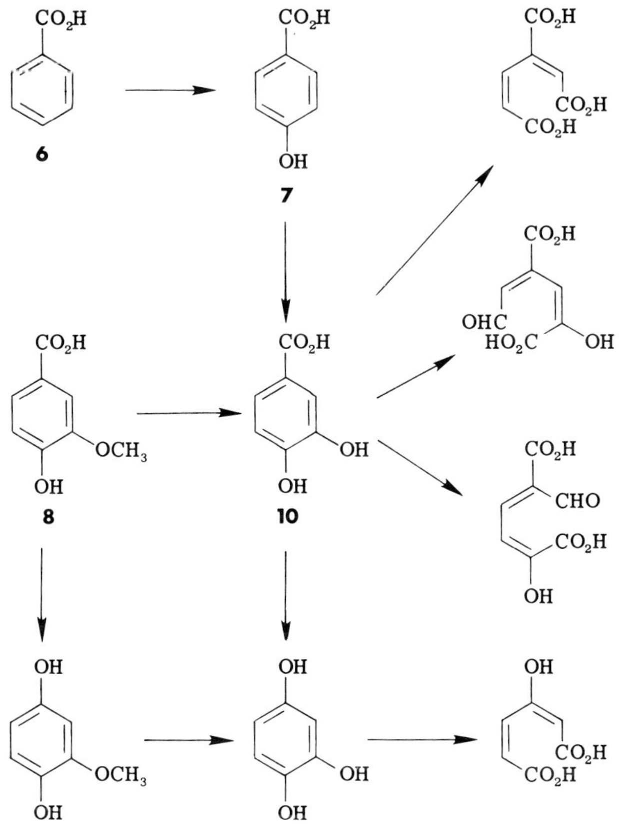

Fig. 8. Alternative metabolic pathways for the disintegration of benzoate (**6**), 4-hydroxybenzoic acid (**7**), vanillic acid (**8**), and protocatechuic acid (**10**) by microorganisms.

The broad literature has been repeatedly summarized (Dagley, 1971, 1977, 1978; Kieslich, 1976; Smith and Rosazza, 1982). These reviews also elaborate on the bewildering complexity of benzoic acid catabolism, because most substrates can be degraded along alternative pathways depending on the organism and the growth conditions applied (see Crawford and Crawford, 1980).

Figures 8 and 9 present an overview of well-established catabolic routes for benzoate (**6**), 4-hydroxybenzoic acid (**7**), vanillic acid (**8**), and salicylic acid (**9**). The pathways demonstrate the great importance of protocatechuic acid (**10**), gentisic acid (**11**), and catechol (**12**) as intermediates and substrates for ring fission reactions. In addition to the various position-specific dioxygenase ring-cleavage reactions, Figs. 8 and 9 further show

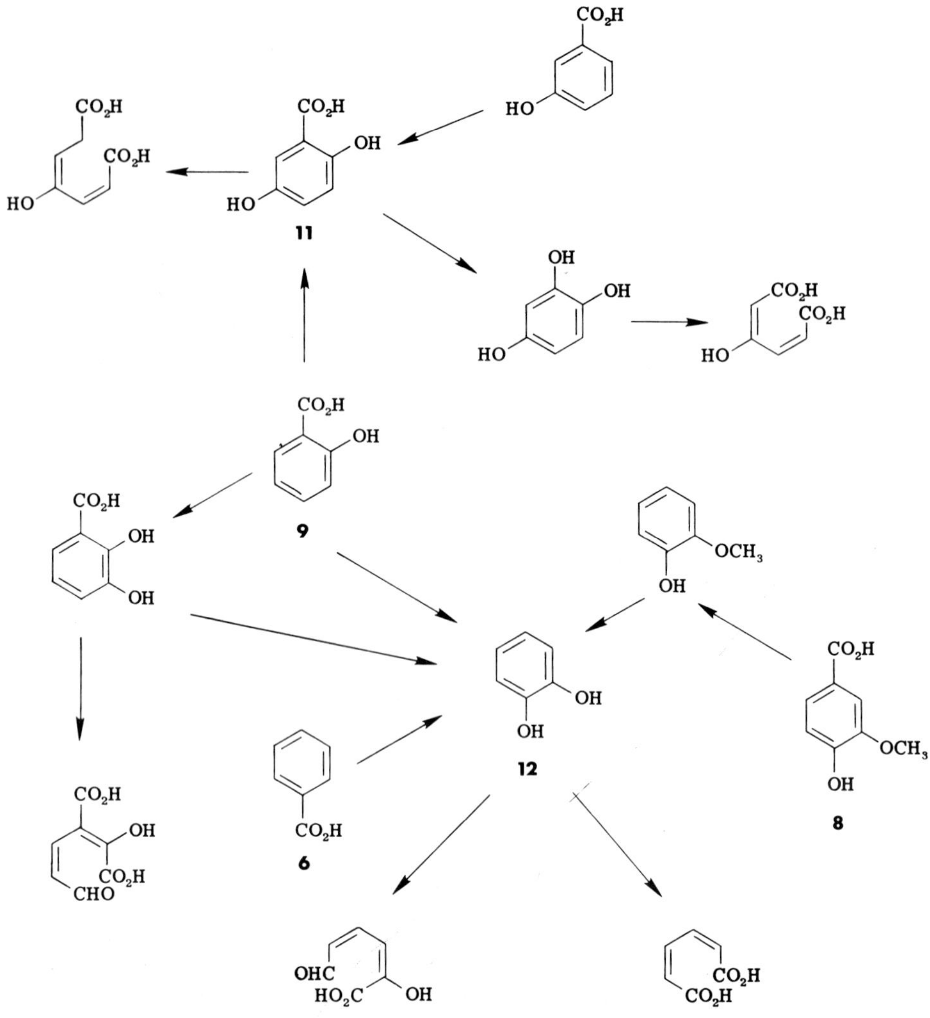

Fig. 9. Alternative metabolic pathways for the disintegration of benzoate (**6**), salicylic acid (**9**), vanillic acid (**8**), and gentisic acid (**11**). Catechol (**12**) is shown as a key intermediate of microbial catabolism.

that the ring-fission substrates are obtained by ring hydroxylation, O-demethylation, and oxidative as well as nonoxidative decarboxylation reactions (Crawford and Crawford, 1980; Anderson and Dagley, 1980; Sutherland *et al.*, 1981; Kieslich, 1976). In view of the great number of potential microbial degraders and the various alternative pathways, each

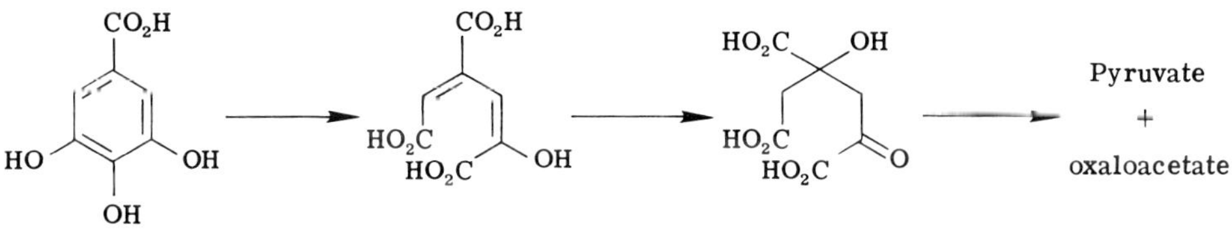

Fig. 10. Degradation of gallic acid by *Pseudomonas* sp.

microorganism must carefully be examined to determine the exact mode of aromatic degradation.

Gallic acid (3,4,5-trihydroxybenzoic acid), which is a building unit of hydrolyzable tannins, has also been repeatedly investigated in the context of its degradation by *Pseudomonas* species (Donnelly and Dagley, 1981; Dagley, 1978; Trevors and Basaraba, 1980). Figure 10 shows that an ortho cleavage reaction finally gives rise to pyruvate and oxaloacetate. If 3-*O*-methylgallic acid is subjected to this ring cleavage reaction, oxaloacetate monomethyl ester is obtained, which indicates that *O*-methyl groups are not necessarily oxidatively removed prior to ring fission.

Benzaldehydes and benzylalcohols are first converted to the corresponding benzoic acids before degradation can occur. The dehydrogenases involved in these oxidation reactions have been characterized (Dagley, 1977).

Inspection of Figs. 6, 8, and 9 demonstrates how alkyl-substituted C_6–C_1 compounds are degraded.

3. C_6–C_2 Compounds

Phenylacetic acids occur widely in the plant kingdom and have also been detected as intermediates in the catabolism of several classes of aromatic compounds (see Sections II,B,7 and IV,C). Numerous investigations on phenylacetic acid catabolism by bacteria and fungi have elucidated the basic pathways of ring-fission reactions for these substrates. It has clearly been found that several catabolic pathways seem to exist for a given hydroxyl compound, depending on the organism and the growth conditions (Dagley, 1977). These results are documented (Fig. 11) by the pathways found for the metabolism of 4-hydroxyphenylacetic acid (**13**) and 3,4-dihydroxyphenylacetic acid (**14**).

In some species, degradation of 4-hydroxyphenylacetic acid is initiated by hydroxylating the benzene nucleus at C_1, the side chain being displaced simultaneously, so that 2,5-dihydroxyphenylacetic acid (**15**) is formed. This acid is also involved in the degradation of L-tyrosine, where acetoacetate and fumarate are the final catabolites (see Section IV,C).

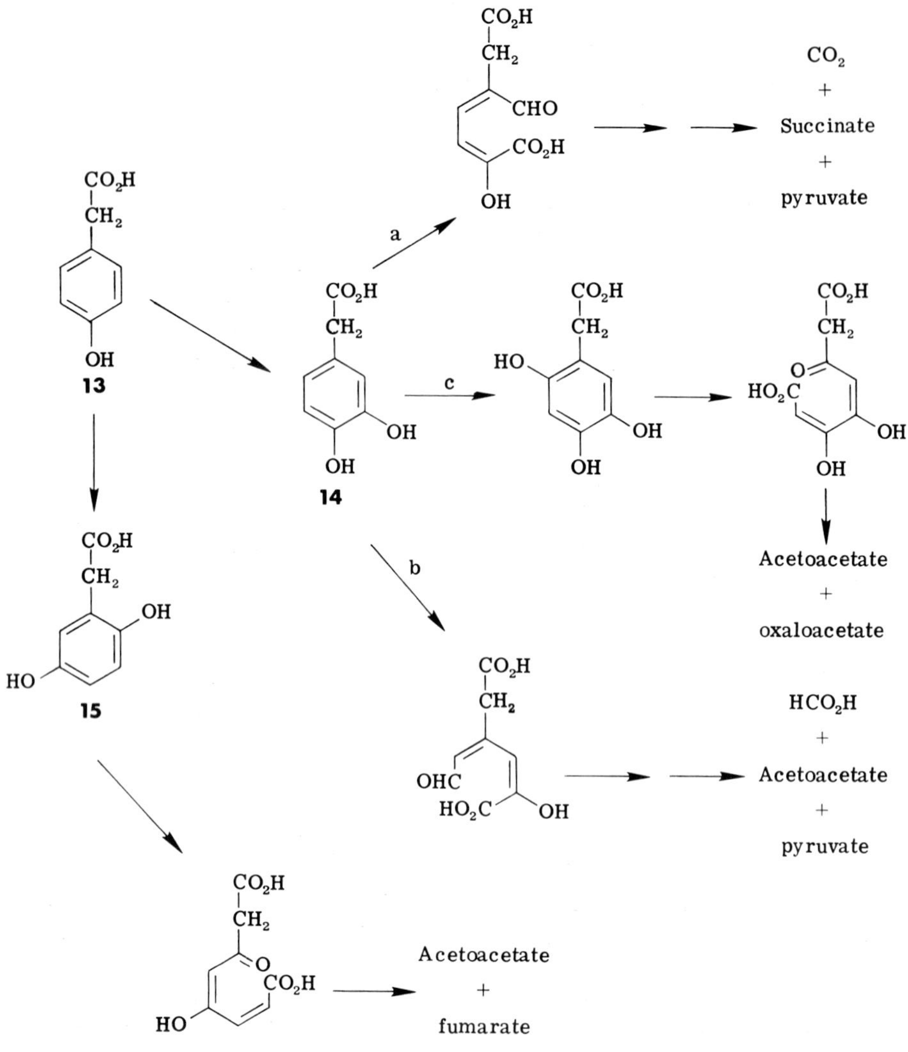

Fig. 11. Alternative microbial pathways for the degradation of 4-hydroxyphenylacetic acid (**13**) and 3,4-dihydroxyphenylacetic acid (**14**). For details, see text.

In other organisms, (**13**) is hydroxylated at C_3, giving 3,4-dihydroxyphenylacetic acid (Hareland *et al.*, 1975; Sparnins *et al.*, 1974). Figure 11 also shows that this acid may be ring-cleaved by a 2,3-dioxygenase (pathway a) or a 4,5-dioxygenase (route b) (Sparnins *et al.*, 1974; Seidman *et al.*, 1969). Furthermore, 3,4-dihydroxyphenylacetic acid may be catabolized through the stage of 2,4,5-trihydroxyphenylacetic acid (route c, Fig. 11), which leads to acetoacetate and oxaloacetate (Sparnins *et al.*, 1978). Similarly, 2-hydroxyphenylacetic acid or 3-hydroxyphenylacetic acid, after a second hydroxylation reaction, may be funneled into the pathways shown in Fig. 11.

The bacterial degradation of 3,4,5-trimethoxyphenylacetic acid is started by two O-demethylation steps to 3,4-dihydroxy-5-methoxyphenylacetic acid. This acid is the ring-fission substrate, and it is dissimilated to methanol, pyruvate, and 3-ketoglutarate (Donnelly *et al.*, 1981).

Intensive work has been carried out on the microbial catabolism of mandelic acid and its derivatives. In principle, mandelate is dehydrogenated to phenylglyoxylate and further degraded through benzaldehyde and benzoate (see Hills and Fewson, 1983).

4. C_6–C_3 Compounds

Cinnamic acids and their aldehyde and alcohol derivatives are the most important natural C_6–C_3 compounds because they represent the building units of lignins. Several species of bacteria and fungi have been isolated as degraders of C_6–C_3 compounds.

The most important pathway for the degradation of cinnamic acids comprises shortening of the unsaturated side chain to yield the equivalent benzoic acid. Thus cinnamic acid, *p*-coumaric acid, ferulic acid, sinapic acid, and 3,4,5-trimethoxycinnamate are converted to benzoate, *p*-hydroxybenzoate, vanillic acid, syringic acid, and 3,4,5-trimethoxybenzoate, respectively (Donnelley and Dagley, 1981; Sutherland *et al.*, 1983; Kieslich, 1976). This step of side-chain degradation could involve a process of β-oxidation or a nonoxidative shortening of the side chain to give the corresponding aldehyde derivative, which is oxidized to the acid (see Iyayi and Dart, 1982). The benzoic acids are further metabolized by O-demethylation and/or ring hydroxylation reactions to yield catechol, protocatechuic acid, and gallic acid, respectively, as suitable substrates for ring fission reactions (see Section II,B,2). Other reports, however, show that ring hydroxylation may precede side-chain degradation, because *Streptomyces* species hydroxylate *p*-coumaric acid to caffeic acid, which is then catabolized through protocatechuic acid (Nambudiri *et al.*, 1972).

Cinnamyl alcohols such as *p*-coumaryl alcohol or coniferyl alcohol are degraded through the corresponding aldehydes to the cinnamic acids, which are further metabolized as outlined above (Iyayi and Dart, 1982; Eggeling and Sahm, 1980). Oxygenated cinnamic acids bearing a hydroxyl group in the para position are converted by many microorganisms to the

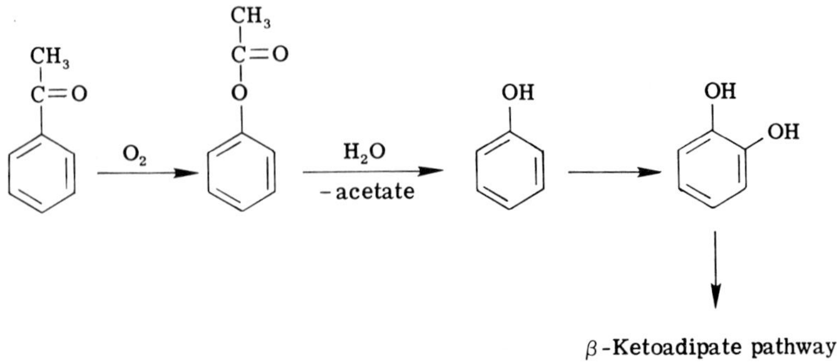

Fig. 12. Proposed pathway of acetophenone metabolism by *Arthrobacter* sp. Redrawn from Cripps (1975).

corresponding styrenes through a nonoxidative decarboxylation (Grammatica *et al.*, 1981). These styrenes and their reduction products *p*-ethylphenols may, however, further be degraded to benzoic acids (Kieslich, 1976).

White-rot fungi such as *Trametes* (Nishida and Fukuzumi, 1978) and *Sporotrichum* (Gupta *et al.*, 1981), when cultivated on ferulic acid and an additional carbon source (glucose, ethanol), transform this cinnamic acid to coniferyl alcohol, coniferylaldehyde, vanillic acid, vanillyl alcohol, 2-methoxyhydroquinone, and 2-methoxyquinone. Other cinnamic acids are similarly metabolized by reductive processes.

5. Acetophenone

Although methyl phenyl ketones such as acetophenone will easily be metabolized via benzoic acid or hydroxylation-dependent ring-cleavage reactions (Kieslich, 1976), an example of the biological equivalent of the chemical Baeyer–Villiger reaction should also be described. Whereas ketones are oxidized chemically to esters by peracids, microbial species metabolize ketones by a sequence of reactions initiated by the insertion of an oxygen atom between the carbon atom of the oxy group and an adjacent carbon atom to form either an ester or a lactone. The enzymes responsible are mixed-function oxygenases. This principle is adequately documented by the metabolism of acetophenone by an *Arthrobacter* sp. (Fig. 12), where the ketone is degraded via phenyl acetate, phenol, and catechol (Cripps, 1975). Other examples of this principle are available (Kieslich, 1976; Cripps, 1975).

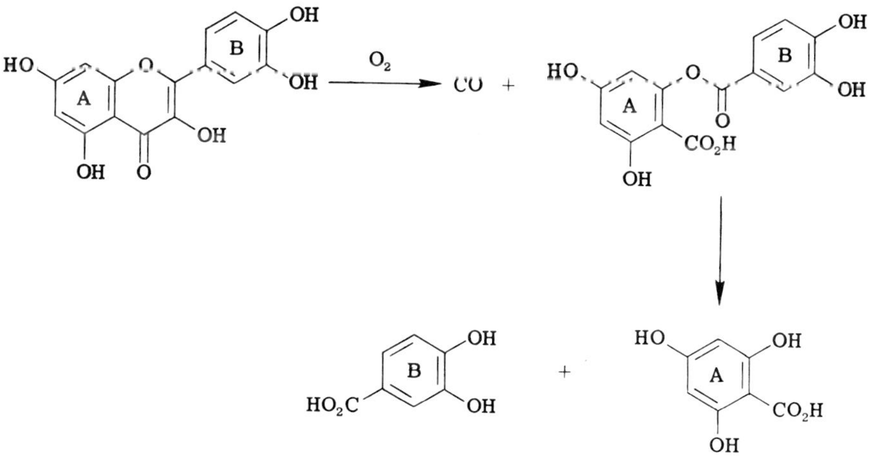

Fig. 13. Degradative pathway of the flavanol quercetin as occurring in fungi.

6. Flavonoids

Numerous bacteria and fungi have been shown to modify or to degrade the various classes of natural flavonoid compounds (review Barz and Hösel, 1975). Many of these flavonoids occur as glycosides or acylglycosides. The hydrolytic removal of acyl groups and the hydrolysis of glycosidic moieties are readily carried out by various microbial species. The principle of microbial flavonoid degradation is best understood by comparison of the pathways shown in Figs. 13 and 14.

The flavonol quercetin (Fig. 13) is degraded by *Aspergillus* and other fungi to two benzoic acids and CO derived from C-3 (Simspon *et al.*, 1960). This oxygenation is catalyzed by the enzyme quercetinase, which has unequivocally been shown to incorporate the two oxygen atoms into the depside moiety, essentially as required for a true dioxygenase (Brown *et al.*, 1982). Phloroglucinol and protocatechuic acid as cleavage products of the depside are dissimilated via well-established routes.

A different mechanism for flavonol disintegration is verified by *Pseudomonas* sp. (Fig. 14). These organisms hydroxylate the substrate at C-8 (ring A) and carry out a dioxygenase-catalyzed meta-type cleavage of ring A. The pathway finally leads to two molecules of oxaloacetate and a substituted benzoic acid from ring B (Schultz *et al.*, 1974). The introductory hydroxylation at C-8 is also found in flavone, dihydroflavonol, and catechin metabolisms (see Barz and Hösel, 1975).

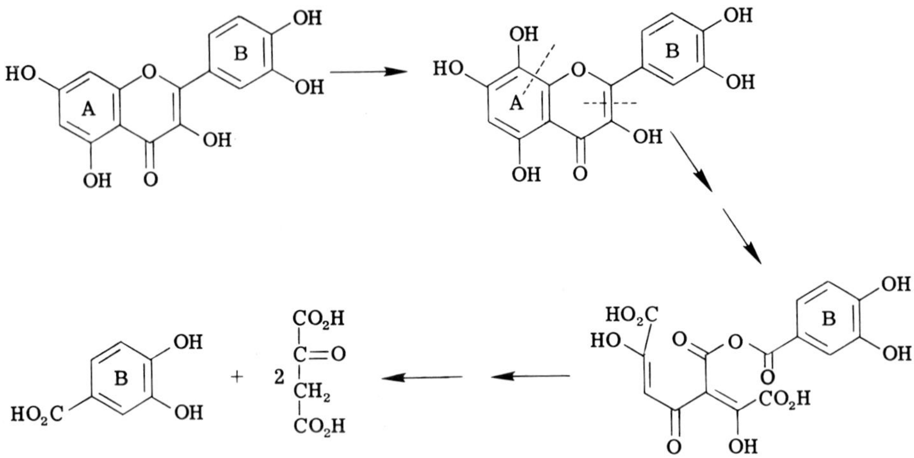

Fig. 14. Degradation of the flavanol quercetin by *Pseudomonas* species via meta cleavage of an 8-hydroxyflavonol.

When degraded by *Pseudomonas* species, catechins are first oxidized at C-4 to yield the dihydroflavonol taxifolin, which is then subject to a degradative mechanism essentially shown in Fig. 14. Thus, catechins and dihydroflavonols yield one mole each of oxaloacetate and 5-(3,4-dihydroxyphenyl)-4-hydroxy-3-oxovalero-δ-lactone. Some other aspects of catechin degradation as carried out by gut microflora are mentioned in Section III,B,8.

7. Isoflavonoids

Isoflavonoids are characteristic tissue and wood constituents of members of *Fabaceae* (Ingham, 1983). These compounds have only recently been investigated more intensively for microbial dissimilation. Most of the earlier work was done in connection with metabolism of forage estrogens in an attempt to clarify their role in the infertility syndrome of mammals (review Barz and Hösel, 1975). Most of the results obtained with animals are essentially due to the action of the intestinal microflora (Section III,B,9).

Recent interest in isoflavonoid degradation, mainly studied in phytopathogenic fungi (*Fusarium, Colletotrichum, Ascochyta, Botrytis, Stemphylium, Helminthosporium*), results from the fact that isoflavones and isoflavanones may act as plant defense compounds (preinfectional inhibitors and phytoalexins) toward fungal pathogens (Barz *et al.*, 1980). Fungal degradation of isoflavonoids may thus be regarded as a detoxication pro-

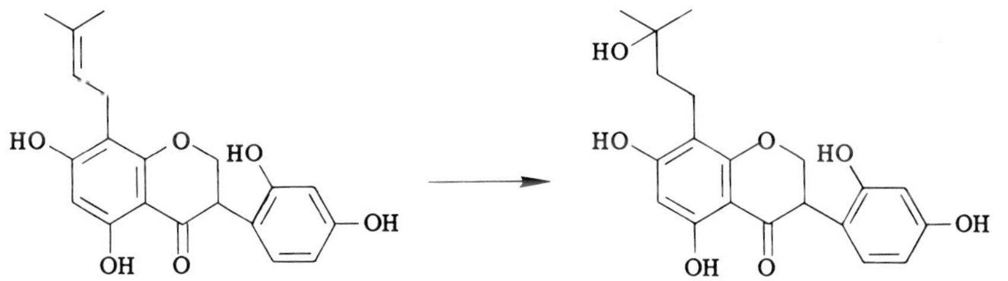

Fig. 15. Hydration of kievitone to kievitone hydrate by *Fusarium solani*. This reaction converts the biologically active phytoalexin to a noninhibitory compound. Redrawn from Smith *et al.* (1980).

cess (VanEtten *et al.*, 1982). A suitable example is the conversion of the biologically active isoflavanone kievitone to the noninhibitory kievitone hydrate (Fig. 15) by *Fusarium solani* (Smith *et al.*, 1980). The responsible enzyme has also been characterized (Smith *et al.*, 1981).

Disintegration of an isoflavone skeleton has recently been elucidated for biochanin A using *Fusarium javanicum* (Willeke and Barz, 1982a). The side-chain phenyl ring B is converted to a phenylacetic acid, with the carboxyl group being derived from C-2 of the γ-pyrone ring. The substrate for cleavage of ring A is presumably 8-hydroxydihydrobiochanin A (Mackenbrock and W. Barz, unpublished). Thus, the C_4 unit of ring A in Fig. 16 most likely is oxaloacetate; a similar mechanism of ring cleavage has been documented for flavonoid catabolism by *Pseudomonas* sp. (compare Fig. 14 in Section II,B,6). Essentially, the same pathway as in Fig. 16 has been shown to occur with several other *Fusarium* fungi for biochanin A (Willeke *et al.*, 1983) and the 4′-hydroxyderivative genistein (Willeke and Barz, 1982b).

The 5-deoxyisoflavones (formononetin, daidzein) are much less efficiently metabolized by microorganisms (Weltring *et al.*, 1982); except for O-demethylation, O-methylation, and 3′-hydroxylation, no degradative reactions have been observed. The remarkable difference between 5-hydroxy- and 5-deoxyisoflavones remains to be determined.

8. Pterocarpans

Pterocarpans, which are isoflavonoids from a biosynthetic point of view (Stoessl, 1982), occur in the heartwood of tropical trees (Ingham, 1983) and have attracted attention due to their fungitoxic properties as phytoalexins (Barz *et al.*, 1980).

Numerous investigations with phytopathogenic fungi have found simple conversion as well as substantial degradative reactions to occur with

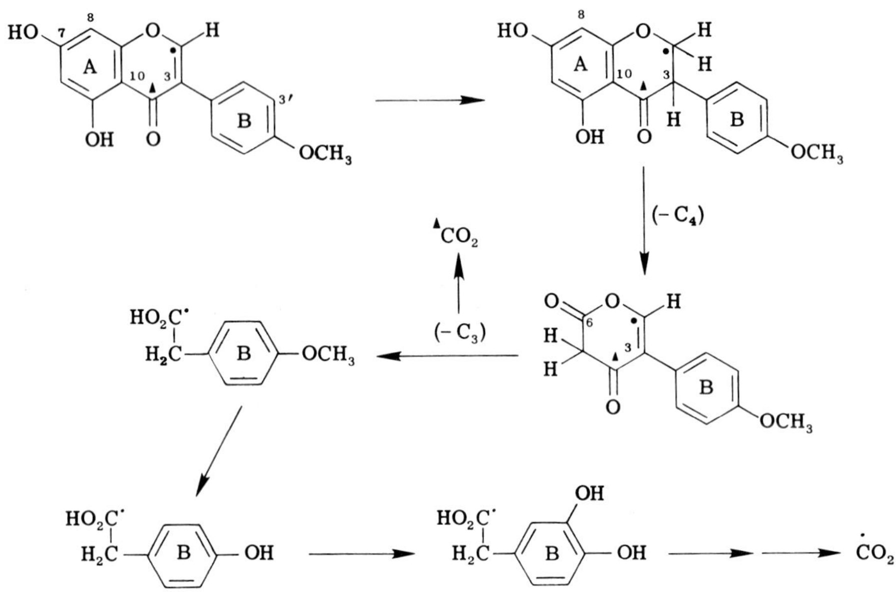

Fig. 16. Degradative pathway of the isoflavone biochanin A as found with *F. javanicum*. Symbols represent position-specific ^{14}C-labeled substrates.

phytoalexin pterocarpans (VanEtten *et al.*, 1982; Barz *et al.*, 1980). Using the most widespread phytoalexin, medicarpin, Fig. 17 shows the various fungal conversion reactions known so far. Reactions b, c, d, e, and g represent characteristic O-demethylation and hydroxylations in various positions. Similar reactions of these types have been found with other phytoalexins (VanEtten *et al.*, 1982). Pathway a represents a remarkable and so far unique oxidative cleavage of the dihydrofuran ring (Denny and VanEtten, 1982). Much more common than reaction a is the reductive cleavage (pathway h) of the benzyl phenyl ether bond, which results in the formation of an isoflavan (VanEtten *et al.*, 1982). Reaction f represents an O-methylation that has also been found with other isoflavonoids (Weltring *et al.*, 1982).

Although several introductory conversion reactions of pterocarpans are well known (Fig. 17), no complete degradative pathway has so far been elucidated. An example (Weltring *et al.*, 1983) on medicarpin degradation by *Fusarium oxysporum* f. sp. *lycopersici* (Fig. 18) at least signals how the isoflavan skeleton may further be degraded. It is postulated that the 3-phenylcoumarin is hydrolytically opened and decarboxylated to yield a stilbene, which may be split to two benzoic acids.

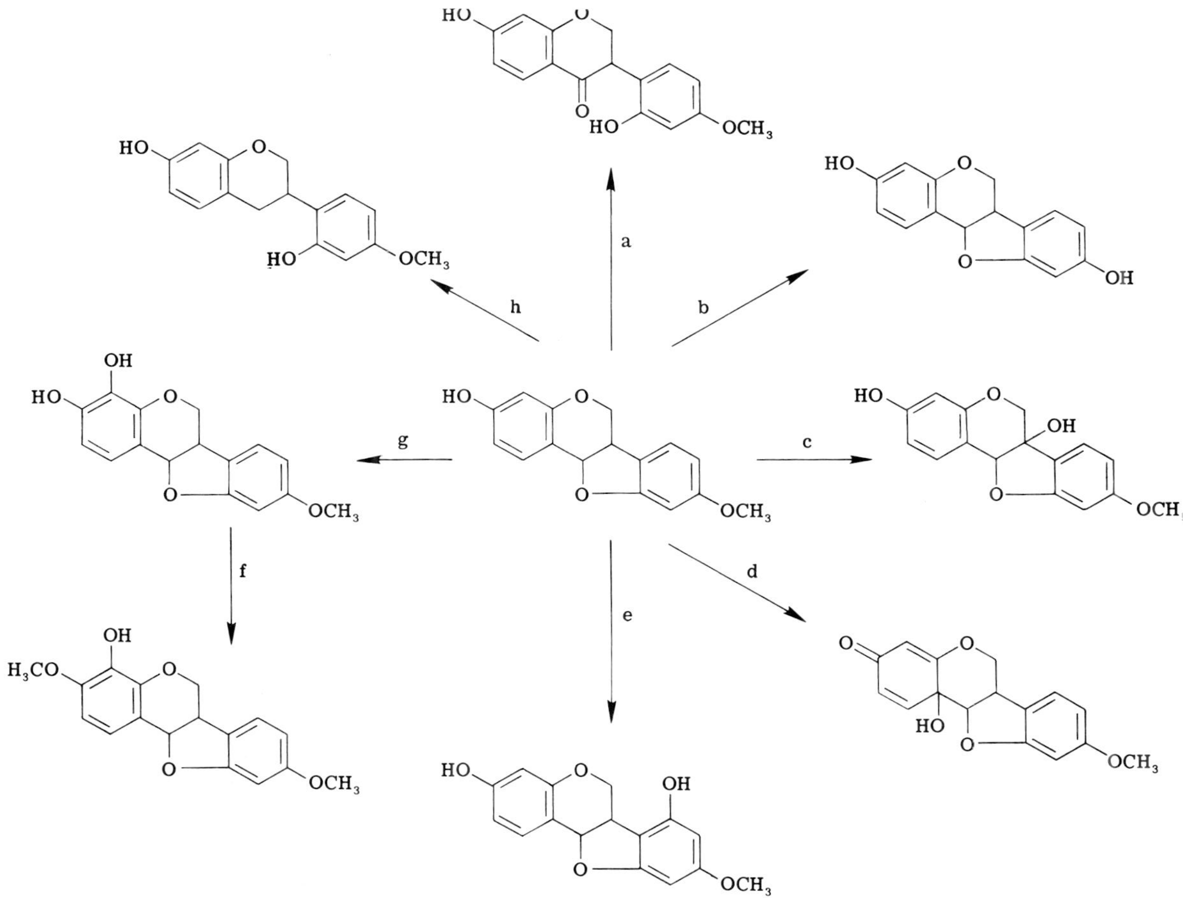

Fig. 17. Presently known fungal conversion reactions of the pterocarpan phytoalexin medicarpin (see VanEtten *et al* , 1982).

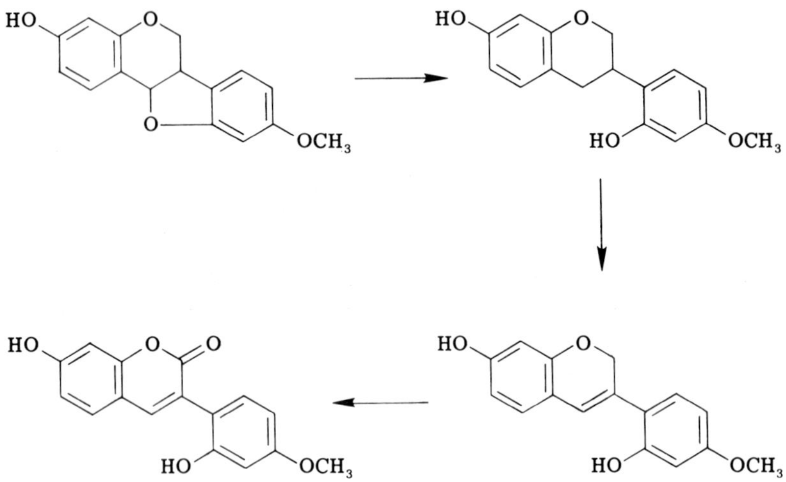

Fig. 18. Degradation of the phytoalexin pterocarpan medicarpin to a 3-phenylcoumarin by *Fusarium oxysporum* f.sp. *lycopersici*. Redrawn from Weltring *et al.* (1983).

9. Rotenoids

Rotenoids are components of the heartwood, roots, stems, and leaves of many leguminous species and have been widely used as nonpersistent pesticides. Despite their apparent low environmental hazard, a well-understood microbial catabolic pathway is not available. Recent investigations by Sariaslani and Rosazza (1983) found that various fungal species transformed rotenone to several metabolites. The two biotransformation products of rotenone (**16**) by *Cunningnamella blakesleeana* were identified as hydroxylation products **17** and **18** of the isopropylene side chain. Further investigations on rotenoid catabolism are, therefore, highly necessary.

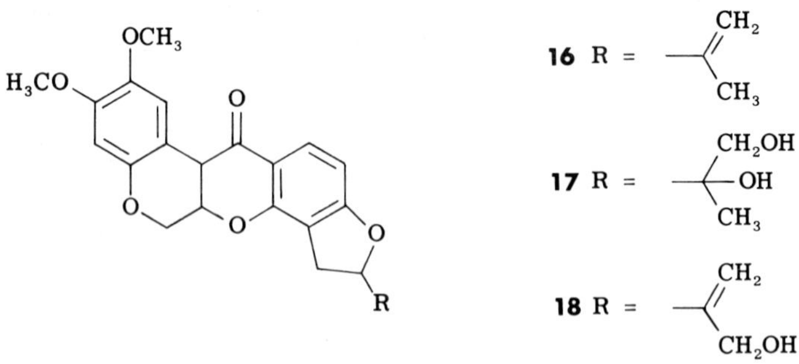

10. Stilbenes

Although several stilbenes (i.e., pinosylvin, pinosylvin monomethyl ether) have long been recognized as fungitoxic heartwood constituents of many conifers and broad-leaved trees, they appear to be readily detoxified by fungi (Loman, 1970). The exact nature of this detoxification process has not been elucidated, and catabolites of stilbenes were not reported. The data are best interpreted by assuming that wood-degrading ascomycetes and basidiomycetes inactivate stilbenes by including, with the aid of laccases, tyrosinases, or peroxidases, polymerization processes of these highly reactive phenols. Stilbene microbial degradation is thus a still open field for future studies; oxidative cleavage of the double bond with formation of two benzoic acids can be expected as a reasonable mode of catabolism.

11. Quinones

Quinones are widely distributed in Nature, and many of them have been found in plant species and as wood extractives. Despite their abundant occurrence, very little is known about the microbial metabolism of these compounds, and no complete catabolic route is presently available. Two recent reports, however, present results on how quinone structures may be cleaved.

Degradation of the walnut naphthoquinone juglone, which exhibits strong biocidal properties, by *Pseudomonas putida* (Rettenmaier *et al.*, 1983) was shown to involve: juglone(5-hydroxynaphthoquinone) → 3-hydroxyjuglone → 2,3-dihydroxybenzoate and, due to meta cleavage, 2-hydroxymuconic acid semialdehyde. The latter compound originated by action of 2,3-dihydroxybenzoate 3,4-dioxygenase with concomitant release of CO_2.

Lapachol is an antitumor naphthoquinone found in the heartwood of several plant species. Among the several microbial metabolites of this quinone (Otten and Rosazza, 1978, 1979), oxidative ring cleavage is a major pathway of microbial conversion. This reaction (Fig. 19) is car-

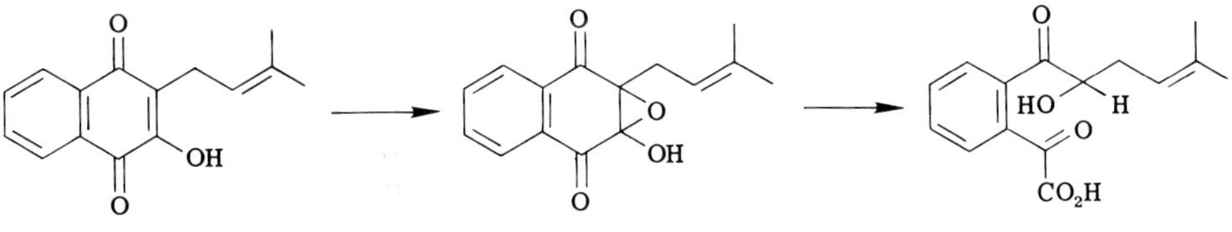

Fig. 19. Conversion of the naphthoquinone lapachol by *Penicillium notatum* to a ketol via an epoxide intermediate. Redrawn from Otten and Rosazza (1983).

ried out by *Penicillium notatum* and, based on $^{18}O_2$ studies, undoubtedly involves an epoxide intermediate (Otten and Rosazza, 1983).

III. Metabolism and Degradation of Aromatic Compounds in Animals

Decomposition of wood, lignified plant tissue, and other plant litter in the biosphere, such as in soil or aqueous environments, is a very complex process that requires the cooperative action of various groups of microorganisms and animals each acting to facilitate further breakdown by the other (Dickinson and Pugh, 1974).

In the animal kingdom, protozoa, nematodes, oligochaetes, micro- and macroarthropods, invertebrates, and vertebrates have been shown to contribute at various stages to the decomposition process. Dissimilatory reactions of organic compounds (sugar, cellulose, lignin, secondary plant constituents) are also encountered with each group of animals mentioned. Information on the biochemical pathways of aromatic degradative processes in lower animals is comparatively meager and rather widespread, so that this field is in many aspects still an underdeveloped area of research. A unifying picture is not yet available.

In some insect–plant interactions, type and concentrations of secondary plant/wood constituents have been shown to play a vital role (Brand *et al.*, 1979). In addition to degradative reactions (Brunet, 1980), the partial modification of wood/plant monomeric constituents (alkaloids, terpenoids, phenolics) by insects could be demonstrated as a biologically very important process. Such insecticidal modification reactions or plant compounds are involved in the biosynthesis of several insect pheromones (Brand *et al.*, 1979).

Aspects of phenolic metabolism and conjugation reactions in fish have been summarized (Call *et al.*, 1980; Franklin *et al.*, 1980).

By far the largest body of information is available on the metabolism and degradation of wood extractives and related plant compounds in mammals. Such studies have long been executed (Williams, 1959), and the available knowledge results from two major areas of investigations.

1. The pronounced physiological effects of many plant constituents (toxins, narcotics, analgesics, anti-inflammatories) on the mammalian body have greatly stimulated investigations on the mammalian metabolism of numerous plant products and other xenobiotics (Hutson, 1975, 1977; Williams, 1959, Mannering, 1971; Mandel, 1971). These pharmacokinetic and pharmacological studies are of great significance in understanding detoxification mechanisms.

2. Analyses of the digestive processes of food dissimilation in herbivores have also significantly contributed to our present knowledge of aromatic metabolism and degradation in mammalian systems.

In the following sections of this chapter, the basic principles of the metabolism and degradation of aromatic wood-derived compounds in mammals are outlined. Furthermore, detailed catabolic pathways of relevant classes of plant products will also be given in addition to some remarks on the contribution of intestinal microorganisms to mammalian metabolism.

A. *The Rationale of Aromatic Metabolism in Mammals*

In mammalian organisms, aromatic wood extractives and other plant-derived aromatic or heterocyclic compounds are in most cases not accepted as nutrients but rather are metabolized as xenobiotics (i.e., physiologically active compounds not synthesized by, and therefore foreign to, the mammalian body). Such compounds must be inactivated or detoxified by biotransformation reactions; subsequently, they are excreted in urine or feces in the form of highly hydrophilic conjugates. In this biphasic process, the conversion of the incorporated often lipophilic compounds to more polar derivatives is summarized as "phase I" whereas the formation of the water-soluble conjugates is called "phase II" (Williams, 1959; Scheline, 1978).

1. Reactions in Phase I

In general, phase I reactions comprise oxidations, reductions, and hydrolyses. With such basic reactions catalyzed by a comparatively small number of enzymes, a tremendous variety of xenobiotics can be modified in preparation for conjugation processes to occur. The most important reactions observed in phase I are summarized in Tables II and III. Comparison of these examples clearly shows that oxidation reactions (Table II) are by far most important biotransformations. They all aim at generating in the xenobiotics OH, NH, or SH groups for conjugation reactions to proceed.

Several of these oxidations, such as aliphatic and aromatic hydroxylation, epoxidations, O-, N-, and S-dealkylations, N- and S-oxidations, and deaminations, are carried out by mixed-function oxygenases (monooxygenases), which were found to be especially associated with the hepatic endoplasmic reticulum. These monooxygenases form an enzyme complex consisting of a hemoprotein, cytochrome *P*-450, and an

Table II

Classification of Oxidation Reactions (Phase I) Involved in the Metabolism of Xenobiotics in Mammals

		Examples	
Reaction	Enzyme involved	Substrate	Products
Oxidations			
Aliphatic hydroxylation	Monooxygenase	Hydrocarbon	Primary alcohol Secondary alcohol Tertiary alcohol
Benzylic hydroxylation	Monooxygenase	Toluene	Benzyl alcohol
Allylic hydroxylation	Monooxygenase	Allylbenzene	1′-Hydroxyallylbenzene
Alicyclic hydroxylation	Monooxygenase	Bornane; cyclohexane	Borneol, cyclohexanol
α-Hydroxylations	Monooxygenase	Nicotine Coumarin	5′-Hydroxynicotine 3-Hydroxycoumarin
β-Hydroxylations	Monooxygenase	Coumarin	4-Hydroxycoumarin
Epoxidation	Monooxygenase Monooxygenase	Styrene Benzene	Styrene epoxide (→Styrene) Arene epoxide (→Phenol or catechol via dihydrodiol)
Alcohol oxidation	NAD-oxidoreductase	Alcohol	Aldehyde
Aldehyde oxidation	NAD-oxidoreductase	Aldehyde	Carboxylic acid
Dehydrogenation	Dehydrogenase	Dihydrodiol Phenylpropionic acid	Catechol Cinnamic acid
β-Oxidation	Chain-shortening system of four enzymes	Cinnamic acid	Benzoic acid
Aromatization	Dehydrogenase	Cyclohexanecarboxylic acid	Benzoic acid
Dealkylation			
N-, O-, S-Demethylation	Monooxygenase	$R—X—CH_3$	R—XH + HCHO
N-, O-, S-Dealkylation	Monooxygenase	$R—X—CH_2—R'$	R—XH + OCH—R′
N-Oxidation	Monooxygenase	Primary/secondary amines Tertiary amine	Hydroxylamines Amineoxide
S-Oxidation	Monooxygenase	Sulfide	Sulfoxide/sulfone
Oxidative deamination	Monooxygenase	Primary amine	Aldehyde + NH_3

Table III

Classification of Reduction and Hydrolysis Reactions (Phase I) Involved in the Metabolism of Xenobiotics in Mammals

		Examples	
Reaction	Enzyme involved	Substrate	Product(s)
Reductions			
Aldehyde reduction	NAD-Oxidoreductase	Aldehyde	Primary alcohol
Ketone reductio	NAD-Oxidoreductase	Ketone	Secondary alcohol
Nitro group reduction	Reductase	Nitrobenzene	Aniline
Sulfur compound reduction	Reductase	Disulfide	Thiol
Hydrolyses			
Ester hydrolysis	Esterase	Ester	Carboxylic acid and alcohol
Amide hydrolysis	Amidase	Amide	Carboxylic acid and NH_3
Epoxide hydrolysis	Hydratase	Arene oxide	Dihydrodiol
		Aliphatic epoxide	Glycol
Glycoside hydrolysis	Glycohydrolase	Glycoside	Sugar and aglycone

NADPH–cytochrome *P*-450 reductase. They introduce one oxygen atom of an oxygen molecule into the substrate and reduce the other oxygen atom to water. Cytochrome *P*-450 catalyzes the oxidation reaction, whereas the NADPH–cytochrome *P*-450 reductase mobilizes the reducing equivalents of the NADPH. Great progress has been made in our understanding of these mixed-function oxygenases (Hutson, 1975, 1977; Gunsalus *et al.*, 1975) which, in general, are substrate-induced and quite often show a remarkably broad substrate specificity.

The reductions (Table III) are catalized by microsomal reductases that require either NADPH or NADH. Of special interest is the reduction of nitro group to yield an amino function. The hydrolytic reactions (Table III) have been intensively studied in mammalian tissues, so that esterases, amidases, and β-glucuronidases together with numerous other glycosidases (glycohydrolases) (La Du *et al.*, 1971) are well understood.

2. Reactions in Phase II

Conjugation reactions in phase II are anabolic processes that generally result in metabolites of reduced biologial activity. In these reactions, a functional group (OH, NH, SH) of the xenobiotic, as modified in phase I, is combined with an endogenous substrate, such as glucuronic acid, glucose, sulfate, glycine, acetate, cysteine, or methyl groups, to form a conjugate. This metabolism leads from lipophilic or weakly hydrophilic compounds to highly water-soluble conjugates that are readily excreted. In most cases formation of a conjugate requires that either the conjugating moiety or the xenobiotic are introduced into the reaction in the form of a nucleotide coenzyme or an activated substrate. Essential data on conjugate formation are summarized in Table IV.

Greatest importance must be placed on glucuronide formation, because this represents the most versatile type of mammalian conjugation reaction. The functional groups listed in Table IV as necessary prerequisites for glucuronylation indicate that with some xenobotics, the formation of more than one glucuronide must be possible. Thus, phenolic benzoic acids are converted to both the ether and ester glucuronides. Details of this and other types of conjugation reactions will become evident in subsequent sections of this chapter. It will also become evident how frequently the various conjugation reactions are employed.

3. Reactions of the Gastrointestinal Microflora

Numerous investigations on mammalian metabolism of plant xenobiotics have shown that the gastrointestinal microflora may greatly contribute to the degradative reactions. More importantly, certain reactions, such as aromatic dehydroxylations, appear to be unique to the gut microflora. Table V presents a survey of degradative reactions contributing to the body microflora. It should be emphasized that several of these modifications are the result of anaerobic metabolic conditions. Gastrointestinal reactions are also of importance for the enterohepatic circulation of xenobiotics (mammalian glucuronide conjugates are excreted in the bile and will then be further metabolized by the microflora of the lower gut) for the metabolic differences observed between various species or strains and in the metabolic adaptation to compounds incorporated repeatedly.

Comparison of Tables II–V reveals that the metabolic pathways as carried out by mammalian tissue enzymes, and gut microorganisms overlap so that a clear differentiation may be difficult to achieve. In such cases the use of antibiotic-treated or germ-free animals, as well as suspensions of gastrointestinal microflora, can be advantageous. Comprehensive treatment of the gut microfloral metabolism of plant xenobiotics is available in reviews (Scheline, 1973, 1978).

Table IV

Classification of Conjugation Reactions (Phase II) Involved in the Metabolism of Xenobiotics in Mammals

Conjugation reaction	Functional group of xenobiotic	Coenzyme or type of substrate activation	Enzyme involved or cosubstrate	Product
Glucuronic acid conjugation	—OH —NH_2 —SH^2 >C—H —COOH	UDP-Glucuronic acid	UDP-Glucuronyl-transferase	—O—, —NH—, —S—, >C— } Glucuronide Glucuronic acid ester
Glucoside conjugation	Phenolic —OH	UDP-Glucose	Glucosyltransferase	Phenolic glucoside
Sulfate conjugation	Aryl or heterocyclic —OH —NH_2	3′-Phosphoadenosine-5′-phosphosulfate	Sulfotransferase	Aryl or heterocyclic sulfate ester, sulfate
Methylation	Aryl OH Primary, secondary, tertiary NH_2 —SH	*S*-Adenosyl-methionine	Methyltransferase	O-Methylether Methylamine Methylsulfide
Amide formation	Aryl or hetero-cyclic COOH	Carboxyl coenzyme A thioester	Glycine Glutamine Taurine	Amide with amino acid
Gluthathione conjugation	Halogen, nitro, epoxi, aryl, alkyl alkene compound with electrophilic carbon atom	None	Glutathione *S*-transferase and enzymes of mer-capturic acid pathway	Mercapturic acid (*N*-acetylcysteine derivative)
Acetylation	Primary NH_2	Acetyl coenzyme A	*N*-Acetyltransferase	*N*-Acetyl derivative

Table V

Classification of Degradative Reactions Carried Out with Plant Xenobiotics by Gastrointestinal Microflora and Relevant Examples

Reaction	Examples: Substrate	Examples: Products
Glycoside hydrolysis	Glycoside	Aglycone and sugar
Ester hydrolysis	Ester	Acid and alcohol
Dehydroxylation of catechol acids	3,4-Dihydroxyphenylacetic acid	*m*-Hydroxyphenyl-acetic acid
Decarboxylation of phenolic acids	Ferulic acid	4-Vinylguaiacol
O-Demethylation	Aromatic methyl ether	Phenol
Heterocyclic ring fission	(+)-Catechin	*m*-Hydroxyphenyl-propionic acid
Double-bond reduction	3,4-Dihydroxycinnamic acid	3,4-Dihydroxyphenyl-propionic acid
Alcohol reduction	Substituted benzyl alcohol	Substituted toluene
Aldehyde reduction	Salicylaldehyde	Salicyl alcohol
Ketone reduction	Isoflavone daidzein	Isoflavan equol
Aromatization	Quinic acid	Benzoic acid, protocatechuic acid

B. *Mammalian Degradation of Different Classes of Aromatic Compounds*

Earlier investigations of mammalian metabolism of plant xenobiotics were largely concerned with the identification of their urinary metabolites. More recent studies have centered on the role of the intestinal microflora in the catabolic processes, on the significance of the biliary–enteric route of excretion, and on the disposition of plant xenobiotics in mammalian tissues following oral and parenteral administration.

In the following subsections, the most important findings with several prominent classes of wood extractives and aromatic plant xenobiotics are given. A complete survey of all experimental observations has not been intended. The reader is also referred to several reviews (Scheline, 1978; Goodwin, 1976).

1. Phenols

Although mono-, di-, and trihydric phenols of the simpler types are of restricted occurrence in the plant kingdom, the metabolism of these compounds elegantly illustrates the principles of xenobiotic mammalian metabolism.

Phenol itself may undergo conjugation reactions directly by virtue of the presence of a free hydroxyl group, or it may undergo oxidation that gives rise to hydroquinone, this in turn being subject to conjugation. The formation of glucuronide or ether sulfate conjugates of both phenolic compounds has largely been observed; however, a phosphate conjugate of phenol has also been identified as an urinary metabolite in cats (Capel *et al.,* 1974).

The same metabolic principles can be seen in the metabolism of catechol, hydroquinone, resorcinol, and orcinol (5-methylresorcinol), which were excreted from various mammalian species as monoglucuronide or monosulfate conjugates. Hydroxylation reactions of such dihydric phenols (i.e., catechol-yielding, 1,2,4-trihydroxybenzene) were also detected. Phloroglucinol (1,3,5-trihydroxybenzene) when given to humans or rats mainly led to phloroglucinol glucuronide (Takaji, 1971).

In addition to conjugation reactions, polyhydric phenols may undergo oxidation to quinone derivatives (Miller *et al.,* 1973), and in the case of dihydric compounds (catechol, pyrogallol), O-methylation is also possible. Such 1,2-dihydroxyaromatic compounds are excellent substrates of the *O*-methyltransferase system of mammalian liver (see Table IV) (Caldwell and Jacoby, 1983). Consequently, following the oral administration of catechol or pyrogallol to rats, the mono-*O*-methyl derivatives have been isolated (Scheline, 1978).

2. Aromatic Acids

Information on the mammalian metabolism of aromatic acids, such as benzoic acids, phenylacetic acids, and cinnamic acids, has long been accumulated. Numerous animals under various nutrition conditions have been tested so that a very clear correlation of the extent of aromatic acid metabolism with the animal species used has been established. Extensive summaries of this field of metabolism have appeared (Williams, 1959; Scheline, 1978; Davison, 1971).

An essential point of aromatic acid metabolism is conjugation reactions that may occur at the carboxyl group or at other suitable functional groups. The leading ideas of these reactions have been discussed in Section III,A,2 (see Table IV). Conjugation with glucuronic acid, amino acids such as glycine or glutamine, the amino compound taurine, or the sulfate group are the most important reactions. The enzymology of such conjugation reactions has been studied very extensively (Jacoby, 1980; Caldwell and Jacoby, 1983).

Table VI

Representative Examples of Main Metabolites of the Most Abundant Aromatic Acids as Formed by Mammals

Aromatic compound	Main metabolites excreted
Benzoic acid	Hippuric acid; benzoyl glucuronide
Salicylic acid	Salicyluric acid, glucuronide ether, glucuronide ester, 2,5-dihydroxibenzoic acid
4-Hydrobenzoic acid	4-Hydroxyhippuric acid, glucuronide ether, glucuronide ester, ester sulfate, 3,4-dihydroxybenzoic acid, phenol
2,5-Dihydroxy benzoic acid	Acid 5-*O*-sulfate, acid 5-*O*-glucuronide, 5-methoxysalicylic acid
3,4-Dihydroxybenzoic acid	Glycine conjugate, 3-methoxy-4-hydroxybenzoic acid, 3-hydroxy-, 4-hydroxy-, and 3-methoxybenzoic acid, catechol
3,4,5-Trihydroxybenzoic acid	4-*O*-Methyl acid, 2-*O*-methylpyrogallol, glucuronide ester
3-Methoxy-4-hydroxy-benzoic acid	Glucuronide ester, ethereal sulfate, glycine conjugate, guaiacol, catechol
3,4-Dimethoxybenzoic acid	Glucuronide ester, 3-hydroxy-4-methoxy- and 3-methoxy-4-hydroxybenzoic acid
Phenylacetic acid	Phenylacetoyl glutamine (monkeys, humans), glycine (rodents, carnivores), taurine, and glucuronide
4-Hydroxyphenylacetic acid	4-Hydroxy-3-methoxyphenylacetic acid, *p*-cresol
2,5-Dihydroxyphenyl-acetic acid	Substrate for ring-fission reaction in homogentisic acid pathway
3,4-Dihydroxyphenyl-acetic acid	3-Hydroxy-4-methoxy- and 3-methoxy-4-hydroxyphenyl acetic acid, 3-hydroxyphenylacetic acid, 4-hydroxy-phenylacetic acid
2-Hydroxycinnamic acid	2-*O*-Glucuronide, glycine conjugate, 2-hydroxydihydrocinnamic acid and its glycine conjugate 4-hydroxycoumarin, 2-hydroxyphenylacetic acid
4-Hydroxycinnamic acid	Glycine conjugate, 4-hydrocinnamic acid, 4-hydrobenzoic acid and its ether sulfate
3,4-Dihydroxycinnamic acid/3-methoxy-4-hydroxycinnamic acid	3,4-Dihydroxystyrene, *p*-ethylcatechol, 3-methoxy-4-hydroxycinnamic acid and its glycine conjugate, 3,4-dihydroxy- and 3-hydroxydihydrocinnamic acid
3,5-Dimethoxy-4-hydroxycinnamic acid	3,5-Dimethoxy-4-hydroxy- 3,4-dihydroxy-5-methoxy- 3,4,5-trihydroxy- and 3,5-dihydroxydihydrocinnamic acids and glycine conjugates

The most important cases of aromatic acid conjugation and metabolism are summarized in Table VI, which presents the main metabolites of several abundant aromatic acids. Basic differences with regard to species variation of metabolite excretion are partly presented, although details must be looked for in suitable reviews (Scheline, 1978).

An interesting feature of aromatic acid metabolism is the excretion of substantial amounts of free acids; this is a function of the acid's pH value, with low pK values leading to decreasing degree of conjugation. In addition to conjugation, several other noteworthy reactions of aromatic acids are found in Table VI.

Decarboxylation of *p*-hydroxy aromatic acids has been found repeatedly. This nonoxidative reaction, which yields a 1-hydroxy or -methoxy compound, has unequivocally been shown to be catalyzed by the gastrointestinal microflora of mammals. Similarly, dehydroxylation reactions that mostly eliminate a parahydroxy substituent are also due to the gut microflora (see Table V) (Scheline, 1978). Reductions of unsaturated side chains as shown to occur with cinnamic acids are likewise of microbial origin (Scheline, 1978).

O-Methylation of one hydroxyl group in a catechol moiety is catalyzed by tissue-located catechol *O*-methyltransferase, which is of great importance in mammalian adrenalin metabolism. The ability of this enzyme system to O-methylate at the 3 or 4 positions was extensively studied by Creveling *et al.* (1972), who found a 7 : 1 ratio between 3-O- and 4-O-methylated product.

Various esters of the aromatic acids shown in Table VI are easily hydrolyzed in the body, and their metabolic fate is largely that of their acid and alcohol components. An example of this is seen with benzyl benzoate, which is completely metabolized via benzyl alcohol to benzoic acid in humans.

Chlorogenic acid [3-(3,4-dihydroxycinnamoyl) quinic acid] is a commonly occurring plant ester. The cinnamic acid moiety leads to 3-hydroxydihydrocinnamic acid as the main urinary metabolite; other metabolites of this cinnamic acid as shown in Table VI were also found.

3. Acyclic Acids

This section deals with the metabolism of (−)-quinic acid (**19**) and shikimic acid (**20**), which are widely distributed plant constituents. These closely related acids both undergo metabolic aromatization to benzoic acid, and this aromatic metabolite is subsequently excreted in the urine as its glycine conjugate (hippuric acid). Catechol has also been detected as an

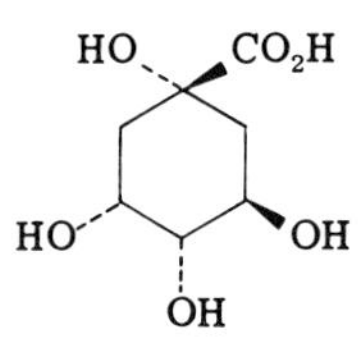

19

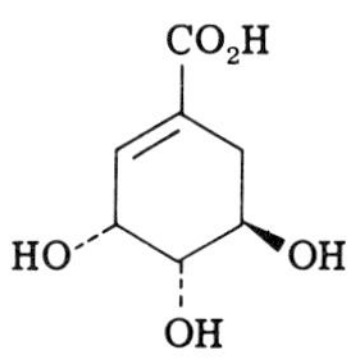

20

alternative minor aromatic catabolite of both alicyclic acids. Several investigations on the aromatization of quinic and shikimic acids have demonstrated that the metabolic activity of the intestinal microorganisms is responsible for the formation of benzoic acid. Species variation in the degree of benzoic acid formation can thus also be explained. For detailed information, the reader is referred to a summary by Scheline (1978).

4. Aromatic Aldehydes, Alcohols, Esters, and Ketones

Simple aromatic alcohols (benzyl alcohol, salicyl alcohol, or vanillyl alcohol) and aldehydes (benzaldehyde, salicyclaldehyde, vanillin, or 3,4-dihydroxybenzaldehyde) are mainly metabolized by oxidation to the respective aromatic acids. The latter are excreted either as the free acids or as conjugates (see Table VI). Excretion of unchanged alcohols or aldehydes in both free and conjugated form has been recorded at low concentrations only (Scheline, 1978; Williams, 1959).

Cinnamyl aldehydes follow the expected pathway of oxidation to the corresponding acid, although the presence of the three-carbon side chain allows for shortening by β-oxidation (see Table II) with subsequent excretion of benzoic acid conjugates.

Alkyl phenyl ethers such as methyl, ethyl, or methylene dioxy ether groups are readily cleaved by action of liver and tissue cytochrome *P*-450 monooxygenase, which catalyzes the introductory hydroxylation reaction. The resulting hemiacetal is readily hydrolyzed to yield formaldehyde and a free hydroxyl group.

Mammalian metabolism (rats, dogs) of safrole (**21**) 1-allyl-3,4,-methylene dioxybenzene) illustrates the metabolic principles of the methylene dioxy moiety and the allyl side chain. As shown in Fig. 20, cleavage of the CH_2 group with subsequent monomethylation of the catechol structure and various oxidation reactions of the side chain are the main metabolic routes (Borchert *et al.,* 1973; Padieu and Maume, 1977). The available studies on other allylbenzene derivatives have been summarized by Scheline (1978).

Alkyl phenyl ketones (such as acetophenone, acetovanillone, resacetophenone, or phloracetophenone) are excreted as glucuronide conjugates and ether sulfates if phenolic hydroxyl groups are available for conjugation reactions to occur. The unsubstituted acetophenone itself was shown to be metabolized in rabbits and dogs both by reduction to methyl phenyl carbinol and by oxidation to benzoic acid and mandelic acid.

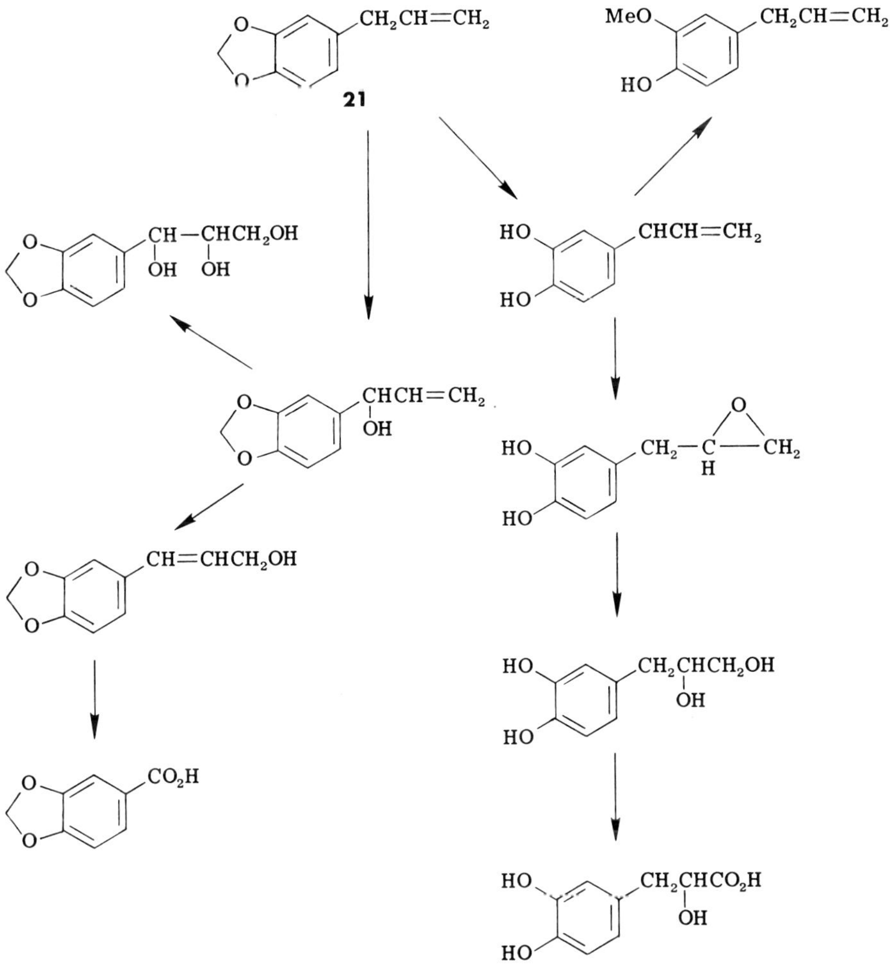

Fig. 20. Main metabolic pathways of safrole (**21**) in rats and dogs to illustrate cleavage of the methylene dioxy group and oxidation of the allyl side chain.

The significance of reduction of carboxyl groups and side-chain oxidation is clearly demonstrated by the results of recent studies on zingerone [4-(4-hydroxy-3-methoxyphenyl)butan-2-one] metabolism in rats (Monge *et al.*, 1976). This compound, a pungent principle of ginger, was largely excreted in the urine as glucuronide and/or sulfate conjugates of zingerone itself and of the corresponding carbinol. Among the various side-chain oxidations, formation of the phenylacetic acids predominated.

5. Coumarins

Extensive investigations on the metabolism of unsubstituted coumarin and various hydroxy- and methoxycoumarins in several mammalian species (dogs, rats, rabbits, guinea pigs, humans) have been carried out. The metabolic routes lead to a considerable number of catabolites (reviewed by Scheline, 1978), which are excreted either as the free compounds or as glucuronides or ether sulfates. Acids derived from coumarins such as *o*-coumaric acid or melilotic acid have also been found as glycine conjugates.

The essential features of mammalian metabolism of coumarins can be summarized as shown in Fig. 21.

1. Hydroxylation reactions may occur at all possible positions, but substitutions in positions 3, 4, and 7 are especially noteworthy.
2. Formation of 3-hydroxycoumarin has been demonstrated as the introductory step on the way to the prominent catabolites *o*-hydroxyphenylacetic acids, which also involves *o*-hydroxyphenylpyruvic acids (Kaighen and Williams, 1961).
3. Reduction of coumarins at the Δ^3 double bond leads to dihydrocoumarins, which are hydrolyzed to melilotic acids (*o*-hydroxyphenylpropionic acid). The reduction step to dihydrocoumarins may be regarded as an intestinal microbial reaction. The *o*-hydroxyphenylpropionic acids are further converted to *o*-hydroxyphenyllactic acids or *o*-hydroxyphenylhydracrylic acids with *o*-coumaric acids as intermediates.
4. Methoxy substituents in position 6 or 7 are O-demethylated with a pronounced preference for the 6 position.

A combination of these metabolic routes may lead to a considerable variety of catabolites, although not all possible compounds have been detected in all of the studies. Polyhydroxylated derivatives are encountered only as minor components, and species variation may further lead to drastic qualitative and quantitative differences in the spectrum of coumarin catabolites.

When 7-hydroxylated 3-phenyl- or 4-phenylcoumarins are fed to rabbits, essentially the same results as described for coumarins are obtained (Krishnaswamy *et al.*, 1971). Catabolites were excreted as the corresponding glucuronide and sulfate conjugates.

6. Stilbene

The majority of the natural phenolic C_6–C_2–C_6 compounds possess fungicidal activity, but very little information on the mammalian metabolism of these wood constituents has been published. Stilbene, an unnatu-

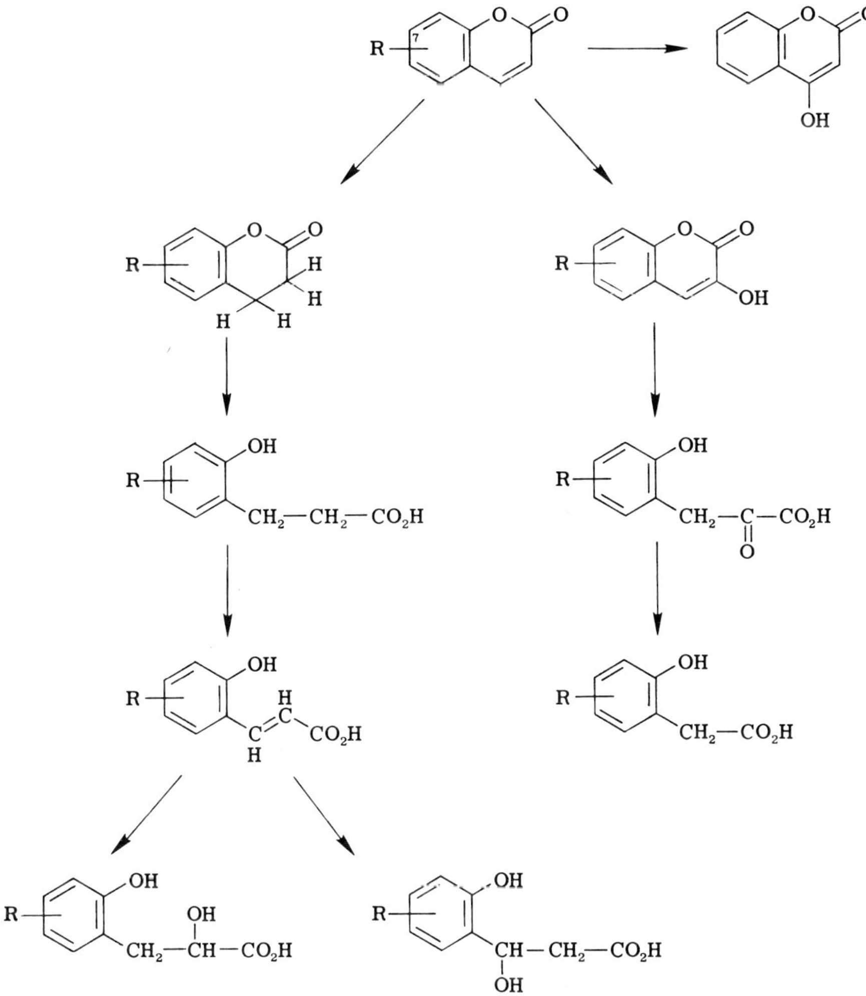

Fig. 21. Metabolites of coumarin and substituted coumarins (R = OH or OCH_3) in mammals (see Scheline, 1978).

ral hydrocarbon, has been the subject of several metabolic investigations leading to various mono- and dihydroxy derivatives (see Tay and Sinsheimer, 1976). Stilbene oxide, also an unnatural compound, has successfully been used as an inducer for both hepatic cytochrome *P*-450 isoenzymes and glutathione reductase (Carlberg *et al.*, 1981) and as a substrate for hepatic epoxide hydrolase (Oesch, 1980). In rabbits, *cis*-stilbene oxide gives rise to *d*-threo-1,2-diphenylethanediol, whereas the *trans*-stilbene oxide leads to *meso*-1,2-diphenylethanediol (Watanabe *et al.*, 1971).

Good insight into the mammalian metabolism of stilbene is derived from the rat and rabbit metabolic pathway of 4-hydroxystilbene (Fig. 22). Several di-, tri-, and tetrahydroxy derivatives as well as reduction of the double bond of the stilbene metabolites to give dibenzyl compounds have been detected (Scheline, 1974). The latter reaction is carried out by intestinal bacteria following the biliary excretion of glucuronide conjugates of the hydroxylated stilbenes. The oxidative cleavage of stilbenes to benzoic acids has been investigated by Watanabe and Akamatsu (1975) in rabbit liver microsomes. The postulated metabolic sequence was given as: stilbene → stilbene oxide → hydrolysis to the diol → oxidation via the benzoin to the benzil derivative, which is *cleaved* to yield two benzoic acids.

7. Lignans

Lignans have been investigated extensively by natural-product chemists and pharmacologists (Rao, 1978) because they have been classified as potential anticancer agents. Several lignans have meanwhile been used in clinical trials to treat tumors (Dombernowski *et al.*, 1972).

The identification in the urine of humans, rats, baboons, and vervet monkeys of glucuronides (Axelson and Setchell, 1980) of several lignans (Setchell *et al.*, 1981) created much interest because these compounds were found to have a cyclic pattern of excretion during the menstrual cycle of women and female animals (Setchell *et al.*, 1980).

As shown in Fig. 23 *trans*-2,3-bis(3-hydroxybenzyl)-γ-butyrolactone (enterolactone, **22**) and 2,3-bis(3-hydroxybenzyl)butane-1,4-diol (enterodiol, **23**) were found as the major urinary lignans in all species investigated, whereas compounds **24–27** have so far only been identified in vervet monkey urine (Setchell *et al.*, 1980).

Compounds **22–27** have been shown to originate from plant dietary sources (Axelson *et al.*, 1982); with regard to enterolactone and enterodiol, the natural 2,3-bis(2-methoxy-4-hydroxybenzyl)butane-1,4-diol (**28**), occurring as a glycoside in linseed, has been postulated as precursor. The intestinal microflora is clearly required for the formation of **22** and **23** from **28** (Axelson and Setchell, 1981). The required reactions, such as glycoside hydrolysis, aromatic dehydroxyilation, O-demethylation, and alcohol side-chain oxidation, are well known from microorganisms (see Table V), although the exact catabolic sequence has still to be elucidated.

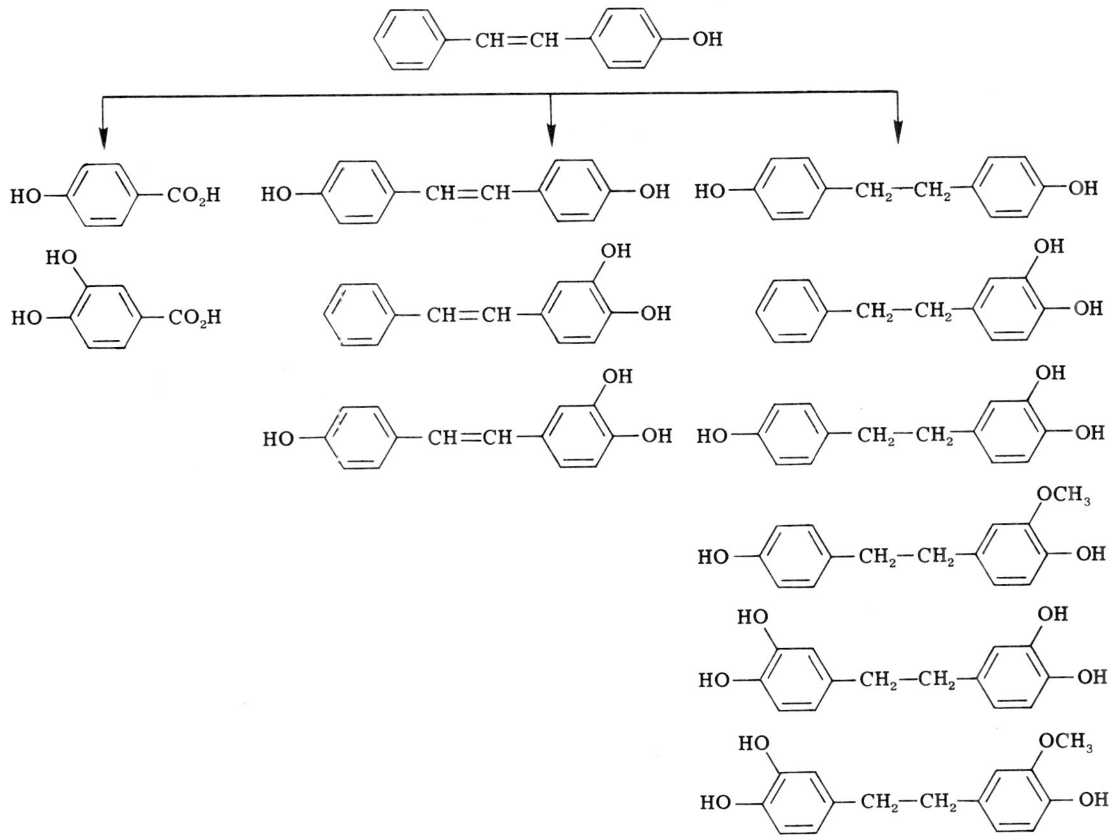

Fig. 22. Metabolites of 4-hydroxystilbene as isolated in studies with rats and rabbits.

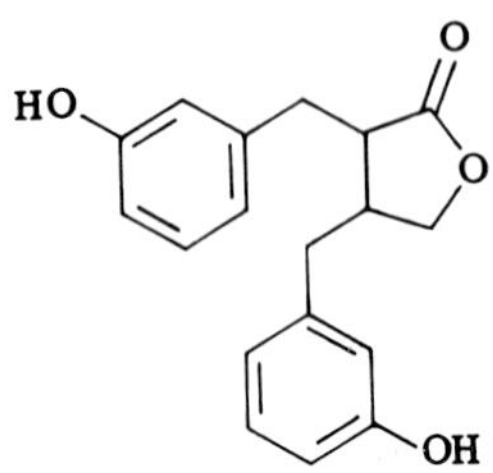

22

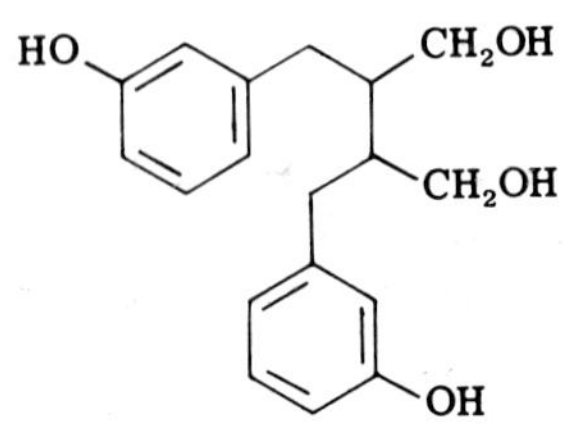

23

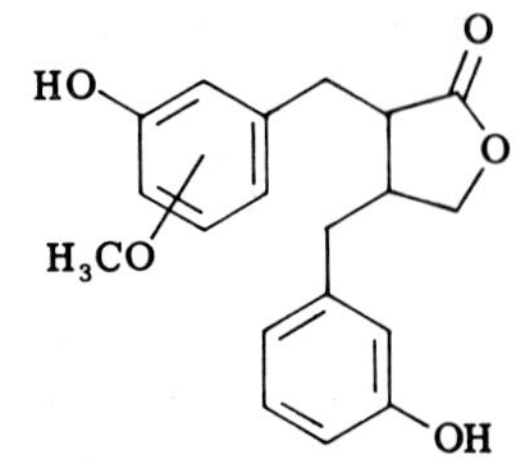

24

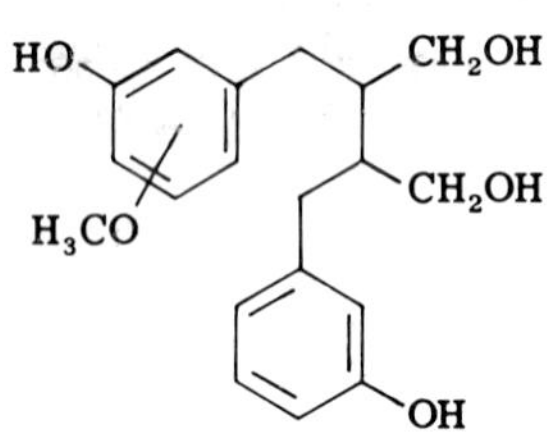

25

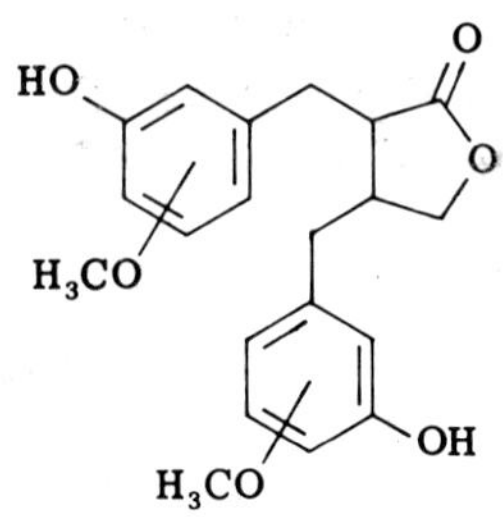

26

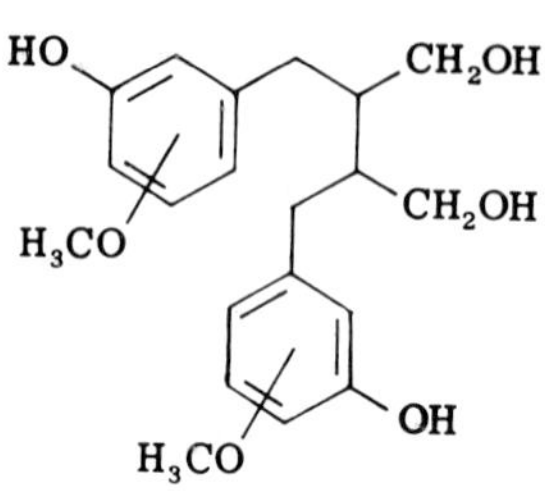

27

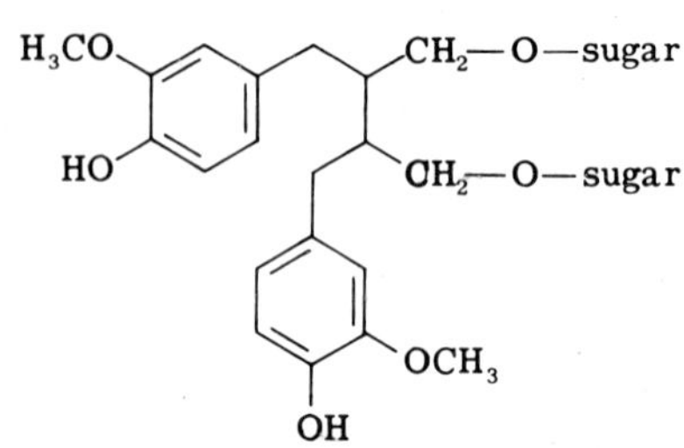

28

Fig. 23. Lignans isolated from the urine of humans, rats, and vervet monkeys [(**22**) and (**23**)] together with the structure (**28**) of the dietary precursor *seco*-isolariciresinol glycoside.

8. Flavonoids

The widespread occurrence of different classes of flavonoid in edible plants and the reported biological activity of certain flavonoids in mammalian systems has long led to numerous investigations on mammalian metabolism of flavonoids (Griffiths, 1975; Gabor, 1975).

The hydrolysis of glycosidic linkages of orally administered flavonoid glycosides is widely reported for flavanone, dihydrochalcone, flavone, and flavonol glycosides (Griffiths and Smith, 1972a,b). The resulting aglycones are in part excreted in feces. Studies on flavonoid glycoside metabolism with preparations of gastrointestinal microflora always resulted in glycoside hydrolysis. On the other hand, in studies with germ-free rats, flavonoid glycosides were isolated from feces, and parenteral administration of flavonoid glucosides to biliary cannulated rats, in which contact between the administered flavonoid glycoside and the gut microflora is precluded, resulted in the excretion of conjugates of the intact glucosides (Hackett *et al.*, 1979). Such data seem to indicate that glycosidases for the hydrolysis of flavonoid glycosides are absent from mammalian tissues.

Studies with germ-free rats have also led to the observation that flavonoid glucuronides such as catechin conjugates, which are synthesized in the bile, are again degraded by the intestinal microflora. It is also known that large amounts of β-glucuronidase of microfloral origin are present in the intestinal contents.

Degradative reactions, including ring fission of flavonoids such as (+)-catechin or the flavonol quercetin, in mammalian systems result in various urinary catabolites that are derived from the side-chain phenyl substituent of the heterocyclic ring. The structures in Fig. 24 demonstrate the degradative sequence of catechin as shown to occur in rats, rabbits, or guinea pigs. The metabolism of (−)-epicatechin, a stereoisomer of (+)-catechin, led to the same catabolites (for review, see Griffiths, 1982).

Similarly, catabolites of the flavonol quercetin in humans were found to be 3-hydroxylphenylacetic acid, 3-methoxy-4-hydroxyphenylacetic acid, 3,4-dihydroxyphenylacetic acid, 3,4-dihydroxytoluene, and β-*m*-hydroxyphenylhydracrylic acid. All of these metabolites were formed via reduction of the γ-pyranone ring of quercetin (Baba *et al.*, 1981; Hiraoka *et al.*, 1981).

Comparable studies with other flavonols, flavones, and flavanones have also led to recognition that ring fission to phenolic acids was a major route of flavonoid degradation (Griffiths, 1975). It was also shown that only flavonoids with a 5,7-dihydroxy substitution pattern in ring A underwent ring fission, as shown for catechin in Fig. 24. Methylation of both or either one of these hydroxy groups greatly reduced degradability. Such data appear to demonstrate that ring A of flavonoids is the primary site for ring fission to occur. So far, catabolites of ring A have not been isolated in such studies (see, however, Section IV). Efforts by various laboratories, especially by Griffiths and associates (see the review by Griffiths, 1982), have collected substantial evidence that these degradative reactions of flavonoids (see Fig. 24) are mediated by the intestinal microflora. Although mammalian tissue enzymes may have contributed to cleavage of flavonoid molecules, the isolation of diagnostic ring fission products from mammalian tissues still has to be proved.

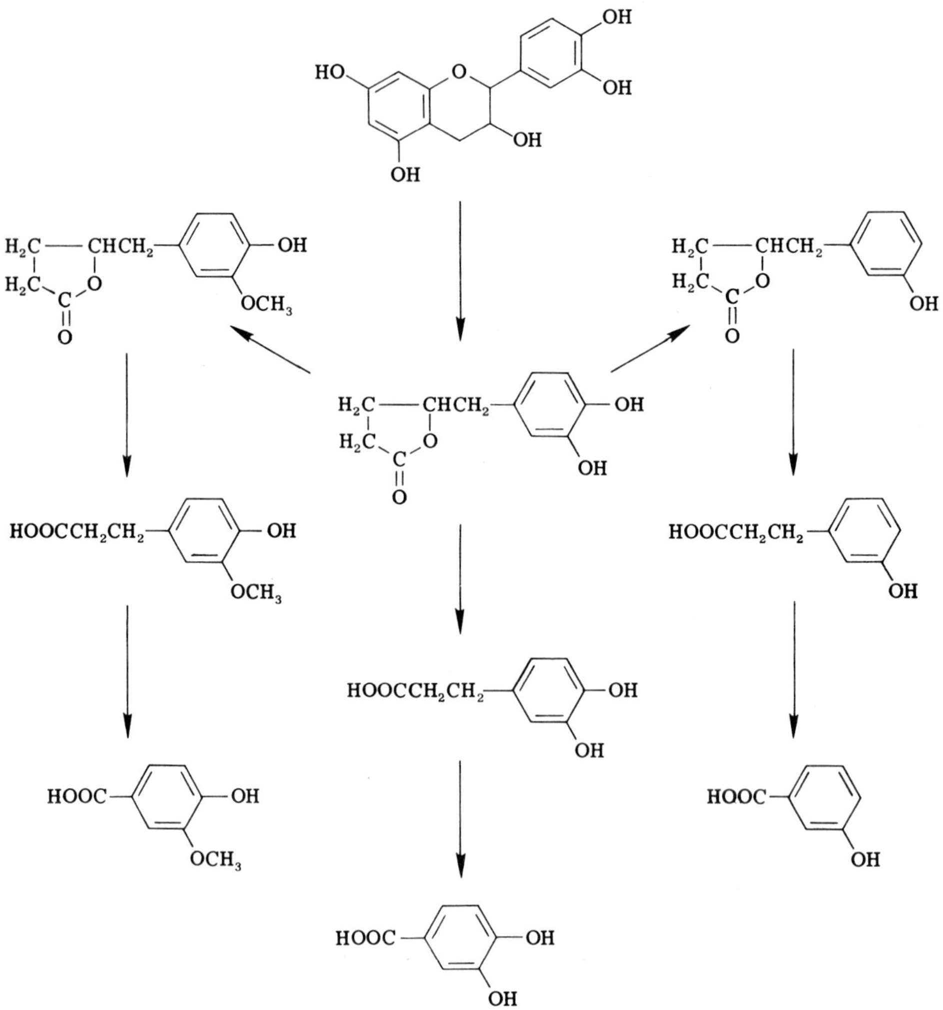

Fig. 24. Urinary catabolites of (+)-catechin isolated from rats, rabbits, and guinea pigs. Redrawn from Griffiths (1982).

In addition to studies of various hydroxylation and reduction reactions of flavonoid aglycones (hydroxylations at position 6, 3′, or 4′; reduction of carbonyl at C-4 to 4α- or 4β-ol) (Griffiths and Smith, 1972a; Buset and Scheline, 1979), greatest efforts have been placed on the elucidation of mammalian conjugation reactions of flavonoids. Such conjugates are formed in the liver and excreted by bile and urine. Methylation of 3′,4′-dihydroxyflavonoids (quercetin, taxifolin, or catechin) at position 3′

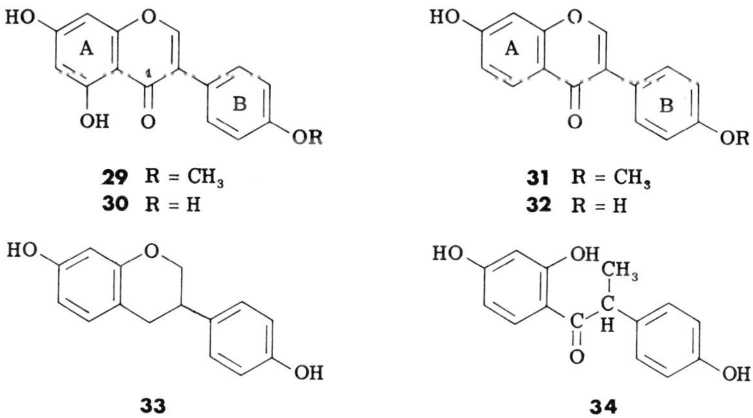

Fig. 25. Structures of naturally occurring isoflavones [(**29**)–(**32**)] and of catabolites equol (**33**) and *o*-demethylangolensin (**34**).

(Brown and Griffiths, 1983a) and formation of glucuronides and sulfates have been observed (Hackett *et al.*, 1979; Griffiths, 1982). Thus, in mouse, rat, human, and marmoset, (+)-catechin and the synthetic 3-*O*-methylcatechin are excreted as 3′-*O*-methyl derivatives combined with glucuronic acid (Hackett and Griffiths, 1983) and sulfate (Hackett and Griffiths, 1983; Hackett *et al.*, 1982). Based on earlier studies on the excretion of flavonoid conjugates (Griffiths, 1975; Gabor, 1975), the various pharmacological and metabolic aspects of catechin conversion and conjugation in several mammalian species have been extensively studied by Griffiths and associates (Griffiths, 1982). Methylation and conjugation have also been shown to be of importance in mammalian chalcone metabolism (Brown and Griffiths, 1983b). Thus, 2′3,4,4′-tetrahydroxychalcone and 2′,3,4-trihydroxychalcone, following administration to bile-cannulated rats, gave rise to the corresponding 3-*O*-methyl ethers and trace amounts of the 4-*O*-methyl derivative. All of these chalcones were selectively excreted in bile and urine as sulfate conjugates. The O-methylation must be considered a tissue-located reaction.

In general, with regard to mammalian metabolism, flavonoids must be considered as the best understood classes of plant xenobiotics.

9. Isoflavonoids

Isoflavonoids, which are mostly restricted to the Fabaceae (Ingham, 1983), have the phenyl ring (B ring) attached at the 3 position of the heterocyclic ring. Most metabolic studies have been performed with the isoflavones (Fig. 25) biochanin A (**29**), genistein (**30**), formononetin (**31**), and daidzein (**32**), which occur in most leguminous fodder plants and in some varieties of wood (Ingham, 1983). O-Demethylation of both biochanin A and formononetin to yield the equivalent 4′-hydroxyl compounds has been obtained with rumen fluid of sheep and cattle (Nilsson *et al.*, 1967; Braden *et al.*, 1967) but not with microorganisms from rat intestine (Griffiths and Smith, 1972a). On the other hand, strong O-demethylation activity for **29** and **31** has been detected in the liver of rats, mice, and rabbits (Nilsson, 1963). Much lower activities were found in the liver of swine, cattle, and sheep. When other isoflavones methoxylated in various positions of ring A or B were applied in studies with rabbit liver, all of these substrates were readily O-demethylated regardless of the exact position of the methyl substituent.

5-Hydroxyisoflavones such as **29** or **30** when given intraruminally to sheep were degraded to *p*-ethyl phenol, which was excreted in the urine (Braden *et al.*, 1967). Experiments with [4-^{14}C]biochanin A (Batterham *et al.*, 1971) indicate that *p*-ethyl phenol is derived from ring B and C-2 and C-3 of the heterocyclic ring (compare biochanin A degradation by *Fusarium* fungi, Section II,B,7). Later studies by Griffiths and Smith (1972a) demonstrated that the degradation of biochanin A to *p*-ethyl phenol was carried out by rat intestinal bacteria.

5-Deoxyisoflavones **31** and **32** (Fig. 25) are not as profoundly degraded as the 5-hydroxy derivatives, because in studies with sheep and guinea pigs (Braden *et al.*, 1967; Batterham *et al.*, 1971; Shutt and Braden, 1968), the isoflavan equol (**33**, Fig. 25) and O-demethyl angolensin (**34**) were the only urinary catabolites. Incubations of formononetin and daidzein with sheep rumen fluid (Nilsson *et al.*, 1967) or rat intestinal microorganisms (Griffiths and Smith, 1972a) resulted in the formation of equol. Therefore, the production of the isoflavan from formononetin and daidzein must be considered as a microbial conversion reaction.

Biliary excretion products of the isoflavones **29–32** and of equol are, chiefly, glucuronides and sulfates. Labow and Layne (1972) demonstrated that the isoflavones form monoglucuronides when incubated with rabbit liver microsomal fractions and UDP-glucuronic acid.

Pterocarpans, which are isoflavonoids with regard to their biosynthesis and which widely occur as constituents of several kinds of wood (Ingham, 1983), have so far not been investigated for their mammalian metabolism. Aspects of microbial degradation of pterocarpans are found in Section II,B,8.

10. Rotenoids

Rotenone (**16**) and various other rotenoids are well known for their insecticidal activity and their toxicity to fish. Major urinary metabolites of rotenone in mice are 6′,7′-dihydro-6′,7′-dihydroxyrotenone and 6′,7′-dihydro-6′,7′,12α-trihydroxyrotenone (Fukami *et al.*, 1969; Yamamoto, 1969). Feeding experiments with natural rotenone in mice and rats demonstrated that O-demethylation at position 3 also occurred to a high degree (Unai *et al.*, 1973). Investigations with microsomal preparations from rat or mouse liver have further shown that hydroxylation reactions in positions 12α and 8′ are also possible. In combination with these substitution reactions, a great variety of conversion reactions of rotenone are thus possible.

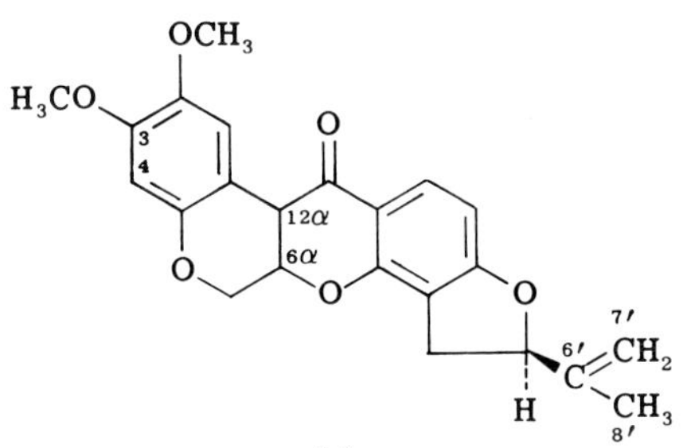

16

11. Quinones

Although quinones are widely distributed in the plant kingdom, only limited information is available on their mammalian metabolism. Quinones are in part detoxified by both glucuronydation and sulfation (Nemoto *et al.*, 1978; Nemoto and Gelboin, 1976). If suitable functional groups for conjugation reactions are not available in the molecule, the actual substrate is either the semiquinone or the hydroquinone. An intermediate reduction step is thus required that is catalyzed by a quinone reductase. The integrated reduction–conjugation process in hepatocytes proceeds with great effectiveness, as shown with several synthetic quinones (Benson *et al.*, 1980).

Hydroquinone is excreted in urine by several mammalian species in form of sulfate and glucuronic acid conjugates (Miller *et al.*, 1976).

Studies on naphthoquinones mainly focused on phylloquinone (vitamin K_1) and its synthetic homolog menadione (2-methyl-1,4-naphthoquinone). With regard to the latter compound, several reports described the excretion of a monoglucuronide and of 4-hydroxy-2-methyl-1-naphthyl sulfate from several species (see Thompson *et al.*, 1972). Phthalic acid has also been shown to be urinary catabolite of menadione in dogs and humans.

The most notable aspect of phylloquinone (**35**) metabolism in humans and rats (Fig. 26) is the oxidation of the side chain of **35** to yield catabolite

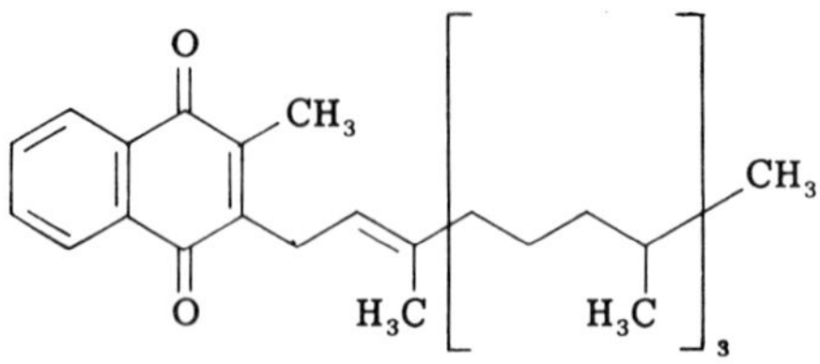

35

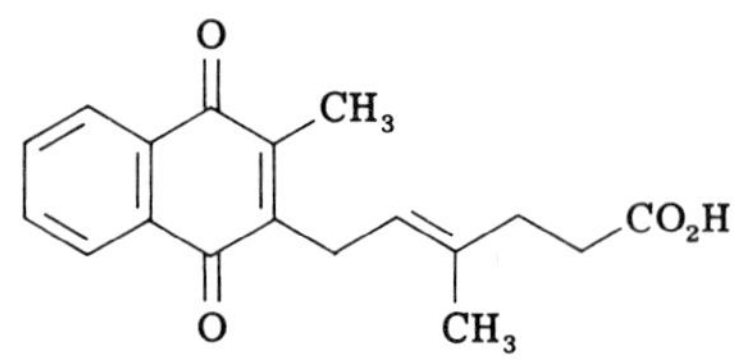

36

Fig. 26. Structure of phylloquinone (**35**) and its main mammalian catabolite (**36**). From Shearer and Barkhan (1973).

36, which occurs in urine as a glucuronide conjugate (Shearer and Barkhan, 1973).

Anthraquinones, many of which are used as dyes and as carthartics, mostly occur as hydroxy- or methoxyanthraquinones. The latter compounds are normally O-demethylated in mammals (Fujita *et al.,* 1961a), whereas 1-hydroxy- or 2-hydroxyanthraquinones are hydroxylated to give the 1,2-dihydroxy compounds. Formation of glucuronides and, to lesser extent, sulfates are the prevailing excretion mechanisms (Fujita *et al.,* 1961b).

Due to the chemical instability of hydroxyanthraquinones, a considerable portion of the applied substrates always appeared to be lost into so far unidentified products. Thus, extensive studies on the mammalian metabolism of various synthetic or natural (chrysophanic acid, rhein, emodin, aloeemodin) 1,8-dihydroxyanthraquinones have not been able to present a complete balance of product conversion and excretion in bile, urine, and feces, respectively. Only 30–40% or less of the applied dose could be accounted for (Breimer and Baars, 1976). Therefore, elucidation of quinone catabolites will be an important future task.

IV. Metabolism and Degradation of Aromatic Compounds in Plants

Numerous investigations in plants and plant cell cultures have amply demonstrated that endogenous (both primary and secondary constituents) and exogenous aromatic compounds are extensively metabolized and often degraded in plants. This potential for catabolic reactions is much larger than previously expected. The various aspects of this new field of plant biochemistry have repeatedly been summarized (Robinson, 1974; Barz and Nicolas, 1978; Barz and Hösel, 1975, 1978; Barz and Köster, 1981). Some essential features will here be outlined.

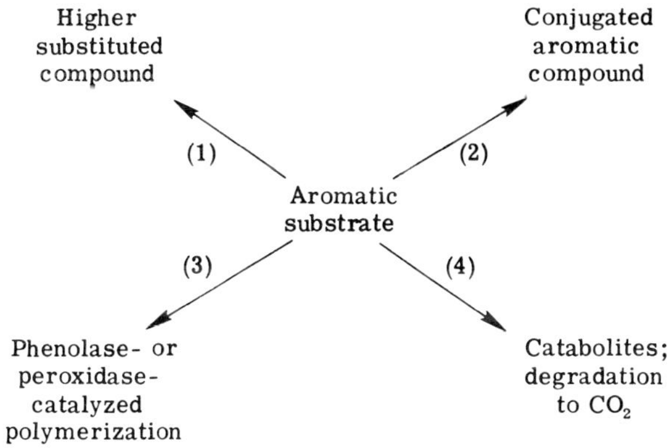

Fig. 27. Metabolic grid showing the four general possibilities for the metabolism of an aromatic substrate in plants. (1) Anabolic conversion; (2) conjugation reaction; (3) oxidative polymerization reaction; and (4) complete degradation.

A. *General Aspects*

Although certain differences between the metabolism of endogenous and exogenous compounds have been observed (Barz and Köster, 1981), the overall metabolic fate of aromatic compounds in plants can be depicted as shown in Fig. 27. Disappearance of a compound from the accumulating tissue or turnover (for definition, see Barz and Köster, 1981) of a particular aromatic product may involve any one of four types of reactions:

1. Interconversion reactions as part of a biosynthetic sequence.
2. Conjugation reactions involving other compounds to which the product under investigation is attached (i.e., formation of glucosides).
3. Oxidative polymerization reactions leading to insoluble structures of high molecular weight.
4. Degradative reactions where the product is converted to a structure of primary metabolism and finally CO_2.

Due to the complex pattern of aromatic plant metabolism, these four metabolic alternatives may occur simultaneously with one particular compound. Furthermore, since these metabolic routes depend on the availability of specific enzymes, the extent of these reactions may greatly vary with plant age, organ differentiation, and various physiological conditions (Barz and Köster, 1981). Suitable examples to substantiate a metabolic grid, as shown in Fig. 27, can be found in investigations on isoflavones and flavonols (Barz, 1980), on the protoalkaloid hordenin (Meyer and Barz, 1978; Meyer, 1983), and on benzoic and cinnamic acids (Barz and Köster, 1981).

B. *Oxidative Polymerization*

Lignified plant material is chemically a very complex material because it consists not only of the well-known lignin-building units polymerized into a cellulose matrix but also a variety of other covalently bound compounds. Thus, aromatic acids have been found to occur in an ester-bound manner. Even more important are the observations that a considerable number of endogenously synthesized and/or foreign compounds were found to be oxidatively incorporated into lignin-like ethanol-insoluble, nonhydrolyzable material. Such oxidative polymerization reactions are catalyzed by peroxidases and constitute for the higher plants an important mechanism to metabolize or detoxify aromatic and heterocyclic compounds (Barz and Hösel, 1975, 1978; Lamoureux *et al.,* 1973; Meyer, 1983; Von der Trenck and Sandermann, 1981; Von der Trenck *et al.,* 1981). A few structural requirements, such as formation of a phenoxy radical or an orthoquinoid structure, have to be met in order to render a compound suitable for peroxidative polymerization (Barz and Köster,

C. *Degradative Reactions*

In the context of this chapter, the degradative reactions of aromatic compounds are of special interest. As shown both for endogenous and foreign compounds, plants seem to be capable of degrading a great variety of aromatic and heterocyclic compounds (Barz and Köster, 1981). The introductory reactions involved in such catabolic pathways are essentially the same as known from microbial systems. The most important are O- and N-demethylation, decarboxylation, aliphatic and aromatic monohydroxylation, N-hydroxylation, aromatic ortho hydroxylation, cleavage of aliphatic and aromatic C—C double bonds, side-chain shortening, oxidation and reduction of oxygen function, and hydrolytic reactions.

While microorganisms use degradable compounds as substrates for growth, plants do not rely on aromatic compounds as energy sources or as growth sustainers. Furthermore, the enzymes of higher plants involved in catabolic pathways appear to be constitutive, in contrast to the substrate-induced enzymes of microorganisms.

Very much against previous belief (Barz, 1980; Barz and Köster, 1981), numerous plants have been shown to possess the capability for the cleav-

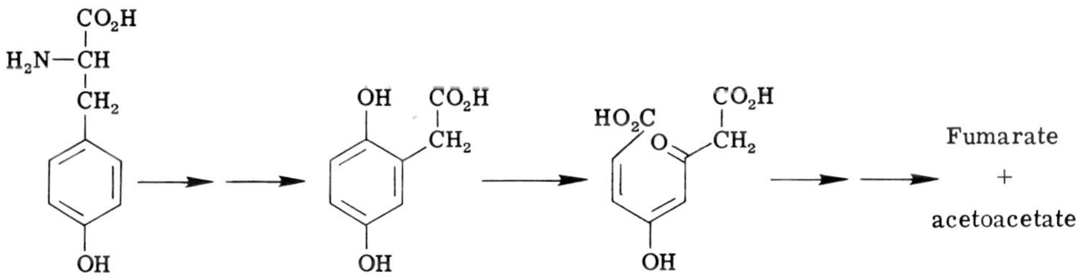

Fig. 28. The homogentisic acid pathway as found in higher plants for cleavage of a 1,4-dihydroxy aromatic compound. Redrawn from Durand and Zenk (1974a,b).

age of aromatic rings. Again, as documented for microbial systems (see Fig. 2), meta or ortho fission of 1,2-dihydroxy aromatic compounds and the cleavage of 1,4-dihydroxy aromatic structures seem to be the basic mechanism for the degradation of such ring systems. Catabolism of L-tyrosine via homogentisic acid (2,5-dihydroxyphenylacetic acid) to aliphatic acids (Fig. 28) as shown to occur in various plants and plant cell cultures (Durand and Zenk, 1974a,b) may serve as a suitable example for this important type of plant dioxygenase for ring cleavage.

Similarly, investigations with 2,3-dihydroxybenzoic acid (Sharma and Vaidyanathan, 1975) 3,4-dihydroxybenzoic acid (protocatechuate) (Mohan *et al.*, 1979), L-DOPA (Barz and Köster, 1981), and indole (Kunapuli and Vaidyanathan, 1983) presented unequivocal evidence that ortho and meta cleavage of aromatic dihydroxy or heterocyclic structures are functioning in plant systems. However, the extent to which dioxygenase-catalyzed ring-cleavage reactions are used in plant systems is presently unknown and requires intensive further studies. Furthermore, our knowledge of these plant dioxygenases is comparatively meager, so that generalizations must be made with great caution.

Another characteristic type of degradative reaction in plants applicable to hydroxyaromatic compounds has been detected in peroxidative cleavage reactions (Patzlaff and Barz, 1978; Barz and Hösel, 1975; Ellis, 1974; Sawahata and Neal, 1982). In accord with chemical analogies on the fission of *ortho*-diphenols with peracids (Schulz and Hecker, 1973; Walter and Hecker, 1973), numerous cases of peroxidase, H_2O_2-dependent cleavage of hydroxyaromatic compounds have been reported (see Barz and Köster, 1981). Investigations with aromatic substrates representing the catechol, resorcinol, or phloroglucinol types of substitution pattern (i.e., phenol, protocatechuate, catechol, 5-hydroxy- or 5-deoxyisoflavones, flavanones, aurones, flavonols) resulted in the observation that plant peroxidases may be responsible for ring-cleavage reactions. Such reac-

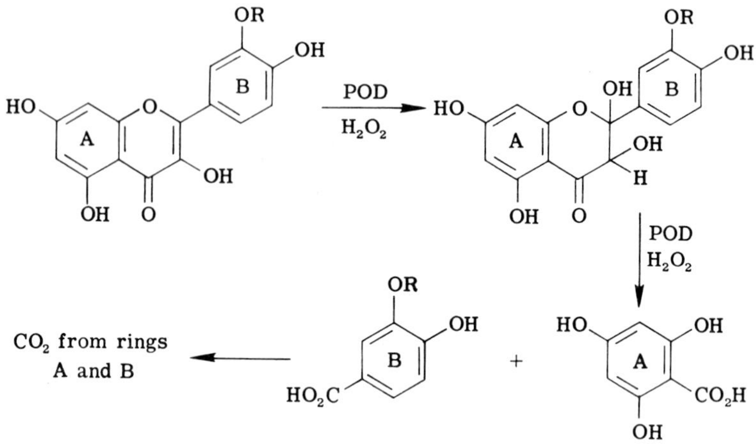

Fig. 29. Catabolites of peroxidative (POD) degradation of flavonols. Redrawn from Barz and Hösel (1978).

occurs. The enzyme mechanisms for these peroxidase-catalyzed ring-fission reactions remain to be elucidated. Figures 29 and 30 show two examples of such peroxidase-catalyzed reactions with the flavanone naringenin and a flavonol (Patzlaff and Barz, 1978; Barz, 1980). It is expected that more cases of this type will be reported and that peroxidases will thus be further recognized as a very potent class of degradative enzymes acting on a large number of aromatic and heterocyclic structures.

With regard to cytological and histological aspects, peroxidase-catalyzed degradation reactions are found especially at peroxidase-rich sites such as the rhizodermis, endodermis, cell walls, xylem, or at wounded tissues (Barz and Hösel, 1978). This enzyme localization is especially suitable for the metabolism of foreign substrates penetrating into the plant body, i.e., polyphenols taken up from soil into roots.

V. Conclusions

Microorganisms, animals, and plants all possess the ability to metabolize and degrade a large variety of exogenous and endogenous compounds. The characteristic differences in the metabolic capabilities of these three large groups of organisms are especially visible when looking at aromatic compounds.

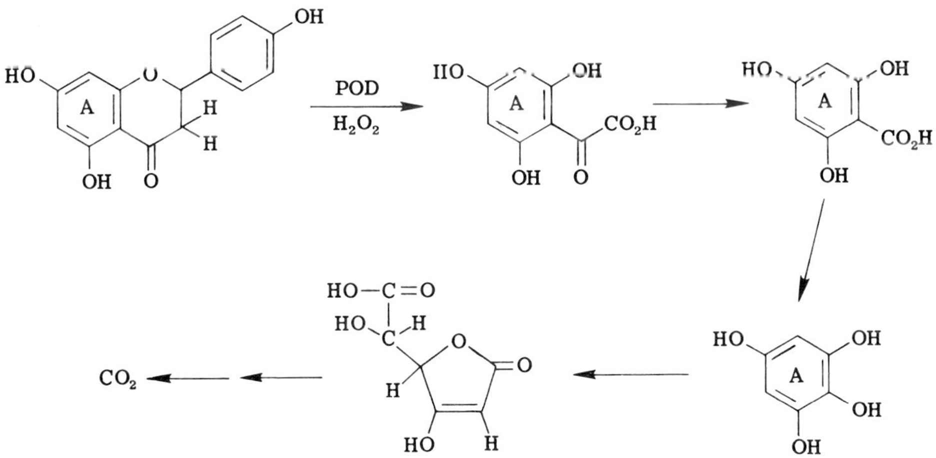

Fig. 30. Catabolites of ring A of the flavanone naringenin isolated from reaction catalyzed by H_2O_2–peroxidase (POD). Redrawn from Patzlaff and Barz (1978).

Microorganisms show an impressive flexibility in their ability to degrade aromatics to carbon dioxide and water. Animals mostly modify aromatics with the aim of excreting suitable conjugates, but they lack the ability to cleave aromatic structures. Reports on degradative reactions in studies with mammals probably result from the associated microflora. Plants, although less intensively studied for the metabolism of aromatic compounds, quite obviously possess the ability for aromatic ring-fission reactions. More important, however, are the peroxidative polymerization reactions, which appear to be very effective with numerous substrates.

In recent years, impressive results on the metabolism and degradation of a wide variety of aromatic products in the biosphere have been obtained. On the other hand, the metabolic fate of many plant- and animal-derived compounds remains to be elucidated. Numerous catabolic pathways are still incomplete and await enzymatic studies. This lack of information is especially obvious in the study of oligomeric or polymeric structures. Since the basic principles of aromatic catabolism are known, future studies will soon reveal the routes of degradation of more complex compounds.

Acknowledgment

The investigations of the authors were supported by Deutsche Forschungsgemeinschaft.

References

Anderson, J. J., and Dagley, S. (1980). *J. Bacteriol.* **141,** 534–543.

Axelson, M., and Setchell, K. D. R. (1980). *FEBS Lett.* **122,** 49–53.

Axelson, M., and Setchell, K. D. R. (1981). *FEBS Lett.* **123,** 337–341.

Axelson, M., Sjövall, J., Gustafsson, B. E., and Setchell, K. D. R. (1982). *Nature* (*London*) **298,** 659–660.

Baba, S., Furuta, T., Horie, M., and Nakagawa, H. (1981). *J. Pharm. Sci.* **70,** 780–785.

Barz, W. (1980). *In* "Pigments in Plants" (F.-C. Czygan, ed.), pp. 210–213. Fischer, Stuttgart.

Barz, W., and Hösel, W. (1975). *In* "The Flavonoids" (J. B. Harborne, T. J. Mabry, and H. Mabry, eds.), pp. 916–969. Chapman & Hall, London.

Barz, W., and Hösel, W. (1978). *Rec. Adv. Phytochem.* **12,** 339–369.

Barz, W., and Nicolas, H.-A. (1978). *In* "Frontiers of Plant Tissue Culture 1978" (T. A. Thorpe, ed.), pp. 345–352. Univ. Calgary Printing Service, Calgary, Alberta.

Barz, W., and Köster, J. (1981). In "The Biochemistry of Plants" (E. E. Conn, and P. K. Stumpf, eds.), Vol. 7, pp. 35–84. Academic Press, New York.

Barz, W., Willeke, U., and Weltring, K.-M. (1980). *Ann. Phytopathol.* **12,** 435–452.

Batterham, T.-J., Shutt, D. J., Hart, N. K., Braden, A. W. H., and Tweeddale, H. J. (1971). *Aust. J. Agric. Res.* **22,** 131–138.

Baumann, P., Doudoroff, M., and Stanier, R. Y. (1968). *J. Bacteriol.* **95,** 1520–1541.

Benson, A. M., Hunkeler, M. J., and Talalay, P. (1980). *Proc. Natl. Acad. Sci. U.S.A.* **77,** 5216–5220.

Blackwood, A. C., Hang, Y. D., Robern, H., and Mathur, D. K. (1970). *Bacteriol. Proc.* **70,** 124.

Borchert, P., Wislocki, P. G., Miller, J. A., and Miller, E. C. (1973). *Cancer Res.* **33,** 575–589.

Braden, A. W. H., Hart, N. K., and Lamberton, J. A. (1967). *Aust. J. Agric. Res.* **18,** 335–348.

Brand, J. M., Young, J. C., and Silverstein, R. M. (1979). *Fortschr. Chem. Org. Naturst.* **37,** 1–190.

Breimer, D. D., and Baars, A. J. (1976). *Pharmacology* **14,** Suppl. 1, 30–47.

Brown, S., and Griffiths, L. A. (1983a). *Experientia* **39,** 198–200.

Brown, S., and Griffiths, L. A. (1983b). *Xenobiotica* **13,** 669–682.

Brown, S. B., Rajananda, V., Holroyd, J. A., and Evans, E. G. V. (1982). *Biochem. J.* **205,** 239–245.

Brunet, P. C. J. (1980). *Insect Biochem.* **10,** 467–500.

Buset, H., and Scheline, R. R. (1979). *Biomed. Mass Spectrom.* **6,** 212–217.

Buswell, J. A., and Eriksson, K.-E. (1979). *FEBS Lett.* **104,** 258–260.

Caldwell, J., and Jacoby, W. B. (1983). "Biological Basis of Detoxication." Academic Press, New York.

Call, D. J., Brooke, L. T., and Lu, P.-Y. (1980). *Arch. Environ. Contam. Toxicol.* **9,** 699–714.

Capel, I. D., Millburn, P., and Williams, R. T. (1974). *Biochem. Soc. Trans.* **2,** 305–306.

Carlberg, I., De Pierre, J. W., and Mannervik, B. (1981). *Biochim. Biophys. Acta* **677,** 140–145.

Chapman, P. J., and Ribbons, D. W. (1976a). *J. Bacteriol.* **125,** 975–984.

Chapman, P. J., and Ribbons, D. W. (1976b). *J. Bacteriol.* **125,** 975–984.

Crawford, D. L., and Crawford, R. L. (1980). *Enzyme Microb. Technol.* **2,** 11–22.

Creveling, C. R., Morris, N., Shimizu, H., Ong, H. H., and Daly, J. (1972). *Molec. Pharmacol.* **8,** 398–409.

Cripps, R. C. (1975). *Biochem. J.* **152,** 233–241.

Crutcher, S. E., and Geary, P. J. (1979). *Biochem. J.* **177,** 393–400.

Dagley, S. (1971). *Adv. Microbial Physiol.* **6,** 1–46.

Dagley, S. (1975). *Am. Scientist* **63,** 681–689.

Dagley, S. (1977). *Surv. Prog. Chem.* **8,** 121–170.

Dagley, S. (1978). *Naturwissenschaften* **65,** 85–95.

Davison, C. (1971). *Ann. N.Y. Acad. Sci.* **179,** 249–268.

Denny, T. P., and VanEtten, H. D. (1982). *Phytochemistry* **21,** 1023–1028.

Dickinson, C. H., and Pugh, G. J. F. (1974). "Biology of Plant Litter Decomposition," Vols. 1–3. Academic Press, London.

Donnelly, M. I., and Dagley, S. (1981). *J. Bacteriol.* **147,** 471–476.

Donnelly, M. I., Chapman, P. J., and Dagley, S. (1981). *J. Bacteriol.* **147,** 477–481.

Dombernowski, P., Nissen, N. J., and Larsen, V. (1972). *Cancer Chemother. Rep.* **56,** 71–87.

Durand, R., and Zenk, M. H. (1974a). *Phytochemistry* **13,** 1483–1492.

Durand, R., and Zenk, M. H. (1974b). *FEBS Lett.* **39,** 218–220.

Eggeling, L., and Sahm, H. (1980). *Arch. Microbiol.* **126,** 141–148.

Ellis, B. E. (1974). *Lloydia* **37,** 168–184.

Franklin, R. B., Elcombe, C. R., Vodicnik, M. J., and Lech, J. J. (1980). *Fed. Proc., Fed. Am. Soc. Exp. Biol.* **39,** 3144–3149.

Fujita, M., Furuya, T., and Matsuo, M. (1961a). *Chem. Pharm. Bull.* **9,** 967–970.

Fujita, M., Furuya, T., and Matsuo, M. (1961b). *Chem. Pharm. Bull.* **9,** 962–966.

Fukami, I.-I., Shishido, T., Fukunaga, K., and Casida, I. E. (1969). *J. Agric. Food Chem.* **17,** 1217–1226.

Gaal, A., and Neujahr, H. Y. (1979). *J. Bacteriol.* **137,** 13–21.

Gabor, M. (1975). "Abriss der Pharmacology von Flavonoiden." Akademiai Kiado, Budapest.

Goodhue, C. T. (1982). *In* "Microbial Transformations of Bioactive Compounds" (J. P. Rosazza, ed.), Vol. I, pp. 9–44, CRC Press, Boca Raton, Florida.

Goodwin, B. L. (1976). "Handbook of Intermediary Metabolism of Aromatic Compounds." Chapman & Hall, London.

Grammatica, P., Ranzi, B. M., and Manitto, P. (1981). *Bioorg. Chem.* **10,** 14–21.

Griffiths, L. A. (1975). *In* "Topics in Flavonoid Chemistry and Biochemistry" (L. Farkas, M. Gabor, and F. Kallay, ed.), p. 325. Elsevier, Amsterdam.

Griffiths, L. A. (1982). *In* "The Flavonoids: Advances in Research" (J. B. Harborne and T. J. Mabry, eds.), pp. 681–718. Chapman & Hall, London.

Griffiths, L. A., and Smith, G. E. (1972a). *Biochem. J.* **128,** 901–911.

Griffiths, L. A., and Smith, G. E. (1972b). *Biochem. J.* **130,** 141–151.

Groseclose, E. E., and Ribbons, D. W. (1981). *J. Bacteriol.* **146,** 460–466.

Gunsalus, I. C., Pederson, T. C., and Sligar, S. G. (1975). *Ann. Rev. Biochem.* **44,** 377–407.

Gupta, J. K., Hamp, S. G., Buswell, J. A., and Eriksson, K.-E. (1981). *Arch. Microbiol.* **128,** 349–354.

Hackett, A. M., and Griffiths, L. A. (1983). *Eur. J. Drug Metab. Pharmacokinet.* **8,** 35–42.

Hackett, A. M., Marsh, I., Barrow, A., and Griffiths, L. A. (1979). *Xenobiotica* **9,** 491–496.

Hackett, A. M., Shaw, I. C., and Griffiths, L. A. (1982). *Experientia* **38,** 538–540.

Hareland, W. A., Crawford, R. L., Chapman, P. J., and Dagley, S. (1975). *J. Bacteriol.* **121,** 272–285.

Hayaishi, O. (1966). *Pharmacol. Rev.* **18,** 71–75.
Hills, C. A., and Fewson, Ch. A. (1983). *J. Gen. Microbiol.* **129,** 2009–2015.
Hiraoka, K., Miyamoto, T., Baba, S., and Furuta, T. (1981). *J. Labelled Compd. Radiopharm.* **18,** 613–617.
Hutson, D. H. (1975). *In* "Foreign Compound Metabolism in Mammals" (D. E. Hathway, ed.), Vol. 3, pp. 449–549. Chemical Society, London.
Hutson, D. H. (1977). *In* "Foreign Compound Metabolism in Mammals" (D. E. Hathway, ed.). Vol. 4, pp. 259–346. Chemical Society, London.
Ingham, J. L. (1983). *Progr. Chem. Org. Nat. Prod.* **43,** 3–266.
Iyayi, C. B., and Dart, R. K. (1982). *J. Gen. Microbiol.* **128,** 1473–1482.
Jacoby, W. B. (1980). "Enzymatic Basis of Detoxication," Vol. I and II, Academic Press, New York.
Kaighen, M., and Williams, R. T. (1961). *J. Med. Chem.* **3,** 25–43.
Kieslich, K. (1976). "Microbial Transformations of Nonsteroid Cyclic Compounds." Georg Thieme Publisher, Stuttgart.
Krishnaswamy, N. R., Seshadri, T. R., and Tahir, P. J. (1971). *Indian J. Expl. Biol.* **9,** 458–461.
Kunapuli, S. P., and Vaidyanathan, C. S. (1983). *Biochem. Int.* **7,** 647–654.
Laborde, A. L., and Gibson, D. T. (1976). *Abstr. Annu. Meet. Am. Soc. Microbiol.* 185.
Labow, R. S., and Layne, D. S. (1972). *Biochem. J.* **128,** 491–497.
La Du, B. N., Mandel, H. G., and Way, E. L. (1971). "Fundamentals of Drug Metabolism and Drug Disposition." Williams & Wilkins, Baltimore.
Lamoureux, G. L., Stafford, L. E., Shimabukuro, R. H., and Zaylskie, R. G. (1973). *J. Agric. Food Chem.* **21,** 1020–1030.
Leisinger, T., Hütter, R., Cook, A. M., and Nüesch, J. (1981). "Microbial Degradation of Xenobiotics and Recalcitrant Compounds." Academic Press, London.
Loewus, F. A., and Runeckles, V. C. (1977). *Recent Adv. Phytochem.,* **11**.
Loman, A. A. (1970). *Can. J. Bot.,* **48,** 737–747.
Mandel, H. G. (1971). Pathways of drug biotransformation: Biochemical conjugations, *In* "Fundamentals of Drug Metabolism and Drug Disposition" (B. N. La Du, H. G. Mandel, and E. L. Way, eds.), pp. 149–186. Williams & Wilkins, Baltimore.
Mannering, G. I. (1971). Microsomal enzyme systems which catalyze drug metabolism, *In* "Fundamentals of Drug Metabolism and Drug Disposition" (B. N. La Du, H. G. Mandel, and E. L. Way, eds.), pp. 206–252. Williams & Wilkins, Baltimore.
Meyer, E. (1983). *Phytochemistry* **22,** 1381–1385.
Meyer, E., and Barz, W. (1978). *Planta Med.* **33,** 336–344.
Miller, J. J., Powell, G. M., Olavesen, A. H., and Curtis, C. G. (1973). *Biochem. Soc. Trans.* **1,** 1163–1165.
Miller, J. J., Powell, G. M., Olaveren, A. H., and Curtis, C. G. (1976). *Toxicol. Appl. Pharmacol.*
Mohan, V. P., Kishore, G., Sugumaran, M., and Vaidyanathan, C. S. (1979). *Plant Sci. Lett.* **16,** 267–272.
Monge, P., Scheline, R., and Solheim, E. (1976). *Xenobiotica* **6,** 411–423.
Nambudiri, A. M. D., Bhat, J. V., and Subba Rao, P. V. (1972). *Biochem. J.* **130,** 425–433.
Nemoto, N., and Gelboin, H. V. (1976). *Biochem. Pharmacol.* **25,** 1221–1226.
Nemoto, N., Takayama, S., and Gelboin, H. V. (1978). *Chem. Biol. Interact.* **23,** 19–30.
Nilsson, A. (1963). *Ark. Kemi* **21,** 97–121.
Nilsson, A., Hill, J. L., and Lloyd Davies, H. (1976). *Biochim. Biophys. Acta* **148,** 92–98.
Nishida, A., and Fukuzumi, T. (1978). *Phytochemistry* **17,** 417–419.

Oesch, F. (1980). Microsomal Epoxide Hydrolase, *In* "Enzymatic Basis of Detoxication." (W. B. Jacoby, ed.), Vol. 2, pp. 277–290. Academic Press, New York.
Otten, S., and Rosazza, J. P. (1978). *Appl. Environ. Microbiol.* **35,** 554–557.
Otten, S., and Rosazza, J. P. (1979). *Appl. Environ. Microbiol.* **38,** 311–313.
Otten, S., and Rosazza, J. P. (1983). *J. Biol. Chem.* **258,** 1610–1613.
Padieu, P., and Maune, B. F. (1977). *In* "Quantitative Mass Spectrometry in Life Sciences." (A. P. Leenheer and R. R. Roncucci, eds.), pp. 49–82. Elsevier Scientific Publishing Co., Amsterdam.
Patzlaff, M., and Barz, W. (1978). *Z. Naturforsch., C: Biosci.* **33,** 675–684.
Rao, C. B. S. (1978). "The Chemistry of Lignans." Audhra University Press, Audhra, India.
Rettenmaier, H., Kupas, U., and Lingens, F. (1983). *FEMS Microbiol. Lett.* **19,** 193–195.
Robinson, T. (1974). Science **184,** 430–435.
Rosazza, J. P., and Smith, R. V. (1979). *Adv. Appl. Microbiol.* **25,** 169–215.
Sahm, H. (1984). *Adv. Biochem. Eng. Biotechnol.* **29,** 83–116.
Sariaslani, F. S., and Rosazza, J. P. (1983). *Appl. Environ. Microbiol.* **45,** 616–21.
Sawahata, T., and Neal, R. A. (1982). *Biochem. Biophys. Res. Commun.* **109,** 988–994.
Scheline, R. R. (1973). *Pharmacol. Rev.* **25,** 451–523.
Scheline, R. R. (1974). *Experientia* **30,** 880–881.
Scheline, R. R. (1978). "Mammalian Metabolism of Plant Xenobiotics." Academic Press, London.
Schoemer, U., and Martin, C. K. A. (1980). *Biotechnol. Bioeng.* **22,** Suppl. 1, 11–25.
Schulz, G., and Hecker, E. (1973). *Z. Naturforsch., C: Biosci.* **28,** 662–674.
Schultz, E., Engle, F. E., and Wood, J. M. (1974). *Biochemistry* **13,** 1768–1771.
Seidman, M. M., Toms, A., and Wood, J. M. (1969). *J. Bacteriol.* **97,** 1192–1197.
Setchell, K. D. R., Lawson, A. M., Mitchell, F. L., Adlercreutz, H., Kirk, D. N., and Axelson, M. (1980). *Nature (London)* **287,** 740–742.
Setchell, K. D. R., Lawson, A. M., Conway, E., Taylor, N. F., Kirk, D. N., Cooley, G., Farrant, R. D., Wynn, St., and Axelson, M. (1981). *Biochem. J.* **197,** 447–458.
Sharma, H. K., and Vaidyanathan, S. (1975). *Eur. J. Biochem.* **56,** 163–171.
Shearer, M. J., and Barkhan, P. (1973). *Biochim. Biophys. Acta* **297,** 300–312.
Shutt, D. A., and Braden, A. W. H. (1968). *Aust. J. Agric. Res.* **19,** 545–553.
Simpson, F. J., Narasimhadari, N., and Westlake, D. W. S. (1960). *Biochem. Biophys. Res. Commun.* **2,** 15–18.
Smith, D. A., Kuhn, P. J., Bailey, J. A., and Burden, R. S. (1980). *Phytochemistry* **19,** 1673–1675.
Smith, D. A., Harrer, J. M., and Cleveland, T. E. (1981). *Phytopathology* **71,** 1212–1215.
Smith, R. V., and Rosazza, J. P. (1975). *Biotechnol. Bioeng.* **17,** 785–814.
Smith, R. V., and Rosazza, J. P. (1982). *In* "Microbial Transformations of Bioactive Compounds" (J. P. Rosazza, ed.), Vol. II, pp. 2–42. CRC Press, Boca Raton, Florida.
Sparnins, V. L., Chapman, P. J., and Dagley, S. (1974). *J. Bacteriol.* **120,** 159–167.
Sparnins, V. L., Anderson, J. J., Omans, J., and Dagley, S. (1978). *J. Bacteriol.* **136,** 449–451.
Stanier, R. Y., and Ornston, L. N. (1973). *Adv. Microb. Physiol.* **9,** 89–151.
Stanier, R. Y., Palleroni, N. J., and Doudoroff, M. (1966). *J. Gen. Microbiol.* **43,** 159–271.
Stoessl, A. (1982). *In* "Phytoalexins" (J. A. Bailey and J. W. Mansfield, eds.), pp. 133–180, Blackie, Glasgow.
Sutherland, J. B., Crawford, D. L., and Pometto, A. L. (1981). *Appl. Environ. Microbiol.* **41,** 442–448.
Sutherland, J. B., Crawford, D. L., and Pometto, A. L. (1983). *Can. J. Microbiol.* **29,** 1253–1257.

Takaji, K., Suzuki, T., Saitoh, Y., and Nishihara, K. (1971). *Yakuzaigaku* **31,** 213–219.
Tay, L. K., and Sinsheimer, J. E. (1976). *Drug Metab. Dispos.* **4,** 154–163.
Thompson, R. M., Gerber, N., Seibert, R. A., and Desiderio, D. M. (1972). *Res. Commun. Chem. Pathol. Pharmacol.* **4,** 543–552.
Trevors, J. T., and Basaraba, J. (1980). *FEMS Microbiol. Lett.* **7,** 307–309.
Unai, T., Cheng, H.-M., Yamamoto, J., and Casida, J. E. (1973). *Agric. Biol. Chem.* **37,** 1937–1944.
VanEtten, H. D., Matthews, D. E., and Smith, D. A. (1982). *In* "Phytoalexins" (J. A. Bailey and J. W. Mansfield, eds.), pp. 181–217. Blackie, Glasgow.
Von der Trenck, Th., and Sandermann, H., Jr. (1981). *FEBS Lett.* **125,** 72–74.
Von der Trenck, Th., Hunkler, D., and Sandermann, H., Jr. (1981). *Z. Naturforsch., C: Biosci.* **36,** 714–719.
Walker, J. R. L., and Taylor, B. G. (1983). *Arch. Microbiol.* **134,** 123–126.
Walsh, T. A., and Ballou, D. P. (1983). *J. Biol. Chem.* **258,** 14413–14421.
Walter, G., and Hecker, E. (1973). *Z. Naturforsch., C: Biosci.* **28,** 675–684.
Watanabe, T. and Akamatsu, K. (1975). *Biochem. Pharmacol.* **24,** 442–447.
Watanabe, T., Akamutsu, K. and Kiyonaga, K. (1971). *Biochem. Biophys. Res. Commun.* **44,** 199–204.
Weltring, K.-M., Mackenbrock, K., and Barz, W. (1982). *Z. Naturforsch.* **37c,** 570–574.
Weltring, K.-M., Barz, W., and Dewick, P. M. (1983). *Phytochemistry* **22,** 2883–2884.
Willeke, U., and Barz, W. (1982a). *Z. Naturforsch.* **37c,** 861–869.
Willeke, U., and Barz, W. (1982b). *Arch. Microbiol.* **132,** 266–269.
Willeke, U., Weltring, K.-M., Barz, W., and VanEtten, H. D. (1983). *Phytochemistry* **22,** 1539–1541.
Williams, R. T. (1959). "Detoxication Mechanisms." Chapman & Hall, London.
Yamaguchi, M., and Fujisawa, H. (1980). *J. Biol. Chem.* **255,** 5058–5063.
Yamamoto, I. (1969). *Residue Rev.* **25,** 161–174.
Yeh, W. K., Gibson, D. T., and Liu, T. N. (1977). *Biochem. Biophys. Res. Commun.* **78,** 401–410.

Index

B

C

M

N

O

P

Q

R

S

T

X

Y